Springer Tracts in Advanced Robotics

Volume 146

The Springer Tracts in Advanced Robotics (STAR) publish new developments and advances in the fields of robotics research, rapidly and informally but with a high quality. The intent is to cover all the technical contents, applications, and multidisciplinary aspects of robotics, embedded in the fields of Mechanical Engineering, Computer Science, Electrical Engineering, Mechatronics, Control, and Life Sciences, as well as the methodologies behind them. Within the scope of the series are monographs, lecture notes, selected contributions from specialized conferences and workshops, as well as selected PhD theses.

Special offer: For all clients with a print standing order we offer free access to the electronic volumes of the Series published in the current year.

Indexed by SCOPUS, DBLP, EI Compendex, zbMATH, SCImago.

All books published in the series are submitted for consideration in Web of Science.

Peter Corke

# Robotics, Vision and Control

Fundamental Algorithms in Python

3rd edition 2023

 Springer

Peter Corke
Queensland University of Technology
Brisbane, QLD, Australia

ISSN 1610-7438 ISSN 1610-742X (electronic)
Springer Tracts in Advanced Robotics
ISBN 978-3-031-06468-5 ISBN 978-3-031-06469-2 (eBook)
https://doi.org/10.1007/978-3-031-06469-2

This Springer imprint is published by the registered company Springer Nature Switzerland AG
The registered company address is: Gewerbestrasse 11, 6330 Cham, Switzerland

*To my family Phillipa, Lucy and Madeline for their indulgence and support; my parents Margaret and David for kindling my curiosity; and to Richard (Lou) Paul who planted the seed that became this book.*

# Foreword

Once upon a time, a very thick document of a dissertation from a faraway land came to me for evaluation. *Visual robot control* was the thesis theme and *Peter Corke* was its author. Here, I am reminded of an excerpt of my comments, which reads, *this is a masterful document, a quality of thesis one would like all of one's students to strive for, knowing very few could attain – very well considered and executed.*

The connection between robotics and vision has been, for over two decades, the central thread of Peter Corke's productive investigations and successful developments and implementations. This rare experience is bearing fruit in this second edition of his book on *Robotics, Vision, and Control.* In its melding of theory and application, this second edition has considerably benefited from the author's unique mix of academic and real-world application influences through his many years of work in robotic mining, flying, underwater, and field robotics.

There have been numerous textbooks in robotics and vision, but few have reached the level of integration, analysis, dissection, and practical illustrations evidenced in this book. The discussion is thorough, the narrative is remarkably informative and accessible, and the overall impression is of a significant contribution for researchers and future investigators in our field. Most every element that could be considered as relevant to the task seems to have been analyzed and incorporated, and the effective use of Toolbox software echoes this thoroughness.

The reader is taken on a realistic walk-through the fundamentals of mobile robots, navigation, localization, manipulator-arm kinematics, dynamics, and joint-level control, as well as camera modeling, image processing, feature extraction, and multi-view geometry. These areas are finally brought together through extensive discussion of visual servo system. In the process, the author provides insights into how complex problems can be decomposed and solved using powerful numerical tools and effective software.

The *Springer Tracts in Advanced Robotics (STAR)* is devoted to bringing to the research community the latest advances in the robotics field on the basis of their significance and quality. Through a wide and timely dissemination of critical research developments in robotics, our objective with this series is to promote more exchanges and collaborations among the researchers in the community and contribute to further advancements in this rapidly growing field.

Peter Corke brings a great addition to our STAR series with an authoritative book, reaching across fields, thoughtfully conceived and brilliantly accomplished.

Oussama Khatib
Stanford, California
October 2016

# Preface

» *Tell me and I will forget.*
*Show me and I will remember.*
*Involve me and I will understand.*
– Chinese proverb

*Simple things should be simple,*
*complex things should be possible.*
– Alan Kay

These are exciting times for robotics. Since the first edition of this book was published over ten years ago we have seen great progress: the actuality of self-driving cars on public roads, multiple robots on Mars (including one that flies), robotic asteroid and comet sampling, the rise of robot-enabled businesses like Amazon, and the DARPA Subterranean Challenge where teams of ground and aerial robots autonomously mapped underground spaces. We have witnessed the drone revolution – flying machines that were once the domain of the aerospace giants can now be bought for just tens of dollars. All of this has been powered by the ongoing improvement in computer power and tremendous advances in low-cost inertial sensors and cameras – driven largely by consumer demand for better mobile phones and gaming experiences. It's getting easier for individuals to create robots – 3D printing is now very affordable, the Robot Operating System (ROS) is capable and widely used, and powerful hobby technologies such as the Arduino, Raspberry Pi, and Dynamixel servo motors are available at low cost. This in turn has enabled the growth of the global maker community, and empowered individuals, working at home, and small startups to do what would once have been done by major corporations.

Robots are machines which acquire data, process it, and take action based on it. The data comes from a variety of sensors that measure, for example, the velocity of a wheel, the angle of a robot arm's joint, or the intensities of millions of pixels that comprise an image of the world. For many robotic applications the amount of data that needs to be processed, in real-time, is massive. For a vision sensor it can be on the order of tens to hundreds of megabytes per second. Progress in robots and machine vision has been, and continues to be, driven by more effective ways to process sensory data.

One axis of progress has been driven by the relentless increase in affordable computational power. ▶ Moore's law predicts that the number of transistors on a chip will double every two years, and this enables ever-increasing amounts of memory, and parallel processing with multiple cores and graphical processing units (GPUs). Concomitantly the size of transistors has shrunk and clock speed has increased.

The other axis of progress is algorithmic, exploiting this abundance of computation and memory to solve robotics problems. Over decades, the research community has developed many solutions for important problems in perception, localization, planning, and control. However, for any particular problem there is a wide choice of algorithms, and each of them may have several implementations. These will be written in a variety of languages, with a variety of API styles and conventions, and with variable code quality, documentation, support and license conditions. This is a significant challenge for robotics today, and "cobbling together" disparate pieces of software has become an essential skill for roboticists. The ROS framework ▶ has helped by standardizing interfaces and allowing common functions to be composed to solve a particular problem. Nevertheless, the software side of robotics is still harder and more time-consuming than it should

When I started in robotics and vision in the mid 1980s, the IBM PC had been recently released – it had a 4.77 MHz 16-bit microprocessor and 16 kbytes (expandable to 256 k) of memory.

See ▶ https://ros.org.

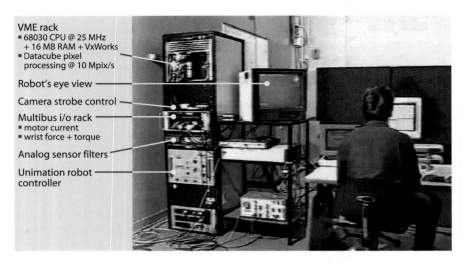

VME rack
- 68030 CPU @ 25 MHz + 16 MB RAM + VxWorks
- Datacube pixel processing @ 10 Mpix/s

Robot's eye view

Camera strobe control

Multibus i/o rack
- motor current
- wrist force + torque

Analog sensor filters

Unimation robot controller

**Fig. 1** Once upon a time a lot of equipment was needed to do vision-based robot control. The author with a large rack full of real-time image processing and robot control equipment and a PUMA 560 robot (1992). Over the intervening 30 years, the number of transistors on a chip has increased by a factor of $2^{30/2} \approx 30,000$ according to Moore's law

be. This unfortunate complexity, and the sheer range of choice, presents a very real barrier to somebody new entering the field.

**»** the software tools used in this book aim to reduce complexity for the reader

The software tools that are used in this book aim to reduce complexity for the reader by providing a coherent and complete set of functionality. We use Python, a popular open-source language and several open-source Toolboxes that provide functionality for robotics and computer vision. This makes common algorithms tangible and accessible. You can read all the code, you can apply it to your own problems, and you can extend it or rewrite it. It gives you a "leg up" as you begin your journey into robotics.

**»** allow the user to work with real problems, not just trivial examples

This book uses that software to illustrate each topic, and this has a number of benefits. Firstly, the software allows the reader to work with real problems, not just trivial examples. For real robots, those with more than two links, or real images with millions of pixels the computation required is beyond unaided human ability. Secondly, these software tools help us gain insight which can otherwise get lost in the complexity. We can rapidly and easily experiment, play *what if* games, and depict the results graphically using the powerful 2D and 3D graphical display tools of Python and Matplotlib.

**»** a cohesive narrative that covers robotics and computer vision – both separately and together

The book takes a conversational approach, weaving text, mathematics and lines of code into a cohesive narrative that covers robotics and computer vision – separately, and together as *robotic vision*. It shows how complex problems can be decomposed and solved using just a few simple lines of code. More formally this is an inductive learning approach, going from specific and concrete examples to the more general.

**»** show how complex problems can be decomposed and solved

The topics covered in this book are guided by real problems that I observed over many years as a practitioner of both robotics and computer vision. Consider the book as a grand tasting menu and I hope that by the end of this book you will share my enthusiasm for these topics.

» consider the book as a grand tasting menu

I was particularly motivated to present a solid introduction to computer vision for roboticists. The treatment of vision in robotics textbooks tends to concentrate on simple binary vision techniques. In this book we will cover a broad range of topics including color vision, advanced segmentation techniques, image warping, image retrieval, stereo vision, pose estimation, deep learning, bundle adjustment, and visual odometry. We also cover nonperspective imaging using fisheye lenses, catadioptric optics, and the emerging area of light-field cameras. These topics are growing in importance for robotics but are not commonly covered. Vision is a powerful sensing modality, and roboticists should have a solid grounding in modern fundamentals. The last part of the book shows how vision can be used as the primary sensor for robot control.

This book is unlike other text books, and deliberately so. While there are already a number of excellent text books that cover robotics and computer vision separately and in depth, there are few that cover both in an integrated fashion. Achieving such integration was a principal goal of the book.

» software is a first-class citizen in this book

Software is a tangible instantiation of the algorithms described – it can be read and it can be pulled apart, modified and put back together again – software is a first-class citizen in this book. There are a number of classic books that use software in an illustrative fashion and have influenced my approach, for example *LaTeX: A document preparation system* (Lamport 1994), *Numerical Recipes in C* (Press et al. 2007), *The Little Lisper* (Friedman et al. 1987) and *Structure and Interpretation of Classical Mechanics* (Sussman et al. 2001). In this book, over 2000 lines of code across over 1600 examples illustrate how the Toolbox software can be used and generally provide *instant gratification* in just a couple of lines of Python code.

» instant gratification in just a couple of lines
of Python code

Thirdly, building the book around Python and the Toolboxes means that we are able to tackle more realistic and more complex problems than other books.

» this book provides a complementary approach

A key motivation of this book is to make robotics, vision and control algorithms *accessible* to a wide audience. Accessibility has a number of dimensions including conceptual, geographical and financial. Conceptually, the mathematics that underpins robotics is inescapable, but the theoretical complexity has been minimized and the book assumes no more than an undergraduate-engineering level of mathematical knowledge. The software-based examples also help to ground the abstract concepts and make them tangible. Using open-source software throughout reduces financial cost for the reader – that software is available at zero cost anywhere in the world. This approach is complementary to the many other excellent textbooks that cover these same topics but which take a stronger, and more traditional, theoretical approach. This book is best read in conjunction with those other texts, and the end of each chapter has a section on further reading that provides pointers to relevant textbooks and key papers. This approach is complementary to the many other excellent textbooks that cover these same topics but which take a stronger, and more traditional, theoretical approach. This book is best read in conjunction with those other texts, and the end of each chapter has a section on further reading that provides pointers to relevant textbooks and key papers.

The fields of robotics and computer vision are underpinned by theories developed by mathematicians, scientists, and engineers over many hundreds of years. Some of their names have become adjectives like Coriolis, Gaussian, Laplacian, or Cartesian; nouns like Jacobian, or units like Newton and Coulomb. They are interesting characters from a distant era when science was a hobby and their day jobs were as doctors, alchemists, gamblers, astrologers, philosophers or mercenaries. In order to know whose shoulders we are standing on, the book includes small vignettes about the lives of some of these people – a smattering of history as a back story.

Creating a Python version of the venerable Robotics Toolbox for MATLAB®, and of this book has been a long-standing aspiration of mine. There were several false starts, and it took a pandemic to finally make it happen. Jesse Haviland Pythonized and extended the ETS notation (▶ Sect. 7.1.1.1) with his package `ropy`, went on to Pythonize the entire Robotics Toolbox, and then developed the Swift simulation environment. Dorian Tsai started the Pythonization of the Machine Vision Toolbox. The graphical editor for the bdsim block diagram package was completed by Daniel Petkov. I thank them deeply for helping me to finally realize this goal.

This book is built on that open-source software, which is in turn is built on other open-source software from the wider community. I'd like to acknowledge the teams behind Python, Matplotlib, NumPy, SciPy, SymPy, pybullet, Python Robotics, Jupyter, OpenCV, Open3D, and PyTorch. Developing text and code was accelerated by awesome open-source or free tools like Visual Studio Code, git, GitHub, meld, TexStudio, xelatex and the whole LaTeX community. The breadth and quality of what the open-source community creates continues to amaze me.

Many people have helped me with critical comments over previous editions – this edition is the better for their input and I thank: Paul Newman, Daniela Rus, Cédric Pradalier, Tim Barfoot, Dmitry Bratanov, Duncan Campbell, Donald Dansereau, Tom Drummond, Malcolm Good, Peter Kujala, Obadiah Lam, Jörn Malzahn, Felipe Nascimento Martins, Ajay Pandey, Dan Richards, Sareh Shirazi, Surya Singh, Ryan Smith, Ben Talbot, Dorian Tsai, Ben Upcroft, François Chaumette, Donald Dansereau, Kevin Lynch, Robert Mahony and Frank Park. For this third edition, my coauthors on the MATLAB version, Witold Jachimczyk and Remo Pillat, have provided a fresh perspective and constructive criticism that has benefited both versions. I am grateful to my colleagues who have provided detailed and insightful feedback on the latest chapter drafts: Christina Kazantzidou (who helped to polish words and mathematical notation), Tobias Fischer, Will Browne, Rob Mahony, Jesse Haviland, Feras Dayoub, Dorian Tsai, Alessandro De Luca, Renaud Detry, Brian Fanous, and Hannes Daepp.

I have tried my hardest to eliminate errors but inevitably some will remain. Please email bug reports to me at rvc@petercorke.com as well as suggestions for improvements and extensions.

This work was partly financially supported by the 2017 Australian University Teacher (AAUT) of the Year Award from the Australian Government Department of Education and Training. I thank Queensland University of Technology for making time available to complete this project. Over all editions I have enjoyed wonderful support from the folk at Springer-Nature, Thomas Ditzinger who has supported this project since before the first edition, and Wilma McHugh. Special thanks to Yvonne Schlatter and the team at le-tex for their wonderful support with typesetting.

Finally, my deepest thanks of all are to Phillipa who has supported me and my book obsession with grace and patience for more than a decade – without her this book could never have been written.

## Notes on the Third Edition

The first two editions of this book ▶ were based on MATLAB® in conjunction with open-source MATLAB Toolboxes that are now thirty years old – that's a long time for any piece of software. Much has happened in the last decade that motivate a change to the software foundations of the book, and that has led to *two* third editions:

The first edition (2011), the second edition as a single volume (2017) and then as a two-volume set (2022).

- The version you are reading, is based on Python which is a popular open-source language with massive third party support. The old MATLAB Toolboxes have been redesigned and reimplemented in Python, taking advantage of popular open-source packages and resources to provide platform portability, fast browser-based 3D graphics, online documentation, fast numerical and symbolic operations, shareable and web-browseable notebooks all powered by GitHub and the open-source community (Corke 2021).
- The alternative version, rewritten with colleagues Witold Jachimczyk and Remo Pillat from MathWorks®, is based on MATLAB, and state-of-the-art toolboxes developed, licensed, and supported by MathWorks.

In addition to changing the software underpinnings of the book, this third edition also provides an opportunity to fix errors, improve mathematical notation, and clarify the narrative throughout. ▶ Chaps. 2 and 7 have been extensively rewritten. This edition also includes new topics such as graph-based path planning, Dubins and Reeds-Shepp paths, branched robots, URDF models, collision checking, task-space control, deep learning for object detection and semantic segmentation, fiducial markers, and point clouds.

Peter Corke
Brisbane, Queensland
March 2022

# Contents

## II    Mobile Robotics

# Nomenclature

The notation used in robotics and computer vision varies considerably across books and research papers. The symbols used in this book, and their units where appropriate, are listed below. Some symbols have multiple meanings and their context must be used to disambiguate them.

| Notation | Description |
|---|---|
| $\boldsymbol{v}$ | a vector |
| $\hat{\boldsymbol{v}}$ | a unit vector parallel to $\boldsymbol{v}$ |
| $\tilde{\boldsymbol{v}}$ | homogeneous representation of the vector $\boldsymbol{v}$ |
| $v_i$ | $i^{\text{th}}$ element of vector $\boldsymbol{v}$ |
| $\tilde{v}_i$ | $i^{\text{th}}$ element of the homogeneous vector $\tilde{\boldsymbol{v}}$ |
| $v_x, v_y, v_z$ | elements of the coordinate vector $\boldsymbol{v}$ |
| $\mathbf{A}$ | a matrix |
| $a_{i,j}$ | the element of $\mathbf{A}$ at row $i$ and column $j$ |
| $\mathbf{0}_{m \times n}$ | an $m \times n$ matrix of zeros |
| $\mathbf{1}_{m \times n}$ | an $m \times n$ identity matrix |
| $\check{q}$ | a quaternion, $\check{q} \in \mathbb{H}$ |
| $\mathring{q}$ | a unit quaternion, $\mathring{q} \in \mathrm{S}^3$ |
| $f(x)$ | a function of $x$ |
| $F_x(x)$ | the first derivative of $f(x)$, $\partial f / \partial x$ |
| $F_{xy}(x, y)$ | the second derivative of $f(x, y)$, $\partial^2 f / \partial x \partial y$ |
| $\{F\}$ | coordinate frame $F$ |

Vectors are generally lower-case Roman or Greek letters, while matrices are generally upper-case Roman or Greek letters. Various decorators are applied to symbols:

| Decorators | Description |
|---|---|
| $x^*$ | desired value of $x$ |
| $x^+$ | predicted value of $x$ |
| $x^\#$ | measured, or observed, value of $x$ |
| $\hat{x}$ | estimated value of $x$, also a unit vector as above |
| $\bar{x}$ | mean of $x$ or relative value |
| $x\langle k \rangle$ | $k^{\text{th}}$ element of a time series |

where $x$ could be a scalar, vector, or matrix.

| Symbol | Description | Unit |
|---|---|---|
| $B$ | viscous friction coefficient | $\mathrm{N\,m\,s\,rad^{-1}}$ |
| $B$ | magnetic flux density | T |
| $\mathbf{C}$ | camera matrix, $\mathbf{C} \in \mathbb{R}^{3\times 4}$ | |
| $\mathbf{C}(\boldsymbol{q}, \dot{\boldsymbol{q}})$ | manipulator centripetal and Coriolis term | $\mathrm{kg\,m^2\,s^{-1}}$ |
| $C$ | configuration space of a robot with $N$ joints: $C \subset \mathbb{R}^N$ | |
| e | mathematical constant, base of natural logarithms, $e = 2.71828\ldots$ | |
| $E$ | illuminance (in lux) | lx |
| $f$ | focal length | m |
| $f$ | force | N |
| $\boldsymbol{f}(\dot{\boldsymbol{q}})$ | manipulator friction torque | Nm |
| g | gravitation acceleration, see ◘ Fig. 3.11 | $\mathrm{ms^{-2}}$ |
| $\boldsymbol{g}(\boldsymbol{q})$ | manipulator gravity term | Nm |
| $\mathbb{H}$ | the set of all quaternions (H for Hamilton) | |
| $J$ | inertia | $\mathrm{kg\,m^2}$ |
| $\mathbf{J}$ | inertia tensor, $\mathbf{J} \in \mathbb{R}^{3\times 3}$ | $\mathrm{kg\,m^2}$ |
| $\mathbf{J}$ | Jacobian matrix | |
| ${}^A\mathbf{J}_B$ | Jacobian transforming velocities in frame {B} to frame {A} | |
| $k, K$ | constant | |
| $\mathbf{K}$ | camera intrinsic matrix, $\mathbf{K} \in \mathbb{R}^{3\times 3}$ | |
| $K_i$ | amplifier gain (transconductance) | $\mathrm{A\,V^{-1}}$ |
| $K_m$ | motor torque constant | $\mathrm{Nm\,A^{-1}}$ |
| $L$ | luminance (in nit) | nt |
| $m$ | mass | kg |
| $\mathbf{M}(\boldsymbol{q})$ | manipulator inertia matrix | $\mathrm{kg\,m^2}$ |
| $\mathbb{N}$ | the set of natural numbers $\{1, 2, 3, \ldots\}$ | |
| $\mathbb{N}_0$ | the set of natural numbers including zero $\{0, 1, 2, \ldots\}$ | |
| $N(\mu, \sigma^2)$ | a normal (Gaussian) distribution with mean $\mu$ and standard deviation $\sigma$ | |
| p | an image plane point | |
| P | a world point | |
| $\mathbf{p}$ | coordinate vector of an image plane point, $\mathbf{p} \in \mathbb{R}^2$ | |
| $\mathbf{P}$ | coordinate vector of a world point, $\mathbf{P} \in \mathbb{R}^3$ | |
| $\mathbb{P}^2$ | projective space of all 2D points, elements are a 3-tuple | |
| $\mathbb{P}^3$ | projective space of all 3D points, elements are a 4-tuple | |
| $\boldsymbol{q}$ | generalized coordinates, robot configuration $\boldsymbol{q} \in C$ | m, rad |
| $\boldsymbol{Q}$ | generalized force $\boldsymbol{Q} \in \mathbb{R}^N$ | N, Nm |
| $\mathbf{R}$ | an orthonormal rotation matrix, $\mathbf{R} \in \mathbf{SO}(2)$ or $\mathbf{SO}(3)$ | |
| $\mathbb{R}$ | set of real numbers | |
| $\mathbb{R}_{>0}$ | set of positive real numbers | |

Nomenclature

| Symbol | Description | Unit |
|---|---|---|
| $\mathbb{R}_{\geq 0}$ | set of non-negative real numbers | |
| $\mathbb{R}^2$ | set of all 2D points, elements are a 2-tuple | |
| $\mathbb{R}^3$ | set of all 3D points, elements are a 3-tuple | |
| $s$ | distance along a path or trajectory, $s \in [0, 1]$ | |
| $s$ | Laplace transform operator | |
| $\mathbf{S}^1$ | set of points on the unit circle $\sim$ set of angles $[0, 2\pi)$ | |
| $\mathbf{S}^n$ | unit sphere embedded in $\mathbb{R}^{n+1}$ | |
| $\mathbf{se}(n)$ | Lie algebra for $\mathbf{SE}(n)$, a vector space of $\mathbb{R}^{(n+1)\times(n+1)}$ augmented skew-symmetric matrices | |
| $\mathbf{so}(n)$ | Lie algebra for $\mathbf{SO}(n)$, a vector space of $\mathbb{R}^{n\times n}$ skew-symmetric matrices | |
| $\mathbf{SE}(n)$ | special Euclidean group of matrices, $\mathbf{T} \in \mathbf{SE}(n)$, $\mathbf{T} \subset \mathbb{R}^{(n+1)\times(n+1)}$, represents pose in $n$ dimensions, aka homogeneous transformation matrix, aka rigid-body transformation | |
| $\mathbf{SO}(n)$ | special orthogonal group of matrices, $\mathbf{R} \in \mathbf{SO}(n)$, $\mathbf{R} \subset \mathbb{R}^{n\times n}$, $\mathbf{R}^\top = \mathbf{R}^{-1}$, $\det(\mathbf{R}) = 1$, represents orientation in $n$ dimensions, aka rotation matrix | |
| $S$ | twist in 2 or 3 dimensions, $S \in \mathbb{R}^3$ or $\mathbb{R}^6$ | |
| $t$ | time | s |
| $\mathcal{T}$ | task space of robot, $\mathcal{T} \subset \mathbf{SE}(3)$ | K |
| $T$ | sample interval | s |
| $T$ | temperature | K |
| $T$ | optical transmission | $m^{-1}$ |
| $\mathbf{T}$ | a homogeneous transformation matrix $\mathbf{T} \in \mathbf{SE}(2)$ or $\mathbf{SE}(3)$ | |
| $^A\mathbf{T}_B$ | a homogeneous transformation matrix representing the pose of frame $\{B\}$ with respect to frame $\{A\}$. If $A$ is not given then assumed relative to world coordinate frame $\{0\}$. Note that $^A\mathbf{T}_B = (^B\mathbf{T}_A)^{-1}$ | |
| $u, v$ | coordinates of point on a camera's image plane | pixels |
| $u_0, v_0$ | coordinates of a camera's principal point | pixels |
| $\bar{u}, \bar{v}$ | normalized image plane coordinates, relative to the principal point | pixels |
| $v$ | velocity | $m\,s^{-1}$ |
| $\boldsymbol{v}$ | velocity vector | $m\,s^{-1}$ |
| $\boldsymbol{w}$ | wrench, a vector of forces and moments $(f_x, f_y, f_z, m_x, m_y, m_z) \in \mathbb{R}^6$ | N, Nm |
| $\hat{\boldsymbol{x}}, \hat{\boldsymbol{y}}, \hat{\boldsymbol{z}}$ | unit vectors aligned with the $x$-, $y$- and $z$-axes, 3D basis vectors | |
| $X, Y, Z$ | Cartesian coordinates of a point | |
| $\bar{x}, \bar{y}$ | retinal image-plane coordinates | m |
| $\mathbb{Z}$ | set of all integers $\{\ldots, -2, -1, 0, 1, 2, \ldots\}$ | |
| $\varnothing$ | null motion | |
| $\alpha$ | roll angle | rad |
| $\beta$ | pitch angle | rad |
| $\gamma$ | steered wheel angle | rad |

| Symbol | Description | Unit |
|---|---|---|
| $\delta$ | an increment | |
| $\epsilon$ | an error or residual, ideally zero | |
| $\gamma$ | yaw angle | rad |
| $\gamma$ | steered-wheel angle for mobile robot | rad |
| $\boldsymbol{\Gamma}$ | 3-angle representation of rotation, $\boldsymbol{\Gamma} \in \mathbb{R}^3$ | rad |
| $\boldsymbol{\Gamma}$ | body torque, $\boldsymbol{\Gamma} \in \mathbb{R}^3$ | Nm |
| $\theta$ | an angle, first Euler angle, heading angle | rad |
| $\lambda$ | wavelength | m |
| $\lambda$ | an eigenvalue | |
| $\boldsymbol{\nu}$ | innovation | |
| $\boldsymbol{\nu}$ | spatial velocity, $(v_x, v_y, v_z, \omega_x, \omega_y, \omega_z) \in \mathbb{R}^6$ | $\mathrm{m\,s^{-1}}$ $\mathrm{rad\,s^{-1}}$ |
| $\boldsymbol{\xi}$ | abstract representation of pose (pronounced ksi) | |
| $^A\boldsymbol{\xi}_B$ | abstract representation of relative pose, frame {B} with respect to frame {A} or rigid-body motion from frame {A} to {B} | |
| $\boldsymbol{\xi}^{t_i}(d)$ | abstract pose that is pure translation of $d$ along axis $i \in \{x, y, z\}$ | |
| $\boldsymbol{\xi}^{r_i}(\theta)$ | abstract pose that is pure rotation a pure rotation of $\theta$ about axis $i \in \{x, y, z\}$ | |
| $\pi$ | mathematical constant $\pi = 3.14159\ldots$ | |
| $\boldsymbol{\pi}$ | a plane | |
| $\rho_{\mathrm{w}}, \rho_{\mathrm{h}}$ | pixel width and height | m |
| $\sigma$ | standard deviation | |
| $\sigma$ | robot joint type, $\sigma = $ R for revolute and $\sigma = $ P for prismatic | |
| $\boldsymbol{\Sigma}$ | element of the Lie algebra of $\mathbf{SE}(3)$, $\boldsymbol{\Sigma} = [\boldsymbol{v}] \in \mathbf{se}(3)$, $\boldsymbol{v} \in \mathbb{R}^6$ | |
| $\tau$ | torque | $\mathrm{N\,m}$ |
| $\tau_{\mathrm{C}}$ | Coulomb friction torque | $\mathrm{N\,m}$ |
| $\phi$ | angle, second Euler angle | rad |
| $\psi$ | third Euler angle | rad |
| $\omega$ | rotational rate | $\mathrm{rad\,s^{-1}}$ |
| $\boldsymbol{\omega}$ | angular velocity vector | $\mathrm{rad\,s^{-1}}$ |
| $\varpi$ | rotational speed of a motor shaft, wheel or propeller | $\mathrm{rad\,s^{-1}}$ |
| $\boldsymbol{\Omega}$ | element of the Lie algebra of $\mathbf{SO}(3)$, $\boldsymbol{\Omega} = [\boldsymbol{v}]_\times \in \mathbf{so}(3)$, $\boldsymbol{v} \in \mathbb{R}^3$ | |
| $\boldsymbol{v}_1 \cdot \boldsymbol{v}_2$ | dot, or inner, product, also $\boldsymbol{v}_1^\top \boldsymbol{v}_2$: $\mathbb{R}^n \times \mathbb{R}^n \mapsto \mathbb{R}$ | |
| $\|\cdot\|$ | norm, or length, of vector, also $\sqrt{\boldsymbol{v} \cdot \boldsymbol{v}}$: $\mathbb{R}^n \mapsto \mathbb{R}_{\geq 0}$ | |
| $\boldsymbol{v}_1 \times \boldsymbol{v}_2$ | cross, or vector, product: $\mathbb{R}^3 \times \mathbb{R}^3 \mapsto \mathbb{R}^3$ | |
| $\tilde{\cdot}$ | $\mathbb{R}^n \mapsto \mathbb{P}^n$ | |
| $\epsilon(\cdot)$ | $\mathbb{P}^n \mapsto \mathbb{R}^n$ | |
| $\mathbf{A}^{-1}$ | inverse of $\mathbf{A}$: $\mathbb{R}^{n \times n} \mapsto \mathbb{R}^{n \times n}$ | |
| $\mathbf{A}^{+}$ | pseudo-inverse of $\mathbf{A}$: $\mathbb{R}^{n \times m} \mapsto \mathbb{R}^{m \times n}$ | |
| $\mathbf{A}^{*}$ | adjugate of $\mathbf{A}$, also $\det(\mathbf{A})\mathbf{A}^{-1}$: $\mathbb{R}^{n \times n} \mapsto \mathbb{R}^{n \times n}$ | |
| $\mathbf{A}^\top$ | transpose of $\mathbf{A}$: $\mathbb{R}^{n \times m} \mapsto \mathbb{R}^{m \times n}$ | |

| Operator | Description |
|---|---|
| $\mathbf{A}^{-\top}$ | transpose of inverse of $\mathbf{A}$, $(\mathbf{A}^\top)^{-1} \equiv (\mathbf{A}^{-1})^\top$: $\mathbb{R}^{n\times n} \mapsto \mathbb{R}^{n\times n}$ |
| $\oplus$ | composition of abstract pose: ${}^x\boldsymbol{\xi}_y \oplus {}^y\boldsymbol{\xi}_z \mapsto {}^x\boldsymbol{\xi}_z$ |
| $\ominus$ | composition of abstract pose with inverse: ${}^x\boldsymbol{\xi}_y \ominus {}^z\boldsymbol{\xi}_y \equiv {}^x\boldsymbol{\xi}_y \oplus {}^y\boldsymbol{\xi}_z \mapsto {}^x\boldsymbol{\xi}_z$ |
| $\ominus$ | inverse of abstract pose: $\ominus\, {}^x\boldsymbol{\xi}_y \equiv {}^y\boldsymbol{\xi}_x$ |
| $\cdot$ | transform a point (coordinate vector) by abstract relative pose: ${}^x\boldsymbol{\xi}_y \cdot {}^y\boldsymbol{p} \mapsto {}^x\boldsymbol{p}$ |
| $\Delta(\cdot)$ | maps incremental relative pose to differential motion: $\mathbf{SE}(3) \mapsto \mathbb{R}^6$ |
| $\Delta^{-1}(\cdot)$ | maps differential motion to incremental relative pose: $\mathbb{R}^6 \mapsto \mathbf{SE}(3)$ |
| $[\cdot]_t$ | translational component of pose: $\mathbf{SE}(n) \mapsto \mathbb{R}^n$ |
| $[\cdot]_R$ | rotational component of pose: $\mathbf{SE}(n) \mapsto \mathbf{SO}(n)$ |
| $[\cdot]_{xy\theta}$ | 2D pose to configuration: $\mathbf{SE}(2) \mapsto \mathbb{R}^2 \times \mathbb{S}^1$ |
| $[\cdot]_\times$ | skew-symmetric matrix: $\mathbb{R} \mapsto \mathbf{so}(2)$, $\mathbb{R}^3 \mapsto \mathbf{so}(3)$ |
| $[\cdot]$ | augmented skew-symmetric matrix: $\mathbb{R}^3 \mapsto \mathbf{se}(2)$, $\mathbb{R}^6 \mapsto \mathbf{se}(3)$ |
| $\vee_\times(\cdot)$ | *unpack* skew-symmetric matrix: $\mathbf{so}(2) \mapsto \mathbb{R}$, $\mathbf{so}(3) \mapsto \mathbb{R}^3$ |
| $\vee(\cdot)$ | *unpack* augmented skew-symmetric matrix: $\mathbf{se}(2) \mapsto \mathbb{R}^3$, $\mathbf{se}(3) \mapsto \mathbb{R}^6$ |
| $\mathrm{Ad}(\cdot)$ | adjoint representation: $\mathbf{SE}(3) \mapsto \mathbb{R}^{6\times 6}$ |
| $\mathrm{ad}(\cdot)$ | logarithm of adjoint representation: $\mathbf{SE}(3) \mapsto \mathbb{R}^{6\times 6}$ |
| $\circ$ | quaternion (Hamiltonian) multiplication: $\mathbb{H} \times \mathbb{H} \mapsto \mathbb{H}$ |
| $\check{v}$ | pure quaternion: $\mathbb{R}^3 \mapsto \mathbb{H}$ |
| $\sim$ | equivalence of representation |
| $\simeq$ | homogeneous coordinate equivalence |
| $\ominus$ | smallest angular difference between two angles: $\mathbb{S}^1 \times \mathbb{S}^1 \mapsto [-\pi, \pi)$ |
| $\mathcal{K}(\cdot)$ | forward kinematics: $C \mapsto \mathcal{T}$ |
| $\mathcal{K}^{-1}(\cdot)$ | inverse kinematics: $\mathcal{T} \mapsto C$ |
| $\mathcal{D}^{-1}(\cdot)$ | manipulator inverse dynamics function: $C, \mathbb{R}^N, \mathbb{R}^N \mapsto \mathbb{R}^N$ |
| $\mathcal{P}(\cdot)$ | camera projection function: $\mathbb{R}^3 \mapsto \mathbb{R}^2$ |
| $*$ | image convolution |
| $\otimes$ | image correlation |
| $\equiv$ | colormetric equivalence |
| $\oplus$ | morphological dilation |
| $\ominus$ | morphological erosion |
| $\circ$ | morphological opening |
| $\bullet$ | morphological closing |
| $[a, b]$ | interval $a$ to $b$ inclusive |
| $(a, b)$ | interval $a$ to $b$ exclusive, not including $a$ or $b$ |
| $[a, b)$ | interval $a$ to $b$, not including $b$ |
| $(a, b]$ | interval $a$ to $b$, not including $a$ |

## Toolbox Conventions

- A Cartesian point, a coordinate vector, is expressed as a 1D NumPy array.
- A set of points is expressed as a 2D NumPy array, with columns representing the coordinate vectors of individual points.
- A robot configuration, a set of joint angles, is expressed as a 1D NumPy array.
- Time series data is expressed as a 2D NumPy array with rows representing time steps.
- A 2D NumPy array has subscripts $[i, j]$ which represent row and column indices respectively. Image coordinates are written $(u, v)$ so an image represented by a NumPy array `I` is indexed as `I[v,u]`.

## Zero-Based Indexing

In Python, lists and arrays are indexed starting from zero. As much as possible, algorithms in this book have been adjusted to follow suit, for example, the joints of a robot manipulator are numbered from zero whereas in other references they are numbered from one. Whenever the book refers to something by number, for example joint one or column two, it implies zero-based indexing. The English words first, second, and third have their normal meaning and apply to indexes zero, one and two. For example: the first column of a matrix is also referred to as column zero; the third joint of a robot is also referred to as joint two. The word zeroth is never used.

## Common Abbreviations

1D  –  1-dimensional

2D  –  2-dimensional

3D  –  3-dimensional

CoM  –  Center of mass

DoF  –  Degrees of freedom

$n$-tuple  –  A group of $n$ numbers, it can represent a point or a vector

# Introduction

**Contents**

© The Author(s), under exclusive license to Springer Nature Switzerland AG 2023
P. Corke, *Robotics, Vision and Control*, Springer Tracts in Advanced Robotics 146,
https://doi.org/10.1007/978-3-031-06469-2_1

## 1.1  A Brief History of Robots

The word *robot* means different things to different people. Science fiction books and movies have strongly influenced what many people expect a robot to be or what it can do – sadly the practice of robotics is far behind this popular conception. The word *robot* is also emotive and some people are genuinely fearful of a future with robots, while others are concerned about what robots mean for jobs, privacy, or warfare. One thing is certain though – robotics will be an important technology in this century. Products such as vacuum cleaning robots have already been with us for over two decades, and self-driving cars are now on the roads with us. These are the vanguard of a wave of smart machines that will appear in our homes and workplaces in the near to medium future.

In the eighteenth century, Europeans were fascinated by automata such as Vaucanson's duck shown in ◻ Fig. 1.1a. These machines, complex by the standards of the day, demonstrated what then seemed *life-like* behavior. The duck used a cam mechanism to sequence its movements and Vaucanson went on to explore the mechanization of silk weaving. Jacquard extended these ideas and developed a loom, shown in ◻ Fig. 1.1b, that was essentially a programmable weaving machine. The pattern to be woven was encoded as a series of holes on punched cards. ◄ This machine has many hallmarks of a modern robot: it performed a physical task and was reprogrammable.

The term *robot* first appeared in a 1920 Czech science fiction play "Rossum's Universal Robots" by Karel Čapek (pronounced chapek). The word *roboti* was suggested by his brother Josef, and in the Czech language has the same linguistic roots

This in turn influenced Sir Charles Babbage in his quest to mechanize computation, which in turn influenced Countess Ada Lovelace to formalize computation and create the first algorithm.

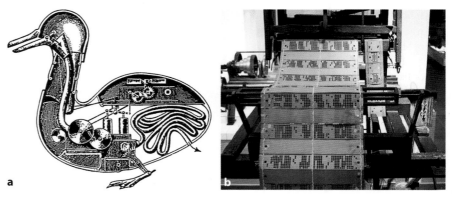

◻ **Fig. 1.1**  Early programmable machines. **a** Vaucanson's duck (1739) was an automaton that could flap its wings, eat grain and defecate. It was driven by a clockwork mechanism and executed a single program; **b** The Jacquard loom (1801) was a reprogrammable machine and the program was held on punched cards (image by George P. Landow from ► https://www.victorianweb.org)

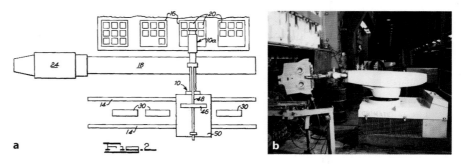

◻ **Fig. 1.2**  Universal automation. **a** A plan view of the machine from Devol's patent; **b** the first Unimation robot – the Unimate – working at a General Motors factory (image courtesy of George C. Devol)

**Fig. 1.3** Manufacturing robots, technological descendants of the Unimate shown in Fig. 1.2. **a** A modern six-axis robot designed for high accuracy and throughput (image courtesy ABB robotics); **b** Universal Robotics collaborative robot can work safely with a human co-worker (© 2022 Universal Robots A/S. All Rights Reserved. This image has been used with permission from Universal Robots A/S)

---

**Excurse 1.1: George C. Devol**

Devol (1912–2011) was a prolific American inventor, born in Louisville, Kentucky. In 1932, he founded United Cinephone Corp. which manufactured phonograph arms and amplifiers, registration controls for printing presses and packaging machines. In 1954, he applied for US patent 2,988,237 for Programmed Article Transfer which introduced the concept of Universal Automation or "Unimation". Specifically it described a track-mounted polar-coordinate arm mechanism with a gripper and a programmable controller – the precursor of all modern robots.

In 2011 he was inducted into the National Inventors Hall of Fame. (image courtesy of George C. Devol)

---

as servitude and slavery. The robots in the play were artificial people or androids and, as in so many robot stories that follow this one, the robots rebel and it ends badly for humanity. Isaac Asimov's robot series, comprising many books and short stories written between 1950 and 1985, explored issues of human and robot interaction, and morality. The robots in these stories are equipped with "positronic

---

**Excurse 1.2: Unimation Inc.**

Devol sought financing to develop his unimation technology and at a cocktail party in 1954 he met Joseph Engelberger who was then an engineer with Manning, Maxwell and Moore. In 1956 they jointly established Unimation (1956–1982), the first robotics company, in Danbury Connecticut. The company was acquired by Consolidated Diesel Corp. (Condec) and became Unimate Inc. a division of Condec. Their first robot went to work in 1961 at a General Motors die-casting plant in New Jersey. In 1968 they licensed technology to Kawasaki Heavy Industries which produced the first Japanese industrial robot. Engelberger served as chief executive until it was acquired by Westinghouse in 1982. People and technologies from this company have gone on to be very influential across the whole field of robotics.

**1**

brains" in which the "Three Laws of Robotics" are encoded. These stories have influenced subsequent books and movies, which in turn have shaped the public perception of what robots are. The mid-twentieth century also saw the advent of the field of *cybernetics* – an uncommon term today but then an exciting science at the frontiers of understanding life and creating intelligent machines.

The first patent for what we would now consider a robot manipulator was filed in 1954 by George C. Devol and issued in 1961. The device comprised a mechanical arm with a gripper that was mounted on a track and the sequence of motions was encoded as magnetic patterns stored on a rotating drum. The first robotics company, Unimation, was founded by Devol and Joseph Engelberger in 1956 and their first industrial robot, shown in ◲ Fig. 1.2b, was installed in 1961. The original vision of Devol and Engelberger for robotic automation has subsequently become a reality. Many millions of robot manipulators, such as shown in ◲ Fig. 1.3, have been built and put to work at tasks such as welding, painting, machine loading and unloading, assembly, sorting, packaging, and palletizing. The use of robots has led to increased productivity and improved product quality. Today, many products that we buy have been assembled or handled by a robot.

## 1.2   Types of Robots

The *first generation* of robots were fixed in place and could not move about the factory. By contrast, *mobile robots* shown in ◲ Figs. 1.4 and 1.5 can move through the world using various forms of mobility. They can locomote over the ground using wheels or legs, fly through the air using fixed wings or multiple rotors, move through the water or sail over it.

An alternative taxonomy is based on the function that the robot performs. *Manufacturing* robots operate in factories and are the technological descendants of the first generation robots created by Unimation Inc. *Service robots* provide services to people such as cleaning, personal care, medical rehabilitation or fetching and carrying as shown in ◲ Fig. 1.5b. *Field robots*, such as those shown in ◲ Fig. 1.4,

◼ **Fig. 1.4** Non-land-based mobile robots. **a** Small autonomous underwater vehicle (Todd Walsh © 2013 MBARI); **b** Global Hawk uncrewed aerial vehicle (UAV) (image courtesy of NASA)

◼ **Fig. 1.5** Mobile robots. **a** Perseverance rover on Mars, self portrait. The mast contains many cameras including stereo camera pairs from which the robot can compute the 3-dimensional structure of its environment (image courtesy of NASA/JPL-Caltech/MSSS); **b** Savioke Relay delivery robot (image courtesy Savioke); **c** self driving car (image courtesy Dept. Information Engineering, Oxford Univ.); **d** Boston Dynamics Spot legged robot (image courtesy Dorian Tsai)

work outdoors on tasks such as environmental monitoring, agriculture, mining, construction and forestry. *Humanoid robots* such as shown in ◼ Fig. 1.6 have the physical form of a human being – they are both mobile robots and service robots.

**1**

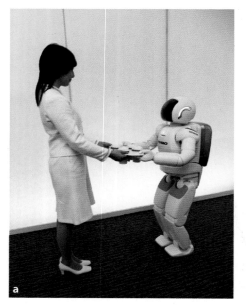

☐ **Fig. 1.6** Humanoid robots. **a** Honda's Asimo humanoid robot (image courtesy Honda Motor Co. Japan); **b** Hubo robot that won the DARPA Robotics Challenge in 2015 (image courtesy KAIST, Korea)

A *manufacturing robot* is typically an arm-type manipulator on a fixed base, such as shown in ☐ Fig. 1.3a, that performs repetitive tasks within a local work cell. Parts are presented to the robot in an orderly fashion which maximizes the advantage of the robot's high speed and precision. High-speed robots are hazardous and safety is achieved by excluding people from robotic work places – typically the robot is placed inside a cage. In contrast, collaborative robots such as shown in ☐ Fig. 1.3b, are human safe – they operate at low speed and stop moving when they encounter an obstruction.

**Excurse 1.5: Telerobots**

The Manhattan Project in World War 2 developed the first nuclear weapons and this required handling of radioactive material. Remotely controlled arms were developed by Ray Goertz at Argonne National Laboratory to exploit the manual dexterity of human operators, while keeping them away from the hazards of the material they were handling. The operators viewed the work space through thick lead-glass windows or via a television link and manipulated the leader arm (on the left). The follower arm (on the right) followed the motion, and forces felt by the follower arm were reflected back to the leader arm, allowing the operator to feel weight and interference force. Telerobotics is still important today for many tasks where people cannot work but which are too complex for a machine to perform by itself, for instance the underwater robots that surveyed the wreck of the Titanic. (image courtesy of Argonne National Laboratory)

*Field and service robots* face specific and significant challenges. The first challenge is that the robot must operate and move in a complex, cluttered and changing environment. For example, a delivery robot in a hospital must operate despite crowds of people and a time-varying configuration of parked carts and trolleys. A Mars rover, as shown in ◻ Fig. 1.5a, must navigate rocks and small craters despite not having an accurate local map in advance of its travel. Robotic, or self-driving cars, such as shown in ◻ Fig. 1.5c, must follow roads, avoid obstacles, and obey traffic signals and the rules of the road. The second challenge for these types of robots is that they must operate safely in the presence of people. The hospital delivery robot operates amongst people, the robotic car contains people, and a robotic surgical device operates *inside* people.

*Telerobots* are robot-like machines that are remotely controlled by a human operator. Perhaps the earliest example was a radio-controlled boat demonstrated by Nikola Tesla in 1898 and which he called a teleautomaton. Such machines were an important precursor to robots, and are still important today for tasks conducted in environments where people cannot work, but which are too complex for a machine to perform by itself. For example the "underwater robots" that surveyed the wreck of the Titanic were technically remotely operated vehicles (ROVs). A modern surgical robot as shown in ◻ Fig. 1.7 is also teleoperated – the motion of the small tools inside the patient are remotely controlled by the surgeon. The patient benefits because the procedure is carried out using a much smaller incision than the old-fashioned approach where the surgeon works inside the body with their hands.

The various Mars rovers autonomously navigate the surface of Mars but human operators provide the high-level goals. That is, the operators tell the robot where to go and the robot itself determines the details of the route. Local decision making on Mars is essential given that the communications delay is several minutes. Some telerobots are hybrids, and the control task is shared or traded with a human operator. In traded control, the control function is passed back and forth between the human operator and the computer. For example an aircraft pilot can pass control to an autopilot and take control back. In shared control, the control function is performed by the human operator and the computer working together. For example an autonomous passenger car might have the computer keeping the car safely in the lane while the human driver controls speed.

**1**

### Excurse 1.6: Cybernetics, Artificial Intelligence and Robotics

Cybernetics flourished as a research field from the 1930s until the 1960s and was fueled by a heady mix of new ideas and results from neurology, control theory, and information theory. Research in neurology had shown that the brain was an electrical network of neurons. Harold Black, Henrik Bode and Harry Nyquist at Bell Labs were researching negative feedback and the stability of electrical networks, Claude Shannon's information theory described digital signals, and Alan Turing was exploring the fundamentals of computation. Walter Pitts and Warren McCulloch proposed an artificial neuron in 1943 and showed how it might perform simple logical functions. In 1951 Marvin Minsky built SNARC (from a B24 autopilot and comprising 3000 vacuum tubes) which was perhaps the first neural-network-based learning machine as his graduate project. William Grey Walter's robotic tortoises showed life-like behavior (see ▶ Sect. 5.1). Maybe an electronic brain could be built!

An important early book was Norbert Wiener's *Cybernetics or Control and Communication in the Animal and the Machine* (Wiener 1965). A characteristic of a cybernetic system is the use of feedback which is common in engineering and biological systems. The ideas were later applied to evolutionary biology, psychology and economics.

In 1956 a watershed conference was hosted by John McCarthy at Dartmouth College and attended by Minsky, Shannon, Herbert Simon, Allen Newell and others. This meeting defined the term artificial intelligence (AI) as we know it today, with an emphasis on digital computers and symbolic manipulation and led to new research in robotics, vision, natural language, semantics and reasoning. McCarthy and Minsky formed the AI group at MIT, and McCarthy left in 1962 to form the Stanford AI Laboratory. Minsky focused on artificially simple "blocks world". Simon, and his student Newell, were influential in AI research at Carnegie-Mellon University from which the Robotics Institute was spawned in 1979. These AI groups were to be very influential in the development of robotics and computer vision in the USA. Societies and publications focusing on cybernetics are still active today.

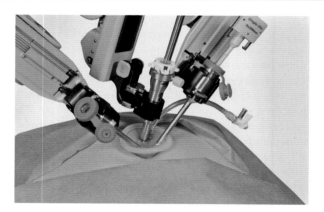

**◻ Fig. 1.7**  The working end of a surgical robot, multiple tools work inside the patient but pass through a single small incision (image © 2015 Intuitive Surgical, Inc)

## 1.3    Definition of a Robot

>> *I can't define a robot, but I know one when I see one*
    – Joseph Engelberger

So what is a robot? There are many definitions and not all of them are particularly helpful. A definition that will serve us well in this book is

>> *a goal oriented machine that can sense, plan, and act.*

A robot *senses* its environment and uses that information, together with a goal, to *plan* some *action*. The action might be to move the tool of a robot manipulator arm to grasp an object, or it might be to drive a mobile robot to some place.

Sensing is critical to robots. *Proprioceptive* sensors measure the state of the robot itself: the angle of the joints on a robot arm, the number of wheel revolutions on a mobile robot, or the current drawn by an electric motor. *Exteroceptive* sensors measure the state of the world with respect to the robot. The sensor might be a simple bump sensor on a robot vacuum cleaner to detect collision. It might be a GPS receiver that measures distances to an orbiting satellite constellation, or a compass that measures the direction of the Earth's magnetic field vector relative to the robot. It might also be an active sensor that emits acoustic, optical or radio pulses in order to measure the distance to points in the world based on the time taken for a reflection to return to the sensor.

## 1.4  Robotic Vision

Another way to sense the world is to capture and interpret patterns of ambient light reflected from the scene. This is what our eyes and brain do, giving us the sense of vision. Our own experience is that vision is a very effective sensor for most things that we do, including recognition, navigation, obstacle avoidance and manipulation. We are not alone in this, and almost all animal species use eyes — in fact evolution has *invented* the eye many times over. ◘ Fig. 1.8 shows some of the diversity of eyes found in nature.

It is interesting to note that even very simple animals, bees for example, with brains comprising just $10^6$ neurons (compared to our $10^{11}$) are able to perform complex and life-critical tasks such as finding food and returning it to the hive using only vision. For more complex animals such as ourselves, the benefits of vision outweigh the very high biological cost of owning an eye: the complex eye itself, muscles to move it, eyelids and tear ducts to protect it, and a large visual cortex (one third of our brain) to process the data it produces.

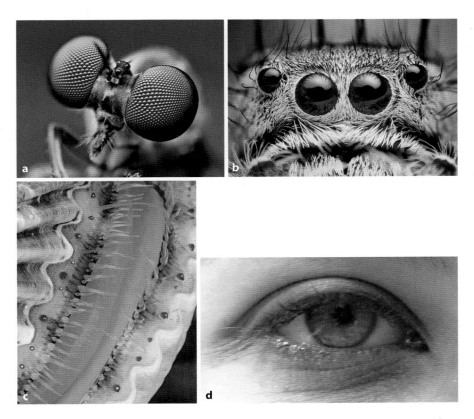

◘ **Fig. 1.8**  The diversity of eyes. **a** Robber fly, Holocephala fusca; **b** jumping spider, Phidippus putnami (images **a** and **b** courtesy Thomas Shahan,
► https://thomasshahan.com). **c** scallop (image courtesy of Sönke Johnsen), each of the small blue spheres is an eye; **d** human eye

**◻ Fig. 1.9**  A cluster of cameras on an outdoor mobile robot: forward looking stereo pair, side looking wide-angle camera, overhead panoramic camera mirror (image courtesy of CSIRO)

The sense of vision has long been of interest to robotics researchers – cameras can mimic the function of an eye and create images, and computer vision algorithms can extract meaning from the images. Combined, they could create powerful vision-based competencies for robots such as recognizing and manipulating objects and navigating within the world. For example, a soccer playing robot could determine the coordinate of a round red object in the scene, while a drone could estimate its motion in the world based on how the world appears to move relative to the drone. ◻ Fig. 1.9 shows a robot with a number of different types of cameras to enable outdoor navigation.

Technological developments have made it increasingly feasible for robots to use cameras as eyes and computers as brains. For much of the history of computer vision, dating back to the 1960s, electronic cameras were cumbersome and expensive, and computer power was inadequate. Today we have CMOS cameras developed for cell phones that cost just a few dollars each, and personal computers come standard with massive parallel computing power. New algorithms, cheap sensors, and plentiful computing power make vision a practical sensor today, and that is a strong focus of this book.

An important limitation of a camera, or an eye, is that the 3-dimensional structure of the scene is lost in the resulting 2-dimensional image. Despite this, humans are particularly good at inferring the 3-dimensional nature of a scene using a number of visual cues. One approach, used by humans and robots, is stereo vision where information from two eyes is combined to estimate the 3-dimensional structure of the scene. The Mars rover shown in ◻ Fig. 1.5a has a stereo camera on its mast, and the robot in ◻ Fig. 1.9 has a stereo camera on its turret.

If the robot's environment is unchanging it could make do with an accurate map and do away with sensing the state of the world, apart from determining its own position. Imagine driving a car with the front window completely covered over and just looking at the GPS navigation system. If you had the road to yourself, you could probably drive successfully from A to B although it might be quite stressful. However, if there were other cars, pedestrians, traffic signals or roadworks this approach would not work. To deal with this realistic scenario you need to look outwards – to sense the world and plan your actions accordingly. For humans this is easy, done without conscious thought, but it is not yet easy to program a machine to do the same – this is the challenge of *robotic vision*.

## 1.5 Ethical Considerations

A number of ethical issues arise from the advent of robotics. Perhaps the greatest concern to the wider public is "robots taking jobs from people". Already today, artificial intelligence systems are encroaching on many information handling tasks including image analysis, decision making, writing sports and finance reports, and credit assessment. People fear that robots will soon encroach on physical tasks, but currently the skill of robots for everyday tasks is poor, reliability and speed is low and the cost is high. However, it is highly likely that, over time, these challenges will be overcome – we cannot shy away from the fact that many jobs now done by people could, in the future, be performed by robots.

The issue of robots and jobs, even today, is complex. Clearly there are jobs which people should not do, for example working in unhealthy or hazardous environments. There are many low-skilled jobs where human labor is increasingly hard to source, for instance in jobs like fruit picking. In many developed countries people no longer aspire to hard physical outdoor work in remote locations. What are the alternatives if people don't want to do the work? Consider again the fruit picking example, and in the absence of available human labor – do we stop eating fruit? do we dramatically raise the wages of fruit pickers (and increase the cost of fruit)? do we import fruit from other places (and increase the cost of fruit as well as its environmental footprint)? or do we use robots to pick locally grown fruit?

In areas like manufacturing, particularly car manufacturing, the adoption of robotic automation has been critical in raising productivity which has allowed that industry to be economically viable in high-wage countries like Europe, Japan and the USA. Without robots, these industries could not exist; they would not employ any people, not pay any taxes, and not consume products and services from other parts of the economy. Automated industry might employ fewer people, but it still

1

makes an important contribution to society. Rather than taking jobs, we could argue that robotics and automation has helped to keep manufacturing industries viable in high-labor cost countries. How do we balance the good of the society with the good of the individual?

There are other issues besides jobs. Consider self-driving cars. We are surprisingly accepting of human-driven cars even though they kill more than one million people every year, yet many are uncomfortable with the idea of self-driving cars even though they will dramatically reduce this loss of life. We worry about who to blame if a robotic car makes a mistake while the carnage caused by human drivers continues. Similar concerns are raised when talking about robotic healthcare and surgery – human surgeons are not perfect but robots are seemingly held to a much higher account. There is a lot of talk about using robots to look after elderly people, but does this detract from their quality of life by removing human contact, conversation and companionship? Should we use robots to look after our children, and even teach them? What do we think of armies of robots fighting and killing human beings?

Robotic fruit picking, cars, health care, elder care and child care might bring economic benefits to our society but is it the right thing to do? Is it a direction that we want our society to go? Once again, how do we balance the good of the society with the good of the individual? These are deep ethical questions that cannot and should not be decided by roboticists alone. But neither should roboticists ignore them. We must not sleepwalk into a technological future just "because we can". It is our responsibility to help shape the future in a deliberate way and ensure that it is good for all. This is an important discussion for all of society, and roboticists have a responsibility to be active participants in this debate.

## 1.6   About the Book

This is a book about robotics and computer vision – separately, and together as robotic vision. These are big topics and the combined coverage is necessarily broad. The goals of the book are:

- to provide a broad and solid base of understanding through theory and the use of examples to make abstract concepts tangible;
- to tackle more complex problems than other more specialized textbooks by virtue of the powerful numerical tools and software that underpins it;
- to provide instant gratification by solving complex problems with relatively little code;
- to complement the many excellent texts in robotics and computer vision;
- to encourage intuition through hands on numerical experimentation.

The approach used is to present background, theory, and examples in an integrated fashion. Code and examples are first-class citizens in this book and are not relegated to the end of the chapter or an associated web site. The examples are woven into the discussion like this

```
>>> R = rot2(0.3)
array([[ 0.9553, -0.2955],
       [ 0.2955,  0.9553]])
```

where the code illuminates the topic being discussed and generally results in a crisp numerical result or a graph in a figure that is then discussed. The examples illustrate how to use the software tools and that knowledge can then be applied to other problems.

### 1.6.1 Python and the Book

> ❯❯ *To do good work, one must first have good tools.*
> – Chinese proverb

The computational foundation of this book is Python 3, a powerful and popular programming language that is likely to be familiar to students, researchers and hobbyists. The core functionality of Python can be extended with packages that can be downloaded, installed and then imported. Python, combined with popular packages such as NumPy, SciPy and Matplotlib ▶ provides a powerful interactive mathematical software environment that makes linear algebra, data analysis and high-quality graphics a breeze. Functionality for robotics and computer vision is provided by additional easily-installable packages:

*Respectively import* `numpy`, `scipy` *and* `matplotlib.pyplot`.

- The *Spatial Maths Toolbox for Python* includes functions for manipulating and converting between datatypes such as vectors, rotation matrices, homogeneous transformations, 3-angle representations, twists and unit quaternions which are necessary to represent 3-dimensional position, orientation and pose.
- The *Robotics Toolbox for Python* provides a diverse range of functions for simulating mobile and arm-type robots. The Toolbox supports a very general method of representing the structure of serial-link manipulators and provides functions for forward and inverse kinematics, differential kinematics, dynamics, visualization and animation. The Toolbox also includes functionality for simulating mobile robots and includes models of wheeled vehicles and quadrotors and controllers for these vehicles. It also provides standard algorithms for robot path planning, localization, map making and SLAM.
- The *Machine Vision Toolbox for Python* provides a rich collection of functions for camera modeling, image processing, image feature extraction, multi-view geometry and vision-based control. The Toolbox also contains functions for image acquisition and display; filtering; blob, point and line feature extraction; mathematical morphology; image warping; stereo vision; homography and fundamental matrix estimation; robust estimation; bundle adjustment; visual Jacobians; geometric camera models; camera calibration and color space operations. Many of the underlying operations are implemented using the popular, efficient and mature OpenCV package.
- The *Block Diagram Simulator* (bdsim) allows block diagrams to be modeled and simulated using Python. It provides over 70 block types, allows the signals on wires between blocks to be any valid Python datatype, and supports subsystems. Complex models can be expressed using concise Python code, but a graphical editor is also included.

The first three packages are reimplementations of the venerable, similarly named, Toolboxes for MATLAB®.

If you're starting out in robotics or vision then the Toolboxes are a significant initial base of code on which to build your project. The Toolboxes mentioned above are provided in source code form. In general the Toolbox code is written in a straightforward manner to facilitate understanding, perhaps at the expense of computational efficiency. ▶ App. A provides details about how to install the Toolboxes, and how to obtain other book resources such as examples and figures.

This book provides examples of how to use many Toolbox functions in the context of solving specific problems but it is not a reference manual. Comprehensive documentation of all Toolbox functions is available through the online documentation generated from the code itself.

**1**

> **Excurse 1.8: Python**
>
> Python is an interpreted high-level programming language that is consistently ranked as one of the most popular programming languages. Python was developed by Guido van Rossum, initially as a Christmas hobby coding project in 1989. It is named after the British TV series Monty Python's Flying Circus, and the documentation has many other similarly inspired references to spam and cheese. The first version was released in 1991, Python 2 followed in 2000 (and was discontinued in 2020 with version 2.7.18), and Python 3 in 2008. Python inherits ideas from many previous programming languages, and supports procedural, object-oriented and functional programming paradigms. It has a clear and simple syntax, dynamic typing, and garbage collection.
>
> A key feature of Python is that it is highly extensible using packages. The main repository for third-party Python software, the Python Package Index (PyPI) at ▶ https://pypi.org, lists over 300,000 packages that can be simply installed using the `pip` command.
>
> Guido van Rossum (1956–) is a Dutch mathematician and computer scientist. He worked at the Centrum Wiskunde & Informatica (CWI) on the ABC programming language which was a strong influence on Python. Until 2018 he was Python's "benevolent dictator for life", and is now a member of the Steering Council.

### 1.6.2 Notation, Conventions and Organization

The mathematical notation used in the book is summarized in the ▶ Nomenclature section. Since the coverage of the book is broad there are just not enough good symbols to go around, and unavoidably some symbols have different meanings in different parts of the book.

There is a lot of Python code in the book and this is indicated in black fixed-width font such as

```
>>> a = 6 * 7
a =
42
```

The Python command prompt is >>> and what follows is the command issued to Python by the user. Subsequent lines, without the prompt, are Python's response. Long commands require continuation lines which begin with an ellipsis ( . . . ). ◄ All functions, classes and methods mentioned in the text or in code segments are cross-referenced and have their own index at the end of the book, allowing you to find different ways that particular functions can be used.

Various boxes are used to organize and differentiate parts of the content.

Code blocks with the prompt characters can be copy and pasted into IPython and will be interpreted correctly.

> These highlight boxes indicate content that is particularly important.

❶ These warning boxes highlight points that are often traps for those starting out.

They are placed as marginal notes near the corresponding marker.

As an author there is a tension between completeness, clarity and conciseness. For this reason a lot of detail has been pushed into side notes ◄ and information boxes, and on a first reading these can be skipped. Some chapters have an *Advanced Topics* section at the end that can also be skipped on a first reading. For clarity, references are cited sparingly in the text but are included at the end of each chapter. However if you are trying to understand a particular algorithm and apply it to your

> **Excurse 1.9: Information**
>
> These information boxes provide technical, historical or biographical information that augment the main text, but which is not critical to its understanding.

own problem, then understanding the details and nuances can be important and the sidenotes, references and *Advanced Topics* will be helpful.

Each chapter ends with a *Wrapping Up* section that summarizes the important lessons from the chapter, discusses some suggested further reading, and provides some exercises. The *Further Reading* subsection discusses prior work and provides extensive references to prior work and more complete description of the algorithms. *Resources* provides links to relevant online code and datasets. *Exercises* extend the concepts discussed within the chapter and are generally related to specific code examples discussed in the chapter. The exercises vary in difficulty from straightforward extension of the code examples to more challenging problems.

Many of the code examples include a plot command which generates a figure. The figure that is included in the book is generally embellished with axis labels, grid, legends and adjustments to line thickness. The embellishment requires many extra lines of code which would clutter the examples, so the GitHub repository provides the detailed Python scripts that generate all of the published figures.

Finally, the book contains links to online resources which can be simply accessed through a web browser. These have short URLs (clickable in the e-book version) and QR codes, which are placed in the margin. Each chapter has a link to a Jupyter notebook on Google CoLaboratory™ (Colab), allowing you to run that chapter's code with zero installation on your own computer, or even run it on a tablet. ▶ Some figures depict three-dimensional scenes and the included two-dimensional image do not do them justice – the links connect to web-based interactive 3D viewers of those scenes.

`chap2.ipynb`

▶ go.sn.pub/ntdGFo

CoLaboratory is a trademark of Google LLC and this book is not endorsed by or affiliated with Google in any way. Some animations and 3D visualizations are not (yet) runnable in CoLaboratory.

### 1.6.3  Audience and Prerequisites

The book is intended primarily for third or fourth year engineering undergraduate students, Masters students and first year Ph.D. students. For undergraduates the book will serve as a companion text for a robotics, mechatronics, or computer vision course or support a major project in robotics or vision. Students should study Part I and the appendices for foundational concepts, and then the relevant part of the book: mobile robotics, arm robots, computer vision or vision-based control. The Toolboxes provide a solid set of tools for problem solving, and the exercises at the end of each chapter provide additional problems beyond the worked examples in the book.

For students commencing graduate study in robotics, and who have previously studied engineering or computer science, the book will help fill the gaps between what you learned as an undergraduate, and what will be required to underpin your deeper study of robotics and computer vision. The book's working code base can help bootstrap your research, enable you to get started quickly, and work productively on your own problems and ideas. Since the source code is available you can reshape it to suit your need, and when the time comes (as it usually does) to code your algorithms in some other language then the Toolboxes can be used to cross-check your implementation.

For those who are no longer students, the researcher or industry practitioner, the book will serve as a useful companion for your own reference to a wide range

**1**

of topics in robotics and computer vision, as well as a handbook and guide for the Toolboxes.

The book assumes undergraduate-level knowledge of linear algebra (matrices, vectors, eigenvalues), basic set theory, basic graph theory, probability, calculus, dynamics (forces, torques, inertia) and control theory. The appendices provide a concise reiteration of key concepts. Computer science students may be unfamiliar with concepts in ▶ Chaps. 4 and 9 such as the Laplace transform, transfer functions, linear control (proportional control, proportional-derivative control, proportional-integral control) and block diagram notation. That material could be skimmed over on a first reading, and Albertos and Mareels (2010) may be a useful introduction to some of these topics. The book also assumes the reader is familiar with using, and programming in, Python and also familiar with object-oriented programming techniques.

### 1.6.4  Learning with the Book

The best way to learn is by doing. Although the book shows the Python commands and the response, there is something special about doing it for yourself. Consider the book as an invitation to tinker. By running the code examples yourself you can look at the results in ways that you prefer, plot the results in a different way, or try the algorithm on different data or with different parameters. The paper edition of the book is especially designed to stay open which enables you to type in commands as you read. You can also look at the online documentation for the Toolbox functions, discover additional features and options, and experiment with those, or read the code to see how it really works and perhaps modify it.

Most of the code examples are quite short so typing them in to Python is not too onerous, however there are a lot of them – more than 1600 code examples. The code for each chapter is available as a Jupyter notebook, see ▶ App. A, which allows the code to be executed cell by cell, or it can be copy and pasted into your own scripts.

A command line tool called `rvctool` is also provided. This is a wrapper for IPython which preloads all the toolboxes using `import *` into the current namespace. While this is perhaps not best Python practice, it does allow you to run examples exactly as they are written in the book.

The Robot Academy (▶ https://robotacademy.net.au) is a free online resource with over 200 short video lessons that cover much, but not all, of the content of this book. Many lessons use code to illustrate concepts, and these examples use MATLAB ◀ rather than Python. Similarly named functions do exist in the Python versions of the Toolboxes.

And version 9 of the open-source Robotics Toolbox for MATLAB.

### 1.6.5  Teaching with the Book

This book can be used to support courses in robotics, mechatronics and computer vision. All courses should include the introduction to position, orientation, pose, pose composition, and coordinate frames that is discussed in ▶ Chap. 2. For a mobile robotics or computer vision course it is sufficient to teach only the 2-dimensional case. For robotic manipulators or multi-view geometry the 2D and 3-dimensional cases should be taught.

All code and most figures in this book are available from the book's web site – you are free to use them with attribution. Code examples in this book are available as Jupyter notebooks and could be used as the basis for demonstrations in lectures or tutorials. Line drawings are provided as PDF files and Python-generated figures are provided as scripts that will recreate the figures. See ▶ App. A for details.

The exercises at the end of each chapter can be used as the basis of assignments, or as examples to be worked in class or in tutorials. Most of the questions are rather open ended in order to encourage exploration and discovery of the effects of parameters and the limits of performance of algorithms. This exploration should be supported by discussion and debate about performance measures and what *best* means. True understanding of algorithms involves an appreciation of the effects of parameters, how algorithms fail and under what circumstances.

The teaching approach could also be inverted, by diving head first into a particular problem and then teaching the appropriate prerequisite material. Suitable problems could be chosen from the Application sections of chapters (see *applications* in the main index), or from any of the exercises. Particularly challenging exercises are so marked.

If you wanted to consider a flipped learning approach then the Robot Academy (▶ https://robotacademy.net.au) could be used in conjunction with your class. Students would watch the video lessons and undertake some formative assessment out of the classroom, and you could use classroom time to work through problem sets.

For graduate-level teaching the papers and textbooks mentioned in the *Further Reading* could form the basis of a student's reading list. They could also serve as candidate papers for a reading group or journal club.

### 1.6.6 Outline

I promised a book with instant gratification but before we can get started in robotics there are some fundamental concepts that we absolutely need to understand, and understand well. Part I introduces the concepts of pose and coordinate frames – how we represent the position and orientation of a robot, a camera or the objects that the robot needs to work with. We discuss how motion between two poses can be *decomposed* into a sequence of elementary translations and rotations, and how elementary motions can be *composed* into more complex motions. ▶ Chap. 2 discusses how pose can be represented in a computer, and ▶ Chap. 3 discusses the relationship between velocity and the derivative of pose, generating a sequence of poses that smoothly follow some path in space and time, and estimating motion from sensors.

With these formalities out of the way we move on to the first main event – robots. There are two important classes of robot: mobile robots and manipulator arms and these are covered in Parts II and III respectively. ▶

Part II begins, in ▶ Chap. 4, with motion models for several types of wheeled vehicles and a multi-rotor flying vehicle. Various control laws are discussed for wheeled vehicles such as moving to a point, following a path and moving to a specific pose. ▶ Chap. 5 is concerned with navigation, that is, how a robot finds a path between points A and B in the world. Two important cases, with and without a map, are discussed. Most navigation techniques require knowledge of the robot's position and ▶ Chap. 6 discusses various approaches to this problem based on dead-reckoning, or landmark observation and a map. We also show how a robot can make a map, and even determine its location while simultaneously mapping an unknown region – the SLAM problem.

Part III is concerned with arm-type robots, or more precisely serial-link manipulators. Manipulator arms are used for tasks such as assembly, welding, material handling and even surgery. ▶ Chap. 7 introduces the topic of kinematics which relates the angles of the robot's joints to the 3-dimensional pose of the robot's tool. Techniques to generate smooth paths for the tool are discussed and two examples show how an arm-robot can draw a letter on a surface, and how multiple arms (acting as legs) can be used to create a model for a simple walking robot. ▶ Chap. 8 discusses the relationships between the rates of change of joint angles

Although robot arms came first chronologically, mobile robotics is mostly a 2-dimensional problem and easier to understand than the 3-dimensional arm-robot case.

1

and tool pose. It introduces the Jacobian matrix and concepts such as singularities, manipulability, null-space motion, and resolved-rate motion control. It also discusses under- and overactuated robots and the general numerical solution to inverse kinematics. ▶ Chap. 9 introduces the design of joint control systems, the dynamic equations of motion for a serial-link manipulator, and the relationship between joint forces and joint motion. It discusses important topics such as variation in inertia, the effect of payload, flexible transmissions, independent-joint versus nonlinear control strategies, and task-space control.

Computer vision is a large field concerned with processing images in order to enhance them for human benefit, interpret the contents of the scene or create a 3D model corresponding to the scene. Part IV is concerned with machine vision, a subset of computer vision, and defined here as the extraction of numerical features from images to provide input for control of a robot. The discussion starts in ▶ Chap. 10 with the fundamentals of light, illumination and color. ▶ Chap. 11 discusses *image processing* which is a domain of 2-dimensional signal processing that transforms one image into another image. The discussion starts with acquiring real-world images and then covers various arithmetic and logical operations that can be performed on images. We then introduce spatial operators such as convolution, segmentation, morphological filtering and finally image shape and size changing. These operations underpin the discussion in ▶ Chap. 12 which describe how numerical features are extracted from images. The features describe homogeneous regions (blobs), lines or distinct points in the scene and are the basis for vision-based robot control. The application of deep-learning approaches for object detection is also introduced. ▶ Chap. 13 describes the geometric model of perspective image creation using lenses and discusses topics such as camera calibration and pose estimation. We introduce nonperspective imaging using wide-angle lenses and mirror systems, camera arrays and light-field cameras. ▶ Chap. 14 is concerned with estimating the underlying three-dimensional geometry of a scene using classical methods such as structured lighting and also combining features found in different views of the same scene to provide information about the geometry and the spatial relationship between the camera views which is encoded in fundamental, essential and homography matrices. This leads to the topic of bundle adjustment and structure from motion, and applications including perspective correction, mosaicing, image retrieval and visual odometry.

Part V discusses how visual features extracted from the camera's view can be used to control arm-type and mobile robots – an approach known as vision-based control or visual servoing. This part pulls together concepts introduced in the earlier parts of the book. ▶ Chap. 15 introduces the classical approaches to visual servoing known as position-based and image-based visual servoing and discusses their respective limitations. ▶ Chap. 16 discusses more recent approaches that address these limitations and also covers the use of nonperspective cameras, underactuated robots and mobile robots.

This is a big book but any one of the parts can be read standalone, with more or less frequent visits to the required earlier material. ▶ Chap. 2 is the only essential reading. Parts II, III or IV could be used respectively for an introduction to mobile robots, arm robots or computer vision class. An alternative approach, following the instant gratification theme, is to jump straight into any chapter and start exploring – visiting the earlier material as required.

### 1.6.7  Further Reading

The Handbook of Robotics (Siciliano and Khatib 2016) provides encyclopedic coverage of the field of robotics today, covering theory, technology and the different types of robot such as telerobots, service robots, field robots, aerial robots, underwater robots and so on. The classic work by Sheridan (2003) discusses the spectrum of autonomy from remote control, through shared and traded control, to full autonomy.

A comprehensive coverage of computer vision is the book by Szeliski (2022). Fascinating reading about eyes in nature can be found in Srinivasan and Venkatesh (1997), Land and Nilsson (2002), Ings (2008), Frisby and Stone (2010), and Stone (2012). A solid introduction to artificial intelligence is the text by Russell and Norvig (2020).

A number of very readable books discuss the future impacts of robotics and artificial intelligence on society, for example Ford (2015), Brynjolfsson and McAfee (2014), and Bostrom (2016). The YouTube video Grey (2014) makes some powerful points about the future of work and is always a great discussion starter.

# Foundations

Contents

# Representing Position and Orientation

**Contents**

© The Author(s), under exclusive license to Springer Nature Switzerland AG 2023
P. Corke, *Robotics, Vision and Control*, Springer Tracts in Advanced Robotics 146,
https://doi.org/10.1007/978-3-031-06469-2_2

We are familiar with numbers for counting and measuring, and tuples of numbers (coordinates) to describe the position of points in 2- or 3-dimensions. In robotics, we are particularly interested in where *objects* are in the world – objects such as mobile robots, the links and tool of a robotic manipulator arm, cameras or work pieces. ► Sect. 2.1 introduces foundational concepts and then, in ► Sect. 2.2 and 2.3 respectively, we apply them to the 2-dimensional and 3-dimensional scenarios that we encounter in robotics. ► Sect. 2.4 covers some advanced topics that could be omitted on a first reading, and ► Sect. 2.5 has additional details about the Python Toolbox we will use throughout the rest of the book.

## 2.1 Foundations

We start by introducing important general concepts that underpin our ability to describe where objects are in the world, both graphically and algebraically.

### 2.1.1 Relative Pose

Orientation is often referred to as *attitude*.

For the 2D and 3D examples in ◻ Fig. 2.1 we see red and blue versions of an object that have a different position and orientation. The combination of position and orientation ◄ of an object is its *pose*, so the red and blue versions of the object have different poses.

The object's initial pose, shown in blue, has been *transformed* into the pose shown in red. The object's shape has not changed, so we call this a *rigid-body transformation*. It is natural to think of this transformation as a motion – the object has moved in some complex way. Its position has changed (it has *translated*) and its orientation has changed (it has *rotated*).

We will use the symbol $\boldsymbol{\xi}$ (pronounced ksigh) to denote motion and depict it graphically as a thick curved arrow. Importantly, as shown in ◻ Fig. 2.2, the motion is always defined with respect to an initial pose.

Next, we will consider the very simple 2D example shown in ◻ Fig. 2.3. We have elaborated our notation to make it clear where the motion is from and to. The leading superscript indicates the initial pose and the trailing subscript indicates the resulting pose. Therefore $^{x}\boldsymbol{\xi}_{y}$ means the motion from pose $x$ to pose $y$.

Motions can be *joined up* to create arbitrary motions – a process called *composition* or *compounding*. We can write this algebraically as

$$^{0}\boldsymbol{\xi}_{2} = \, ^{0}\boldsymbol{\xi}_{1} \oplus \, ^{1}\boldsymbol{\xi}_{2} \tag{2.1}$$

which says that the motion $^{0}\boldsymbol{\xi}_{2}$ is the same as the motion $^{0}\boldsymbol{\xi}_{1}$ followed by the motion $^{1}\boldsymbol{\xi}_{2}$. We use the special symbol $\oplus$ to remind ourselves that we are *adding*

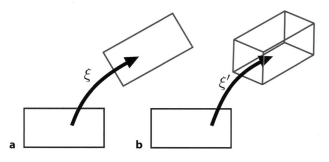

◻ **Fig. 2.1**  Rigid motion of a shape **a** rectangular object in 2D; **b** cuboid object in 3D, the initial blue view is a "side on" view of a 3D cuboid

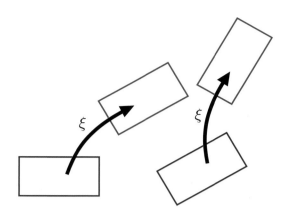

**Fig. 2.2** The motion is always defined relative to the starting pose shown in blue. $\xi$ is the same in both cases

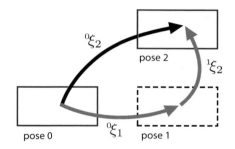

**Fig. 2.3** The motion from pose 0 to pose 2 is equivalent to the motion from pose 0 to pose 1 and then to pose 2

motions not regular numbers. The motions are consecutive, and this means that the pose names on each side of the $\oplus$ must be the same. Although **Fig. 2.3** only shows translations, the motions that we compose can include rotations. For the 3D case, there is a richer set of motions to choose from – there are more directions to translate in, and more axes to rotate about.

It is important to note that the order of the motions is important. For example, if we walk forward 1 m and then turn clockwise by 90°, we have a different pose than if we turned 90° clockwise and then walked forward 1 m. Composition of motions is not commutative.

The corollary is that any motion can be *decomposed* into a number of smaller or simpler motions. For the 2D case in **Fig. 2.1**, one possible decomposition of $\xi$ is a horizontal translation, followed by a vertical translation followed by a rotation about its center.

For any motion there is an *inverse* or *opposite* motion – the motion that gets us back to where we started. For example, for a translation of 2 m to the right, the inverse motion is a translation of 2 m to the left, while for a rotation of 30° clockwise, the inverse is a rotation of 30° counter-clockwise. In general, for a motion from pose $X$ to pose $Y$ the inverse motion is from pose $Y$ to pose $X$, that is

$$^{X}\xi_{Y} \oplus {}^{Y}\xi_{X} = {}^{X}\xi_{X} = \varnothing \tag{2.2}$$

where $\varnothing$ is the *null motion* and means no motion.

We will introduce another operator $\ominus$ that turns any motion into its inverse

$$^{Y}\xi_{X} \equiv \ominus\, {}^{X}\xi_{Y}\,,$$

and now we can write (2.2) as

$$^{X}\xi_{Y} \ominus {}^{X}\xi_{Y} = \varnothing\,.$$

While this looks a lot like subtraction, we use the special symbol $\ominus$ to remind ourselves that we are *subtracting* motions not regular numbers. It follows, that adding or subtracting a *null motion* to a motion leaves the motion unchanged, that is

$$^X\xi_Y \oplus \varnothing = {}^X\xi_Y, \ {}^X\xi_Y \ominus \varnothing = {}^X\xi_Y$$

For the example in ■ Fig. 2.3, we could also write

$$^0\xi_1 = {}^0\xi_2 \oplus {}^2\xi_1$$

and while there is no arrow in ■ Fig. 2.3 for $^2\xi_1$, it would simply be the reverse of the arrow shown for $^1\xi_2$. Since $^2\xi_1 = \ominus\,^1\xi_2$, we can write

$$^0\xi_1 = {}^0\xi_2 \ominus {}^1\xi_2$$

which is very similar to (2.1) except that we have effectively *subtracted* $^1\xi_2$ from both sides – to be very precise, we have subtracted it from the right of *both* sides of the equation since order is critically important. Taking this one step at a time

$$^0\xi_2 = {}^0\xi_1 \oplus {}^1\xi_2$$
$$^0\xi_2 \ominus {}^1\xi_2 = {}^0\xi_1 \oplus {}^1\xi_2 \ominus {}^1\xi_2$$
$$^0\xi_2 \ominus {}^1\xi_2 = {}^0\xi_1 \oplus \varnothing$$
$$= {}^0\xi_1 \ .$$

$$\xrightarrow{\oplus \to \ominus}$$
$$^0\xi_2 = {}^0\xi_1 \left(\oplus {}^1\xi_2\right)$$

Alternatively, just as we do in regular algebra, we could do this in a single step by "*taking* $^1\xi_2$ *across to the other side*" and "*negating*" it, as shown above, to the right.

At the outset, we talked about the pose of the objects in ■ Fig. 2.1. There is no absolute way to describe the pose of an object, we can only describe its pose with respect to some other pose, and we introduce a *reference pose* for this purpose. Any pose is always a *relative pose*, described by the motion required to get there from the reference pose.

### 2.1.2 Coordinate Frames

To describe relative pose, we need to describe its two components: translation and rotation. To achieve this, we rigidly attach a right-handed coordinate frame to the object, as shown in ■ Fig. 2.4. Coordinate frames are a familiar concept from mathematics and comprise two or three orthogonal axes which intersect at a point called the *origin*.

▶ go.sn.pub/VH3R8d

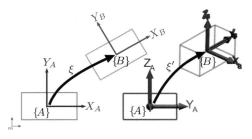

■ **Fig. 2.4** An object with attached coordinate frame shown in two different poses. The axes of 3D coordinate frames are frequently colored red, green and blue for the $x$-, $y$- and $z$-axes respectively. The reference frame is shown in black

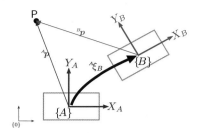

**◘ Fig. 2.5** The point P can be described by coordinate vectors expressed in frame {A} or {B}. The pose of {B} relative to {A} is $^A\xi_B$

The position and orientation of the coordinate frame is sufficient to fully describe the position and orientation of the object it is attached to. Each frame is named, in this case {A} or {B}, and the name is also reflected in the subscripts on the axis labels. Translation is the distance between the origins of the two coordinate frames, and rotation can be described by the set of angles between corresponding axes of the coordinate frames. Together, these represent the change in pose, the relative pose, or the motion described by $\xi$.

The coordinate frame associated with the reference pose is denoted by {0} and called the *reference frame*, the *world frame* and denoted by {W}, or the *global frame*. If the leading superscript is omitted, the reference frame is assumed. The frame associated with a moving body, such as a vehicle, is often called the *body frame* or *body-fixed frame* and denoted by {B}. Any point on the moving body is defined by a constant coordinate vector with respect to the body frame.

◘ Fig. 2.5 shows that a point P can be described by two different coordinate vectors: $^A p$ with respect to frame {A}, or $^B p$ with respect to frame {B}. The *basis* of the coordinate frames are unit frames in the directions of the frame's axes. These two coordinate vectors are related by

$$^A p = {}^A\xi_B \cdot {}^B p$$

where the · operator *transforms* the coordinate vector, resulting in a new coordinate vector that describes the same point but with respect to the coordinate frame resulting from the motion $^A\xi_B$.

We can think of the right-hand side as a motion from {A} to {B} and then to P. We indicate this *composition* of a motion and a vector with the · operator to distinguish it from proper motion composition where we use the ⊕ operator.

Coordinate frames are an extremely useful way to think about real-world robotics problems such as shown in ◘ Fig. 2.6. We have attached coordinate frames to the key objects in the problem and, next, we will look at the relationships between the poses that they represent.

We have informally developed a useful set of rules for dealing with motions. We can compose motions

$$^X\xi_Z = {}^X\xi_Y \oplus {}^Y\xi_Z \tag{2.3}$$

which we read as the motion from X to Z is equivalent to a motion from X to Y and then from Y to Z. The pose names on either side of the ⊕ operator must match.

There is an operator for inverse motion

$$^Y\xi_X = \ominus {}^X\xi_Y \tag{2.4}$$

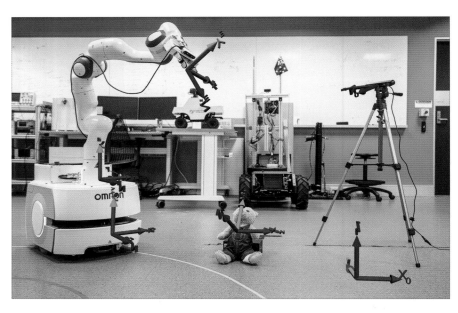

**Fig. 2.6** Examples of coordinate frames for a real-world scenario. Frames include the mobile robot {M}, arm base {B}, arm end effector {E}, camera {C}, workpiece {P} and the reference frame {0} (image by John Skinner and Dorian Tsai)

and a null motion denoted by $\varnothing$ such that

$$
{}^{X}\boldsymbol{\xi}_X = \varnothing, \ \ {}^{X}\boldsymbol{\xi}_Y \ominus {}^{X}\boldsymbol{\xi}_Y = \varnothing, \ \ \ominus {}^{X}\boldsymbol{\xi}_Y \oplus {}^{X}\boldsymbol{\xi}_Y = \varnothing,
$$
$$
{}^{X}\boldsymbol{\xi}_Y \oplus \varnothing = {}^{X}\boldsymbol{\xi}_Y, \ \ {}^{X}\boldsymbol{\xi}_Y \ominus \varnothing = {}^{X}\boldsymbol{\xi}_Y, \ \ \varnothing \oplus {}^{X}\boldsymbol{\xi}_Y = {}^{X}\boldsymbol{\xi}_Y \tag{2.5}
$$

A motion transforms the coordinate vector of a point to reflect the change in coordinate frame due to that motion

$$
{}^{X}\boldsymbol{p} = {}^{X}\boldsymbol{\xi}_Y \cdot {}^{Y}\boldsymbol{p} \ . \tag{2.6}
$$

While this looks like regular algebra it is important to remember that $\boldsymbol{\xi}$ is not a number, it is a relative pose or motion. In particular, the order of operations matters, especially when rotations are involved. We cannot assume commutativity as we do with regular numbers. Formally, $\boldsymbol{\xi}$ belongs to a non-abelian group with the operator $\oplus$, an inverse $\ominus$ and an identity element $\varnothing$.

In the literature, it is common to use a symbol like $*$, $\cdot$ or nothing instead of $\oplus$, and $X^{-1}$ to indicate inverse instead of $\ominus X$.

**Excurse 2.1: The Reference Frame Rule**
In relative pose composition, we can check that we have our reference frames correct by ensuring that the subscript and superscript on each side of the $\oplus$ operator are matched. We can then *cancel out* the intermediate subscripts and superscripts, leaving just the outermost subscript and superscript.

**Excurse 2.2: Groups**

Groups are a useful and important mathematical concept and are directly related to our discussion about motion and relative pose. Formally, a *group* is an abstract system represented by an ordered pair $G = (G, \diamond)$ which comprises a set of objects $G$ equipped with a single *binary operation* denoted by $\diamond$ which combines any two elements of the group to form a third element, also belonging to the group. A group satisfies the following axioms:

| | | |
|---|---|---|
| Closure | $a \diamond b \in G$ | $\forall a, b \in G$ |
| Associativity | $(a \diamond b) \diamond c =$ $a \diamond (b \diamond c)$ | $\forall a, b, c \in G$ |
| Identity element | $a \diamond e = e \diamond a = a$ | $\forall a \in G, \exists e \in G$ |
| Inverse element | $a \diamond a^{-1} =$ $a^{-1} \diamond a = e$ | $\forall a \in G, \exists a^{-1} \in G$ |

A group does *not* satisfy the axiom of commutativity which means that the result of applying the group operation depends on the order in which they are written – an abelian group is one that does satisfy the axiom of commutativity.

For the group of relative motions, the operator is composition. The axioms state that the result of composing two motions is another motion, that the order of motions being composed is significant, and that there exists an inverse motion, and an identity (null) motion.

We will shortly introduce mathematical objects to represent relative motion, for example, rotation matrices, homogeneous transformation matrices, quaternions, and twists – these all form groups under the operation of composition.

### 2.1.3 Pose Graphs

◼ Fig. 2.7 is a pose graph – a directed graph ▶ which comprises vertices (blue circles) representing poses and edges (arrows) representing relative poses or motions. This is a clear, but abstract, representation of the spatial relationships present in the problem shown in ◼ Fig. 2.6. Pose graphs can be used for 2D or 3D problems – that just depends on how we *implement* $\xi$.

See ▶ App. I for more details about graphs.

The black arrows represent known relative poses, and the gray arrows are unknown relative poses that we wish to determine. In order for the robot to grasp the workpiece, we need to know its pose relative to the robot's end effector, that is, $^E\xi_P$. We start by looking for a pair of equivalent paths – two different paths that have the same start and end pose, one of which includes the unknown. We choose the paths shown as red dashed lines. Each path has the same start and end pose, they are equivalent motions, so we can equate them

$$^0\xi_M \oplus {}^M\xi_B \oplus {}^B\xi_E \oplus \underline{{}^E\xi_P} = {}^0\xi_C \oplus {}^C\xi_P .$$

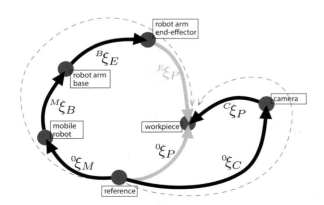

◼ **Fig. 2.7** A pose graph for the scenario shown in ◼ Fig. 2.6. Blue circles are vertices of the graph and represent poses. while the arrows are edges of the graph and represent relative poses or motions

**2**

Now we can perform the algebraic operations discussed earlier and manipulate the expression to isolate the unknown motion

$$^{E}\boldsymbol{\xi}_{P} = \ominus\,^{B}\boldsymbol{\xi}_{E} \ominus\,^{M}\boldsymbol{\xi}_{B} \ominus\,^{0}\boldsymbol{\xi}_{M} \oplus\,^{0}\boldsymbol{\xi}_{C} \oplus\,^{C}\boldsymbol{\xi}_{P}$$

It is easy to write such an expression by inspection. To determine $^{X}\boldsymbol{\xi}_{Y}$, find any path through the pose graph from {X} to {Y} and write down the relative poses of the edges in a left to right order – if we traverse the edge in the direction of its arrow, precede it with the $\oplus$ operator, otherwise use $\ominus$.

### 2.1.4  Summary

□ Fig. 2.8 provides a graphical summary of the key concepts we have just covered. The key points to remember are:

- The position and orientation of an object is referred to as its pose.
- A motion, denoted by $\boldsymbol{\xi}$, causes a change in pose – it is a relative pose defined with respect to the initial pose.
- There is no absolute pose, a pose is always relative to a reference pose.
- We can perform algebraic manipulation of expressions written in terms of relative poses using the operators $\oplus$ and $\ominus$, and the concept of a null motion $\varnothing$.
- We can represent a set of poses, with known relative poses, as a pose graph.
- To quantify pose, we rigidly attach a coordinate frame to an object. The origin of that frame is the object's position, and the directions of the frame's axes describe its orientation.
- The constituent points of a rigid object are described by constant coordinate vectors relative to its coordinate frame.
- Any point can be described by a coordinate vector with respect to any coordinate frame. A coordinate vector can be transformed between frames by applying the relative pose of those frames to the vector using the · operator.

The details of what $\boldsymbol{\xi}$ *really is* have been ignored up to this point. This was a deliberate strategy to get us thinking generally about the pose of objects and motions between poses, rather than immediately getting into the details of implementation. The true nature of $\boldsymbol{\xi}$ depends on the problem. Is it a 2D or 3D problem? Does it involve translations, rotations, or both? The implementation of $\boldsymbol{\xi}$ could be a vector, a matrix or something more exotic like a twist, a unit quaternion or a unit dual quaternion.

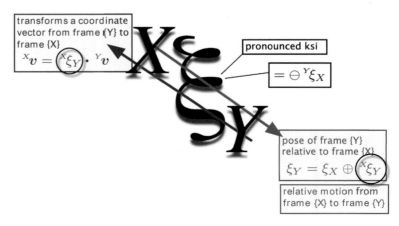

□ **Fig. 2.8**  Everything you need to know about relative pose

### Excurse 2.3: Euclid of Alexandria

Euclid (325–265 BCE) was a Greek mathematician, who was born and lived in Alexandria, Egypt, and is considered the "father of geometry", His great work *Elements*, comprising 13 books, captured and systematized much early knowledge about geometry and numbers. It deduces the properties of planar and solid geometric shapes from a set of 5 axioms and 5 postulates.

*Elements* is probably the most successful book in the history of mathematics. It describes plane geometry and is the basis for most people's first introduction to geometry and formal proof, and is the basis of what we now call Euclidean geometry. Euclidean distance is simply the distance between two points on a plane. Euclid also wrote *Optics* which describes geometric vision and perspective.

### Excurse 2.4: Euclidean versus Cartesian geometry

Euclidean geometry is concerned with points and lines in the Euclidean plane (2D) or Euclidean space (3D). It is based entirely on a set of axioms and makes no use of arithmetic. Descartes added a coordinate system (2D or 3D) and was then able to describe points, lines and other curves in terms of algebraic equations. The study of such equations is called analytic geometry and is the basis of all modern geometry. The Cartesian plane (or space) is the Euclidean plane (or space) with all its axioms and postulates *plus* the extra facilities afforded by the added coordinate system. The term Euclidean geometry is often used to mean that Euclid's fifth postulate (parallel lines never intersect) holds, which is the case for a planar surface but not for a curved surface.

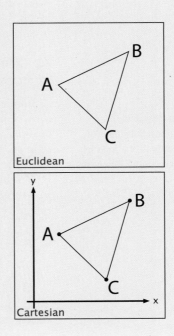

In the rest of this chapter, we will explore a number of concrete representations of rotation, translation and pose, and ways to convert between them. To make it real, we introduce software tools for Python that can create and manipulate these mathematical objects, and will support everything we do in the rest of the book.

**2**

Descartes (1596–1650) was a French philosopher, mathematician and part-time mercenary. He is famous for the philosophical statement "*Cogito, ergo sum*" or "*I am thinking, therefore I exist*" or "*I think, therefore I am*". He was a sickly child and developed a life-long habit of lying in bed and thinking until late morning. A possibly apocryphal story is that during one such morning he was watching a fly walk across the ceiling and realized that he could describe its position in terms of its distance from the two edges of the ceiling. This is the basis of the *Cartesian* coordinate system and modern (analytic) geometry, which he described in his 1637 book *La Géométrie*. For the first time, mathematics and geometry were connected, and modern calculus was built on this foundation by Newton and Leibniz. Living in Sweden, at the invitation of Queen Christina, he was obliged to rise at 5 A.M., breaking his lifetime habit – he caught pneumonia and died. His remains were later moved to Paris, and are now lost apart from his skull which is in the Musée de l'Homme. After his death,

the Roman Catholic Church placed his works on the Index of Prohibited Books – the index was abolished in 1966.

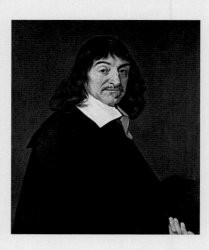

## 2.2  Working in Two Dimensions (2D)

The relative orientation of the $x$- and $y$-axes obey the right-hand rule. For the 2D case the $y$-axis is obtained by rotating the $x$-axis counter-clockwise by 90°.

A 2-dimensional world, or plane, is familiar to us from high-school geometry. We use a right-handed ◄ Cartesian coordinate system or coordinate frame with orthogonal axes, denoted $x$ and $y$, and typically drawn with the $x$-axis pointing to the right and the $y$-axis pointing upwards. The point of intersection is called the origin.

The basis vectors are unit vectors parallel to the axes and are denoted by $\hat{x}$ and $\hat{y}$. A point is represented by its $x$- and $y$-coordinates $(x, y)$ or as a coordinate vector from the origin to the point

$$\boldsymbol{p} = x\hat{\boldsymbol{x}} + y\hat{\boldsymbol{y}} \tag{2.7}$$

which is a linear combination of the basis vectors.

◘ Fig. 2.9 shows a red coordinate frame {B} that we wish to describe with respect to the blue frame {A}. We see clearly that the origin of {B} has been displaced, and the frame rotated counter-clockwise. We will consider the problem in two parts: pure rotation and then rotation plus translation.

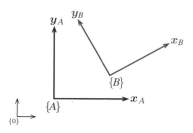

◘ **Fig. 2.9**  Two 2D coordinate frames {A} and {B}, defined with respect to the reference frame {0}. {B} is rotated and translated with respect to {A}

### 2.2.1 Orientation in Two Dimensions

#### 2.2.1.1 2D Rotation Matrix

❏ Fig. 2.10 shows two coordinate frames {A} and {B} with a common origin but different orientation. Frame {B} is obtained by rotating frame {A} by $\theta$ in the positive (counter-clockwise) direction about the origin.

Frame {A} is described by the directions of its axes, and these are parallel to the basis vectors $\hat{\boldsymbol{x}}_A$ and $\hat{\boldsymbol{y}}_A$ which are defined with respect to the reference frame {0}. In this 2D example, the basis vectors have two elements which we write as column vectors and stack side-by-side to form a $2 \times 2$ matrix $(\hat{\boldsymbol{x}}_A \quad \hat{\boldsymbol{y}}_A)$. This matrix *completely* describes frame {A}.

Similarly, frame {B} is completely described by its basis vectors $\hat{\boldsymbol{x}}_B$ and $\hat{\boldsymbol{y}}_B$ and these can be expressed in terms of the basis vectors of frame {A}

$$\hat{\boldsymbol{x}}_B = \hat{\boldsymbol{x}}_A \cos\theta + \hat{\boldsymbol{y}}_A \sin\theta$$
$$\hat{\boldsymbol{y}}_B = -\hat{\boldsymbol{x}}_A \sin\theta + \hat{\boldsymbol{y}}_A \cos\theta$$

or in matrix form as

$$(\hat{\boldsymbol{x}}_B \quad \hat{\boldsymbol{y}}_B) = (\hat{\boldsymbol{x}}_A \quad \hat{\boldsymbol{y}}_A) \underbrace{\begin{pmatrix} \cos\theta & -\sin\theta \\ \sin\theta & \cos\theta \end{pmatrix}}_{^A\mathbf{R}_B(\theta)} \qquad (2.8)$$

where

$$^A\mathbf{R}_B(\theta) = \begin{pmatrix} \cos\theta & -\sin\theta \\ \sin\theta & \cos\theta \end{pmatrix}$$

is a special type of matrix called a *rotation matrix* that transforms frame {A}, described by $(\hat{\boldsymbol{x}}_A \quad \hat{\boldsymbol{y}}_A)$, into frame {B} described by $(\hat{\boldsymbol{x}}_B \quad \hat{\boldsymbol{y}}_B)$. Rotation matrices have some special properties:

- The columns are the basis vectors that define the axes of the rotated 2D coordinate frame, and therefore have unit length and are orthogonal.
- It is an orthogonal (also called orthonormal) matrix ▶ and therefore its inverse is the same as its transpose, that is, $\mathbf{R}^{-1} = \mathbf{R}^\top$.
- The matrix-vector product $\mathbf{R}\boldsymbol{v}$ preserves the length and relative orientation of vectors $\boldsymbol{v}$ and therefore its determinant is $+1$.
- It is a member of the Special Orthogonal (SO) group of dimension 2 which we write as $\mathbf{R} \in \mathbf{SO}(2) \subset \mathbb{R}^{2 \times 2}$. Being a group under the operation of matrix multiplication means that the product of any two matrices belongs to the group, as does its inverse.

▶ App. B is a refresher on vectors, matrices and linear algebra.

❏ Fig. 2.5 shows a point and two coordinate frames, {A} and {B}. The point can be described by the coordinate vector $\left({}^A p_x, {}^A p_y\right)^\top$ with respect to frame {A} or the

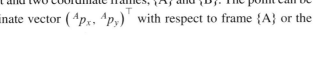

❏ **Fig. 2.10** Rotated coordinate frames in 2D. Basis vectors are shown as thick arrows

**2**

coordinate vector $\left( {}^{B}p_x, \; {}^{B}p_y \right)^{\top}$ with respect to frame {B}. From (2.7), we can write a coordinate vector as a linear combination of the basis vectors of the reference frame which in matrix form is

$$\underline{{}^{A}\boldsymbol{p}} = \begin{pmatrix} \hat{\boldsymbol{x}}_A & \hat{\boldsymbol{y}}_A \end{pmatrix} \begin{pmatrix} {}^{A}p_x \\ {}^{A}p_y \end{pmatrix}, \tag{2.9}$$

$$ {}^{B}\boldsymbol{p} = \begin{pmatrix} \hat{\boldsymbol{x}}_B & \hat{\boldsymbol{y}}_B \end{pmatrix} \begin{pmatrix} {}^{B}p_x \\ {}^{B}p_y \end{pmatrix}. \tag{2.10}$$

Substituting (2.8) into (2.10) yields

$$\underline{{}^{A}\boldsymbol{p}} = \begin{pmatrix} \hat{\boldsymbol{x}}_A & \hat{\boldsymbol{y}}_A \end{pmatrix} {}^{A}\mathbf{R}_B(\theta) \begin{pmatrix} {}^{B}p_x \\ {}^{B}p_y \end{pmatrix} \tag{2.11}$$

and we have changed the left-hand side's reference frame, since it is now expressed in terms of the basis vectors of {A} – we write ${}^{A}\boldsymbol{p}$ instead of ${}^{B}\boldsymbol{p}$. Equating the two definitions of ${}^{A}\boldsymbol{p}$ in (2.9) and (2.11), we can write

$$\begin{pmatrix} \hat{\boldsymbol{x}}_A & \hat{\boldsymbol{y}}_A \end{pmatrix} \underline{\begin{pmatrix} {}^{A}p_x \\ {}^{A}p_y \end{pmatrix}} = \begin{pmatrix} \hat{\boldsymbol{x}}_A & \hat{\boldsymbol{y}}_A \end{pmatrix} \underline{{}^{A}\mathbf{R}_B(\theta) \begin{pmatrix} {}^{B}p_x \\ {}^{B}p_y \end{pmatrix}} \tag{2.12}$$

and then, equating the underlined coefficients, leads to

$$\begin{pmatrix} {}^{A}p_x \\ {}^{A}p_y \end{pmatrix} = {}^{A}\mathbf{R}_B(\theta) \begin{pmatrix} {}^{B}p_x \\ {}^{B}p_y \end{pmatrix} \tag{2.13}$$

which shows that the rotation matrix ${}^{A}\mathbf{R}_B(\theta)$ transforms a coordinate vector from frame {B} to {A}.

A rotation matrix has all the characteristics of relative pose $\boldsymbol{\xi}$ described by (2.3) to (2.6).

---

**$\xi$ as an SO(2) matrix**

For the case of pure rotation in 2D, $\boldsymbol{\xi}$ can be implemented by a rotation matrix $\mathbf{R} \in \mathbf{SO}(2)$. Its implementation is:

| composition | $\xi_1 \oplus \xi_2$ | $\mapsto \mathbf{R}_1\mathbf{R}_2$, matrix multiplication |
|---|---|---|
| inverse | $\ominus\xi$ | $\mapsto \mathbf{R}^{-1} = \mathbf{R}^{\top}$, matrix transpose |
| identity | $\varnothing$ | $\mapsto \mathbf{R}(0) = \mathbf{1}_{2\times 2}$, identity matrix |
| vector-transform | $\xi \cdot \boldsymbol{v}$ | $\mapsto \mathbf{R}\boldsymbol{v}$, matrix-vector product |

Composition is commutative, that is, $\mathbf{R}_1\mathbf{R}_2 = \mathbf{R}_2\mathbf{R}_1$, and $\mathbf{R}(-\theta) = \mathbf{R}^{\top}(\theta)$.

---

To make this tangible, we will create an **SO**(2) rotation matrix using the Toolbox software

```
>>> R = rot2(0.3)
array([[ 0.9553,  -0.2955],
       [ 0.2955,   0.9553]])
```

where the angle is specified in radians. The orientation represented by a rotation matrix can be visualized as a coordinate frame

```
>>> trplot2(R);
```

We can observe some of the properties just mentioned, for example, the determinant is equal to one

```
>>> np.linalg.det(R)
1
```

and that the product of two rotation matrices is also a rotation matrix

```
>>> np.linalg.det(R @ R)
1
```

and note that we use the operator @, rather than *, to indicate matrix multiplication of NumPy arrays.

The Toolbox also supports symbolic mathematics, ▶ for example

```
>>> from sympy import Symbol, Matrix, simplify
>>> theta = Symbol('theta')
theta
>>> R = Matrix(rot2(theta))  # convert to SymPy matrix
Matrix([
[cos(theta), -sin(theta)],
[sin(theta),  cos(theta)]])
>>> simplify(R * R)
Matrix([
[cos(2*theta), -sin(2*theta)],
[sin(2*theta),  cos(2*theta)]])
>>> R.det()
sin(theta)**2 + cos(theta)**2
>>> R.det().simplify()
1
```

> You will need to have SymPy installed. Note that SymPy uses the * operator for multiplication of scalar and arrays.

### 2.2.1.2 Matrix Exponential for Rotation

There is a fascinating, and very useful, connection between a rotation matrix and the exponential of a skew-symmetric matrix. We can easily demonstrate this by considering, again, a pure rotation of 0.3 radians expressed as a rotation matrix

```
>>> R = rot2(0.3);
```

We can take the matrix logarithm using the SciPy function `logm` ▶

```
>>> L = linalg.logm(R)
array([[      0,    -0.3],
       [    0.3,       0]])
```

> `logm` is different to the function `np.log` which computes the logarithm of each element of the matrix. A logarithm can be computed using a power series with a matrix, rather than scalar, argument. The logarithm of a matrix is not unique and `logm` computes the principal logarithm.

and the result is a simple matrix with just two nonzero elements – they have a magnitude of 0.3 which is the rotation angle. This matrix is an example of a $2 \times 2$ skew-symmetric matrix. It has only one unique element

```
>>> S = vex(L)
array([0.3])
```

which is the rotation angle as a single-element array, not a scalar. This is our first encounter with Lie (pronounced lee) group theory, and we will continue to explore this topic in the rest of this chapter.

Exponentiating the logarithm of the rotation matrix L, using the matrix exponential function `expm`, ▶ yields the original rotation matrix

```
>>> linalg.expm(L)
array([[ 0.9553,  -0.2955],
       [ 0.2955,   0.9553]])
```

> `expm` is different to the function `exp` which computes the exponential of each element of the matrix. The matrix exponential can be computed using a power series with a matrix, rather than scalar, argument: $expm(\mathbf{A}) = \mathbf{1} + \mathbf{A} + \mathbf{A}^2/2! + \mathbf{A}^3/3! + \cdots$

The exponential of a skew-symmetric matrix is always a rotation matrix, with all the special properties outlined earlier. The skew-symmetric matrix can be reconstructed from the single element of S, so we can also write

```
>>> linalg.expm(skew(S))
array([[ 0.9553,  -0.2955],
       [ 0.2955,   0.9553]])
```

**2**

---

**Excurse 2.6: 2D Skew-Symmetric Matrix**

In 2 dimensions, the skew- or anti-symmetric matrix is

$$[\omega]_\times = \begin{pmatrix} 0 & -\omega \\ \omega & 0 \end{pmatrix} \in \mathbf{so}(2) \tag{2.14}$$

which has a distinct structure with a zero diagonal and only one unique value $\omega \in \mathbb{R}$, and $[\omega]_\times^\top = -[\omega]_\times$. The vector space of 2D skew-symmetric matrices is denoted $\mathbf{so}(2)$ and is the Lie algebra of $\mathbf{SO}(2)$. The Toolbox implements the $[\cdot]_\times$ operator as

```
>>> X = skew(2)
array([[ 0, -2],
       [ 2,  0]])
```

and the inverse operator $\vee_\times(\cdot)$ as

```
>>> vex(X)
array([        2])
```

---

In general, we can write

$$\mathbf{R} = e^{[\theta]_\times} \in \mathbf{SO}(2)$$

where $\theta$ is the rotation angle, and $[\cdot]_\times : \mathbb{R} \mapsto \mathbf{so}(2) \subset \mathbb{R}^{2\times2}$ is a mapping from a scalar to a skew-symmetric matrix.

### 2.2.2 Pose in Two Dimensions

To describe the relative pose of the frames shown in ◻ Fig. 2.9, we need to account for the translation between the origins of the frames as well as the rotation. The frames are shown in more detail in ◻ Fig. 2.11.

#### 2.2.2.1 2D Homogeneous Transformation Matrix

The first step is to transform the coordinate vector $^B\boldsymbol{p} = (^Bx, \,^By)$, with respect to frame {B}, to $^{A'}\boldsymbol{p} = (^{A'}x, \,^{A'}y)$ with respect to frame {A'} using the rotation matrix $^{A'}\mathbf{R}_B(\theta)$ which is a function of the orientation $\theta$. Since frames {A'} and {A} are parallel, the coordinate vector $^A\boldsymbol{p}$ is obtained by adding $^A\boldsymbol{t}_B = (t_x, t_y)^\top$ to $^{A'}\boldsymbol{p}$

$$\begin{pmatrix} ^Ax \\ ^Ay \end{pmatrix} = \begin{pmatrix} ^{A'}x \\ ^{A'}y \end{pmatrix} + \begin{pmatrix} t_x \\ t_y \end{pmatrix} \tag{2.15}$$

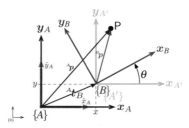

◻ **Fig. 2.11** Rotated and translated coordinate frames where the axes of frame {A'} are parallel to the axes of frame {A}

$$= \begin{pmatrix} \cos\theta & -\sin\theta \\ \sin\theta & \cos\theta \end{pmatrix} \begin{pmatrix} {}^B x \\ {}^B y \end{pmatrix} + \begin{pmatrix} t_x \\ t_y \end{pmatrix} \tag{2.16}$$

$$= \begin{pmatrix} \cos\theta & -\sin\theta & t_x \\ \sin\theta & \cos\theta & t_y \end{pmatrix} \begin{pmatrix} {}^B x \\ {}^B y \\ 1 \end{pmatrix} \tag{2.17}$$

or more compactly as

$$\begin{pmatrix} {}^A x \\ {}^A y \\ 1 \end{pmatrix} = \begin{pmatrix} {}^A\mathbf{R}_B(\theta) & {}^A\boldsymbol{t}_B \\ \mathbf{0}_{1\times2} & 1 \end{pmatrix} \begin{pmatrix} {}^B x \\ {}^B y \\ 1 \end{pmatrix} \tag{2.18}$$

where ${}^A\boldsymbol{t}_B$ is the translation of the origin of frame {B} with respect to frame {A}, and ${}^A\mathbf{R}_B(\theta)$ is the orientation of frame {B} with respect to frame {A}.

> If we consider the homogeneous transformation as a relative pose or rigid-body motion, this corresponds to the coordinate frame being first translated by ${}^A\boldsymbol{t}_B$ with respect to frame {A}, and *then* rotated by ${}^A\mathbf{R}_B(\theta)$.

The coordinate vectors for point P are now expressed in *homogeneous form* and we denote that with a tilde. Now we can write

$$\begin{aligned} {}^A\tilde{\boldsymbol{p}} &= \begin{pmatrix} {}^A\mathbf{R}_B(\theta) & {}^A\boldsymbol{t}_B \\ \mathbf{0}_{1\times2} & 1 \end{pmatrix} {}^B\tilde{\boldsymbol{p}} \\ &= {}^A\mathbf{T}_B\, {}^B\tilde{\boldsymbol{p}} \end{aligned}$$

and ${}^A\mathbf{T}_B$ is referred to as a *homogeneous transformation* – it transforms homogeneous vectors. The matrix has a very specific structure and belongs to the Special Euclidean (SE) group of dimension 2 which we write as $\mathbf{T} \in \mathbf{SE}(2) \subset \mathbb{R}^{3\times3}$.

The matrix ${}^A\mathbf{T}_B$ represents translation and orientation or relative pose and has all the characteristics of relative pose $\boldsymbol{\xi}$ described by (2.3) to (2.6).

> **$\xi$ as an SE(2) matrix**
> For the case of rotation and translation in 2D, $\boldsymbol{\xi}$ can be implemented by a homogeneous transformation matrix $\mathbf{T} \in \mathbf{SE}(2)$ which is sometimes written as an ordered pair $(\mathbf{R}, \boldsymbol{t}) \in \mathbf{SO}(2) \times \mathbb{R}^2$. The implementation is:
>
> composition $\quad \boldsymbol{\xi}_1 \oplus \boldsymbol{\xi}_2 \quad \mapsto \mathbf{T}_1\mathbf{T}_2 = \begin{pmatrix} \mathbf{R}_1\mathbf{R}_2 & \boldsymbol{t}_1 + \mathbf{R}_1\boldsymbol{t}_2 \\ \mathbf{0}_{1\times2} & 1 \end{pmatrix}$, matrix multiplication
>
> inverse $\quad \ominus\boldsymbol{\xi} \quad \mapsto \mathbf{T}^{-1} = \begin{pmatrix} \mathbf{R}^\top & -\mathbf{R}^\top\boldsymbol{t} \\ \mathbf{0}_{1\times2} & 1 \end{pmatrix}$, matrix inverse
>
> identity $\quad \varnothing \quad \mapsto \mathbf{1}_{3\times3}$, identity matrix
>
> vector-transform $\quad \boldsymbol{\xi} \cdot \boldsymbol{v} \quad \mapsto \epsilon(\mathbf{T}\tilde{\boldsymbol{v}})$, matrix-vector product
>
> where $\tilde{\phantom{i}} : \mathbb{R}^2 \mapsto \mathbb{P}^2$ and $\epsilon(\cdot) : \mathbb{P}^2 \mapsto \mathbb{R}^2$. Composition is not commutative, that is, $\mathbf{T}_1\mathbf{T}_2 \neq \mathbf{T}_2\mathbf{T}_1$.

**2**

> **Excurse 2.7: Homogeneous Vectors**
>
> A vector $\boldsymbol{p} = (x, y)$ is written in homogeneous form as $\tilde{\boldsymbol{p}} = (x_1, x_2, x_3) \in \mathbb{P}^2$ where $\mathbb{P}^2$ is the 2-dimensional projective space, and the tilde indicates the vector is homogeneous.
>
> Homogeneous vectors have the important property that $\tilde{\boldsymbol{p}}$ is equivalent to $\lambda \tilde{\boldsymbol{p}}$ for all $\lambda \neq 0$ which we write as $\tilde{\boldsymbol{p}} \simeq \lambda \tilde{\boldsymbol{p}}$. That is, $\tilde{\boldsymbol{p}}$ represents the same point in the plane irrespective of the overall scaling factor. Homogeneous representation is also used in computer vision which we discuss in Part IV. Additional details are provided in ▶ App. C.2.
>
> To convert a point to homogeneous form we typically append an element equal to one, for the 2D case this is $\tilde{\boldsymbol{p}} = (x, y, 1)$. The dimension of the vector has been increased by one, and a point on a plane is now represented by a 3-vector. The Euclidean or nonhomogeneous coordinates are related by $x = x_1/x_3$, $y = x_2/x_3$ and $x_3 \neq 0$.

To recap, the rotation matrix for a rotation of 0.3 rad is

```
>>> rot2(0.3)
array([[  0.9553,   -0.2955],
       [  0.2955,    0.9553]])
```

which can be composed with other rotation matrices, or used to transform coordinate vectors. The homogeneous transformation matrix for a rotation of 0.3 rad is

```
>>> trot2(0.3)
array([[  0.9553,   -0.2955,          0],
       [  0.2955,    0.9553,          0],
       [       0,         0,          1]])
```

and we see the rotation matrix in the top-left corner, and zeros and a one on the bottom row. The top two elements of the right-hand column are zero indicating zero translation. This matrix represents relative pose, it can be composed with other relative poses, or used to transform *homogeneous* coordinate vectors.

Next, we will compose two relative poses: a translation of $(1, 2)$ followed by a rotation of $30°$

```
>>> TA = transl2(1, 2) @ trot2(30, "deg")
array([[   0.866,       -0.5,          1],
       [     0.5,      0.866,          2],
       [       0,          0,          1]])
```

The function `transl2` creates a 2D relative pose with a finite translation but zero rotation, while `trot2` creates a relative pose with a finite rotation but zero translation. ◄ We can plot a coordinate frame representing this pose, relative to the reference frame, by

For describing rotations, the Toolbox has functions to create a rotation matrix (`rot2`) or a homogeneous transformation with zero translation (`trot2`).

```
>>> plotvol2([0, 5]); # new plot with both axes from 0 to 5
>>> trplot2(TA, frame="A", color="b");
```

The options to `trplot` specify that the label for the frame is {A} and it is colored blue and this is shown in ◼ Fig. 2.12. For completeness, we will add the reference frame to the plot

```
>>> T0 = transl2(0, 0);
>>> trplot2(T0, frame="0", color="k");  # reference frame
```

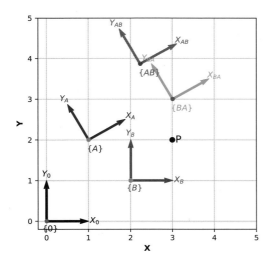

■ **Fig. 2.12** Coordinate frames drawn using the Toolbox function `trplot2`

We create another relative pose which is a displacement of $(2, 1)$ and zero rotation

```
>>> TB = transl2(2, 1)
array([[    1,         0,         2],
       [    0,         1,         1],
       [    0,         0,         1]])
```

which we plot in red

```
>>> trplot2(TB, frame="B", color="r");
```

Now we can compose the two relative poses

```
>>> TAB = TA @ TB
array([[  0.866,      -0.5,      2.232],
       [    0.5,     0.866,      3.866],
       [      0,         0,          1]])
```

and plot the result as a green coordinate frame

```
>>> trplot2(TAB, frame="AB", color="g");
```

We see that the displacement of $(2, 1)$ has been applied with respect to frame {A}. It is important to note that our final displacement is not $(3, 3)$ because the displacement is with respect to the rotated coordinate frame. The noncommutativity of composition is clearly demonstrated by reversing the order of multiplication

```
>>> TBA = TB @ TA;
>>> trplot2(TBA, frame="BA", color="c");
```

and we see that frame {BA} is different to frame {AB}.

Now we define a point $(3,2)$ relative to the world frame

```
>>> P = np.array([3, 2]);
```

which is a 1D array, and add it to the plot

```
>>> plot_point(P, "ko", text="P");
```

To determine the coordinate of the point with respect to {A} we write

$$^{0}\boldsymbol{p} = {}^{0}\boldsymbol{\xi}_{A} \cdot {}^{A}\boldsymbol{p}$$

and then rearrange as

$$^{A}\boldsymbol{p} = (\ominus \, {}^{0}\boldsymbol{\xi}_{A}) \cdot {}^{0}\boldsymbol{p} \, .$$

**2**

Substituting numerical values we obtain

```
>>> np.linalg.inv(TA) @ np.hstack([P, 1])
array([  1.732,      -1,        1])
```

where we first converted the Euclidean point coordinates to homogeneous form by appending a one. The result is also in homogeneous form and has a negative *y*-coordinate in frame {A}. Using the Toolbox, we could also have expressed this as

```
>>> h2e(np.linalg.inv(TA) @ e2h(P))
array([[  1.732],
       [     -1]])
```

where the result is in Euclidean coordinates. The function `e2h` converts Euclidean coordinates to homogeneous, and `h2e` performs the inverse conversion. Even more concisely, we can write

```
>>> homtrans(np.linalg.inv(TA), P)
array([[  1.732],
       [     -1]])
```

which handles conversion of the coordinate vectors to and from homogeneous form.

### 2.2.2.2   Rotating a Coordinate Frame

The pose of a coordinate frame is fully described by an $\mathbf{SE}(2)$ matrix and here we will explore rotation of coordinate frames. First, we create and plot a reference coordinate frame {0} and a target frame {X}

```
>>> plotvol2([-5, 4, -1, 5]);
>>> T0 = transl2(0, 0);
>>> trplot2(T0, frame="0", color="k");
>>> TX = transl2(2, 3);
>>> trplot2(TX, frame="X", color="b");
```

Next, we create an $\mathbf{SE}(2)$ matrix representing a rotation of 2 radians (nearly 120°)

```
>>> TR = trot2(2);
```

and plot the effect of the two possible orders of composition

```
>>> trplot2(TR @ TX, frame="RX", color="g");
>>> trplot2(TX @ TR, frame="XR", color="g");
```

The results are shown as green coordinate frames in ◘ Fig. 2.13. We see that the frame {RX} has been rotated about the origin, while frame {XR} has been rotated about the origin of {X}.

What if we wished to rotate a coordinate frame about some arbitrary point C? The first step is to define the coordinate of the point and display it

```
>>> C = np.array([3, 2]);
>>> plot_point(C, "ko", text="C");
```

and then compute a transform to rotate about point C

```
>>> TC = transl2(C) @ TR @ transl2(-C)
array([[ -0.4161,  -0.9093,    6.067],
       [  0.9093,  -0.4161,   0.1044],
       [       0,        0,       1]])
```

We apply it to frame {X}

```
>>> trplot2(TC @ TX, frame="XC", color="r");
```

and see that the red frame has been rotated about point C.

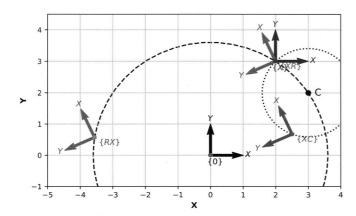

**◘ Fig. 2.13** The frame {X} is rotated by 2 radians about {0} to give frame {RX}, about {X} to give {XR}, and about point C to give frame {XC}

To understand how this works we can unpack what happens when we apply TC to TX. Since $\mathbf{T}_C$ premultiplies $\mathbf{T}_X$, the first transform that is applied to {X} is the right-most one, `trans12(-C)`, which performs an origin shift that places C at the origin of the reference frame. Then we apply $\mathbf{T}_R$, a pure rotation which rotates the shifted version of {X} about the origin, which is where C now is – therefore {X} is rotated about C. Finally, we apply `trans12(C)` which is the inverse origin shift, that places C back at its original position and {X} to its final pose. This transformation is an example of a conjugation but its construction was somewhat elaborate. Premultiplication of matrices is discussed in ► Sect. 2.4.1. A more intuitive way to achieve this same result is by using twists which we introduce in ► Sect. 2.2.2.4.

### 2.2.2.3 Matrix Exponential for Pose

We can take the logarithm of the **SE**(2) matrix TC from the previous example

```
>>> L = linalg.logm(TC)
array([[      0,      -2,       4],
       [      2,       0,      -6],
       [      0,       0,       0]])
```

and the result is an augmented skew-symmetric matrix: the upper-left corner is a $2 \times 2$ skew-symmetric matrix: the upper right column is a 2-vector, and the bottom row is zero. The three unique elements can be unpacked

```
>>> S = vexa(L)
array([      4,      -6,       2])
```

and the last element is the rotation angle.

Exponentiating L yields the original **SE**(2) matrix, and L can be reconstructed from just the three elements of s

```
>>> linalg.expm(skewa(S))
array([[ -0.4161,  -0.9093,    6.067],
       [  0.9093,  -0.4161,   0.1044],
       [      0,       0,        1]])
```

In general, we can write

$$\mathbf{T} = e^{[S]} \in \mathbf{SE}(2)$$

where $S \in \mathbb{R}^3$ and $[\cdot] : \mathbb{R}^3 \mapsto \mathbf{se}(2) \subset \mathbb{R}^{3 \times 3}$. The logarithm and exponential can be computed efficiently using the Toolbox functions `trlog2` and `trexp2` respectively. The vector $S$ is a *twist vector* which we will discuss next.

**2**

---

**Excurse 2.8: 2D Augmented Skew-Symmetric Matrix**

In 2 dimensions, the augmented skew-symmetric matrix corresponding to the vector $S = (v_x, v_y, \omega)$ is

$$[S] = \begin{pmatrix} 0 & -\omega & v_x \\ \omega & 0 & v_y \\ \hline 0 & 0 & 0 \end{pmatrix} \in \mathbf{se}(2) \tag{2.19}$$

which has a distinct structure with a zero diagonal and bottom row, and a skew-symmetric matrix in the top-left corner. The vector space of 2D augmented skew-symmetric matrices is denoted $\mathbf{se}(2)$ and is the Lie algebra of $\mathbf{SE}(2)$. The Toolbox implements the [·] operator as

```
>>> X = skewa([1, 2, 3])
array([[      0,      -3,       1],
       [      3,       0,       2],
       [      0,       0,       0]])
```

and the inverse operator $\vee(\cdot)$ as

```
>>> vexa(X)
array([       1,       2,       3])
```

---

#### 2.2.2.4  2D Twists

In ▶ Sect. 2.2.2.2, we transformed a coordinate frame by rotating it about a specified point. The corollary is that, given any two frames, we can find a rotational center and rotation angle that will *rotate* the first frame into the second. This is the key concept behind what is called a twist.

A rotational, or revolute, twist about the point specified by the coordinate c is created by

```
>>> S = Twist2.UnitRevolute(C)
(2 -3; 1)
```

where the class method `UnitRevolute` constructs a rotational twist. The result is a `Twist2` object that encapsulates a 2D twist vector $(v, \omega) \in \mathbb{R}^3$ comprising a *moment* $v \in \mathbb{R}^2$ and a scalar $\omega$. This particular twist is a *unit twist* that describes a rotation of 1 rad about the point c.

For the example in ▶ Sect. 2.2.2.2, we require a rotation of 2 rad about the point C, so we scale the unit twist and exponentiate it

```
>>> linalg.expm(skewa(2 * S.S))
array([[ -0.4161,  -0.9093,    6.067],
       [  0.9093,  -0.4161,   0.1044],
       [       0,        0,        1]])
```

where s.s is the twist vector as a NumPy array. The `Twist2` object has a shorthand method for this

```
>>> S.exp(2)
  -0.4161   -0.9093    6.067
   0.9093   -0.4161   0.1044
        0         0        1
```

The result is not actually a 3 × 3 NumPy array even though it looks like one. It is an SE2 object which encapsulates a NumPy array, this is discussed further in ▶ Sect. 2.5.

This has the same value as the transformation computed in the previous section, ◀ but more concisely specified in terms of the center and magnitude of rotation. The center of rotation, called the pole, is encoded in the twist

```
>>> S.pole
array([       3,       2])
```

For the case of pure translational motion, the rotational center is at infinity. A translational, or prismatic, unit twist is therefore specified only by the direction of motion. For example, motion in the $y$-direction is created by

```
>>> S = Twist2.UnitPrismatic([0, 1])
(0 1; 0)
```

which represents a displacement of 1 in the $y$-direction. To create an **SE**(2) transformation for a specific translation, we scale and exponentiate it

```
>>> S.exp(2)
    1          0          0
    0          1          2
    0          0          1
```

and the result has a zero rotation and a translation of 2 in the $y$-direction.

For an arbitrary 2D homogeneous transformation

```
>>> T = transl2(3, 4) @ trot2(0.5)
array([[  0.8776,   -0.4794,         3],
       [  0.4794,    0.8776,         4],
       [       0,         0,         1]])
```

the twist is

```
>>> S = Twist2(T)
(3.9372 3.1663; 0.5)
```

which describes a rotation of

```
>>> S.w
0.5
```

about the point

```
>>> S.pole
array([  -3.166,     3.937])
```

and exponentiating this twist

```
>>> S.exp(1)
    0.8776     -0.4794      3
    0.4794      0.8776      4
    0           0           1
```

yields the original **SE**(2) transform – completely described by the three elements of the twist vector.

---

**$\xi$ as a 2D twist**

For the case of rotation and translation in 2D, $\boldsymbol{\xi}$ can be implemented by a twist $\boldsymbol{S} \in \mathbb{R}^3$. The implementation is:

| | | |
|---|---|---|
| composition | $\boldsymbol{\xi}_1 \oplus \boldsymbol{\xi}_2$ | $\mapsto \log\!\left(e^{[S_1]}e^{[S_2]}\right)$, product of exponential |
| inverse | $\ominus\boldsymbol{\xi}$ | $\mapsto -\boldsymbol{S}$, negation |
| identity | $\varnothing$ | $\mapsto \mathbf{0}_{1\times 3}$, zero vector |
| vector transform | $\boldsymbol{\xi} \cdot \boldsymbol{v}$ | $\mapsto \epsilon(e^{[S]}\tilde{\boldsymbol{v}})$, matrix-vector product using homogeneous vectors |

where $\tilde{\cdot} : \mathbb{R}^2 \mapsto \mathbb{P}^2$ and $\epsilon(\cdot) : \mathbb{P}^2 \mapsto \mathbb{R}^2$. Composition is not commutative, that is, $e^{[S_1]}e^{[S_2]} \neq e^{[S_2]}e^{[S_1]}$. Note that log and exp have efficient closed form, rather than transcendental, solutions which makes composition relatively inexpensive.

**2**

## 2.3  **Working in Three Dimensions (3D)**

The 3-dimensional case is an extension of the 2-dimensional case discussed in ▶ Sect. 2.2. We add an extra coordinate axis, denoted by $z$, that is orthogonal to both the $x$- and $y$-axes. The direction of the $z$-axis obeys the *right-hand rule* and forms a *right-handed coordinate frame*. The basis vectors are unit vectors parallel to the axes, denoted by $\hat{x}$, $\hat{y}$ and $\hat{z}$ ◄ and related such that ◄

These basis vectors are sometimes denoted $e_1$, $e_2$, $e_3$.

$$\hat{z} = \hat{x} \times \hat{y}, \quad \hat{x} = \hat{y} \times \hat{z}, \quad \hat{y} = \hat{z} \times \hat{x} \ . \tag{2.20}$$

In all these identities, the symbols from left to right (across the equals sign) are a cyclic rotation of the sequence $\ldots z, x, y, z, x, y, \ldots$.

A point P is represented by its $x$-, $y$- and $z$-coordinates $(x, y, z)$ or as a coordinate vector from the origin to the point

$$p = x\hat{x} + y\hat{y} + z\hat{z}$$

which is a linear combination of the basis vectors.

▢ Fig. 2.14 shows two 3-dimensional coordinate frames and we wish to describe the red frame {B} with respect to the blue frame {A}. We can see clearly that the origin of {B} has been displaced by the 3D vector $^A t_B$ and then rotated in some complex fashion. We will again consider the problem in two parts: pure rotation and then rotation plus translation. Rotation is surprisingly complex for the 3-dimensional case and the next section will explore some of the many ways to describe it.

▶ go.sn.pub/ATBZR0

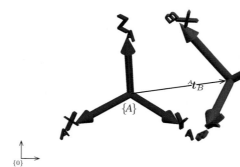

▢ **Fig. 2.14**   Two 3D coordinate frames {A} and {B} defined with respect to the reference frame. {B} is rotated and translated with respect to {A}. There is an ambiguity when viewing 3D coordinate frames on a page. Your brain may flip between two interpretations, with the origin either toward or away from you – knowing the right-hand rule resolves the ambiguity

### Excurse 2.9: Right-hand rule
A right-handed coordinate frame is defined by the first three fingers of your right hand which indicate the relative directions of the $x$-, $y$- and $z$-axes respectively.

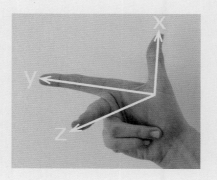

### 2.3.1 Orientation in Three Dimensions

The two frames shown in ◘ Fig. 2.14 clearly have different orientations, and we want some way to describe the orientation of one with respect to the other. We can imagine picking up frame {A} in our hand and rotating it until it looked just like frame {B}.

We start by considering rotation about a single coordinate frame axis. ◘ Fig. 2.15 shows a right-handed coordinate frame, and that same frame after it has been rotated by various angles about different axes.

Rotation in 3D has some subtleties which are illustrated in ◘ Fig. 2.16, where a sequence of two rotations are applied in different orders. We see that the final orientation depends on the order in which the rotations are applied – this is a confounding characteristic of the 3-dimensional world. It is important to remember:

> **!** In three dimensions, rotation is not commutative – the result depends on the order in which rotations are applied.

Mathematicians have developed many ways to represent rotation and we will introduce those that are most commonly encountered in robotics: rotation matrices, Euler and Cardan angles, rotation axis and angle, exponential coordinates, and unit

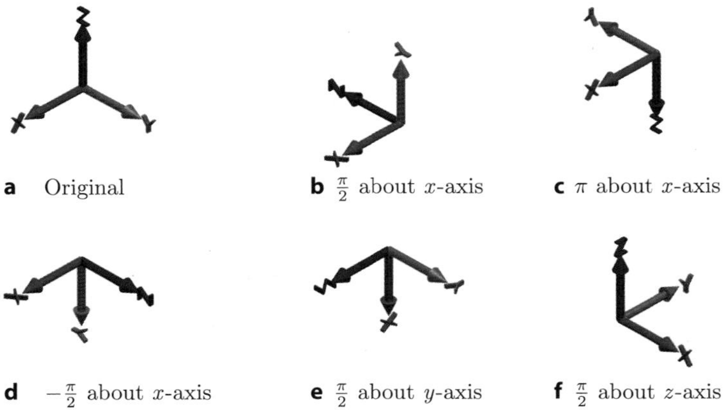
**a** Original
**b** $\frac{\pi}{2}$ about $x$-axis
**c** $\pi$ about $x$-axis
**d** $-\frac{\pi}{2}$ about $x$-axis
**e** $\frac{\pi}{2}$ about $y$-axis
**f** $\frac{\pi}{2}$ about $z$-axis

◘ **Fig. 2.15** Rotation of a 3D coordinate frame. **a** The original coordinate frame, **b–f** frame **a** after various rotations (in units of radians) as indicated

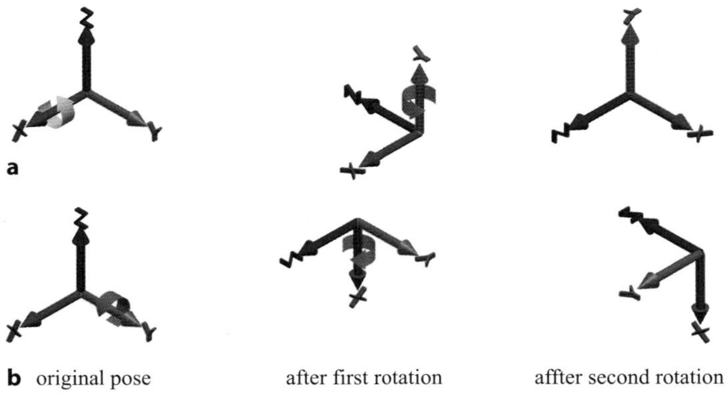
**a**
**b** original pose
after first rotation
affter second rotation

◘ **Fig. 2.16** Example showing the noncommutativity of rotation. In the top row the coordinate frame is rotated by $\frac{\pi}{2}$ about the $x$-axis and then $\frac{\pi}{2}$ about the $y$-axis. In the bottom row the order of rotations is reversed. The results are clearly different

► go.sn.pub/XU7D6b

**2**

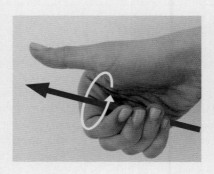

quaternions. All can be represented as NumPy arrays or as Python classes. The Toolbox provides many ways to create and convert between these representations.

#### 2.3.1.1  **3D Rotation Matrix**

Just as for the 2-dimensional case, we can represent the orientation of a coordinate frame by its basis vectors expressed in terms of the reference coordinate frame. Each basis vector has three elements and they form the columns of a $3 \times 3$ *orthogonal matrix* $^A\mathbf{R}_B$

$$
\begin{pmatrix} ^Ap_x \\ ^Ap_y \\ ^Ap_z \end{pmatrix} = {}^A\mathbf{R}_B \begin{pmatrix} ^Bp_x \\ ^Bp_y \\ ^Bp_z \end{pmatrix}
\tag{2.21}
$$

which transforms a coordinate vector defined with respect to frame {B} to a coordinate vector with respect to frame {A}. A 3-dimensional rotation matrix $\mathbf{R}$ has the same special properties as its 2D counterpart:

- The columns are the basis vectors that define the axes of the rotated 3D coordinate frame, and therefore have unit length and are orthogonal.
- It is orthogonal (also called orthonormal) ◄ and therefore its inverse is the same as its transpose, that is, $\mathbf{R}^{-1} = \mathbf{R}^{\top}$.

*See ► App. B for a refresher on vectors, matrices and linear algebra.*

- The matrix-vector product $\mathbf{R}\boldsymbol{v}$ preserves the length and relative orientation of vectors $\boldsymbol{v}$ and therefore its determinant is $+1$.
- It is a member of the Special Orthogonal (SO) group of dimension 3 which we write as $\mathbf{R} \in \mathbf{SO}(3) \subset \mathbb{R}^{3\times3}$. Being a group under the operation of matrix multiplication means that the product of any two matrices belongs to the group, as does its inverse.

The rotation matrices that correspond to a coordinate frame rotation of $\theta$ about the $x$-, $y$- and $z$-axes are

$$
\mathbf{R}_x(\theta) = \begin{pmatrix} 1 & 0 & 0 \\ 0 & \cos\theta & -\sin\theta \\ 0 & \sin\theta & \cos\theta \end{pmatrix}
$$

$$
\mathbf{R}_y(\theta) = \begin{pmatrix} \cos\theta & 0 & \sin\theta \\ 0 & 1 & 0 \\ -\sin\theta & 0 & \cos\theta \end{pmatrix}
$$

$$\mathbf{R}_z(\theta) = \begin{pmatrix} \cos\theta & -\sin\theta & 0 \\ \sin\theta & \cos\theta & 0 \\ 0 & 0 & 1 \end{pmatrix}$$

The Toolbox provides functions to compute these elementary rotation matrices, for example $\mathbf{R}_x\left(\frac{\pi}{2}\right)$ is

```
>>> R = rotx(pi / 2)
array([[      1,         0,         0],
       [      0,         0,        -1],
       [      0,         1,         0]])
```

and the functions `roty` and `rotz` compute $\mathbf{R}_y(\theta)$ and $\mathbf{R}_z(\theta)$ respectively.

The orientation represented by a rotation matrix can be visualized as a coordinate frame – rotated with respect to the reference coordinate frame

```
>>> trplot(R);
```

which is shown in ◘ Fig. 2.17a. The interpretation as a motion or relative pose, can be made explicit by visualizing the rotation matrix as an animation

```
>>> tranimate(R)
```

which shows the reference frame moving to the specified relative pose. If you have a pair of anaglyph stereo glasses ▶you can see this in more vivid 3D by either of

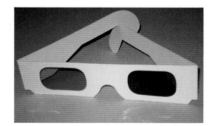

```
>>> trplot(R, anaglyph=True)
>>> tranimate(R, anaglyph=True);
```

Now we will apply another rotation, this time about the $y$-axis resulting from the first rotation

```
>>> R = rotx(pi / 2) @ roty(pi / 2)
array([[      0,         0,         1],
       [      1,         0,         0],
       [      0,         1,         0]])
>>> trplot(R);
```

which gives the frame shown in ◘ Fig. 2.17b and the $x$-axis now points in the direction of the world $y$-axis. This frame is the same as the rightmost frame in ◘ Fig. 2.16a.

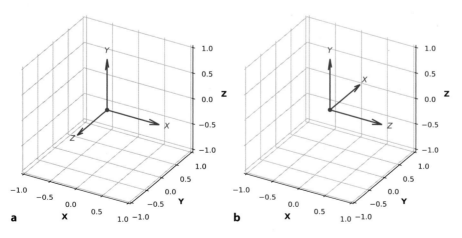

**a** **b**

◘ **Fig. 2.17** 3D coordinate frames displayed using `trplot`. **a** Reference frame rotated by $\frac{\pi}{2}$ about the $x$-axis, **b** frame **a** rotated by $\frac{\pi}{2}$ about the $y$-axis

**Excurse 2.11: Reading a Rotation Matrix**
The columns from left to right tell us the directions of the new frame's axes in terms of the current coordinate frame.

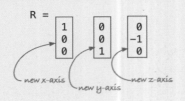

In this case, the new frame has its $x$-axis in the old $x$-direction $(1, 0, 0)$, its $y$-axis in the old $z$-direction $(0, 0, 1)$, and the new $z$-axis in the old negative $y$-direction $(0, -1, 0)$. In this case, the $x$-axis was unchanged, since this is the axis around which the rotation occurred. The rows are the converse – the current frame axes in terms of the new frame axes.

The noncommutativity of rotation is clearly shown by reversing the order of the rotations

```
>>> roty(pi / 2) @ rotx(pi / 2)
array([[        0,        1,        0],
       [        0,        0,       -1],
       [       -1,        0,        0]])
```

which has a very different value, and this orientation is shown in ◘ Fig. 2.16b.

$\xi$ **as an SO(3) matrix**
For the case of pure rotation in 3D, $\xi$ can be implemented by a rotation matrix $\mathbf{R} \in \mathbf{SO}(3)$. The implementation is:

| | | |
|---|---|---|
| composition | $\xi_1 \oplus \xi_2$ | $\mapsto \mathbf{R}_1\mathbf{R}_2$, matrix multiplication |
| inverse | $\ominus\xi$ | $\mapsto \mathbf{R}^{-1} = \mathbf{R}^\top$, matrix transpose |
| identity | $\varnothing$ | $\mapsto \mathbf{1}_{3\times3}$, identity matrix |
| vector-transform | $\xi \cdot v$ | $\mapsto \mathbf{R}v$, matrix-vector product |

Composition is not commutative, that is, $\mathbf{R}_1\mathbf{R}_2 \neq \mathbf{R}_2\mathbf{R}_1$.

#### 2.3.1.2 Three-Angle Representations

» Euler's rotation theorem:
*Any two independent orthonormal coordinate frames can be related by a sequence of rotations (not more than three) about coordinate axes, where no two successive rotations may be about the same axis.* (Kuipers 1999).

Euler's rotation theorem means that a rotation between any two coordinate frames in 3D can be represented by a sequence of three rotation angles each associated with a particular coordinate frame axis. The rotations are applied consecutively: the first rotates the world frame {0} around some axis to create a new coordinate frame {1}; then a rotation about some axis of {1} results in frame {2}; and finally a rotation about some axis of {2} results in frame {3}.

There are a total of twelve unique rotation sequences. Six involve repetition, but not successive, of rotations about one particular axis: XYX, XZX, YXY, YZY, ZXZ, or ZYZ. Another six are characterized by rotations about all three axes: XYZ, XZY, YZX, YXZ, ZXY, or ZYX.

**Excurse 2.12: Leonhard Euler**

Euler (1707–1783) was a Swiss mathematician and physicist who dominated eighteenth century mathematics. He was a student of Johann Bernoulli and applied new mathematical techniques such as calculus to many problems in mechanics and optics. He developed the functional notation, $y = f(x)$, and in robotics we use his rotation theorem and his equations of motion in rotational dynamics.

He was prolific and his collected works fill 75 volumes. Almost half of this was produced during the last seventeen years of his life when he was completely blind.

**❗ "Euler angles" is an ambiguous term**

It is common practice to refer to all possible three-angle representations as Euler angles but this is insufficiently precise since there are twelve rotation sequences. The particular angle sequence needs to be specified, but it is often an implicit convention within a particular technological field.

In mechanical dynamics, the ZYZ sequence is commonly used

$$\mathbf{R}(\phi, \theta, \psi) = \mathbf{R}_z(\phi)\,\mathbf{R}_y(\theta)\,\mathbf{R}_z(\psi) \tag{2.22}$$

and the Euler angles are written as the 3-vector $\boldsymbol{\Gamma} = (\phi, \theta, \psi) \in (\mathbf{S}^1)^3$. ▶ To compute the equivalent rotation matrix for $\boldsymbol{\Gamma} = (0.1, 0.2, 0.3)$, we write

$(\mathbf{S}^1)^3$ or $\mathbf{S}^1 \times \mathbf{S}^1 \times \mathbf{S}^1$ denotes a 3-tuple of angles, whereas $\mathbf{S}^3$ is an angle in 4D space.

```
>>> R = rotz(0.1) @ roty(0.2) @ rotz(0.3);
```

or more conveniently

```
>>> R = eul2r(0.1, 0.2, 0.3)
array([[  0.9021,  -0.3836,   0.1977],
       [  0.3875,   0.9216,   0.01983],
       [ -0.1898,   0.05871,  0.9801]])
```

The inverse problem is finding the ZYZ Euler angles that correspond to a given rotation matrix

```
>>> gamma = tr2eul(R)
array([    0.1,      0.2,      0.3])
```

If $\theta$ is negative, the rotation matrix is

```
>>> R = eul2r(0.1, -0.2, 0.3)
array([[  0.9021,  -0.3836,  -0.1977],
       [  0.3875,   0.9216,  -0.01983],
       [  0.1898,  -0.05871,  0.9801]])
```

and the inverse function

```
>>> gamma = tr2eul(R)
array([   -3.042,        0.2,    -2.842])
```

returns a set of different Euler angles where the sign of $\theta$ has changed, and the values of $\phi$ and $\psi$ have been offset by $-\pi$. However, the corresponding rotation matrix

```
>>> eul2r(gamma)
array([[   0.9021,    -0.3836,    -0.1977],
       [   0.3875,     0.9216,   -0.01983],
       [   0.1898,   -0.05871,     0.9801]])
```

is the same. This means that there are two different sets of Euler angles that generate the same rotation matrix – the mapping from a rotation matrix to Euler angles is not unique. The Toolbox *always* returns a positive value for $\theta$.

For the case where $\theta = 0$

```
>>> R = eul2r(0.1, 0, 0.3)
array([[   0.9211,    -0.3894,          0],
       [   0.3894,     0.9211,          0],
       [        0,          0,          1]])
```

the inverse function returns

```
>>> tr2eul(R)
array([        0,          0,        0.4])
```

which is again quite different, but these Euler angles will generate the same rotation matrix. The explanation is that if $\theta = 0$ then $\mathbf{R}_y = \mathbf{1}_{3\times3}$ and (2.22) becomes

$$\mathbf{R} = \mathbf{R}_z(\phi)\mathbf{R}_z(\psi) = \mathbf{R}_z(\phi + \psi)$$

which is a function of the sum $\phi + \psi$. The inverse operation can determine this sum and arbitrarily split it between $\phi$ and $\psi$ – by convention we choose $\phi = 0$. The case $\theta = 0$ is a *singularity* and will be discussed in more detail in the next section.

The other widely used convention are the Cardan angles: roll, pitch and yaw which we denote as $\alpha$, $\beta$ and $\gamma$ respectively. ◀

Roll-pitch-yaw angles are also known as Tait-Bryan angles, after Peter Tait a Scottish physicist and quaternion supporter, and George Bryan a pioneering Welsh aerodynamicist. They are also known as nautical angles. For aeronautical applications, the angles are called bank, attitude and heading respectively.

**❶ Roll-pitch-yaw angle ambiguity**

Confusingly, there are two different roll-pitch-yaw sequences in common use: ZYX or XYZ, depending on whether the topic is mobile robots or robot manipulators respectively.

When describing the attitude of vehicles such as ships, aircraft and cars, the convention is that the $x$-axis of the body frame points in the forward direction and its $z$-axis points either up or down. We start with the world reference frame and in order:

- rotate about the world $z$-axis by the yaw angle, $\gamma$, so that the $x$-axis points in the direction of travel, then
- rotate about the $y$-axis of the frame above by the pitch angle, $\beta$, which sets the angle of the vehicle's longitudinal axis relative to the horizontal plane (pitching the nose up or nose down), and then finally
- rotate about the $x$-axis of the frame above by the roll angle, $\alpha$, so that the vehicle's body rolls about its longitudinal axis.

which leads to the ZYX angle sequence

$$\mathbf{R}(\alpha, \beta, \gamma) = \mathbf{R}_z(\gamma)\,\mathbf{R}_y(\beta)\,\mathbf{R}_x(\alpha) \tag{2.23}$$

### Excurse 2.13: Gerolamo Cardano

Cardano (1501–1576) was an Italian Renaissance mathematician, physician, astrologer, and gambler. He was born in Pavia, Italy, the illegitimate child of a mathematically gifted lawyer. He studied medicine at the University of Padua and later was the first to describe typhoid fever. He partly supported himself through gambling and his book about games of chance *Liber de ludo aleae* contains the first systematic treatment of probability as well as effective cheating methods. His family life was problematic: his eldest son was executed for poisoning his wife, and his daughter was a prostitute who died from syphilis (about which he wrote a treatise). He computed and published the horoscope of Jesus, was accused of heresy, and spent time in prison until he abjured and gave up his professorship.

He published the solutions to the cubic and quartic equations in his book *Ars magna* in 1545, and also invented the combination lock, the gimbal consisting of three concentric rings allowing a compass or gyroscope to rotate freely (see

Fig. 2.19), and the Cardan shaft with universal joints – the drive shaft used in motor vehicles today.

and the roll, pitch and yaw angles are written as the 3-vector $\Gamma = (\alpha, \beta, \gamma) \in (\mathbf{S}^1)^3$. For example

```
>>> R = rpy2r(0.1, 0.2, 0.3, order="zyx")
array([[ 0.9363,  -0.2751,   0.2184],
       [ 0.2896,   0.9564,  -0.03696],
       [ -0.1987,  0.09784,  0.9752]])
```

where the arguments are given in the order roll, pitch, then yaw which is opposite to the order they appear in (2.23).

The inverse is

```
>>> gamma = tr2rpy(R, order="zyx")
array([    0.1,      0.2,      0.3])
```

When describing the orientation of a robot gripper, as shown in ◻ Fig. 2.20, the convention is that its coordinate frame has the $z$-axis pointing forward and the $y$-axis is parallel to a line between the finger tips. This leads to the XYZ angle sequence

$$\mathbf{R}(\alpha, \beta, \gamma) = \mathbf{R}_x(\gamma)\, \mathbf{R}_y(\beta)\, \mathbf{R}_z(\alpha) \tag{2.24}$$

for example

```
>>> R = rpy2r(0.1, 0.2, 0.3, order="xyz")
array([[ 0.9752, -0.09784,   0.1987],
       [ 0.1538,   0.9447,  -0.2896],
       [ -0.1593,   0.313,   0.9363]])
```

and the inverse is

```
>>> gamma = tr2rpy(R, order="xyz")
array([    0.1,      0.2,      0.3])
```

The roll-pitch-yaw sequence allows all angles to have arbitrary sign and it has a singularity when $\beta = \pm\frac{\pi}{2}$.

**2**

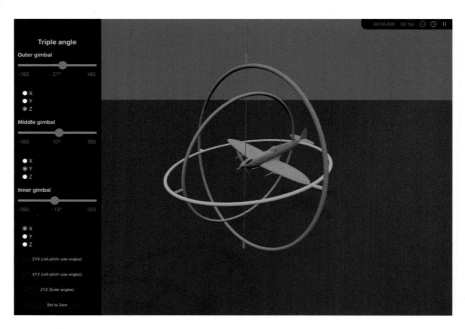

◻ **Fig. 2.18** The Robotics Toolbox demo `tripleangledemo` allows you to explore Euler angles and roll-pitch-yaw angles and see how they change the orientation of a body in 3D space. This displays in a browser tab using the Swift visualizer

This can also be run from the operating system command line as `tripleangledemo`.

*"The LM (Lunar Module) Body coordinate system is right-handed, with the +X axis pointing up through the thrust axis, the +Y axis pointing right when facing forward which is along the +Z axis. The rotational transformation matrix is constructed by a 2-3-1 (YZX) Euler sequence, that is: Pitch about Y, then Roll about Z and, finally, Yaw about X. Positive rotations are pitch up, roll right, yaw left."* (Hoag 1963).

Operationally, this was a significant limiting factor with this particular gyroscope (Hoag 1963) and could have been alleviated by adding a fourth gimbal, as was used on other spacecraft. It was omitted on the Lunar Module for reasons of weight and space.

Rotations obey the cyclic rotation rules

$$\mathbf{R}_x(\tfrac{\pi}{2})\mathbf{R}_y(\theta)\mathbf{R}_x^\top(\tfrac{\pi}{2}) \equiv \mathbf{R}_z(\theta)$$
$$\mathbf{R}_y(\tfrac{\pi}{2})\mathbf{R}_z(\theta)\mathbf{R}_y^\top(\tfrac{\pi}{2}) \equiv \mathbf{R}_x(\theta)$$
$$\mathbf{R}_z(\tfrac{\pi}{2})\mathbf{R}_x(\theta)\mathbf{R}_z^\top(\tfrac{\pi}{2}) \equiv \mathbf{R}_y(\theta)$$

and anti-cyclic rotation rules

$$\mathbf{R}_y^\top(\tfrac{\pi}{2})\mathbf{R}_x(\theta)\mathbf{R}_y(\tfrac{\pi}{2}) \equiv \mathbf{R}_z(\theta)$$
$$\mathbf{R}_z^\top(\tfrac{\pi}{2})\mathbf{R}_y(\theta)\mathbf{R}_z(\tfrac{\pi}{2}) \equiv \mathbf{R}_x(\theta)$$
$$\mathbf{R}_x^\top(\tfrac{\pi}{2})\mathbf{R}_z(\theta)\mathbf{R}_x(\tfrac{\pi}{2}) \equiv \mathbf{R}_y(\theta)$$

The Toolbox includes an interactive graphical tool

```
>>> %run -m tripleangledemo
```

that allows you to experiment with Euler angles or roll-pitch-yaw angles and visualize their effect on the orientation of an object, as shown in ◻ Fig. 2.18. ◀

### 2.3.1.3  Singularities and Gimbal Lock

A fundamental problem with all the three-angle representations just described is singularity. This is also known as gimbal lock, a geeky term made famous in the movie Apollo 13. The term is related to mechanical gimbal systems.

One example is the mechanical gyroscope, shown in ◻ Fig. 2.19, that was used for spacecraft navigation. The innermost assembly is the *stable member* which has three orthogonal gyroscopes that hold it at a constant orientation with respect to the universe. It is mechanically connected to the spacecraft via a gimbal mechanism which allows the spacecraft to rotate around the stable platform without exerting any torque on it. The attitude of the spacecraft is determined directly by measuring the angles of the gimbal axes with respect to the stable platform – in this design the gimbals form a Cardanian YZX sequence giving yaw-roll-pitch angles. ◀ The orientation of the spacecraft's body-fixed frame {B} with respect to the stable platform frame {S} is

$$^S\mathbf{R}_B = \mathbf{R}_y(\beta)\mathbf{R}_z(\alpha)\mathbf{R}_x(\gamma) \ . \tag{2.25}$$

Consider the situation when the rotation angle of the middle gimbal ($\alpha$, roll about the spacecraft's $z$-axis) is 90° – the axes of the inner ($\beta$) and outer ($\gamma$) gimbals are aligned and they share the *same* rotation axis. Instead of the original three rotational axes, since two are parallel, there are now only two effective rotational axes – we say that one degree of freedom has been lost. ◀ Substituting the identity ◀

$$\mathbf{R}_y(\theta)\,\mathbf{R}_z\!\left(\tfrac{\pi}{2}\right) \equiv \mathbf{R}_z\!\left(\tfrac{\pi}{2}\right)\mathbf{R}_x(\theta)$$

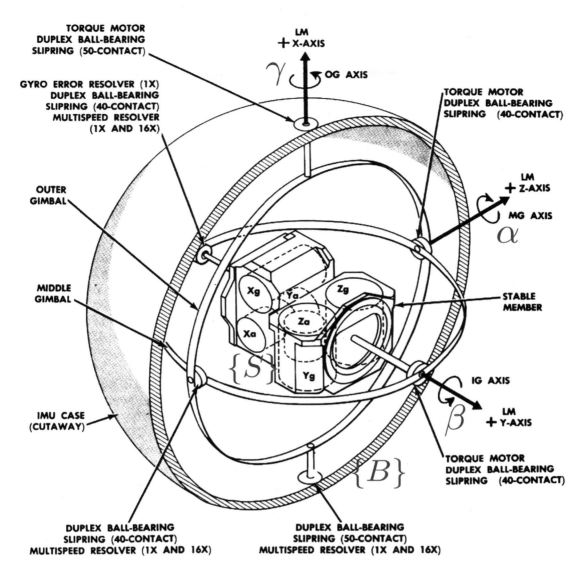

**◘ Fig. 2.19** Schematic of Apollo Lunar Module (LM) inertial measurement unit (IMU). The vehicle's coordinate system has the $x$-axis pointing up through the thrust axis, the $z$-axis forward, and the $y$-axis pointing right. Starting at the stable platform {S} and working outwards toward the spacecraft's body frame {B}, the rotation angle sequence is *YZX*. The components labeled $X_g$, $Y_g$ and $Z_g$ are the $x$-, $y$- and $z$-axis gyroscopes and those labeled $X_a$, $Y_a$ and $Z_a$ are the $x$-, $y$- and $z$-axis accelerometers (redrawn after Apollo Operations Handbook, LMA790-3-LM)

for the first two terms of (2.25), leads to

$$^S\mathbf{R}_B = \mathbf{R}_z(\tfrac{\pi}{2})\,\mathbf{R}_x(\beta)\,\mathbf{R}_x(\gamma) = \mathbf{R}_z(\tfrac{\pi}{2})\,\mathbf{R}_x(\beta + \gamma)$$

which is unable to represent any rotation about the $y$-axis. This is dangerous because any spacecraft rotation about the $y$-axis would also rotate the stable element and thus ruin its precise alignment with the stars: hence the anxiety onboard Apollo 13.

The loss of a degree of freedom means that mathematically we cannot invert the transformation, we can only establish a linear relationship between two of the angles. In this case, the best we can do is determine the sum of the pitch and yaw angles. We observed a similar phenomenon with the ZYZ-Euler angle singularity in the previous section.

All three-angle representations of orientation, whether Eulerian or Cardanian, suffer the problem of gimbal lock when two axes become aligned. For ZYZ-Euler

**2**

angles, this occurs when $\theta = k\pi$, $k \in \mathbb{Z}$; and for roll-pitch-yaw angles when pitch $\beta = \pm(2k + 1)\frac{\pi}{2}$. The best we can do is carefully choose the angle sequence and coordinate system to ensure that the singularity occurs for an orientation outside the normal operating envelope of the system. ◀

For an aircraft or submarine, normal operation involves a pitch angle around zero, and a singularity only occurs for impossible orientations such as nose straight up or straight down.

Singularities are an unfortunate consequence of using a minimal representation – in this case, using just three parameters to represent orientation. To eliminate this problem, we need to adopt different representations of orientation. Many in the Apollo LM team would have preferred a four gimbal system and the clue to success, as we shall see shortly in ▶ Sect. 2.3.1.7, is to introduce a fourth parameter.

### 2.3.1.4 **Two-Vector Representation**

For arm-type robots, it is useful to consider a coordinate frame {E} attached to the end effector, as shown in ◻ Fig. 2.20. By convention, the axis of the tool is associated with the $z$-axis and is called the *approach vector* and denoted $\hat{\boldsymbol{a}} = (a_x, a_y, a_z)$. For some applications, it is more convenient to specify the approach vector than to specify ZYZ-Euler or roll-pitch-yaw angles.

However, specifying the direction of the $z$-axis is insufficient to describe the coordinate frame – we also need to specify the direction of the $x$- or $y$-axes. For robot grippers, this is given by the *orientation vector*, $\hat{\boldsymbol{o}} = (o_x, o_y, o_z)$ which is parallel to the gripper's $y$-axis and is typically a vector between the finger tips. These two unit vectors are sufficient to completely define the rotation matrix

$$\mathbf{R} = \begin{pmatrix} n_x & o_x & a_x \\ n_y & o_y & a_y \\ n_z & o_z & a_z \end{pmatrix} \tag{2.26}$$

since the remaining column, the normal vector $\hat{\boldsymbol{n}} = (n_x, n_y, n_z)$, can be computed using (2.20) as $\hat{\boldsymbol{n}} = \hat{\boldsymbol{o}} \times \hat{\boldsymbol{a}}$. Consider an example where the robot's gripper is facing downward toward a work table

```
>>> a = [0, 0, -1];
```

and its finger tips lie on a line parallel to the vector $(1, 1, 0)$

```
>>> o = [1, 1, 0];
```

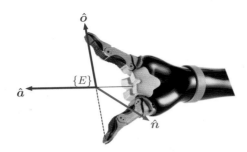

◻ **Fig. 2.20** Robot end-effector coordinate system defines the pose in terms of an *approach* vector $\hat{a}$ and an *orientation* vector $\hat{o}$ parallel to a line joining the finger tips. The $\hat{n}, \hat{o}$ and $\hat{a}$ vectors correspond to the $x$-, $y$- and $z$-axes respectively of the end-effector coordinate frame (image of robot gripper courtesy of Kinova Robotics)

Using the Toolbox, the rotation matrix is ▶

In this case, o is not a unit vector but oa2r will unitize it.

```
>>> R = oa2r(o, a)
array([[ -0.7071,    0.7071,        0],
       [  0.7071,    0.7071,        0],
       [       0,         0,       -1]])
```

Any two nonparallel vectors are sufficient to define a coordinate frame. Even if the two vectors $\hat{a}$ and $\hat{o}$ are not orthogonal, they still define a plane, and the computed $\hat{n}$ is normal to that plane. In this case, we need to compute a new value for $\hat{o}' = \hat{a} \times \hat{n}$ which lies in the plane but is orthogonal to each of $\hat{a}$ and $\hat{n}$.

There are many other situations where we know the orientation in terms of two vectors. For a camera, we might use the optical axis (by convention the $z$-axis) as $\hat{a}$, and the right-side of the camera (by convention the $x$-axis) as $\hat{n}$. For a mobile robot, we might use the gravitational acceleration vector measured with accelerometers (by convention the $z$-axis) as $\hat{a}$, and the heading direction measured with an electronic compass (by convention the $x$-axis) as $\hat{n}$.

### 2.3.1.5 Rotation About an Arbitrary Vector

Two coordinate frames of arbitrary orientation are related by a *single* rotation about some axis in space. For the example rotation used earlier

```
>>> R = rpy2r(0.1, 0.2, 0.3);
```

we can determine such an angle and axis by

```
>>> theta, v = tr2angvec(R)
```

where

```
>>> theta
0.3655
```

is the angle of rotation and

```
>>> v
array([ 0.1886,    0.5834,      0.79])
```

is a unit vector parallel to the rotation axis. Note that this is not unique since a rotation of $-\theta$ about the vector $-v$ results in the same orientation as a rotation of $\theta$ about the vector $v$ – this is referred to as a double mapping.

This information is encoded in the eigenvalues and eigenvectors of **R**

```
>>> e, x = np.linalg.eig(R)
```

where the eigenvalues are

```
>>> e
array([0.934+0.357j, 0.934-0.357j, 1.   +0.j   ])
```

**2**

The matrices e and x are complex, and NumPy denotes the imaginary part with $j = \sqrt{-1}$. Some elements are real, that is they have a zero imaginary part).

and the associated eigenvectors are the corresponding columns of ◄

```
>>> x
array([[-0.694+0.j  , -0.694-0.j  ,  0.189+0.j  ],
       [ 0.079+0.569j,  0.079-0.569j,  0.583+0.j  ],
       [ 0.107-0.42j ,  0.107+0.42j ,  0.79 +0.j  ]])
```

From the definition of eigenvalues and eigenvectors, we recall that

$$\mathbf{R}\boldsymbol{v} = \lambda \boldsymbol{v}$$

where $\boldsymbol{v}$ is the eigenvector corresponding to the eigenvalue $\lambda$. For the case $\lambda = 1$, we can write

$$\mathbf{R}\boldsymbol{v} = \boldsymbol{v}$$

which implies that the corresponding eigenvector $\boldsymbol{v}$ is *unchanged* by the rotation. There is only one such vector, and that is the vector *about which* the rotation occurs. In the example above, the third eigenvalue is equal to one, so the rotation axis $\boldsymbol{v}$ is the third column of x.

A rotation matrix will always have one real eigenvalue at $\lambda = 1$ and in general a complex pair $\lambda = \cos\theta \pm j\sin\theta$, where $\theta$ is the rotation angle. The angle of rotation in this case is ◄

The rotation angle is also related to the trace of R as discussed in ▶ App. B.2.1 but is limited to solutions in quadrants 1 and 2 only. The angle function finds a solution in any quadrant.

```
>>> theta = np.angle(e[0])
0.3655
```

The inverse problem, converting from angle and vector to a rotation matrix, is achieved using Rodrigues' rotation formula

$$\mathbf{R} = \mathbf{1}_{3\times 3} + \sin\theta [\hat{\boldsymbol{v}}]_\times + (1 - \cos\theta)[\hat{\boldsymbol{v}}]_\times^2 \tag{2.27}$$

where $[\hat{v}]_\times$ is a skew-symmetric matrix. We can use this formula to determine the rotation matrix for a rotation of 0.3 rad about the $x$-axis

```
>>> R = angvec2r(0.3, [1, 0, 0])
array([[      1,        0,        0],
       [      0,   0.9553,  -0.2955],
       [      0,   0.2955,   0.9553]])
```

which is equivalent to `rotx(0.3)`.

The Euler-Rodrigues formula is a variation that directly transforms a coordinate vector $x$

$$x' = \mathbf{R}(\theta, \hat{v})x = x + 2s(\omega \times x) + 2(\omega \times (\omega \times x))$$

where $s = \cos\frac{\theta}{2}$ and $\omega = \hat{v}\sin\frac{\theta}{2}$. Collectively, $(s, \omega) \in \mathbb{R}^4$ are known as the Euler parameters of the rotation. ▶

This is actually a unit quaternion, which is introduced in ▶ Sect. 2.3.1.7.

The angle-axis representation is parameterized by four numbers: one for the angle of rotation, and three for the unit vector defining the axis. We can pack these four numbers into a single vector $v \in \mathbb{R}^3$ – the direction of the rotational axis is given by the unit vector $\hat{v}$ and the rotation angle is some function of the vector magnitude $\|v\|$. ▶ This leads to some common 3-parameter representations such as:

- $\hat{v}\theta$, the Euler vector or exponential coordinates,
- $\hat{v}\tan\frac{\theta}{2}$, the Rodrigues' vector,
- $\hat{v}\sin\frac{\theta}{2}$, the unit quaternion vector component, or
- $\hat{v}\tan\theta$.

These forms are minimal and efficient in terms of data storage but the direction $\hat{v}$ is not defined when $\theta = 0$. That case corresponds to zero rotation and needs to be detected and dealt with explicitly.

A 3D unit vector contains redundant information and can be described using just two numbers, the unit-norm constraint means that any element can be computed from the other two. Another way to think about this is to consider a unit sphere, then all possible unit vectors from the center can be described by the latitude and longitude of the point at which they touch the surface of the sphere.

### 2.3.1.6 Matrix Exponential for Rotation

Consider an $x$-axis rotation expressed as a rotation matrix

```
>>> R = rotx(0.3)
array([[      1,        0,        0],
       [      0,   0.9553,  -0.2955],
       [      0,   0.2955,   0.9553]])
```

As we did for the 2-dimensional case, we can take the logarithm using the function `logm` ▶

```
>>> L = linalg.logm(R)
array([[      0,        0,        0],
       [      0,        0,     -0.3],
       [      0,      0.3,        0]])
```

`logm` is different to the function `np.log` which computes the logarithm of each element of the matrix. A logarithm can be computed using a power series with a matrix, rather than scalar, argument. The logarithm of a matrix is not unique and `logm` computes the principal logarithm.

and the result is a sparse matrix with two elements that have a magnitude of 0.3, which is the rotation angle. This matrix has a zero diagonal and is another example of a skew-symmetric matrix, in this case $3 \times 3$.

*Unpacking* the skew-symmetric matrix gives

```
>>> S = vex(L)
array([    0.3,        0,        0])
```

and we find the original rotation angle is in the first element, corresponding to the $x$-axis about which the rotation occurred. This is a computational alternative to the eigenvector approach introduced in ▶ Sect. 2.3.1.5. A computationally efficient way to compute the logarithm of an **SO**(3) matrix is

```
>>> L = trlog(R);
```

**2**

expm is different to the function exp which computes the exponential of each element of the matrix. The matrix exponential can be computed using a power series with a matrix, rather than scalar, argument: $\text{expm}(\mathbf{A}) = \mathbf{1} + \mathbf{A} + \mathbf{A}^2/2! + \mathbf{A}^3/3! + \cdots$

Exponentiating the logarithm of the rotation matrix L using the matrix exponential function expm, ◀ yields the original rotation matrix

```
>>> linalg.expm(L)
array([[      1,       0,        0],
       [      0,  0.9553,  -0.2955],
       [      0,  0.2955,   0.9553]])
```

A computationally efficient way to compute the exponential of an $\mathbf{SO}(3)$ matrix is

```
>>> trexp(L);
```

The exponential of a skew-symmetric matrix is always a rotation matrix, with all the special properties outlined earlier. The skew-symmetric matrix can be reconstructed from the 3-vector s, so we can also write

```
>>> linalg.expm(skew(S))
array([[      1,       0,        0],
       [      0,  0.9553,  -0.2955],
       [      0,  0.2955,   0.9553]])
```

In fact, the function

```
>>> R = rotx(0.3);
```

is equivalent to

```
>>> R = linalg.expm(0.3 * skew([1, 0, 0]));
```

where we have specified the rotation in terms of a rotation angle and a rotation axis (as a unit vector).

In general, we can write

$$\mathbf{R} = e^{\theta[\hat{\boldsymbol{\omega}}]_\times} \in \mathbf{SO}(3)$$

where $\theta$ is the rotation angle, $\hat{\boldsymbol{\omega}}$ is a unit vector parallel to the rotation axis, and the notation $[\cdot]_\times : \mathbb{R}^3 \mapsto \mathbf{so}(3) \subset \mathbb{R}^{3\times3}$ indicates a mapping from a vector to a skew-symmetric matrix. Since $\theta[\hat{\boldsymbol{\omega}}]_\times = [\theta\hat{\boldsymbol{\omega}}]_\times$, we can completely describe the rotation by the vector $\theta\hat{\boldsymbol{\omega}} \in \mathbb{R}^3$ – a rotational representation known as exponential coordinates. Rodrigues' rotation formula (2.27) is a computationally efficient means of computing the matrix exponential for the special case where the argument is a skew-symmetric matrix. ▶

### 2.3.1.7  Unit Quaternions

**»** *Quaternions came from Hamilton after his really good work had been done; and, though beautifully ingenious, have been an unmixed evil to those who have touched them in any way, including Clark Maxwell.*
– Lord Kelvin, 1892

Unit quaternions are widely used today for robotics, computer vision, computer graphics and aerospace navigation systems. The quaternion was invented nearly 200 years ago as an extension of the complex number – a hypercomplex number – with a real part and three complex parts

$$\breve{q} = s + v_x i + v_y j + v_z k \in \mathbb{H} \tag{2.29}$$

where the orthogonal complex numbers $i, j$ and $k$ are defined such that

$$i^2 = j^2 = k^2 = ijk = -1 \ . \tag{2.30}$$

Today, it is more common to consider a quaternion as an ordered pair $\breve{q} = (s, \boldsymbol{v})$, sometimes written as $s + \boldsymbol{v}$, where $s \in \mathbb{R}$ is the scalar part and $\boldsymbol{v} \in \mathbb{R}^3$ is the vector part. In this book, we will also write it in component form as

$$\breve{q} = s\langle v_x, v_y, v_z \rangle \ .$$

Quatenions form a vector space $\breve{q} \in \mathbb{H}$ and therefore allow operations such as addition and subtraction, performed element-wise, and multiplication by a scalar. Quaternions also allow conjugation

$$\breve{q}^* = s\langle -v_x, -v_y, -v_z \rangle \in \mathbb{H} \ , \tag{2.31}$$

and quaternion multiplication

$$\breve{q}_q \circ \breve{q}_2 = s_1 s_2 - \boldsymbol{v}_1 \cdot \boldsymbol{v}_2 \langle s_1 \boldsymbol{v}_2 + s_2 \boldsymbol{v}_1 + \boldsymbol{v}_1 \times \boldsymbol{v}_2 \rangle \in \mathbb{H} \tag{2.32}$$

which is known as the quaternion or Hamilton product and is not commutative. ▶ The inner product is a scalar

$$\breve{q}_1 \cdot \breve{q}_2 = s_1 s_2 + v_{x_1} v_{x_2} + v_{y_1} v_{y_2} + v_{z_1} v_{z_2} \in \mathbb{R} \tag{2.33}$$

and the magnitude of a quaternion is

$$\|\breve{q}\| = \sqrt{\breve{q} \cdot \breve{q}} = \sqrt{s^2 + v_x^2 + v_y^2 + v_z^2} \in \mathbb{R} \ . \tag{2.34}$$

A pure quaternion is one whose scalar part is zero.

To represent rotations, we use unit quaternions or versors, quaternions with unit magnitude $\|\breve{q}\| = 1$ and denoted by $\mathring{q}$. ▶ They can be considered as a rotation of $\theta$ about the unit vector $\hat{\boldsymbol{v}}$ which are related to the components of the unit quaternion by

$$\mathring{q} = \cos\frac{\theta}{2}\langle \hat{\boldsymbol{v}} \sin\frac{\theta}{2} \rangle \ \in S^3 \tag{2.35}$$

which has similarities to the angle-axis representation of ▶ Sect. 2.3.1.5. It also has a double mapping which means that $\mathring{q}$ and $-\mathring{q}$ represent the same orientation.

In the Toolbox, these are implemented by the `UnitQuaternion` class and the constructor converts a passed argument such as a rotation matrix to a unit quaternion, for example

This is implemented by the Toolbox function `trexp` which is equivalent to `expm` but is more efficient.

If we write the quaternion as a 4-vector $(s, v_x, v_y, v_z)$, then multiplication can be expressed as a matrix-vector product where

$$\breve{q} \circ \breve{q}' = \begin{pmatrix} s & -v_x & -v_y & -v_z \\ v_x & s & v_z & -v_y \\ v_y & v_z & s & v_x \\ v_z & v_y & -v_x & s \end{pmatrix}\begin{pmatrix} s' \\ v'_x \\ v'_y \\ v'_z \end{pmatrix}$$

The $4 \times 4$ matrix is given by the `matrix` property of the `Quaternion` class.

The four elements of the unit quaternion form a vector and are known as the Euler parameters of the rotation.

### Excurse 2.17: Sir William Rowan Hamilton

Hamilton (1805–1865) was an Irish mathematician, physicist, and astronomer. He was a child prodigy with a gift for languages and by age thirteen knew classical and modern European languages as well as Persian, Arabic, Hindustani, Sanskrit, and Malay. Hamilton taught himself mathematics at age 17, and discovered an error in Laplace's Celestial Mechanics. He spent his life at Trinity College, Dublin, and was appointed Professor of Astronomy and Royal Astronomer of Ireland while still an undergraduate. In addition to quaternions, he contributed to the development of optics, dynamics, and algebra. He also wrote poetry and corresponded with Wordsworth who advised him to devote his energy to mathematics.

According to legend, the key quaternion equation, (2.30), occurred to Hamilton in 1843 while walking along the Royal Canal in Dublin with his wife, and this is commemorated by a plaque on Broome bridge:

» Here as he walked by on the 16th of October 1843 Sir William Rowan Hamilton in a flash of genius discovered the fundamental formula for quaternion

multiplication $i^2 = j^2 = k^2 = ijk = -1$ & cut it on a stone of this bridge.

His original carving is no longer visible, but the bridge is a pilgrimage site for mathematicians and physicists.

```
>>> q = UnitQuaternion(rpy2r(0.1, 0.2, 0.3))
 0.9833 <<  0.0343,  0.1060,  0.1436 >>
```

The class supports a number of standard operators and methods, for example, quaternion multiplication ◄ is achieved by the overloaded multiplication operator

```
>>> q = q * q;
```

and inversion, the conjugate of a unit quaternion, is

```
>>> q.inv()
 0.9339 << -0.0674, -0.2085, -0.2824 >>
```

Multiplying a unit quaternion by its inverse, yields the identity quaternion

```
>>> q * q.inv()
 1.0000 <<  0.0000,  0.0000,  0.0000 >>
```

which represents a null rotation. This can be expressed more succinctly as

```
>>> q / q
 1.0000 <<  0.0000,  0.0000,  0.0000 >>
```

The unit quaternion $\mathring{q}$ can be converted to a rotation matrix by

```
>>> q.R
array([[  0.7536,  -0.4993,   0.4275],
       [  0.5555,   0.8315, -0.008145],
       [ -0.3514,   0.2436,   0.904]])
```

A coordinate vector can be rotated by a quaternion using the overloaded multiplication operator

```
>>> q * [1, 0, 0]
array([  0.7536,   0.5555,  -0.3514])
```

and we can also plot the orientation represented by the unit quaternion

```
>>> q.plot();
```

Compounding two orthonormal rotation matrices requires 27 multiplications and 18 additions. The quaternion form requires 16 multiplications and 12 additions. This saving can be particularly important for embedded systems.

as a coordinate frame in a similar style to that shown in ◘ Fig. 2.17. The `UnitQuaternion` class has many methods and operators which are described fully in the online documentation.

---

**$\xi$ as a unit quaternion**

For the case of pure rotation in 3D, $\xi$ can be implemented with a unit quaternion $\mathring{q} \in S^3$. The implementation is:

| | | |
|---|---|---|
| composition | $\xi_1 \oplus \xi_2$ | $\mapsto \mathring{q}_1 \circ \mathring{q}_2$, Hamilton product |
| inverse | $\ominus \xi$ | $\mapsto \mathring{q}^* = s\langle -v \rangle$, quaternion conjugation |
| identity | $\varnothing$ | $\mapsto 1\langle 0 \rangle$ |
| vector-transform | $\xi \cdot v$ | $\mapsto \mathring{q} \circ \mathring{v} \circ \mathring{q}^*$, where $\mathring{v} = 0\langle v \rangle$ is a pure quaternion |

Composition is not commutative, that is, $\mathring{q}_1 \circ \mathring{q}_2 \neq \mathring{q}_2 \circ \mathring{q}_1$.

---

It is possible to store only the vector part of a unit quaternion, since the scalar part can be recovered by $s = \pm(1 - v_x^2 - v_y^2 - v_z^2)^{1/2}$. The sign ambiguity can be resolved by ensuring that the scalar part is positive, negating the quaternion if required, before taking the vector part. This 3-vector form of a unit quaternion is frequently used for optimization problems such as posegraph relaxation or bundle adjustment.

**❗ Beware the left-handed quaternion**

Hamilton defined the complex numbers such that $ijk = -1$ but in the aerospace community it is common to use quaternions where $ijk = 1$. These are known as JPL, flipped or left-handed quaternions. This can be confusing, and you need to understand the convention adopted by any software libraries or packages you might use. This book and the Toolbox software adopt Hamilton's convention.

**❗ Quaternion element ordering**

We have described quaternions with the components implicitly ordered as $s, v_x, v_y, v_z$, and the quaternion functions in the `spatialmath.base` package consider the quaternion as a 4-element 1D NumPy array with this ordering. Some software tools, notably ROS, consider the quaternion as a vector ordered as $v_x, v_y, v_z, s$.

## 2.3.2 Pose in Three Dimensions

To describe the relative pose of the frames shown in ◘ Fig. 2.14, we need to account for the translation between the origins of the frames, as well as the rotation. We will combine particular methods of representing rotation with representations of translation, to create tangible representations of relative pose in 3D.

### 2.3.2.1 Homogeneous Transformation Matrix

The derivation for the 3D homogeneous transformation matrix is similar to the 2D case of (2.18) but extended to account for the $z$-dimension

$$\begin{pmatrix} {}^A x \\ {}^A y \\ {}^A z \\ 1 \end{pmatrix} = \begin{pmatrix} {}^A \mathbf{R}_B & {}^A t_B \\ \mathbf{0}_{1 \times 3} & 1 \end{pmatrix} \begin{pmatrix} {}^B x \\ {}^B y \\ {}^B z \\ 1 \end{pmatrix}$$

**2**

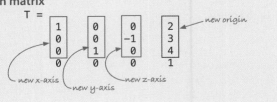

where ${}^A t_B \in \mathbb{R}^3$, as shown in ◙ Fig. 2.14, is a vector defining the origin of frame {B} with respect to frame {A}, and ${}^A \mathbf{R}_B \in \mathbf{SO}(3)$ is the rotation matrix which describes the orientation of frame {B} with respect to frame {A}. If points are represented by homogeneous coordinate vectors, then

$$
\begin{aligned}
{}^A \tilde{p} &= \begin{pmatrix} {}^A\mathbf{R}_B & {}^A t_B \\ \mathbf{0}_{1\times 3} & 1 \end{pmatrix} {}^B \tilde{p} \\
&= {}^A\mathbf{T}_B \, {}^B \tilde{p}
\end{aligned}
\tag{2.36}
$$

where ${}^A\mathbf{T}_B$ is a $4 \times 4$ homogeneous transformation matrix which is commonly called a "transform". This matrix has a very specific structure and belongs to the Special Euclidean group of dimension 3 which we write as $\mathbf{T} \in \mathbf{SE}(3) \subset \mathbb{R}^{4\times 4}$.

The $4 \times 4$ homogeneous transformation matrix is very commonly used in robotics, computer graphics and computer vision. It is supported by the Toolbox and will be used throughout this book as a concrete representation of $\xi$ for 3-dimensional relative pose.

The Toolbox has many functions to create 3D homogeneous transformations. We can demonstrate composition of transforms by

```
>>> T = transl(2, 0, 0) @ trotx(pi / 2) @ transl(0, 1, 0)
array([[      1,        0,        0,        2],
       [      0,        0,       -1,        0],
       [      0,        1,        0,        1],
       [      0,        0,        0,        1]])
```

For describing rotation, the Toolbox has functions that create a rotation matrix (eg. `rotx`, `rpy2r`) or a homogeneous transformation with zero translation (eg. `trotx`, `rpy2tr`. Some Toolbox functions accept a rotation matrix or a homogeneous transformation and ignore the translational component, for example, `tr2rpy`.

The function `transl` creates a relative pose with a finite translation but zero rotation, while `trotx` creates a relative pose corresponding to a rotation of $\frac{\pi}{2}$ about the $x$-axis with zero translation. ◄ We can think of this expression as representing a walk along the $x$-axis for 2 units, *then* a rotation by 90° about the $x$-axis, *then* a walk of 1 unit along the new $y$-axis. The result, as shown in the last column of the resulting matrix is a translation of 2 units along the original $x$-axis and 1 unit along the original $z$-axis. We can plot the corresponding coordinate frame by

```
>>> trplot(T);
```

The rotation matrix component of T is

```
>>> t2r(T)
array([[      1,        0,        0],
       [      0,        0,       -1],
       [      0,        1,        0]])
```

For historical reasons, `transl` converts $\mathbb{R}^3 \mapsto \mathbf{SE}(3)$ as well as $\mathbf{SE}(3) \mapsto \mathbb{R}^3$, depending on the arguments provided. `transl2` behaves in a similar way.

and the translation component is a column vector ◄

```
>>> transl(T)
array([      2,        0,        1])
```

**ξ as an SE(3) matrix**

For the case of rotation and translation in 3D, $\boldsymbol{\xi}$ can be implemented by a homogeneous transformation matrix $\mathbf{T} \in \mathbf{SE}(3)$ which is sometimes written as an ordered pair $(\mathbf{R}, t) \in \mathbf{SO}(3) \times \mathbb{R}^3$. The implementation is:

composition $\quad \boldsymbol{\xi}_1 \oplus \boldsymbol{\xi}_2 \quad \mapsto \mathbf{T}_1\mathbf{T}_2 = \begin{pmatrix} \mathbf{R}_1\mathbf{R}_2 & t_1 + \mathbf{R}_1 t_2 \\ \mathbf{0}_{1\times 3} & 1 \end{pmatrix}$, matrix multiplication

inverse $\quad \ominus\boldsymbol{\xi} \quad \mapsto \mathbf{T}^{-1} = \begin{pmatrix} \mathbf{R}^\top & -\mathbf{R}^\top t \\ \mathbf{0}_{1\times 3} & 1 \end{pmatrix}$, matrix inverse

identity $\quad \varnothing \quad \mapsto \mathbf{1}_{4\times 4}$, identity matrix

vector-transform $\quad \boldsymbol{\xi} \cdot \boldsymbol{v} \quad \mapsto \epsilon(\mathbf{T}\tilde{\boldsymbol{v}})$, matrix-vector product using homogeneous vectors

where $\tilde{\cdot} : \mathbb{R}^3 \mapsto \mathbb{P}^3$ and $\epsilon(\cdot) : \mathbb{P}^3 \mapsto \mathbb{R}^3$. Composition is not commutative, that is, $\mathbf{T}_1\mathbf{T}_2 \neq \mathbf{T}_2\mathbf{T}_1$.

### 2.3.2.2 Matrix Exponential for Pose

Consider the $\mathbf{SE}(3)$ matrix

```
>>> T = transl(2, 3, 4) @ trotx(0.3)
array([[      1,        0,        0,        2],
       [      0,   0.9553,  -0.2955,        3],
       [      0,   0.2955,   0.9553,        4],
       [      0,        0,        0,        1]])
```

and its logarithm

```
>>> L = linalg.logm(T)
array([[      0,        0,        0,        2],
       [      0,        0,     -0.3,    3.577],
       [      0,      0.3,        0,     3.52],
       [      0,        0,        0,        0]])
```

is a sparse matrix and the rotation magnitude, 0.3, is clearly evident. The structure of this matrix is an augmented skew-symmetric matrix: the upper-left corner is a $3 \times 3$ skew-symmetric matrix, the upper right column is a 3-vector, and the bottom row is zero. We can unpack the six unique elements by

```
>>> S = vexa(L)
array([      2,    3.577,     3.52,      0.3,        0,        0])
```

and the last three elements describe rotation, and the value 0.3 is in the first "slot" indicating a rotation about the first axis – the $x$-axis.

Exponentiating the logarithm matrix yields the original $\mathbf{SE}(3)$ matrix, and the logarithm can be reconstructed from just the six elements of s

```
>>> linalg.expm(skewa(S))
array([[      1,        0,        0,        2],
       [      0,   0.9553,  -0.2955,        3],
       [      0,   0.2955,   0.9553,        4],
       [      0,        0,        0,        1]])
```

In general, we can write

$$\mathbf{T} = e^{[S]} \in \mathbf{SE}(3)$$

where $S \in \mathbb{R}^6$ and $[\cdot] : \mathbb{R}^6 \mapsto \mathbf{se}(3) \subset \mathbb{R}^{4\times 4}$. The vector $S$ is a *twist vector* as we will discuss next.

**2**

**Excurse 2.19: 3D Augmented Skew-Symmetric Matrix**

In 3 dimensions, the augmented skew-symmetric matrix, corresponding to the vector $S = (v_x, v_y, v_z, \omega_x, \omega_y, \omega_z)$, is

$$[S] = \left( \begin{array}{ccc|c} 0 & -\omega_z & \omega_y & v_x \\ \omega_z & 0 & -\omega_x & v_y \\ -\omega_y & \omega_x & 0 & v_z \\ \hline 0 & 0 & 0 & 0 \end{array} \right) \in \mathbf{se}(3) \tag{2.37}$$

which has a distinct structure with a zero diagonal and bottom row, and a skew-symmetric matrix in the top-left corner. The vector space of 3D augmented skew-symmetric matrices is denoted $\mathbf{se}(3)$ and is the Lie algebra of $\mathbf{SE}(3)$. The Toolbox implements the $[\cdot]$ operator as

```
>>> X = skewa([1, 2, 3, 4, 5, 6])
array([[      0,      -6,       5,       1],
       [      6,       0,      -4,       2],
       [     -5,       4,       0,       3],
       [      0,       0,       0,       0]])
```

and the inverse operator $\vee(\cdot)$ as

```
>>> vexa(X)
array([      1,       2,       3,       4,       5,       6])
```

### 2.3.2.3  3D Twists

In 3D, a twist is equivalent to screw or *helicoidal* motion – motion about and along some line in space as shown in ◘ Fig. 2.24. Rigid-body motion between any two poses can be represented by such screw motion. For pure translation, the rotation is about a screw axis at infinity.

We represent a screw as a twist vector $(\boldsymbol{v}, \boldsymbol{\omega}) \in \mathbb{R}^6$ comprising a moment $\boldsymbol{v} \in \mathbb{R}^3$ and a vector $\boldsymbol{\omega} \in \mathbb{R}^3$ parallel to the screw axis. The moment encodes a point lying on the twist axis, as well as the screw pitch which is the ratio of the distance along the screw axis to the rotation about the screw axis.

Consider the example of a rotation about the $x$-axis which has a screw axis parallel to the $x$-axis and passing through the origin

```
>>> S = Twist3.UnitRevolute([1, 0, 0], [0, 0, 0])
(0 0 0; 1 0 0)
```

This particular twist is a *unit twist* that describes a rotation of 1 rad about the $x$-axis.

To create an $\mathbf{SE}(3)$ transformation for a specific rotation, we scale it and exponentiate it

```
>>> linalg.expm(0.3 * skewa(S.S));
```

where the `.S` attribute is the twist vector as a NumPy array, or more compactly

```
>>> S.exp(0.3)
   1         0         0         0
   0         0.9553   -0.2955    0
   0         0.2955    0.9553    0
   0         0         0         1
```

The result is an `SE3` object which encapsulates a $4 \times 4$ NumPy array. The is discussed further in ▶ Sect. 2.5.

which is identical to the value we would obtain using `trotx(0.3)`. ◄ However, the twist allows us to express rotation about an axis passing through any point, not just the origin.

To illustrate the underlying screw motion, we will progressively rotate a coordinate frame about a screw. We define a screw parallel to the $z$-axis that passes

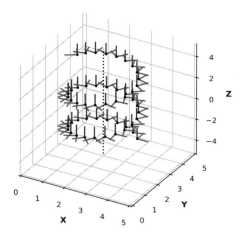

**◻ Fig. 2.21** A coordinate frame displayed for different values of $\theta$ about a screw parallel to the $z$-axis and passing through the point $(2, 3, 2)$. The $x$-, $y$- and $z$-axes are indicated by red, green and blue lines respectively

through the point $(2, 3, 2)$ and has a pitch of 0.5

```
>>> S = Twist3.UnitRevolute([0, 0, 1], [2, 3, 2], 0.5)
(3 -2 0.5; 0 0 1)
```

and the coordinate frame to be rotated is described by an **SE**(3) matrix

```
>>> X = transl(3, 4, -4);
```

For values of $\theta$ in the range 0 to 15 rad, we evaluate the twist for each value of $\theta$, ▶ apply it to the frame {X} and plot the result

The `.A` attribute of the resulting `SE3` instance is the **SE**(3) matrix as a NumPy array.

```
>>> for theta in np.arange(0, 15, 0.3):
...     trplot(S.exp(theta).A @ X, style="rgb", width=2)
```

The plot in ◻ Fig. 2.21 clearly shows the screw motion in the successive poses of the frame {X} as it is rotated about the screw axis. The screw axis is a 3D line

```
>>>  L = S.line()
{ -3 2 -1; 0 0 1}
>>> L.plot("k:", linewidth=2);
```

which is shown in the plot as a black dotted line. ▶

A translational, or prismatic, unit twist in the $y$-direction is created by

The 3D line is represented by a `Line3` object in terms of Plücker coordinates which are covered in ▶ App. C.1.2.2.

```
>>> S = Twist3.UnitPrismatic([0, 1, 0])
(0 1 0; 0 0 0)
```

and describes a displacement of 1 in the specified direction. To create an **SE**(3) transformation for a specific translation, we scale and exponentiate the unit twist

```
>>> S.exp(2)
  1        0        0        0
  0        1        0        2
  0        0        1        0
  0        0        0        1
```

which has a zero rotation and a translation of 2 in the $y$-direction.

We can also convert an arbitrary **SE**(3) matrix to a twist. For the transform from ▶ Sect. 2.3.2.2, the twist vector is

```
>>> T = transl(1, 2, 3) @ eul2tr(0.3, 0.4, 0.5);
>>> S = Twist3(T)
(1.1204 1.6446 3.1778; 0.041006 0.4087 0.78907)
```

2

and we see that the twist vector is the unique elements of the logarithm of the **SE**(3) matrix. The screw in this case is parallel to

```
>>> S.w
array([ 0.04101,    0.4087,    0.7891])
```

passes through the point

```
>>> S.pole
array([0.001138,    0.8473,    -0.4389])
```

has a pitch of

```
>>> S.pitch
3.226
```

and the rigid-body motion requires a rotation about the screw of

```
>>> S.theta
0.8896
```

radians, which is about 1/7th of a turn.

---

**$\xi$ as a 3D twist**

For the case of rotation and translation in 3D, $\xi$ can be implemented by a twist $S \in \mathbb{R}^6$. The implementation is:

| | | |
|---|---|---|
| composition | $\xi_1 \oplus \xi_2$ | $\mapsto \log\big(e^{[S_1]}e^{[S_2]}\big)$, product of exponential |
| inverse | $\ominus \xi$ | $\mapsto -S$, negation |
| identity | $\varnothing$ | $\mapsto \mathbf{0}_{1\times 6}$, zero vector |
| vector-transform | $\xi \cdot v$ | $\mapsto \epsilon(e^{[S]}\tilde{v})$, matrix-vector product using homogeneous vectors |

where $\tilde{\cdot} : \mathbb{R}^3 \mapsto \mathbb{P}^3$ and $\epsilon(\cdot) : \mathbb{P}^3 \mapsto \mathbb{R}^3$. Composition is not commutative, that is, $e^{[S_1]}e^{[S_2]} \neq e^{[S_2]}e^{[S_1]}$. Note that log and exp have efficient closed form, rather than transcendental, solutions which makes composition relatively inexpensive.

---

### 2.3.2.4 Vector-Quaternion Pair

Another way to represent pose in 3D is to combine a translation vector and a unit quaternion. It represents pose using just 7 numbers, is easy to compound, and singularity free.

---

**$\xi$ as a vector-quaternion pair**

For the case of rotation and translation in 3D, $\xi$ can be implemented by a vector and a unit quaternion written as an ordered pair $(t, \mathring{q}) \in \mathbb{R}^3 \times S^3$. The implementation is:

| | | |
|---|---|---|
| composition | $\xi_1 \oplus \xi_2$ | $\mapsto (t_1 + \mathring{q}_1 \cdot t_2,\ \mathring{q}_1 \circ \mathring{q}_2)$ |
| inverse | $\ominus \xi$ | $\mapsto (-\mathring{q}^* \cdot t,\ \mathring{q}^*)$ |
| identity | $\varnothing$ | $\mapsto (\mathbf{0}_3, 1\langle 0 \rangle)$ |
| vector-transform | $\xi \cdot v$ | $\mapsto \mathring{q} \cdot v + t$ |

Composition is not commutative, that is, $\xi_1 \xi_2 \neq \xi_2 \xi_1$. The vector transformation operator $\cdot : S^3 \times \mathbb{R}^3 \mapsto \mathbb{R}^3$ can be found in ▶ Sect. 2.3.1.7.

---

### 2.3.2.5  Unit Dual Quaternion

An early approach to representing pose with quaternions was Hamilton's bi-quaternion where the quaternion coefficients were themselves complex numbers. Later, William Clifford ▶ developed the dual number which is defined as an ordered pair $d=(x,y)$ that can be written as $d = x + y\varepsilon$, where $\varepsilon^2 = 0$. Clifford created a quaternion dual number with $x, y \in \mathbb{H}$ which he also called a bi-quaternion but is today called a dual quaternion.

William Clifford (1845–1879) was an English mathematician and geometer. He made contributions to geometric algebra, and Clifford Algebra is named in his honor.

> **$\xi$ as a unit dual quaternion**
>
> For the case of rotation and translation in 3D, $\boldsymbol{\xi}$ can be implemented by a unit dual quaternion $(\mathring{q}_r, \breve{q}_d) \in S^3 \times \mathbb{H}$ which comprises a unit quaternion $\mathring{q}_r$ (the real part) representing rotation, and a quaternion $\breve{q}_d = \frac{1}{2}\breve{t} \circ \mathring{q}_r$ (the dual part) where $\breve{t}$ is a pure quaternion representing translation. The implementation is:
>
> | | | |
> |---|---|---|
> | composition | $\boldsymbol{\xi}_1 \oplus \boldsymbol{\xi}_2$ | $\mapsto (\mathring{q}_{r_1} \circ \mathring{q}_{r_2}, \mathring{q}_{r_1} \circ \breve{q}_{d_2} + \breve{q}_{d_1} \circ \mathring{q}_{r_2})$ |
> | inverse | $\ominus\boldsymbol{\xi}$ | $\mapsto (\mathring{q}_r, \breve{q}_d)^* = (\mathring{q}_r^*, \breve{q}_d^*)$, conjugation |
> | identity | $\varnothing$ | $\mapsto (1\langle 0\rangle, 1\langle 0\rangle)$ |
> | vector-transform | $\boldsymbol{\xi} \cdot \boldsymbol{v}$ | $\mapsto (\mathring{q}_r, \breve{q}_d) \circ (1\langle\mathbf{0}\rangle, \breve{v}) \circ (\mathring{q}_r, \breve{q}_d)^*$, where the middle term is a pure dual quaternion. |
>
> Composition is not commutative, that is, $\boldsymbol{\xi}_1\boldsymbol{\xi}_2 \neq \boldsymbol{\xi}_2\boldsymbol{\xi}_1$. Note that, in general, the Hamilton product of a general quaternion $\breve{p}$ and a unit quaternion $\mathring{q}$ is a non-unit quaternion.

This representation is quite compact, requiring just 8 numbers; it is easy to compound using a special multiplication table; and it is easy to normalize to eliminate the effect of finite precision arithmetic. However, it has no real useful computational advantage over matrix methods.

## 2.4  Advanced Topics

### 2.4.1  Pre- and Post-Multiplication of Transforms

Many texts make a distinction between *pre-multiplying* and *post-multiplying* relative transformations. For the case of a series of rotation matrices, the former rotates with respect to the world reference frame, whereas the latter rotates with respect to the current or rotated frame. The difference is best understood using the pose graph shown in ◘ Fig. 2.22. The pose {A} is defined with respect to the reference pose {0}.

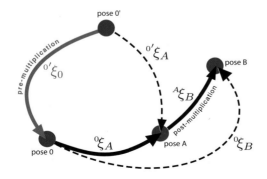

◘ **Fig. 2.22**  Pre- and post-multiplication shown as a pose graph

**2**

Post-multiplication should be familiar by now, and is the motion from {A} to {B} expressed in frame {A} – the rotated frame. The result, $^0\boldsymbol{\xi}_B$, is the motion from {0} to {A} and then to {B}.

Pre-multiplication is shown by the red arrow. It requires that we introduce a new pose for the start of the arrow which we will call {0′}. We can compose these motions – the super- and sub-scripts match – as $^{0'}\boldsymbol{\xi}_0 \oplus {}^0\boldsymbol{\xi}_A = {}^{0'}\boldsymbol{\xi}_A$. Effectively this is a change in the reference frame – from {0} to {0′}. The transformation is therefore with respect to the reference frame, albeit a new one.

A three-angle rotation sequence such as ZYX roll-pitch-yaw angles from (2.23)

$$^0\mathbf{R}_B = \mathbf{R}_z(\gamma)\,\mathbf{R}_y(\beta)\,\mathbf{R}_x(\alpha)$$

can be interpreted in two different, but equivalent, ways. Firstly, using pre-multiplication, we can consider this right to left as consecutive rotations about the reference frame by $\mathbf{R}_x(\alpha)$, then by $\mathbf{R}_y(\beta)$ and then by $\mathbf{R}_z(\gamma)$ – these are referred to as extrinsic rotations. Alternatively, from left to right as sequential rotations, each with respect to the previous frame, by $\mathbf{R}_z(\gamma)$, then by $\mathbf{R}_y(\beta)$, and then by $\mathbf{R}_x(\alpha)$ – these are referred to as intrinsic rotations.

### 2.4.2  Rotating a Frame Versus Rotating a Point

Many sources distinguish between transforming the coordinate vector of a point $\boldsymbol{p}$ by *rotating the reference frame* (passive rotation)

$$\boldsymbol{q} = \mathbf{R}^p\,\boldsymbol{p}$$

and *rotating the point* (active rotation)

$$\boldsymbol{q} = \mathbf{R}^a\,\boldsymbol{p}\ \ .$$

This becomes less confusing if we rewrite using our super- and sub-script notation. The passive rotation

$$^0\boldsymbol{q} = {}^0\mathbf{R}_B^p\,{}^B\boldsymbol{p}$$

transforms a coordinate vector from the rotated body frame to the world frame, while the active rotation

$$^B\boldsymbol{q} = {}^B\mathbf{R}_0^a\,{}^0\boldsymbol{p}$$

transforms a coordinate vector from the world frame to the rotated body frame. The active and passive rotations are inverses and $\mathbf{R}^a = (\mathbf{R}^p)^\top$.

In this book, we will not make such distinctions. It is simpler to consider that points are fixed to objects and defined by coordinate vectors relative to the object's coordinate frame. We use our established mechanisms to describe motion between frames, and how such motion transforms these coordinate vectors.

### 2.4.3  Direction Cosine Matrix

Consider two coordinate frames {A} and {B}. The rotation matrix $^A\mathbf{R}_B$ can be written as

$$^A\mathbf{R}_B = \begin{pmatrix} \cos(\hat{\boldsymbol{x}}_A, \hat{\boldsymbol{x}}_B) & \cos(\hat{\boldsymbol{x}}_A, \hat{\boldsymbol{y}}_B) & \cos(\hat{\boldsymbol{x}}_A, \hat{\boldsymbol{z}}_B) \\ \cos(\hat{\boldsymbol{y}}_A, \hat{\boldsymbol{x}}_B) & \cos(\hat{\boldsymbol{y}}_A, \hat{\boldsymbol{y}}_B) & \cos(\hat{\boldsymbol{y}}_A, \hat{\boldsymbol{z}}_B) \\ \cos(\hat{\boldsymbol{z}}_A, \hat{\boldsymbol{x}}_B) & \cos(\hat{\boldsymbol{z}}_A, \hat{\boldsymbol{y}}_B) & \cos(\hat{\boldsymbol{z}}_A, \hat{\boldsymbol{z}}_B) \end{pmatrix} \tag{2.38}$$

where $\cos(\hat{\boldsymbol{u}}_A, \hat{\boldsymbol{v}}_B)$, $u, v \in \{x, y, z\}$ is the cosine of the angle between the unit vector $\hat{\boldsymbol{u}}_A$ in {A}, and the unit vector $\hat{\boldsymbol{v}}_B$ in {B}. Since $\hat{\boldsymbol{u}}_A$ and $\hat{\boldsymbol{v}}_B$ are unit vectors $\cos(\hat{\boldsymbol{u}}_A, \hat{\boldsymbol{v}}_B) = \hat{\boldsymbol{u}}_A \cdot \hat{\boldsymbol{v}}_B$. This matrix is called the direction cosine matrix (DCM) and is frequently used in aerospace applications.

### 2.4.4 Efficiency of Representation

Many of the representations discussed in this chapter have more parameters than strictly necessary. For instance, an **SO**(2) matrix is $2 \times 2$ and has four elements that represent a single rotation angle. This redundancy can be explained in terms of constraints: the columns are each unit-length ($\|\boldsymbol{c}_1\| = \|\boldsymbol{c}_2\| = 1$), which provides two constraints; and the columns are orthogonal ($\boldsymbol{c}_1 \cdot \boldsymbol{c}_2 = 0$), which adds another constraint. Four matrix elements with three constraints is effectively one independent value.

⬛ Table 2.1 shows the number of parameters $N$ for various representations, as well as the minimum number of parameters $N_{\min}$ required. The matrix representations are the least efficient, they have the lowest $N_{\min}/N$, which has a real cost in terms of computer memory. However, this is offset by the simplicity of composition which is simply matrix multiplication and that can be executed very efficiently on modern computer hardware. If space is an issue, then we can use simple tricks like not storing the bottom row of matrices which is equivalent to representing the homogeneous transformation matrix as an ordered pair $(\mathbf{R}, \boldsymbol{t})$ which is common in computer vision.

### 2.4.5 Distance Between Orientations

The distance between two Euclidean points is the length of the line between them, that is, $\|\boldsymbol{p}_1 - \boldsymbol{p}_2\|$, $\boldsymbol{p}_i \in \mathbb{R}^n$. The *distance* between two orientations is more complex and, unlike translational distance, it has an upper bound. Some common metrics for

⬛ **Table 2.1** Number of parameters required for different pose representations. The † variant does not store the constant bottom row of the matrix

| Type | $N$ | $N_{\min}$ | $N_{\min}/N$ |
| --- | --- | --- | --- |
| **SO**(2) matrix | 4 | 1 | 25% |
| **SE**(2) matrix | 9 | 3 | 33% |
| **SE**(2) as $2 \times 3$ matrix † | 6 | 3 | 50% |
| 2D twist vector | 3 | 3 | **100%** |
| **SO**(3) matrix | 9 | 3 | 33% |
| Unit quaternion | 4 | 3 | 75% |
| Unit quaternion vector part | 3 | 3 | **100%** |
| Euler parameters | 3 | 3 | **100%** |
| **SE**(3) matrix | 16 | 6 | 38% |
| **SE**(3) as $3 \times 4$ matrix † | 12 | 6 | 50% |
| 3D twist vector | 6 | 6 | **100%** |
| Vector + unit quaternion | 7 | 6 | 86% |
| Unit dual quaternion | 8 | 6 | 75% |

rotation matrices and quaternions are

$$d(\mathbf{R}_1, \mathbf{R}_2) = \| \mathbf{1}_{3 \times 3} - \mathbf{R}_1 \mathbf{R}_2^\top \| \in [0, 2] \tag{2.39}$$

$$d(\mathbf{R}_1, \mathbf{R}_2) = \| \log \mathbf{R}_1 \mathbf{R}_2^\top \| \in [0, \pi] \tag{2.40}$$

$$d(\mathring{q}_1, \mathring{q}_2) = 1 - |\mathring{q}_1 \cdot \mathring{q}_2| \in [0, 1] \tag{2.41}$$

$$d(\mathring{q}_1, \mathring{q}_2) = \cos^{-1} |\mathring{q}_1 \cdot \mathring{q}_2| \in [0, \pi/2] \tag{2.42}$$

$$d(\mathring{q}_1, \mathring{q}_2) = 2 \tan^{-1} \frac{\|\mathring{q}_1 - \mathring{q}_2\|}{\|\mathring{q}_1 + \mathring{q}_2\|} \in [0, \pi/2] \tag{2.43}$$

$$d(\mathring{q}_1, \mathring{q}_2) = 2 \cos^{-1} |\mathring{q}_1 \cdot \mathring{q}_2| \in [0, \pi] \tag{2.44}$$

where the $\cdot$ operator for unit quaternions is the inner product given by (2.33). These measures are all proper distance metrics which means that $d(x, x) = 0$, $d(x, y) = d(y, x)$, and the triangle inequality holds $d(x, z) \le d(x, y) + d(y, z)$. For a metric, $d(x, y) = 0$ implies that $x = y$ but for unit quaternions this is not true since, due to the double mapping, it is also true that $d(\mathring{q}, -\mathring{q}) = 0$. This problem can be avoided by ensuring that unit quaternions always have a non-negative scalar part. (2.42) and (2.43) are equivalent but the latter is numerically more robust.

These metrics can be computed by the `angdist` method of the `SO3` and `UnitQuaternion` classes which, by default, use (2.43). For example

```
>>> UnitQuaternion.Rx(pi / 2).angdist(UnitQuaternion.Rz(-pi / 2))
1.047
```

### 2.4.6 Normalization

The IEEE-754 standard for double precision (64-bit) floating point arithmetic has around 16 decimal digits of precision.

Floating-point arithmetic has finite precision ◄ and each operation introduces some very small error. Consecutive operations will accumulate that error. A rotation matrix has, by definition, a determinant of one

```
>>> R = np.eye(3,3);
>>> np.linalg.det(R) - 1
0
```

but if we repeatedly multiply by a valid rotation matrix the result

```
>>> for i in range(100):
...     R = R @ rpy2r(0.2, 0.3, 0.4);
>>> np.linalg.det(R) - 1
5.773e-15
```

This may vary between computers.

has a small error ◄ – the determinant is no longer equal to one and the matrix is no longer a proper orthonormal rotation matrix. To fix this, we need to normalize the matrix, a process which enforces the constraints on the columns $c_i$ of the rotation matrix $\mathbf{R} = [c_0, c_1, c_2]$. We assume that the third column has the correct direction

$$c'_2 = c_2$$

then the first column is made orthogonal to the last two

$$c'_0 = c_1 \times c'_2 \ .$$

However, the last two columns may not have been orthogonal so

$$c'_1 = c'_2 \times c'_0 \ .$$

Finally, the columns are all normalized to unit magnitude

$$c''_i = \frac{c'_i}{\|c'_i\|}, \ i = 0, 1, 2$$

In the Toolbox, normalization is implemented by

```
>>> R = trnorm(R);
```

and the determinant is now much closer to one ▶

```
>>> np.linalg.det(R) - 1
2.22e-16
```

This error is now at the limit of IEEE-754 standard 64-bit (double precision) floating point arithmetic which is $2.2204 \times 10^{-16}$ and given by the NumPy function `np.finfo.eps`.

Normalization can also be applied to an **SE**(3) matrix, in which case only the **SO**(3) rotation submatrix is normalized.

Arithmetic issues also occur when multiplying unit quaternions – the norm, or magnitude, of the unit quaternion diverges from one. However, this is much easier to fix since normalizing the quaternion simply involves dividing all elements by the norm

$$\mathring{q}' = \frac{\mathring{q}}{\|\mathring{q}\|}$$

which is implemented by

```
>>> q = q.unit();
```

The `SO2`, `SE2`, `SO3`, `SE3` and `UnitQuaternion` classes also support a variant of multiplication using the `@` operator

```
>>> T = T1 @ T2
>>> q = q1 @ q2
```

which performs an explicit normalization after the multiplication.

Normalization does not need to be performed after every multiplication since it is an expensive operation. However it is advisable for situations like the example above, where one transform is being repeatedly updated.

### 2.4.7 Understanding the Exponential Mapping

In this chapter, we have glimpsed some connection between rotation matrices, homogeneous transformation matrices, skew-symmetric matrices, matrix logarithms and matrix exponentiation. The roots of this connection are in the mathematics of Lie groups which are covered in text books on algebraic topology. This is advanced mathematics and many people starting out in robotics will find their content somewhat inaccessible. An introduction to the essentials of this topic is given in ▶ App. D. In this section, we will use an intuitive approach, based on undergraduate engineering mathematics, to shed some light on the connection.

■ Fig. 2.23 shows a point P, defined by a coordinate vector $p$, being rotated about an axis. The axis is parallel to the vector $\omega$ whose magnitude $\|\omega\|$ specifies the rate of rotation about the axis. ▶ We wish to rotate the point by an angle $\theta$ about this axis and the velocity of the point is known from mechanics to be

Such a vector is an angular velocity vector which will be properly introduced in the next chapter.

$$\dot{p} = \omega \times p$$

and we can replace the cross product with a skew-symmetric matrix-vector product

$$\dot{p} = [\omega]_\times p \ . \tag{2.45}$$

We can find the solution to this first-order differential equation by analogy to the simple scalar case

$$\dot{x} = ax$$

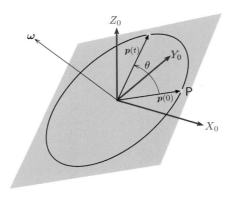

■ **Fig. 2.23**   The point P follows a circular path in the plane normal to the axis $\boldsymbol{\omega}$

whose solution is

$$x(t) = e^{at} x(0)$$

which implies that the solution to (2.45) is

$$\boldsymbol{p}(t) = e^{[\boldsymbol{\omega}]_\times t}\, \boldsymbol{p}(0) \;.$$

If $\|\boldsymbol{\omega}\| = 1$, then after $t$ seconds, the coordinate vector will have rotated by $t$ radians. We require a rotation by $\theta$, so we can set $t = \theta$ to give

$$\boldsymbol{p}(\theta) = e^{[\hat{\boldsymbol{\omega}}]_\times \theta}\, \boldsymbol{p}(0)$$

which describes the vector $\boldsymbol{p}(0)$ being rotated to $\boldsymbol{p}(\theta)$. A matrix that rotates a coordinate vector is a rotation matrix, and this implies that our matrix exponential is a rotation matrix

$$\mathbf{R}(\theta,\, \hat{\boldsymbol{\omega}}) = e^{[\hat{\boldsymbol{\omega}}]_\times \theta} \in \mathbf{SO}(3) \;.$$

Now consider the more general case of rotational and translational motion. We can write

$$\dot{\boldsymbol{p}} = [\boldsymbol{\omega}]_\times \boldsymbol{p} + \boldsymbol{v}$$

and, rearranging into matrix form

$$\begin{pmatrix} \dot{\boldsymbol{p}} \\ 0 \end{pmatrix} = \begin{pmatrix} [\boldsymbol{\omega}]_\times & \boldsymbol{v} \\ 0 & 0 \end{pmatrix} \begin{pmatrix} \boldsymbol{p} \\ 1 \end{pmatrix}$$

and introducing homogeneous coordinates, this becomes

$$\begin{aligned} \dot{\tilde{\boldsymbol{p}}} &= \begin{pmatrix} [\boldsymbol{\omega}]_\times & \boldsymbol{v} \\ 0 & 0 \end{pmatrix} \tilde{\boldsymbol{p}} \\ &= \boldsymbol{\Sigma}(\boldsymbol{v}, \boldsymbol{\omega}) \tilde{\boldsymbol{p}} \end{aligned}$$

where $\boldsymbol{\Sigma}(\boldsymbol{v}, \boldsymbol{\omega})$ is a $4 \times 4$ augmented skew-symmetric matrix. Again, by analogy with the scalar case, we can write the solution as

$$\tilde{\boldsymbol{p}}(\theta) = e^{\boldsymbol{\Sigma}(\boldsymbol{v}, \boldsymbol{\omega})\theta}\, \tilde{\boldsymbol{p}}(0) \;.$$

A matrix that rotates and translates a point in homogeneous coordinates is a homogeneous transformation matrix, and this implies that our matrix exponential is a homogeneous transformation matrix

$$\mathbf{T}(\theta, \hat{\boldsymbol{\omega}}, \boldsymbol{v}) = e^{\Sigma(v,\omega)\,\theta} = e^{\begin{pmatrix} [\hat{\boldsymbol{\omega}}]_\times & \boldsymbol{v} \\ 0 & 0 \end{pmatrix}\theta} \in \mathbf{SE}(3)$$

where $[\hat{\boldsymbol{\omega}}]_\times \theta$ defines the magnitude and axis of rotation and $\boldsymbol{v}\theta$ is the translation.

### 2.4.8 More About Twists

» Chasles' theorem:
  *Any displacement of a body in space can be accomplished by means of a rotation of the body about a unique line in space accompanied by a translation of the body parallel to that line.*

In this chapter, we have introduced and demonstrated twists in 2D and 3D. Here, we will more formally define them and discuss the relationship between twists and homogeneous transformation matrices via the exponential mapping.

The motion described by Chasles' theorem is that of a body, rigidly attached to a nut, rotating about a screw as illustrated in ◘ Fig. 2.24. As the nut rotates, a frame attached to the nut rotates and translates in space. Simplistically, this is the essence of screw theory and its mathematics was developed by Sir Robert Ball in the late 19th century for the analysis of mechanisms.

The general displacement of a rigid body in 3D can be represented by a twist vector

$$\boldsymbol{S} = (\boldsymbol{v}, \ \boldsymbol{\omega}) \in \mathbb{R}^6$$

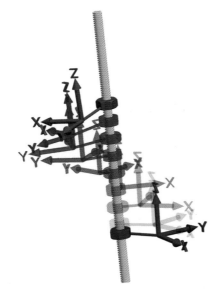

▶ go.sn.pub/tyNyom

◘ **Fig. 2.24** Conceptual depiction of a screw. The blue coordinate frame is attached to a nut by a rigid rod. As the nut rotates around the screw, the pose of the frame changes as shown in red. The corollary is that, given any two frames, we can determine a screw axis, pitch and amount of rotation that will transform one frame into the other (the screw thread shown is illustrative only, and does not reflect the pitch of the screw)

where $\boldsymbol{v} \in \mathbb{R}^3$ is referred to as the moment and encodes the position of the screw axis in space and the pitch of the screw, and $\boldsymbol{\omega} \in \mathbb{R}^3$ is the direction of the screw axis.

We begin with unit twists. For the case of pure rotation, the unit twist parallel to the unit vector $\hat{\boldsymbol{a}}$ and passing through the point Q defined by coordinate vector $\boldsymbol{q}$ is

$$\hat{S} = (\boldsymbol{q} \times \hat{\boldsymbol{a}}, \hat{\boldsymbol{a}})$$

and corresponds to unit rotation, 1 radian, about the screw axis.

The screw pitch is the ratio of the distance along the screw axis to the rotation about the axis. For pure rotation, the pitch is zero. For the more general case, where the screw axis is parallel to the vector $\hat{\boldsymbol{a}}$ and passes through the point defined by $\boldsymbol{q}$, and has a pitch of $p$, the unit twist is

$$\hat{S} = (\boldsymbol{q} \times \hat{\boldsymbol{a}} + p\hat{\boldsymbol{a}}, \hat{\boldsymbol{a}})$$

and the screw pitch is

$$p = \hat{\boldsymbol{w}}^\top \boldsymbol{v} \ .$$

At one extreme, we have a pure rotation, where the screw pitch is zero. At the other extreme, we have a pure translation, which is equivalent to an infinite pitch, and, for translation parallel to the vector $\hat{\boldsymbol{a}}$, the unit twist is

$$\hat{S} = (\hat{\boldsymbol{a}}, 0)$$

and corresponds to unit motion along the screw axis.

A unit twist describes the motion induced by a rotation of 1 rad about the screw axis. A unit rotational twist has $\|\boldsymbol{\omega}\| = 1$, while a unit translational twist has $\boldsymbol{\omega} = 0$ and $\|\boldsymbol{v}\| = 1$. We can also consider a unit twist as defining a family of motions

$$S = \theta \hat{S}$$

where $\hat{S}$ completely defines the screw, and the scalar parameter $\theta$ describes the amount of rotation about the screw.

A twist vector can be written in a non-compact form as an augmented skew-symmetric matrix, which for the 3D case is

$$[S] = \left( \begin{array}{ccc|c} 0 & -\omega_z & \omega_y & v_x \\ \omega_z & 0 & -\omega_x & v_y \\ -\omega_y & \omega_x & 0 & v_z \\ \hline 0 & 0 & 0 & 0 \end{array} \right) \in \mathbf{se}(3) \ .$$

The matrix belongs to the vector space $\mathbf{se}(3)$, the Lie algebra of $\mathbf{SE}(3)$, and is the *generator* of a rigid-body displacement. The displacement as an $\mathbf{SE}(3)$ matrix is obtained by the exponential mapping

$$\mathbf{T} = \mathrm{e}^{[S]} \in \mathbf{SE}(3) \ .$$

If the motion is expressed in terms of a unit twist $\theta \hat{S}$ we can write

$$\mathbf{T}(\theta, \hat{S}) = \mathrm{e}^{\theta[\hat{S}]} \in \mathbf{SE}(3)$$

and the matrix exponential has an efficient closed-form

$$\mathbf{T}(\theta, \hat{S}) = \begin{pmatrix} \mathbf{R}(\theta, \hat{\boldsymbol{\omega}}) & \left( \theta \mathbf{1}_{3\times3} + (1 - \cos\theta)[\hat{\boldsymbol{\omega}}]_\times + (\theta - \sin\theta)[\hat{\boldsymbol{\omega}}]_\times^2 \right) \hat{\boldsymbol{v}} \\ \mathbf{0}_{1\times3} & 1 \end{pmatrix} \tag{2.46}$$

**Excurse 2.20: Michel Chasles**
Chasles (1793–1880) was a French mathematician born at Épernon. He studied at the École Polytechnique in Paris under Poisson and in 1814 was drafted to defend Paris in the War of the Sixth Coalition. In 1837, he published a work on the origin and development of methods in geometry, which gained him considerable fame and he was appointed as professor at the École Polytechnique in 1841, and at the Sorbonne in 1846.

He was an avid collector and purchased over 27,000 forged letters purporting to be from Newton, Pascal and other historical figures – all written in French! One from Pascal claimed he had discovered the laws of gravity before Newton. In 1867, Chasles took this to the French Academy of Science where scholars recognized the fraud. Eventually, Chasles admitted he had been deceived and revealed he had spent nearly 150,000 francs on the letters. He is buried in Cimetière du Père Lachaise in Paris.

where $\mathbf{R}(\theta, \hat{\boldsymbol{\omega}})$ can be computed cheaply using Rodrigues' rotation formula (2.27). This is the algorithm implemented by `trexp` and `trexp2`.

The Toolbox implements twists as classes: `Twist3` for the 3D case and `Twist2` for the 2D case – and they have many properties and methods. A 3D rotational unit twist is created by

```
>>> S = Twist3.UnitRevolute([1, 0, 0], [0, 0, 0])
(0 0 0; 1 0 0)
```

which describes unit rotation about the $x$-axis. The twist vector, and its components, can be accessed

```
>>> S.S
array([      0,        0,        0,        1,        0,        0])
>>> S.v
array([      0,        0,        0])
>>> S.w
array([      1,        0,        0])
```

We can display the Lie algebra of the twist

```
>>> S.se3()
array([[      0,        0,        0,        0],
       [      0,        0,       -1,        0],
       [      0,        1,        0,        0],
       [      0,        0,        0,        0]])
```

which is an **se**(3) matrix. If we multiply this by 0.3 and exponentiate it

```
>>> trexp(0.3 * S.se3())
array([[      1,        0,        0,        0],
       [      0,   0.9553,  -0.2955,        0],
       [      0,   0.2955,   0.9553,        0],
       [      0,        0,        0,        1]])
```

the result is the same as `trotx(0.3)`. The class provides a shorthand way to achieve this result

```
>>> S.exp(0.3)
   1         0         0         0
   0    0.9553   -0.2955         0
   0    0.2955    0.9553         0
   0         0         0         1
```

The class supports composition using the overloaded multiplication operator

```
>>> S2 = S * S
(0 0 0; 2 0 0)
>>> S2.printline(orient="angvec", unit="rad")
t = 0, 0, 0; angvec = (2 | 1, 0, 0)
```

and the result in this case is a twist of two units, or 2 rad, about the $x$-axis.

The `line` method returns the screw's line of action as a `Line3` object that represents a line in 3-dimensional space expressed in Plücker coordinates

```
>>> line = S.line()
{ 0 0 0; 1 0 0}
```

which can be plotted by

```
>>> line.plot("k:", linewidth=2);
```

Any rigid-body motion described by $\mathbf{T} \in \mathbf{SE}(n)$ can also be described by a twist

$$S = \vee(\log \mathbf{T}) \ .$$

The screw is defined by the unit twist

$$\hat{S} = S/\theta$$

and the motion parameter is

$$\theta = \begin{cases} \|\boldsymbol{\omega}\|, & \text{if } \|\boldsymbol{\omega}\| > 0 \\ \|\boldsymbol{v}\|, & \text{if } \|\boldsymbol{\omega}\| = 0 \end{cases} \ .$$

Consider the example

```
>>> T = transl(1, 2, 3) @ eul2tr(0.3, 0.4, 0.5);
>>> S = Twist3(T)
(1.1204 1.6446 3.1778; 0.041006 0.4087 0.78907)
```

The unit twist is

```
>>> S / S.theta
(1.2594 1.8488 3.5722; 0.046096 0.45943 0.88702)
```

which is the same result as

```
>>> S.unit();
```

Scaling the twist by zero is the null motion expressed as an **SE**(3) matrix

```
>>> S.exp(0)
   1        0        0        0
   0        1        0        0
   0        0        1        0
   0        0        0        1
```

and scaling by one is the original rigid-body displacement

```
>>> S.exp(1)
   0.6305   -0.6812    0.372     1
   0.6969    0.7079    0.1151    2
  -0.3417    0.1867    0.9211    3
   0         0         0         1
```

If we took just half the required rotation

```
>>> S.exp(0.5)
   0.9029   -0.3796    0.2017    0.5447
   0.3837    0.9232    0.01982   0.9153
  -0.1937    0.05949   0.9793    1.542
   0         0         0         1
```

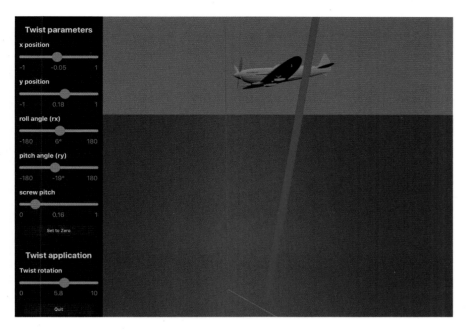

◨ **Fig. 2.25** The Robotics Toolbox example `twistdemo` allows you to explore twists and see how they change the orientation of a body in 3D space. The aircraft is centered at the origin and the twist parameters such as its intersection point with the $xy$-plane, direction and pitch can be set using the sliders. The twist rotation slider rotates the aircraft about the twist axis (thin blue cylinder). This displays in a browser tab using the Swift visualizer

we would have an intermediate displacement. This is not a linear interpolation but rather a *helicoidal* interpolation.

The Toolbox includes an interactive graphical tool

```
>>> %run -m twistdemo
```

that allows you to experiment with twists and visualize the effect of twist parameters and exponent on the pose of an object, as shown in ◨ Fig. 2.25. ▶

This can also be run from the operating system command line as `twistdemo`.

**❗ The proper way to compound twists**

Since twists are the logarithms of rigid-body transformations, there is a temptation to compound the transformations by *adding* the twists rather than multiplying $\mathbf{SE}(3)$ matrices. This works of course for real numbers, if $x = \log X$ and $y = \log Y$, then

$$Z = XY = e^x e^y = e^{x+y}$$

but for the matrix case this would *only be true* if the matrices commute, and rotation matrices do not, therefore

$$Z = XY = e^x e^y \neq e^{x+y} \quad \text{if} \quad x, y \in \mathbf{so}(n) \quad \text{or} \quad \mathbf{se}(n) \, .$$

The bottom line is that there is no shortcut to compounding poses, we must compute $z = \log(e^x e^y)$ not $z = x + y$. Fortunately, these matrices have a special structure and efficient means to compute the logarithm and exponential (2.46).

## 2.4.9 Configuration Space

So far, we have considered the pose of objects in terms of the position and orientation of a coordinate frame affixed to them. For an arm-type robot, we might affix

**Excurse 2.21: Sir Robert Ball**

Ball (1840–1913) was an Irish astronomer born in Dublin. He became Professor of Applied Mathematics at the Royal College of Science in Dublin in 1867, and in 1874 became Royal Astronomer of Ireland and Andrews Professor of Astronomy at the University of Dublin. In 1892, he was appointed Lowndean Professor of Astronomy and Geometry at Cambridge University and became director of the Cambridge Observatory. He was a Fellow of the Royal Society and in 1900 became the first president of the Quaternion Society.

He is best known for his contributions to the science of kinematics described in his treatise *The Theory of Screws* (1876), but he also published *A Treatise on Spherical Astronomy* (1908) and a number of popular articles on astronomy. He is buried at the Parish of the Ascension Burial Ground in Cambridge.

---

a coordinate frame to its end effector, while for a mobile robot or UAV we might affix a frame to its body – its body-fixed frame. This is sufficient to describe the state of the robot in the familiar 2D or 3D Euclidean space which is referred to as the task space or operational space, since it is where the robot performs tasks or operates. However, it says nothing about the joints and the shape of the arm that positions the end effector.

An alternative way of thinking about this comes from classical mechanics and is referred to as the *configuration* of a system. The configuration is the smallest set of parameters, called generalized coordinates, that are required to fully describe the position of *every* particle in the system. This is not as daunting as it may appear since real-world objects are generally rigid, and, within each of these, the particles have a constant coordinate vector with respect to the object's coordinate frame.

If the *system* is a train moving along a track, then all the particles comprising the train move together and we need only a single generalized coordinate $q$, the distance along the track from some datum, to describe their location. A robot arm with a fixed base and two rigid links, connected by two rotational joints has a configuration that is completely described by two generalized coordinates – the two joint angles $(q_1, q_2)$. The generalized coordinates can, as their name implies, represent displacements or rotations.

The number of independent ◀ generalized coordinates $N$ is known as the number of degrees of freedom of the system. Any configuration of the system is represented by a point $q$ in its $N$-dimensional configuration space, or C-space, denoted by $C$ and $q \in C$. We can also say that $\dim C = N$. For the train example $C \subset \mathbb{R}$, which says that the displacement is a bounded real number. For the 2-joint robot, the generalized coordinates are both angles, so $C \subset \mathbf{S}^1 \times \mathbf{S}^1$.

Consider again the train moving along its rail. We might be interested to describe the train in terms of its position on a plane which we refer to as its task space – in this case, the task space is $\mathcal{T} \subset \mathbb{R}^2$. We might also be interested in the latitude and longitude of the train, in which case the task space would be $\mathcal{T} \subset \mathbf{S}^2$. We could also choose the task space to be our everyday 3D world where $\mathcal{T} \subset \mathbb{R}^3 \times \mathbf{S}^3$ which accounts for height changes as the train moves up and down hills and its orientation changes as it moves around curves and vertical gradients. Any point in the configuration space can be mapped to a point in the task space $q \in C \mapsto \tau \in \mathcal{T}$ and the mapping depends on the particular task space that we choose. However, in general, not all points in the task space can be mapped to the configuration space

That is, there are no holonomic constraints on the system.

– some points in the task space are not accessible. While every point along the rail line can be mapped to a point in the task space, most points in the task space will not map to a point on the rail line. The train is constrained by its fixed rails to move in a subset of the task space. If the dimension of the task space exceeds the dimension of the configuration space, $\dim \mathcal{T} > \dim C$, the system can only access a lower-dimensional subspace of the entire task space.

The simple 2-joint robot arm can access a subset of points in a plane, so a useful task space might be $\mathcal{T} \subset \mathbb{R}^2$. The dimension of the task space equals the dimension of the configuration space, $\dim \mathcal{T} = \dim C$, and this means that the mapping between task and configuration spaces is bi-directional but it is not necessarily unique – for this type of robot, in general, two different configurations map to a single point in task space. Points in the task space beyond the physical reach of the robot are not mapped to the configuration space. If we chose a task space with more dimensions such as 2D or 3D pose, then $\dim \mathcal{T} > \dim C$ and the robot would only be able to access points within a subset of that space.

Now consider a snake-robot arm, such as shown in ◾ Fig. 8.10, with 20 joints and $C \subset (\mathbf{S}^1)^{20}$ and $\dim \mathcal{T} < \dim C$. In this case, an infinite number of configurations in a $20 - 6 = 14$-dimensional subspace of the 20-dimensional configuration space will map to the same point in task space. This means that, in addition to the task of positioning the robot's end effector, we can *simultaneously* perform motion in the configuration subspace to control the shape of the arm to avoid obstacles in the environment. Such a robot is referred to as overactuated or redundant and this topic is covered in ▶ Sect. 8.3.4.2.

The airframe of a quadrotor, such as shown in ◾ Fig. 4.22d, is a single rigid-body whose configuration is completely described by six generalized coordinates, its position and orientation in 3D space $C \subset \mathbb{R}^3 \times (\mathbf{S}^1)^3$ where the orientation is expressed in some three-angle representation. For such a robot the most logical task space would be $\mathbb{R}^3 \times \mathbf{S}^3$ which is equivalent to the configuration space and $\dim \mathcal{T} = \dim C$. However, the quadrotor has only four actuators which means it cannot *directly* access all the points in its configuration space and hence its task space. Such a robot is referred to as an underactuated system and we will revisit this important topic in ▶ Sect. 4.2.

## 2.5 Using the Toolbox

This book leverages a large amount of open-source software as shown in ◾ Fig. 2.26. This chapter has introduced the functionality of the Spatial Maths Toolbox for Python which supports the Robotics Toolbox for Python and the Machine Vision Toolbox for Python that we will use in later chapters.

The Spatial Maths Toolbox has two levels. The *base* level is accessed by

```
>>> from spatialmath.base import *
```

which imports the functions that were introduced in this chapter. It provides a rich set of functions that operate on NumPy arrays to implement 2D and 3D rotation matrices and homogeneous transformation matrices, for example, functions such as `trans12`, `rotx` and `trexp`. It implements many quaternion functions where the quaternion is represented by a 4-element 1D NumPy array. It also provides a rich collection of graphics functions like `trplot` and `plot_point`, and many utility functions. Many of these functions have the same names as functions in the open-source Spatial Math Toolbox for MATLAB®.

The *top* level is accessed by

```
>>> from spatialmath import *
```

and provides a rich set of classes, some of which we have already used, for example, `UnitQuaternion`, `Twist2`, `Twist3` and `Line3`. The top level also provides

**2**

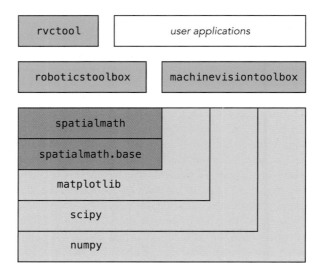

■ **Fig. 2.26** Open-source software stack. Components shown in orange are the subject of this chapter and used by the blue components which underpin the rest of this book. Those shown in gray are standard packages in the Python ecosystem

classes that represent rotation matrices and homogeneous transformation matrices, named respectively SO2 and SE2 for two dimensions, and SO3 and SE3 for three dimensions.

We can highlight the difference with a simple example

```
>>> R = rotx(0.3)  # create SO(3) matrix as NumPy array
array([[      1,        0,        0],
       [      0,   0.9553,  -0.2955],
       [      0,   0.2955,   0.9553]])
>>> type(R)
numpy.ndarray
>>> R = SO3.Rx(0.3)  # create SO3 object
   1          0          0
   0     0.9553    -0.2955
   0     0.2955     0.9553
>>> type(R)
spatialmath.pose3d.SO3
```

The visual appearance of the displayed value of the SO3 object is different to the standard NumPy representation of 3 × 3 matrix.

Instances of these classes *encapsulate* a NumPy array which is the actual rotation or homogeneous transformation matrix. We can access that encapsulated NumPy array using its *array* attribute ◄

All spatial maths objects provide this property.

```
>>> R.A
array([[      1,        0,        0],
       [      0,   0.9553,  -0.2955],
       [      0,   0.2955,   0.9553]])
```

These are Python class methods and the Toolbox convention is that they are capitalized.

There are many constructor-like methods to create an SO3 object, for example ◄

```
>>> R = SO3(rotx(0.3));                  # convert an SO(3) matrix
>>> R = SO3.Rz(0.3);                     # rotation about z-axis
>>> R = SO3.RPY(10, 20, 30, unit="deg"); # from roll-pitch-yaw ang.
>>> R = SO3.AngleAxis(0.3, (1, 0, 0));   # from angle and axis
>>> R = SO3.EulerVec((0.3, 0, 0));       # from an Euler vector
```

as well as conversion methods

```
>>> R.rpy();        # convert to roll-pitch-yaw angles
>>> R.eul();        # convert to Euler angles
>>> R.printline();  # compact single-line print
rpy/zyx = 10°, 20°, 30°
```

The classes are polymorphic, which means that they have many methods with the same name that perform the equivalent function. For example

```
>>> R = SO3.RPY(10, 20, 30, unit="deg");           # SO(3) matrix
>>> T = SE3.RPY(10, 20, 30, unit="deg");           # SE(3) matrix
>>> S = Twist3.RPY(10, 20, 30, unit="deg");        # 3D twist
>>> q = UnitQuaternion.RPY(10, 20, 30, unit="deg"); # unit quat.
```

all represent a rotation described in terms of ZYX roll-pitch-angles.

To illustrate the use of these classes, we revisit the example from the end of ► Sect. 2.2.2.1, which becomes

```
>>> TA = SE2(1, 2) * SE2(30, unit="deg");
>>> type(TA)
spatialmath.pose2d.SE2
```

and the value of the encapsulated **SE**(2) matrix is

```
>>> TA
   0.866    -0.5      1
   0.5       0.866    2
   0         0        1
```

In some environments, the elements are color coded: the rotation submatrix is red, the translation vector is blue, the other elements gray. We could write this more concisely as

```
>>> TA = SE2(1, 2, 30, unit="deg");
```

We can access the encapsulated **SE**(2) array using the .A attribute as already mentioned, or access the rotation and translation components

```
>>> TA.R
array([[   0.866,      -0.5],
       [    0.5,    0.866]])
>>> TA.t
array([     1,        2])
```

as NumPy arrays. We can plot the coordinate frame represented by this pose

```
>>> TA.plot(frame="A", color="b");
```

or display it in a compact single line form

```
>>> TA.printline()
t = 1, 2; 30°
```

Returning to that earlier example, we can easily transform a vector by the inverse transformation

```
>>> P = [3, 2];
>>> TA.inv() * P
array([[   1.732],
       [      -1]])
```

and the class overloads the Python * operator and handles the details of converting the vector between Euclidean and homogeneous forms.

The advantages of using classes rather than functions are:

- They improve code readability and can be used in an almost identical fashion to NumPy arrays.
- They provide type safety. NumPy allows you to add two **SO**(3) matrices even though that is not a defined operation. Adding two so3 objects results in a NumPy array, not an so3 object.

**2**

- They are polymorphic, which makes it easy to switch between using, say, rotation matrices and quaternions, or lifting a solution from two to three dimensions. We only need to change the object constructor.
- That all objects of the same type can be compounded using the overloaded $\star$ operator.
- These classes inherit from the Python list class. This means that they can hold multiple values, and support operations like indexing, `append` and `len`. This is useful to represent a set of related poses or orientations, for example, poses of all the link frames in a robot manipulator arm or poses along a trajectory. For example

```
>>> R = SO3.Rx(np.linspace(0, 1, 5));
>>> len(R)
5
>>> R[3]
   1            0           0
   0         0.7317     -0.6816
   0         0.6816      0.7317
```

is an `SO3` instance that contains five rotation matrices, rotations about the $x$-axis by angles from 0 to 1 rad in 5 steps. The class handles broadcasting, when compounding with other objects of the same type or even in the case of vector transformation

```
>>> R * [1, 2, 3]
array([[      1,         1,         1,         1,         1],
       [      2,      1.196,    0.3169,   -0.5815,    -1.444],
       [      3,      3.402,     3.592,     3.558,     3.304]])
```

where the result has columns that are the vector $(1, 2, 3)^\top$ rotated by the consecutive elements of `R`.

The Python classes incur some performance penalty, the overhead of object construction and method invocation. If computational performance is important, consider using the base functions instead. In general there is a one-to-one mapping between class methods and base functions.

For the rest of this book, we will use classes exclusively. ◄

## 2.6 Wrapping Up

In this chapter, we introduced relative pose which describes the position and orientation of one object with respect to another. We can think of relative pose as the motion of a rigid body. We discussed how such motions can be composed and decomposed, developed some algebraic rules for manipulating poses, and showed how sets of related poses can be represented as a pose graph. It is important to remember that composition is noncommutative – the order in which relative poses are applied is significant.

To quantify the pose of an object, we affix a 2D or 3D coordinate frame to it. Relative pose is the distance between the origins of the frames and the relative angles between the axes of the coordinate frames. Points within the object are represented by constant coordinate vectors relative to a coordinate frame.

We have discussed a variety of mathematical objects to tangibly represent pose. We have used orthonormal rotation matrices for the 2- and 3-dimensional case to represent orientation and shown how it can rotate a point's coordinate vector from one coordinate frame to another. Its extension, the homogeneous transformation matrix, can be used to represent both orientation and translation, and we have shown how it can rotate and translate a point expressed as a homogeneous coordinate vector from one frame to another. Rotation in 3 dimensions has subtlety and complexity and we have looked at various parameterizations such as ZYZ-Euler angles, roll-pitch-yaw angles, and unit quaternions. Touching on Lie group theory, we showed that rotation matrices, from the group $\mathbf{SO}(2)$ or $\mathbf{SO}(3)$, are the result of exponentiating skew-symmetric generator matrices. Similarly, homogeneous transformation matrices, from the group $\mathbf{SE}(2)$ or $\mathbf{SE}(3)$, are the result of exponentiating

**Fig. 2.27** Physical coordinate frames are very helpful when thinking about robotics problems. For information about how build your own coordinate frames visit ► https://petercorke.com/resources/3d-frame. There are frames you can 2D print and fold, or 3D print (image by Dorian Tsai)

augmented skew-symmetric generator matrices. We have also introduced twists as a concise way of describing relative pose in terms of rotation around a screw axis, a notion that comes to us from screw theory, and these twists are the unique elements of the generator matrices.

A simple graphical summary of key concepts was given in ■ Fig. 2.8. There are two important lessons from this chapter. The first is that there are *many* mathematical objects that can be used to represent pose. There is no right or wrong – each has strengths and weaknesses, and we typically choose the representation to suit the problem at hand. Sometimes we wish for a vectorial representation, perhaps for interpolation, in which case $(x, y, \theta)$ or $(x, y, z, \Gamma)$ might be appropriate, where $\Gamma$ is a 3-angle sequence, but this representation cannot be easily compounded. Sometimes we may only need to describe 3D rotation in which case $\Gamma$ or $\mathring{q}$ is appropriate.

The second lesson is that coordinate frames are your friend. The essential first step in many vision and robotics problems is to assign coordinate frames to all objects of interest as shown in ■ Fig. 2.6. Then add the important relative poses, known and unknown, in a pose graph, write down equations for the loops and solve for the unknowns. ■ Fig. 2.27 shows how to build your own coordinate frames – you can pick them up and rotate them to make these ideas tangible. Don't be shy, embrace the coordinate frame.

We now have solid foundations for moving forward. The notation has been defined and illustrated, and we have started our hands-on work with Python and the Toolboxes. The next chapter discusses motion and coordinate frames that change with time, and after that we are ready to move on and discuss robots.

## 2.6.1 Further Reading

The treatment in this chapter is a hybrid mathematical and graphical approach that covers the 2D and 3D cases by means of abstract representations and operators which are later made tangible. The standard robotics textbooks such as Kelly

**2**

(2013), Siciliano et al. (2009), Spong et al. (2006), Craig (2005), and Paul (1981) all introduce homogeneous transformation matrices for the 3-dimensional case but differ in their approach. These books also provide good discussion of the other representations, such as angle-vector and three-angle vectors. Spong et al. (2006, § 2.5.1) have a good discussion of singularities. Siegwart et al. (2011) explicitly cover the 2D case in the context of mobile robotics.

The book by Lynch and Park (2017) is a comprehensive introduction to twists and screws as well as covering the standard matrix approaches. Solà et al. (2018) presents the essentials of Lie-group theory for robotics. Algebraic topology is a challenging topic but Selig (2005) provides a readable way into the subject. Grassia (1998) provides an introduction to exponential mapping for rotations.

Quaternions are discussed in Kelly (2013) and briefly in Siciliano et al. (2009). The books by Vince (2011) and Kuipers (1999) are readable and comprehensive introductions to quaternions. Quaternion interpolation is widely used in computer graphics and animation and the classic paper by Shoemake (1985) is very readable introduction to this topic. Solà (2017) provides a comprehensive discussion of quaternion concepts, including JPL quaternions, and Kenwright (2012) describes the fundamentals of quaternions and dual quaternions. Dual quaternions, and a library with Python bindings, are presented by Adorno and Marhinho (2021). The first publication about quaternions for robotics is probably Taylor (1979), and followed up in subsequent work by Funda et al. (1990). Terzakis et al. (2018) describe and compare a number of common rotational representations, while Huynh (2009) introduces and compares a number of distance metrics for 3D rotations.

You will encounter a wide variety of different notation for rotations and transformations in textbooks and research articles. This book uses $^A\mathbf{T}_B$ to denote an **SE**(3) representation of a rigid-body motion from frame {A} to frame {B}. A common alternative notation is $\mathbf{T}_B^A$ or even $^A_B\mathbf{T}$. To describe points, this book uses $^A\boldsymbol{p}_B$ to denote a coordinate vector from the origin of frame {A} to the point B whereas others use $\boldsymbol{p}_B^A$, or even $^C\boldsymbol{p}_B^A$ to denote a vector from the origin of frame {A} to the point **B** but with respect to coordinate frame {C}. Twists can be written as either $(\boldsymbol{v}, \boldsymbol{\omega})$ as in this book, or as $(\boldsymbol{\omega}, \boldsymbol{v})$.

#### ■■ Tools and Resources

This chapter has introduced a pure Python toolbox for creating and manipulating objects such as rotation matrices, homogeneous transform matrices, quaternions and twists. There are numerous other packages that perform similar functions, but have different design objectives. Manif is a Lie theory library written in C++ with Python bindings, ► https://artivis.github.io/manif/python. Sophus is a C++ implementation of Lie groups for 2D and 3D geometric problems, ► https://github.com/strasdat/Sophus. tf2 is part of the ROS ecosystem and keeps track of multiple coordinate frames and maintains the relationship between them in a tree structure which is similar to the pose graph introduced in this chapter. tf2 can manipulate vectors, quaternions and homogeneous transformations, and read and write those types as ROS geometry messages, ► http://wiki.ros.org/tf2.

#### ■■ Historical and General

Quaternions had a tempestuous beginning. Hamilton and his supporters, including Peter Tait, were vigorous in defending Hamilton's precedence in inventing quaternions and promoted quaternions as an alternative to vectors. Vectors are familiar to us today, but at that time were just being developed by Josiah Gibbs. The paper by Altmann (1989) is an interesting description on this tussle of ideas.

Rodrigues developed his eponymous formula in 1840 although Gauss discovered it in 1819 but, as usual, did not publish it. It was published in 1900. The article by Pujol (2012) revisits the history of mathematical ideas for representing rotation. Quaternions have even been woven into fiction (Pynchon 2006).

## 2.6.2 Exercises

1. Create a 2D rotation matrix. Visualize the rotation using `trplot2`. Use it to transform a vector. Invert it and multiply it by the original matrix; what is the result? Reverse the order of multiplication; what is the result? What is the determinant of the matrix and its inverse?
2. Build a coordinate frame as shown in ◘ Fig. 2.27 and reproduce the 3D rotation shown in ◘ Figs. 2.15 and 2.16.
3. Create a 3D rotation matrix. Visualize the rotation using `trplot` or `tranimate`. Use it to transform a vector. Invert it and multiply it by the original matrix; what is the result? Reverse the order of multiplication; what is the result? What is the determinant of the matrix and its inverse?
4. Explore the many options associated with `trplot`.
5. Animate a rotating cube:
   a) Write a function to plot the edges of a cube centered at the origin.
   b) Modify the function to accept an argument which is a homogeneous transformation that is applied to the cube vertices before plotting.
   c) Animate rotation about the $x$-axis.
   d) Animate rotation about all axes.
6. Create a vector-quaternion class to describe pose and which supports composition, inverse and point transformation.
7. Compute the matrix exponential using the power series. How many terms are required to match the result shown to standard 64-bit floating point precision?
8. Generate the sequence of plots shown in ◘ Fig. 2.16.
9. Write the function `na2r` which is analogous to `oa2r` but creates a rotation matrix from the normal and approach vectors.
10. For the 3-dimensional rotation about the vector $[2, 3, 4]$ by $0.5$ rad compute an **SO**(3) rotation matrix using: the matrix exponential functions `scipy.linalg.expm` and `trexp`, Rodrigues' rotation formula (code this yourself), and the Toolbox function `angvec2tr`. Compute the equivalent unit quaternion.
11. Create two different rotation matrices, in 2D or 3D, representing frames {A} and {B}. Determine the rotation matrix $^A\mathbf{R}_B$ and $^B\mathbf{R}_A$. Express these as a rotation axis and angle, and compare the results. Express these as a twist.
12. Create a 2D or 3D homogeneous transformation matrix. Visualize the rigid-body displacement using `tranimate`. Use it to transform a vector. Invert it and multiply it by the original matrix, what is the result? Reverse the order of multiplication; what happens?
13. Create three symbolic variables to represent roll, pitch and yaw angles, then use these to compute a rotation matrix using `rpy2r`. You may want to symbolically simplify the result. Use this to transform a unit vector in the $z$-direction. Looking at the elements of the rotation matrix, devise an algorithm to determine the roll, pitch and yaw angles. Hint: find the pitch angle first.
14. Experiment with the `tripleangle` application in the Toolbox. Explore roll, pitch and yaw motions about the nominal attitude and at singularities.
15. Using the composition rule for **SE**(3) matrices from ▶ Sect. 2.3.2.1, show that $\mathbf{TT}^{-1} = \mathbf{1}$.
16. Is the inverse of a homogeneous transformation matrix equal to its transpose?
17. In ▶ Sect. 2.2.2.2, we rotated a frame about an arbitrary point. Derive the expression for computing TC that was given.
18. Explore the effect of negative roll, pitch or yaw angles. Does transforming from RPY angles to a rotation matrix then back to RPY angles give a different result to the starting value as it does for Euler angles?
19. Show that $e^x e^y \neq e^{x+y}$ for the case of matrices. Hint: expand the first few terms of the exponential series.

20. A camera has its $z$-axis parallel to the vector $[0, 1, 0]$ in the world frame, and its $y$-axis parallel to the vector $[0, 0, -1]$. What is the attitude of the camera with respect to the world frame expressed as a rotation matrix and as a unit quaternion?

21. Pick a random $\mathbf{SE}(3)$ matrix, and plot it and the reference frame. Compute and display the screw axis that rotates the reference frame to the chosen frame. Hint: convert the chosen frame to a twist.

# Time and Motion

» *The only reason for time is so that everything doesn't happen at once.*
— Albert Einstein

## Contents

© The Author(s), under exclusive license to Springer Nature Switzerland AG 2023
P. Corke, *Robotics, Vision and Control*, Springer Tracts in Advanced Robotics 146,
https://doi.org/10.1007/978-3-031-06469-2_3

**3**

The previous chapter was concerned with describing the pose of objects in two or three-dimensional space. This chapter extends those concepts to poses that change as a function of time. ▶ Sect. 3.1 introduces the derivative of time-varying position, orientation and pose and relates these to concepts from classical mechanics such as velocity and angular velocity. We also cover discrete-time approximations to the derivatives which are useful for computer implementation of algorithms such as inertial navigation. ▶ Sect. 3.2 is a brief introduction to dynamics, the motion of objects under the influence of forces and torques. We also discuss the important difference between inertial and noninertial reference frames.

▶ Sect. 3.3 discusses how to generate a temporal sequence of poses, a trajectory, that smoothly changes from an initial pose to a final pose. This could describe the path followed by a robot gripper moving to grasp an object, or the flight path of an aerial robot. ▶ Sect. 3.4 brings many of these topics together for the important application of inertial navigation. We introduce three common types of inertial sensor and learn how to use their measurements to update the estimate of pose for a moving object such as a robot.

## 3.1  Time-Varying Pose

This section considers poses that change with time, and we discuss how to describe the rate of change of 3D pose which has both a translational and rotational velocity component. The translational velocity is straightforward: it is the rate of change of the position of the origin of the coordinate frame. Rotational velocity is a little more complex.

### 3.1.1  Rate of Change of Orientation

We consider the time-varying orientation of the body frame {B} with respect to a fixed frame {A}. There are many ways to represent orientation but the rotation matrix and exponential form, introduced in ▶ Sect. 2.3.1.6, is convenient

$$^A\mathbf{R}_B(t) = \mathrm{e}^{\theta(t)\left[^A\hat{\boldsymbol{\omega}}_B(t)\right]_\times} \in \mathbf{SO}(3)$$

where the rotation is described by a rotational axis $^A\hat{\boldsymbol{\omega}}_B(t)$ defined with respect to frame {A} and a rotational angle $\theta(t)$, and where $[\cdot]_\times$ is a skew-symmetric, or anti-symmetric, matrix.

At an instant in time $t$, we will assume that the axis has a fixed direction and the frame is rotating around that axis. The derivative with respect to time is

$$^A\dot{\mathbf{R}}_B(t) = \dot{\theta}(t)\left[^A\hat{\boldsymbol{\omega}}_B\right]_\times \mathrm{e}^{\theta(t)\left[^A\hat{\boldsymbol{\omega}}_B\right]_\times} \in \mathbb{R}^{3\times3}$$
$$= \dot{\theta}(t)\left[^A\hat{\boldsymbol{\omega}}_B\right]_\times {}^A\mathbf{R}_B(t)$$

which is a general $3 \times 3$ matrix, not a rotation matrix. We can write this succinctly as

$$^A\dot{\mathbf{R}}_B = \left[^A\boldsymbol{\omega}_B\right]_\times {}^A\mathbf{R}_B \in \mathbb{R}^{3\times3} \tag{3.1}$$

where $^A\omega_B = \dot{\theta}\,{}^A\hat{\boldsymbol{\omega}}_B$ is the *angular velocity* of frame {B} with respect to frame {A}. This is a vector quantity $^A\boldsymbol{\omega}_B = (\omega_x, \omega_y, \omega_z)$ that defines the *instantaneous* axis and rate of rotation. The unit vector $^A\hat{\boldsymbol{\omega}}_B$ is parallel to the axis about which the coordinate frame is rotating at a particular instant of time, and the magnitude $\| {^A}\boldsymbol{\omega}_B \|$ is the rate of rotation $\dot{\theta}$ about that axis.

Consider now that angular velocity is measured in the moving frame {B}, for example, it is measured by gyroscope sensors onboard a moving vehicle. We know that

$${^A}\boldsymbol{\omega}_B = {^A}\mathbf{R}_B \, {^B}\boldsymbol{\omega}_B$$

and, using the identity $[\mathbf{R}\boldsymbol{v}]_\times = \mathbf{R}[\boldsymbol{v}]_\times \mathbf{R}^\top$, it follows that

$$ {^A}\dot{\mathbf{R}}_B = {^A}\mathbf{R}_B \left[ {^B}\boldsymbol{\omega}_B \right]_\times \in \mathbb{R}^{3\times 3} \tag{3.2}$$

and we see that the order of the rotation matrix and the skew-symmetric matrix have been swapped.

The derivatives of a unit quaternion, the quaternion equivalent of (3.1) and (3.2), are

$$ {^A}\mathring{\dot{q}}_B = \tfrac{1}{2} \, {^A}\breve{\omega}_B \circ {^A}\mathring{q}_B = \tfrac{1}{2} \, {^A}\mathring{q}_B \circ {^B}\breve{\omega}_B \in \mathbb{H} \tag{3.3}$$

which is a regular quaternion, not a unit quaternion, and $\breve{\omega}$ is a pure quaternion formed from the angular velocity vector. These equations implemented by the `UnitQuaternion` methods `dot` and `dotb` respectively.

### 3.1.2 Rate of Change of Pose

The derivative of pose can be determined by expressing pose as an **SE**(3) matrix

$${^A}\mathbf{T}_B = \begin{pmatrix} {^A}\mathbf{R}_B & {^A}\boldsymbol{t}_B \\ \mathbf{0}_{1\times 3} & 1 \end{pmatrix} \in \mathbf{SE}(3)$$

and taking the derivative with respect to time, then substituting (3.1) gives

$${^A}\dot{\mathbf{T}}_B = \begin{pmatrix} {^A}\dot{\mathbf{R}}_B & {^A}\dot{\boldsymbol{t}}_B \\ \mathbf{0}_{1\times 3} & 0 \end{pmatrix} = \begin{pmatrix} \left[ {^A}\boldsymbol{\omega}_B \right]_\times {^A}\mathbf{R}_B & {^A}\dot{\boldsymbol{t}}_B \\ \mathbf{0}_{1\times 3} & 0 \end{pmatrix} \in \mathbb{R}^{4\times 4} \; . \tag{3.4}$$

The rate of change can be described in terms of the current orientation ${^A}\mathbf{R}_B$ and *two* velocities. The linear, or translational, velocity ${^A}\boldsymbol{v}_B = {^A}\dot{\boldsymbol{t}}_B \in \mathbb{R}^3$ is the velocity of the origin of {B} with respect to {A}. The angular velocity ${^A}\boldsymbol{\omega}_B \in \mathbb{R}^3$ has already been introduced. We can combine these two velocity vectors to create the spatial velocity vector

$${^A}\boldsymbol{\nu}_B = \left( {^A}\boldsymbol{v}_B, \; {^A}\boldsymbol{\omega}_B \right) \in \mathbb{R}^6$$

which is the instantaneous velocity of frame {B} with respect to {A}.

Every point in the body has the same angular velocity, but the translational velocity of a point depends on its position within the body. It is common to place {B} at the body's center of mass.

**3**

> **Spatial velocity convention**
>
> There is no firm convention for the order of the velocities in the spatial velocity vector. This book always uses $v = (v, \omega)$ but other sources, including the MATLAB third edition of this book, use $v = (\omega, v)$.

### 3.1.3 Transforming Spatial Velocities

■ Fig. 3.1a shows two fixed frames and a moving coordinate frame. An observer on fixed frame {B} observes an object moving with a spatial velocity $^B v$ with respect to frame {B}. For an observer on frame {A}, that frame's spatial velocity is

$$
^A v = \begin{pmatrix} ^A \mathbf{R}_B & \mathbf{0}_{3\times3} \\ \mathbf{0}_{3\times3} & ^A \mathbf{R}_B \end{pmatrix} \, ^B v = \, ^A \mathbf{J}_B (^A \mathbf{T}_B) \, ^B v \tag{3.5}
$$

where $^A \mathbf{J}_B(\cdot)$ is a Jacobian or interaction matrix which is a function of the relative orientation of the two frames, and independent of the translation between them. We can investigate this numerically, mirroring the setup of ■ Fig. 3.1a ◀

> We use `aTb` to denote $^A\mathbf{T}_B$ in the code in a readable and valid Python way.

```
>>> aTb = SE3.Tx(-2) * SE3.Rz(-pi/2) * SE3.Rx(pi/2);
```

If the spatial velocity in frame {B} is

```
>>> bV = [1, 2, 3, 4, 5, 6];
```

then the spatial velocity in frame {A} is

```
>>> aJb = aTb.jacob();
>>> aJb.shape
(6, 6)
>>> aV = aJb @ bV
array([ -3,   -1,    2,   -6,   -4,    5])
```

where the `jacob` method has returned a $6 \times 6$ NumPy array. We see that the velocities have been transposed, and sometimes negated due to the different orientation of the frames. The $x$-axis translational and rotational velocities in frame {B} have been mapped to the $-y$-axis translational and rotational velocities in frame {A}, the $y$-axis velocities have been mapped to the $z$-axis, and the $z$-axis velocities have been mapped to the $-x$-axis.

Next, we consider the case where the two frames are rigidly attached, and moving together, as shown in ■ Fig. 3.1b. There is an observer in each frame who is

▶ go.sn.pub/bKBuTA

▶ go.sn.pub/rp7jDE

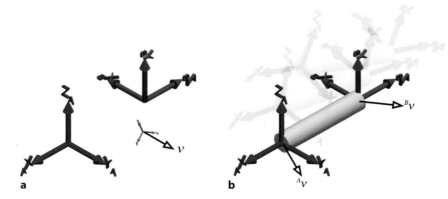

■ **Fig. 3.1** Spatial velocity can be expressed with respect to frame {A} or frame {B}. **a** Both frames are stationary and the object frame is moving with velocity $v$, **b** both frames are moving together

able to estimate their velocity with respect to their own frame. The velocities they observe are related by

$$
{}^A\boldsymbol{v} = \begin{pmatrix} {}^A\mathbf{R}_B & \left[{}^A\boldsymbol{t}_B\right]_\times {}^A\mathbf{R}_B \\ \mathbf{0}_{3\times3} & {}^A\mathbf{R}_B \end{pmatrix} {}^B\boldsymbol{v} = \text{Ad}({}^A\mathbf{T}_B) \; {}^B\boldsymbol{v}
$$

which uses the adjoint matrix of the relative pose and this does depend on the translation between the frames. For example, using the relative pose from the previous example, and with frame {B} experiencing pure translation, the velocity experienced by the observer in frame {A}

```
>>> aV = aTb.Ad() @ [1, 2, 3, 0, 0, 0]
array([ -3,  -1,   2,   0,   0,   0])
```

is also pure translation, but the axes have been transposed due to the relative orientation of the two frames. If the velocity in frame B is pure angular velocity about the $x$-axis

```
>>> aV = aTb.Ad() @ [0, 0, 0, 1, 0, 0]
array([  0,   0,   2,   0,  -1,   0])
```

then the observer in frame {A} measures that same angular velocity magnitude, but now transposed to be about the $-y$-axis. There is also a translational velocity in the $z$-direction which is due to the origin of {A} following a circular path around the $x$-axis of frame B. If we combine these translational and rotational velocities in frame {B}, then the spatial velocity in frame {A} will be

```
>>> aV = aTb.Ad() @ [1, 2, 3, 1, 0, 0]
array([ -3,  -1,   4,   0,  -1,   0])
```

> ⊘ **Twist Confusion**
>
> Many sources (including ROS) refer to spatial velocity, as defined above, as a *twist*. This is not to be confused with the twists introduced in ▶ Sects. 2.2.2.4 and 2.3.2.3. The textbook Lynch and Park (2017) uses the term *velocity twist* $\bar{V}$ which has important differences to the spatial velocity introduced above (the bar is introduced here to clearly differentiate the notation). The velocity twist of a body-fixed frame {B} is ${}^B\bar{V} = ({}^B\bar{\boldsymbol{\omega}}, {}^B\bar{\boldsymbol{v}})$ which has a rotational and translational velocity component, but ${}^B\bar{\boldsymbol{v}}$ is the body-frame velocity of an imaginary point rigidly attached to the body but located at the world frame origin. The body- and world-frame velocity twists are related by the adjoint matrix rather than (3.5). Confusingly, the older book by Murray et al. (1994) refers to the velocity twist as spatial velocity.

### 3.1.4 Incremental Rotation

In robotics, there are many applications where we need to integrate angular velocity, measured by sensors in the body frame, in order to estimate the orientation of the body frame. This is a key operation in inertial navigation systems which we will discuss in greater detail in ▶ Sect. 3.4. If we assume that the rotational axis is constant over the sample interval $\delta_t$, then the change in orientation over the interval can be expressed in angle-axis form where angle is $\|\boldsymbol{\omega}\delta_t\|$. The orientation update is

$$
\mathbf{R}_{B\langle t+\delta_t\rangle} = \mathbf{R}_{B\langle t\rangle}\, e^{\delta_t[\boldsymbol{\omega}]_\times} \tag{3.6}
$$

and involves the matrix exponential which is expensive to compute.

We introduced $\dot{\mathbf{R}}$ in ▶ Sect. 3.1.1 which describes the rate of change of rotation matrix elements with time. We can write this as a first-order approximation to the

The only valid operators for the group **SO**(n) are composition ⊕ and inverse ⊖, so the result of subtraction cannot belong to the group. The result is a 3 × 3 matrix of element-wise differences. Groups are introduced in ▶ Sect. 2.1.2 and ▶ App. D.

derivative ◀

$$\dot{\mathbf{R}} \approx \frac{\mathbf{R}_{\langle t+\delta_t \rangle} - \mathbf{R}_{\langle t \rangle}}{\delta_t} \in \mathbb{R}^{3\times3} \tag{3.7}$$

where $\mathbf{R}_{\langle t \rangle}$ is the rotation matrix at time $t$ and $\delta_t$ is an infinitesimal time step. Consider an object whose body frames {B} at two consecutive time steps $\mathbf{R}_{B\langle t \rangle}$ and $\mathbf{R}_{B\langle t+\delta_t \rangle}$ are related

$$\mathbf{R}_{B\langle t+\delta_t \rangle} = \mathbf{R}_{B\langle t \rangle}{}^{B\langle t \rangle}\mathbf{R}_{B\langle t+\delta_t \rangle} = \mathbf{R}_{B\langle t \rangle}\mathbf{R}^{\Delta}$$

by a small rotation $\mathbf{R}^{\Delta} \in \mathbf{SO}(3)$ expressed in the first body frame. We substitute (3.2) into (3.7) and rearrange to obtain

$$\mathbf{R}^{\Delta} \approx \delta_t \left[ {}^{B}\boldsymbol{\omega} \right]_{\times} + \mathbf{1}_{3\times3} \tag{3.8}$$

This is the first two terms of the Rodrigues' rotation formula (2.27) when $\theta = \delta_t \omega$.

which says that a small rotation can be approximated by the sum of a skew-symmetric matrix and an identity matrix. ◀ We can see this structure clearly for a small rotation about the $x$-axis

```
>>> rotx(0.001)
array([[    1,        0,        0],
       [    0,        1,   -0.001],
       [    0,    0.001,        1]])
```

Rearranging (3.8) allows us to approximate the angular velocity vector

$$\boldsymbol{\omega} \approx \tfrac{1}{\delta_t} \vee_{\times} \left( \mathbf{R}_{B\langle t \rangle}^{\top} \mathbf{R}_{B\langle t+\delta_t \rangle} - \mathbf{1}_{3\times3} \right)$$

from two consecutive rotation matrices where the operator $\vee_{\times}(\cdot)$ unpacks the unique elements of a skew-symmetric matrix into a vector. The exact value can be found by $\vee_{\times} \log(\mathbf{R}_{B\langle t \rangle}^{\top} \mathbf{R}_{B\langle t+\delta_t \rangle})$.

Returning now to the rotation integration problem, we can substitute (3.2) into (3.7) and rearrange as

$$\mathbf{R}_{B\langle t+\delta_t \rangle} \approx \mathbf{R}_{B\langle t \rangle}(\mathbf{1}_{3\times3} + \delta_t[\boldsymbol{\omega}]_{\times}) \tag{3.9}$$

which is cheap to compute and involves no trigonometric operations, but is an approximation. We can numerically explore the tradeoffs in this approximation

```
>>> import time
>>> Rexact = np.eye(3); Rapprox = np.eye(3);  # null rotation
>>> w = np.array([1, 0, 0]);   # rotation of 1rad/s about x-axis
>>> dt = 0.01;            # time step
>>> t0 = time.process_time();
>>> for i in range(100):  # exact integration over 100 time steps
...     Rexact = Rexact @ trexp(skew(w*dt));  # update by composition
>>> print(time.process_time() - t0)
0.008975
>>> t0 = time.process_time();
>>> for i in range(100): # approx. integration over 100 steps
...     Rapprox += Rapprox @ skew(w*dt); # update by addition
>>> print(time.process_time() - t0)
0.000851
```

This is a very crude way to measure code execution time.

The approximate solution is ten times faster than the exact one, ◀ but the repeated non-group addition operations will result in an improper rotation matrix. We can check this by examining its determinant, and the difference from the correct value of +1

```
>>> np.linalg.det(Rapprox) - 1
0.01
```

which is a significantly error. Even the exact solution has an improper rotation matrix

```
>>> np.linalg.det(Rexact) - 1
-2.89e-15
```

though the error is very small, and is due to accumulated finite-precision arithmetic error as discussed in ▶ Sect. 2.4.6.

We can normalize the rotation matrices in both cases, and then check the overall result which should be a rotation around the $x$-axis by 1 radian

```
>>> tr2angvec(trnorm(Rexact))
(1, array([        1,        0,        0]))
>>> tr2angvec(trnorm(Rapprox))
(1, array([        1,        0,        0]))
```

We see that the approximate solution is good to the default printing precision of four significant figures. The sample time used in this example is rather high, but was chosen for the purpose of illustration.

We can also approximate the unit quaternion derivative by a first-order difference

$$\dot{\mathring{q}} \approx \frac{\mathring{q}\langle k+1 \rangle - \mathring{q}\langle k \rangle}{\delta_t} \in \mathbb{H}$$

which combined with (3.3) gives us the approximation

$$\mathring{q}\langle k+1 \rangle \approx \mathring{q}\langle k \rangle + \frac{\delta_t}{2}\mathring{q}\langle k \rangle \circ \breve{\omega} \tag{3.10}$$

where $\breve{\omega}$ is a pure quaternion. This is even cheaper to compute than the rotation matrix approach, and while the result will not be a proper unit quaternion it can be normalized cheaply as discussed in ▶ Sect. 2.4.6.

### ❗ Addition is Not a Group Operator for Rotation Matrices

In this section we have updated rotation matrices by addition, but addition is not a group operation for **SO**(3). The resulting matrix will not belong to **SO**(3) – its determinant will not equal one and its columns will not be orthogonal unit vectors. Similarly, we have updated unit quaternions by addition which is not a group operation for $\mathbf{S}^3$. The result will not be a proper unit quaternion belonging to $\mathbf{S}^3$ – its magnitude will not be one.

However, if the values we add are *small*, then this problem is minor, and can be largely corrected by *normalization* as discussed in ▶ Sect. 2.4.6. We can ensure that those values are small by ensuring that $\delta_t$ is small which implies a high sample rate. For inertial navigation systems operating on low-end computing hardware, there is a tradeoff between sample rate and accuracy when using approximate update methods.

**3**

### 3.1.5 Incremental Rigid-Body Motion

Consider two poses $\boldsymbol{\xi}_1$ and $\boldsymbol{\xi}_2$ which differ infinitesimally and are related by

$$\boldsymbol{\xi}_2 = \boldsymbol{\xi}_1 \oplus \boldsymbol{\xi}^{\Delta}$$

where $\boldsymbol{\xi}^{\Delta} = \ominus \boldsymbol{\xi}_1 \oplus \boldsymbol{\xi}_2$. If $\boldsymbol{\xi}$ is represented by an $\mathbf{SE}(3)$ matrix we can write

$$\mathbf{T}^{\Delta} = \mathbf{T}_1^{-1} \mathbf{T}_2 = \begin{pmatrix} \mathbf{R}^{\Delta} & \boldsymbol{t}^{\Delta} \\ \mathbf{0}_{1\times 3} & 1 \end{pmatrix}$$

then $\boldsymbol{t}^{\Delta} \in \mathbb{R}^3$ is an incremental displacement. $\mathbf{R}^{\Delta} \in \mathbf{SO}(3)$ is an incremental rotation matrix which, by (3.8), will be an identity matrix plus a skew-symmetric matrix which has only three unique elements $\boldsymbol{r}^{\Delta} = \vee_{\times}(\mathbf{R}_{\Delta} - \mathbf{1}_{3\times 3})$. The incremental rigid-body motion can therefore be described by just six parameters

$$\Delta(\boldsymbol{\xi}_1, \boldsymbol{\xi}_2) \mapsto \boldsymbol{\delta} = (\boldsymbol{t}^{\Delta}, \boldsymbol{r}^{\Delta}) \in \mathbb{R}^6 \tag{3.11}$$

This is useful in optimization procedures that seek to minimize the error between two poses: we can choose the cost function $e = \|\Delta(\xi_1, \xi_2)\|$ which is equal to zero when $\xi_1 \equiv \xi_2$. This is very approximate when the poses are significantly different, but becomes ever more accurate as $\xi_1 \to \xi_2$.

which is a *spatial displacement*. ◄ A body with constant spatial velocity $\boldsymbol{v}$ for $\delta_t$ seconds undergoes a spatial displacement of $\boldsymbol{\delta} = \delta_t \boldsymbol{v}$. This is an approximation to the more expensive operation $\vee(\log \mathbf{T}^{\Delta})$.

The inverse of this operator is

$$\Delta^{-1}(\boldsymbol{\delta}) \mapsto \boldsymbol{\xi}^{\Delta} \tag{3.12}$$

and if $\boldsymbol{\xi}$ is represented by an $\mathbf{SE}(3)$ matrix then

$$\mathbf{T}^{\Delta} = \begin{pmatrix} \left[\boldsymbol{r}^{\Delta}\right]_{\times} + \mathbf{1}_{3\times 3} & \boldsymbol{t}^{\Delta} \\ \mathbf{0}_{1\times 3} & 1 \end{pmatrix}$$

which is an approximation to the more expensive operation $e^{[\boldsymbol{\delta}]}$.

The spatial displacement operator and its inverse are implemented by the Toolbox functions `tr2delta` and `delta2tr` respectively. These functions are computationally cheap compared to the exact operations using `logm` and `expm`, but do assume that the displacements are infinitesimal – they become increasingly inaccurate with displacement magnitude.

---

**Excurse 3.1: Sir Isaac Newton**

Newton (1642–1727) was an English natural philosopher and alchemist. He was Lucasian professor of mathematics at Cambridge, Master of the Royal Mint, and the thirteenth president of the Royal Society. His achievements include the three laws of motion, the mathematics of gravitational attraction, the motion of celestial objects, the theory of light and color (see ► Exc. 10.1), and building the first reflecting telescope.

Many of these results were published in 1687 in his great 3-volume work *The Philosophiae Naturalis Principia Mathematica* (Mathematical principles of natural philosophy). In 1704 he published *Opticks*, which was a study of the nature of light and color and the phenomena of diffraction. The SI unit of force is named in his honor. He is buried in Westminster Abbey, London.

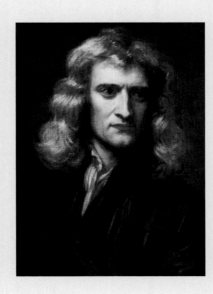

## 3.2 Accelerating Bodies and Reference Frames

So far we have considered only the first derivative, the velocity of a coordinate frame. However, all motion is ultimately caused by a force or a torque which leads to acceleration and this is the domain of dynamics.

### 3.2.1 Dynamics of Moving Bodies

For translational motion, Newton's second law describes, in the inertial frame {0}, the acceleration of a particle with position $x$ and mass $m$

$$m\,^0\ddot{x}_B = \,^0f_B \tag{3.13}$$

due to the applied force $^0f_B$.

Rotational motion in three dimensions is described by Euler's equations of motion which relates the angular acceleration of the body in the body frame

$$^B\mathbf{J}_B\,^B\dot{\boldsymbol{\omega}}_B + \,^B\boldsymbol{\omega}_B \times \left(^B\mathbf{J}_B\,^B\boldsymbol{\omega}_B\right) = \,^B\boldsymbol{\tau}_B \tag{3.14}$$

to the applied torque or moment $^B\boldsymbol{\tau}_B$ and a positive-definite rotational inertia tensor $^B\mathbf{J}_B \in \mathbb{R}^{3\times3}$. Nonzero angular acceleration implies that angular velocity, the axis and angle of rotation, evolves over time. ▶

We will simulate the rotational motion of an object moving in space. We define the inertia tensor as

```
>>> J = np.array([[ 2, -1, 0],
...               [-1,  4, 0],
...               [ 0,  0, 3]]);
```

which is positive definite ▶ and the nonzero off-diagonal terms will cause it to *tumble*. The initial conditions for orientation and angular velocity are

```
>>> orientation = UnitQuaternion();  # identity quaternion
>>> w = 0.2 * np.array([1, 2, 2]);
```

The simulation is implemented as a Python generator which computes angular acceleration by (3.14), and uses rectangular integration to update the angular velocity and orientation.

```
>>> dt = 0.05;  # time step
>>> def update():
...     global orientation, w
...     for t in np.arange(0, 10, dt):
...         wd = -np.linalg.inv(J) @ (np.cross(w, J @ w))  # (3.14)
...         w += wd * dt
...         orientation *= UnitQuaternion.EulerVec(w * dt)
...         yield orientation.R
>>> tranimate(update())
```

The `EulerVec` method acts like a constructor to create a new `UnitQuaternion` from a 3-element Euler vector. `tranimate` takes successive values from the generator and animates a graphical coordinate frame.

In the absence of torque, the angular velocity a body is not necessarily constant, for example, the angular velocity of a satellite tumbling in space is not constant. This is quite different to the linear velocity case where, in the absence of force, velocity remains constant. For rotational motion, it is angular momentum $\boldsymbol{h} = \mathbf{J}\boldsymbol{\omega}$ in the inertial frame that is constant.

Off-diagonal elements of a positive-definite matrix can be negative, it is the eigenvalues that must be positive.

**3**

---

**Excurse 3.2: Rotational Inertia Tensor**

The rotational inertia of a body that moves in three dimensions is described by a rank-2 tensor whose elements depend on the choice of the coordinate frame. An inertia tensor can be written and used like a $3 \times 3$ matrix.

$$\mathbf{J} = \begin{pmatrix} J_{xx} & J_{xy} & J_{xz} \\ J_{xy} & J_{yy} & J_{yz} \\ J_{xz} & J_{yz} & J_{zz} \end{pmatrix}$$

which is symmetric and positive definite, that is, it has positive eigenvalues. The eigenvalues also satisfy the triangle inequality: the sum of any two eigenvalues is always greater than, or equal to, the other eigenvalue.

The diagonal elements of $\mathbf{J}$ are the positive moments of inertia, and the off-diagonal elements are products of inertia. Only six of these nine elements are unique: three moments and three products of inertia. The products of inertia are all zero if the object's mass distribution is symmetrical with respect to the coordinate frame.

---

### 3.2.2 Transforming Forces and Torques

The translational force $\boldsymbol{f} = (f_x, f_y, f_z)$ and rotational torque, or moment, $\boldsymbol{m} = (m_x, m_y, m_z)$ applied to a body can be combined into a 6-vector that is called a wrench $\boldsymbol{w} = (f_x, f_y, f_z, m_x, m_y, m_z) \in \mathbb{R}^6$.

A wrench $^B\boldsymbol{w}$ is defined with respect to the coordinate frame {B} and applied at the origin of that frame. For coordinate frame {A} attached to the same body, the wrench $^A\boldsymbol{w}$ is equivalent if it causes the same motion of the body when applied to the origin of frame {A}. The wrenches are related by

$$^A\boldsymbol{w} = \begin{pmatrix} {}^B\mathbf{R}_A & \left[{}^B t_A\right]_\times {}^B\mathbf{R}_A \\ \mathbf{0}_{3\times3} & {}^B\mathbf{R}_A \end{pmatrix}^\top {}^B\boldsymbol{w} = \mathrm{Ad}^\top\!\left({}^B\mathbf{T}_A\right) {}^B\boldsymbol{w} \tag{3.15}$$

which is similar to the spatial velocity transform of (3.5) but uses the transpose of the adjoint of the *inverse* relative pose.

Continuing the example from ◻ Fig. 3.1b, we define a wrench with respect to frame {B} that exerts force along each axis

```
>>> bW = [1, 2, 3, 0, 0, 0];
```

The equivalent wrench in frame {A} would be

```
>>> aW = aTb.inv().Ad().T @ bW
array([ -3,  -1,   2,   0,   4,   2])
```

which has the same force components as applied at {B}, but transposed and negated to reflect the different orientations of the frames. The forces applied to the origin of {B} will also exert a moment on the body, so the equivalent wrench at {A} must include this: a torque of 4 Nm about the $y$-axis and 2 Nm about the $z$-axis. Note that there is no moment about the $x$-axis of frame {A}.

**❗ Wrench convention**

There is no firm convention for the order of the forces and torques in the wrench vector. This book always uses $\boldsymbol{w} = (\boldsymbol{f}, \boldsymbol{m})$ but other sources, including the MATLAB third edition of this book, use $\boldsymbol{w} = (\boldsymbol{m}, \boldsymbol{f})$.

### 3.2.3  Inertial Reference Frame

In robotics it is important to distinguish between an *inertial reference frame* and a non-inertial reference frame. An inertial reference frame is crisply defined as:

» *a reference frame that is not accelerating or rotating.*

Consider a particle P at rest with respect to a stationary reference frame {0}. Frame {B} is moving with constant velocity $^0v_B$ relative to frame {0}. From the perspective of frame {B}, the particle would be moving at constant velocity, in fact $^Bv_P = -^0v_B$. The particle is not accelerating and obeys Newton's first law:

» *in the absence of an applied force, a particle moves at a constant velocity*

and therefore frame {B} is an inertial reference frame.

Now consider that frame {B} is accelerating, at a constant acceleration $^0a_B$ with respect to {0}. From the perspective of frame {B}, the particle appears to be accelerating without an applied force, in fact $^Ba_P = -^0a_B$. This violates Newton's first law and an observer in frame {B} would have to invoke some intangible force to explain what they observe. We call such a force a fictitious, apparent, pseudo, inertial or d'Alembert force – they only exist in an accelerating or noninertial reference frame. This accelerating frame {B} is *not* an inertial reference frame.

Gravity could be considered to be an intangible force since it causes objects to accelerate with respect to a frame that is not accelerating. However, in Newtonian mechanics, gravity is considered a real body force $mg$ and a free object will accelerate relative to the inertial frame. ▶

An everyday example of a noninertial reference frame is an accelerating car or airplane. Inside an accelerating vehicle we observe fictitious forces pushing objects, including ourselves, around in a way that is not explained by Newton's first law.

For a rotating reference frame, things are more complex still. Imagine two people standing on a rotating turntable, and throwing a ball back and forth. They would observe that the ball follows a curved path in space, and they would also have to invoke an intangible force to explain what they observe.

If the reference frame {B} is rotating with angular velocity $\omega$ about its origin, then Newton's second law (3.13) becomes

$$m\left( {^B\dot{v}} + \underbrace{\omega \times \left(\omega \times {^B p}\right)}_{\text{centripetal}} + \underbrace{2\omega \times {^B v}}_{\text{Coriolis}} + \underbrace{\dot{\omega} \times {^B p}}_{\text{Euler}} \right) = {^0 f}$$

with three *new* acceleration terms. Centripetal acceleration always acts inward toward the origin. If the point is moving, then Coriolis acceleration will be normal to its velocity. If the rotational velocity varies with time, then Euler acceleration will be normal to the position vector. The centripetal term can be moved to the right-hand side, in which case it becomes a fictitious outward centrifugal force. This complexity is symptomatic of being in a noninertial reference frame, and another definition of an inertial frame is:

» *one in which the physical laws hold good in their simplest form*
   – Albert Einstein: The foundation of the general theory of relativity.

In robotics, the term inertial frame and world coordinate frame tend to be used loosely and interchangeably to indicate a frame fixed to some point on the Earth. This is to distinguish it from the body frame attached to the robot or vehicle. The surface of the Earth is an approximation of an inertial reference frame – the effect of the Earth's rotation is a finite acceleration less than 0.03 m s$^{-2}$ due to centripetal acceleration. From the perspective of an Earth-bound observer, a moving body will

Einstein's equivalence principle is that "*we assume the complete physical equivalence of a gravitational field and a corresponding acceleration of the reference system*" – we are unable to distinguish between gravity and being on a rocket accelerating at 1 *g* far from the gravitational influence of any celestial object.

**3**

Coriolis acceleration is significant for large-scale weather systems and therefore meteorological prediction, but is below the sensitivity of low-cost sensors such as those found in smart phones.

experience Coriolis acceleration. ◄ These effects are small, dependent on latitude, and typically ignored.

## 3.3 Creating Time-Varying Pose

In robotics, we often need to generate a pose that varies smoothly with time from $\xi_0$ to $\xi_1$, for example, the desired motion of a robot end effector or a drone. We require position and orientation to vary smoothly with time. In simple terms, this is moving smoothly along a path that is itself smooth. Smoothness in this context means that its first few temporal derivatives of position and orientation are continuous. Typically, velocity and acceleration are required to be continuous and sometimes also the derivative of acceleration or jerk.

The first problem is to construct a smooth path from $\xi_0$ to $\xi_1$ which could be defined by some function $\xi(s)$, $s \in [0, 1]$. The second problem is to construct a smooth trajectory, that is smooth motion along the path which requires that $s(t)$ is a smooth function of time.

We start by discussing how to generate smooth trajectories in one dimension. We then extend that to the multi-dimensional case and then to piecewise-linear trajectories that visit a number of intermediate points without stopping.

### 3.3.1 Smooth One-Dimensional Trajectories

We start our discussion with a scalar function of time that has a specified initial and final value. An obvious candidate for such a smooth function is a polynomial function of time. Polynomials are simple to compute and can easily provide the required smoothness and boundary conditions. A quintic (fifth-order) polynomial is commonly used

$$q(t) = At^5 + Bt^4 + Ct^3 + Dt^2 + Et + F \tag{3.16}$$

where time $t \in [0, T]$. The first- and second-derivatives are also polynomials

$$\dot{q}(t) = 5At^4 + 4Bt^3 + 3Ct^2 + 2Dt + E \tag{3.17}$$

$$\ddot{q}(t) = 20At^3 + 12Bt^2 + 6Ct + 2D \tag{3.18}$$

and therefore smooth. The third-derivative, jerk, will be a quadratic.

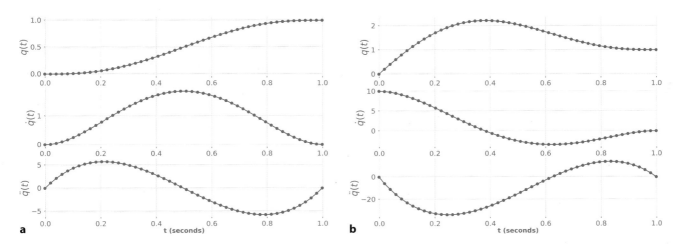

**Fig. 3.2** Quintic polynomial trajectory. From top to bottom is position, velocity and acceleration versus time step. **a** With zero initial and final velocity, **b** initial velocity of 10 and a final velocity of 0. The discrete-time points are indicated by dots

The trajectory has defined boundary conditions for position ($q_0$, $q_T$), velocity ($\dot{q}_0$, $\dot{q}_T$) and acceleration ($\ddot{q}_0$, $\ddot{q}_T$). Writing (3.16) to (3.18) for the boundary conditions $t = 0$ and $t = T$ gives six equations which we can write in matrix form as

$$
\begin{pmatrix} q_0 \\ q_T \\ \dot{q}_0 \\ \dot{q}_T \\ \ddot{q}_0 \\ \ddot{q}_T \end{pmatrix} = \begin{pmatrix} 0 & 0 & 0 & 0 & 0 & 1 \\ T^5 & T^4 & T^3 & T^2 & T & 1 \\ 0 & 0 & 0 & 0 & 1 & 0 \\ 5T^4 & 4T^3 & 3T^2 & 2T & 1 & 0 \\ 0 & 0 & 0 & 2 & 0 & 0 \\ 20T^3 & 12T^2 & 6T & 2 & 0 & 0 \end{pmatrix} \begin{pmatrix} A \\ B \\ C \\ D \\ E \\ F \end{pmatrix}.
$$

The matrix is square ▶ and, if $T \neq 0$, we can invert it to solve for the coefficient vector ($A, B, C, D, E, F$).

The Toolbox provides a function to generate the trajectory described by (3.16)

This is the reason for choice of quintic polynomial. It has six coefficients that enable it to meet the six boundary conditions on initial and final position, velocity and acceleration.

```
>>> traj = quintic(0, 1, np.linspace(0, 1, 50));
```

where the arguments are respectively the initial position, the final position and the time which varies from 0 to 1 in 50 steps. The function returns an object that contains the trajectory data and its plot method

```
>>> traj.plot();
```

produces the graph shown in ◘ Fig. 3.2a. We can access the position, velocity and acceleration data as `traj.q`, `traj.qd` and `traj.qdd` – each is a 50-element NumPy array. We observe that the initial and final velocity and acceleration are all zero – the default value. ▶

The initial and final velocities can be set to nonzero values, for example, an initial velocity of 10 and a final velocity of 0

For this simple case, with no waypoints this is also the minimum jerk trajectory.

```
>>> quintic(0, 1, np.linspace(0, 1, 50), qd0=10, qdf=0);
```

creates the trajectory shown in ◘ Fig. 3.2b. This illustrates an important problem with polynomials – nonzero initial velocity causes the polynomial to overshoot the terminal value, in this example, peaking at over 2 on a trajectory from 0 to 1.

**3**

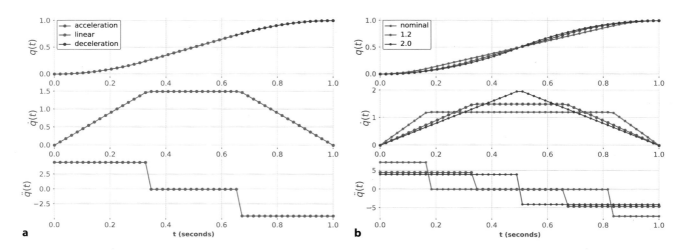

□ **Fig. 3.3**    Trapezoidal trajectory. **a** trajectory for default linear segment velocity; **b** trajectories for specified linear segment velocities

Another problem with polynomials, a very practical one, can be seen in the middle graph of □ Fig. 3.2a. The velocity peaks when $t = 0.5$ s which means that, for most of the time, the velocity is far less than the maximum. The mean velocity

```
>>> qd = traj.qd;
>>> qd.mean() / qd.max()
0.5231
```

is only 52% of the peak so we are not using the motor as fully as we could. A real robot joint has a well-defined maximum speed and, for minimum-time motion, we want to be operating at that maximum for as much of the time as possible. We would like the velocity curve to be *flatter* on top.

A well-known alternative is a trapezoidal hybrid trajectory

```
>>> traj = trapezoidal(0, 1, np.linspace(0, 1, 50));
>>> traj.plot();
```

where the arguments have the same meaning as for `quintic` and the trajectory, plotted in a similar way, is shown in □ Fig. 3.3a. As with `quintic`, the velocity and acceleration trajectories are also stored within the object.

The velocity trajectory has a trapezoidal shape, hence the name, comprising three linear segments. The corresponding segments of the position trajectory are a straight line (constant velocity segment) with parabolic blends. The term blend refers to a trajectory segment that smoothly joins linear segments. This type of trajectory is commonly used in industrial motor drives. It is continuous in position and velocity, but not in acceleration.

The function `trapezoidal` has *chosen* the velocity of the linear segment to be

```
>>> traj.qd.max()
1.5
```

but this can be overridden by specifying an additional argument

```
>>> traj1_2 = trapezoidal(0, 1, np.linspace(0, 1, 50), V=1.2);
>>> traj2_0 = trapezoidal(0, 1, np.linspace(0, 1, 50), V=2);
```

The system has one design degree of freedom. There are six degrees of freedom (blend time, three parabolic coefficients and two linear coefficients) and five constraints (total time, initial and final position and velocity).

The trajectories for these different cases are overlaid in □ Fig. 3.3b. We see that as the velocity of the linear segment increases, its duration decreases and ultimately its duration would be zero. In fact, the velocity cannot be chosen arbitrarily – too high or too low a value for the maximum velocity will result in an infeasible trajectory and the function will return an error. ◄

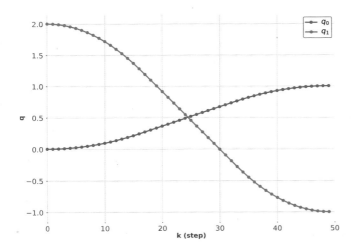

**◘ Fig. 3.4** Multi-axis motion. $q_0$ varies from $0 \rightarrow 1$ and $q_1$ varies from $2 \rightarrow -1$

### 3.3.2 Multi-Axis Trajectories

Most useful robots have more than one axis of motion and it is quite straightforward to extend the smooth scalar trajectory to the vector case. In terms of configuration space (► Sect. 2.4.9), these axes of motion correspond to the dimensions of the robot's configuration space – to its degrees of freedom. We represent the robot's configuration as a vector $q \in \mathbb{R}^N$ where $N$ is the number of degrees of freedom. The configuration of a 3-joint robot would be its joint coordinates $q = (q_0, q_1, q_2)$. The configuration vector of wheeled mobile robot might be its position $q = (x, y)$ or its position and heading angle $q = (x, y, \theta)$. For a 3-dimensional body that had an orientation in **SO**(3) we would use roll-pitch-yaw angles $q = (\alpha, \beta, \gamma)$, or for a pose in **SE**(3) we would use $q = (x, y, z, \alpha, \beta, \gamma)$. ► In all these cases, we would require smooth multi-dimensional motion from an initial configuration vector to a final configuration vector.

Any 3-parameter representation could be used: roll-pitch-yaw angles, Euler angles or exponential coordinates.

Consider a 2-axis $xy$-type robot moving from configuration $(0, 2)$ to $(1, -1)$ in 50 steps with a trapezoidal profile. Using the Toolbox, this is achieved using the function `mtraj` and we write

```
>>> traj = mtraj(trapezoidal, [0, 2], [1, -1], 50);
```

where `traj.q` is a $50 \times 2$ NumPy array with one row per time step and one column per axis. The first argument is a function object that generates a smooth *scalar* trajectory as a function of time, in this case `trapezoidal` but `quintic` could also be used. The result is again a trajectory object and the 2-axis trajectory

```
>>> traj.plot();
```

is shown in ◘ Fig. 3.4.

If we wished to create a trajectory for 3-dimensional pose, we could convert an `SE3` pose instance `T` to a 6-vector by a command like

```
>>> q = np.array([T.t, T.rpy()])
```

though as we shall see later, interpolation of 3-angle representations has some complexities.

**3**

### 3.3.3 **Multi-Segment Trajectories**

In robotics applications, there is often a need to move smoothly along a path through one or more intermediate or *via* points or *waypoints* without stopping. This might be to avoid obstacles in the workplace, or to perform a task that involves following a piecewise-linear trajectory such as welding a seam or applying a bead of sealant in a manufacturing application.

To formalize the problem, consider that the trajectory is defined by $M$ configurations $q_k$, $k = 0, \ldots, M-1$ so there will be $M-1$ motion segments. As in the previous section, $q_k \in \mathbb{R}^N$ is a configuration vector.

The robot starts from $q_0$ at rest and finishes at $q_{M-1}$ at rest, but moves through (or close to) the intermediate configurations without stopping. The problem is over constrained and, in order to attain continuous velocity, we surrender the ability to reach each intermediate configuration. This is easiest to understand for the 1-dimensional case shown in ◘ Fig. 3.5. The motion comprises linear motion segments with polynomial blends, like `trapezoidal`, but here we choose quintic polynomials because they are able to match boundary conditions on position, velocity and acceleration at their start and end points.

The first segment of the trajectory accelerates from the initial configuration $q_0$ and zero velocity, and joins the line heading toward the second configuration $q_1$. The blend time is set to be a constant $t_{\mathrm{acc}}$ and $t_{\mathrm{acc}}/2$ before reaching $q_1$, the trajectory executes a polynomial blend, of duration $t_{\mathrm{acc}}$, onto the line from $q_1$ to $q_2$, and the process repeats. The constant velocity $\dot{q}_k$ can be specified for each segment. The average acceleration during the blend is

$$\ddot{q} = \frac{\dot{q}_{k+1} - \dot{q}_k}{t_{\mathrm{acc}}} \ .$$

If the maximum acceleration capability of the axis is known, then the minimum blend time can be computed. ◄

The axes may have different maximum motor speeds and, on a particular motion segment, the axes will generally have different distances to travel. The first step in planning a segment is to determine which axis will be the last to complete the segment, and the speed of the other axes is reduced to ensure that the motion is coordinated, that is, all axes reach the next target $q_k$ at the *same time*.

Consider again a 2-axis $xy$-type robot following a path defined by the corners of a rotated square. The trajectory can be generated by

```
>>> via = SO2(30, unit="deg") * np.array([
...     [-1, 1, 1, -1, -1], [1, 1, -1, -1, 1]]);
>>> traj0 = mstraj(via.T, dt=0.2, tacc=0, qdmax=[2, 1]);
```

The real limit of the axis will be its peak, rather than average, acceleration. The peak acceleration for the blend can be determined from (3.18) once the quintic coefficients are known.

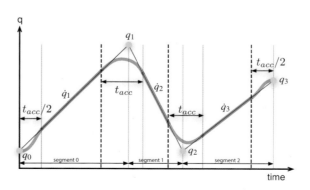

◘ **Fig. 3.5** Notation for multi-segment trajectory showing four points and three motion segments. Blue indicates constant velocity motion, red indicates the blends where acceleration occurs

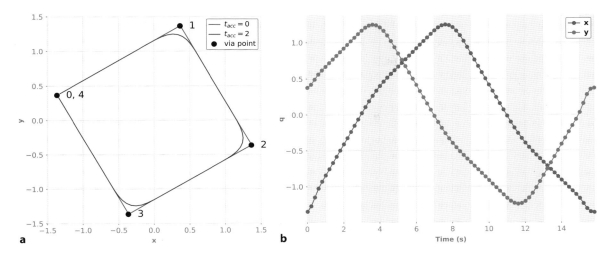

**◘ Fig. 3.6** Multi-segment multi-axis trajectories. **a** configuration of robot (tool position) for different acceleration times; **b** configuration versus time for the case $t_{acc} = 2$ s with segment blends indicated by gray bars

where the arguments are: the array of via points, each column is the coordinate of a waypoint; the sample time; the acceleration time; ▶ and a vector of maximum speeds for each axis. ▶ The function returns an object that holds the trajectory data as in the earlier examples. We can plot $y$ against $x$ to see the path of the robot

```
>>> xplot(traj0.q[:, 0], traj0.q[:, 1], color="red");
```

as shown by the red line in ◘ Fig. 3.6a. If we increase the acceleration time to two seconds

```
>>> traj2 = mstraj(via.T, dt=0.2, tacc=2, qdmax=[2, 1]);
```

the trajectory shown in blue is more rounded as the polynomial blending functions do their work. The smoother trajectory also takes more time to complete

```
>>> len(traj0), len(traj2)
(28, 80)
```

The configuration variables, as a function of time, are shown in ◘ Fig. 3.6b and the linear segments and blends can be clearly seen. This function also accepts optional initial and final velocity arguments, and $t_{acc}$ can be a vector specifying different acceleration times for each of the $M$ blends.

Keep in mind that this function simply interpolates configuration represented as a vector. In this example, the vector was assumed to be Cartesian coordinates. For rotation, we could consider applying this function to Euler or roll-pitch-yaw angles but this is not an ideal way to interpolate orientation as we will discuss in the next section.

Any times are rounded up internally to a multiple of the time step.

Alternatively, the time per segment can be specified using the argument `tsegment`.

### 3.3.4 Interpolation of Orientation in 3D

In robotics, we often need to interpolate orientation, for example, we require the end effector of a robot to smoothly change from orientation $\boldsymbol{\xi}_0$ to $\boldsymbol{\xi}_1$ in $\mathbf{S}^3$. We require some function $\boldsymbol{\xi}(s) = \sigma(\boldsymbol{\xi}_0, \boldsymbol{\xi}_1, s)$ where $s \in [0, 1]$, with boundary conditions $\sigma(\boldsymbol{\xi}_0, \boldsymbol{\xi}_1, 0) = \boldsymbol{\xi}_0$ and $\sigma(\boldsymbol{\xi}_0, \boldsymbol{\xi}_1, 1) = \boldsymbol{\xi}_1$ and where $\sigma(\boldsymbol{\xi}_0, \boldsymbol{\xi}_1, s)$ varies *smoothly* for intermediate values of $s$. How we implement this depends very much on our concrete representation of $\boldsymbol{\xi}$.

A workable and commonly used approach is to consider a 3-parameter representation such as Euler or roll-pitch-yaw angles, $\boldsymbol{\Gamma} \in (\mathbf{S}^1)^3$, or exponential coordi-

**3**

nates and use linear interpolation

$$\sigma(\boldsymbol{\Gamma}_0,\ \boldsymbol{\Gamma}_1,\ s) = (1-s)\boldsymbol{\Gamma}_0 + s\boldsymbol{\Gamma}_1, s \in [0, 1]$$

and converting the interpolated vector back to a rotation matrix. For example, we define two orientations

```
>>> R0 = SO3.Rz(-1) * SO3.Ry(-1);
>>> R1 = SO3.Rz(1) * SO3.Ry(1);
```

and find the equivalent ZYX roll-pitch-yaw angles

```
>>> rpy0 = R0.rpy(); rpy1 = R1.rpy();
```

and create a trajectory between them with 50 uniform steps

```
>>> traj = mtraj(quintic, rpy0, rpy1, 50);
```

and `traj.q` is a $50 \times 3$ array of roll-pitch-yaw angles. This is most easily visualized as an animation, so we will first convert those angles back to $\mathbf{SO}(3)$ matrices

```
>>> pose = SO3.RPY(traj.q);
>>> len(pose)
50
```

which has created a single SO3 object with 50 values which we can animate

```
>>> pose.animate();
```

For large orientation change we see that the axis, around which the coordinate frame rotates, changes along the trajectory. The motion, while smooth, can look a bit uncoordinated. There will also be problems if either $\boldsymbol{\xi}_0$ or $\boldsymbol{\xi}_1$ is close to a singularity in the particular vector representation of orientation.

Interpolation of unit quaternions is only a little more complex than for 3-angle vectors and results in a rotation about a *fixed* axis in space. Using the Toolbox, we first find the two equivalent unit quaternions

```
>>> q0 = UnitQuaternion(R0); q1 = UnitQuaternion(R1);
```

and then interpolate them in 50 uniform steps

```
>>> qtraj = q0.interp(q1, 50);
>>> len(qtraj)
50
```

where the object is the initial orientation and the arguments to the `interp` method are the final orientation and the number of steps. The result is a `UnitQuaternion` object that holds 50 values, which we can animate by

```
>>> qtraj.animate()
```

Quaternion interpolation is achieved using spherical linear interpolation (*slerp*) in which the unit quaternions follow a great circle ◄ path on a 4-dimensional hypersphere. The result in three dimensions is rotation about a fixed axis in space.

A great circle on a sphere is the intersection of the sphere and a plane that passes through its center. On Earth, the equator and all lines of longitude are great circles. Ships and aircraft prefer to follow great circles because they represent the shortest path between two points on the surface of a sphere.

> **🛑 We cannot linearly interpolate a rotation matrix**
>
> If pose is represented by an orthogonal rotation matrix $\mathbf{R} \in \mathbf{SO}(3)$, we *cannot* use linear interpolation $\sigma(\mathbf{R}_0, \mathbf{R}_1, s) = (1-s)\mathbf{R}_0 + s\mathbf{R}_1$. Scalar multiplication and addition are not valid operations for the group of $\mathbf{SO}(3)$ matrices, and that means that the resulting $\mathbf{R}$ would not be an $\mathbf{SO}(3)$ matrix – the column norm and inter-column orthogonality constraints would be violated.

### 3.3.4.1  Direction of Rotation

When moving between two points on a circle, we can travel clockwise or counterclockwise – the result is the same but the distance traveled may be different. The same choices exist when we move on a great circle. In this example, we animate a rotation about the $z$-axis, from an angle of $-2$ radians to $+2$ radians

```
>>> q0 = UnitQuaternion.Rz(-2); q1 = UnitQuaternion.Rz(2);
>>> q = q0.interp(q1, 50);
>>> q.animate()
```

which is a path that takes the *long way* around the circle, moving 4 radians when we could travel just $2\pi - 4 \approx 2.28$ radians in the opposite direction. We can request that the shortest path be taken

```
>>> q = q0.interp(q1, 50, shortest=True);
>>> q.animate()
```

and the animation clearly shows the difference.

### 3.3.5 Cartesian Motion in 3D

Another common requirement in robotics is to find a smooth path between two 3D poses in $\mathbb{R}^3 \times \mathbf{S}^3$ which involves change in position as well as in orientation. In robotics, this is often referred to as Cartesian motion.

We represent the initial and final poses as $\mathbf{SE}(3)$ matrices

```
>>> T0 = SE3.Trans([0.4, 0.2, 0]) * SE3.RPY(0, 0, 3);
>>> T1 = SE3.Trans([-0.4, -0.2, 0.3]) * SE3.RPY(-pi/4, pi/4, -pi/2);
```

The SE3 object has a method interp that interpolates between two poses for normalized distance $s \in [0, 1]$ along the path, for example the midway pose between T0 and T1 is

```
>>> T0.interp(T1, 0.5)
   0.09754   -0.702     0.7055     0
   0.702      0.551     0.4512     0
  -0.7055     0.4512    0.5465     0.15
   0          0         0          1
```

where the object is the initial pose and the arguments to the interp method are the final pose and the normalized distance. The translational component is linearly interpolated, while rotation is interpolated using unit quaternion spherical linear interpolation as introduced above. A trajectory between the two poses in 51 steps is created by

```
>>> Ts = T0.interp(T1, 51);
```

and the result is an SE3 object with 51 values

```
>>> len(Ts)
51
```

representing the pose at each time step, and once again the easiest way to visualize this is by animation

```
>>> Ts.animate()
```

which shows the coordinate frame moving and rotating from pose T0 to pose T1. The $\mathbf{SE}(3)$ value at the mid-point on the path is

```
>>> Ts[25]
   0.09754   -0.702     0.7055     0
   0.702      0.551     0.4512     0
  -0.7055     0.4512    0.5465     0.15
   0          0         0          1
```

and is the same as the result given above.

The translational part of this trajectory is obtained by

```
>>> P = Ts.t;
>>> P.shape
(51, 3)
```

**3**

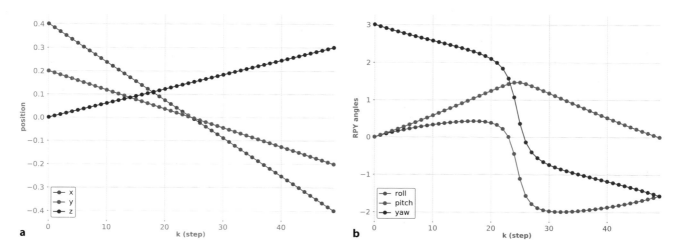

**Fig. 3.7**    Cartesian motion. **a** Cartesian position versus time; **b** roll-pitch-yaw angles versus time

which is an array where the rows represent position at consecutive time steps. This is plotted

```
>>> xplot(P, labels="x y z");
```

in  Fig. 3.7 along with the orientation as roll-pitch-yaw angles

```
>>> rpy = Ts.rpy();
>>> xplot(rpy, labels="roll pitch yaw");
```

We see that the position coordinates vary linearly with time, but that the roll-pitch-yaw angles are nonlinear with time and that has two causes. Firstly, roll-pitch-yaw angles are a nonlinear transformation of the linearly-varying quaternion orientation. Secondly, this particular trajectory passes very close to the roll-pitch-yaw singularity, ◄ at around steps 24 and 25, and a symptom of this is the rapid rate of change of roll-pitch-yaw angles around this point. The coordinate frame is not rotating faster at this point – you can verify that in the animation – the rotational parameters are changing very quickly, and this is a consequence of the particular representation.

However, the motion has a velocity and acceleration *discontinuity* at the first and last points. While the path is smooth in space, the distance *s* along the path is not smooth in time. Speed along the path jumps from zero to some finite value and then

Covered in ► Sect. 2.3.1.3.

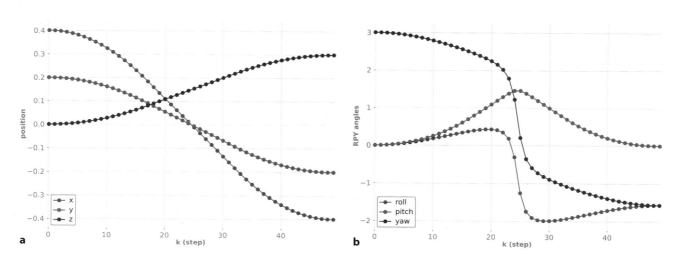

**Fig. 3.8**    Cartesian motion with trapezoidal path distance profile. **a** Cartesian position versus time, **b** roll-pitch-yaw angles versus time

drops to zero at the end – there is no initial acceleration or final deceleration. The scalar functions `quintic` and `trapezoidal` discussed earlier can be used to generate the interpolation variable $s$ so that motion *along* the path is smooth. We can pass a vector of normalized distances along the path as the second argument to `interp`

```
>>> Ts = T0.interp(T1, trapezoidal(0, 1, 50).q);
```

The path is unchanged, but the coordinate frame now accelerates to a constant speed along the path and decelerates at the end, resulting in the smoother trajectories shown in ◘ Fig. 3.8. The Toolbox provides a convenient shorthand `ctraj` function for this

```
>>> Ts = ctraj(T0, T1, 50);
```

where the arguments are the initial and final pose and the number of time steps.

## 3.4 Application: Inertial Navigation

An inertial navigation system or INS is a self contained unit that estimates its velocity, orientation and position by measuring accelerations and angular velocities and integrating them over time. Importantly, it has no external inputs such as radio signals from satellites. This makes it well suited to applications such as submarine, spacecraft and missile guidance where it is not possible to communicate with radio navigation aids, or which must be immune to radio jamming. These particular applications drove development of the technology during the cold war and space race of the 1950s and 1960s. Those early systems were large, see ◘ Fig. 3.9a, extremely expensive and the technical details were national secrets. Today, INSs are considerably cheaper and smaller as shown in ◘ Fig. 3.9b; the sensor chips shown in ◘ Fig. 3.9c can cost as little as a few dollars and they are built into every smart phone.

An INS estimates its pose with respect to an inertial reference frame which is typically denoted as {0}. ▶ The frame typically has its $z$-axis upward or downward and the $x$- and $y$-axes establish a locally horizontal tangent plane. Two common conventions have the $x$-, $y$- and $z$-axes respectively parallel to north-east-down (NED) or east-north-up (ENU) directions. The coordinate frame {B} is attached to the moving vehicle or robot and is known as the body- or body-fixed frame.

As discussed in ▶ Sect. 3.2.3, the Earth's surface is not an inertial reference frame but, for most robots with nonmilitary grade sensors, this is a valid assumption.

◘ **Fig. 3.9** Inertial sensors. **a** SPIRE (Space Inertial Reference Equipment) from 1953 was 1.5 m in diameter and weighed 1200 kg; **b** a modern inertial navigation system the LORD Micro-Strain 3DM-GX4-25 has triaxial gyroscopes, accelerometers and magnetometer, a pressure altimeter, is only 36×24×11 mm and weighs 16 g (image courtesy of LORD MicroStrain); **c** 9 Degrees of Freedom IMU Breakout (LSM9DS1-SEN-13284 from Spark-Fun Electronics), the chip itself is only 3.5 × 3 mm

**3**

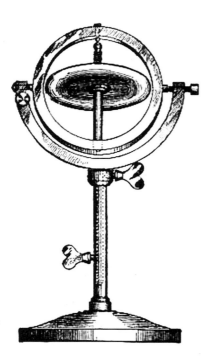

The engineering challenge was to create a mechanism that allowed the vehicle to rotate around the stable platform without exerting any torque on the gyroscopes. This required exquisitely engineered low-friction gimbals and bearing systems.

Typically by strain gauges attached to the bearings of the rotor shaft.

### 3.4.1 Gyroscopes

Any sensor that measures the rate of change of orientation is known, for historical reasons, as a gyroscope.

#### 3.4.1.1 How Gyroscopes Work

The term gyroscope conjures up an image of a childhood toy – a spinning disk in a round frame that can balance on the end of a pencil. Gyroscopes are confounding devices – you try to turn them one way but they resist and turn (precess) in a different direction. This unruly behavior is described by a simplified version of (3.14)

$$\tau = \omega \times h \tag{3.19}$$

where $h$ is the angular momentum of the gyroscope, a vector parallel to the rotor's axis of spin and with magnitude $\|h\| = J\varpi$, where $J$ is the rotor's inertia and $\varpi$ its rotational speed. It is the cross product in (3.19) that makes the gyroscope move in a contrary way.

If no torque is applied to the gyroscope, its angular momentum remains constant in the inertial reference frame which implies that the axis will maintain a *constant direction* in that frame. Two gyroscopes with orthogonal axes form a stable platform that will maintain a *constant orientation* with respect to the inertial reference frame – fixed with respect to the universe. This was the principle of many early spacecraft navigation systems such as that shown in ◘ Fig. 2.19 – the gimbals allowed the spacecraft to rotate about the stable platform, and the relative orientation was determined by measuring the gimbal angles. ◄

Alternatively, we can fix the gyroscope to the vehicle in the strapdown configuration as shown in ◘ Fig. 3.10. If the vehicle rotates with an angular velocity $\omega$, the attached gyroscope will *precess* and exert an orthogonal torque $\tau$ which can be measured. ◄ If the magnitude of $h$ is high, then this kind of sensor is very sensitive – a very small angular velocity leads to an easily measurable torque.

Over the last few decades, this rotating disk technology has been eclipsed by sensors based on optical principles such as the ring-laser gyroscope (RLG) and the fiber-optic gyroscope (FOG). These are highly accurate sensors but expensive and bulky. The low-cost sensors used in smart phones and drones are based on micro-electro-mechanical systems (MEMS) fabricated on silicon chips. Details of the designs vary but all contain a mass which vibrates at many kHz in a plane. Rotation about an axis normal to the plane causes an orthogonal displacement within the plane that is measured capacitively.

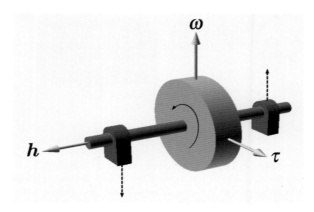

◘ **Fig. 3.10** Gyroscope in strapdown configuration. Angular velocity $\omega$ induces a torque $\tau$ which can be sensed as forces at the bearings shown in red

Gyroscopic angular velocity sensors measure rotation about a single axis. Typically, three gyroscopes are packaged together and arranged so that their sensitive axes are orthogonal. The three outputs of such a triaxial gyroscope are the components of the angular velocity vector $^B\boldsymbol{\omega}_B^{\#}$ measured in the body frame {B}, and we introduce the notation $\boldsymbol{x}^{\#}$ to explicitly indicate a value of $\boldsymbol{x}$ measured by a sensor.

Interestingly, nature has invented gyroscopic sensors. All vertebrates have angular velocity sensors as part of their vestibular system. In each inner ear, we have three semi-circular canals – fluid-filled organs that measure angular velocity. They are arranged orthogonally, just like a triaxial gyroscope, with two measurement axes in a vertical plane and one diagonally across the head.

### 3.4.1.2  Estimating Orientation

If the orientation of the sensor frame is initially $\boldsymbol{\xi}_B$, then the evolution of estimated orientation can be written in discrete-time form as

$$\hat{\boldsymbol{\xi}}_{B\langle k+1\rangle} \leftarrow \hat{\boldsymbol{\xi}}_{B\langle k\rangle} \oplus {}^{B\langle k\rangle}\boldsymbol{\xi}_{B\langle k+1\rangle} \tag{3.20}$$

where we use the hat notation to explicitly indicate an estimate of orientation, and $k \in \mathbb{N}_0$ is the index of the time step. $^{B\langle k\rangle}\boldsymbol{\xi}_{B\langle k+1\rangle}$ is the incremental rotation over the time step which can be computed from measured angular velocity.

It is common to assume that $^B\boldsymbol{\omega}_B$ is constant over the time interval $\delta_t$. In terms of $\mathbf{SO}(3)$ rotation matrices, the orientation update is

$$\hat{\mathbf{R}}_{B\langle k+1\rangle} \leftarrow \hat{\mathbf{R}}_{B\langle k\rangle}\, e^{\delta_t\left[{}^B\boldsymbol{\omega}_B^{\#}\right]_\times} \tag{3.21}$$

and the result should be periodically normalized to eliminate accumulated numerical error, as discussed in ▶ Sect. 2.4.6.

We will demonstrate this integration using unit quaternions and simulated angular velocity data for a tumbling body. The Toolbox example function `IMU`

```
>>> from imu_data import IMU
>>> true, _ = IMU()
```

returns an object which describes the true motion of the body. The rows of the array `true.omega` represent consecutive body-frame angular velocity measurements with corresponding times given by elements of the vector `true.t`. We choose the initial orientation to be the null rotation

```
>>> orientation = UnitQuaternion();  # identity quaternion
```

---

**Excurse 3.4: MIT Instrumentation Laboratory**

Important development of inertial navigation technology took place at the MIT Instrumentation Laboratory which was led by Charles Stark Draper. In 1953, the feasibility of inertial navigation for aircraft was demonstrated in a series of flight tests with a system called SPIRE (Space Inertial Reference Equipment) shown in ◼ Fig. 3.9a. It was 1.5 m in diameter and weighed 1200 kg. SPIRE guided a B-29 bomber on a 12-hour trip from Massachusetts to Los Angeles without the aid of a pilot and with Draper aboard. In 1954, the first self-contained submarine inertial navigation system (SINS) was introduced to service. The Instrumentation Lab also developed the Apollo Guidance Computer, a one-cubic-foot computer that guided the Apollo Lunar Module to the surface of the Moon in 1969.

Today, high-performance inertial navigation systems based on fiber-optic gyroscopes are widely available and weigh around 1 kg while low-cost systems based on MEMS technology can weigh just a few grams and cost a few dollars.

and then for each time step we update the orientation and keep the orientation history by exploiting the list-like properties of the `UnitQuaternion` class

```
>>> for w in true.omega[:-1]:
...    next = orientation[-1] @ UnitQuaternion.EulerVec(w * true.dt);
...    orientation.append(next);
>>> len(orientation)
400
```

The orientation is updated by a unit quaternion, returned by the `EulerVec` constructor, that corresponds to a rotation angle and axis given by the magnitude and direction of its argument. The `@` operator performs unit-quaternion multiplication and normalizes the product, ensuring the result has a unit norm. ◄ We can animate the changing orientation of the body frame

> The `.increment` method of the UnitQuaternion class does this in a single call.

```
>>> orientation.animate(time=true.t)
```

or view the roll-pitch-yaw angles as a function of time

```
>>> xplot(true.t, orientation.rpy(), labels="roll pitch yaw");
```

### 3.4.2  Accelerometers

Accelerometers are sensors that measure acceleration. Even when not moving, they sense the acceleration due to gravity which defines the direction we know as *downward*. Gravitational acceleration is a function of our distance from the Earth's center, and the material in the Earth beneath us. The Earth is not a perfect sphere ◄ and points in the equatorial region are further from the center. Gravitational acceleration can be approximated by

> The technical term is an oblate spheroid, it bulges out at the equator because of centrifugal acceleration due to the Earth's rotation. The equatorial diameter is around 40 km greater than the polar diameter.

$$g \approx 9.780327\left(1 + 0.0053024\sin^2\theta - 0.0000058\sin^2 2\theta\right) - 0.000003086h$$

where $\theta$ is the angle of latitude and $h$ is height above sea level. A map of gravity showing the effect of latitude and topography is shown in ◘ Fig. 3.11.

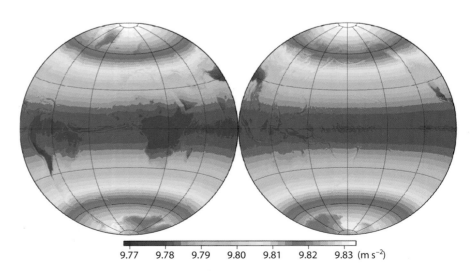

9.77    9.78    9.79    9.80    9.81    9.82    9.83  (m s$^{-2}$)

◘ **Fig. 3.11** Variation in Earth's gravitational acceleration shows the imprint of continents and mountain ranges. The hemispheres shown are centered on the prime (left) and anti (right) meridian respectively (from Hirt et al. 2013)

**Excurse 3.5: Charles Stark (Doc) Draper**

Draper (1901–1987) was an American scientist and engineer, often referred to as "the father of inertial navigation." Born in Windsor, Missouri, he studied at the University of Missouri then Stanford where he earned a B.A. in psychology in 1922, then at MIT an S.B. in electro-chemical engineering and an S.M. and Sc.D. in physics in 1928 and 1938 respectively. He started teaching while at MIT and became a full professor in aeronautical engineering in 1939. He was the founder and director of the MIT Instrumentation Laboratory which made important contributions to the theory and practice of inertial navigation to meet the needs of the cold war and the space program.

Draper was named one of Time magazine's Men of the Year in 1961 and inducted to the National Inventors Hall of Fame in 1981. The Instrumentation lab was renamed Charles Stark Draper Laboratory (CSDL) in his honor. (image courtesy of The Charles Stark Draper Laboratory Inc.)

### 3.4.2.1  How Accelerometers Work

An accelerometer is conceptually a very simple device comprising a mass, known as the proof mass, supported by a spring as shown in ◻ Fig. 3.12. In the inertial reference frame, Newton's second law for the proof mass is

$$m\ddot{x}_m = F_s - mg \tag{3.22}$$

and, for a spring with natural length $l_0$, the relationship between force and extension $d$ is

$$F_s = kd \ .$$

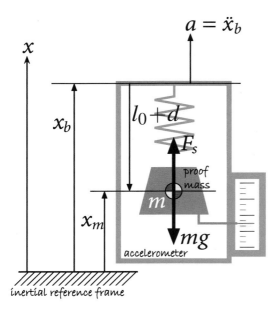

◻ **Fig. 3.12**  The essential elements of an accelerometer and notation

The various displacements are related by

$$x_b - (l_0 + d) = x_m$$

and taking the double derivative then substituting (3.22) gives

$$\ddot{x}_b - \ddot{d} = \frac{1}{m}kd - g \ .$$

We assume that $\ddot{d} = 0$ in steady state. Typically, there is a frictional element to damp out oscillation of the proof mass. This adds a term $-B\dot{x}_m$ to the right-hand side of (3.22).

The quantity we wish to measure is the acceleration of the accelerometer $a = \ddot{x}_b$ and the relative displacement of the proof mass ◄

$$d = \frac{m}{k}(a + g)$$

is linearly related to that acceleration. In an accelerometer, the displacement is measured and scaled by $k/m$ so that the output of the sensor is

$$a^{\#} = a + g \ .$$

If this accelerometer is stationary, then $a = 0$ yet the measured acceleration would be $a^{\#} = 0 + g = g$ in the upward direction. This is because our model has included the Newtonian gravity force $m\boldsymbol{g}$, as discussed in ► Sect. 3.2.3. Accelerometer output is sometimes referred to as specific, inertial or proper acceleration.

🛑 **Accelerometer readings are nonintuitive**

It is quite unintuitive that a stationary accelerometer indicates an upward acceleration of $1\,g$ since it is clearly not accelerating. Intuition would suggest that, if anything, the acceleration should be in the downward direction which is how the device would accelerate if dropped. Adding to the confusion, some smart phone sensor apps incorrectly report positive acceleration in the *downward* direction when the phone is stationary.

Accelerometers measure acceleration along a single axis. Typically, three accelerometers are packaged together and arranged so that their sensitive axes are orthogonal. The three outputs of such a triaxial accelerometer are the components of the acceleration vector ${}^B\boldsymbol{a}_B^{\#}$ measured in the body frame {B}.

Inconsistency between motion sensed in our ears and motion perceived by our eyes is the root cause of motion sickness.

Nature has also invented the accelerometer. All vertebrates have acceleration sensors called ampullae as part of their vestibular system. We have two in each inner ear to help us maintain balance: ◄ the saccule measures vertical acceleration, and the utricle measures front-to-back acceleration. The proof mass in the ampullae is a collection of calcium carbonate crystals called otoliths, literally ear stones, on a gelatinous substrate which serves as the spring and damper. Hair cells embedded in the substrate measure the displacement of the otoliths due to acceleration.

### 3.4.2.2 Estimating Pose and Body Acceleration

◘ Fig. 3.13 shows frame {0} with its $z$-axis vertically upward, and the acceleration is

$$
{}^0\boldsymbol{a}_B = \begin{pmatrix} 0 \\ 0 \\ g \end{pmatrix}
$$

We could use any 3-angle sequence.

where $g$ is the local gravitational acceleration from ◘ Fig. 3.11. In a body-fixed frame {B} at an arbitrary orientation expressed in terms of ZYX roll-pitch-yaw angles ◄

$$
{}^0\boldsymbol{\xi}_B = \boldsymbol{\xi}^{r_z}(\gamma) \oplus \boldsymbol{\xi}^{r_y}(\beta) \oplus \boldsymbol{\xi}^{r_x}(\alpha)
$$

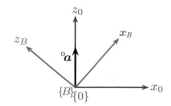

**◻ Fig. 3.13** Gravitational acceleration is defined as the $z$-axis of the fixed frame {0} and is measured by an accelerometer in the rotated body frame {B}

the gravitational acceleration will be

$$
{}^B\boldsymbol{a}_B = \left(\ominus \, {}^0\boldsymbol{\xi}_B\right) \cdot {}^0\boldsymbol{a}_B = \begin{pmatrix} -g \sin\beta \\ g \cos\beta \sin\alpha \\ g \cos\beta \cos\alpha \end{pmatrix} . \tag{3.23}
$$

The *measured* acceleration vector from the sensor in frame {B} is

$$
{}^B\boldsymbol{a}_B^{\#} = (a_x^{\#}, a_y^{\#}, a_z^{\#})^{\top}
$$

and equating this with (3.23) we can solve for the roll and pitch angles

$$
\sin\hat{\beta} = \frac{-a_x^{\#}}{g}
$$

$$
\tan\hat{\alpha} = \frac{a_y^{\#}}{a_z^{\#}}, \ \hat{\beta} \neq \pm\frac{\pi}{2} \tag{3.24}
$$

and we use the hat notation to indicate that these are estimates of the angles. ▶ Notice that there is no solution for the yaw angle and in fact $\gamma$ does not even appear in (3.23). The gravity vector is parallel to the vertical axis and rotating around that axis, yaw rotation, will not change the measured value at all. ▶ Also note that $\hat{\beta}$ depends on the local value of gravitational acceleration whose variation is shown in ◻ Fig. 3.11.

These angles are sufficient to determine whether a phone, tablet or camera is in portrait or landscape orientation.

**🛈 Accelerometers measure gravity *and* body motion**

We have made a very strong assumption that the measured acceleration ${}^B\boldsymbol{a}_B^{\#}$ is only due to gravity. On a real robot, the sensor will experience additional acceleration as the robot moves and this will introduce an error in the estimated orientation.

Frequently, we want to estimate the motion of the vehicle in the inertial frame, and the total measured acceleration in {0} is due to gravity *and* motion

$$
{}^0\boldsymbol{a}^{\#} = {}^0\boldsymbol{g} + {}^0\boldsymbol{a}_B .
$$

We observe acceleration in the body frame so the acceleration in the world frame is

$$
{}^0\hat{\boldsymbol{a}}_B = {}^0\hat{\mathbf{R}}_B \, {}^B\boldsymbol{a}_B^{\#} - {}^0\boldsymbol{g} \tag{3.25}
$$

and we assume that ${}^0\hat{\mathbf{R}}_B$ and $g$ are both known. ▶ Integrating acceleration with respect to time

$$
{}^0\hat{\boldsymbol{v}}_B(t) = \int_0^t {}^0\hat{\boldsymbol{a}}_B(t) \, dt \tag{3.26}
$$

Another way to consider this is that we are essentially measuring the direction of the gravity vector with respect to the frame {B} and a vector provides only two unique *pieces* of directional information, since one component of a unit vector can always be written in terms of the other two.

The first assumption is a strong one and problematic in practice. Any error in the rotation matrix results in incorrect cancellation of the gravity component of $\boldsymbol{a}^{\#}$ which leads to an error in the estimated body acceleration.

**3**

gives the velocity of the body frame, and integrating again

$$
{}^{0}\hat{\boldsymbol{p}}_{B}(t) = \int_{0}^{t} {}^{0}\hat{\boldsymbol{v}}_{B}(t) \, \mathrm{d}t \tag{3.27}
$$

gives its position. We can assume vehicle acceleration is zero and estimate orientation, or assume orientation and estimate vehicle acceleration. We cannot estimate both since there are more unknowns than measurements.

### 3.4.3 Magnetometers

The Earth is a massive but weak magnet. The poles of this geomagnet are the Earth's north and south magnetic poles which are constantly moving and located quite some distance from the planet's rotational axis. At any point on the planet, the magnetic flux lines can be considered a vector $\boldsymbol{m}$ whose magnitude and direction can be accurately predicted and mapped as shown in ▢ Fig. 3.14.

The direction of the Earth's north rotational pole, where the rotational axis intersects the surface of the northern hemisphere.

▢ Fig. 3.14b, c shows the vector's direction in terms of two angles: declination and inclination. A horizontal projection of the vector $\boldsymbol{m}$ points in the direction of magnetic north and the declination angle $D$ is measured from true north ◄ clockwise to that projection. The inclination or dip angle $I$ of the vector is measured in a vertical plane downward ◄ from horizontal to $\boldsymbol{m}$. The length of the vector, the magnetic flux density, is measured by a magnetometer in units of Tesla (T) and for the Earth this varies from 25–65 μT ◄ as shown in ▢ Fig. 3.14a.

In the Northern hemisphere, inclination is positive, that is, the vector points into the ground.

#### 3.4.3.1 How Magnetometers Work

The key element of most modern magnetometers is a Hall-effect sensor, a semiconductor device which produces a voltage proportional to the magnetic flux density in a direction normal to the current flow. Typically, three Hall-effect sensors are packaged together, and arranged so that their sensitive axes are orthogonal. The three outputs of such a triaxial magnetometer are the components of the Earth's magnetic flux density vector ${}^{B}\boldsymbol{m}^{\#}$ measured in the body frame {B}.

By comparison, a modern MRI machine has a magnetic field strength of 4–8 T.

Yet again nature leads, and creatures from bacteria to turtles and birds are known to sense magnetic fields. The effect is particularly well known in pigeons and there is debate about whether or not humans have this sense. The actual biological sensing mechanism has not yet been discovered.

#### 3.4.3.2 Estimating Heading

▢ Fig. 3.16 shows an inertial coordinate frame {0} with its $z$-axis vertically upward and its $x$-axis in the horizontal plane and pointing toward *magnetic* north. The magnetic field vector therefore lies in the $xz$-plane

$$
{}^{0}\boldsymbol{m} = B \begin{pmatrix} \cos I \\ 0 \\ \sin I \end{pmatrix}
$$

where $B$ is the magnetic flux density and $I$ the inclination angle which are both known from ▢ Fig. 3.14. In a body-fixed frame {B} at an arbitrary orientation expressed in terms of ZYX roll-pitch-yaw angles ◄

We could use any 3-angle sequence.

$$
{}^{0}\boldsymbol{\xi}_{B} = \boldsymbol{\xi}^{r_{z}}(\gamma) \oplus \boldsymbol{\xi}^{r_{y}}(\beta) \oplus \boldsymbol{\xi}^{r_{x}}(\alpha)
$$

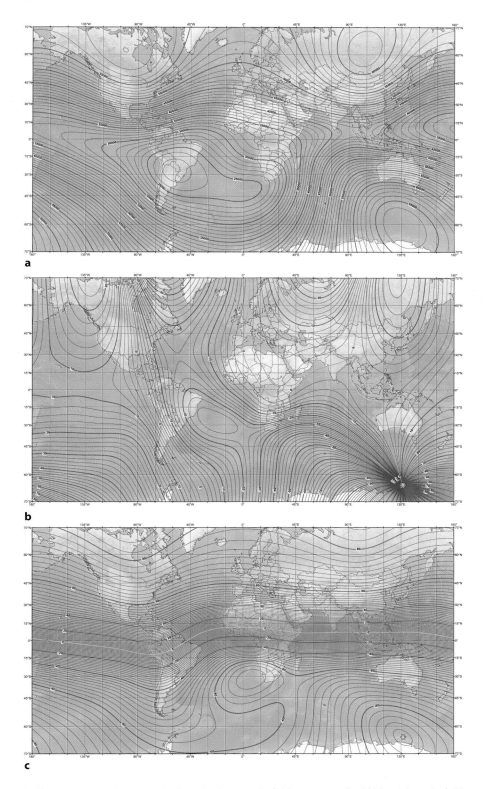

□ **Fig. 3.14**   A predicted model of the Earth magnetic field parameters for 2015. **a** Magnetic field flux density (nT); **b** magnetic declination (degrees); **c** magnetic inclination (degrees). Magnetic poles indicated by *asterisk* (maps by NOAA/NGDC and CIRES ► http://ngdc.noaa.gov/geomag/WMM, published Dec 2014)

3

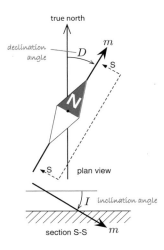

**Fig. 3.15** A compass needle aligns with the horizontal component of the Earth's magnetic field vector which is offset from true north by the declination angle $D$. The magnetic field vector has a dip or inclination angle of $I$ with respect to horizontal

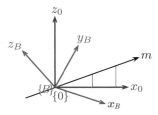

**Fig. 3.16** The reference frame {0} has its $x$-axis aligned with the horizontal projection of $\boldsymbol{m}$ and its $z$-axis vertical. Magnetic field is measured by a magnetometer in the rotated body frame {B}

### Excurse 3.6: Edwin Hall

Hall (1855–1938) was an American physicist born in Maine. His Ph.D. research in physics at the Johns Hopkins University in 1880 discovered that a magnetic field exerts a force on a current in a conductor. He passed current through thin gold leaf and, in the presence of a magnetic field normal to the leaf, was able to measure a very small potential difference between the sides of the leaf. This is now known as the Hall effect. While it was then known that a magnetic field exerted a force on a current carrying conductor, it was believed the force acted on the conductor not the current itself – electrons were yet to be discovered. He was appointed as professor of physics at Harvard in 1895 where he worked on thermoelectric effects.

the magnetic flux density will be

$$
{}^B\boldsymbol{m} = \left(\ominus\, {}^0\boldsymbol{\xi}_B\right) \cdot {}^0\boldsymbol{m}
$$

$$
= \begin{pmatrix} B \sin I \sin \alpha \sin \gamma + \cos \alpha \cos \gamma \sin \beta + B \cos I \cos \beta \cos \gamma \\ B \cos I \cos \beta \sin \gamma - B \sin I \cos \gamma \sin \alpha - \cos \alpha \sin \beta \sin \gamma \\ B \sin I \cos \alpha \cos \beta - B \cos I \sin \beta \end{pmatrix} .
$$

$$
(3.28)
$$

The measured magnetic flux density vector from the sensor in frame {B} is

$$
{}^B\boldsymbol{m}^{\#} = (m_x^{\#}, m_y^{\#}, m_z^{\#})^{\top}
$$

and equating this with (3.28) we can solve for the yaw angle

$$
\tan \hat{\gamma} = \frac{\cos \beta \left( m_z^{\#} \sin \alpha - m_y^{\#} \cos \alpha \right)}{m_x^{\#} + B \sin I \sin \beta}
$$

which is the magnetic heading and related to the true heading by

$$
{}^{\text{tn}}\hat{\gamma} = \hat{\gamma} - D \ .
$$

This assumes that we know the local declination angle as well as the roll and pitch angles $\alpha$ and $\beta$. The latter could be estimated from acceleration measurements using (3.24), and many triaxial Hall-effect sensor chips also include a triaxial accelerometer for just this purpose.

Magnetometers are great in theory but problematic in practice. Firstly, our modern world is full of magnets and electromagnets. Buildings contain electrical wiring and robots themselves are full of electric motors, batteries and electronics. These all add to, or overwhelm, the local geomagnetic field. Secondly, many objects in our world contain ferromagnetic materials such as the reinforcing steel in buildings or the steel bodies of cars, ships or robots. These distort the geomagnetic field leading to local changes in its direction. These effects are referred to respectively as hard- and soft-iron distortion of the magnetic field. ▶

These can be calibrated out but the process requires that the sensor is physically rotated. The compass app in your phone might sometimes ask you to do that.

### 3.4.4 Inertial Sensor Fusion

An inertial navigation system uses the sensors we have just discussed to determine the pose of a vehicle – its position and its orientation. Early inertial navigation systems, such as shown in ◻ Fig. 2.19, used mechanical gimbals to keep the accelerometers at a constant orientation with respect to the stars using a gyro-stabilized platform. The acceleration measured on this platform is by definition referred to the inertial frame and simply needs to be integrated to obtain the velocity of the platform, and integrated again to obtain its position. In order to achieve accurate position estimates over periods of hours or days, the gimbals and gyroscopes had to be of extremely high quality so that the stable platform did not drift. The acceleration sensors also needed to be extremely accurate.

The modern strapdown inertial measurement configuration uses no gimbals. The angular velocity, acceleration and magnetic field sensors are rigidly attached to the vehicle. The collection of inertial sensors is referred to as an inertial measurement unit or IMU. A 6-axis IMU comprises triaxial gyroscopes and accelerometers

Increasingly, these sensor packages also include a barometric pressure sensor to measure changes in altitude.

while a 9-axis IMU comprises triaxial gyroscopes, accelerometers and magnetometers. ◄ A system that only determines orientation is called an attitude and heading reference system or AHRS.

The sensors we use, particularly the low-cost ones in smart phones and drones, are far from perfect. The output of any sensor $x^{\#}$ – gyroscope, accelerometer or magnetometer – is a corrupted version of the true value $x$ and a common model is

$$x^{\#} = sx + b + \varepsilon$$

Some sensors may not be perfectly linear and a sensor datasheet will characterize this.

Some sensors also exhibit cross-sensitivity. They may give a weak response to a signal in an orthogonal direction or from a different mode, quite commonly low-cost gyroscopes respond to vibration and acceleration as well as rotation.

The effect of an orientation error is dangerous on something like a quadrotor. For example, if the estimated pitch angle is too high then the vehicle control system will pitch down by the same amount to keep the craft "level", and this will cause it to accelerate forward.

where $s$ is the scale factor, $b$ is the offset or bias, and $\varepsilon$ is random error or "noise". $s$ is usually specified by the manufacturer to some tolerance, perhaps $\pm 1\%$, and for a particular sensor this can be determined by some calibration procedure. ◄ Bias $b$ is ideally equal to zero but will vary from device to device. Bias that varies over time is often called sensor drift. Scale factor and bias are typically both a function of temperature. ◄

In practice, bias is the biggest problem because it varies with time and temperature and has a very deleterious effect on the estimated position and orientation. Consider a positive bias on the output of a gyroscopic sensor – the output is higher than it should be. At each time step in (3.20), the incremental rotation will be bigger than it should be, which means that the orientation error will grow linearly with time. ◄

If we use (3.25) to estimate the vehicle's acceleration, then the error in orientation means that the measured gravitational acceleration is incorrectly canceled out and will be indistinguishable from *actual* vehicle acceleration. This offset in acceleration becomes a linear-time error in velocity and a quadratic-time error in position. Given that the orientation error is already linear in time, we end up with a cubic-time error in position, and this is ignoring the effects of accelerometer bias. Sensor bias is problematic! A rule of thumb is that gyroscopes with bias stability of $0.01$ $\deg h^{-1}$ will lead to position error growing at a rate of 1 $\text{nmi} \, h^{-1}$ ($1.85$ $\text{km} \, h^{-1}$). Military grade systems have very impressive stability, for missiles it is less than $0.000\,02$ $\deg h^{-1}$ which is in stark contrast to consumer grade devices which are in the range $0.01$–$0.2$ deg per *second*.

To see the effect of bias on the estimated orientation, we will use the Toolbox example function IMU

```
>>> from imu_data import IMU
>>> true, imu = IMU()
```

Not completely unknown, you can look at the source code of `imu_data.py`.

which returns two objects that are respectively the true motion of a rotating body and simulated IMU data from that body. The latter contains "measured" gyroscope, accelerometer and magnetometer data which include a fixed but unknown bias. ◄ The data is organized as rows of the arrays `imu.gyro`, `imu.accel` and `imu.magno` respectively, with one row per time step, and the corresponding times are given by elements of the vector `imu.t`. We repeat the example from ► Sect. 3.4.1.2, but now with sensor bias

```
>>> q = UnitQuaternion();
>>> for wm in imu.gyro[:-1]:
...     q.append(q[-1] @ UnitQuaternion.EulerVec(wm * imu.dt))
```

As discussed in ► Sect. 2.4.5.

The angular error ◄ between the estimated and true orientation

```
>>> xplot(true.t, q.angdist(true.orientation), color="red");
```

is shown as the red line in ◘ Fig. 3.17a. We can clearly see growth in angular error over time due to bias on the sensor signals.

If we knew the bias, we could subtract it from the sensor measurement before integration. A simple way of estimating bias is to leave the IMU stationary for a few seconds and compute the average value of all the sensor outputs. ◄ This is really only valid over a short time period because the bias is not constant.

Many hobby drones do this just before they take off.

A more sophisticated approach is to estimate the bias online ▶ but, to do this, we need to combine information from different sensors – an approach known as sensor fusion. We rely on the fact that different sensors have complementary characteristics. Bias on angular rate sensors causes the orientation estimate error to grow with time, but for accelerometers it will only cause an orientation offset. Accelerometers respond to translational motion while good gyroscopes do not. Magnetometers provide partial information about roll, pitch and yaw, are immune to acceleration, but do respond to stray magnetic fields and other distortions. There are many ways to achieve this kind of fusion. A common approach is to use an estimation tool called an extended Kalman filter described in ▶ App. H. Given a full nonlinear mathematical model that relates the sensor signals and their biases to the vehicle pose and knowledge about the noise (uncertainty) on the sensor signals, the filter gives an optimal estimate of the pose and bias that best explain the sensor signals.

Here we will consider a simple, but effective, alternative called the explicit complementary filter. The rotation update step is performed using an augmented version of (3.21)

$$ {}^B\boldsymbol{\xi}_{\Delta\langle k\rangle} = e^{\delta_t \left[ {}^B\boldsymbol{\omega}_B^{\#}\langle k\rangle - \hat{\boldsymbol{b}}\langle k\rangle + k_P\boldsymbol{\sigma}_R\langle k\rangle \right]_\times} . \tag{3.29} $$

The key differences are that the estimated bias $\hat{\boldsymbol{b}}$ is subtracted from the sensor measurement and a term based on the orientation error $\boldsymbol{\sigma}_R$ is added. The estimated bias changes with time according to

$$ \hat{\boldsymbol{b}}\langle k+1\rangle \leftarrow \hat{\boldsymbol{b}}\langle k\rangle - k_I\boldsymbol{\sigma}_R\langle k\rangle \tag{3.30} $$

and also depends on the orientation error $\boldsymbol{\sigma}_R$. $k_P > 0$ and $k_I > 0$ are both well-chosen constants. The orientation error is derived from *N vector measurements* ${}^B\boldsymbol{v}_i^{\#}$

$$ \boldsymbol{\sigma}_R\langle k\rangle = \sum_{i=0}^{N-1} k_i \, {}^B\boldsymbol{v}_i^{\#}\langle k\rangle \times {}^B\boldsymbol{v}_i\langle k\rangle $$

where

$$ {}^B\boldsymbol{v}_i\langle k\rangle = \left( \ominus\, {}^0\hat{\boldsymbol{\xi}}_{B\langle k\rangle} \right) \cdot {}^0\boldsymbol{v}_i $$

is a vector signal, known in the inertial frame (for example gravitational acceleration), and rotated into the body-fixed frame by the inverse of the estimated orientation ${}^0\hat{\boldsymbol{\xi}}_B$. Any error in direction between these vectors will yield a nonzero cross-product which is the axis around which to rotate one vector into the other. The filter uses this difference – the innovation – to improve the orientation estimate by feeding it back into (3.29). This filter allows an unlimited number of body-frame vectorial measurements ${}^B\boldsymbol{v}_i$ to be fused together, for example, we could include magnetic field or any other kind of direction data such as the altitude and azimuth of visual landmarks, stars or planets.

Now we implement the explicit complementary filter. It has just a few extra lines of code compared to the example above

```
>>> kI = 0.2; kP = 1;
>>> b = np.zeros(imu.gyro.shape);
>>> qcf = UnitQuaternion();
>>> data = zip(imu.gyro[:-1], imu.accel[:-1], imu.magno[:-1]);
>>> for k, (wm, am, mm) in enumerate(data):
...     qi = qcf[-1].inv()
...     sR = np.cross(am, qi * true.g) + np.cross(mm, qi * true.B)
...     wp = wm - b[k,:] + kP * sR
...     qcf.append(qcf[k] @ UnitQuaternion.EulerVec(wp * imu.dt))
...     b[k+1,:] = b[k,:] - kI * sR * imu.dt
```

Our brain has an online mechanism to cancel out the bias in our vestibular gyroscopes. It uses the recent average rotation as the bias, based on the reasonable assumption that we do not undergo prolonged rotation. If we do, then that angular velocity becomes the new normal so that when we stop rotating we perceive rotation in the opposite direction. We call this dizziness.

3

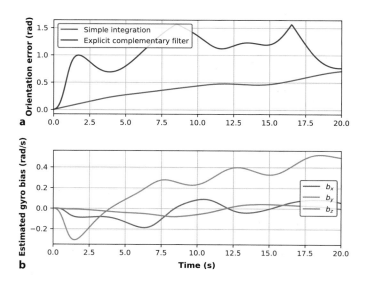

**◘ Fig. 3.17** **a** Effect of gyroscope bias on simple gyro integration and explicit complementary filter; **b** estimated gyroscope bias from the explicit complementary filter

and plot the angular difference between the estimated and true orientation

```
>>> xplot(true.t, qcf.angdist(true.orientation), color="blue");
```

as the blue line in ◘ Fig. 3.17a.

Bringing together information from multiple sensors has checked the growth in orientation error, despite all the sensors having a bias. The estimated gyroscope bias is shown in ◘ Fig. 3.17b and we can see the bias estimates converging on their true value.

## 3.5  Wrapping Up

This chapter has discussed the important issue of time-varying pose, from several different perspectives. The first was from the perspective of differential calculus, and we showed that the temporal derivative of a rotation matrix or a quaternion is a function of the angular velocity of the body. The skew-symmetric matrix appears in the rotation matrix case, and we should no longer be surprised about this given its intimate connection to rotation via Lie group theory as discussed in the previous chapter. We then looked at a finite-time difference as an approximation to the derivative and showed how this leads to computationally cheap methods to update rotation matrices and quaternions given knowledge of angular velocity. We also discussed the dynamics of moving bodies that translate and rotate under the influence of forces and torques, inertial and noninertial reference frames and the notion of fictitious forces.

The second perspective was creating motion – a sequence of poses, a trajectory, that a robot can follow. An important characteristic of a trajectory is that it is smooth with respect to time – position and orientation have a continuous first, and possibly higher, derivative. We started by discussing how to generate smooth trajectories in one dimension and then extended that to the multi-dimensional case and then to piecewise-linear trajectories that visit a number of intermediate points. Smoothly-varying rotation was achieved by interpolating roll-pitch-yaw angles and unit quaternions.

The third perspective was to use measured angular velocity, acceleration and magnetic field data to estimate orientation and position. We used our knowledge to investigate the important problem of inertial navigation, estimating the pose of a moving body given imperfect measurements from sensors on the moving body.

### 3.5.1 Further Reading

The earliest work on manipulator Cartesian trajectory generation was by Paul (1972, 1979) and Taylor (1979). The multi-segment trajectory is discussed by Paul (1979, 1981) and the concept of segment transitions or blends is discussed by Lloyd and Hayward (1991). These early papers, and others, are included in the compilation on Robot Motion (Brady et al. 1982). Polynomial and trapezoidal trajectories are described in detail by Spong et al. (2006) and multi-segment trajectories are covered at length in Siciliano et al. (2009) and Craig (2005). We have discussed only a simple multisegment trajectory generator which is inspired by Paul (1981) and more sophisticated approaches based on splines are possible. A Python package to generate smooth paths subject to kinematic and dynamic constraints can be found at ▶ https://hungpham2511.github.io/toppra.

There is a lot of literature related to the theory and practice of inertial navigation systems. The thesis of Achtelik (2014) describes a sophisticated extended Kalman filter for estimating the pose, velocity and sensor bias for a small aerial robot. The explicit complementary filter used in this chapter is described by Hua et al. (2014). The book by Groves (2013) covers inertial and terrestrial radio and satellite navigation and has a good coverage of Kalman filter state estimation techniques. Titterton and Weston (2005) provides a clear and concise description of the principles underlying inertial navigation with a focus on the sensors themselves but is perhaps a little dated with respect to modern low-cost sensors. Data sheets on many low-cost inertial and magnetic field sensing chips can be found at ▶ https://www.sparkfun.com in the Sensors category.

The book *Digital Apollo* (Mindell 2008) is a very readable story of the development of the inertial navigation system for the Apollo Moon landings. The article by Corke et al. (2007) describes the principles of inertial sensors and the functionally equivalent sensors located in the inner ear of mammals that play a key role in maintaining balance.

### 3.5.2 Exercises

1. Express the incremental rotation $\mathbf{R}^\Delta$ as an exponential series and verify (3.9).
2. Derive the unit quaternion update equation (3.10).
3. Derive the expression for fictitious forces in a rotating reference frame from ▶ Sect. 3.2.3.
4. Make a simulation that includes a particle moving at constant velocity and a rotating reference frame. Plot the position of the particle in the inertial and the rotating reference frame and observe how the motion in the latter changes as a function of the particle's translational velocity and the reference frame angular velocity.
5. Compute the centripetal acceleration due to the Earth's orbital rotation about the Sun. How does this compare to that due to the Earth's axial rotation?
6. For the rotational integration example (▶ Sect. 3.4.1.2):
   a) experiment by changing the sample interval and adding rotation matrix normalization at each step, or every Nth step.
   b) use the `timeit` package to measure the execution time of the approximate solution with normalization, and the exact solution using the matrix exponential.

**3**

c) Redo the example using quaternions, and experiment by changing the sample interval and adding normalization at each step, or every Nth step.

7. Redo the quaternion-based angular velocity integration (▶ Sect. 3.4.1.2) using rotation matrices.

8. At your location, determine the magnitude and direction of centripetal acceleration you would experience. If you drove at $100 \, \text{km} \, \text{h}^{-1}$ due east what is the magnitude and direction of the Coriolis acceleration you would experience? What about at $100 \, \text{km} \, \text{h}^{-1}$ due north? The vertical component is called the Eötvös effect, how much lighter does it make you?

9. For a `quintic` trajectory from 0 to 1 in 50 steps explore the effects of different initial and final velocities, both positive and negative. Under what circumstances does the quintic polynomial overshoot and why?

10. For a `trapezoidal` trajectory from 0 to 1 in 50 steps explore the effects of specifying the velocity for the constant velocity segment. What are the minimum and maximum bounds possible?

11. For a trajectory from 0 to 1 in $T$ seconds with a maximum possible velocity of 0.5, compare the total trajectory times required for each of the `quintic` and `trapezoidal` trajectories.

12. Use the `animate` method to compare rotational interpolation using Euler angles, roll-pitch-yaw angles, quaternions, and twists. Hint: use the quaternion `interp` method and `mtraj`.

13. Develop a method to quantitatively compare the performance of the different orientation interpolation methods. Hint: plot the locus followed by $\hat{z}$ on a unit sphere.

14. Repeat the example of ◘ Fig. 3.7 for the case where:
    a) the interpolation does *not* pass through a singularity. Hint – change the start or goal pitch angle. What happens?
    b) the final orientation is at a singularity. What happens?

15. Convert an **SE**(3) matrix to a twist and exponentiate the twist with values in the interval [0,1]. How does this form of interpolation compare with others covered in this chapter?

16. For the `mstraj` example (▶ Sect. 3.3.3)
    a) Repeat with different initial and final velocity.
    b) Investigate the effect of increasing the acceleration time. Plot total time as a function of acceleration time.
    c) The function has a third output argument that provides parameters for each segment, investigate these.

17. Modify `mstraj` so that acceleration limits are taken into account when determining the segment time.

18. There are a number of iOS and Android apps that display and record sensor data from the device's gyros, accelerometers and magnetometers. Alternatively you could use an inertial sensor shield for a RaspberryPi or Arduino and write your own logging software. Run one of these and explore how the sensor signals change with orientation and movement. If the app supports logging, what acceleration is recorded when you throw the phone into the air?

19. Consider a gyroscope with a 20 mm diameter steel rotor that is 4 mm thick and rotating at 10,000 rpm. What is the magnitude of $h$? For an angular velocity of $5 \, \text{deg} \, \text{s}^{-1}$, what is the generated torque?

20. Using (3.19), can you explain how a toy gyroscope is able to balance on a single point with its spin axis horizontal? What holds it upright?

21. An accelerometer has fallen off the table. Assuming no air resistance and no change in orientation, what value does it return as it falls?

22. Implement the algorithm to determine roll and pitch angles from accelerometer measurements.
    a) Devise an algorithm to determine if it is in portrait or landscape orientation.

b) Create a trajectory for the accelerometer using `quintic` to generate motion in either the $x$- or $y$-direction. What effect does the acceleration along the path have on the estimated angle?

c) Calculate the orientation using quaternions rather than roll-pitch-yaw angles.

23. Determine the Euler angles as a function of the measured acceleration. You could install SymPy to help you with the algebra.

24. You are in an aircraft flying at 30,000 feet over your current location. How much lighter are you?

25. Determine the magnetic field strength, declination and inclination at your location. Visit the website ▶ http://www.ngdc.noaa.gov/geomag-web.

26. Using the sensor reading app from above, orient the phone so that the magnetic field vector has only a $z$-axis component. Where is the magnetic field vector with respect to your phone?

27. Using the sensor reading app from above, log some inertial sensor data from a phone while moving it around. Use that data to estimate the changing orientation or full pose of the phone. Can you do this in real time?

28. Experiment with varying the parameters of the explicit complementary filter (▶ Sect. 3.4.4). Change the bias or add Gaussian noise to the simulated sensor readings.

# Mobile Robotics

**Contents**

In this part we discuss mobile robots, a class of robots that are able to move through the environment. Mobile robots are a more recent technology than robot manipulator arms, which we cover in Part III, but since the most common mobile robots operate in two dimensions they allow for a conceptually more gentle introduction.

Today, there is intense interest in mobile robots – the next frontier in robotics – with advances in self-driving cars, planetary rovers, humanoids and drones, and new application areas such as warehouse logistics and aerial taxis. The first commercial applications of mobile robots came in the 1980s when automated guided vehicles (AGVs) were developed for transporting material around factories and these have since become a mature technology. Those early free-ranging mobile wheeled vehicles typically used fixed infrastructure for guidance, for example, a painted line on the floor, a buried cable that emits a radio-frequency signal, or wall-mounted fiducial markers. The last decade has seen significant achievements in mobile robots that can operate without navigational infrastructure.

The chapters in this part of the book cover the fundamentals of mobile robotics. ▶ Chap. 4 discusses the motion and control of two exemplar robot platforms: wheeled vehicles that operate on a planar surface, and aerial robots that move in 3-dimensional space – specifically quadrotor aerial robots. ▶ Chap. 5 is concerned with navigation, how a robot finds its way to a goal in an environment that contains obstacles using direct sensory information or a map. Most navigation strategies require knowledge of the robot's position and this is the topic of ▶ Chap. 6 which examines techniques such as dead reckoning, and the use of maps combined with observations of landmarks. We also show how a robot can make a map, and even determine its location while simultaneously navigating an unknown region. To achieve all this, we will leverage our knowledge from ▶ Sect. 2.2 about representing position, orientation and pose in two dimensions.

# Mobile Robot Vehicles

## Contents

© The Author(s), under exclusive license to Springer Nature Switzerland AG 2023
P. Corke, *Robotics, Vision and Control*, Springer Tracts in Advanced Robotics 146,
https://doi.org/10.1007/978-3-031-06469-2_4

**4**

■ Figs. 4.1 to 4.3 show examples of mobile robots that are diverse in form and function. Mobile robots are not just limited to operations on the ground, and ■ Fig. 4.2 shows examples of aerial robots known as uncrewed aerial vehicles (UAVs), underwater robots known as autonomous underwater vehicles (AUVs), and robotic boats which are known as autonomous or uncrewed surface vehicles (ASVs or USVs). Collectively these are often referred to as Autonomous Uncrewed Systems (AUSs). Field robotic systems such as trucks in mines, container transport vehicles in shipping ports, and self-driving tractors for broad-acre agriculture are shown in ■ Fig. 4.3. While some of these are very specialized or research prototypes, commercial field robots are available for a growing range of applications.

The diversity we see is in the robotic platform – the robot's physical embodiment, its means of locomotion and its sensors. The robot's sensors are clearly visible in ■ Fig. 4.1b, ■ Fig. 4.2a, d, and ■ Fig. 4.3a. Internally, these robots would have strong similarity in their sensor-processing pipeline and navigation algorithms. The robotic vacuum cleaner of ■ Fig. 4.1a has few sensors and uses reactive strategies to clean the floor. By contrast, the 2007-era self-driving vehicle

■ **Fig. 4.1**  Some mobile ground robots: **a** The Roomba® robotic vacuum cleaner, 2008 (image courtesy iRobot®). **b** Boss, Tartan Racing team's autonomous car that won the DARPA Urban Challenge, 2007 (© Carnegie-Mellon University)

■ **Fig. 4.2**  Some mobile air and water robots: **a** Yamaha RMAX® helicopter with 3 m blade diameter (image by Sanjiv Singh); **b** ScanEagle, fixed-wing robotic aircraft (image courtesy of Insitu); **c** Ranger-Bot, a 6-thruster underwater robot (2020) (image courtesy Biopixel); **d** Autonomous Surface Vehicle (image by Matthew Dunbabin)

**Fig. 4.3** **a** Exploration: Mars Science Laboratory (MSL) rover, known as Curiosity, undergoing testing (image courtesy NASA/Frankie Martin); **b** Logistics: an automated straddle carrier that moves containers; Port of Brisbane, 2006 (image courtesy of Port of Brisbane Pty Ltd); **c** Mining: autonomous haul truck (Copyright © 2015 Rio Tinto); **d** Agriculture: QUT's AgBot-2 broad-acre weeding robot (image courtesy Owen Bawden)

shown in ☐ Fig. 4.1b displays a multitude of sensors that provide the vehicle with awareness of its surroundings.

This chapter discusses how a robot platform moves, that is, how its pose changes with time as a function of its control inputs. There are many different types of robot platform but, in this chapter, we will consider only four important exemplars. ▶ Sect. 4.1 covers three different types of wheeled vehicles that operate in a 2-dimensional world. They can be propelled forwards or backwards and their heading direction is controlled by some steering mechanism. ▶ Sect. 4.2 describes a quadrotor, a flying vehicle, which is an example of a robot that moves in 3-dimensional space. Quadrotors or "drones" are becoming increasingly popular as a robot platform since they are low-cost and can be easily modeled and controlled.

▶ Sect. 4.3 revisits the concept of configuration space and dives more deeply into important issues of underactuation and nonholonomy.

## 4.1 Wheeled Mobile Robots

Wheeled locomotion is one of humanity's great innovations. The wheel was invented around 3000 BCE and the two-wheeled cart around 2000 BCE. Today, four-wheeled vehicles are ubiquitous, and the total automobile population of the planet is well over one billion. ▶ The effectiveness of cars, and our familiarity with them, makes them a natural choice for robot platforms that move across the ground.

We know from our everyday experience with cars that there are limitations on how they move. It is not possible to drive sideways, but with some practice we can learn to follow a path that results in the vehicle being to one side of its initial position – this is parallel parking. Neither can a car rotate on the spot, but we can follow a path that results in the vehicle being at the same position but rotated by 180° – a three-point turn. The necessity to perform such maneuvers is the hallmark of a system that is nonholonomic – an important concept, which is discussed further in ▶ Sect. 4.3. Despite these minor limitations, the car is the simplest and most effec-

There are also around one billion bicycles, but robotic bicycles are not common. While balancing is not too hard to automate, stopping and starting requires additional mechanical complexity.

**4**

From Sharp 1896

Often, perhaps incorrectly, called the Ackermann model.

tive means of moving in a planar world that we have yet found. The car's motion model and the challenges it raises for control will be discussed in ▶ Sect. 4.1.1.

In ▶ Sect. 4.1.2 we will introduce differentially-steered vehicles, which are mechanically simpler than cars and do not have steered wheels. This is a common configuration for small mobile robots and also for larger machines such as bulldozers. ▶ Sect. 4.1.3 introduces novel types of wheels that *are* capable of omnidirectional motion, and then models a vehicle based on these wheels.

## 4.1.1 Car-Like Vehicle

Cars with steerable wheels are a very effective class of vehicles, and the archetype for most ground robots such as those shown in ◘ Fig. 4.3. In this section we will create a model for a car-like vehicle, and develop controllers that can drive the car to a point, along a line, follow an arbitrary path, and finally, drive to a specific pose. ◀

A commonly used model for the low-speed behavior of a four-wheeled car-like vehicle is the kinematic bicycle model ◀ shown in ◘ Fig. 4.4. The bicycle has a rear wheel fixed to the body and the plane of the front wheel rotates about the vertical axis to steer the vehicle.

The pose of the vehicle is represented by its body coordinate frame {B} shown in ◘ Fig. 4.4, which has its $x$-axis in the vehicle's forward direction and its origin at the center of the rear axle. The orientation of its $x$-axis with respect to the world frame is the vehicle's heading angle (yaw angle or simply *heading*). The *configuration* of the vehicle is represented by the generalized coordinates $q = (x, y, \theta) \in C$ where $C \subset \mathbb{R}^2 \times \mathbf{S}^1$ is the configuration space. We assume that the velocity of each wheel is in the plane of the wheel, and that the wheels roll without skidding or slipping sideways

$$^B v = (v, 0) \ . \tag{4.1}$$

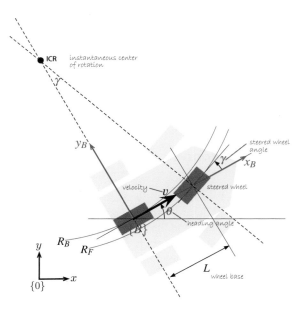

◘ **Fig. 4.4** Bicycle model of a car-like vehicle. The 4-wheeled car is shown in light gray, and the equivalent 2-wheeled bicycle model in dark gray. The vehicle's body frame is shown in red, and the world coordinate frame in black. The steered wheel angle is $\gamma$ with respect to the body frame, and the velocity of the back wheel, in the $x$-direction, is $v$. The two wheel axes are extended as dashed lines and intersect at the Instantaneous Center of Rotation (ICR). The back and front wheels follow circular arcs around the ICR of radius $R_B$ and $R_F$ respectively

The dashed lines show the direction along which the wheels cannot move, the lines of no motion, and these intersect at a point known as the Instantaneous Center of Rotation (ICR). The reference point of the vehicle, the origin of frame {B}, thus follows a circular path and its angular velocity is

$$\dot{\theta} = \frac{v}{R_B} \qquad (4.2)$$

and by simple geometry the turning radius is

$$R_{\mathrm{B}} = \frac{L}{\tan \gamma} \qquad (4.3)$$

where $L$ is the length of the vehicle or the *wheel base*. As we would expect, the turning circle increases with vehicle length – a bus has a bigger turning circle than a car. The angle of the steered wheel, $\gamma$, is typically limited mechanically and its maximum value dictates the minimum value of $R_{\mathrm{B}}$.

In a car, we use the *steering wheel* to control the angle of the steered wheels on the road, and these two angles are linearly related by the steering ratio. ▶ Curves on roads are circular arcs or clothoids so that road following is easy for a driver – requiring constant or smoothly varying steering-wheel angle. Since $R_{\mathrm{F}} > R_{\mathrm{B}}$, the front wheel must follow a longer path and therefore rotate more quickly than the back wheel.

When a four-wheeled vehicle goes around a corner, the two steered wheels follow circular paths of slightly different radii and therefore the angles of the steered wheels $\gamma_L$ and $\gamma_R$ should be very slightly different. This is achieved by the commonly used Ackermann steering *space* mechanism that results in lower wear and tear on the tyres. The driven wheels must rotate at different speeds on corners so a differential gearbox is required between the motor and the driven wheels.

The velocity of the robot in the world frame is described by

> Steering ratio is the ratio of the steering-wheel angle to the steered-wheel angle. For cars, it is typically in the range 12 : 1 to 20 : 1. Some cars have "variable ratio" steering where the ratio is higher (less sensitive) around the zero angle.

$$\dot{x} = v \cos \theta$$
$$\dot{y} = v \sin \theta$$
$$\dot{\theta} = \frac{v}{L} \tan \gamma \qquad (4.4)$$

which are the equations of motion, or the motion model. This is a kinematic model since it describes the velocities of the vehicle but not the forces or torques that cause the velocity. The rate of change of heading $\dot{\theta}$ is referred to as turn rate, heading rate or yaw rate and can be measured by a gyroscope. It can also be deduced from the angular velocity of the wheels on the left- and right-hand sides of the vehicle which follow arcs of different radii, and therefore rotate at different speeds.

Equation (4.4) captures some other important characteristics of a car-like vehicle. When $v = 0$ then $\dot{\theta} = 0$; that is, it is not possible to change the vehicle's orientation when it is not moving. As we know from driving, we must be moving

---

**Excurse 4.1: Vehicle Coordinate System**

The coordinate system that we will use, and a common one for vehicles of all sorts, is that the $x$-axis is forward (longitudinal motion), the $y$-axis is to the left side (lateral motion), which implies that the $z$-axis is upward. For aerospace and underwater applications, the $x$-axis is forward and the $z$-axis is often downward.

**4**

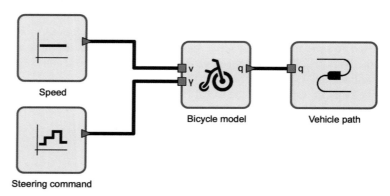

**Fig. 4.5** Block diagram model `models/lanechange` that demonstrates a simple lane changing maneuver. The piecewise-constant block controls the angle of the steered wheel, left then right (the waveform shown in the block is generic). The vehicle has a default wheelbase $L = 1$

Details about installing `bdsim` and using it to run the block diagram models shown in this book, are given in ▶ App. A

The model also includes a maximum velocity limit, a velocity rate limiter to model finite acceleration, and a limiter on the steered-wheel angle to model the finite range of the steered wheel. These can be changed by modifying parameters at the top of the file `lanechange.py`.

in order to turn. When the steered-wheel angle $\gamma = \frac{\pi}{2}$, the front wheel is orthogonal to the back wheel, the vehicle cannot move forward, and the model enters an undefined region.

Expressing (4.1) in the world coordinate frame, the vehicle's velocity in its lateral or body-frame $y$-direction is

$$\dot{y}\cos\theta - \dot{x}\sin\theta \equiv 0 \tag{4.5}$$

which is called a nonholonomic constraint and will be discussed further in ▶ Sect. 4.3.1. This equation cannot be integrated to form a relationship between $x$, $y$ and $\theta$.

We can explore the behavior of the vehicle for simple steering input using the Python-based block-diagram simulation package `bdsim`. ◀. The model shown in ■ Fig. 4.5 contains a `Bicycle` block that implements (4.4). ◀ The velocity input is a constant, and the steered-wheel angle is a positive pulse followed by a negative pulse.

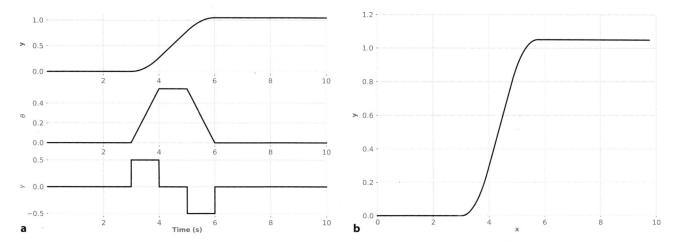

**◻ Fig. 4.6** Simple lane-changing maneuver using the model `models/lanechange` shown in ◻ Fig. 4.5. **a** Vehicle response as a function of time, **b** motion in the $xy$-plane, the vehicle moves in the positive $x$-direction

Running the model simulates the motion of the vehicle ▶

```
>>> %run -m lanechange -H
```

This assumes the `models` folder is on the Python module search path, which is the case when running the example using `rvctool`.

where the option `-H` tells the block diagram model to not block at the end of its execution – it normally blocks until all simulation windows have been dismissed. Running the model in this way adds a simulation results object `out` to the global namespace

```
>>> out
results:
t        | ndarray (115,)
x        | ndarray (115, 3)
xnames   | list = ['x', 'y', '$\\theta$']
ynames   | list = []
```

This object has a number of attributes such as simulation time `out.t` and system state `out.x`, in this case, the vehicle configuration $(x, y, \theta)$ with one row per time step. The configuration can be plotted against time

```
>>> plt.plot(out.t, out.x);  # q vs time
```

as shown in ◻ Fig. 4.6a, and the vehicle's position in the $xy$-plane

```
>>> plt.plot(out.x[:,0], out.x[:,1]);  # x vs y
```

is shown in ◻ Fig. 4.6b and demonstrates a simple lane-changing trajectory.

### 4.1.1.1 Driving to a Point

Consider the problem of moving toward a goal point $(x^*, y^*)$ on the $xy$-plane. We will control the robot's velocity to be proportional to its distance from the goal

$$v^* = K_v \sqrt{(x^* - x)^2 + (y^* - y)^2}$$

and to steer toward the goal, which is at the vehicle-relative angle in the world frame of ▶

This angle is in the interval $[-\pi, \pi)$ and is computed using the `atan2` function.

$$\theta^* = \tan^{-1} \frac{y^* - y}{x^* - x} , \tag{4.6}$$

a simple proportional controller

$$\gamma = K_h(\theta^* \ominus \theta), \quad K_h > 0 \tag{4.7}$$

**4**

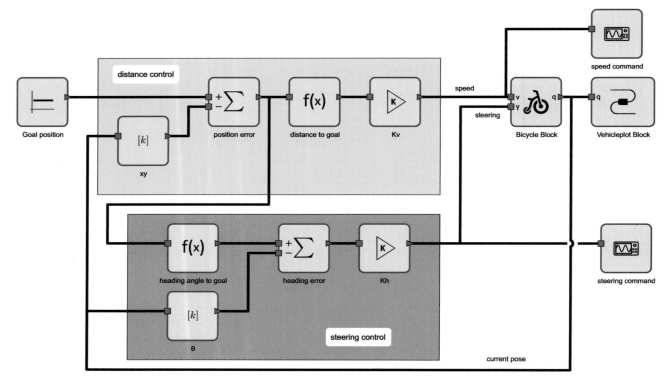

◻ **Fig. 4.7**   Block diagram model `models/drivepoint` that drives a vehicle to a point

The Toolbox function `angdiff` computes the difference between two angles and returns a difference in the interval $[-\pi, \pi]$. The function `wrap_mpi_pi` ensures that an angle is in the interval $[-\pi, \pi]$. This is also the shortest distance around the circle, as discussed in ▶ Sect. 3.3.4.1. To add or subtract angles in a blockdiagram, set the `mode` option of a $\boxed{\Sigma}$ block to c (for circle).

is sufficient – it turns the steered wheel toward the target. We use the operator $\ominus$ since $\theta^*, \theta \in \mathbf{S}^1$ are angles, not real numbers, and the result is always in the interval $[-\pi, \pi)$. ◄

A block diagram model is shown in ◻ Fig. 4.7. We specify a goal coordinate

```
>>> pgoal = (5, 5);
```

and an initial vehicle configuration

```
>>> qs = (8, 5, pi / 2);
```

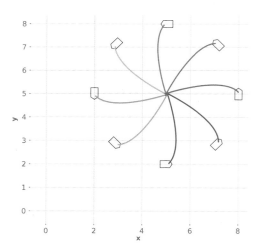

◻ **Fig. 4.8**   Vehicle motion to a point from different initial configurations, using the block diagram model `models/drivepoint` shown in ◻ Fig. 4.7. The goal $(x^*, y^*)$ is $(5, 5)$, and the controller parameters are $K_v = 0.5$ and $K_h = 1.5$

**Excurse 4.3: atan vs atan2**

In robotics, we frequently need to find $\theta$ given an expression for one or more of $\sin \theta$, $\cos \theta$ or $\tan \theta$. The inverse functions are not one-to-one and have ranges of $[-\frac{\pi}{2}, \frac{\pi}{2}]$, $[0, \pi]$ and $[-\frac{\pi}{2}, \frac{\pi}{2}]$ respectively which are all subsets of the full possible range of $\theta$ which is $[-\pi, \pi)$.

In Python, these inverse functions are implemented by `math.asin, math.acos, math.atan` and `math.atan2`. The figure shows the trigonometric quadrants associated with the sign of the argument to the inverse functions: blue for positive and red for negative. Using NumPy, the functions are `np.arcsin, np.arccos, np.arctan` and `np.arctan2`.

Consider the example for $\theta$ in quadrant 3 (Q3), where $\sin \theta = s = \frac{-1}{\sqrt{2}}$ and $\cos \theta = c = \frac{-1}{\sqrt{2}}$. Then `asin(s)` $\rightarrow -\pi/4$, `acos(c)` $\rightarrow 3\pi/4$ and `atan(s/c)` $\rightarrow \pi/4$, three valid but quite different results. For the `atan(s/c)` case, the quotient of the two negative numbers is positive, placing the angle in quadrant 1. In contrast, the `math.atan2` or `np.arctan2`

function is passed the numerator and denominator separately, and takes the individual signs into account to determine the angle in any quadrant. In this case, `atan2(s, c)` $\rightarrow -3\pi/4$, which is in quadrant 3.

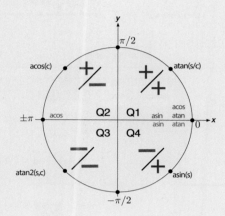

and then simulate the motion

```
>>> %run -i -m drivepoint -H
```

where the `-i` option gives the model access to the Python namespace where `qs` and `pgoal` are defined.

Once again, the simulation results are saved in the object `out` from which we can extract the configuration as a function of time. The vehicle's path on the $xy$-plane is

```
>>> q = out.x;  # configuration vs time
>>> plt.plot(q[:, 0], q[:, 1]);
```

which is shown in ◼ Fig. 4.8 for a number of starting configurations. In each case the vehicle has moved forward and turned onto a path toward the goal point. The final part of each path approaches a straight line, and the final orientation therefore depends on the starting point.

### 4.1.1.2 Driving Along a Line

Another useful task for a mobile robot is to follow a line on the $xy$-plane ▶ defined by $ax + by + c = 0$. We achieve this using two steering controllers. The first controller

2-dimensional lines in homogeneous form are discussed in ▶ App. C.2.1.

$$\gamma_d = K_d d, \quad K_d > 0$$

turns the robot toward the line to minimize the robot's normal distance from the line

$$d = \frac{(a, b, c) \cdot (x, y, 1)}{\sqrt{a^2 + b^2}}$$

where $d > 0$ if the robot is to the right of the line. The second controller adjusts the heading angle, or orientation, of the vehicle to be parallel to the line

$$\theta^* = \tan^{-1} \frac{a}{-b}$$

**4**

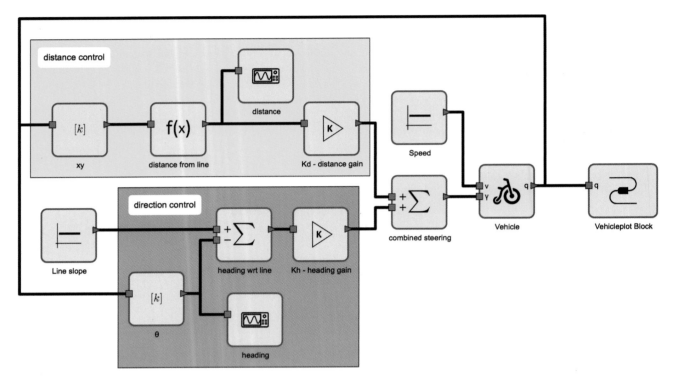

**◘ Fig. 4.9** Block diagram model `models/driveline` drives a vehicle along a line. The line parameters $(a, b, c)$ are set in the workspace variable L

using the proportional controller

$$\gamma_h = K_h(\theta^* \ominus \theta), \quad K_h > 0 \ .$$

The combined control law

$$\gamma = \gamma_d + \gamma_h = K_d d + K_h(\theta^* \ominus \theta)$$

turns the steered wheel so as to drive the robot toward the line and move along it.

The simulation model is shown in ◘ Fig. 4.9 and we specify the target line as a 3-vector $(a, b, c)$

```
>>> L = (1, -2, 4);
```

and an initial configuration

```
>>> qs = (8, 5, pi / 2);
```

and then simulate the motion

```
>>> %run -i -m driveline -H
```

The vehicle's path for a number of different starting configurations is shown in ◘ Fig. 4.10.

### 4.1.1.3 Driving Along a Path

Instead of following a straight line, we might wish to follow an arbitrary path on the $xy$-plane. The path might come from one of the path planners that are introduced in ▶ Chap. 5.

A simple and effective algorithm for path following is pure pursuit in which we move at constant velocity and *chase* a point that is a constant distance ahead on the path. The path is defined by a set of $N$ points $\boldsymbol{p}_i, i = 0, \ldots, N-1$ and the robot's

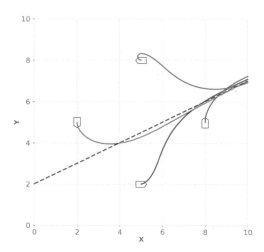

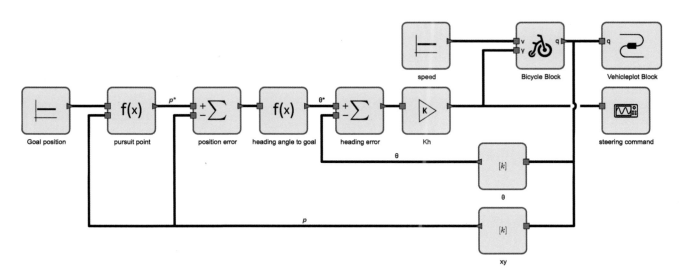

**Fig. 4.10** Vehicle motion along a line from different initial configurations, using the simulation model `models/driveline` shown in **Fig. 4.9. The line $(1, -2, 4)$ is shown dashed, and the controller parameters are $K_d = 0.5$ and $K_h = 1$

current position is $\boldsymbol{p}_r$. The first step is to find the index of the point on the path that is closest to the robot

$$
j = \arg \min_{i=0}^{N-1} \| \boldsymbol{p}_r - \boldsymbol{p}_i \| \ .
$$

Next, we find the first point along the path that is at least a distance $d$ ahead of that point

$$
k = \arg \min_{i=j+1}^{N-1} | \| \boldsymbol{p}_i - \boldsymbol{p}_j \| - d |
$$

and then the robot drives toward the point $\boldsymbol{p}_k$. In terms of the old fable about using a carrot to lead a donkey, what we are doing is always keeping the carrot a distance $d$ in front of the donkey and moving it from side to side to steer the donkey.

This problem is now the same as the one we tackled in ▶ Sect. 4.1.1.1, moving to a point, except that this time the point is moving. Robot speed is constant so

**Fig. 4.11** Block diagram model `models/drivepursuit` drives the vehicle along a piecewise-linear trajectory

**4**

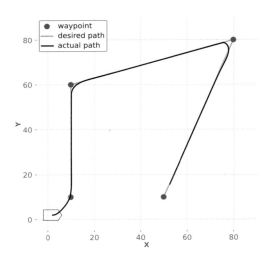

**Fig. 4.12** Pure pursuit motion along a path defined by waypoints, using the simulation model `models/drivepursuit` shown in Fig. 4.11. The robot starts at the origin, the look ahead distance $d = 5$, and the heading control gain is $K_h = 0.5$

only heading control, (4.6) and (4.7), is required. The block diagram model shown in Fig. 4.11 defines a piecewise-linear path defined by a number of waypoints. It can be simulated by

```
>>> %run -m drivepursuit -H
```

and the results are shown in Fig. 4.12. The robot starts at the origin, turns toward the start of the path and then smoothly follows it. The robot will *cut the corners* and the amount of corner cutting is proportional to the pursuit distance $d$.

#### 4.1.1.4 Driving to a Configuration

The final control problem we discuss is driving to a specific configuration $(x^*, y^*, \theta^*)$. The controller of Fig. 4.7 can drive the robot to a goal position, but the final orientation depends on the starting position. For tasks like parking the orientation of the vehicle is important.

In order to control the final orientation, we first rewrite (4.4) in matrix form

$$\begin{pmatrix} \dot{x} \\ \dot{y} \\ \dot{\theta} \end{pmatrix} = \begin{pmatrix} \cos\theta & 0 \\ \sin\theta & 0 \\ 0 & 1 \end{pmatrix} \begin{pmatrix} v \\ \omega \end{pmatrix}$$

We have effectively converted the bicycle kinematic model to a unicycle model which we discuss in ▶ Sect. 4.1.2.

where the inputs to the vehicle model are the speed $v$ and the turning rate $\omega$ which can be achieved by choosing the steered-wheel angle as ◄

$$\gamma = \tan^{-1} \frac{\omega L}{v}$$

We then transform the equations into polar coordinate form using the notation shown in Fig. 4.13 and apply a change of variables

$$\rho = \sqrt{\Delta_x^2 + \Delta_y^2}$$

$$\alpha = \tan^{-1} \frac{\Delta_y}{\Delta_x} - \theta$$

$$\beta = -\theta - \alpha$$

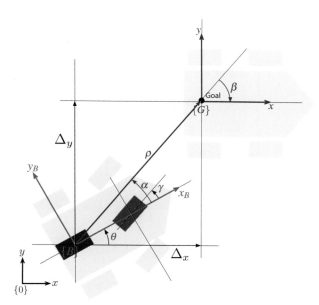

◻ **Fig. 4.13** Polar coordinate notation for the bicycle model vehicle moving toward a goal configuration: $\rho$ is the distance to the goal, $\beta$ is the angle of the goal vector with respect to the world frame, and $\alpha$ is the angle of the goal vector with respect to the vehicle frame

which results in

$$
\begin{pmatrix} \dot{\rho} \\ \dot{\alpha} \\ \dot{\beta} \end{pmatrix} = \begin{pmatrix} -\cos\alpha & 0 \\ \dfrac{\sin\alpha}{\rho} & -1 \\ -\dfrac{\sin\alpha}{\rho} & 0 \end{pmatrix} \begin{pmatrix} v \\ \omega \end{pmatrix}, \quad \text{if } \alpha \in \left(-\tfrac{\pi}{2}, \tfrac{\pi}{2}\right] \tag{4.8}
$$

and assumes that the goal frame {G} is in front of the vehicle. The linear control law

$$
v = k_\rho \rho
$$
$$
\omega = k_\alpha \alpha + k_\beta \beta
$$

drives the robot to a unique equilibrium at $(\rho, \alpha, \beta) = (0, 0, 0)$. The intuition behind this controller is that the terms $k_\rho \rho$ and $k_\alpha \alpha$ drive the robot along a line toward {G} while the term $k_\beta \beta$ rotates that line so that $\beta \rightarrow 0$. The closed-loop system

$$
\begin{pmatrix} \dot{\rho} \\ \dot{\alpha} \\ \dot{\beta} \end{pmatrix} = \begin{pmatrix} -k_\rho \rho \cos\alpha \\ k_\rho \sin\alpha - k_\alpha \alpha - k_\beta \beta \\ -k_\rho \sin\alpha \end{pmatrix}
$$

is stable so long as

$$
k_\rho > 0, \quad k_\beta < 0, \quad k_\alpha - k_\rho > 0 .
$$

To implement this in practice, the distance and bearing angle to the goal $(\rho, \alpha)$ could be measured by a camera or laser range finder, and the angle $\beta$ could be derived from $\alpha$ and vehicle heading $\theta$ as measured by a compass.

For the case where the goal is behind the robot, that is $\alpha \notin (-\tfrac{\pi}{2}, \tfrac{\pi}{2}]$, we reverse the vehicle by negating $v$ and $\gamma$ in the control law. This negation depends on the initial value of $\alpha$ and remains constant throughout the motion.

**4**

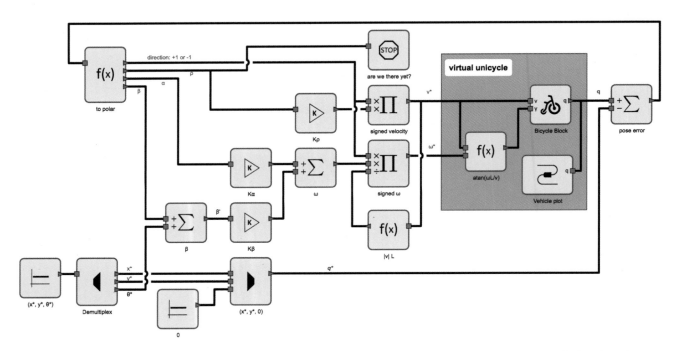

**Fig. 4.14** Block diagram model `models/driveconfig` drives the vehicle to a specified configuration. The initial and final configurations are set by the workspace variable `qs` and `qg` respectively

So far we have described a *regulator* that drives the vehicle to the configuration $(0, 0, 0)$. To move the robot to an arbitrary configuration $(x^*, y^*, \theta^*)$, we perform a change of coordinates

$$x' = x - x^*, \quad y' = y - y^*, \quad \theta' = \theta, \quad \beta = \beta' + \theta^* \ .$$

This configuration controller is implemented by the block diagram model shown in Fig. 4.14 and the transformation from bicycle to unicycle kinematics is clearly shown, mapping angular velocity $\omega$ to steered-wheel angle $\gamma$. We specify a goal configuration

```
>>> qg = (5, 5, pi / 2);
```

and an initial configuration

```
>>> qs = (9, 5, 0);
```

and then simulate the motion

```
>>> %run -i -m driveconfig -H
```

As before, the simulation results are saved in the variable `out` and can be plotted

```
>>> q = out.x;  # configuration vs time
>>> plt.plot(q[:, 0], q[:, 1]);
```

to show the vehicle's path in the $xy$-plane. The vehicle's path for a number of starting configurations is shown in Fig. 4.15. The vehicle moves forwards or backwards, and takes a smooth path to the goal configuration. ◄

The controller is based on the bicycle model of (4.4), but the vehicle model implemented by the `Bicycle` class has additional hard nonlinearities including steered-wheel angle limits and velocity rate limiting. If those limits are violated, the configuration controller may fail.

## 4.1.2  Differentially-Steered Vehicle

A car-like vehicle has steerable wheels that are mechanically complex. Differential steering eliminates this complexity and steers by independently controlling the

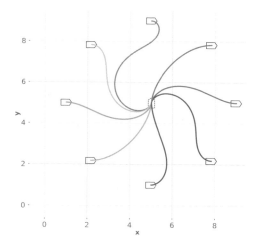

**Fig. 4.15** Vehicle motion to a configuration from different initial configurations, using the simulation model `models/driveconfig` shown in **◘** Fig. 4.14. The goal configuration $(5, 5, \frac{\pi}{2})$ is shown dashed, and the controller parameters are $K_\rho = 15$, $K_\alpha = 5$ and $K_\beta = -2$. Note that in some cases the robot has *backed into* the goal configuration

**◘ Fig. 4.16** Clearpath Robotics™ Husky™ robot with differential steering (image by Tim Barfoot)

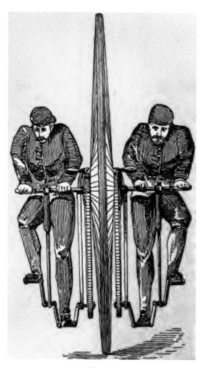

Fɪɢ 171.

From Sharp 1896

speed of the wheels on each side of the vehicle – if the speeds are not equal, the vehicle will turn. Very simple differentially-steered robots have two driven wheels and a front and/or back castor to provide stability. Larger differentially-steered vehicles such as the one shown in **◘** Fig. 4.16 employ a pair of wheels on each side, with each pair sharing a drive motor via some mechanical transmission. Very large differentially-steered vehicles such as bulldozers and tanks sometimes employ caterpillar tracks instead of wheels.

A commonly used model for the low-speed behavior of a differentially-steered vehicle is the kinematic unicycle model shown in **◘** Fig. 4.17. The unicycle has a single wheel located at the centroid of the wheels. ▶

The pose of the vehicle is represented by the body coordinate frame {B} shown in **◘** Fig. 4.17, with its *x*-axis in the vehicle's forward direction and its origin at the centroid of the unicycle wheel. The configuration of the vehicle is represented by the generalized coordinates $\boldsymbol{q} = (x, y, \theta) \in C$ where $C \subset \mathbb{R}^2 \times \mathbf{S}^1$ is the configuration

**4**

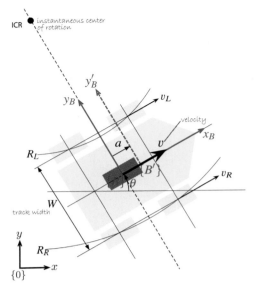

space. The vehicle's velocity is $v$ in the vehicle's $x$-direction, and we again assume that the wheels roll without slipping sideways

$$^B\boldsymbol{v} = (v, 0) \ .$$

The vehicle follows a curved path centered on the Instantaneous Center of Rotation (ICR). The left-hand wheels roll at a speed of $v_L$ but move along an arc of radius $R_L$. Similarly, the right-hand wheels roll at a speed of $v_R$ but move along an arc of radius $R_R$. The angular velocity of $\{B\}$ is

$$\dot{\theta} = \frac{v_L}{R_L} = \frac{v_R}{R_R}$$

and since $R_R = R_L + W$ we can write the turn rate

$$\dot{\theta} = \frac{v_R - v_L}{W} \tag{4.9}$$

in terms of the differential velocity and wheel separation $W$. The equations of motion are therefore

$$
\begin{aligned}
\dot{x} &= v \cos\theta \\
\dot{y} &= v \sin\theta \\
\dot{\theta} &= \frac{v_\Delta}{W}
\end{aligned}
\tag{4.10}
$$

where $v = \frac{1}{2}(v_R + v_L)$ and $v_\Delta = v_R - v_L$ are the average and differential velocities respectively. For a desired speed $v$ and turn rate $\dot{\theta}$, the wheel speeds are

$$v_R = v + \frac{W}{2}\dot{\theta}$$
$$v_L = v - \frac{W}{2}\dot{\theta} \tag{4.11}$$

There are similarities and differences to the bicycle model of (4.4). The turn rate for this vehicle is directly proportional to $v_\Delta$ and is independent of speed – the vehicle can turn even when not moving forward. For the 4-wheeled case shown in ▪ Fig. 4.17, the axes of the wheels do not intersect the ICR so, when the vehicle is turning, the wheel velocity vectors $v_L$ and $v_R$ are not tangential to the path – there is a component in the lateral direction that violates the no-slip constraint. This causes skidding or scuffing which is extreme when the vehicle is turning on the spot – hence differential steering is also called skid steering. Similar to the car-like vehicle, we can write an expression for velocity in the vehicle's $y$-direction expressed in the world coordinate frame

$$\dot{y}\cos\theta - \dot{x}\sin\theta \equiv 0 \tag{4.12}$$

which is the nonholonomic constraint. It is important to note that the ability to turn on the spot does not make the vehicle holonomic and is fundamentally different to the ability to move in an arbitrary direction which we will discuss next.

If we move the vehicle's reference frame to {B'} as shown in ▪ Fig. 4.17 and ignore orientation, we can rewrite (4.10) in matrix form as

$$\begin{pmatrix} \dot{x} \\ \dot{y} \end{pmatrix} = \begin{pmatrix} \cos\theta & -a\sin\theta \\ \sin\theta & a\cos\theta \end{pmatrix} \begin{pmatrix} v \\ \omega \end{pmatrix}$$

and if $a \neq 0$ this can be inverted

$$\begin{pmatrix} v \\ \omega \end{pmatrix} = \begin{pmatrix} \cos\theta & \sin\theta \\ -\frac{1}{a}\sin\theta & \frac{1}{a}\cos\theta \end{pmatrix} \begin{pmatrix} \dot{x} \\ \dot{y} \end{pmatrix} \tag{4.13}$$

to give the required forward speed and turn rate to achieve an arbitrary velocity $(\dot{x}, \dot{y})$ for the origin of frame {B'}.

### 4.1.3 Omnidirectional Vehicle

The vehicles we have discussed so far have a constraint on lateral motion, the nonholonomic constraint, which necessitates complex maneuvers in order to achieve some goal configurations. Alternative wheel designs, such as shown in ▪ Fig. 4.18, remove this constraint and allow omnidirectional motion. Even more radical is the spherical wheel shown in ▪ Fig. 4.20.

In this section, we will discuss the mecanum or *Swedish* wheel ▶ shown in ▪ Fig. 4.18b and schematically in ▪ Fig. 4.19. It comprises a number of rollers set around the circumference of the wheel with their axes at an angle of $\alpha$ relative to the axle of the wheel. The dark roller is the one on the bottom of the wheel and currently in contact with the ground. The rollers have a barrel shape so that only one point on the roller is in contact with the ground at any time.

▪ Fig. 4.19 shows the wheel coordinate frame {W} with its $x$-axis pointing in the direction of wheel motion. Rotation of the wheel will cause forward velocity of $R\varpi\hat{x}_w$ where $R$ is the wheel radius and $\varpi$ is the wheel's rotational rate. However, because the roller is free to roll in the direction indicated by the green line, normal to the roller's axis, there is potentially arbitrary velocity in that direction. A desired

Mecanum was a Swedish company where the wheel was invented by Bengt Ilon in 1973. It is described in US patent 3876255 (expired).

**4**

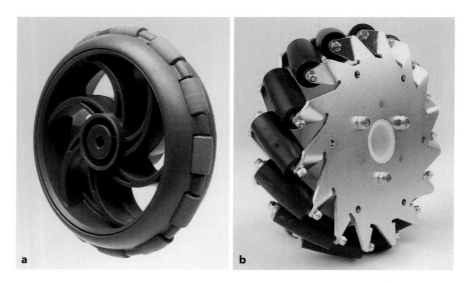

**◪ Fig. 4.18** Two types of omnidirectional wheel, note the different roller orientation. **a** Allows the wheel to *roll* sideways (courtesy VEX Robotics); **b** allows the wheel to *drive* sideways (courtesy of Nexus Robotics)

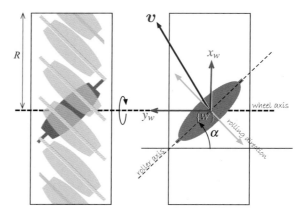

**◪ Fig. 4.19** Schematic of a mecanum wheel in plan view. The rollers are shown in gray and the dark gray roller is in contact with the ground. The green arrow indicates the rolling direction

velocity $v$ can be resolved into two components, one parallel to the direction of wheel motion and one parallel to the rolling direction

$$
v = \underbrace{v_w \hat{\boldsymbol{x}}_w}_{\text{driven}} + \underbrace{v_r (\cos\alpha\, \hat{\boldsymbol{x}}_w + \sin\alpha\, \hat{\boldsymbol{y}}_w)}_{\text{rolling}}
$$
$$
= (v_w + v_r \cos\alpha)\hat{\boldsymbol{x}}_w + v_r \sin\alpha\, \hat{\boldsymbol{y}}_w \tag{4.14}
$$

where $v_w$ is the speed due to wheel rotation and $v_r$ is the rolling speed. Expressing $v = v_x \hat{\boldsymbol{x}}_w + v_y \hat{\boldsymbol{y}}_w$ in component form allows us to solve for the rolling speed $v_r = v_y / \sin\alpha$ and substituting this into the first term we can solve for the required wheel velocity

$$
v_w = v_x - v_y \cot\alpha \ . \tag{4.15}
$$

The required wheel rotation rate is then $\varpi = v_w / R$. If $\alpha = 0$ then $v_w$ is undefined since the roller axes are parallel to the wheel axis and the wheel can provide no traction. If $\alpha = \frac{\pi}{2}$ as in ◪ Fig. 4.18a, the wheel allows sideways rolling but not sideways driving since there is zero coupling from $v_w$ to $v_y$.

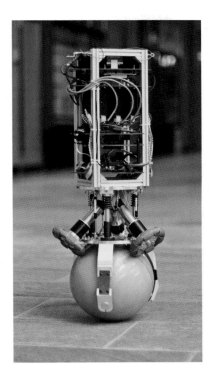

**Fig. 4.20** The Rezero ballbot developed at ETH Zürich (image by Péter Fankhauser)

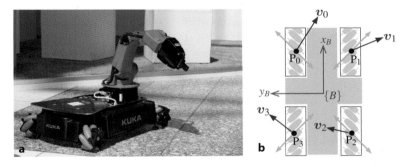

**Fig. 4.21** **a** KUKA youBot®, which has four mecanum wheels (image courtesy youBot Store); **b** schematic of a vehicle with four mecanum wheels in the youBot configuration

A single mecanum wheel does not allow any control in the rolling direction but for three or more mecanum wheels, suitably arranged, the motion in the rolling direction of any one wheel will be driven by the other wheels. A vehicle with four mecanum wheels is shown in ☐ Fig. 4.21. Its pose is represented by the body frame {B} with its $x$-axis in the vehicle's forward direction and its origin at the centroid of the four wheels. The configuration of the vehicle is represented by the generalized coordinates $q = (x, y, \theta) \in C$ where $C \subset \mathbb{R}^2 \times \mathbf{S}^1$. The rolling axes of the wheels are orthogonal, and when the wheels are not rotating, the vehicle cannot roll in any direction or rotate – it is effectively locked.

The four wheel-contact points indicated by black dots have coordinate vectors ${}^B p_i, i = 0, 1, 2, 3$. For a desired body velocity ${}^B v_B$ and turn rate ${}^B \omega_B$, the velocity at each wheel-contact point is

$$
{}^B v_i = {}^B v_B + {}^B \omega_B \hat{z}_B \times {}^B p_i
$$

and we then apply (4.14) and (4.15) to determine wheel rotational rates $\varpi_i$, while noting that $\alpha$ has the opposite sign for wheels 1 and 3 in (4.14).

Mecanum-wheeled robots exist, for example ◘ Fig. 4.21, but they are not common. The wheels themselves are more expensive than conventional wheels, and their additional complexity means more points of wear and failure. Mecanum wheels work best on a hard smooth floor, making them well suited to indoor applications such as warehouse logistics. Conversely, they are ill-suited to outdoor applications where the ground might be soft or uneven, and the rollers could become fouled with debris. To create an omnidirectional base, the minimum requirement is three mecanum wheels placed at the vertices of an equilateral triangle and with their rolling axes intersecting at the center.

## 4.2 Aerial Robots

» *In order to fly, all one must do is simply miss the ground.*
–Douglas Adams

Aerial robots or uncrewed aerial vehicles (UAV) are becoming increasingly common and span a great range of sizes and shapes as shown in ◘ Fig. 4.22. Applications include military operations, surveillance, meteorological observation, robotics research, commercial photography and increasingly hobbyist and personal use. A growing class of flying machines are known as micro air vehicles or MAVs, which are smaller than 15 cm in all dimensions. Fixed-wing UAVs are similar in principle to passenger aircraft with wings to provide lift, a propeller or jet to provide forward thrust and control surfaces for maneuvering. Rotorcraft UAVs have a variety of configurations that include the conventional *helicopter* design with a main and tail rotor, a *coax* with counter-rotating coaxial rotors and *quadrotors*. Rotorcraft UAVs have the advantage of being able to take off vertically and to hover.

Aerial robots differ from ground robots in some important ways. Firstly, they have 6 degrees of freedom and their configuration $\boldsymbol{q} \in C$ where $C \subset \mathbb{R}^3 \times (\mathbf{S}^1)^3$ is

◘ **Fig. 4.22** Aerial robots. **a** Global Hawk uncrewed aerial vehicle (UAV) (image courtesy of NASA), **b** a micro air vehicle (MAV) (image courtesy of AeroVironment, Inc.); **c** Ingenuity on Mars, uses two counter-rotating coaxial propellers (image courtesy of NASA); **d** a quadrotor that has four rotors and a block of sensing and control electronics in the middle (image courtesy of 3D Robotics)

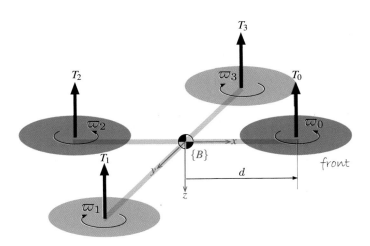

■ **Fig. 4.23** Quadrotor notation showing the four rotors, their thrust vectors and directions of rotation. The body frame {B} is attached to the vehicle and has its origin at the vehicle's center of mass. Rotors 0 and 2 shown in blue rotate counterclockwise (viewed from above) while rotors 1 and 3 shown in red rotate clockwise

the 3-dimensional configuration space. Secondly, they are actuated by forces; that is, their motion model is expressed in terms of forces, torques and accelerations rather than velocities as was the case for the ground vehicle models – we will use a dynamic rather than a kinematic model. Underwater robots have many similarities to aerial robots and can be considered as vehicles that *fly through water* and there are underwater equivalents to fixed-wing aircraft and rotorcraft. The principal differences underwater are an upward buoyancy force, drag forces that are much more significant than in air, and added mass.

In this section, we will create a model for a quadrotor flying vehicle such as shown in ■ Fig. 4.22d. Quadrotors are now widely available, both as commercial products and as open-source projects. Compared to fixed-wing aircraft, they are highly maneuverable and can be flown safely indoors, which makes them well suited for laboratory or hobbyist use. Compared to conventional helicopters, with a large main rotor and tail rotor, the quadrotor is easier to fly, does not have the complex swash plate mechanism and is easier to model and control.

The notation for the quadrotor model is shown in ■ Fig. 4.23. The body coordinate frame {B} has its $z$-axis downward following the aerospace convention. The quadrotor has four rotors, labeled 0 to 3, mounted at the end of each cross arm. Hex- and octo-rotors are also popular, with the extra rotors providing greater payload lift capability. The approach described here can be generalized to N rotors, where N is even.

The rotors are driven by electric motors powered by electronic speed controllers. Some low-cost quadrotors use small motors and reduction gearing to achieve sufficient torque. The rotor speed is $\varpi_i$ and the thrust is an upward vector

$$T_i = b\varpi_i^2, \quad i \in \{0, 1, 2, 3\} \tag{4.16}$$

in the vehicle's $-z$-direction, where $b > 0$ is the lift constant that depends on air density, the cube of the radius of the *rotor disk*, ▶ the number of rotor blades, and the chord length of the blade. ▶

The translational dynamics of the vehicle in world coordinates is given by Newton's second law

$$m\dot{\boldsymbol{v}} = \begin{pmatrix} 0 \\ 0 \\ mg \end{pmatrix} - {}^0\mathbf{R}_B \begin{pmatrix} 0 \\ 0 \\ T \end{pmatrix} - B\boldsymbol{v} \tag{4.17}$$

The disk described by the tips of the rotor blades. Its radius is the rotor blade length.

When flying close to the ground a rotorcraft experiences *ground effect* – increased lift due to a cushion of air trapped beneath it. For helicopters this effect occurs for heights up to 2–3 rotor radii, while for multirotors it occurs for height up to 5–6 rotor radii (Sharf et al 2014).

**4**

> **Excurse 4.4: Rotor Flapping**
>
> The propeller blades on a rotor craft have fascinating dynamics. When flying into the wind, the blade tip coming forward experiences greater lift while the receding blade has less lift. This is equivalent to a torque about an axis pointing into the wind and the rotor blades behave like a gyroscope (see ▶ Sect. 3.4.1.1) so the net effect is that the plane of the rotor disk pitches up by an amount proportional to the apparent or net wind speed, countered by the blade's bending stiffness and the change in lift as a function of blade bending. The pitched rotor disk causes a component of the thrust vector to retard the vehicle's forward motion, and this velocity-dependent force acts like a friction force. This is known as blade flapping and is an important characteristic of blades on all types of rotorcraft.

where $\boldsymbol{v}$ is the velocity of the vehicle's center of mass in the world frame, $g$ is gravitational acceleration, $m$ is the total mass of the vehicle, $B$ is aerodynamic friction and $T$ is the total upward thrust

$$T = \sum_{i=0}^{N-1} T_i \; . \tag{4.18}$$

The first term in (4.17) is the weight force due to gravity which acts downward in the world frame, the second term is the total thrust in the vehicle frame rotated into the world coordinate frame, and the third term is aerodynamic drag.

Pairwise differences in rotor thrusts cause the vehicle to rotate. The torque about the vehicle's $x$-axis, the *rolling* torque, is generated by the moments

$$\tau_x = dT_3 - dT_1$$

where $d$ is the distance from the rotor axis to the center of mass. We can write this in terms of rotor speeds by substituting (4.16)

$$\tau_x = db\big(\varpi_3^2 - \varpi_1^2\big) \tag{4.19}$$

and similarly for the $y$-axis, the *pitching* torque is

$$\tau_y = db\big(\varpi_0^2 - \varpi_2^2\big) \; . \tag{4.20}$$

The torque applied to each propeller by the motor is opposed by aerodynamic drag

$$Q_i = k\varpi_i^2$$

where $k > 0$ depends on the same factors as $b$. This torque exerts a reaction torque on the airframe which acts to rotate the airframe about the propeller shaft in the opposite direction to its rotation. The total reaction torque about the $z$-axis is

$$\begin{aligned}\tau_z &= Q_0 - Q_1 + Q_2 - Q_3 \\ &= k\big(\varpi_0^2 - \varpi_1^2 + \varpi_2^2 - \varpi_3^2\big)\end{aligned} \tag{4.21}$$

where the different signs are due to the different rotation directions of the rotors. A yaw torque can be created simply by appropriate coordinated control of all four rotor speeds.

The total torque applied to the airframe according to (4.19) to (4.21) is $\boldsymbol{\tau} = (\tau_x, \tau_y, \tau_z)^\top$ and the rotational acceleration is given by Euler's equation of motion from (3.14)

$$\mathbf{J}\dot{\boldsymbol{\omega}} = -\boldsymbol{\omega} \times \mathbf{J}\boldsymbol{\omega} + \boldsymbol{\tau} \tag{4.22}$$

where $\mathbf{J}$ is the $3 \times 3$ inertia tensor of the vehicle and $\boldsymbol{\omega}$ is the angular velocity vector. The motion of the quadrotor is obtained by integrating the forward dynamics equations (4.17) and (4.22).

The forces and moments on the airframe, given by (4.18) to (4.21), can be written in matrix form

$$\begin{pmatrix} \boldsymbol{\tau} \\ T \end{pmatrix} = \begin{pmatrix} 0 & -db & 0 & db \\ db & 0 & -db & 0 \\ k & -k & k & -k \\ -b & -b & -b & -b \end{pmatrix} \begin{pmatrix} \varpi_0^2 \\ \varpi_1^2 \\ \varpi_2^2 \\ \varpi_3^2 \end{pmatrix} = \mathbf{M} \begin{pmatrix} \varpi_0^2 \\ \varpi_1^2 \\ \varpi_2^2 \\ \varpi_3^2 \end{pmatrix} \tag{4.23}$$

and are functions of the squared rotor speeds. The mixing matrix $\mathbf{M}$ is constant, and full rank since $b, k, d > 0$ and can be inverted

$$\begin{pmatrix} \varpi_0^2 \\ \varpi_1^2 \\ \varpi_2^2 \\ \varpi_3^2 \end{pmatrix} = \mathbf{M}^{-1} \begin{pmatrix} \tau_x \\ \tau_y \\ \tau_z \\ T \end{pmatrix} \tag{4.24}$$

to solve for the rotor speeds required to apply a specified moment $\boldsymbol{\tau}$ and thrust $T$ to the airframe.

To control the vehicle, we will employ a nested control structure that we describe for pitch and $x$-translational motion. The innermost loop uses a proportional and derivative controller ▶ to compute the required pitching torque on the airframe

$$\tau_y^* = K_{\tau,p}\left(\beta^* - \hat{\beta}\right) + K_{\tau,d}\left(\dot{\beta}^* - \dot{\hat{\beta}}\right) \tag{4.25}$$

based on the error between the desired and estimated pitch angle, as well as their rates of change. ▶ The gains $K_{\tau,p}$ and $K_{\tau,d}$ are determined by classical control design approaches based on an approximate dynamic model and then tuned to achieve good performance. The estimated vehicle pitch angle $\hat{\beta}$ would come from an inertial navigation system as discussed in ▶ Sect. 3.4 and $\dot{\hat{\beta}}$ would be derived from gyroscopic sensors. The required rotor speeds are then determined using (4.24).

Consider a coordinate frame $\{B'\}$ attached to the vehicle and with the same origin as $\{B\}$ but with its $x$- and $y$-axes in the horizontal plane and parallel to the ground. The thrust vector is parallel to the $-z$-axis of frame $\{B\}$ and pitching the nose down, rotating about the $y$-axis by $\beta$, generates a force

$${}^{B'}\!f = \boldsymbol{\xi}^{r_y}(\beta) \cdot \begin{pmatrix} 0 \\ 0 \\ -T \end{pmatrix} = \begin{pmatrix} -T \sin \beta \\ 0 \\ -T \cos \beta \, , \end{pmatrix}$$

which has a component

$${}^{B'}\!f_x = -T \sin \beta \approx -T\beta$$

that accelerates the vehicle in the $-{}^{B'}\!x$-direction, and we have assumed that $\beta$ is small. We can control the velocity in this direction with a proportional control law

$${}^{B'}\!f_x^* = m K_f\left({}^{B'}\!v_x^* - {}^{B'}\!\hat{v}_x\right)$$

where $K_f > 0$ is a gain. ▶ Combining these two equations, we obtain the desired pitch angle

$$\beta^* \approx -\frac{m}{T} K_f\left({}^{B'}\!v_x^* - {}^{B'}\!\hat{v}_x\right) \tag{4.26}$$

The rotational dynamics has a second-order transfer function of $\beta(s)/\tau_y(s) = 1/(Js^2 + Bs)$ where $J$ is rotational inertia, $B$ is aerodynamic damping which is generally quite small, and $s$ is the Laplace operator. To regulate a second-order system requires a proportional-derivative controller, as the rate of change as well as the value of the signal needs to be taken into account.

The term $\dot{\beta}^*$ is commonly ignored.

The negative sign is because a positive pitch creates a thrust in the $-x$-direction. For the roll axis, there is no negative sign, positive roll angle leads to positive force and velocity in the $y$-direction.

**4**

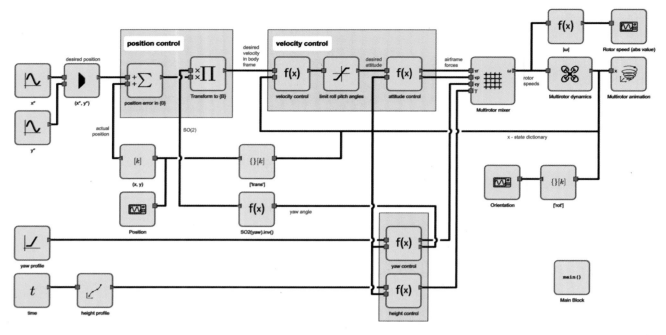

**◻ Fig. 4.24** Block diagram model `models/quadrotor` for a closed-loop simulation of a quadrotor. The signal x is the quadrotor state: position, orientation, velocity and orientation rate. The vehicle lifts off and flies in a circle at constant altitude while yawing about its own center

required to achieve the desired forward velocity. Using (4.25), we compute the required pitching torque and then, using (4.24), the required rotor speeds. For a vehicle in vertical equilibrium, the total thrust equals the weight force, so $m/T \approx 1/g$.

The estimated vehicle velocity $^B\hat{v}_x$ could come from an inertial navigation system as discussed in ▶ Sect. 3.4 or a GPS receiver. If the desired position of the vehicle in the $xy$-plane of the world frame is $^0p \in \mathbb{R}^2$, then the desired velocity is given by the proportional control law

$$^0v^* = K_p\left(^0p^* - {}^0\hat{p}\right) \tag{4.27}$$

based on the error between the desired and actual position. The desired velocity in the $xy$-plane of frame {B′} is

$$^{B'}v = \ominus \, ^0\xi^r_{B'}(\gamma) \cdot {}^0v$$

which is a function of the yaw angle $\gamma$

$$\begin{pmatrix} ^{B'}v_x \\ ^{B'}v_y \end{pmatrix} = \begin{pmatrix} \cos\gamma & -\sin\gamma \\ \sin\gamma & \cos\gamma \end{pmatrix}^\top \begin{pmatrix} v_x \\ v_y \end{pmatrix}$$

◻ Fig. 4.24 shows a block diagram model of the complete control system for a quadrotor. ◀ Working our way left to right and starting at the top, we have the desired position of the quadrotor in world coordinates. The position error is rotated from the world frame to the body frame and becomes the desired velocity. The velocity controller implements (4.26) and its equivalent for the roll axis, and outputs the desired pitch and roll angles of the quadrotor. The attitude controller is a proportional-derivative controller that determines the appropriate pitch and roll torques to achieve these angles based on feedback of current attitude and attitude rate. ◀ The yaw control block determines the error in heading angle and implements a proportional-derivative controller to compute the required yaw torque that is achieved by speeding up one pair of rotors and slowing the other pair.

This model is hierarchical and organized in terms of subsystems.

Using the coordinate conventions shown in ◻ Fig. 4.23, $x$-direction motion requires a negative pitch angle, and $y$-direction motion requires a positive roll angle, so the gains have different signs for the roll and pitch loops.

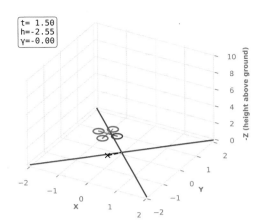

**Fig. 4.25** One frame from the quadrotor model `models/quadrotor` shown in **Fig. 4.24**. The rotor disks are shown as circles and their orientation includes the effect of blade flapping. The ×-marker on the ground plane is a projection of the vehicle's centroid

Altitude is controlled by a proportional-derivative controller

$$T = K_p(z^* - \hat{z}) + K_d\left(\dot{z}^* - \hat{\dot{z}}\right) + T_w$$

which determines the average rotor speed. $T_w = mg$ is the weight of the vehicle and this is an example of feedforward control – used here to counter the effect of gravity that is otherwise a constant disturbance to the altitude control loop. The alternatives to feedforward control would be to have very high value of $K_p$ for the altitude loop which might lead to instability, or a proportional-integral-derivative (PID) controller that might require a long time for the integral term to remove the steady-state error, and then lead to overshoot. We will revisit gravity compensation in ► Chap. 9 in the context of robot manipulator arms.

The mixer block combines the three torque demands and the vertical thrust demand, and implements (4.24) to determine the appropriate rotor speeds. Rotor speed limits are applied here. These are input to the quadrotor block ► which implements the forward dynamics, integrating (4.17) to (4.22) to determine the position, orientation, velocity and orientation rate. The output of this block is the 12-element state vector $x = ({}^0p, {}^0\Gamma, {}^B\dot{p}, {}^B\dot{\Gamma})$. As is common in aerospace applications, we represent orientation $\Gamma \in (\mathbf{S}^1)^3$ and orientation rate $\dot{\Gamma} \in \mathbb{R}^3$ in terms of ZYX roll-pitch-yaw angles. Note that position and attitude are in the world frame, while the rates are expressed in the body frame. ►

The simulation can be run by

```
>>> %run -m quadrotor -H
```

which loads the default set of quadrotor parameters and displays an animation and various time histories in separate windows. The vehicle lifts off and flies around a circle while slowly yawing about its own $z$-axis. A snapshot is shown in ■ Fig. 4.25. The simulation saves the time history in the variable `out`

```
>>> t = out.t; x = out.x;
>>> x.shape
(213, 12)
```

and the state $x$ has one row per time step, and each row is a 12-element state vector. We can plot 2D position versus time by

```
>>> plt.plot(t, x[:, 0], t, x[:, 1]);
```

To recap on control of the quadrotor. A position error leads to a required translational velocity to correct the error. To achieve that velocity, we adjust the pitch

This block can simulate an $N$-rotor vehicle, the default is $N = 4$.

The block output is actually a dictionary containing items: `"x"` $= ({}^0p, {}^0\Gamma, {}^B\dot{p}, {}^B\dot{\Gamma})$, `"rot"` $= ({}^0\Gamma, {}^B\dot{\Gamma})$, and `"trans"` $= ({}^0p, {}^B\dot{p})$.

**4**

and roll of the vehicle so that a component of the vehicle's thrust acts in the horizontal plane and generates a force to accelerate it. To achieve the required pitch and roll angles, we apply moments to the airframe by means of differential-propeller thrust which, in turn, requires rotor speed control. When the vehicle is performing this motion, the total thrust must be increased so that the vertical component still balances gravity. As the vehicle approaches its goal, the airframe must be rotated in the opposite direction so that a component of thrust decelerates the motion.

This indirection from translational motion to rotational motion is a consequence of the vehicle being underactuated – we have just four rotor speeds to adjust, but the vehicle's configuration space is 6-dimensional. In the configuration space, we cannot move in the $x$- or $y$-directions, but we can move in the pitch- or roll-direction which *leads* to motion in the $x$- or $y$-directions. The cost of underactuation is once again a maneuver. The pitch and roll angles are a means to achieve translational control but cannot be independently set if we wish to hold a constant position.

## 4.3 Advanced Topics

### 4.3.1 Nonholonomic and Underactuated Systems

We introduced the notion of configuration space in ▶ Sect. 2.4.9, and it is useful to revisit it now that we have discussed several different types of mobile robot platform. Common vehicles – as diverse as cars, hovercrafts, ships and aircraft – are all able to move forward effectively but are unable to instantaneously move sideways. This is a very sensible tradeoff that simplifies design and caters to the motion we most commonly require of the vehicle. Lateral motion for occasional tasks such as parking a car, docking a ship, or landing an aircraft are possible, albeit with some complex maneuvering but humans can learn this skill.

The configuration space of a train was introduced in ▶ Sect. 2.4.9. To fully describe all its constituent particles, we need to specify just one generalized coordinate: its distance $q$ along the track from some datum. It has one degree of freedom, and its configuration space is $C \subset \mathbb{R}$. The train has one motor, or actuator, to move it along the track and this is equal to the number of degrees of freedom. We say the train is fully actuated.

Consider a hovercraft that moves over a planar surface. To describe its configuration, we need to specify three generalized coordinates: its position in the $xy$-plane and its heading angle. It has three degrees of freedom, and its configuration space is $C \subset \mathbb{R}^2 \times \mathbf{S}^1$. This hovercraft has two propellers whose axes are parallel but not collinear. The sum of their thrusts provide a forward force and the difference in thrusts generates a yawing torque for steering. The number of actuators, two, is less than its degrees of freedom $\dim C = 3$, and we call this an underactuated system. This imposes significant limitations on the way in which it can move. At any point in time, we can control the forward (parallel to the thrust vectors) acceleration and the rotational acceleration of the hovercraft, but there is zero sideways (or lateral) acceleration since it cannot generate any lateral thrust. Nevertheless, with some clever maneuvering, like with a car, the hovercraft can follow a path that will take it to a place to one side of where it started. In the hovercraft's 3-dimensional configuration space, this means that at any point there are certain directions in which *acceleration* is not possible. We can reach points in those directions but not directly, only by following some circuitous path.

All aerial and underwater vehicles have a configuration that is completely described by six generalized coordinates – their position and orientation in 3D space. The configuration space is $C \subset \mathbb{R}^3 \times (\mathbf{S}^1)^3$ where the orientation is expressed in some three-angle representation – since $\dim C = 6$, the vehicles have six degrees of freedom. A quadrotor has four actuators, four thrust-generating propellers, and this is fewer than its degrees of freedom making it underactuated. Controlling the four propellers causes motion in the up/down, roll, pitch and yaw directions of the configuration space, but not in the forward/backward or left/right directions. To access those degrees of freedom, it is necessary to perform a *maneuver*: pitch down so that the thrust vector provides a horizontal force component, accelerate forward, pitch up so that the thrust vector provides a horizontal force component to decelerate, and then level out.

For a helicopter, only four of the six degrees of freedom are practically useful: up/down, forward/backward, left/right and yaw. Therefore, a helicopter requires a minimum of four actuators: the main rotor generates a thrust vector whose magnitude is controlled by the collective pitch and whose direction is controlled by the lateral and longitudinal cyclic pitch. The tail rotor provides a yawing moment. This leaves two degrees of freedom unactuated, roll and pitch angles, but clever design ensures that gravity actuates them and keeps them close to zero like the keel of a boat – without gravity a helicopter cannot work. A fixed-wing aircraft moves forward very efficiently and also has four actuators: engine thrust provides acceleration in the forward direction and the ailerons, elevator and rudder exert respectively roll, pitch and yaw moments on the aircraft. ▶ To access the missing degrees of freedom such as up/down and left/right translation, the aircraft must pitch or yaw while moving forward.

Some low-cost hobby aircraft have no rudder and rely only on ailerons to bank and turn the aircraft. Even cheaper hobby aircraft have no elevator and rely on motor speed to control height.

The advantage of underactuation is having fewer actuators. In practice, this means real savings in terms of cost, complexity and weight. The consequence is that, at any point in its configuration space, there are certain directions in which the vehicle cannot move. Fully-actuated designs are possible but not common, for example, the RangerBot underwater robot shown in ◘ Fig. 4.2c has six degrees of freedom and six actuators. These can exert an arbitrary force and torque on the vehicle, allowing it to accelerate in any direction or about any axis.

A 4-wheeled car has many similarities to the hovercraft discussed above. It moves over a planar surface and its configuration is fully described by its position in the $xy$-plane and a rotation angle. It has three degrees of freedom and its configuration space is $\mathcal{C} \subset \mathbb{R}^2 \times \mathbf{S}^1$. A car has two actuators, one to move forwards or backwards and one to change the heading direction. A car, like a hovercraft, is underactuated. We know from our experience with cars that we cannot move directly in certain directions and sometimes needs to perform a maneuver to reach our goal as shown in ◘ Fig. 4.26.

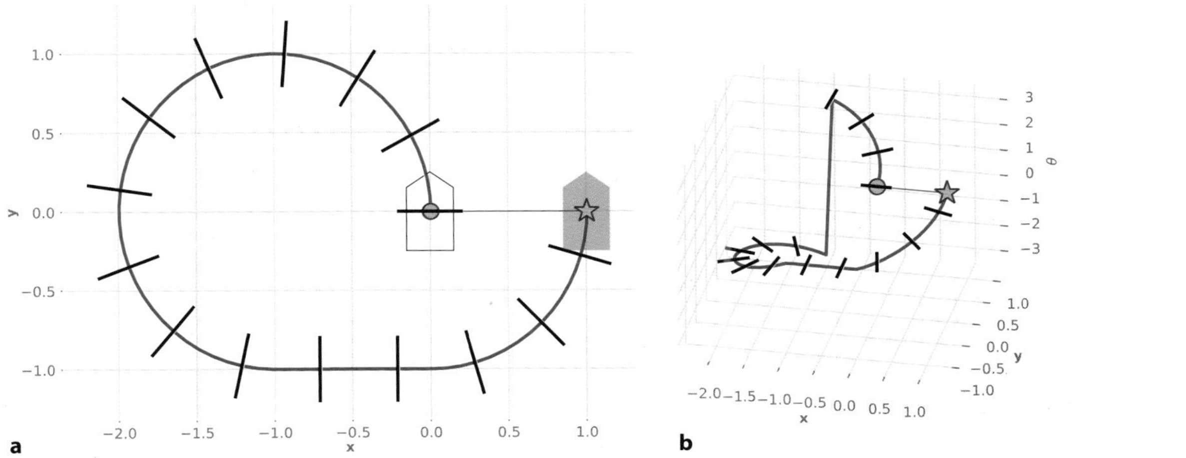


◘ **Fig. 4.26** Sideways motion of a car-like vehicle. The car cannot move directly from the circle to the star due to the nonholonomic constraint. **a** Motion in the plane and the black line segments indicate the directions in which motion is not possible; **b** the same motion shown in configuration space, note that the heading angle has wrapped around

**4**

The hovercraft, aerial, and underwater vehicles are controlled by forces, so in this case the constraints are on vehicle acceleration in configuration space, not velocity.

For example, fixing the end of a 10-joint robot arm introduces six holonomic constraints (position and orientation) so the arm would have only 4 degrees of freedom.

The constraint cannot be integrated to a constraint in terms of configuration variables, so such systems are also known as nonintegrable systems.

A differentially- or skid-steered vehicle, such as a tank, is also underactuated – it has only two actuators, one for each track. While this type of vehicle can turn on the spot, it cannot move sideways. To do that it has to turn, proceed, stop then turn – this need to maneuver is the clear signature of an underactuated system.

We might often wish for an ability to drive our car sideways, but the standard wheel provides real benefit when cornering – lateral friction between the wheels and the road provides, for free, the centripetal force which would otherwise require an extra actuator to provide. The hovercraft has many similarities to a car but we can push a hovercraft sideways – we cannot do that with a car. This lateral friction is a distinguishing feature of the car.

The inability to slip sideways is a constraint, the *rolling* constraint, on the velocity ◄ of the vehicle just as underactuation is. A vehicle with one or more *velocity constraints*, due to underactuation or a rolling constraint, is referred to as a nonholonomic system. A key characteristic of these systems is that they cannot always move *directly* from one configuration to another. Sometimes, they must take a longer indirect path which we recognize as a maneuver. A car has a velocity constraint due to its wheels and is also underactuated.

In contrast, a holonomic constraint restricts the possible configurations that the system can achieve – it can be expressed as an equation written in terms of the configuration variables. ◄ A nonholonomic constraint, such as (4.5) and (4.12), is one that restricts the *velocity* (or acceleration) of a system in configuration space – it can only be expressed in terms of the derivatives of the configuration variables. ◄ The nonholonomic constraint does not restrict the possible configurations the system can achieve, but it does preclude instantaneous velocity or acceleration in certain directions as shown by the black line segments in ◘ Fig. 4.26.

In control theoretic terms, Brockett's theorem states that nonholonomic systems are controllable, but they cannot be stabilized to a desired state using a differentiable, or even continuous, pure state-feedback controller. A time-varying or nonlinear control strategy is required which means that the robot follows some generally nonlinear path. For example, the controller (4.8) introduces a discontinuity at $\rho = 0$. One exception is an underactuated system moving in a 3-dimensional space within a force field, for example, a gravity field – gravity acts like an additional actuator and makes the system linearly controllable (but not holonomic), as we showed for the quadrotor example in ▶ Sect. 4.2.

Mobility parameters for the various robots that we have discussed here, and earlier in ▶ Sect. 2.4.9, are tabulated in ◘ Table 4.1. We will discuss under and overactuation in the context of arm robots in ▶ Sect. 8.3.4.

◘ **Table 4.1**  Summary of configuration space characteristics for various robots. A nonholonomic system is underactuated and/or has a rolling constraint

|  | dim $C$ | Degrees of freedom | Number of actuators | Actuation | Rolling constraints | Holonomic |
|---|---|---|---|---|---|---|
| Train | 1 | 1 | 1 | Full |  | ✓ |
| 2-joint robot arm | 2 | 2 | 2 | Full |  | ✓ |
| 6-joint robot arm | 6 | 6 | 6 | Full |  | ✓ |
| 10-joint robot arm | 10 | 10 | 10 | Over |  | ✓ |
| Hovercraft | 3 | 3 | 2 | Under |  |  |
| Car | 3 | 2 | 2 | Under | ✓ |  |
| Helicopter | 6 | 6 | 4 | Under |  |  |
| Fixed-wing aircraft | 6 | 6 | 4 | Under |  |  |
| RangerBot | 6 | 6 | 6 | Full |  | ✓ |

## 4.4  Wrapping Up

In this chapter, we have created and discussed models and controllers for a number of common, but quite different, robot platforms. We first discussed wheeled robots. For car-like vehicles, we developed a kinematic model that we used to develop a number of different controllers in order that the platform could perform useful tasks such as driving to a point, driving along a line, following a path or driving to a configuration. We then discussed differentially-steered vehicles on which many robots are based, and omnidirectional robots based on novel wheel types. Then we discussed a simple but common aerial vehicle, the quadrotor, and developed a dynamic model and a hierarchical control system that allowed the quadrotor to fly a circuit. This hierarchical or nested control approach is described in more detail in ▶ Sect. 9.1.7 in the context of robot arms.

We also extended our earlier discussion about configuration space to include the velocity constraints due to underactuation and rolling constraints from wheels.

The next chapters in this Part will discuss how to plan paths for robots through complex environments that contain obstacles and then how to determine the location of a robot.

### 4.4.1  Further Reading

Comprehensive modeling of mobile ground robots is provided in the book by Siegwart et al. (2011). In addition to the models covered here, it presents in-depth discussion of a variety of wheel configurations with different combinations of driven wheels, steered wheels and passive castors. The book by Kelly (2013) also covers vehicle modeling and control. Both books also provide a good introduction to perception, localization and navigation, which we will discuss in the coming chapters.

The paper by Martins et al. (2008) discusses kinematics, dynamics and control of differentially-steered robots. Astolfi (1999) develops pose control for a car-like vehicle. The Handbook of Robotics (Siciliano and Khatib 2016, part E) covers modeling and control of various vehicle types including aerial and underwater. The theory of helicopters with an emphasis on robotics is provided by Mettler (2003), but the definitive reference for helicopter dynamics is the very large book by Prouty (2002). The book by Antonelli (2014) provides comprehensive coverage of modeling and control of underwater robots.

Some of the earliest papers on quadrotor modeling and control are by Pounds, Mahony and colleagues (Hamel et al. 2002; Pounds et al. 2004, 2006). The thesis by Pounds (2007) presents comprehensive aerodynamic modeling of a quadrotor with a particular focus on blade flapping, a phenomenon well known in conventional helicopters but largely ignored for quadrotors. A tutorial introduction to the control of multi-rotor aerial robots is given by Mahony, Kumar, and Corke (2012).

Mobile ground robots are now a mature technology for transporting parts around manufacturing plants. The research frontier is now for vehicles that operate autonomously in outdoor environments (Siciliano and Khatib 2016, part F). Research into the automation of passenger cars has been ongoing since the 1980s, and increasing autonomous driving capability is becoming available in new cars.

■ ■ Historical and Interesting

The Navlab project at Carnegie-Mellon University started in 1984 and a series of autonomous vehicles, Navlabs, were built and a large body of research has resulted. All vehicles made strong use of computer vision for navigation. In 1995, the supervised autonomous Navlab 5 made a 3000-mile journey, dubbed "No Hands Across

America" (Pomerleau and Jochem 1995, 1996). The vehicle steered itself 98% of the time largely by visual sensing of the white lines at the edge of the road.

In Europe, Ernst Dickmanns and his team at Universität der Bundeswehr München demonstrated autonomous control of vehicles. In 1988, the VaMoRs system, a 5-tonne Mercedes-Benz van, could drive itself at speeds over $90 \, \text{kmh}^{-1}$ (Dickmanns and Graefe 1988b; Dickmanns and Zapp 1987; Dickmanns 2007). The European Prometheus Project ran from 1987 to 1995 and, in 1994, the robot vehicles VaMP and VITA-2 drove more than $1000 \, \text{km}$ on a Paris multi-lane highway in standard heavy traffic at speeds up to $130 \, \text{kmh}^{-1}$. They demonstrated autonomous driving in free lanes, convoy driving, automatic tracking of other vehicles, and lane changes with autonomous passing of other cars. In 1995, an autonomous S-Class Mercedes-Benz made a $1600 \, \text{km}$ trip from Munich to Copenhagen and back. On the German Autobahn, speeds exceeded $175 \, \text{kmh}^{-1}$ and the vehicle executed traffic maneuvers such as overtaking. The mean time between human interventions was $9 \, \text{km}$ and it drove up to $158 \, \text{km}$ without any human intervention. The UK part of the project demonstrated autonomous driving of an XJ6 Jaguar with vision (Matthews et al. 1995) and radar-based sensing for lane keeping and collision avoidance. In the USA in the 2000s, DARPA ran a series of Grand Challenges for autonomous cars. The 2005 desert and 2007 urban challenges are comprehensively described in compilations of papers from the various teams in Buehler et al. (2007, 2010). More recent demonstrations of self-driving vehicles are a journey along the fabled silk road described by Bertozzi et al. (2011) and a classic road trip through Germany by Ziegler et al. (2014).

Ackermann's magazine can be found online at ▶ http://smithandgosling. wordpress.com/2009/12/02/ackermanns-repository-of-arts, and the carriage steering mechanism is published in the March and April issues of 1818 (pages 162 and 234). King-Hele (2002) provides a comprehensive discussion about the prior work on steering geometry and Darwin's earlier invention.

**▪▪ Toolbox Notes**

In addition to the block diagram blocks used in this chapter, the Toolbox provides Python classes that implement the kinematic equations and that we will use in ▶ Chap. 6. For example, we can create a vehicle model with steered-wheel angle and speed limits

```
>>> veh = Bicycle(speed_max=1, steer_max=np.deg2rad(30));
>>> veh.q
array([     0,      0,      0])
```

which has an initial configuration of $(0, 0, 0)$. We can apply $(v, \gamma)$ and after one time step the configuration is

```
>>> veh.step([0.3, 0.2])
array([   0.03, 0.006081])
>>> veh.q
array([   0.03,      0, 0.006081])
```

We can evaluate (4.4) for a particular configuration and set of control inputs $(v, \gamma)$

```
>>> veh.deriv(veh.q, [0.3, 0.2])
array([   0.3, 0.001824,  0.06081])
```

This class inherits from the abstract base class `VehicleBase` as do other vehicle classes, for example, the `Unicycle` class implements the unicycle kinematics with inputs of $(v, \omega)$ and the `DiffSteer` class implements a differentially-steered robot with inputs of $(\omega_L, \omega_R)$ – all have many common methods.

## 4.4.2 Exercises

1. For a 4-wheeled vehicle with $L = 2$ m and width between wheel centers of 1.5 m
   a) What steered-wheel angle is needed for a turn rate of 10 degs$^{-1}$ at a forward speed of 20 kmh$^{-1}$?
   b) Compute the difference in angle for the left and right steered-wheels, in an Ackermann steering system, when driving around arcs of radius 10, 50 and 100 m.
   c) If the vehicle is moving at 80 kmh$^{-1}$, compute the difference in back wheel rotation rates for curves of radius 10, 50 and 100 m.
2. Write an expression for turn rate in terms of the angular rotation rate of the two back wheels. Investigate the effect of errors in wheel radius and vehicle width.
3. Consider a car and bus with $L = 4$ and $L = 12$ m respectively. To follow an arc with radius of 10, 20 and 50 m, determine the respective steered-wheel angles.
4. For a number of steered-wheel angles in the range $-45$ to $45°$ and a velocity of $2$ ms$^{-1}$, overlay plots of the vehicle's trajectory in the $xy$-plane.
5. Implement the $\ominus$ operator used in ▶ Sect. 4.1.1.1 and check against the code for `angdiff`.
6. Driving to a point (▶ Sect. 4.1.1.1), plot $x$, $y$ and $\theta$ against time.
7. Pure pursuit example (▶ Sect. 4.1.1.3)
   a) Investigate what happens if you vary the look-ahead distance.
   b) Modify the pure pursuit example so that the robot follows a slalom course.
   c) Modify the pure pursuit example to follow a target moving back and forth along a line.
8. Driving to a configuration (▶ Sect. 4.1.1.4)
   a) Repeat the example with a different initial orientation.
   b) Implement a parallel parking maneuver. Is the resulting path practical?
   c) Experiment with different controller parameters.
9. Create a GUI interface with a simple steering wheel and velocity control, and use this to create a very simple driving simulator. Alternatively, interface a gaming steering wheel and pedal to Python.
10. Adapt the various controllers in ▶ Sect. 4.1.1 to the differentially-steered robot.
11. Derive (4.9) from the preceding equation.
12. For constant forward velocity, plot $v_L$ and $v_R$ as a function of ICR position along the $y$-axis. Under what conditions do $v_L$ and $v_R$ have a different sign?
13. Implement a controller using (4.13) that moves a unicycle robot in its $y$-direction. How does the robot's orientation change as it moves?
14. Sketch the design for a robot with three mecanum wheels. Ensure that it cannot roll freely and that it can drive in any direction. Write code to convert from desired vehicle translational and rotational velocity to wheel rotation rates.
15. For the 4-wheeled omnidirectional robot of ▶ Sect. 4.1.3, write an algorithm that will allow it to move in a circle of radius 0.5 m around a point with its nose always pointed towards the center of the circle.
16. Quadrotor control (▶ Sect. 4.2)
    a) At equilibrium, compute the speed of all the propellers.
    b) Experiment with different control gains. What happens if you reduce the the gains applied to rates of change of quantities?
    c) Remove the gravity feedforward constant and experiment with large altitude gain or a PI controller.
    d) When the vehicle has nonzero roll and pitch angles, the magnitude of the vertical thrust is reduced and the vehicle will slowly descend. Add compensation to the vertical thrust to correct this.
    e) Simulate the quadrotor flying inverted, that is, its $z$-axis is pointing upwards.

**4**

f) Program a ballistic motion. Have the quadrotor take off at 45 deg to horizontal, then remove all thrust.

g) Program a smooth landing. What does smooth mean?

h) Program a barrel roll maneuver. Have the quadrotor fly horizontally in its $x$-direction and then increase the roll angle from 0 to $2\pi$.

i) Program a flip maneuver. Have the quadrotor fly horizontally in its $x$-direction and then increase the pitch angle from 0 to $2\pi$.

j) Repeat the exercise but for a 6-rotor configuration.

k) Use the function `mstraj` to create a trajectory through ten via points $(X_i, Y_i, Z_i, \theta_y)$ and modify the controller of ◻ Fig. 4.24 for smooth pursuit of this trajectory.

l) Create a GUI interface with a simple joystick control, and use this to create a very simple flying simulator. Alternatively, interface a gaming joystick to Python.

# Navigation

## Contents

© The Author(s), under exclusive license to Springer Nature Switzerland AG 2023
P. Corke, *Robotics, Vision and Control*, Springer Tracts in Advanced Robotics 146,
https://doi.org/10.1007/978-3-031-06469-2_5

**5**

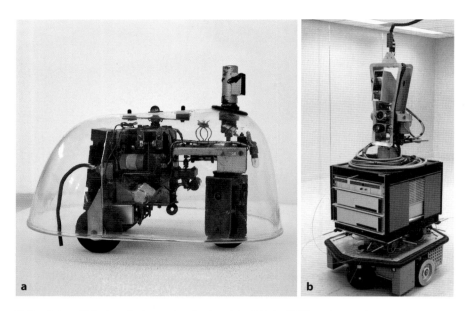

■ **Fig. 5.1  a** Elsie the robotic tortoise developed in 1940 by William Grey Walter. Burden Institute Bristol (1948). Now in the collection of the Smithsonian Institution but not on display (image courtesy Reuben Hoggett collection). **b** Shakey. SRI International (1968). Now in the Computer Museum in Mountain View (image courtesy SRI International)

chap5.ipynb

▶ go.sn.pub/yLoW7Z

Robot navigation is the problem of guiding a robot towards a goal. The goal might be specified in terms of some feature in the environment, for instance, moving toward a light, or in terms of some geometric coordinate or map reference.

The human approach to navigation is to make maps and erect signposts, and at first glance it seems obvious that robots should operate the same way. However, many robotic tasks can be achieved without any map at all, using an approach referred to as *reactive navigation* – sensing the world and reacting to what is sensed. For example, the early robotic tortoise Elsie, shown in ■ Fig. 5.1a, *reacted* to her environment and could seek out a light source and move around obstacles without having any explicit plan, or any knowledge of the position of herself or the light.

■ **Fig. 5.2**  Time lapse photograph of an early iRobot® Roomba robot cleaning a room (image by Chris Bartlett)

Today, tens of millions of robotic vacuum cleaners are cleaning floors and many of them do so without using any map of the rooms in which they work. Instead, they do the job by making random moves and sensing only that they have made contact with an obstacle as shown in ◘ Fig. 5.2.

An alternative approach was embodied in the 1960s robot Shakey, shown in ◘ Fig. 5.1b, which was capable of 3D perception. Shakey created a map of its environment and then reasoned about the map to plan a path to its destination. This is a more human-style *map-based navigation* like that used today in the autopilots of aircraft and ships, self-driving cars and mobile robots. The problem of using a map to find a path to a goal is known as path planing or motion planning. This approach supports more complex tasks but is itself more complex. It requires having a map of the environment and knowing the robot's position with respect to the map at all times – both of these are non-trivial problems that we introduce in ▸ Chap. 6.

Reactive and map-based approaches occupy opposite ends of a spectrum for mobile robot navigation. Reactive systems can be fast and simple, since sensation is connected directly to action. This means there is no need for resources to acquire and hold a representation of the world, nor any capability to reason about that representation. In nature, such strategies are used by simple organisms such as insects. Systems that make maps and reason about them require more resources, but are capable of performing more complex tasks. In nature, such strategies are used by more complex creatures such as mammals.

The remainder of this chapter discusses the reactive and map-based approaches to robot navigation with a focus on wheeled robots operating in a planar environment. Reactive navigation is covered in ▸ Sect. 5.1 and a selection of map-based motion planning approaches are introduced in ▸ Sect. 5.2.

## 5.1 Introduction to Reactive Navigation

Surprisingly complex tasks can be performed by a robot even if it has no map and no "idea" about where it is. The first-generation of Roomba robotic vacuum cleaners used only random motion and information from contact sensors to perform

**Excurse 5.1: William Grey Walter**

William Grey Walter (1910–1977) was a neurophysiologist and pioneering cyberneticist born in Kansas City, Missouri. He studied at King's College Cambridge and, unable to obtain a research fellowship at Cambridge, undertook neurophysiological research at hospitals in London and, from 1939, at the Burden Neurological Institute in Bristol. He developed electro-encephalographic brain topography which used multiple electrodes on the scalp and a triangulation algorithm to determine the amplitude and location of brain activity.

Walter was influential in the then new field of cybernetics. He built robots to study how complex reflex behavior could arise from neural interconnections. His tortoise Elsie (of the species *Machina Speculatrix*) was built in 1948. It was a three-wheeled robot capable of phototaxis that could also find its way to a recharging station. A second generation tortoise (from 1951) is in the collection of the Smithsonian Institution. Walter published popular articles in Scientific American (1950 and 1951) and a book *The Living Brain* (1953). He was badly injured in a car accident in 1970 from which he never fully recovered. (Image courtesy Reuben Hoggett collection)

**5**

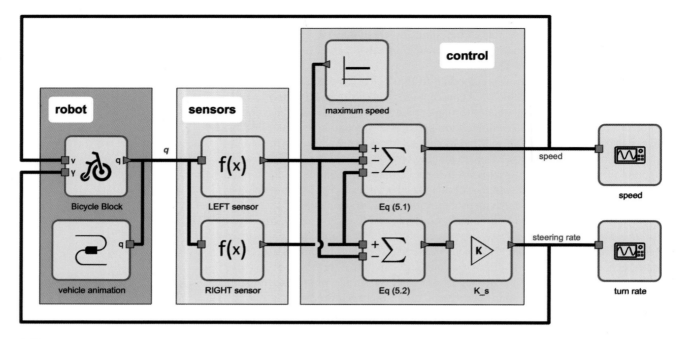

**Fig. 5.3** The block diagram model `models/braitenberg` drives a vehicle toward the maxima of a scalar field, which is defined inside the `f(x)` blocks that model the sensors. The vehicle plus controller is an example of a Braitenberg vehicle

a complex cleaning task, as shown in  Fig. 5.2. Insects such as ants and bees gather food and return it to their nest based on input from their senses, they have far too few neurons to create any kind of mental map of the world and plan paths through it. Even single-celled organisms such as flagellate protozoa exhibit goal-seeking behaviors. In this case, we need to temporarily modify our earlier definition of a robot to be:

> a goal oriented machine that can sense, ~~plan~~ and act.

Walter's robotic tortoise shown in  Fig. 5.1 demonstrated that it could moves toward a light source. It was seen to exhibit "life-like behavior" and was an important result in the then emerging scientific field of Cybernetics. This might seem like an exaggeration from the technologically primitive 1940s but this kind of behavior, known as phototaxis, ◄ is exhibited by simple organisms.

More generally, a *taxis* is the response of an organism to a stimulus gradient.

### 5.1.1  Braitenberg Vehicles

This is a fine philosophical point, the plan could be considered to be implicit in the details of the connections between the motors and sensors.

A very simple class of goal-achieving robots are known as Braitenberg vehicles and are characterized by direct connection between sensors and motors. They have no explicit internal representation of the environment in which they operate and nor do they make explicit plans. ◄

This is similar to the problem of moving to a point discussed in ► Sect. 4.1.1.1.

Consider the problem of a robot, moving in two dimensions, that is seeking the local maxima of a scalar field – the field could be light intensity or the concentration of some chemical. ◄ The block diagram model shown in  Fig. 5.3 achieves this using velocity and steering inputs derived directly from the sensors. ◄

This is similar to Braitenberg's Vehicle 4a.
We can make the measurements simultaneously using two spatially separated sensors or from one sensor over time as the robot moves.

To ascend the gradient, we need to estimate the gradient direction at the robot's location and this requires at least two measurements of the field. ◄ In this example, we use a common trick from nature, bilateral sensing, where two sensors are arranged symmetrically on each side of the robot's body. The sensors are modeled by the function blocks in  Fig. 5.3 and are parameterized by the position of the

**Excurse 5.2: Braitenberg Vehicles**
Valentino Braitenberg (1926–2011) was an Italian-Austrian neuroscientist and cyberneticist, and former director at the Max Planck Institute for Biological Cybernetics in Tübingen, Germany.

A Braitenberg vehicle is an automaton that combines sensors, actuators and their direct interconnection to produce goal-oriented behaviors. They were introduced in his 1986 book *Vehicles: Experiments in Synthetic Psychology* which describes reactive goal-achieving vehicles. The book describes the vehicles conceptually as analog circuits, but today they would typically be implemented using a microcontroller. Walter's tortoise predates Braitenberg's book, but would today be considered as a Braitenberg vehicle. Image courtesy of The MIT Press, © MIT 1984.

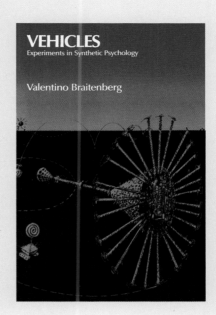

sensor with respect to the robot's body, and the sensing function. The sensors are positioned at ±2 units in the vehicle's lateral or $y$-direction. The field to be sensed is a simple inverse-square field defined by

```
>>> def sensorfield(x, y):
...    xc, yc = (60, 90)
...    return 200 / ((x - xc) ** 2 + (y - yc) ** 2 + 200)
```

that returns the sensor value $s(x, y) \in [0, 1]$ which is a function of the sensor's position in the plane. This particular function has a peak value at the coordinate $(60, 90)$.

The vehicle speed is

$$v = 2 - s_R - s_L \tag{5.1}$$

where $s_R$ and $s_L$ are the right and left sensor readings respectively. At the goal, where $s_R = s_L = 1$, the velocity becomes zero.

Steering angle is based on the difference between the sensor readings

$$\gamma = K_s(s_R - s_L) \tag{5.2}$$

and $K_s$ is the steering gain. When the field is equal in the left- and right-hand sensors, the robot moves straight ahead. ▶

We run the simulation by

```
>>> %run -m braitenberg -H
```

and it shows the robot moving toward the maxima of the scalar field – turning toward the goal and slowing down as it approaches – asymptotically achieving the goal position. The path of the robot is shown in ◑ Fig. 5.4. The initial configuration, steering gain and speed can all be adjusted by changing the constants at the top of the file `models/braitenberg.py`.

This particular sensor-action control law results in a specific robotic *behavior*. An obstacle would block this robot since its only behavior is to steer toward the goal, but an additional behavior could be added to handle this case and drive around

Similar strategies are used by moths whose two antennae are exquisitely sensitive odor detectors. A male moth uses the differential signal to "steer" toward a pheromone-emitting female.

**5**

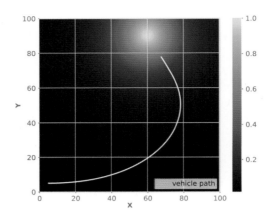

◩ **Fig. 5.4** Path of the Braitenberg vehicle using the model `models/braitenberg` shown in ◩ Fig. 5.3. The vehicle moves toward the maximum of a 2D scalar field, whose magnitude is shown color coded

an obstacle. We could add another behavior to search randomly for the source if none was visible. Walter's tortoise had four behaviors, and switching was based on light level and a touch sensor.

Multiple behaviors and the ability to switch between them leads to an approach known as behavior-based robotics. The subsumption architecture was proposed as a means to formalize the interaction between different behaviors. Complex, some might say *intelligent looking*, behaviors can be manifested by such systems. However, as more behaviors are added, the complexity of the system grows rapidly and interactions between behaviors become more complex to express and debug. Ultimately, the advantage of simplicity in not using a map becomes a limiting factor in achieving efficient and taskable behavior.

### 5.1.2 Simple Automata

Another class of reactive robots are known as *bugs* – simple automata that perform goal seeking in the presence of obstacles. Many *bug* algorithms have been proposed and they share the ability to sense when they are in proximity to an obstacle. In this respect, they are similar to the Braitenberg class vehicle, but the *bug* includes a state machine and other logic between the sensor and the motors. The automata have memory, which the earlier Braitenberg vehicle lacked. ◄ In this section, we will investigate a specific *bug* algorithm known as *bug2*.

We start by loading a model of a house that is included with the Toolbox,

```
>>> house = rtb_load_matfile("data/house.mat");
```

which is a dictionary containing two items. The floorplan

```
>>> floorplan = house["floorplan"];
>>> floorplan.shape
(397, 596)
```

is a 2D NumPy array whose elements are zero or one, representing free space or obstacle respectively. The matrix is displayed as an image in ◩ Fig. 5.5. This is an example of an occupancy grid, which will be properly introduced in ▶ Sect. 5.2. The dictionary also includes the positions of named places within the house, as items of a dictionary-like object

```
>>> places = house["places"];
>>> places.keys()
dict_keys(['kitchen', 'garage', 'br1', 'br2', 'br3', 'nook', ...
```

Braitenberg's book describes a series of increasingly complex vehicles, some of which incorporate memory. However the term *Braitenberg vehicle* has become associated with the simplest vehicles he described.

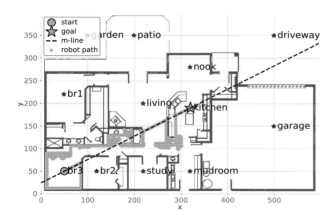

**◘ Fig. 5.5** Driveable space is white, obstacles are red, and named places are indicated in blue. The start location is a solid circle and the goal is a star. Approximate scale is 45 mm per cell

and the center of bedroom 3, for example, has the coordinate

```
>>> places.br3
array([50, 50], dtype=uint8)
```

For the *bug* algorithm we make a number of assumptions:
- the robot operates in a grid world and occupies one grid cell.
- the robot is capable of omnidirectional motion and can move to any of its eight neighboring grid cells.
- the robot can determine its position on the plane. This is a nontrivial problem that will be discussed in detail in ▶ Chap. 6.
- the robot can only sense its goal and whether adjacent cells are occupied using a simulated proximity sensor.

We create an instance of the `Bug2` class

```
>>> bug = Bug2(occgrid=floorplan);
```

and pass in the occupancy grid which is used to generate the sensory inputs for the simulated robot. We can display the robot's environment by

```
>>> bug.plot();
```

and the simulation is run by

```
>>> bug.run(start=places.br3, goal=places.kitchen, animate=True);
```

whose arguments are the start and goal positions $(x, y)$ of the robot within the house. The method displays an animation of the robot moving toward the goal and the path is shown as a series of green dots in ◘ Fig. 5.5.

The strategy of the *bug2* algorithm is quite simple. The first thing it does is compute a straight line from the start position to the goal – the m-line – and attempts to drive along it. If it encounters an obstacle, it turns right and continues until it encounters a point on the m-line that is closer to the goal than when it departed from the m-line. ▶

The `run` method returns the robot's path

```
>>> path = bug.run(start=places.br3, goal=places.kitchen);
>>> path.shape
(1307, 2)
```

as a 2D array with one row per time step, and in this case the path comprises 1307 points.

In this example, the *bug2* algorithm has reached the goal but it has taken a very suboptimal route, traversing the inside of a wardrobe, behind doors and visiting

It could be argued that the m-line represents an explicit plan. Thus, *bug* algorithms occupy a position somewhere between Braitenberg vehicles and map-based planning systems in the spectrum of approaches to navigation.

5

two bathrooms. For this example, it might have been quicker to turn left rather than right at the first obstacle, but that strategy might give a worse outcome somewhere else. Many variants of the *bug* algorithm have been developed – some may improve performance for one type of environment but will generally show worse performance in others. The robot is fundamentally limited by not using a map – without the "big picture" it is doomed to take paths that are locally, rather than globally, optimal.

## 5.2    Introduction to Map-Based Navigation

The key to achieving the *best* path between points A and B, as we know from everyday life, is to use a map. Typically, *best* means the shortest distance but it may also include some penalty term or cost related to traversability which is how easy the terrain is to drive over – it might be quicker to travel further, but faster, over better roads. A more sophisticated planner might also consider the size of the robot, the kinematics and dynamics of the vehicle and avoid paths that involve turns that are tighter than the vehicle can execute. Recalling our earlier definition of a robot as a

» goal oriented machine that can sense, plan and act,

the remaining sections of this chapter concentrate on path planning. We will discuss two types of maps that can be conveniently represented within a computer to solve navigation problems: mathematical graphs and occupancy grids.

The key elements of a graph, shown in ◻ Fig. 5.6a, are vertices or nodes (the dots) and edges (the lines) that respectively represent places and paths between places. Graphs can be directed, where the edges are arrows that indicate a direction of travel; or undirected, where the edges are without arrows and travel is possible in either direction. For mapping, we typically consider that the vertices are placed or *embedded* in a Cartesian coordinate system at the position of the places that they represent – this is called an embedded graph. The edges *represent* driveable routes ◀ between pairs of vertices and have an associated cost to traverse – typically distance or time but it could also be road roughness. Edges could have other attributes relevant to planning such as a speed limit, toll fees, or the number of coffee shops.

The edges of the graph are generally not the actual route between vertices, they simply represent the fact that there is a driveable path between them.

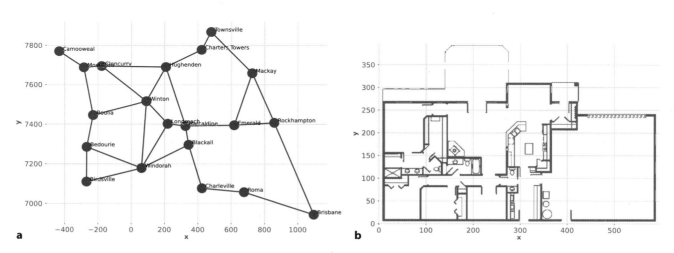

**a**    **b**

◻ **Fig. 5.6** Two common map representations **a** mathematical graph comprising vertices (places) and edges (routes), **b** occupancy grid comprising an array of square cells that have two values shown as red or white to describe whether the cell is occupied or empty respectively

## Excurse 5.3: Graph

A graph is an abstract representation of a set of objects connected by links typically denoted as an ordered pair $(V, E)$. The objects, $V$, are called vertices or nodes, and the links, $E$, that connect some pairs of vertices are called edges or arcs. Edges can be directed (arrows) or undirected as in the case shown here. Edges can have an associated weight or cost associated with moving from one of its vertices to the other. A sequence of edges from one vertex to another is a path. Graphs can be used to represent transport or communications networks and even social relationships, and the associated branch of mathematics is graph theory. The navigation classes use a Python graph package called pgraph, see ▶ App. I.

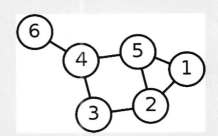

## Excurse 5.4: Humans and Maps

Animals, particularly mammals, have a powerful ability to navigate from one place to another. This ability includes recognition of familiar places or landmarks, planning ahead, reasoning about changes and adapting the route if required. The hippocampus structure in the vertebrate brain holds spatial memory and plays a key role in navigation.

Humans are mammals with a unique ability for written and graphical communications, so it is not surprising that recording and sharing navigation information in the form of maps is an ancient activity.

The oldest surviving map is carved on a mammoth tusk and is over 25,000 years old. Archaeologists believe it may have depicted the landscape around the village of Pavlov in the Czech Republic, showing hunting areas and the Thaya river. Today, maps are ubiquitous and we rely heavily on them for our daily life.

Non-embedded graphs are also common. Subway maps, for example, do not necessarily place the stations at their geographic coordinates, rather they are placed so as to make the map easier to understand. Topological graphs, often used in robotics, comprise vertices, representing places, which can be recognized even if their position in the world is not known, and paths that link those places. For example, a robot may be able to recognize that a place is a kitchen, and that by going through a particular doorway it arrives at a place it can recognize as a living room. This is sufficient to create two vertices and one edge of the topological graph, even though the vertices have no associated Cartesian coordinates.

◘ Fig. 5.6b shows a grid-based map. Here, the environment is considered to be a grid of cells, typically square, and each cell holds information about the traversability of that cell. The simplest case is called an occupancy grid, and a cell has two possible values: one indicates the cell is occupied and we cannot move into it – it is an obstacle; or zero if it is unoccupied. The size of the cell depends on the application. The memory required to hold the occupancy grid increases with the spatial area represented, and inversely with the cell size. For modern computers, this representation is feasible even for very large areas.

5

**Excurse 5.5: The Origin of Graph Theory**

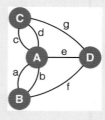

FIGURE 98. *Geographic Map:*
*The Königsberg Bridges.*

The origins of graph theory can be traced back to the Königsberg bridge problem. This city, formerly in Germany but now in Russia and known as Kaliningrad, included an island in the river Pregel and it was linked to the land on either side of the river by seven bridges (of which only three stand today).

A topic of debate was whether it was possible to walk through the city and cross each bridge just once. In 1736, Leonhard Euler proved that it was *not possible*, and his solution was the beginning of graph theory.

---

Universal Transverse Mercator (UTM) is a mapping system that maps the Earth's sphere as sixty north-south strips or zones numbered 1 to 60, each 6° wide. Each zone is divided into bands lettered C to X, each 8° high. The zone and band defines a grid zone, which is represented as a planar grid with its coordinate frame origin at the center.

| 49L | 50L | 51L | 52L | 53L | 54L | 55L | 56L |
| 49K | 50K | 51K | 52K | 53K | 54K | 55K | 56K |
| 49J | 50J | 51J | 52J | 53J | 54J | 55J | 56J |
| 49H | 50H | 51H | 52H | 53H | 54H | 55H | 56H |
| 49G | 50G | 51G | 52G | 53G | 54G | 55G | 56G |

(image created using the MATLAB® Mapping Toolbox™)

## 5.3 Planning with a Graph-Based Map

We will work with embedded graphs where the position of the vertices in the world is known. For this example, we consider the towns and cities shown in ◨ Fig. 5.7, and we import a datafile included with the Toolbox

```
>>> data = rtb_load_jsonfile("data/queensland.json");
```

where `data` is a dictionary containing two items. Firstly, `data["places"]` is a dictionary that maps place names to a dictionary of attributes such as latitude, longitude and UTM map coordinates in units of kilometers. ◄ We can easily plot these places, along with their names

```
>>> for name, info in data["places"].items():
...     plot_point(info["utm"], text=name)
```

Secondly, `data["routes"]` is a list of dictionaries that each define a graph edge: the name of its start and end place, the route length and the driving speed. For example

```
>>> data["routes"][0]
{'start': 'Camooweal', 'end': 'Mount Isa', 'distance': 188,
 'speed': 100}
```

We will use the Python `pgraph` package to create a graph representation of the data. The first step is to create an undirected graph object and then add vertices and edges from the data we just loaded

```
>>> import pgraph
>>> g = pgraph.UGraph()  # create empty undirected graph
>>> for name, info in data["places"].items():
```

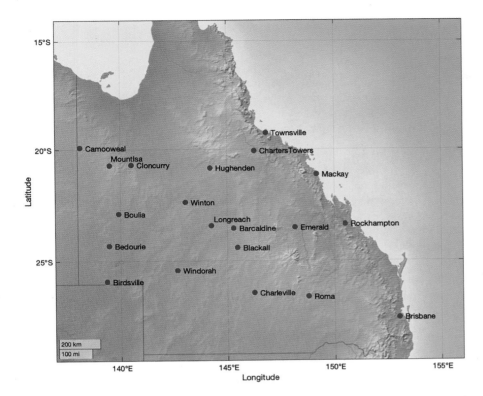

**Fig. 5.7** Cities and towns in Queensland, Australia used for graph-based planning (image created by Remo Pillat using the MATLAB® Mapping Toolbox™)

```
...    g.add_vertex(name=name, coord=info["utm"])  # add a place
>>> for route in data["routes"]:
...    g.add_edge(route["start"], route["end"],
...            cost=route["distance"])  # add a route
```

The `UGraph` object has many methods, for example, we can plot the graph we just created

```
>>> g.plot()
```

which produces the plot shown in ◼ Fig. 5.6a. The graph object has properties that report the number of vertices and edges

```
>>> g.n
20
>>> g.ne
29
```

We can retrieve a vertex by its name

```
>>> g["Brisbane"]
UVertex[Brisbane, coord=(1096, 6947)]
```

which is an object of `UVertex` class which represents a vertex in an undirected graph. Its neighbors, or adjacent vertices, are

```
>>> g["Brisbane"].adjacent()
[UVertex[Rockhampton, coord=(858.9, 7410)],
 UVertex[Roma, coord=(677.7, 7060)]]
```

The degree of a vertex is the number of neighbors it is connected to

```
>>> g["Brisbane"].degree
2
```

and the average degree of the entire graph is

```
>>> g.average_degree()
2.9
```

Average degree indicates how well-connected a graph is. A higher value means that there are, on average, more paths between vertices and hence more options for travel.

A component is a set of connected vertices, and this graph

```
>>> g.nc
 1
```

has just one component and is said to be fully connected – there is a path between every pair of vertices. A disjoint graph has multiple components – we can find a path within a component but not between them.

We can obtain references to the `Edge` objects that represent the edges. A list of the edges from the Brisbane vertex is given by

```
>>> edges = g["Brisbane"].edges()
[Edge{[Rockhampton] -- [Brisbane], cost=682},
 Edge{[Brisbane] -- [Roma], cost=482}]
```

and each edge has references to its endpoints

```
>>> edges[0].endpoints
[UVertex[Rockhampton, coord=(858.9, 7410)],
 UVertex[Brisbane, coord=(1096, 6947)]]
```

Now, we are ready to plan a path, and will consider the specific example of a journey from Hughenden to Brisbane. A simple-minded solution would be to randomly choose one of the neighboring vertices, move there, and repeat the process. We might eventually reach our goal – but there is no guarantee of that – and it is extremely unlikely that the path would be optimal. A more systematic approach is required and we are fortunate this problem has been well-studied – it is the well-known graph search problem.

Starting at Hughenden, there are just four roads we can take to get to a neighboring place. From one of those places, say Winton, we can visit in turn each of its neighbors. However, one of those neighbors is Hughenden, our starting point, and backtracking cannot lead to an optimal path. We need some means of keeping track of where we have been. The frontier (or open set) contains all the vertices that are scheduled for exploration, and the  explored set (or closed set) contains all the vertices that we have explored and are finished with.

We will introduce three algorithms, starting with breadth-first search

```
>>> path, length = g.path_BFS("Hughenden", "Brisbane")
>>> length
1759
>>> ", ".join([p.name for p in path])
'Hughenden, Barcaldine, Emerald, Rockhampton, Brisbane'
```

which has returned two values: a list of vertices defining the path from start to goal, and the total path length in kilometers. We can visualize the path by first displaying the plot, then overlaying the path

```
>>> g.plot()
>>> g.highlight_path(path)
```

which is shown in ◘ Fig. 5.8a. The algorithm is sequential and it is instructive to look at its step-by-step actions

```
 >>> g.path_BFS("Hughenden", "Brisbane", verbose=True)
FRONTIER: Hughenden
EXPLORED:
   expand Hughenden
     add Cloncurry to the frontier
```

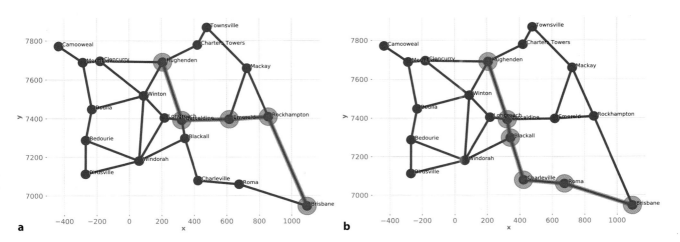

**Fig. 5.8** Path planning results. **a** Breadth-first search (BFS) with a path length of 1759 km; **b** Uniform-cost search (UCS) and A* search with a path length of 1659 km

```
    add Charters Towers to the frontier
    add Barcaldine to the frontier
    add Winton to the frontier
    move Hughenden to the explored list

FRONTIER: Cloncurry, Charters Towers, Barcaldine, Winton
EXPLORED: Hughenden
    expand Cloncurry
        add Mount Isa to the frontier
        move Cloncurry to the explored list

FRONTIER: Charters Towers, Barcaldine, Winton, Mount Isa
EXPLORED: Hughenden, Cloncurry
    expand Charters Towers
        add Townsville to the frontier
        move Charters Towers to the explored list
...
```

The frontier is initialized with the start vertex. At each step, the first vertex from the frontier is *expanded* – its neighbors added to the frontier and that vertex is retired to the explored set. At the last expansion

```
FRONTIER: Rockhampton, Bedourie, Birdsville, Roma
EXPLORED: Hughenden, Cloncurry, Charters Towers, Barcaldine, Winton,
Mount Isa, Townsville, Blackall, Emerald, Longreach, Boulia,
Windorah, Camooweal, Mackay, Charleville
    expand Rockhampton
        goal Brisbane reached
15 vertices explored, 3 remaining on the frontier
```

the goal is encountered and the algorithm stops. There were three vertices on the frontier, still unexpanded, that might have led to a better path.

At this point, it is useful to formalize the problem: at each step we want to choose the next vertex $v$ that minimizes the cost

$$f(v) = g(v) + h(v) \tag{5.3}$$

where $g(v)$ is the cost of the path so far – the *cost to come* (from the start), and $h(v)$ is the remaining distance to the goal – the *cost to go*. However, we do not know $h(v)$ since we have not yet finished planning. For now, we settle for choosing the next vertex $v$ that minimizes $f(v) = g(v)$. That is, we choose

$$v^* = \arg\min_v f(v) \tag{5.4}$$

5

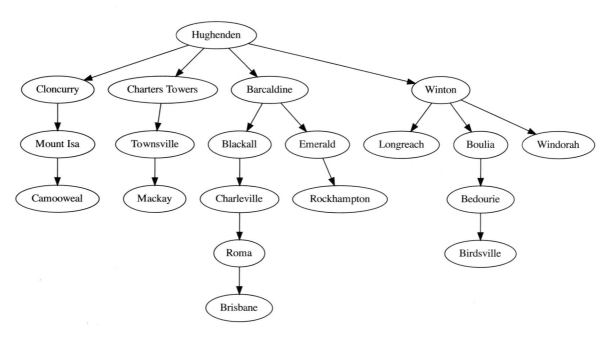

⬛ **Fig. 5.9**  Search tree for uniform-cost search

as the next vertex to expand. The intuition is that in order to achieve the shortest overall path, at each step we should choose the vertex associated with the smallest distance so far.

This leads to another well-known  graph search algorithm called uniform-cost search (UCS). In the previous search, we expanded vertices from the frontier on a first-in first-out basis, but now we will choose the frontier vertex that has the lowest cumulative distance from the start – the lowest cost to come.

Running the UCS algorithm

```
>>> path, length, parents = g.path_UCS("Hughenden", "Brisbane")
>>> length
1659
>>> ", ".join([p.name for p in path])
'Hughenden, Barcaldine, Blackall, Charleville, Roma, Brisbane'
```

we obtain a path that is 100 km shorter and shown in ⬛ Fig. 5.8b. UCS has found the shortest path length whereas BFS found the path with the fewest vertices.

As the algorithm progresses, it records, for every vertex, the vertex that it travelled from – the vertex's parent. This information is returned in `parents` which is a dictionary, for example, the parent of Winton is Hughenden

```
>>> parents["Winton"]
'Hughenden'
```

We can convert this dictionary into a non-embedded directed graph

```
>>> tree = pgraph.DGraph.Dict(parents);
```

`showgraph` uses the GraphViz package and `dot` to create a pretty graph layout.

which represents the search tree and can be displayed graphically ◄

```
>>> tree.showgraph()
```

as shown in ⬛ Fig. 5.9. This is the final version of the search tree, and it is built incrementally as the graph is explored.

Once again, it is instructive to look at the step-by-step operation

```
>>> g.path_UCS("Hughenden", "Brisbane", verbose=True)
FRONTIER: Hughenden(0)
EXPLORED:
```

```
   expand Hughenden
      add Cloncurry to the frontier
      add Charters Towers to the frontier
      add Barcaldine to the frontier
      add Winton to the frontier
   move Hughenden to the explored list

FRONTIER: Cloncurry(401), Charters Towers(248), Barcaldine(500),
Winton(216)
EXPLORED: Hughenden
   expand Winton
      add Longreach to the frontier
      add Boulia to the frontier
      add Windorah to the frontier
   move Winton to the explored list

FRONTIER: Cloncurry(401), Charters Towers(248), Barcaldine(500),
Longreach(396), Boulia(579), Windorah(703)
EXPLORED: Hughenden, Winton
   expand Charters Towers
      add Townsville to the frontier
   move Charters Towers to the explored list
...
```

The value of $g(v)$, the cost to come, is given in parentheses and at each step the vertex with the lowest value is chosen for expansion. ▶

Something interesting happens during expansion of the Bedourie vertex

```
FRONTIER: Emerald(807), Bedourie(773), Charleville(911),
Birdsville(1083), Rockhampton(1105)
EXPLORED: Hughenden, Winton, Charters Towers, Townsville, Longreach,
Cloncurry, Barcaldine, Mount Isa, Boulia, Blackall, Windorah,
Camooweal, Mackay
   expand Bedourie
 reparent Birdsville: cost 966 via Bedourie is less than cost 1083
 via Windorah, change parent from Windorah to Bedourie
      move Bedourie to the explored list
```

One of its neighbors is Birdsville, which is still in the frontier. The cost of travelling to Birdsville from the start via Windorah is 1083 km, whereas the cost of traveling to Bedourie from the start is 773 km and then it is a further

```
>>> g["Bedourie"].edgeto(g["Birdsville"]).cost
193
```

to Birdsville. This means that Birdsville is really only $773 + 193 = 966$ km from the start, and we have discovered a shorter route than the one already computed. The algorithm changes the parent of Birdsville in the search tree, shown in ◘ Fig. 5.9, to reflect that.

At the third-last step Brisbane, our goal, enters the frontier

```
FRONTIER: Rockhampton(1077), Roma(1177)
EXPLORED: Hughenden, Winton, Charters Towers, Townsville, Longreach,
Cloncurry, Barcaldine, Mount Isa, Boulia, Blackall, Windorah,
Camooweal, Mackay, Bedourie, Emerald, Charleville, Birdsville
   expand Rockhampton
      add Brisbane to the frontier
      move Rockhampton  to the explored list

FRONTIER: Roma(1177), Brisbane(1759)
EXPLORED: Hughenden, Winton, Charters Towers, Townsville, Longreach,
Cloncurry, Barcaldine, Mount Isa, Boulia, Blackall, Windorah,
Camooweal, Mackay, Bedourie, Emerald, Charleville, Birdsville,
Rockhampton
   expand Roma
 reparent Brisbane: cost 1659 via Roma is less than cost 1759 via
 Rockhampton, change parent from Rockhampton to Roma
```

In many implementations, the frontier is maintained as an ordered list, or priority queue, from smallest to largest value of $f(v)$.

**5**

**Excurse 5.6: Shakey and the A\* Trick**

Shakey, shown in ◼ Fig. 5.1, was a mobile robot system developed at Stanford Research Institute from 1966 to 1972. It was a testbed for computer vision, planning, navigation, and communication using natural language – artificial intelligence as it was then understood. Many novel algorithms were integrated into a physical system for the first time. The robot was a mobile base with a TV camera, rangefinder and bump sensors connected by radio links to a PDP-10 mainframe computer. Shakey can be seen in the Computer History Museum in Mountain View, CA.

The A\* trick was published in a 1966 paper *A formal basis for the heuristic determination of minimum cost paths* by Shakey team members Peter Hart, Nils Nilsson, and Bertram Raphael.

```
         move Roma to the explored list

FRONTIER: Brisbane(1659)
EXPLORED: Hughenden, Winton, Charters Towers, Townsville, Longreach,
Cloncurry, Barcaldine, Mount Isa, Boulia, Blackall, Windorah,
Camooweal, Mackay, Bedourie, Emerald, Charleville, Birdsville,
Rockhampton, Roma
    expand Brisbane
19 vertices explored, 0 remaining on the frontier
```

but the algorithm does not stop immediately in case there is a better route to Brisbane to be discovered. In fact, at the second-last step we find a better path through Roma rather than Rockhampton. UCS has found the shortest path but it has explored every vertex, and this can be expensive for very large graphs. The next algorithm we introduce can obtain the same result by exploring just a subset of the vertices.

The trick to achieving this is to rewrite (5.3) as

$$f(v) = g(v) + h^*(v) \tag{5.5}$$

where $h^*(v)$ is an *estimate* of the distance to the goal – referred to as the *heuristic* distance. A commonly used heuristic is the distance "as the crow flies" or Euclidean distance – easily computed since our graph is embedded and the position of every vertex is known.

The final algorithm we introduce is the very well-known  A\* search which incorporates this heuristic. We run it in a similar way to UCS

```
>>> path, length, parents = g.path_Astar("Hughenden", "Brisbane",
... summary=True)
12 vertices explored, 3 remaining on the frontier
>>> length
1659
>>> ", ".join([p.name for p in path])
'Hughenden, Barcaldine, Blackall, Charleville, Roma, Brisbane'
```

and it returns the same optimal path as UCS, but has explored only 12 vertices compared to 19 for UCS. We can visualize this by highlighting the vertices that were explored, those that are included in the returned search tree

```
>>> # find the unique vertex names
>>> visited = set(list(parents.keys()) + list(parents.values()));
>>> g.plot()
>>> g.highlight_vertex(visited, color="yellow")
```

and shown in ◼ Fig. 5.10. The vertices Camooweal, Bedourie and Birdsville, far away from the path, were not visited.

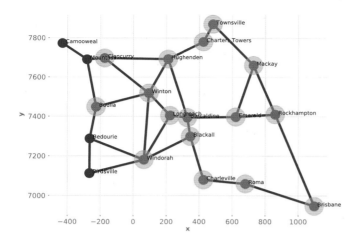

**◻ Fig. 5.10** Vertices explored by A* search

The first two steps of A* are

```
>>> g.path_Astar("Hughenden", "Brisbane", verbose=True)
FRONTIER: Hughenden(0)
EXPLORED:
    expand Hughenden
        add Cloncurry to the frontier
        add Charters Towers to the frontier
        add Barcaldine to the frontier
        add Winton to the frontier
     move Hughenden  to the explored list

FRONTIER: Cloncurry(1878), Charters Towers(1319), Barcaldine(1390),
Winton(1371)
EXPLORED: Hughenden
    expand Charters Towers
        add Townsville to the frontier
      move Charters Towers  to the explored list
```

and the value in parentheses this time is the cost $g(v) + h^*(v)$ – the sum of the cost-to-come (as used by UCS) plus an *estimate* of the cost-to-go.

The performance of A* depends critically on the heuristic. A* is *guaranteed* to return the lowest-cost path only if the heuristic is admissible, that is, it *does not overestimate* the cost of reaching the goal. The Euclidean distance is an admissible heuristic since it is the minimum possible cost and by definition cannot be an overestimate. Of course, $h^*(v) = 0$ is admissible but that is equivalent to UCS and all the vertices will be visited. As the heuristic cost component increases from zero up to its maximum admissible value, the efficiency increases as fewer vertices are explored. A non-admissible heuristic may lead to a non-optimal path.

### 5.3.1   **Minimum-Time Path Planning**

For some problems, we might wish to minimize time rather than distance. Minimum time is important when there is the possibility of trading off a short slow route for a longer but faster route. We will continue the example from the last section but change the edge costs to reflect the road speeds which are either $100\,\mathrm{kmh}^{-1}$ for a major road or $50\,\mathrm{kmh}^{-1}$ for a minor road. We will build a new graph

```
>>> g = pgraph.UGraph()
>>> for name, info in data["places"].items():
...     g.add_vertex(name=name, coord=info["utm"])
>>> for route in data["routes"]:
```

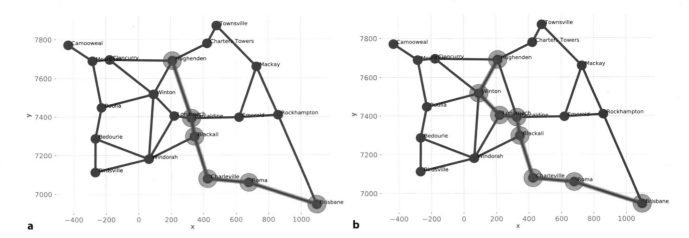

**Fig. 5.11** A* path for **a** minimum distance and **b** minimum time

```
...      g.add_edge(route["start"], route["end"],
...                   cost=route["distance"] / route["speed"])
```

and the edge cost is now in units of hours. Before we run A* we need to define the heuristic cost for this case. For the minimum-distance case, the heuristic was the distance as the crow flies between the vertex and the goal. For the minimum-time case, the heuristic cost will be the time to travel as the crow flies from the vertex to the goal, but the question is at which speed? Remember that the heuristic must not *overestimate* the cost of attaining the goal. If we choose the slow speed, then on a fast road we would overestimate the time, making this an inadmissible metric. However, if we choose the fast speed, then on a slow road, we will underestimate the travel time and that *is* admissible. Therefore our heuristic cost assumes travel at $100\,\mathrm{km}\mathrm{h}^{-1}$ and is

```
>>> g.heuristic = lambda x: np.linalg.norm(x) / 100
```

Now, we can run the A* planner

```
>>> path, time, _ = g.path_Astar("Hughenden", "Brisbane")
>>> time
16.61
>>> ", ".join([p.name for p in path])
'Hughenden, Winton, Longreach, Barcaldine, Blackall, Charleville,
Roma, Brisbane'
```

and the length of the path is in units of hours. The minimum-distance and minimum-time routes are compared in ☐ Fig. 5.11. We see that the minimum-time route has avoided the slow road between Hughenden and Barcaldine and instead chosen the faster, but longer, route through Winton and Longreach.

### 5.3.2  Wrapping Up

In practice, road networks have additional complexity, particularly in urban areas. Travel time varies dynamically over the course of the day, and that requires changing the edge costs and replanning. Some roads are one-way only, and we can represent that using a directed graph, a DGraph instance rather than a UGraph instance. For two-way roads, we would have to add two edges between each pair of vertices and these could have different costs for each direction. For the one-way road, we would add just a single edge.

Planners have a number of formal properties. A planner is *complete* if it reports a solution, or the absence of a solution, in finite time. A planner is *optimal* if it reports the best path with respect to some metric. The *complexity* of a planner is concerned with the computational time and memory resources it requires.

## 5.4 Planning with an Occupancy-Grid Map

The occupancy grid is an array of cells, typically square, and each cell holds information about the traversability of that cell. In the simplest case, a cell has two possible values: one indicates the cell is occupied and we cannot move into it – it is an obstacle; or zero if it is unoccupied.

We want a planner that can return the shortest obstacle-free path through the grid, that is, the path is *optimal* and *admissible*. In this section, we make some assumptions:

- the robot operates in a grid world;
- the robot occupies one grid cell;
- the robot is capable of omnidirectional motion and can move to any of its eight neighboring grid cells.

### 5.4.1 Distance Transform

Consider a 2D array of zeros with just a single nonzero element representing the goal as shown in ◘ Fig. 5.12a. The distance transform of this array is another 2D array, of the same shape, but the value of each element is its distance ▶ from the

The distance between two points $(x_1, y_1)$ and $(x_2, y_2)$ where $\Delta_x = x_2 - x_1$ and $\Delta_y = y_2 - y_1$ can be the Euclidean distance or $L_2$ norm $\sqrt{\Delta_x^2 + \Delta_y^2}$; or Manhattan (also known as city-block) distance or $L_1$ norm $|\Delta_x| + |\Delta_y|$.

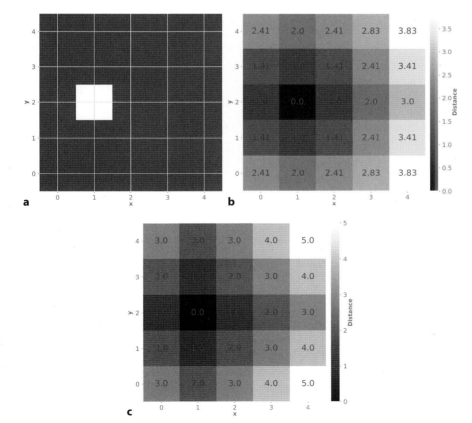

◘ **Fig. 5.12** The distance transform **a** the goal cell is set to one, others are zero; **b** Euclidean distance ($L_2$) transform where cell values shown in red are the Euclidean distance from the goal cell; **c** Manhattan ($L_1$) distance transform

**5**

> **Excurse 5.7: Creating an Occupancy Grid**
>
> There are many ways we can create an occupancy-grid map. For example, we could create a 2D NumPy array filled with zeros (all cells are free space)
>
> ```
> >>> map = np.zeros((100, 100));
> ```
>
> and use NumPy operations such as
>
> ```
> >>> map[40:50, 20:80] = 1;  # set to occupied
> ```
>
> to create obstacles. To set 100 random elements to be occupied is achieved by
>
> ```
> >>> k = np.random.choice(map.size, 100, replace=False);
> >>> map.ravel()[k] = 1;
> ```
>
> We could also use any kind of painting program that produces a black and white image, export an image file, and import that into Python using Pillow, OpenCV or SciPy. For example, the occupancy grid in ◘ Fig. 5.5 was derived from an image file but online buildings plans and street maps could also be used as discussed in ▶ Sect. 11.1.6.

original nonzero pixel. Two approaches to measuring distances are commonly used for robot path planning. The Toolbox default is Euclidean or $L_2$ distance which is shown in ◘ Fig. 5.12b – the distance to horizontal and vertical neighboring cells is one, and to diagonal neighbors is $\sqrt{2}$. The alternative is Manhattan distance, also called city-block or $L_1$ distance, which is shown in ◘ Fig. 5.12c. The distance to horizontal and vertical neighboring cells is one, and motion to diagonal neighbors is not permitted. The distance transform is an image processing technique that we will discuss further in ▶ Sect. 11.6.4.

To demonstrate the distance transform for robot navigation, we first create a simple occupancy grid

```
>>> simplegrid = np.zeros((6, 6));
>>> simplegrid[2:5, 3:5] = 1
>>> simplegrid[3:5, 2] = 1
```

which is shown in ◘ Fig. 5.13a. Next, we create a distance transform motion planner and pass it the occupancy grid

```
>>> dx = DistanceTransformPlanner(occgrid=simplegrid);
```

> ❗ **Occupancy grid coordinates and matrix indices are not the same**
> The occupancy grid is a 2D array whose coordinates are conventionally expressed as `map[row, column]` and the row is the vertical dimension of a matrix. Here, we use the Cartesian convention of a horizontal $x$-coordinate first, followed by the $y$-coordinate therefore the matrix is always indexed as `map[y, x]` in the code.

Then we create a plan to reach a specific goal at cell $(1, 1)$

```
>>> dx.plan(goal=(1, 1))
```

and the distance transform can be visualized

```
>>> dx.plot(labelvalues=True);
```

as shown in ◘ Fig. 5.13b. The numbers in red are the distance of the cell from the goal, taking into account travel *around* obstacles. For Manhattan distance, we use a similar set of commands

```
>>> dx = DistanceTransformPlanner(occgrid=simplegrid,
... distance="manhattan");
>>> dx.plan(goal=(1, 1))
>>> dx.plot(labelvalues=True);
```

and the plan is shown in ◘ Fig. 5.13c.

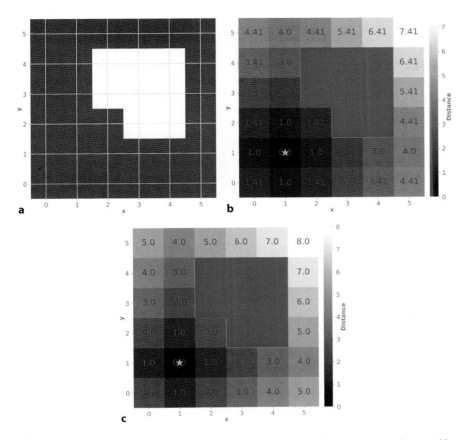

**◘ Fig. 5.13** Distance transform path planning. **a** The occupancy grid displayed as an image, white pixels are nonzero and represent obstacles; **b** Euclidean distance ($L_2$) transform, red pixels are obstacles; **c** Manhattan ($L_1$) distance transform. Numbers in red are the distance of the cell from the goal, which is denoted by a star

The plan to reach the goal is implicit in the distance map. The optimal path to the goal, from *any* cell, is a move to the neighboring cell that has the smallest distance to the goal. The process is repeated until the robot reaches a cell with a distance value of zero which is the goal. If we plot distance from the goal as a 3-dimensional surface

```
>>> dx.plot_3d();
```

as shown in ◘ Fig. 5.14a, then reaching the goal is simply a matter of rolling *down-hill* on the distance function from *any* starting point. We have converted a fairly complex planning problem into one that can now be handled by a Braitenberg-class robot that makes local decisions based on the distance to the goal.

The optimal path from an arbitrary starting point $(5, 4)$ to the goal is obtained by *querying* the plan

```
>>> path = dx.query(start=(5, 4))
array([[5, 4],
       [5, 3],
       [5, 2],
       [4, 1],
       [3, 1],
       [2, 1],
       [1, 1]])
```

**5**

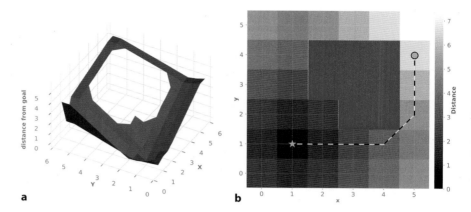

**◘ Fig. 5.14    a** The distance transform as a 3D function, where height is distance from the goal. The path to the goal is simply a downhill run. **b** Distance transform with overlaid path

There is a duality between start and goal with this type of planner. By convention, the plan is based on the goal location and we query for a particular start location, but we could base the plan on the start position and then query for a particular goal.

and the result has one path point per row, from start to goal. ◄ We can overlay this on the occupancy grid

```
>>> dx.plot(path);
```

as shown in ◘ Fig. 5.14b. We can also animate the motion of the robot along the path from start to goal by

```
>>> dx.query(start=(5, 4), animate=True);
```

which displays the robot's path as a series of green dots moving toward the goal.

Using the distance transform for the more realistic house navigation problem is now quite straightforward. We load the house floorplan and named places

```
>>> house = rtb_load_matfile("data/house.mat");
>>> floorplan = house["floorplan"];
>>> places = house["places"];
```

For a goal in the kitchen

```
>>> dx = DistanceTransformPlanner(occgrid=floorplan);
>>> dx.plan(goal=places.kitchen)
>>> dx.plot();
```

the distance transform is shown in ◘ Fig. 5.15. The distance map indicates the distance from any point to the goal, in units of grid cells, taking into account travel *around* obstacles. A path from bedroom three to the goal is

```
>>> path = dx.query(start=places.br3);
>>> path.shape
(338, 2)
```

which we can visualize overlaid on the occupancy grid and distance map

```
>>> dx.plot(path);
```

as shown in ◘ Fig. 5.15.

This navigation algorithm has exploited its global view of the world and has, through exhaustive computation, found the shortest possible path which is 338 steps long. In contrast, the *bug2* algorithm of ▶ Sect. 5.1.2 without the global view has just bumped its way through the world and had a path length of 1307 steps – nearly four times longer. The penalty for achieving the optimal path is upfront computational cost. This particular implementation of the distance transform is iterative. Each iteration has a cost of $O(N^2)$ and the number of iterations is at least $O(N)$, where $N$ is the dimension of the map. In this case, the summary option

```
>>> dx.plan(places.kitchen, summary=True);
571 iterations, 9925 unreachable cells
```

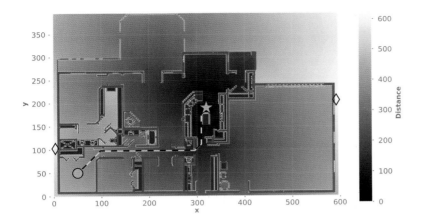

**◻ Fig. 5.15** The distance transform path. The background gray intensity represents the cell's distance from the goal in units of cell size as indicated by the scale on the right-hand side. Obstacles are indicated in red, unreachable cells are blue. Wavefront collisions are indicated by the black diamonds

shows us that 571 iterations were required. Although the plan is expensive to create, once it has been created it can be used to plan a path from *any* initial point to that goal. For large occupancy grids, this approach to planning will become impractical. The roadmap methods that we discuss later in this chapter provide an effective means to find paths in large maps at greatly reduced computational cost.

We can visualize the iterations of the distance transform algorithm

```
>>> dx.plan(places.kitchen, animate=True);
```

that shows the distance values propagating as a wavefront outward from the goal. The wavefront moves outward, spills through doorways into adjacent rooms and outside the house. ▶ Along the left and right edges of the distance map there are discontinuities where two wavefronts collide. The collisions delineate two different paths to the goal. On one side of the collision the optimal path is upwards toward the goal, while on the other side the optimal path is downwards toward the goal.

The way that the distance values spread outward from the goal inspires other names such as the wavefront, brushfire or grassfire algorithm.

Cells that are completely enclosed by obstacles can never be reached by the wavefront. In the example above, there are 9925 of these cells, and they are colored blue in ◻ Fig. 5.15.

The scale associated with this occupancy grid is 45 mm per cell and we have assumed the robot occupies a single grid cell – this is a very small robot. The planner could therefore find paths through gaps that a larger real robot would fail to pass through. A common solution to this problem is to *inflate* the occupancy grid – bigger obstacles and a small robot is equivalent to the actual obstacles and a bigger robot. For example, if the robot fits within a 500 mm circle (11 cells) then we shrink it by 5 cells in all directions, leaving just a single grid cell. To compensate, we expand or inflate all the obstacles by 5 cells in all directions – it is like applying a very thick layer of paint ▶

Obstacle inflation is an image processing technique called morphological dilation, which is discussed in ▶ Sect. 11.6.

```
>>> dx = DistanceTransformPlanner(occgrid=floorplan, inflate=5);
>>> dx.plan(places.kitchen);
>>> p = dx.query(places.br3);
>>> dx.plot(p);
```

and this is shown in ◻ Fig. 5.16. The inflated obstacles are shown in pink and the previous route to the kitchen has become blocked. The robot follows a different route, through the narrower corridors, to reach its goal. ▶

A more sophisticated approach is based on an *inflation ratio* which relates robot speed to inflation radius. At high-speed, a large inflation radius is used to give plenty of clearance from obstacles.

**5**

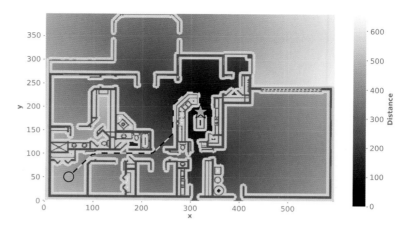

**Fig. 5.16** Distance transform path with obstacles inflated by 5 cells. Original obstacle cells are shown in red and the inflated obstacle cell are shown in pink

### 5.4.2 D*

D* is a well-known and popular algorithm for robot path computation. Similar to the distance transform, it uses an occupancy-grid representation of the world and finds an optimal path. Under the hood, D* converts the occupancy grid to a graph and then finds the optimal path using an A*-like algorithm. ◀ D* has several features that are useful for real-world applications. Firstly, it generalizes the occupancy grid to a cost map that represents the cost $c \in \mathbb{R}_{>0}$ of traversing each cell in the horizontal or vertical direction. The cost of traversing the cell diagonally is $c\sqrt{2}$. For cells corresponding to obstacles, $c = \infty$ (np.inf in NumPy). If we are interested in the shortest time to reach the goal, then cost is the time to drive across the cell. If we are interested in minimizing damage to the vehicle or maximizing passenger comfort, then cost might be related to the roughness of the terrain within the cell. ◀

D* stands for dynamic A* and is an extension of the A* algorithm for finding minimum-cost paths through a graph that supports efficient replanning when route costs change.

Secondly, D* supports incremental replanning. This is important if, while we are moving, we discover that the world is different to our map. If we discover that a cell has a different cost, either higher or lower, then we can incrementally replan to find a better path. The incremental replanning has a lower computational cost than completely replanning as would be required using the distance transform method just discussed.

The costs assigned to cells will also depend on the characteristics of the vehicle as well as the world, for example, a large 4-wheel drive vehicle may have a finite cost to cross a rough area, whereas for a small car this cost might be infinite.

To implement the D* planner using the Toolbox, we use a similar pattern as before and create a D* navigation object

```
>>> dstar = DstarPlanner(occgrid=floorplan);
```

which converts the passed occupancy grid into a cost map, which we can retrieve

```
>>> c = dstar.costmap;
```

The elements of c will be 1 or $\infty$ representing free and occupied cells respectively. These are the edge costs in the internal graph representation of the grid. The D* planner can also be passed a costmap, a CostMap object, rather than an occupancy grid. A plan for moving to a goal in the kitchen is generated by

```
>>> dstar.plan(goal=places.kitchen);
```

which creates a dense directed graph where every cell or graph vertex has a distance to the goal, and a link to the neighboring cell that is closest to the goal. Just like the graph search algorithms discussed in ▶ Sect. 5.3, there is a frontier list and an

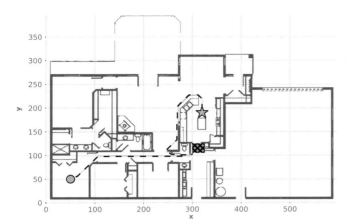

**Fig. 5.17** Path from D* planner with the map being modified dynamically. The high-cost region indicated by the orange hatched rectangle is discovered during the mission, causing the robot to backtrack and find an alternative route to the goal

explored list. The Python implementation is slow, taking tens of seconds for a map of this size and the total number of vertices expanded is

```
>>> nexpand0 = dstar.nexpand
359764
```

The path from an arbitrary starting point to the goal

```
>>> path = dstar.query(start=places.br3);
```

is almost identical to the one given by the distance transform planner shown in ◘ Fig. 5.15.

The real power of D* comes from being able to efficiently change the cost map during the mission. This is actually quite a common requirement in robotics since real sensors have a finite range and a robot discovers more of world as it proceeds. We inform D* about changes using the `sensor` argument, which is a function object that updates the costmap as the robot proceeds. To block off the bottom doorway to the kitchen when the robot is nearly there, we create a sensor function

```
>>> def sensorfunc(pos):
...     if pos[0] == 300:  # near the door?
...         changes = []
...         for x in range(300, 325):
...             for y in range(115,125):
...                 changes.append((x, y, np.inf))
...         return changes
```

where `pos` is the robot's current position. The function returns a list of change tuples comprising the map cell's coordinate and new cost. Now, if we query for a route to the kitchen

```
>>> dstar.query(start=places.br3, sensor=sensorfunc);
```

the path is the optimal route via the bottom doorway to the kitchen until the robot arrives there and determines that the door is shut. The sensor function assigns an infinite cost to a small rectangular region across the doorway which is shown hatched in ◘ Fig. 5.17. The changes invoke D* replanning, the changed cells are added to the frontier and the changes ripple out across the map and vertices are rearranged in the search tree. When completed, the robot's move to its lowest-cost neighbor will be optimal with respect to the changed environment. The robot will back track and follow a path through the top doorway to the kitchen as shown in ◘ Fig. 5.17. The number of vertices expanded to accommodate the change is

```
>>> dstar.nexpand - nexpand0
42792
```

**5**

which is around 12% of the original planning task. D\* allows updates to the map to be made at any time while the robot is moving.

## 5.5 Planning with Roadmaps

In robotic path planning, the analysis of the map is referred to as the *planning phase*. The *query phase* uses the result of the planning phase to find a path from A to B. The algorithms introduced in ▶ Sect. 5.4, the distance transform and D\*, require a significant amount of computation for the planning phase, but the query phase is very cheap. However, in both cases, the plan depends on the goal, so if the goal changes the expensive planning phase must be re-executed. D\* does allow the path to be recomputed during query, but it does not support a changed goal.

The disparity in planning and query costs has led to the development of roadmap methods where the query can include both the start and goal positions. The planning phase performs analysis of the map that supports all starting points and goals. A good analogy is making a journey by train. We first find a local path to the nearest train station, travel through the train network, get off at the station closest to our goal, and then take a local path to the goal. The train network is invariant and planning a path through the train network is straightforward using techniques like those we discussed in ▶ Sect. 5.3. Planning paths to and from the entry and exit stations respectively is also straightforward since they are likely to be short paths and we can use the occupancy-grid techniques introduced in ▶ Sect. 5.4.

The robot navigation problem then becomes one of building a network of obstacle-free paths through the environment that serve the function of the train network. In robot path planning, such a network is referred to as a *roadmap*. The roadmap need only be computed once, and can then be used like the train network to get us from any start location to any goal location.

We will illustrate the principles by creating a roadmap from the occupancy grid using some image processing techniques, as shown in ◻ Fig. 5.18. The first step is to make a copy of the occupancy grid we imported earlier

```
>>> occgrid = floorplan.copy();
```

and add an obstacle perimeter around the outer border

```
>>> occgrid[0, :] = 1
>>> occgrid[-1, :] = 1
>>> occgrid[:, 0] = 1
>>> occgrid[:, -1] = 1
```

Next, we will use the image processing capability of the Toolbox to convert the occupancy grid to an image of the freespace and display it

```
>>> freespace = Image(occgrid == 0)
Image: 596 x 397 (bool)
>>> freespace.disp();
```

as shown in ◻ Fig. 5.18a. Free space cells are depicted as white pixels which have a value of `True`, and occupied cells are depicted as black pixels and have a value of `False`. The topological skeleton of the free space is computed by a morphological image processing algorithm known as thinning ◄ applied to the free space of ◻ Fig. 5.18a

Also known as skeletonization. We will cover this topic in ▶ Sect. 11.6.3.

```
>>> skeleton = freespace.thin().disp();
```

and the result is shown in ◻ Fig. 5.18b where the black and white pixels have values of 0 and 1 respectively. We see that the free space, the white cells, have shrunk to a thin network of connected white cells that are equidistant from the boundaries of the original obstacles.

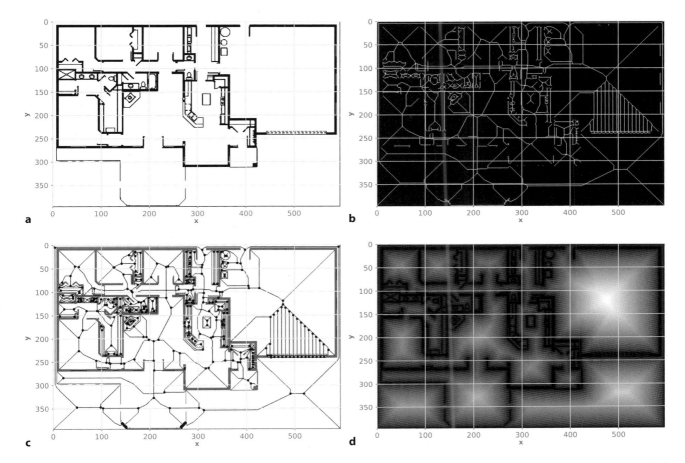

**Fig. 5.18** Steps in the creation of a Voronoi roadmap. **a** Free space is indicated by white cells; **b** the skeleton of the free space is a network of adjacent white cells no more than one cell thick; **c** the skeleton (inverted) with the obstacles overlaid in red and roadmap junction points indicated by black dots; **d** the distance transform of the obstacles, where the brightness of a point increases with its distance from the nearest obstacle

■ Fig. 5.18c shows the free-space network overlaid on the original map. We have created a network of paths that span the free space and which could be used for obstacle-free travel around the map. ▶ These paths are the edges of a generalized Voronoi diagram. We could obtain a similar result by computing the distance transform of the obstacles, ■ Fig. 5.18a, and this is shown in ■ Fig. 5.18d. The value of each pixel is the distance to the nearest obstacle and the bright ridge lines correspond to the skeleton of ■ Fig. 5.18b. Thinning or skeletonization, like the distance transform, is a computationally expensive iterative algorithm but it illustrates the principles of finding paths through free space.

The high computational cost of the distance transform and skeletonization methods makes them infeasible for large maps and has led to the development of probabilistic methods. These methods sparsely sample the map and the most well-known of these methods is the probabilistic roadmap or PRM method.

Finding the path is a two-phase process: planning, and query. We first create a PRM planner object

```
>>> prm = PRMPlanner(occgrid=floorplan, seed=0);
```

and use that to generate the plan ▶

```
>>> prm.plan(npoints=50)
```

which is a roadmap, represented internally as an embedded graph with 50 vertices. Note that the plan – the roadmap – is independent of the start and the goal.

The junctions in the roadmap are indicated by black dots. The junctions, or triple points, are found using the morphological image processing method `triplepoint`.

The argument `seed=0` sets the seed of the private random number generator, used by this planner, to zero. This ensures that the result of this command, based on random numbers, will be repeatable.

5

**Excurse 5.8: Voronoi Tessellation**

The Voronoi tessellation of a set of planar points, known as sites, is a set of polygonal Voronoi cells. Each cell corresponds to a site and consists of all points that are closer to its site than to any other site. The edges of the cells are the points that are equidistant to the two nearest sites. Using Python, we can generate a Voronoi diagram by

```
>>> sites = np.random.uniform(size=(2,9));
>>> from scipy.spatial import Voronoi
>>> vor = Voronoi(sites.T);
```

A generalized Voronoi diagram comprises cells defined by measuring distances to finite-sized objects rather than points.

Georgy Voronoi (1868–1908) was a Russian mathematician, born in what is now Ukraine. He studied at Saint Petersburg University and was a student of Andrey Markov.

One of his students Boris Delaunay defined the eponymous triangulation which has dual properties with the Voronoi diagram.

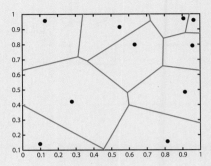

The obstacle-free path is tested by sampling at a fixed number of points along its length.

Random points are chosen, and if they are in freespace, they are added to the graph. Once $N$ vertices have been added we attempt to connect each vertex to other vertices by an obstacle-free straight-line path. ◄ The resulting graph is stored within the PRM object and a summary can be displayed by

```
>>> prm
PRMPlanner:
  BinaryOccupancyGrid: 596 x 397, cell size=1, x = [0.0, 595.0],
    y = [0.0, 396.0], 8.8% occupied
  UGraph: 50 vertices, 222 edges, 12 components
```

which indicates the number of edges and connected components in the graph. The graph can be visualized

```
>>> prm.plot();
```

as shown in ◘ Fig. 5.19a. The dots are the randomly chosen points – the vertices of the embedded graph. The lines are obstacle-free paths between pairs of points – the edges of the graph which have a cost equal to the line length. This graph has 222 edges and contains 12 components. The vertex and edge color indicates which graph component they belong to, and each component is assigned a unique color. There are two large components but several rooms are not connected to either of those, and this means that we cannot plan a path to or from those rooms. The advantage of PRM is that relatively few points need to be tested to ascertain that the points and the paths between them are obstacle free.

To improve connectivity, we can run the planner again – the random points will be different and will perhaps give better connectivity. It is common practice to iterate on the planner until a path is found, noting that there is no guarantee a valid plan will be found. We can increase the chance of success by specifying more points

```
>>> prm.plan(npoints=300)
>>> prm.plot();
>>> prm
PRMPlanner:
  BinaryOccupancyGrid: 596 x 397, cell size=1, x = [0.0, 595.0],
    y = [0.0, 396.0], 8.8% occupied
  UGraph: 300 vertices, 10000 edges, 13 components
```

which gives the result shown in ◘ Fig. 5.19b which connects all rooms in the house with a graph containing 10,000 edges.

The query phase finds a path from the start point to the goal. This is simply a matter of moving to the closest vertex in the roadmap (the start vertex), following a

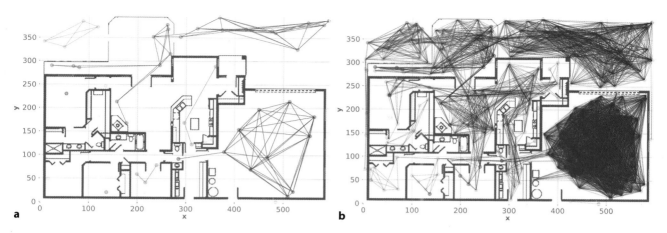

**Fig. 5.19** Probablistic roadmap (PRM) planner and the random graphs produced in the planning phase. **a** Poorly-connected roadmap graph with 50 vertices; **b** well-connected roadmap graph with 300 vertices

---

**Excurse 5.9: Random Numbers**

Creating good random numbers is a non-trivial problem. A random number generator (RNG) emits a very long, but finite length, sequence of numbers that are an excellent approximation to a random sequence – a pseudo-random number (PRN) sequence. The generator maintains an internal state which is effectively the position within the sequence. At startup NumPy obtains a random starting point in this sequence from the operating system. Every time you start a program the first random number will be different.

Many algorithms discussed in this book make use of random numbers and this means that the results can never be replicated. To solve this problem, we can *seed* the random number sequence

```
>>> np.random.rand(5)   # seeded by NumPy (not repeatable)
array([ 0.7512,   0.4447,   0.2153,   0.9116,   0.692])
>>> np.random.seed(42)
>>> np.random.rand(5)   # with seed of 42
array([ 0.3745,   0.9507,   0.732,   0.5987,   0.156])
>>> np.random.seed(42)
>>> np.random.rand(5)   # with seed of 42
array([ 0.3745,   0.9507,   0.732,   0.5987,   0.156])
```

and we see that the random sequence can be restarted.

To make examples in this book repeatable, Toolbox objects that use random numbers have their own "private" RNG and its seed can be explicitly set, typically using the `seed` argument to its constructor.

---

minimum-distance A* route through the roadmap, getting off at the vertex closest to the goal and then traveling to the goal. For the house navigation problem, this is

```
>>> path = prm.query(start=places.br3, goal=places.kitchen);
>>> prm.plot(path)
```

and the path shown in ■ Fig. 5.20 is quite efficient. Note that we provide the start and the goal position to the query phase. An advantage of this planner is that, once the roadmap is created by the planning phase, we can change the goal and starting points very cheaply, only the query phase needs to be repeated. The path

```
>>> path = prm.query(start=places.br2, goal=places.kitchen);
>>> path.shape
(8, 2)
```

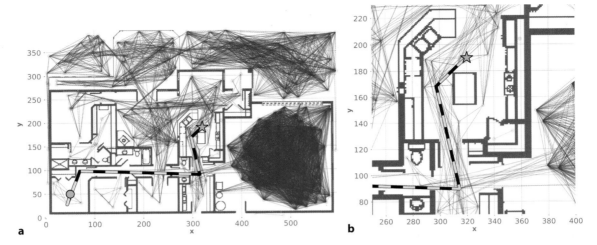

**◘ Fig. 5.20** Probablistic roadmap (PRM) planner **a** showing the path taken by the robot via vertices of the roadmap which are highlighted by a yellow and black striped line; **b** close-up view of goal region where the short path from a roadmap vertex to the goal can be seen

is a list of the vertex coordinates that the robot passes through – a list of waypoints or via points. The distance between these points is variable and there is no guarantee that the first and last path segments – connecting to the roadmap – are obstacle-free. For a robot to follow such a path, we could use the path-following controller discussed in ► Sect. 4.1.1.3.

There are some important tradeoffs in achieving this computational efficiency:

—  the random sampling of the free space means that a different roadmap is created every time the planner is run, resulting in different paths and path lengths.

—  the planner can fail by creating a network consisting of disjoint components as shown in ◘ Fig. 5.19a. If the start and goal vertices are not connected by the roadmap, that is, they are close to different components, the `query` method will report an error. The only solution is to rerun the planner and/or increase the number of vertices. We can iterate on the planner until we find a path between the start and goal.

—  long narrow gaps between obstacles, such as corridors, are unlikely to be exploited since the probability of randomly choosing points that lie along such spaces is very low.

—  the paths generated are not as smooth as those generated by the distance transform and D*, and may involve non-optimal back-tracking or very sharp turns.

The computational advantages of PRM extend into higher-dimensional planning problems such as 3D assembly or protein folding. The application to planning an obstacle-free path in vehicle configuration space will be covered in ► Sect. 5.6.5.

## 5.6 Planning Driveable Paths

The planners introduced so far can produce paths that are optimal and admissible but not necessarily *feasible*, that is, a real vehicle may not actually be able to follow the path. A path-following controller, such as described in ► Sect. 4.1.1.3, will guide a robot along the path but steering constraints mean that we cannot guarantee how closely the vehicle will follow the path. If we cannot precisely follow the collision-free path, then we cannot guarantee that the robot will not collide with an obstacle.

An alternative is to *design* a path from the outset that we know the vehicle can follow. The next two planners that we introduce take into account the motion

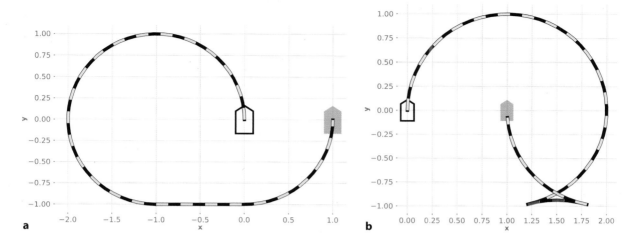

**◻ Fig. 5.21** Continuous path to achieve lateral translation of vehicle: **a** Dubins path, **b** Reeds-Shepp path. The initial configuration is shown by the outline vehicle shape, and the goal is shown by the gray vehicle shape. Forward motion is a yellow and black striped line, backward motion is a red and black striped line

model of the vehicle, and relax the assumption we have so far made that the robot is capable of omnidirectional motion.

In ▶ Chap. 4, we modeled the motion of wheeled vehicles and how their configuration evolves over time as a function of inputs such as steering and velocity. To recap, the bicycle model follows an arc of radius $L/\tan\gamma$ where $\gamma$ is the angle of the steered wheel and $L$ is the wheelbase. In this section, it is useful to consider the curvature $\kappa$ of the path driven by the robot, which is the inverse of the arc radius. For the bicycle model, we can write

$$\kappa = \frac{\tan\gamma}{L} \, . \tag{5.6}$$

When $\gamma = 0$, the curvature is zero, which is a straight line. For a real vehicle, $\gamma \in [-\gamma_m, \gamma_m]$ and the highest-curvature path the vehicle can follow is

$$\kappa_m = \frac{\tan\gamma_m}{L} \, . \tag{5.7}$$

Consider the challenging problem of moving a car-like vehicle sideways, which is similar to the parallel parking problem. The start and goal configurations are

```
>>> qs = (0, 0, pi/2);
>>> qg = (1, 0, pi/2);
```

For the bicycle motion model, the nonholonomic constraint means that the robot cannot always move directly between two configurations. It may have to follow some indirect path, which we call a maneuver. The rest of this section introduces various approaches to solving this problem.

### 5.6.1 Dubins Path Planner

A Dubins path is the shortest curve that connects two planar configurations with a constraint on the curvature of the path and assuming forward motion only. While the restriction to forward motion might seem very limiting, we should remember that many useful vehicles such as ships and aircraft have this constraint.

**5**

We can use formal concepts of continuity here. $C^0$ continuity means that the end of one segment is the same as the start of the next segment. $C^1$ continuity means that the tangents are the same at those two points. $C^2$ continuity means that curvature is the same at those two points. In this case, we have $C^0$ and $C^1$ continuity, but not $C^2$ continuity.

We create an instance of a Dubins path planner for a maximum-path curvature of one

```
>>> dubins = DubinsPlanner(curvature=1)
```

and use this to compute a Dubins path

```
>>> path, status = dubins.query(qs, qg)
>>> path.shape
(74, 3)
```

which has one row per time step and each row is a vehicle configuration $(x, y, \theta)$. We can plot just the translational part of this path

```
>>> dubins.plot(path);
```

which is shown in ◘ Fig. 5.21a. The planner returns additional status information

```
>>> status
DubinsStatus(segments=['L', 'S', 'L'], length=7.283,
seglengths=[4.712, 1.0, 1.5708])
```

which indicates this path comprises a left turn, a straight segment and another left turn. The total path length is 7.283 and the segment lengths are given by `seglengths`. They are 75% of a circle ($3\pi/2$), a straight line of length 1, and 25% of a circle ($\pi/2$). The path is sampled at discrete intervals and that interval, a distance of 0.1 in this case, is a parameter to the `DubinsPath` constructor.

We can also plot the trajectory in 3D configuration space

```
>>> dubins.plot(path, configspace=True);
```

as shown in ◘ Fig. 5.22a. The path is continuous and the slope of the tangents is continuous, but curvature is not continuous. ◄

## 5.6.2 Reeds-Shepp Path Planner

For vehicles that are capable of forward and backward motion, we can use the Reeds-Shepp planner instead

```
>>> rs = ReedsSheppPlanner(curvature=1)
>>> path, status = rs.query(qs, qg)
>>> path.shape
(66, 3)
>>> status
ReedsSheppStatus(segments=['R', 'L', 'R'], length=6.283,
seglengths=[4.460, -0.5054, 1.318],
direction=[1, 1, 1,... -1, -1, -1, -1, -1, ... 1, 1, 1])
```

The `status` object in this case indicates the path comprises a right turn, a left turn but traveling backwards as indicated by the sign of the corresponding segment length, and then another right turn traveling forwards. The direction at each point on the path is given by the `direction` attribute – these values are either 1 for forward or -1 for backward motion. We can plot the path, with color coding for direction by

```
>>> rs.plot(path, direction=status.direction);
```

which is shown in ◘ Fig. 5.21b. The vehicle drives forward around a large arc, makes a short reversing motion, and then drives forward to the goal configuration. The trajectory in 3D configuration space

```
>>> rs.plot(path, configspace=True);
```

is shown in ◘ Fig. 5.22b.

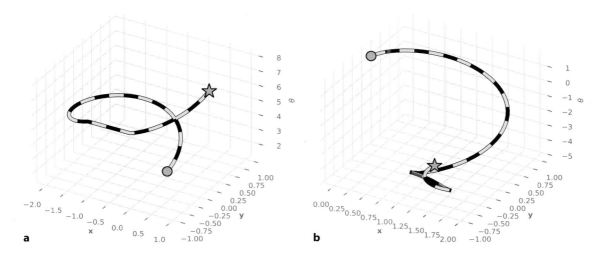

■ **Fig. 5.22** Paths in configuration space: **a** Dubins path, **b** Reeds-Shepp path. The initial configuration is indicated by a circle and the goal is indicated by a star

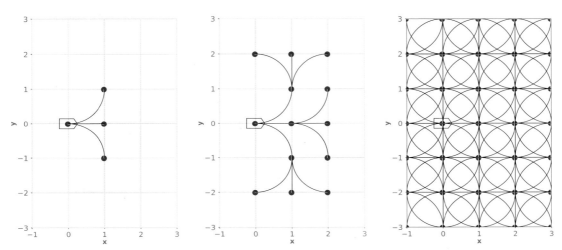

■ **Fig. 5.23** Lattice plan after 1, 2 and 8 iterations. The initial configuration is $(0, 0, 0)$

### 5.6.3 Lattice Planner

The lattice planner computes a path between discrete points in the robot's 3-dimensional configuration space. Consider ■ Fig. 5.23a that shows the robot, initially at the origin, and driving forward to the three points indicated by blue dots. ▶ Similar to the Dubins planner, each path is an arc with a curvature of $\kappa \in \{1, 0, -1\}$ and the robot's heading direction at the end of each arc is $\theta \in \{0, \frac{\pi}{2}, \pi, -\frac{\pi}{2}\}$ radians. We can create this by

```
>>> lp = LatticePlanner();
>>> lp.plan(iterations=1, summary=True)
4 vertices and 3 edges created
>>> lp.plot()
```

which is shown in ■ Fig. 5.23a.

At the end of each branch, we can add the same set of three motions suitably rotated and translated

```
>>> lp.plan(iterations=2, summary=True)
13 vertices and 12 edges created
>>> lp.plot()
```

The pitch of the grid, the horizontal and vertical distance between vertices, is dictated by the turning radius of the vehicle.

5

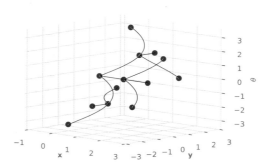

**Fig. 5.24** Lattice plan after 2 iterations shown in 3-dimensional configuration space

and this is shown in ◘ Fig. 5.23b. The graph now contains 13 vertices and represents 9 paths each 2 segments long. Each vertex represents a configuration $(x, y, \theta)$, not just a position, and we can plot the vertices and paths in configuration space

```
>>> lp.plot(configspace=True)
```

which is shown in ◘ Fig. 5.24.

By increasing the number of iterations

```
>>> lp.plan(iterations=8, summary=True)
780 vertices and 1698 edges created
>>> lp.plot()
```

we can fill in more possible paths as shown in ◘ Fig. 5.23c and the paths now extend well beyond the area shown. There are only four possible values for the heading angle $\theta \in \{0, \frac{\pi}{2}, -\pi, -\frac{\pi}{2}\}$ for each 2D grid coordinate. A left-turning arc from $\theta = \frac{\pi}{2}$ will result in a heading angle of $\theta = -\pi$.

Now that we have created the graph, we can compute a path between any two configurations using the `query` method. For the same problem as in ◘ Fig. 5.21

```
>>> path, status = lp.query(qs, qg);
>>> path.shape
(6, 3)
```

the lowest-cost path is a series of waypoints in configuration space, each row represents the configuration-space coordinates $(x, y, \theta)$ of a vertex in the lattice along the path from start to goal configuration. The path, overlaid on the lattice

```
>>> lp.plot(path)
```

is shown in ◘ Fig. 5.25a and is the same as the Dubins curve we saw in ◘ Fig. 5.21a.

Internally A* search is used to find a path through the lattice graph and the edge costs are the arc lengths. The `status` return value

```
>>> status
LatticeStatus(cost=7.283, segments=['L', 'L', 'L', 'S', 'L'],
            edges=[Edge{[0SRRRS] -- [0SLSLLR], cost=1.571},
                Edge{[0SLSLLR] -- [0LLSLS], cost=1.571},
                Edge{[0LLSLS] -- [0LLSLSL], cost=1.571},
                Edge{[0LLSLSL] -- [0LLLSL], cost=1},
                Edge{[0LLLSL] -- [0SSRRRS], cost=1.571}])
```

The vertex names in the edge list are formed from the sequence of motions required to reach that vertex from the origin, during the lattice building process.

contains the motion segments, the edge list ◄ and edge costs. The edge costs are either $\frac{\pi}{2}$ or 1, but we could apply a penalty to a particular type of edge, for example, we could increase the cost associated with turning left

```
>>> lattice = LatticePlanner(costs=[1, 10, 1])  # S, L, R
>>> lp.plan(iterations=8)
>>> path, status = lp.query(qs, qg)
```

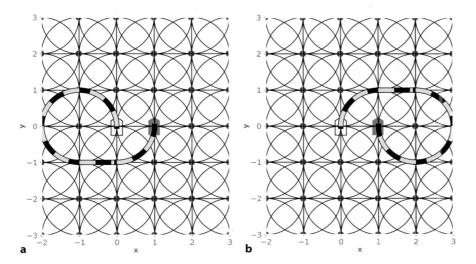

**□ Fig. 5.25** Lattice planner for lateral translation **a** With uniform-cost; **b** with increased penalty for left turns

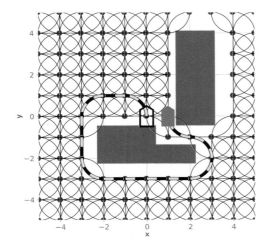

**□ Fig. 5.26** Lattice planner for lateral translation with obstacles

and the result is shown in □ Fig. 5.25b. Controlling the edge costs provides some control over the returned path compared to the Dubins planner. ▶

Another advantage of the lattice planner over the Dubins planner is that it can be integrated with a 2D occupancy grid. We first create a binary occupancy-grid object with unit-sized cells initialized to `False` (not obstacles)

```
>>> og = BinaryOccupancyGrid(workspace=[-5, 5, -5, 5], value=False)
BinaryOccupancyGrid: 11 x 11, cell size=1, x = [-5.0, 5.0],
y = [-5.0, 5.0], 0.0% occupied
```

in a workspace spanning $[-5, 5]$ in the $x$- and $y$-directions. Then we define some obstacle regions

```
>>> og.set([-2, 0, -2, -1], True)
>>> og.set([2, 3, 0, 4], True)
>>> og.set([0, 2, -2, -2], True)
```

where the first argument is an inclusive range of cells in the $x$- and $y$-directions using workspace coordinates. Then we repeat the planning process

```
>>> lattice = LatticePlanner(occgrid=og)
>>> lattice.plan(iterations=None)
```

Internally, the Dubins planner calculates multiple possible paths, but always returns the shortest. If there are several equally short paths, the returned path is arbitrarily chosen.

**5**

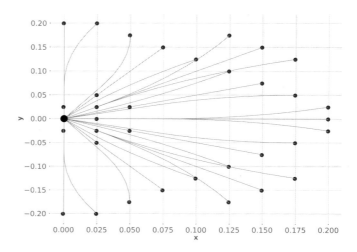

◻ **Fig. 5.27** A more sophisticated lattice generated by the package **sbpl** with 43 paths based on the kinematic model of a unicycle. Note that the paths shown here have been rotated by 90° clockwise for consistency with ◻ Fig. 5.23

```
iteration 1, frontier length 3
iteration 2, frontier length 3
iteration 3, frontier length 6
iteration 4, frontier length 15
...
iteration 23, frontier length 4
iteration 24, frontier length 2
finished after 25 iterations
```

In this case, we did not specify the number of iterations, so the planner will iterate until it can add no more vertices to the free space. We then query for a path

```
>>> path, status = lattice.query(qs, qg)
>>> lattice.plot(path)
```

and the result is shown in ◻ Fig. 5.26.

The lattice planners are similar to PRM in that a local planner is required to move from the start configuration to the nearest discrete configuration in the lattice, and from a different discrete lattice vertex to the goal. This is a very coarse lattice with just three branches at every vertex. More sophisticated lattice planners employ a richer set of motion primitives, such as shown in ◻ Fig. 5.27. In an autonomous driving scenario, this palette of curves can be used for turning corners or lane-changing. To keep the size of the graph manageable, it is typical to generate all the arcs forward from the vehicle's current configuration, for a limited number of iterations. At each iteration, curves that intersect obstacles are discarded. From this set of admissible paths, the vehicle chooses the curve that meets its current objective and takes a step along that path – the process is repeated at every control cycle.

### 5.6.4  Curvature Polynomials

The Dubins, Reeds-Shepp and lattice planners all generate paths that appear smooth, and if the path curvature is achievable by the vehicle, then it can follow the path. However, the curvature is discontinuous – at any point on the path $\kappa \in \{\kappa_m, 0, -\kappa_m\}$ and there are no intermediate values. If you were driving this path in a car, you would have to move the steering wheel instantaneously between the three angles corresponding to hard left turn, straight ahead and hard right turn.

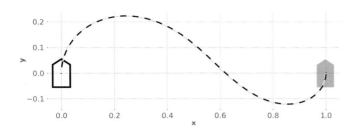

**◻ Fig. 5.28** Continuous path to achieve lateral translation using a curvature polynomial

Instantaneous motion is not possible, and the finite time required to move the steering wheel will lead to errors in path following, which raises the possibility of a collision.

Road designers have known for a long time that path curvature must change continuously with distance. In general, roads follow clothoid curves ▶ where curvature changes linearly with distance. Position, as a function of path length $s$ is given by the Fresnel integrals

Also called Euler or Cornu spirals.

$$x(s) = \int_0^s \cos\frac{at^2}{2}dt, \;\; y(s) = \int_0^s \sin\frac{at^2}{2}dt \tag{5.8}$$

which can be integrated numerically, and $a$ is the tightness of the curve.

To compute a path with continuous curvature between two configurations, we require a third-order cubic polynomial

$$\kappa(s) = \kappa_0 + as + bs^2 + cs^3, \; 0 \le s \le s_f$$

where $s_f$ is the unknown path length and $\kappa_0, a, b$ and $c$ are the unknown polynomial coefficients. Integrating curvature along the path gives the heading angle

$$\theta(s) = \theta_0 + \int_0^s \kappa(t)dt = \theta_0 + \kappa_0 s + \frac{a}{2}s^2 + \frac{b}{3}s^3 + \frac{c}{4}s^4$$

and position along the path is

$$x(s) = x_0 + \int_0^s \cos\theta(t)dt, \;\; y(s) = y_0 + \int_0^s \sin\theta(t)dt \; . \tag{5.9}$$

Given the start and goal configuration, we can solve for the unknowns $\kappa_0, a, b, c$ and $s_f$ using numerical optimization. This requires a cost function which is the distance between the final pose given by (5.9) and the desired goal pose. This is implemented by

```
>>> cpoly = CurvaturePolyPlanner()
>>> path, status = cpoly.query(qs, qg)
>>> status
CurvaturePolyStatus(length=1.299, maxcurvature=9.9549,
poly=array([ 3.059, 3.095, 2.234, -4.865]))
>>> cpoly.plot(path);
```

and the solution is shown in ◻ Fig. 5.28. Compared to the Dubins curve of ◻ Fig. 5.21, it differs in having continuous curvature and therefore it can be followed exactly by a wheeled vehicle. ◻ Fig. 5.29 compares the path curvature of the Dubins and polynomial approaches. We note, firstly, the curvature discontinuity in the Dubins curve. Secondly, the polynomial approach has a continuous curvature but has exceeded the maximum-curvature constraint. Other approaches to this problem include using Bézier splines or B-splines.

**5**

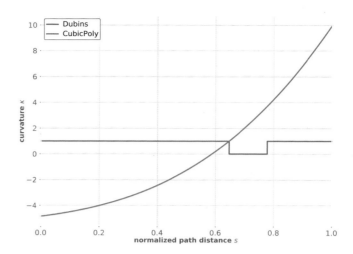

**◻ Fig. 5.29** Curvature versus distance along path for the Dubins and curvature polynomial paths

### 5.6.5 **Planning in Configuration Space**

The final planner that we introduce is the rapidly exploring random tree or RRT. The RRT is able to take into account the motion model of the vehicle, but unlike the lattice planner, which plans over a regular grid, RRT uses a probabilistic approach like the PRM planner. The problem we will consider is the classical *piano mover's problem* shown in ◻ Fig. 5.30a – we have to move the piano from a start to a goal configuration through a narrow gap.

To set up the problem, we create some polygonal obstacles, rather than an occupancy grid, since it is easier to find the intersection between a polygonal vehicle outline and polygonal obstacles. We define a map workspace spanning the range 0 to 10 in both directions and then add the obstacles expressed in terms of their

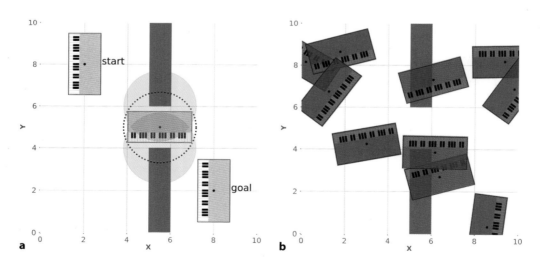

**◻ Fig. 5.30** The piano mover's problem. **a** The start and goal configurations are shown, and we can see that, if properly oriented, the piano will fit through the gap between the obstacles, shown in red. The reference point of the piano is shown by a black dot. The dashed circle is the minimal enclosing circle for the piano and the faint red discs indicate the required obstacle inflation. **b** Ten random piano configurations that are colored red if they collide with obstacles, otherwise they are green

vertices

```
>>> map = PolygonMap(workspace=[0, 10]);
>>> map.add([(5, 50), (5, 6), (6, 6), (6, 50)])
>>> map.add([(5, 4), (5, -50), (6, -50), (6, 4)])
```

Note that the obstacles extend well above and below the working area to prevent the planner from taking shortcuts. We can plot the map by

```
>>> map.plot()
```

which displays with red-colored obstacle polygons shown in ◘ Fig. 5.30a. The start and goal configurations of the piano are on either side of the gap at

```
>>> qs = (2, 8, -pi/2);
>>> qg = (8, 2, -pi/2);
```

and we can visualize these by

```
>>> piano = VehicleIcon("piano", scale=3)
>>> piano.plot(qs);
>>> piano.plot(qg);
```

When moving the piano between the obstacles its orientation is important, and we can see that it fits through the gap only if it is approximately aligned with the $x$-axis. Earlier, in ▶ Sect. 5.4.1, we handled a finite-sized robot by inflating the obstacles. The minimum enclosing circle of the robot is shown as a black dashed circle. If we attempted to inflate the obstacles by the radius of this circle, as indicated by the faint red discs, the gap would be completely closed. We need a more sophisticated approach to this problem.

A key component of the RRT algorithm is to test whether a particular vehicle configuration intersects with obstacles. The vehicle is a rectangle that is $3 \times 1.5$ which we represent by a polygon

```
>>> l, w = 3, 1.5;
>>> vpolygon = Polygon2([(-l/2, w/2), (-l/2, -w/2),
...                      (l/2, -w/2), (l/2, w/2)]);
```

defined symmetrically about the origin. For a particular configuration

```
>>> q = (2, 4, 0);
```

we can test for a collision by

```
>>> map.iscollision(vpolygon.transformed(SE2(q)))
False
```

where the `transformed` method of the `Polygon2` object returns a new `Polygon2` object whose vertices have been transformed by an **SE**(2) rigid body transformation constructed from the configuration $(x, y, \theta)$. ◘ Fig. 5.30b shows ten random configurations and their collision status.

RRT maintains a graph of robot configurations, as shown in ◘ Fig. 5.31, and each vertex is a configuration $q = (x, y, \theta) \in \mathbb{R}^2 \times \mathbf{S}^1$, which is represented by a 3-vector. The goal configuration is added as a vertex.

A random collison-free configuration $q_{new}$ is chosen, achieved by randomly choosing configurations in the workspace until one that does not intersect obstacles is found. The vertex with the closest configuration $q_{near}$ is found. ▶ Next, a path from the configuration $q_{new}$ to $q_{near}$ is computed using the Dubins path planner. If the path is collision-free ▶ then $q_{new}$ is added to the graph, and the edge to $q_{near}$ has a copy of the Dubins path. The process is repeated until the desired number of vertices has been added to the graph.

The result is a set of paths that are collision-free and driveable by this nonholonomic vehicle. ▶ For any desired starting configuration, we can find the closest configuration in the graph, and then use graph search to find a path to the goal.

The distance measure must account for a difference in position and orientation but, from a strict consideration of units, it is not proper to combine units of length (meters) and angles (radians). In some cases, it might be appropriate to weight the terms to account for the different scaling of these quantities.

A number of vehicle configurations along the path are tested for intersection with the obstacles.

As with all planners we can swap the meaning of start and goal. We can create the RRT for a particular start configuration and then query for a path to a goal.

5

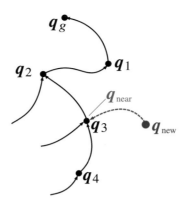

**Fig. 5.31** The RRT is a directed graph of configurations $q_i$ where $q_g$ is the goal. Edges represent admissible (obstacle free) and feasible (driveable) paths

To use RRT, we must first create a model to describe the vehicle kinematics

```
>>> vehicle = Bicycle(steer_max=1, L=2, polygon=vpolygon);
```

which uses the car-like bicycle model with a maximum steered-wheel angle of 1 rad, from which the planner can determine the maximum permissible path curvature

```
>>> vehicle.curvature_max
0.7787
```

The vehicle polygon is also passed as a parameter and is used by RRT for collision checking.

In familiar fashion, we create an RRT planner object

```
>>> rrt = RRTPlanner(map=map, vehicle=vehicle, npoints=50,
...                  showsamples=True, seed=0)
```

which has a number of methods, for example

```
>>> q = rrt.qrandom()
array([    6.37,    2.698,   -2.884])
>>> rrt.iscollision(q)
True
```

to create a random configuration in the workspace, and test for a collision at a specific vehicle configuration.

We create an RRT with its root at the goal

```
>>> rrt.plan(goal=qg, animate=True)
```

and the result is shown in ◼Fig. 5.32. To query a path to the goal from the start configuration

```
>>> path, status = rrt.query(start=qs);
```

returns a path in configuration space as well as additional status information. We can overlay the RRT graph by

```
>>> rrt.g.plot()
```

and the path through the graph can be highlighted by

```
>>> rrt.g.highlight_path(status.vertices, color="red")
```

as shown in ◼Fig. 5.32a. The path to the right of the gap has a loop which is inefficient – rerunning the RRT planner *may* lead to a better solution.

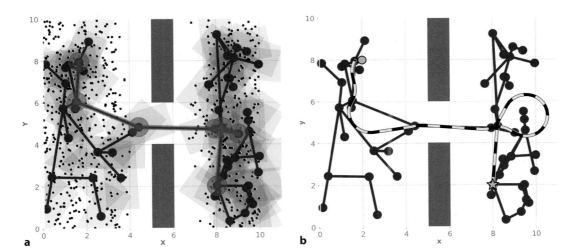

**⬛ Fig. 5.32  a** The RRT vertices and edges are shown in dark blue. The black dots are the random samples in free space that were not added to the graph. The faint blue rectangles are collision-free piano configurations. The path through the graph is highlighted in pink. **b** The path taken by the vehicle is the concatenation of a number of Dubins curves

The path is the concatenation of a series of Dubins paths between configurations in the graph and we can plot its projection in the plane

```
>>> rrt.plot(path);
```

which is shown in ⬛ Fig. 5.32b.

This example illustrates some important points about RRTs. Firstly, as for the PRM planner, there may be some distance (and orientation) between the start configuration and the nearest vertex in the RRT, in this case

```
>>> status.initial_d
0.5996
```

Minimizing this requires tuning RRT parameters, such as the number of vertices. Secondly, the path is feasible but not optimal. In this case, the vehicle has made a large and unnecessary loop on the right-hand side of the gap. Since the planner is probabilistic, rerunning it or increasing the number of vertices may help. Techniques have been proposed to optimize the paths generated by RRT while still retaining the curvature limits and collision avoidance. There are many variants of RRT – we have covered only the most basic to illustrate the principle.

## 5.7  Advanced Topics

### 5.7.1  A* vs Dijkstra Search

A very famous approach to graph search is Dijkstra's *shortest-path algorithm* from 1956. It is framed in terms of the shortest path from a source vertex to all other vertices in the graph and produces a shortest-path tree. The source vertex could be the start or the goal. The algorithm is similar to UCS in that it visits every vertex and computes the distance of every vertex from the start. However, UCS does require a start and a goal, whereas Dijkstra's search requires only a start (or a goal). Dijkstra's algorithm maintains two sets of vertices, visited and unvisited, which are analogous to UCS's explored and frontier sets – Dijkstra's unvisited set is initialized with all vertices, whereas the UCS frontier is initialized with only the start vertex. The computed cost expands outwards from the starting vertex and is analogous to the distance transform or wavefront planner.

5

> **Excurse 5.10: Edsger Dijkstra**
>
> Dijkstra (1930-2002) was a pioneering Dutch computer scientist and winner of the 1972 Turing award. A theoretical physicist by training, he worked in industry and academia at the time when computing was transitioning from a craft to a discipline, there were no theoretical underpinnings, and few computer languages. He was an early advocate of new programming styles to improve the quality of programs, coined the phrase "structured programming" and developed languages and compilers to support this. In the 1960s, he worked in concurrent and distributed computing on topics such as mutual exclusion, semaphores, and deadlock. He also developed an algorithm for finding shortest paths in graphs. Many of his contributions to computer science are critical to the implementation of robotic systems today.

## 5.7.2  Converting Grid Maps to Graphs

We have treated grids and graphs as distinct representations, but we can easily convert an occupancy grid to a graph. We place a vertex in every grid cell and then add edges to every neighboring cell that is not occupied. If the vehicle can drive to all neighbors, then a vertex can have at most eight neighbors. If it is a Manhattan-world problem, then a vertex can have at most four neighbors.

## 5.7.3  Converting Between Graphs and Matrices

A graph with $n$ vertices can be represented as an $n \times n$ distance matrix $\mathbf{D}$. For example

```
>>> import pgraph
>>> g = pgraph.UGraph()
>>> for i in range(4):  # add 4 vertices
...    g.add_vertex()
>>> g[0].connect(g[1], cost=1);   # 0 -- 1
>>> g[0].connect(g[3], cost=2);   # 0 -- 3
>>> g[1].connect(g[2], cost=3);   # 1 -- 2
>>> g.distance()
array([[     0,        1,        0,        2],
       [     1,        0,        3,        0],
       [     0,        3,        0,        0],
       [     2,        0,        0,        0]])
```

The elements of the matrix $d_{i,j}$ are the edge cost from vertex $i$ to vertex $j$ or 0 if vertices $i$ and $j$ are not connected. For an undirected graph, as shown here, the matrix is symmetric.

## 5.7.4  Local and Global Planning

The global planners we have discussed are able to produce optimal paths because they have full knowledge of the state of the world, but in reality this is not sufficient. When we drive a car from A to B, we can compute a globally optimal path through the road network, but that is critically reliant on the accuracy and currency of the map – in reality a map is never going to reflect all the road closures and road works, let alone the cars, bicycles and pedestrians that we will encounter. When we drive, we make numerous small adjustments to accommodate those elements of the environment that are unmappable and for this we use our sensory system

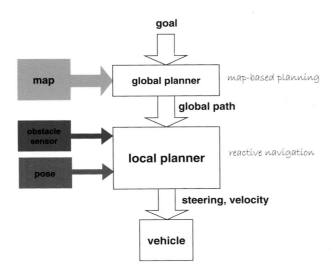

**◻ Fig. 5.33** Global and local path planning architecture

– most importantly, our eyes. In practice, we use both map-based planning and reactive strategies.

This insight is reflected in the common robot planning architecture shown in ◻ Fig. 5.33. The task of the local planner is to keep the vehicle as close as possible to the global optimal path, while accommodating a priori unknowable local conditions. Local path planning is based on sensory input and is particularly challenging since sensors have limited range and therefore global information is not available. For example, in the process of avoiding one local obstacle, we might turn left to avoid it, and encounter another obstacle, which we would not have encountered had we turned right. This was the challenge we encountered with the *bug2* planner in ▶ Sect. 5.1.2.

## 5.8　Wrapping Up

Robot navigation is the problem of guiding a robot towards a goal and we have covered a spectrum of approaches. The simplest was the purely reactive Braitenberg-type vehicle. Then we added limited memory to create state machine based automata such as *bug2*, which can deal with obstacles, however, the paths that it finds can be far from optimal.

A number of different map-based planning algorithms were then introduced. We started with graph-based maps and covered a number of algorithms for finding optimal paths through a road network, minimizing distance or travel time. For grid-based maps, we started with the distance transform to find an optimal path to the goal. D* finds a similar path but allows grid cells to have a continuous traversability measure rather than being considered as only free space or obstacle. D* also supports computationally cheap incremental replanning for small changes in the map. PRM reduces the computational burden by probabilistic sampling but at the expense of somewhat less optimal paths and no guarantee of finding a path even if one exists. To accommodate the motion constraints of real vehicles, we introduced the Dubins and Reeds-Shepp planners that find driveable paths comprising arcs and straight lines, and a lattice planner that also supports planning around obstacles. These paths, while technically driveable, are discontinuous in curvature and curvature polynomials were introduced to overcome that limitation. Finally, we looked at RRT for planning a path through configuration space in the presence of obstacles and with vehicle motion constraints. All the map-based approaches

5

require a map and knowledge of the robot's location, and these are both topics that we will cover in the next chapter.

### 5.8.1  **Further Reading**

Comprehensive coverage of planning for robots is provided by two text books. Choset et al. (2005) covers geometric and probabilistic approaches to planning, as well as the application to robots with dynamic and nonholonomic constraints. LaValle (2006) covers motion planning, planning under uncertainty, sensor-based planning, reinforcement learning, nonlinear systems, trajectory planning, nonholonomic planning, and is available online for free at ▶ http://lavalle.pl/planning. In particular, these books provide a much more sophisticated approach to representing obstacles in configuration space and cover potential-field planning methods that we have not discussed. The powerful planning techniques discussed in these books can be applied beyond robotics to very high-order systems such as vehicles with trailers, robotic arms or even the shape of molecules. LaValle (2011a) and LaValle (2011b) provide a concise two-part tutorial introduction. More succinct coverage of planning is given by Kelly (2013), Siegwart et al. (2011), the Robotics Handbook (Siciliano and Khatib 2016, § 7), and also by Spong et al. (2006) and Siciliano et al. (2009).

The *bug1* and *bug2* algorithms were described by Lumelsky and Stepanov (1986), and Ng and Bräunl (2007) implemented and compared eleven variations of the Bug algorithm in a number of different environments. Graph-based planning is well-covered by Russell and Norvig (2020) and the road network example in ▶ Sect. 5.3 was inspired by their use of a map of Romania. The distance transform is well-described by Borgefors (1986) and its early application to robotic navigation was explored by Jarvis and Byrne (1988). Efficient approaches to implementing the distance transform include the two-pass method of Hirata (1996), fast marching methods, or reframing it as a graph search problem that can be solved using Dijkstra's method; the last two approaches are compared by Alton and Mitchell (2006). The algorithm for finding shortest paths within a graph was by Dijkstra (1959), and the later A* algorithm (Hart et al. 1968) improves the search efficiency. Any occupancy-grid map can be converted to a graph which is the approach used in the D* algorithm by Stentz (1994) which allows cheap replanning when the map changes. There have been many further extensions including, but not limited to, Field D* (Ferguson and Stentz 2006) and D* lite (Koenig and Likhachev 2005). D* is used in many real-world robot systems and many implementations exist, including open source.

The ideas behind PRM started to emerge in the mid 1990s and it was first described by Kavraki et al. (1996). Geraerts and Overmars (2004) compare the efficacy of a number of subsequent variations that have been proposed to the basic PRM algorithm. Approaches to planning that incorporate the vehicle's dynamics include state-space sampling (Howard et al. 2008), and the RRT, which is described in LaValle (1998, 2001). More recently, RRT* has been proposed by Karaman et al. (2011).

The approaches to connecting two poses by means of arcs and line segments were introduced by Dubins (1957) and Reeds and Shepp (1990). Owen et al. (2015) extend the Dubins algorithm to 3D which is useful for aircraft applications. Lattice planners are covered in Pivtoraiko et al. (2009). A good introduction to path curvature, clothoids and curvature polynomials is given by Kelly (2013), and a comprehensive treatment is given by Bertolazzi and Frego (2014).

■ ■ Historical and Interesting

The defining book in cybernetics was written by Wiener in 1948 and updated in 1965 (Wiener 1965). Walter published a number of popular articles (1950, 1951) and a book (1953) based on his theories and experiments with robotic tortoises, while Holland (2003) describes Walter's contributions.

The definitive reference for Braitenberg vehicles is Braitenberg's own book (1986) which is a whimsical, almost poetic, set of thought experiments. Vehicles of increasing complexity (fourteen vehicle families in all) are developed, some including nonlinearities, memory and logic to which he attributes anthropomorphic characteristics such as love, fear, aggression and egotism. The second part of the book outlines the factual basis of these machines in the neural structure of animals.

Early behavior-based robots included the Johns Hopkins Beast, built in the 1960s, and Genghis (Brooks 1989) built in 1989. Behavior-based robotics are described in the early paper by Brooks (1986), the book by Arkin (1999), and the Robotics Handbook (Siciliano and Khatib 2016, § 13). Matarić's Robotics Primer (Matarić 2007) and associated comprehensive web-based resources are also an excellent introduction to reactive control, behavior-based control and robot navigation. A rich collection of archival material about early cybernetic machines, including Walter's tortoise and the Johns Hopkins Beast can be found at the Cybernetic Zoo ▶ http://cyberneticzoo.com.

### 5.8.2 Resources

Python Robotics (▶ https://github.com/AtsushiSakai/PythonRobotics) is a comprehensive and growing collection of algorithms in Python for robotic path planning, as well as localization, mapping and SLAM. OMPL, the Open MotionPlanning Library (▶ https://ompl.kavrakilab.org) written in C++, is a powerful and widely used set of motion planners. It has a Python-based app that provides a convenient means to explore planning problems. Steve LaValle's website ▶ http://lavalle.pl/software.html has many code resources (C++ and Python) related to motion planning. Lattice planners are included in the sbpl package from the Search-Based Planning Lab (▶ https://github.com/sbpl/sbpl) which has MATLAB tools for generating motion primitives and C++ code for planning over the lattice graphs, and pysbpl (▶ https://github.com/poine/pysbpl) provides Python 3 bindings.

### 5.8.3 Exercises

1. Braitenberg vehicles (▶ Sect. 5.1.1)
   a) Experiment with different starting configurations and control gains.
   b) Modify the signs on the steering signal to make the vehicle light-phobic.
   c) Modify the `sensorfield` function so that the peak moves with time.
   d) The vehicle approaches the maxima asymptotically. Add a stopping rule so that the vehicle stops when either sensor detects a value greater than 0.95.
   e) Create a scalar field with two peaks. Can you create a starting pose where the robot gets confused?
2. Occupancy grids. Create some different occupancy grids and test them on the different planners discussed.
   a) Create an occupancy grid that contains a maze and test it with various planners. See ▶ http://rosettacode.org/wiki/Maze_generation.
   b) Create an occupancy grid from a downloaded floor plan.
   c) Create an occupancy grid from a city street map (▶ Sect. 11.1.6), perhaps apply color segmentation (▶ Sect. 12.1.1.2) to segment roads from other features. Can you convert this to a cost map for D* where different roads or intersections have different costs?
   d) Experiment with obstacle inflation.

5

e) At 1 m cell size, how much memory is required to represent the surface of the Earth? How much memory is required to represent just the land area of Earth? What cell size is needed in order for a map of your country to fit in 1 Gbyte of memory?

3. Bug algorithms (▶ Sect. 5.1.2)
   a) Create a map to challenge *bug2*. Try different starting points.
   b) Create an obstacle map that contains a maze. Can *bug2* solve the maze?
   c) Experiment with different start and goal locations.
   d) Create a bug trap. Make a hollow box, and start the bug inside a box with the goal outside. What happens?
   e) Modify *bug2* to change the direction it turns when it hits an obstacle.
   f) Implement other bug algorithms such as *bug1* and *tangent bug*. Do they perform better or worse?

4. Distance transform (▶ Sect. 5.4.1)
   a) Create an obstacle map that contains a maze and solve it using distance transform.

5. D* planner (▶ Sect. 5.4.2)
   a) Add a low-cost region to the living room. Can you make the robot prefer to take this route to the kitchen?
   b) Block additional doorways to challenge the robot.

6. PRM planner (▶ Sect. 5.5)
   a) Run the PRM planner 100 times and gather statistics on the resulting path length.
   b) Vary the value of the distance threshold parameter and observe the effect.
   c) Use the output of the PRM planner as input to a pure pursuit planner as discussed in ▶ Chap. 4.

7. Dubins and Reeds-Shepp planners (▶ Sect. 5.6.1 and ▶ Sect. 5.6.2)
   a) Find a path to implement a 3-point turn.
   b) Investigate the effect of changing the maximum curvature.

8. Lattice planner (▶ Sect. 5.6.3)
   a) How many iterations are required to completely fill the region of interest shown in ◘ Fig. 5.23c?
   b) How does the number of vertices and the spatial extent of the lattice increase with the number of iterations?
   c) Given a car with a wheelbase of 4.5 m and maximum steered-wheel angles of $\pm 30°$, what is the lattice grid size?
   d) Compute and plot curvature as a function of path length for a path through the lattice such as the one shown in ◘ Fig. 5.25a.
   e) Design a controller that will take a unicycle or bicycle model with a finite steered-wheel angle rate (there is a parameter to specify this) that will drive the vehicle along the paths shown in ◘ Fig. 5.25.

9. Curvature polynomial (▶ Sect. 5.6.4)
   a) Investigate Bézier, Hermite and B-splines to create smooth paths between positions or configurations.
   b) Modify the curvature polynomial planner to have an upper bound on curvature (hard).

10. RRT planner (▶ Sect. 5.6.5)
   a) Experiment with RRT parameters such as the number of points and the vehicle steered-wheel angle limits.
   b) Compute and plot the steered-wheel angle required to move from start to goal pose.
   c) Add a local planner to move from initial pose to the closest vertex, and from the final vertex to the goal pose.
   d) Determine a path through the graph that minimizes the number of reversals of direction.

# Localization and Mapping

» *In order to get somewhere, we need to know where we are*

## Contents

© The Author(s), under exclusive license to Springer Nature Switzerland AG 2023
P. Corke, *Robotics, Vision and Control*, Springer Tracts in Advanced Robotics 146,
https://doi.org/10.1007/978-3-031-06469-2_6

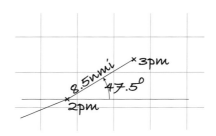

■ **Fig. 6.1** Position estimation by dead reckoning. The ship's position at 3 P.M. is based on its position at 2 P.M., the estimated distance traveled since, and the average compass heading

**6**

`chap6.ipynb`

▶ go.sn.pub/DORGc6

In our discussion of map-based navigation so far, we assumed that the robot had a means of knowing its position. In this chapter we discuss some of the common techniques used to estimate the pose of a robot in the world – a process known as localization.

Today GPS makes outdoor localization so easy that we often take this capability for granted. Unfortunately, GPS isn't the perfect sensor many people imagine it to be. It relies on very weak radio signals received from distant satellites, which means that GPS fails or has error where there is no *line of sight* radio reception, for instance indoors, underwater, underground, in urban canyons or in deep mining pits. GPS signals can also be jammed and for some applications that is not acceptable.

GPS has only been in use since 1995 yet humankind has been navigating the planet and localizing for thousands of years. In this chapter we will introduce the *classical* navigation principles such as dead reckoning and the use of landmarks on which modern robotic navigation is founded.

Dead reckoning is the estimation of position based on estimated speed, direction and time of travel with respect to a previous estimate. ■ Fig. 6.1 shows how a ship's position is updated on a chart. Given the average compass heading over the previous hour and the distance traveled, the position at 3 P.M. can be found using elementary geometry from the position at 2 P.M. However the measurements on which the update is based are subject to both systematic and random error. Modern instruments are quite precise but 500 years ago clocks, compasses and speed measurement were primitive. The recursive nature of the process, each estimate is based on the previous one, means that errors will accumulate over time and for sea voyages of many-years this approach was quite inadequate.

**Excurse 6.1: Measuring Speed at Sea**

A ship's log is an instrument that provides an estimate of the distance traveled. The oldest method of determining the speed of a ship at sea was the Dutchman's log – a floating object was thrown into the water at the ship's bow and the time for it to pass the stern was measured using an hourglass. Later came the chip log, a flat quarter-circle of wood with a lead weight on the circular side causing it to float upright and resist towing. It was tossed overboard and a line with knots at 50 foot intervals was payed out. A special hourglass, called a log glass, ran for 30 s, and each knot on the line over that interval corresponds to approximately 1 nmi h$^{-1}$ or 1 knot. A nautical mile (nmi) is now defined as 1.852 km. (Image modified from Text-Book of Seamanship, Commodore S. B. Luce 1891)

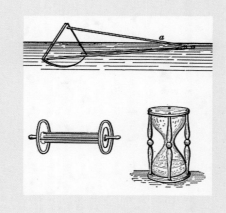

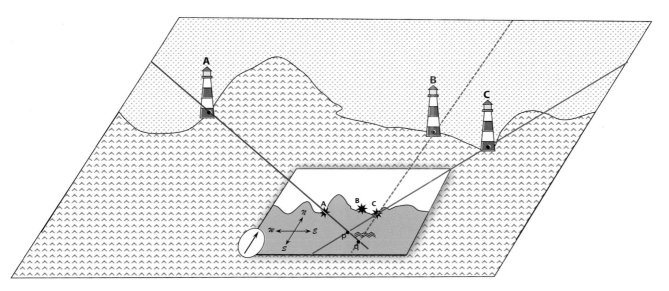

◻ **Fig. 6.2** Position estimation using a map. Lines of sight from two light-houses, **A** and **C**, and their corresponding positions on the map provide an estimate *p* of position. If we mistake lighthouse **B** for **C** then we obtain an incorrect estimate *q*

The Phoenicians were navigating at sea more than 4000 years ago and they did not even have a compass – that was developed 2000 years later in China. The Phoenicians navigated with crude dead reckoning but wherever possible they used *additional information* to correct their position estimate – sightings of islands and headlands, primitive maps and observations of the Sun and the Pole Star.

A landmark is a visible feature in the environment whose position is known with respect to some coordinate frame. ◻ Fig. 6.2 shows schematically a map and a number of lighthouse landmarks. We first of all use a compass to align the north axis of our map with the direction of the north pole. The direction of a single landmark constrains our position to lie along a line on the map. Sighting a second landmark places our position on another constraint line, and our position must be at their intersection – a process known as resectioning. ▶ For example, lighthouse **A** constrains us to lie along the blue line while lighthouse **C** constrains us to lie along the red line – our true position **p** is at the intersection of these two lines.

However, this process is critically reliant on correctly associating the observed landmark with the corresponding feature on the map. If we mistake one lighthouse for another, for example we see **B** but think it is **C** on the map, then the red-dashed line leads to a significant error in estimated position – we would believe we were at **q** instead of **p**. This belief would lead us to overestimate our distance from the coastline. If we decided to sail toward the coast we would run aground on rocks and be surprised since they were not where we expected them to be. This is unfortunately a very common error and countless ships have foundered because of this fundamental data association error. This is why lighthouses flash! In the eighteenth century technological advances enabled lighthouses to emit unique flashing patterns so that the identity of the particular lighthouse could be reliably determined and associated with a point on a navigation chart.

The earliest mariners had no maps, or lighthouses or even compasses. They had to create maps as they navigated by incrementally adding new landmarks to their maps just beyond the boundaries of what was already known. It is perhaps not surprising that so many early explorers came to grief ▶ and that maps were tightly-kept state secrets.

Robots operating today in environments without GPS face *exactly* the same problems as the ancient navigators and, perhaps surprisingly, borrow heavily from navigational strategies that are centuries old. A robot's estimate of distance traveled

Resectioning is the estimation of position by measuring the bearing angles to known landmarks. Triangulation is the estimation of position by measuring the bearing angles to an unknown point from each of the landmarks.

Magellan's 1519 expedition started with 237 men and 5 ships but most, including Magellan, were lost along the way. Only 18 men and 1 ship returned.

**6**

### Excurse 6.2: Celestial Navigation

The position of celestial bodies in the sky is a predictable function of the time and the observer's latitude and longitude. This information can be tabulated and is known as ephemeris (meaning daily) and such data has been published annually in Britain since 1767 as the *The Nautical Almanac* by HM Nautical Almanac Office. The elevation of a celestial body with respect to the horizon can be measured using a sextant, a handheld optical instrument.

Time and longitude are coupled, the star field one hour later is the same as the star field 15° to the east. However the northern Pole Star, *Polaris* or the *North Star*, is very close to the celestial pole and its elevation angle is independent of longitude and time, allowing latitude to be determined very conveniently from a single sextant measurememt.

Solving the longitude problem was the greatest scientific challenge to European governments in the eighteenth century since it was a significant impediment to global navigation and maritime supremacy. The British Longitude Act of 1714 created a prize of £20,000 which spurred the development of nautical chronometers, clocks that could maintain high accuracy onboard ships. More than fifty years later John Harrison developed a suitable chronometer, a copy of which was used by Captain James Cook on his second voyage of 1772–1775. After a three year journey the error in estimated longitude was just 13 km. With accurate knowledge of time, the elevation angle of stars could be used to estimate latitude and longitude. This technological advance enabled global exploration and trade. (Image courtesy archive.org)

THE

NAUTICAL ALMANAC

AND

ASTRONOMICAL EPHEMERIS,

FOR THE YEAR 1781.

Publifhed by Order of the

COMMISSIONERS OF LONGITUDE.

LONDON

PRINTED BY WILLIAM RICHARDSON, PRINTER;
AND SOLD BY
J. NOURSE, in the Strand, and Meff. MOUNT and PAGE on Tower-Hill,
Bookfellers to the faid COMMISSIONERS.
M DCC LXXIX.
[Price Three Shillings and Six Pence.]

will be imperfect and it may have no map, or perhaps an imperfect or incomplete map. Additional information from observed features in the world is critical to minimizing a robot's localization error, but the possibility of data association errors remains.

We can define the localization problem formally: $x$ is the true, but unknown, position or pose of the robot and $\hat{x}$ is our best estimate of that. We also wish to know the *uncertainty* of the estimate that we can consider in statistical terms as the standard deviation associated with the estimate $\hat{x}$.

It is useful to describe the robot's estimated position in terms of a probability density function (PDF) over all possible positions of the robot. Some example PDFs are shown in ◘ Fig. 6.3 where the magnitude of the function $f(x, y)$ at any point $(x, y)$ is the relative likelihood of the robot being at that position. ◄ A Gaussian function is commonly used and can be described succinctly in terms of intuitive concepts such as mean and standard deviation. The robot is most likely to be at the position of the peak (the mean) and increasingly less likely to be at positions further away from the peak. ◘ Fig. 6.3a shows a peak with a small standard deviation that indicates that the robot's position is very well known. There is an almost zero probability that the robot is at the point indicated by the vertical black line. In contrast, the peak in ◘ Fig. 6.3b has a large standard deviation that means that we are less certain about the position of the robot. There is a reasonable probability that the robot is at the point indicated by the vertical line. Using a PDF also allows for multiple hypotheses about the robot's position. For example the PDF of ◘ Fig. 6.3c describes a robot that is quite certain that it is at one of four places. This is more useful than it seems at face value. Consider an indoor robot that has observed a door and there are four doors on the map. In the absence of any other

By extension, for robot pose rather than position, the PDF $f(x, y, \theta)$ will be a 4-dimensional surface. The volume under the surface is always 1.

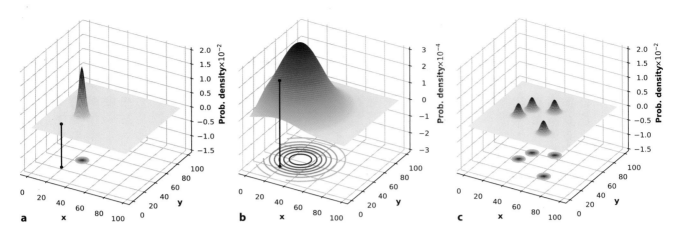

◻ **Fig. 6.3** Notions of robot position and uncertainty in the $xy$-plane, $\theta$ is not considered. The vertical axis is the relative likelihood of the robot being at that position, sometimes referred to as belief or bel$(x, y)$. Contour lines are displayed on the lower plane. **a** The robot has low position uncertainty, $\sigma = 1$; **b** the robot has much higher position uncertainty, $\sigma = 20$; **c** the robot has multiple hypotheses for its position, each $\sigma = 1$

---

### Excurse 6.3: Radio-Based Localization

One of the earliest systems was LORAN, based on the British World War II GEE system. A network of transmitters around the world emitted synchronized radio pulses and a receiver measured the difference in arrival time between pulses from a pair of radio transmitters. Knowing the identity of two transmitters and the time difference (TD) constrains the receiver to lie along a hyperbolic curve shown on navigation charts as *TD lines*. Using a second pair of transmitters (which may include one of the first pair) gives another hyperbolic constraint curve, and the receiver must lie at the intersection of the two curves.

The Global Positioning System (GPS) was proposed in 1973 but did not become fully operational until 1995. It comprises around 30 active satellites orbiting the Earth in six planes at a distance of 20,200 km. A GPS receiver measures the time of travel of radio signals from four or more satellites whose orbital position is encoded in the GPS signal. With four known points in space and four measured time delays it is possible to compute the $(x, y, z)$ position of the receiver and the time. If the GPS signals are received after reflecting off some surface, the distance traveled is longer and this will introduce an error in the position estimate. This effect is known as multi-pathing and is a common problem in large-scale industrial facilities.

Variations in the propagation speed of radio waves through the atmosphere is the main cause of error in the position estimate. However, these errors vary slowly with time and are approximately constant over large areas. This allows the error to be measured at a reference station and transmitted as an augmentation to compatible nearby receivers that can offset the error – this is known as Differential GPS (DGPS). This information can be transmitted via the internet, via coastal radio networks to ships, or by satellite networks such as WAAS, EGNOS or OmniSTAR to aircraft or other users. RTK GPS achieves much higher precision in time measurement by using phase information from the carrier signal. The original GPS system deliberately added error, euphemistically termed selective availability, to reduce its utility to military opponents but this *feature* was disabled in May 2000. Other satellite navigation systems include the European Galileo, the Chinese Beidou, and the Russian GLONASS. The general term for these different systems is Global Navigation Satellite System (GNSS). GNSS made LORAN obsolete and much of the global network has been decommissioned. However recently, concern over the dependence on, and vulnerability of, GNSS had led to work on enhanced LORAN (eLORAN).

---

information the robot is equally likely to be in the vicinity of *any* of the four doors. We will revisit this approach in ▶ Sect. 6.6.

Determining the PDF based on knowledge of how the robot moves and its observations of the world is a problem in *estimation*, which we can define as:

》 *the process of inferring the value of some quantity of interest, $x$, by processing data that is in some way dependent on $x$.*

For example, a ship's navigator or a surveyor estimates position by measuring the bearing angles to known landmarks or celestial objects, and a GPS receiver estimates latitude and longitude by observing the time delay from moving satellites whose positions are known.

For our robot localization problem, the true and estimated state are vector quantities. Uncertainty is represented by a covariance matrix whose diagonal elements are the variance of the corresponding states, and the off-diagonal elements are correlations between states which indicate how uncertainties in the states are connected.

> ❗ **Probability of being at a pose**
>
> The value of a PDF at a point is *not* the probability of being at that position. Instead, consider a small region of the *xy*-plane, the volume under that region of the PDF is the probability of being in that region.

## 6.1 Dead Reckoning Using Odometry

Dead reckoning is the estimation of a robot's pose based on its estimated speed, direction and time of travel with respect to a previous estimate.

An odometer is a sensor that measures distance traveled. For wheeled vehicles this is commonly achieved by measuring the angular rotation of the wheels using a sensor like an encoder. The direction of travel can be measured using an electronic compass, or the change in direction can be measured using a gyroscope or the difference in angular velocity of a left- and right-hand side wheel. These sensors are imperfect due to systematic errors such an incorrect wheel radius or gyroscope bias, and random errors such as slip between wheels and the ground. In robotics, the term odometry generally means measuring distance and direction of travel.

For robots without wheels, such as aerial and underwater robots, there are alternatives. Inertial navigation, introduced in ▶ Sect. 3.4, uses accelerometers and gyroscopes to estimate change in pose. Visual odometry is a computer-vision approach based on observations of the world moving past the robot, and this is covered in ▶ Sect. 14.8.3. Laser odometry is based on changes in observed point clouds measured with lidar sensors, and is covered in ▶ Sect. 6.8.1.

### 6.1.1 Modeling the Robot

The first step in estimating the robot's pose is to write a function, $f(\cdot)$, that describes how the robot's configuration changes from one time step to the next. A vehicle model such as (4.4) or (4.9) describes the progression of the robot's configuration as a function of its control inputs, however for real robots we rarely have direct access to these control inputs. Most robotic platforms have proprietary motion control systems that accept motion commands from the user such as desired position, and report odometry information.

Instead of using (4.4) or (4.9) directly we will write a discrete-time model for the change of configuration based on odometry where $\boldsymbol{\delta}_{\langle k \rangle} = (\delta_d, \delta_\theta)^\top$ is the distance traveled and change in heading over the preceding interval, and $k$ is the time step. The initial pose is represented by an **SE**(2) matrix

$$
\mathbf{T}_{\langle k \rangle} = \begin{pmatrix} \cos \theta_{\langle k \rangle} & -\sin \theta_{\langle k \rangle} & x_{\langle k \rangle} \\ \sin \theta_{\langle k \rangle} & \cos \theta_{\langle k \rangle} & y_{\langle k \rangle} \\ 0 & 0 & 1 \end{pmatrix} .
$$

We make a simplifying assumption that motion over the time interval is *small* so the order of applying the displacements is not significant. We choose to move forward in the robot's body frame $x$-direction, see ◻ Fig. 4.4, by $\delta_d$, and then rotate by $\delta_\theta$ giving the new pose

$$
\begin{aligned}
\mathbf{T}_{\langle k+1 \rangle} &= \begin{pmatrix} \cos\theta_{\langle k \rangle} & -\sin\theta_{\langle k \rangle} & x_{\langle k \rangle} \\ \sin\theta_{\langle k \rangle} & \cos\theta_{\langle k \rangle} & y_{\langle k \rangle} \\ 0 & 0 & 1 \end{pmatrix} \begin{pmatrix} 1 & 0 & \delta_d \\ 0 & 1 & 0 \\ 0 & 0 & 1 \end{pmatrix} \begin{pmatrix} \cos\delta_\theta & -\sin\delta_\theta & 0 \\ \sin\delta_\theta & \cos\delta_\theta & 0 \\ 0 & 0 & 1 \end{pmatrix} \\
&= \begin{pmatrix} \cos(\theta_{\langle k \rangle}+\delta_\theta) & -\sin(\theta_{\langle k \rangle}+\delta_\theta) & x_{\langle k \rangle}+\delta_d\cos\theta_{\langle k \rangle} \\ \sin(\theta_{\langle k \rangle}+\delta_\theta) & \cos(\theta_{\langle k \rangle}+\delta_\theta) & y_{\langle k \rangle}+\delta_d\sin\theta_{\langle k \rangle} \\ 0 & 0 & 1 \end{pmatrix}
\end{aligned}
$$

which we can represent concisely as a configuration vector $\boldsymbol{q} = (x, y, \theta)$

$$
q_{\langle k+1 \rangle} = [\mathbf{T}_{\langle k+1 \rangle}]_{xy\theta} = \begin{pmatrix} x_{\langle k \rangle}+\delta_d\cos\theta_{\langle k \rangle} \\ y_{\langle k \rangle}+\delta_d\sin\theta_{\langle k \rangle} \\ \theta_{\langle k \rangle}+\delta_\theta \end{pmatrix} \tag{6.1}
$$

which gives the new configuration in terms of the previous configuration and the odometry.

In practice, odometry is not perfect and we model the error by imagining a random number generator that corrupts the output of a perfect odometer. The measured output of the real odometer is the perfect, but unknown, odometry $(\delta_d, \delta_\theta)$ plus the output of the random number generator $(v_d, v_\theta)$. Such random errors are often referred to as noise, or more specifically as sensor noise. ▶ The random numbers are not known and cannot be measured, but we assume that we know the distribution from which they are drawn.

We assume that the noise is additive. Another option is multiplicative noise which is appropriate if the noise magnitude is related to the signal magnitude.

The robot's configuration at the next time step, including the odometry error, is

$$
q_{\langle k+1 \rangle} = \boldsymbol{f}(\boldsymbol{q}_{\langle k \rangle}, \boldsymbol{\delta}_{\langle k \rangle}, \boldsymbol{v}_{\langle k \rangle}) = \begin{pmatrix} x_{\langle k \rangle}+(\delta_d+v_d)\cos\theta_{\langle k \rangle} \\ y_{\langle k \rangle}+(\delta_d+v_d)\sin\theta_{\langle k \rangle} \\ \theta_{\langle k \rangle}+\delta_\theta+v_\theta \end{pmatrix} \tag{6.2}
$$

where $k$ is the time step, $\delta_{\langle k \rangle} = (\delta_d, \delta_\theta)^\top \in \mathbb{R}^{2\times 1}$ is the odometry measurement and $\boldsymbol{v}_{\langle k \rangle} = (v_d, v_\theta)^\top \in \mathbb{R}^{2\times 1}$ is the random measurement noise over the preceding interval. ▶

The odometry noise is *inside* the model of our process (robot motion) and is referred to as process noise.

In the absence of any information to the contrary, we model the odometry noise as $\boldsymbol{v} = (v_d, v_\theta)^\top \sim N(0, \mathbf{V})$, a zero-mean multivariate Gaussian process ▶ with covariance

$$
\mathbf{V} = \begin{pmatrix} \sigma_d^2 & 0 \\ 0 & \sigma_\theta^2 \end{pmatrix}.
$$

A normal distribution of angles on a circle is actually not possible since $\theta \in \mathbf{S}^1 \notin \mathbb{R}$, that is angles wrap around $2\pi$. However, if the covariance for angular states is small this problem is minimal. A normal-like distribution of angles on a circle is the von Mises distribution, see `numpy.random.vonmises`.

This odometry covariance matrix is diagonal so errors in distance and heading are *independent*. ▶ Choosing $\mathbf{V}$ is not always easy but we can conduct experiments or make some reasonable engineering assumptions. In the examples which follow, we choose $\sigma_d = 2$ cm and $\sigma_\theta = 0.5°$ per sample interval which leads to a covariance matrix of

In reality this is unlikely to be the case, since odometry distance errors tend to be worse when change of heading is high.

```
>>> V = np.diag([0.02, np.deg2rad(0.5)]) ** 2;
```

Objects derived from the Toolbox `VehicleBase` superclass provide a method `f()` that implements the appropriate configuration update equation. For the case of a robot with bicycle kinematics, with the motion model (4.4) and the configuration update (6.2), we create a `Bicycle` object

```
>>> robot = Bicycle(covar=V, animation="car")
Bicycle: x = [ 0, 0, 0 ]
  L=1, steer_max=1.41372, speed_max=inf, accel_max=inf
```

which shows the default parameters such as the robot's length, speed, steering limit and the sample interval which defaults to 0.1 s. The `animation` option specifies the shape of the polygon used to animate the robot's motion when its `run` method is called. The object provides a method to simulate motion over one time step

```
>>> odo = robot.step((1, 0.3))
array([  0.1025,   0.02978])
```

where we have specified a speed of 1 ms$^{-1}$ and a steered-wheel angle of 0.3 rad. The function updates the robot's true configuration and returns a noise corrupted odometer reading. ◀ With a sample interval of 0.1 s the robot reports that it is moving approximately 0.1 m over the interval and changing its heading by approximately 0.03 rad. The robot's true (but "unknown") configuration can be seen by

> We simulate the odometry noise using random numbers that have zero-mean and a covariance given by V. The random noise means that repeated calls to this function will return different values.

```
>>> robot.q
array([      0.1,          0,   0.03093])
```

Given the reported odometry, we can estimate the configuration of the robot after one time step using (6.2), which is implemented by

```
>>> robot.f([0, 0, 0], odo)
array([  0.1025,          0,   0.02978])
```

where the discrepancy with the exact value is due to the use of a noisy odometry measurement.

For the scenarios that we want to investigate, we require the simulated robot to drive for a long time period within a defined spatial region. The `RandomPath` class is a *driver* that steers the robot to randomly selected waypoints within a specified region. We create an instance of a driver object and connect it to the robot

```
>>> robot.control = RandomPath(workspace=10)
```

where the argument specifies a working region that spans $\pm 10$ m in the $x$- and $y$-directions. We can display an animation of the robot with its driver by

```
>>> robot.run(T=10);
```

which runs the simulation, with animation, for 10 s by repeatedly calling the `step` method and saving a history of the true state of the robot within the `Bicycle` object.

## 6.1.2 Estimating Pose

The problem we face, just like the ship's navigator, is how to estimate our new pose given the previous pose and noisy odometry. The mathematical tool that we will use to solve this problem is the Kalman filter. This is a recursive algorithm that updates, at each time step, the *optimal* estimate of the unknown true configuration and the uncertainty associated with that estimate, based on the previous estimate, the control input, and noisy measurement data.

The Kalman filter is described more completely in ▶ App. H. This filter can be shown to provide the optimal estimate of the system state assuming that the noise is zero-mean and Gaussian. The Kalman filter is formulated for linear systems but our model of the robot's motion (6.2) is nonlinear – for this case we use the extended Kalman filter.

We will start by considering the one-dimensional example shown in ◻ Fig. 6.4a. The initial PDF has a mean value of $\hat{x}_{\langle k \rangle} = 2$ and a variance $P_{\langle k \rangle} = 0.25$. The

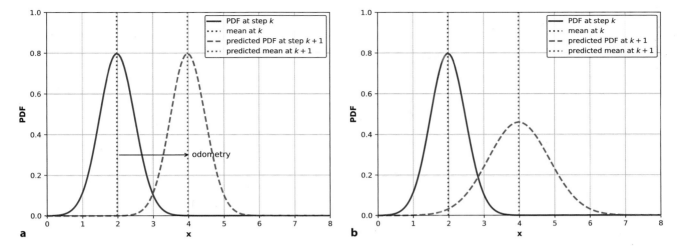

**◻ Fig. 6.4** One-dimensional example of Kalman filter prediction. The initial PDF has a mean of 2 and a variance of 0.25, and the odometry value is 2. **a** Predicted PDF for the case of no odometry noise $V = 0$; **b** Updated PDF for the case $V = 0.5$

odometry is 2 and its variance is $V = 0$ so the predicted PDF at the next time step is simply shifted to a mean value of $\hat{x}_{\langle k+1 \rangle} = 4$. In ◻ Fig. 6.4b the odometry variance is $V = 0.5$ so the predicted PDF still has a mean value of $\hat{x}_{\langle k+1 \rangle} = 4$, but the distribution is now wider, accounting for the uncertainty in odometry.

For the multi-dimensional robot localization problem the state vector is the robot's configuration

$$\boldsymbol{x} = (x_v,\ y_v,\ \theta_v)^\top$$

and the prediction equations ▶

> The Kalman filter, ◻ Fig. 6.8, has two steps: *prediction* based on the model, and *update* based on sensor data. In this dead-reckoning case we use only the prediction equation.

$$\hat{\boldsymbol{x}}^+{}_{\langle k+1 \rangle} = \boldsymbol{f}(\hat{\boldsymbol{x}}_{\langle k \rangle},\ \hat{\boldsymbol{u}}_{\langle k \rangle}) \tag{6.3}$$

$$\hat{\mathbf{P}}^+{}_{\langle k+1 \rangle} = \mathbf{F}_x \hat{\mathbf{P}}_{\langle k \rangle} \mathbf{F}_x^\top + \mathbf{F}_v \hat{\mathbf{V}} \mathbf{F}_v^\top \tag{6.4}$$

describe how the state and covariance evolve with time. The term $\hat{\boldsymbol{x}}^+{}_{\langle k+1 \rangle}$ indicates an estimate of $\boldsymbol{x}$ at time $k + 1$ based on information up to, and including, time $k$. $\hat{\boldsymbol{u}}_{\langle k \rangle}$ is the input to the process, which in this case is the measured odometry, so $\hat{\boldsymbol{u}}_{\langle k \rangle} = \boldsymbol{\delta}_{\langle k \rangle}$. $\hat{\mathbf{V}}$ is our *estimate* of the covariance of the odometry noise which in reality we do not know.

The covariance matrix is symmetric

$$\hat{\mathbf{P}} = \left( \begin{array}{cc|c} \sigma_x^2 & \mathrm{cov}(x, y) & \mathrm{cov}(x, \theta) \\ \mathrm{cov}(x, y) & \sigma_y^2 & \mathrm{cov}(y, \theta) \\ \hline \mathrm{cov}(x, \theta) & \mathrm{cov}(y, \theta) & \sigma_\theta^2 \end{array} \right) \tag{6.5}$$

and represents uncertainty in the estimated robot configuration, as well as dependence between elements of the configuration. Variance ▶ is concerned with a single state and $\mathrm{cov}(u, u) = \sigma_u^2$. Covariance between the states $u$ and $v$ is written as $\mathrm{cov}(u, v) = \mathrm{cov}(v, u)$, and is zero if $u$ and $v$ are independent. The highlighted top-left corner is the covariance of robot *position* rather than configuration.

The two terms in (6.4) have a particular form, a similarity transformation, $\mathbf{F} \boldsymbol{\Sigma} \mathbf{F}^\top$ where $\boldsymbol{\Sigma}$ is a covariance matrix. For a function $\boldsymbol{y} = \boldsymbol{f}(\boldsymbol{x})$ the covariance

> Covariance is defined as
> $\mathrm{cov}(u, v) = \sum_{i=0}^{N-1}(u_i - \bar{u})(v_i - \bar{v})$.

6

### Excurse 6.4: Reverend Thomas Bayes

Bayes (1702–1761) was a nonconformist Presbyterian minister who studied logic and theology at the University of Edinburgh. He lived and worked in Tunbridge-Wells in Kent and through his association with the 2nd Earl Stanhope he became interested in mathematics and was elected to the Royal Society in 1742. After his death his friend Richard Price edited and published his work in 1763 as *An Essay towards solving a Problem in the Doctrine of Chances* which contains a statement of a special case of Bayes' theorem. Bayes is buried in Bunhill Fields Cemetery in London.

Bayes' theorem shows the relation between a conditional probability and its inverse: the probability of a hypothesis given observed evidence and the probability of that evidence given the hypothesis. Consider the hypothesis that the robot is at position X and it makes a sensor observation S of a known landmark. The *posterior* probability that the robot is at X given the observation S is

$$P(X|S) = \frac{P(S|X)P(X)}{P(S)}$$

where $P(X)$ is the *prior* probability that the robot is at X (not accounting for any sensory information), $P(S|X)$ is the

likelihood of the sensor observation S given that the robot is at X, and $P(S)$ is the prior probability of the observation S. The Kalman filter, and the Monte-Carlo estimator we discuss later in this chapter, are essentially two different approaches to solving this inverse problem.

### Excurse 6.5: The Kalman Filter

Rudolf Kálmán (1930-2016) was a mathematical system theorist born in Budapest. He obtained his bachelors and masters degrees in electrical engineering from MIT, and Ph.D. in 1957 from Columbia University. He worked as a Research Mathematician at the Research Institute for Advanced Study, in Baltimore, from 1958–1964 where he developed his ideas on estimation. These were met with some skepticism among his peers and he chose a mechanical (rather than electrical) engineering journal for his paper *A new approach to linear filtering and prediction problems* because "When you fear stepping on hallowed ground with entrenched interests, it is best to go sideways". He received many awards including the IEEE Medal of Honor, the Kyoto Prize and the Charles Stark Draper Prize.

Stanley F. Schmidt (1926–2015) was a research scientist who worked at NASA Ames Research Center and was an early advocate of the Kalman filter. He developed the first implementation as well as the nonlinear version now known as the extended Kalman filter. This led to its incorporation in the Apollo navigation computer for trajectory estimation. (Extract from Kálmán's famous paper (1960) on the right reprinted with permission of ASME)

of $x$ and $y$ are related by $\Sigma_y = \mathbf{F}\,\Sigma_x\,\mathbf{F}^\top$ where $\mathbf{F}$ is a Jacobian matrix – the vector version of a derivative which are revised in ▶ App. E. The Jacobian matrices are obtained by differentiating (6.2) and evaluating the result at $v = 0$ giving ▶

$$\mathbf{F}_x = \left.\frac{\partial f}{\partial x}\right|_{v=0} = \begin{pmatrix} 1 & 0 & -\delta_d \sin\theta_v \\ 0 & 1 & \delta_d \cos\theta_v \\ 0 & 0 & 1 \end{pmatrix} \tag{6.6}$$

$$\mathbf{F}_v = \left.\frac{\partial f}{\partial v}\right|_{v=0} = \begin{pmatrix} \cos\theta_v & 0 \\ \sin\theta_v & 0 \\ 0 & 1 \end{pmatrix} \tag{6.7}$$

The noise value $v$ cannot be measured, so we evaluate the derivative using its mean value $v = 0$.

which are functions of the current robot state and odometry. We drop the time step notation $\langle k \rangle$ to reduce clutter. All objects of the `Vehicle` superclass provide methods `Fx` and `Fv` to compute these Jacobians, for example

```
>>> robot.Fx([0, 0, 0], [0.5, 0.1])
array([[       1,          0,          0],
       [       0,          1,        0.5],
       [       0,          0,          1]])
```

where the first argument is the configuration at which the Jacobian is computed and the second is the odometry.

To simulate the robot and the EKF using the Toolbox, we must define the initial covariance. We choose a diagonal matrix

```
>>> x_sdev = [0.05, 0.05, np.deg2rad(0.5)];
>>> P0 = np.diag(x_sdev) ** 2;
```

written in terms of the standard deviation of position and orientation which are respectively 5 cm and 0.5° in orientation – this implies that we have a good idea of the initial configuration. We pass this to the constructor for an `EKF` object

```
>>> ekf = EKF(robot=(robot, V), P0=P0)
EKF object: 3 states, estimating: vehicle pose, map
  robot: Bicycle: x = [ 0, 0, 0 ]
    L=1, steer_max=1.41372, speed_max=inf, accel_max=inf
  V_est:  [ 0.0004, 0 | 0, 7.62e-05 ]
```

along with the robot kinematic model and the odometry covariance matrix, defined earlier.

Running the filter for 20 seconds

```
>>> ekf.run(T=20);
```

drives the robot along a random path starting at the origin. At each time step the filter updates the state estimate using various methods provided by the vehicle object.

We can plot the true path taken by the robot, stored within the `VehicleBase` superclass object, by

```
>>> robot.plot_xy(color="b")
```

and the filter's estimate of the path stored within the `EKF` object,

```
>>> ekf.plot_xy(color="r")
```

and these are shown in ◼ Fig. 6.5. We see some divergence between the true and estimated robot path.

The covariance at time step 150 is

```
>>> P150 = ekf.get_P(150)
array([[ 0.1666, -0.04335,  0.03476],
       [-0.04335,  0.09977, -0.01525],
       [ 0.03476, -0.01525,  0.01158]])
```

**6**

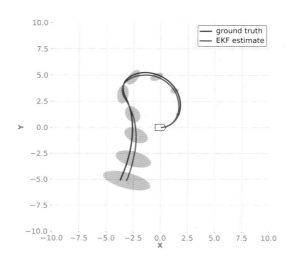

☐ **Fig. 6.5** Dead reckoning using the EKF showing the true path and estimated path of the robot. 95% confidence ellipses are indicated in *green*. The robot starts at the origin

The matrix is symmetric and the diagonal elements are the estimated variance associated with the states, that is $\sigma_x^2$, $\sigma_y^2$ and $\sigma_\theta^2$ respectively. The standard deviation $\sigma_x$ of the PDF associated with the robot's $x$-coordinate is

```
>>> np.sqrt(P150[0, 0])
0.4081
```

There is a 95% chance that the robot's $x$-coordinate is within the $\pm 2\sigma$ bound or $\pm 0.81$ m in this case. We can compute uncertainty for $y$ and $\theta$ similarly.

The off-diagonal terms are correlation coefficients and indicate that the uncertainties between the corresponding variables are related. For example the value $p_{0,2} = p_{2,0} = 0.0348$ indicates that the uncertainties in $x$ and $\theta$ are related – error in heading angle causes error in $x$-position and vice versa. Conversely, new information about $\theta$ can be used to correct $\theta$ as well as $x$. The uncertainty in position is described by the top-left $2 \times 2$ covariance submatrix of $\hat{\mathbf{P}}$, which can be interpreted as an ellipse defining a confidence bound on position. We can overlay such ellipses on the plot by

```
>>> ekf.plot_ellipse(filled=True, facecolor="g", alpha=0.3)
```

as shown in ☐ Fig. 6.5. These correspond to the default 95% confidence bound and are plotted by default every 10 time steps. The robot started at the origin and as it progresses we see that the ellipses become larger as the estimated uncertainty increases. The ellipses only show the uncertainty in $x$- and $y$-position, but uncertainty in $\theta$ also grows.

> The elements of **P** have different units: m² and rad². The uncertainty is therefore a mixture of spatial and angular uncertainty with an implicit weighting. If the range of the position variables $x, y \gg \pi$ then positional uncertainty dominates.

> A positive-definite matrix can be thought of as the matrix equivalent of a positive number.

The estimated uncertainty in position and heading ◄ is given by $\sqrt{\det(\hat{\mathbf{P}})}$, which we can plot as a function of time

```
>>> t = ekf.get_t();
>>> pn = ekf.get_Pnorm();
>>> plt.plot(t, pn);
```

as the solid curve in ☐ Fig. 6.6. We observe that the uncertainty never decreases and this is because in (6.4) we keep adding a positive-definite matrix ◄ (the second term) to the estimated covariance.

### 🛈 V versus $\hat{\mathbf{V}}$

We have used the odometry covariance matrix V twice. The first usage, in the Vehicle constructor, is the covariance **V** of the zero-mean Gaussian noise source that is added to the true odometry to *simulate* odometry error in (6.2). In a real

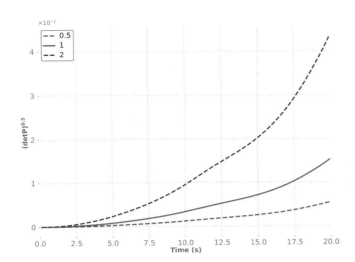

**Fig. 6.6** Overall uncertainty given by $\sqrt{\det\left(\hat{\mathbf{P}}\right)}$ versus time shows monotonically increasing uncertainty. Changing the magnitude of $\hat{\mathbf{V}}$ alters the rate of uncertainty growth. Curves are shown for $\hat{\mathbf{V}} = \alpha\mathbf{V}$ where $\alpha = 0.5, 1, 2$

application this noise is generated by some physical process *hidden inside* the robot and we would not know its parameters. The second usage, in the EKF constructor is $\hat{\mathbf{V}}$, which is our best *estimate* of the odometry covariance and is used in the filter's state covariance update equation (6.4).

The relative values of $\mathbf{V}$ and $\hat{\mathbf{V}}$ control the rate of uncertainty growth as shown in ■ Fig. 6.6. If $\hat{\mathbf{V}} > \mathbf{V}$ then $\hat{\mathbf{P}}$ will be larger than it should be and the filter is pessimistic – it overestimates uncertainty and is less certain than it should be. If $\hat{\mathbf{V}} < \mathbf{V}$ then $\hat{\mathbf{P}}$ will be smaller than it should be and the filter will be *overconfident* of its estimate – the actual uncertainty is greater than the estimated uncertainty. In practice some experimentation is required to determine the appropriate value for the estimated covariance.

---

**Excurse 6.6: Error Ellipses**

We consider the PDF of the robot's position (ignoring orientation) such as shown in ■ Fig. 6.3 to be a 2-dimensional Gaussian probability density function

$$p(\mathbf{x}) = \frac{1}{(2\pi)\det\left(\mathbf{P}_{xy}\right)^{1/2}}e^{\left\{-\frac{1}{2}(\mathbf{x}-\boldsymbol{\mu}_x)^\top\mathbf{P}_{xy}^{-1}(\mathbf{x}-\boldsymbol{\mu}_x)\right\}}$$

where $\mathbf{x} = (x, y)^\top$ is the position of the robot, $\boldsymbol{\mu}_x = (\hat{x}, \hat{y})^\top$ is the estimated mean position and $\mathbf{P}_{xy} \in \mathbb{R}^{2\times 2}$ is the position covariance matrix, the top left of the covariance matrix $\mathbf{P}$ as shown in (6.5). A horizontal cross-section of the PDF is a contour of constant probability which is an ellipse defined by the points $\mathbf{x}$ such that

$$(\mathbf{x} - \boldsymbol{\mu}_x)^\top\mathbf{P}_{xy}^{-1}(\mathbf{x} - \boldsymbol{\mu}_x) = s \ .$$

Such error ellipses are often used to represent positional uncertainty as shown in ■ Fig. 6.5. A large ellipse corresponds to a wider PDF peak and less certainty about position. To obtain a particular confidence contour (eg. 99%) we choose $s$ as the inverse of the $\chi^2$ cumulative distribution function for 2 degrees of freedom. In Python we can use the function chi2inv(C, 2) from the scipy.stats.distributions package, where C $\in [0, 1]$ is the confidence value. Such confidence values can be passed to several EKF methods when specifying error ellipses.

A handy scalar measure of total position uncertainty is the area of the ellipse $\pi r_1 r_2$ where the radii $r_i = \sqrt{\lambda_i}$ and $\lambda_i$ are the eigenvalues of $\mathbf{P}_{xy}$. Since $\det(\mathbf{P}_{xy}) = \prod_i \lambda_i$ the ellipse area – the scalar uncertainty – is proportional to $\sqrt{\det(\mathbf{P}_{xy})}$. See also ▶ App. C.1.4 and ▶ App. G.

## 6.2  Localizing with a Landmark Map

We have seen how uncertainty in position grows without bound using dead reckoning alone. The solution, as the Phoenicians worked out 4000 years ago, is to use additional information from observations of known features or *landmarks* in the world.

Continuing the one-dimensional example of ◘ Fig. 6.4b, we now assume that the robot also has a sensor that can measure the robot's position with respect to some external reference. The PDF of the sensor measurement is shown as the cyan curve in ◘ Fig. 6.7a and has a mean $\hat{x} = 5$ as well as some uncertainty due to its own variance $\hat{W} = 1$. The mean of the sensor measurement is different to the mean of the prediction based on odometry, $\hat{x} = 4$ – perhaps the odometry was underreporting due to wheel slippage. We now have two PDFs for the robot's position, the dashed-red prediction from odometry and the cyan measurement from the sensor. The true PDF is the *product* of these two Gaussians, which is also a Gaussian, and this is shown as the solid-red curve. We have combined two uncertain estimates to produce a new, optimal, estimate – there is a noticeable reduction in the uncertainty associated with the robot's position.

Using a better sensor with a variance of $\hat{W} = 0.3$, as shown in ◘ Fig. 6.7b, the corrected prediction shown by the solid-red curve has less variance and is closer to the mean of the sensor measurement. Given two estimates, one from odometry and one from the sensor, we place more trust in the estimate with lower variance.

For a robotics problem we consider that the robot is able to observe a set of landmarks using an onboard sensor. For the moment we will consider that the position of the landmarks is known, and we will use a map that contains $N$ fixed but randomly located landmarks whose positions are known. The Toolbox provides a `LandmarkMap` object

```
>>> map = LandmarkMap(20, workspace=10)
LandmarkMap object with 20 landmarks,
workspace=(-10.0: 10.0, -10.0: 10.0)
```

that in this case contains $N = 20$ landmarks uniformly randomly spread over a region spanning $\pm 10$ m in the $x$- and $y$-directions and this can be displayed by

```
>>> map.plot()
```

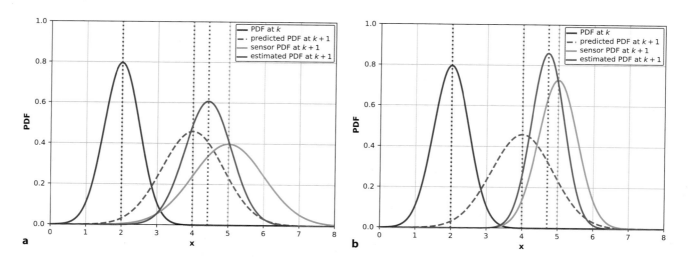

◘ **Fig. 6.7** One-dimensional example of Kalman filter prediction, the blue and dashed-red curves are from ◘ Fig. 6.4b. A sensor now provides a measurement to improve the prediction. **a** for the case of sensor variance $\hat{W} = 1$; **b** for the case of sensor variance $\hat{W} = 0.3$

The robot is equipped with a sensor that provides observations of the landmarks *with respect to the robot* as described by

$$z = h(q, \, p_i) \tag{6.8}$$

where $q = (x_v, y_v, \theta_v)^\top$ is the robot configuration, and $p_i = (x_i, y_i)^\top$ is the *known* position of the $i$th landmark in the world frame.

To make this tangible we will consider a common type of sensor that measures the range and bearing angle to a landmark in the environment, for instance a radar or lidar ▶ such as shown in ◘ Fig. 6.25b-d. The sensor is mounted onboard the robot so the observation of the $i$th landmark is

> Respectively **ra**dio or **l**ight **d**etection **a**nd **r**anging, that measure distance by timing the round trip time for a microwave signal or infrared pulse to be reflected from an object and return to the sensor.

$$z = h(q, \, p_i) = \begin{pmatrix} \sqrt{(y_i - y_v)^2 + (x_i - x_v)^2} \\ \tan^{-1} \frac{y_i - y_v}{x_i - x_v} - \theta_v \end{pmatrix} + \begin{pmatrix} w_r \\ w_\beta \end{pmatrix} \tag{6.9}$$

where $z = (r, \beta)^\top$ and $r$ is the range, $\beta$ the bearing angle, and $w = (w_r, w_\beta)^\top$ is a zero-mean Gaussian random variable that models errors in the sensor

$$\begin{pmatrix} w_r \\ w_\beta \end{pmatrix} \sim N(0, \, \mathbf{W}), \quad \mathbf{W} = \begin{pmatrix} \sigma_r^2 & 0 \\ 0 & \sigma_\beta^2 \end{pmatrix}.$$

The constant diagonal covariance matrix indicates that range and bearing errors are independent. ▶

For this example we set the sensor uncertainty to be $\sigma_r = 0.1\,\text{m}$ and $\sigma_\beta = 1°$ giving a sensor covariance matrix

> It also indicates that covariance is independent of range but in reality covariance may increase with range since the strength of the return signal, laser or radar, drops rapidly $(1/r^4)$ with distance $r$ to the target.

```
>>> W = np.diag([0.1, np.deg2rad(1)]) ** 2;
```

We model this type of sensor with a `RangeBearingSensor` object ▶

> A subclass of `SensorBase`.

```
>>> sensor = RangeBearingSensor(robot=robot, map=map, covar=W,
...             angle=[-pi/2, pi/2], range=4, animate=True)
RangeBearingSensor sensor class
  LandmarkMap object with 20 landmarks,
    workspace=(-10.0: 10.0, -10.0: 10.0)
  W = [ 0.01, 0 | 0, 0.000305 ]
  sampled every 1 samples
  range: (0: 4)
  angle: (-1.57: 1.57)
```

which is connected to the robot and map objects, and we specify the sensor covariance matrix `W` along with the maximum range and the bearing angle limits. The `reading` method provides the range and bearing to a randomly selected visible ▶ landmark along with its identity, for example

> The landmark is chosen randomly from the set of visible landmarks, those that are within the field of view and the minimum and maximum range limits. If no landmark is visible the method returns (None, None).

```
>>> z, i = sensor.reading()
>>> z
array([    3.04,   -1.559])
>>> i
15
```

The identity is an integer $i = 0, \ldots, 19$ since the map was created with 20 landmarks. We have avoided the data association problem by assuming that we know the identity of the sensed landmark. The position of landmark 15 can be looked up in the map

```
>>> map[15]
array([  -6.487,   -3.795])
```

Using (6.9) the robot can estimate the range and bearing angle to the landmark based on its own estimated position and the known position of the landmark from the map. The difference between the observation $z$ and the estimated observation

$$\boldsymbol{v}_{\langle k \rangle} = \boldsymbol{z}_{\langle k \rangle} - \boldsymbol{h}\left(\hat{\boldsymbol{x}}^+_{\langle k \rangle}, \boldsymbol{p}_i\right) \tag{6.10}$$

indicates an error in the robot's pose estimate $\hat{\boldsymbol{x}}^+$ – the robot isn't where it *thought* it was. This difference is called the *innovation* since it represents valuable *new* information, and is key to the operation of the Kalman filter.

The second step of the Kalman filter updates the *predicted* state and covariance computed using (6.3) and (6.4), and denoted by the superscript $^+$. The state estimate update step

$$\hat{\boldsymbol{x}}_{\langle k+1 \rangle} = \hat{\boldsymbol{x}}^+_{\langle k+1 \rangle} + \mathbf{K}\boldsymbol{v}_{\langle k+1 \rangle} \tag{6.11}$$

uses a gain term $\mathbf{K}$ to add some *amount* of the innovation to the predicted state estimate so as to reduce the estimation error. We could choose some constant value of $\mathbf{K}$ to drive the innovation toward zero, but there is a better way

$$\mathbf{K} = \mathbf{P}^+_{\langle k+1 \rangle}\mathbf{H}_x^\top\left(\mathbf{H}_x\mathbf{P}^+_{\langle k+1 \rangle}\mathbf{H}_x^\top + \mathbf{H}_w\hat{\mathbf{W}}\mathbf{H}_w^\top\right)^{-1} \tag{6.12}$$

This can also be written as

$$\mathbf{S} = \mathbf{H}_x\mathbf{P}^+_{\langle k+1 \rangle}\mathbf{H}_x^\top + \mathbf{H}_w\hat{\mathbf{W}}\mathbf{H}_w^\top$$
$$\mathbf{K} = \mathbf{P}^+_{\langle k+1 \rangle}\mathbf{H}_x^\top\mathbf{S}^{-1}$$

where $\mathbf{S}$ is the estimated covariance of the innovation.

where $\hat{\mathbf{W}}$ is the estimated covariance of the sensor noise and $\mathbf{H}_x$ and $\mathbf{H}_w$ are Jacobians. ◄ $\mathbf{K}$ is called the Kalman gain matrix and it *distributes* the innovation from the landmark observation, a 2-vector, to update every element of the state vector – the position and orientation of the robot. The off-diagonal elements of the covariance matrix are the error correlations between the states, and the Kalman filter uses these so that innovation in one state optimally updates the other states it is correlated with. The size of the state estimate update depends on the relevant sensor covariance – if uncertainty is low then the state estimate has a larger update than if the uncertainty was high.

The covariance update also depends on the Kalman gain

$$\hat{\mathbf{P}}_{\langle k+1 \rangle} = \hat{\mathbf{P}}^+_{\langle k+1 \rangle} - \mathbf{K}\mathbf{H}_x\hat{\mathbf{P}}^+_{\langle k+1 \rangle} \tag{6.13}$$

and it is important to note that the second term in (6.13) is *subtracted* from the estimated covariance. This provides a means for covariance to decrease that was not possible for the dead-reckoning case of (6.4).

The Jacobians $\mathbf{H}_x$ and $\mathbf{H}_w$ are obtained by differentiating (6.9) yielding

$$\mathbf{H}_x = \left.\frac{\partial \boldsymbol{h}}{\partial \boldsymbol{x}}\right|_{\boldsymbol{w}=0} = \begin{pmatrix} -\frac{x_i - x_v}{r} & -\frac{y_i - y_v}{r} & 0 \\ \frac{y_i - y_v}{r^2} & -\frac{x_i - x_v}{r^2} & -1 \end{pmatrix} \tag{6.14}$$

which is a function of landmark and robot position, and landmark range; and

$$\mathbf{H}_w = \left.\frac{\partial \boldsymbol{h}}{\partial \boldsymbol{w}}\right|_{\boldsymbol{w}=0} = \begin{pmatrix} 1 & 0 \\ 0 & 1 \end{pmatrix}. \tag{6.15}$$

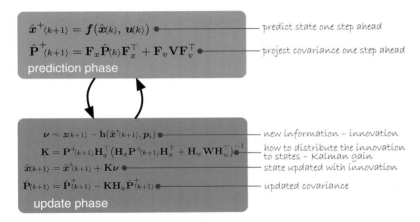

The `RangeBearingSensor` object above includes methods `h` to implement (6.9) and `Hx` and `Hw` to compute these Jacobians respectively.

The EKF comprises two phases: prediction and update, and these are summarized in □ Fig. 6.8.

We now have all the pieces to build an estimator that uses odometry and observations of map features. The Toolbox implementation is

```
>>> map = LandmarkMap(20, workspace=10);
>>> V = np.diag([0.02, np.deg2rad(0.5)]) ** 2
array([[ 0.0004,        0],
       [      0, 7.615e-05]])
>>> robot = Bicycle(covar=V, animation="car");
>>> robot.control = RandomPath(workspace=map)
>>> W = np.diag([0.1, np.deg2rad(1)]) ** 2
array([[   0.01,        0],
       [      0, 0.0003046]])
>>> sensor = RangeBearingSensor(robot=robot, map=map, covar=W,
...             angle=[-pi/2, pi/2], range=4, animate=True);
>>> P0 = np.diag([0.05, 0.05, np.deg2rad(0.5)]) ** 2;
>>> ekf = EKF(robot=(robot, V), P0=P0, map=map, sensor=(sensor, W));
```

The `workspace` argument to the `LandmarkMap` constructor sets the map span to ±10 m in each dimension and this information is also used by the `RandomPath` and `RangeBearingSensor` constructors.

Running the simulation for 20 seconds

```
>>> ekf.run(T=20)
```

shows an animation of the robot moving, and observations being made to the landmarks. We plot the saved results

```
>>> map.plot()
>>> robot.plot_xy();
>>> ekf.plot_xy();
>>> ekf.plot_ellipse()
```

which are shown in □ Fig. 6.9a. We see that the error ellipses grow as the robot moves, just as we observed for the dead reckoning case of □ Fig. 6.5, and then shrink due to a landmark observation that corrects the robot's estimated state. The close-up view in □ Fig. 6.9b shows the robot's actual and estimated path, and the latter has a sharp kink at the correction. New information, in addition to odometry, has been used to correct the state in the Kalman filter update phase. The error ellipses are now much smaller and many can hardly be seen.

□ Fig. 6.10a shows that the overall uncertainty is no longer growing monotonically. When the robot observes a landmark the estimated covariance is dramatically

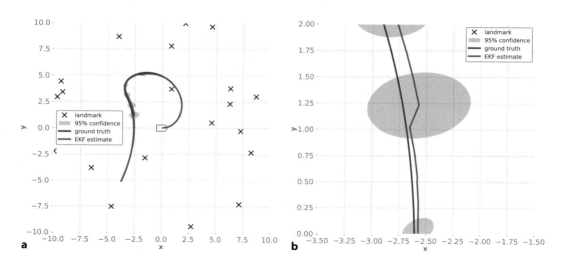

**Fig. 6.9 a** EKF localization with dead reckoning and landmarks, showing the true and estimated path of the robot. Black stars are landmarks and the green ellipses are 95% confidence bounds. The robot starts at the origin. **b** Closeup of the robot's true and estimated path showing the effect of a landmark observation

reduced. ◼ Fig. 6.10b shows the error associated with each component of pose and the pink background is the estimated 95% confidence bound (derived from the covariance matrix) and we see that the error is mostly within this envelope. Below this is plotted the landmark observations and we see that the confidence bounds are tight (indicating low uncertainty) while landmarks are being observed, but that they start to grow during periods with no observations.

The EKF is an extremely powerful tool that allows data from many and varied sensors to update the state, which is why the estimation problem is also referred to as sensor fusion. For example, heading angle from a compass, yaw rate from a gyroscope, target bearing angle from a camera, position from GPS could all be used to update the state. For each sensor we need only to provide the observation function $h(\cdot)$, the Jacobians $\mathbf{H}_x$ and $\mathbf{H}_w$ and some estimate of the sensor covariance $\hat{\mathbf{W}}$. The function $h(\cdot)$ can be nonlinear and even noninvertible – the EKF will do the rest.

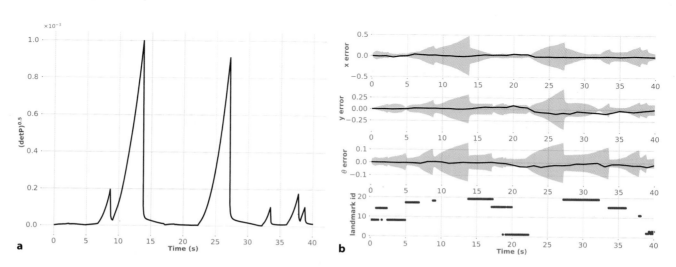

**Fig. 6.10 a** Covariance magnitude as a function of time. Overall uncertainty is given by $\sqrt{\det(\mathbf{P})}$ and shows that uncertainty does not continually increase with time. **b** Top: pose estimation error with 95% confidence bound shown in coral; bottom: scatter plot of observed landmarks and periods of no observations can be clearly seen

### Excurse 6.7: Data Association

So far we have assumed that we can determine the identity of the observed landmark, but in reality this is rarely the case. Instead, we can compare our observation to the predicted position of all currently known landmarks and make a decision as to which landmark it is most likely to be, or whether it is a new landmark. This decision needs to take into account the uncertainty associated with the robot's pose, the sensor measurement and the landmarks in the map. This is the data association problem. Errors in this step are potentially catastrophic – incorrect innovation would introduce error in the state of the robot that increases the chance of an incorrect data association on the next cycle. In practice, filters only use a landmark when there is a very high confidence in its estimated identity – a process that involves Mahalanobis distance and $\chi^2$ confidence tests. If the situation is ambiguous it is best not to use the landmark – a bad update can do more harm than good.

An alternative is to use a multi-hypothesis estimator, such as the particle filter that we will discuss in ▶ Sect. 6.6, which can model the possibility of observing landmark A or landmark B, and future observations will reinforce one hypothesis and weaken the others. The extended Kalman filter uses a Gaussian probability model, with just one peak, which limits it to holding only one hypothesis about the robot's pose. (Picture: the wreck of the Tararua, 1881)

### Excurse 6.8: Artificial Landmarks

For some applications it might be appropriate to use artificial landmarks or fiducial markers that can be cheaply sensed by a camera. These are not only visually distinctive in the environment, but they can also encode a unique identity. Commonly used 2-dimensional bar codes include ArUco markers, ARTags and AprilTags (shown here). If the camera is calibrated we can determine the range to the marker and its surface normal. The Toolbox supports detection and reading of ArUco markers and AprilTags and these are discussed in ▶ Sect. 13.6.1.

### ❗ W versus $\hat{\mathbf{W}}$

As discussed earlier for V, we also use W twice. The first usage, in the constructor for the `RangeBearingSensor` object, is the covariance $\mathbf{W}$ of the zero-mean Gaussian noise that is added to the computed range and bearing to *simulate* sensor error as in (6.9). In a real application this noise is generated by some physical process *hidden inside* the sensor and we would not know its parameters. The second usage, $\hat{\mathbf{W}}$ is our best *estimate* of the sensor covariance which is used by the Kalman filter (6.12).

The relative magnitude of $\mathbf{W}$ and $\hat{\mathbf{W}}$ affects the rate of covariance growth. Large $\hat{\mathbf{W}}$ indicates that less trust should be given to the sensor measurements so the covariance reduction at each observation will be lower.

## 6.3 Creating a Landmark Map

So far we have taken the existence of the map for granted, and this is understandable given that maps today are common and available for free via the internet. Nevertheless somebody, or something, has to create the maps we use. Our next example considers the problem of creating a map – determining the positions of the landmarks – in an environment that the robot is moving through.

As before we have a range and bearing sensor mounted on the robot, which measures, imperfectly, the position of landmarks with respect to the robot. We will assume that the sensor can determine the identity of each observed landmark, and that the robot's pose is known perfectly – it has ideal localization. This is unrealistic but this scenario is an important conceptual stepping stone to the next section. ◀

Since the robot's pose is known perfectly we do not need to estimate it, but we do need to estimate the coordinates of the landmarks. For this problem the state vector comprises the coordinates of the $M$ landmarks that have been observed so far

$$x = (x_0, \ y_0, \ x_1, \ y_1, \ \dots, \ x_{M-1}, \ y_{M-1})^\top \in \mathbb{R}^{2M \times 1} \ .$$

The covariance matrix is symmetric

$$\hat{\mathbf{P}} = \begin{pmatrix} \sigma_{x_1}^2 & \text{cov}(x_1, y_1) & \text{cov}(x_1, x_2) & \text{cov}(x_1, y_2) & \cdots \\ \text{cov}(x_1, y_1) & \sigma_{y_1}^2 & \text{cov}(y_1, x_2) & \text{cov}(y_1, y_2) & \cdots \\ \text{cov}(x_1, x_2) & \text{cov}(x_1, y_2) & \sigma_{x_2}^2 & \text{cov}(x_2, y_2) & \cdots \\ \text{cov}(y_1, x_2) & \text{cov}(y_1, y_2) & \text{cov}(x_2, y_2) & \sigma_{y_2}^2 & \\ \vdots & \vdots & \vdots & \vdots & \ddots \end{pmatrix} \in \mathbb{R}^{2M \times 2M} \tag{6.16}$$

and each $2 \times 2$ block on the diagonal is the covariance of the corresponding landmark position, and the off-diagonal blocks are the covariance between the landmarks.

Initially $M = 0$ and is incremented every time a previously-unseen landmark is observed. The state vector has a variable length since we do not know in advance how many landmarks exist in the environment. The estimated covariance $\hat{\mathbf{P}}$ also has variable size.

The prediction equation is straightforward in this case since the landmarks are assumed to be stationary

$$\hat{x}^+_{\langle k+1 \rangle} = \hat{x}_{\langle k \rangle} \tag{6.17}$$

$$\hat{\mathbf{P}}^+_{\langle k+1 \rangle} = \hat{\mathbf{P}}_{\langle k \rangle} \tag{6.18}$$

We introduce the function $g(\cdot)$, which is the inverse of $h(\cdot)$ and returns the world coordinates of the observed landmark based on the known robot pose $(x_v, y_v, \theta_v)$ and the sensor observation

$$g(x, z) = \begin{pmatrix} x_v + r\cos(\theta_v + \beta) \\ y_v + r\sin(\theta_v + \beta) \end{pmatrix} \tag{6.19}$$

which is also known as the inverse measurement function, or the inverse sensor model. Since $\hat{x}$ has a variable length we need to extend the state vector and the covariance matrix when we encounter a previously-unseen landmark. The state vector is extended by the function $y(\cdot)$

$$x'_{\langle k \rangle} = y(x_{\langle k \rangle}, \ z_{\langle k \rangle}, \ x_{v \langle k \rangle}) \tag{6.20}$$

$$= \begin{pmatrix} x_{\langle k \rangle} \\ g(x_{v \langle k \rangle}, \ z_{\langle k \rangle}) \end{pmatrix} \tag{6.21}$$

A close and realistic approximation would be a high-quality RTK GPS+INS system operating in an outdoor environment with no buildings or hills to obscure satellites, or a motion capture (MoCap) system in an indoor environment.

which appends the sensor-based estimate of the new landmark's coordinates to those already in the map. The order of feature coordinates within $\hat{x}$ therefore depends on the order in which they are observed.

The covariance matrix also needs to be extended when a new landmark is observed and this is achieved by

$$\hat{\mathbf{P}}'_{\langle k \rangle} = \mathbf{Y}_z \begin{pmatrix} \hat{\mathbf{P}}_{\langle k \rangle} & \mathbf{0}_{2 \times M} \\ \mathbf{0}_{M \times 2} & \hat{\mathbf{W}} \end{pmatrix} \mathbf{Y}_z^\top$$

where $\mathbf{Y}_z$ is the insertion Jacobian

$$\mathbf{Y}_z = \frac{\partial \mathbf{y}}{\partial z} = \left( \begin{array}{cc|c} \mathbf{1}_{M \times M} & & \mathbf{0}_{M \times 2} \\ \mathbf{G}_x & \mathbf{0}_{2 \times (M-3)} & \mathbf{G}_z \end{array} \right) \tag{6.22}$$

that relates the rate of change of the extended state vector to the new observation. $M$ is the dimension of $\hat{\mathbf{P}}$ *prior* to it being extended and the Jacobians are

$$\mathbf{G}_x = \frac{\partial \mathbf{g}}{\partial \mathbf{x}} = \begin{pmatrix} 0 & 0 & 0 \\ 0 & 0 & 0 \end{pmatrix} \tag{6.23}$$

$$\mathbf{G}_z = \frac{\partial \mathbf{g}}{\partial z} = \begin{pmatrix} \cos(\theta_v + \beta) & -r \sin(\theta_v + \beta) \\ \sin(\theta_v + \beta) & r \cos(\theta_v + \beta) \end{pmatrix}. \tag{6.24}$$

$\mathbf{G}_x$ is zero since $\mathbf{g}(\cdot)$ is independent of the map and therefore $\mathbf{x}$. An additional Jacobian for $\mathbf{h}(\cdot)$ is

$$\mathbf{H}_{p_i} = \frac{\partial \mathbf{h}}{\partial \mathbf{p}_i} = \begin{pmatrix} \frac{x_i - x_v}{r} & \frac{y_i - y_v}{r} \\ -\frac{y_i - y_v}{r^2} & \frac{x_i - x_v}{r^2} \end{pmatrix} \tag{6.25}$$

which describes how the sensor measurement changes with respect to the position of landmark $i$ for a particular robot pose, and is implemented by the method `Hp`.

For the mapping case, the Jacobian $\mathbf{H}_x$ used in (6.13) describes how the landmark observation changes with respect to the full state vector. However, the observation depends only on the position of the observed landmark, so this Jacobian is mostly zeros

$$\mathbf{H}_x = \frac{\partial \mathbf{h}}{\partial \mathbf{x}} \bigg|_{\mathbf{w}=0} = \begin{pmatrix} \mathbf{0}_{2 \times 2} & \cdots & \mathbf{H}_{p_i} & \cdots & \mathbf{0}_{2 \times 2} \end{pmatrix} \in \mathbb{R}^{2 \times 2M} \tag{6.26}$$

where $\mathbf{H}_{p_i}$ is at the location in the state vector corresponding to landmark $i$. This structure represents the fact that observing a particular landmark provides information to estimate the position of that landmark, but no others.

The Toolbox implementation is

```
>>> map = LandmarkMap(20, workspace=10);
>>> robot = Bicycle(covar=V, animation="car");
>>> robot.control = RandomPath(workspace=map);
>>> W = np.diag([0.1, np.deg2rad(1)]) ** 2
array([[    0.01,        0],
       [       0, 0.0003046]])
>>> sensor = RangeBearingSensor(robot=robot, map=map, covar=W,
...            range=4, angle=[-pi/2, pi/2], animate=True);
>>> ekf = EKF(robot=(robot, None), sensor=(sensor, W));
```

The map object is not passed to `EKF` in this case since it is unknown to the filter. The estimated odometry covariance is given as `None` to indicate that the robot's pose estimate is perfect, and the initial covariance is not given since it is initially a $0 \times 0$ matrix. We run the simulation for 100 seconds

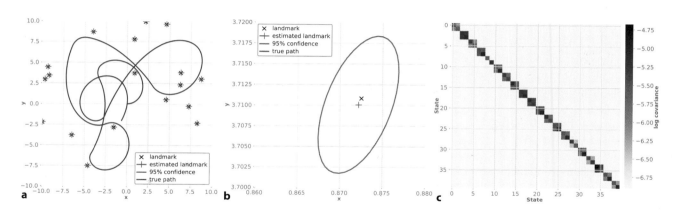

**Fig. 6.11** **a** EKF mapping results, the uncertainty ellipses are too small to see. **b** Enlarged view of landmark 8 and its covariance ellipse. **c** the final covariance matrix has a block diagonal structure

```
>>> ekf.run(T=100);
```

and see an animation of the robot moving and the covariance ellipses associated with the map features evolving over time. The estimated landmark positions

```
>>> map.plot();
>>> ekf.plot_map();
>>> robot.plot_xy();
```

are shown in ◻ Fig. 6.11a as 95% confidence ellipses along with the true landmark positions and the path taken by the robot. ◻ Fig. 6.11b shows a close-up view of landmark 8 and we see that the estimate is close to the true value and well within the confidence ellipse.

The covariance matrix is shown graphically in ◻ Fig. 6.11c and has a block-diagonal structure. The off-diagonal elements are zero, which implies that the landmark estimates are uncorrelated or independent. This is to be expected since observing one landmark provides no new information about any other landmark.

Internally the EKF object uses a dictionary to relate the landmark's identity, returned by the RangeBearingSensor object, to the position of that landmark's coordinates within the state vector. For example, landmark  10

```
>>> ekf.landmark(10)
(12, 22)
```

was seen a total of 22 times during the simulation, was the twelth (base 0) landmark to be seen, and its estimated position can be found in the EKF state vector starting at elements 24 ($12 \times 2$)

```
>>> ekf.x_est[24:26]
array([   6.312,    3.756])
```

Its estimated covariance is a submatrix within $\hat{\mathbf{P}}$

```
>>> ekf.P_est[24:26, 24:26]
array([[0.0001317, 5.822e-06],
       [5.822e-06, 0.0002383]])
```

## 6.4  Simultaneous Localization and Mapping

Finally, we tackle the problem of determining our position and creating a map at the same time. This is an old problem in marine navigation and cartography – incrementally extending maps while also using the map for navigation. ◻ Fig. 6.12 shows what can be done without GPS from a moving ship with poor odometry and

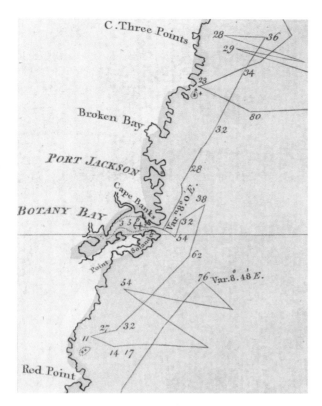

**◘ Fig. 6.12** Map of the Eora territory coast (today, the Sydney region of eastern Australia) created by Captain James Cook in 1770. The path of the ship and the map of the coast were determined at the same time. Numbers indicate depth in fathoms (1.83 m) (National Library of Australia, MAP NK 5557 A)

infrequent celestial position *fixes*. In robotics this problem is known as simultaneous localization and mapping (SLAM) or concurrent mapping and localization (CML). This is a "chicken and egg" problem since we need a map to localize and we need to localize to make the map. However, based on what we have learned in the previous sections this problem is now quite straightforward to solve.

The state vector comprises the robot configuration *and* the coordinates of the $M$ landmarks that have been observed so far

$$\boldsymbol{x} = (x_v, \ y_v, \ \theta_v, \ x_0, \ y_0, \ x_1, \ y_1, \ \dots, \ x_{M-1}, \ y_{M-1})^\top \in \mathbb{R}^{(2M+3)\times 1}$$

and therefore has a variable length. The estimated covariance is a variable-sized symmetric matrix with a block structure

$$\hat{\mathbf{P}} = \begin{pmatrix} \hat{\mathbf{P}}_{vv} & \hat{\mathbf{P}}_{vm} \\ \hat{\mathbf{P}}_{vm}^\top & \hat{\mathbf{P}}_{mm} \end{pmatrix} \in \mathbb{R}^{(2M+3)\times(2M+3)}$$

where $\hat{\mathbf{P}}_{vv} \in \mathbb{R}^{3\times3}$ is the covariance of the robot pose as described by (6.5), $\hat{\mathbf{P}}_{mm} \in \mathbb{R}^{2M\times2M}$ is the covariance of the estimated landmark positions as described by (6.16), and $\hat{\mathbf{P}}_{vm} \in \mathbb{R}^{3\times2M}$ is the correlation between robot and landmark states.

The predicted robot state and covariance are given by (6.3) and (6.4) and the sensor-based update is given by (6.11) to (6.15). When a new feature is observed the state vector is updated using the insertion Jacobian $\mathbf{Y}_z$ given by (6.22) but in this case $\mathbf{G}_x$ is not zero

$$\mathbf{G}_x = \frac{\partial \boldsymbol{g}}{\partial \boldsymbol{x}} = \begin{pmatrix} 1 & 0 & -r\sin(\theta_v + \beta) \\ 0 & 1 & r\cos(\theta_v + \beta) \end{pmatrix} \tag{6.27}$$

**6**

since the estimate of the new landmark depends on the state vector which now contains the robot's configuration.

For the SLAM case, the Jacobian $\mathbf{H}_x$ used in (6.13) describes how the landmark observation changes with respect to the state vector. The observation depends only on two elements of the state vector: the position of the robot and the position of the observed landmark

$$\mathbf{H}_x = \left.\frac{\partial \boldsymbol{h}}{\partial \boldsymbol{x}}\right|_{\boldsymbol{w}=0} = \left(\mathbf{H}_{x_v} \cdots \mathbf{0}_{2\times 2} \cdots \mathbf{H}_{p_i} \cdots \mathbf{0}_{2\times 2}\right) \in \mathbb{R}^{2\times(2M+3)} \quad (6.28)$$

where $\mathbf{H}_{p_i}$ is at the location in the state vector corresponding to landmark $i$. This is similar to (6.26) but with an extra nonzero block $\mathbf{H}_{x_v}$ given by (6.14).

The Kalman gain matrix $\mathbf{K}$ *distributes* innovation from the landmark observation, a 2-vector, to update *every* element of the state vector – the configuration of the robot *and* the position of *every* landmark in the map.

The Toolbox implementation is by now quite familiar

```
>>> map = LandmarkMap(20, workspace=10);
>>> W = np.diag([0.1, np.deg2rad(1)]) ** 2
array([[    0.01,          0],
       [       0, 0.0003046]])
>>> robot = Bicycle(covar=V, x0=(3, 6, np.deg2rad(-45)),
...            animation="car");
>>> robot.control = RandomPath(workspace=map);
>>> W = np.diag([0.1, np.deg2rad(1)]) ** 2
array([[    0.01,          0],
       [       0, 0.0003046]])
>>> sensor = RangeBearingSensor(robot=robot, map=map, covar=W,
...            range=4, angle=[-pi/2, pi/2], animate=True);
>>> P0 = np.diag([0.05, 0.05, np.deg2rad(0.5)]) ** 2;
>>> ekf = EKF(robot=(robot, V), P0=P0, sensor=(sensor, W));
```

and once again the map object is not passed to `EKF` since it is unknown to the filter and `P0` is the initial $3 \times 3$ covariance for the robot state. The initial configuration of the robot is $(3, 6, 45°)$ but the EKF is using the default robot configuration of $(0, 0, 0)$.

We run the simulation for 40 seconds

```
>>> ekf.run(T=40);
```

and as usual an animation of the moving robot is shown. We also see the covariance ellipses associated with the map features evolving over time. We can plot the landmarks and the path taken by the robot

```
>>> map.plot();       # plot true map
>>> robot.plot_xy();  # plot true path
```

which are shown in ◘ Fig. 6.13a. Next we will plot the EKF estimates

```
>>> ekf.plot_map();      # plot estimated landmark position
>>> ekf.plot_ellipse();  # plot estimated covariance
>>> ekf.plot_xy();       # plot estimated robot path
```

which are shown in ◘ Fig. 6.13b and we see that the robot's path is quite different – it is shifted and rotated with respect to the robot path shown in ◘ Fig. 6.13a. The estimated path and landmarks are relative to the map frame. A large part of the difference is because the EKF has used a default initial robot configuration of $(0, 0, 0)$ which is different to the actual initial configuration of the robot $(3, 6, -45°)$. Sensor noise also affects the pose of the map frame: its position depends on the initial estimate of the robot's pose and the odometry error at the first filter step, and the frame's orientation depends on the dead-reckoned orientation and sensor noise when the first measurement was made. Knowing the true and es-

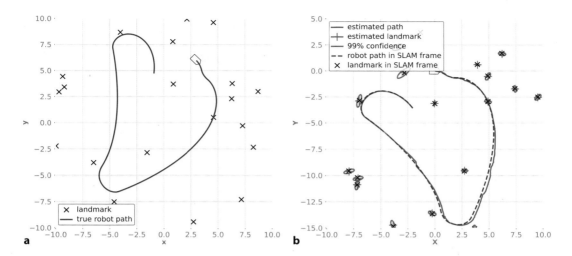

**Fig. 6.13** Simultaneous localization and mapping **a** The true robot path and landmark positions in the world reference frame; **b** The estimated robot path and landmark positions in the map frame, the true robot path and landmarks are overlaid after being transformed into the map frame. Ellipses show the 95% confidence interval

timated landmark positions of at least two landmarks, we can compute an $\mathbf{SE}(2)$ transformation from the map frame to the world frame

```
>>> T = ekf.get_transform(map)
   0.7088    -0.7054     2.101
   0.7054     0.7088    -6.315
   0          0          1
```

which we can use to transform the robot's path and the landmarks from the world frame to the map frame and these are overlaid in ■ Fig. 6.13b. This shows very close agreement once we have brought the true and estimated values into alignment.

■ Fig. 6.14a shows that uncertainty is decreasing over time. The final covariance matrix is shown graphically in ■ Fig. 6.14b and we see a complex structure. Unlike the mapping case of ■ Fig. 6.11, $\hat{\mathbf{P}}_{mm}$ is not block diagonal and the finite off-diagonal terms represent correlations *between* the landmarks in the map. The landmark uncertainties never increase, the position prediction model is that they do not move, but they can never drop below the initial position uncertainty of the robot in $\hat{\mathbf{P}}_0$. The block $\hat{\mathbf{P}}_{vm}$ is the correlation between errors in the robot configuration and the landmark positions.

A landmark's position estimate is a function of the robot's pose, so errors in that pose appear as errors in the landmark position – and vice versa. The correlations are used by the Kalman filter to apply the innovation from a landmark observation to improving the estimate of every other landmark in the map as well as the robot configuration. Conceptually it is as if all the states were connected by springs and the movement of any one affects all the others.

The extended Kalman filter is a powerful tool for estimation but has a number of limitations. Firstly, the size of the matrices involved increases with the number of landmarks and can lead to memory and computational bottlenecks as well as numerical problems. Secondly, the underlying assumption of the Kalman filter is that all errors have a Gaussian distribution and this is far from true for sensors like lidars that we will discuss later in this chapter. Thirdly, it requires good estimates of covariance of the noise sources, which in practice is challenging.

**6**

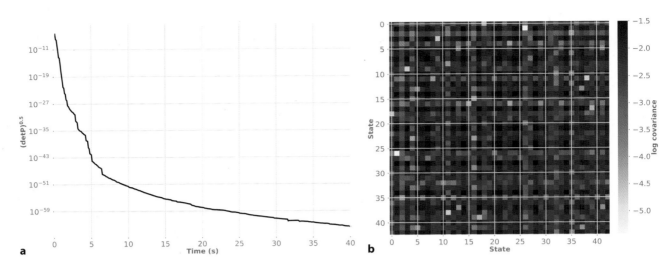

a   b

◻ **Fig. 6.14**   Simultaneous localization and mapping. **a** Covariance versus time; **b** the final covariance matrix

## 6.5   Pose-Graph SLAM

An alternative approach to the SLAM problem is to formulate it as a pose graph as shown in ◻ Fig. 6.15, where the robot's path is considered to be a sequence of distinct poses and the task is to estimate those poses. There are two types of vertex: robot poses denoted by circles and landmark positions denoted by stars. An edge between two vertices represents a spatial constraint between them due to some observation $z_{i,j}$. The constraint between poses is typically derived from odometry – distance travelled and change in heading. The constraint between a pose and a landmark is typically derived from a lidar sensor or a camera – distance and bearing angle from the robot to the landmark.

As the robot progresses counter clockwise around the square, it compounds an increasing number of uncertain relative poses from odometry so that the cumulative error in the estimated pose of the vertices is increasing – the problem with dead reckoning we discussed in ▶ Sect. 6.1. By the time the robot reaches the fourth position the pose error is significant. However, it is able to observe landmark A, which it saw previously and this adds the constraint shown in red, which forms a loop in the pose graph – this type of event is referred to as a *loop closure*.

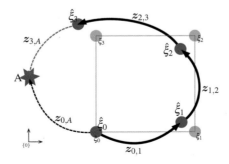

◻ **Fig. 6.15**   Fundamentals of pose-graph SLAM. The robot is moving counter-clockwise around the square. The estimated poses are denoted by blue circles and the solid arrows are observed odometry as the robot moves between poses. The blue star is a landmark point and observations of that point from poses in the graph are shown as dashed lines. $z_{i,j}$ is a sensor observation associated with the edge from vertex $i$ to $j$. We assume that $\hat{\xi}_0 = \xi_0$

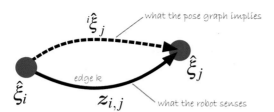

**Fig. 6.16** A single edge of the pose graph showing the estimated robot poses $\hat{\boldsymbol{\xi}}_i$ and $\hat{\boldsymbol{\xi}}_j$, the relative pose based on the sensor measurement $\boldsymbol{z}_{i,j}$, and the implicit relative pose shown as a dashed line

The function $\boldsymbol{g}(\cdot)$, for example (6.19), estimate the landmark's world coordinates given the robot pose and a sensor observation. The estimate $\boldsymbol{g}(\hat{\boldsymbol{\xi}}_0, \boldsymbol{z}_{0,A})$ will be subject to sensor error, but the estimate $\boldsymbol{g}(\hat{\boldsymbol{\xi}}_3, \boldsymbol{z}_{3,A})$ will be subject to sensor error and the accumulated odometry error in $\hat{\boldsymbol{\xi}}_3$. This discrepancy, analogous to the innovation of the Kalman filter, can be used to adjust the vertices – poses and landmarks – so as to minimize the total error. However, there is insufficient information to determine exactly where the error lies, so naively adjusting the $\hat{\boldsymbol{\xi}}_3$ vertex to fit the landmark observation might increase the error in another part of the graph. To minimize the error consistently over the whole graph we will formulate this as a minimization problem, which ensures that an error in one part of the graph is optimally distributed across the whole graph.

We consider first the case without landmarks, and all vertices represent pose. The state vector contains the poses of all the vertices

$$x = \{\boldsymbol{\xi}_0, \boldsymbol{\xi}_1, \ldots, \boldsymbol{\xi}_{N-1}\} \tag{6.29}$$

and we seek an estimate $\hat{x}$ that minimizes the error across all the edges

$$x^* = \arg\min_x \sum_k F_k(x) \tag{6.30}$$

where $F_k(x) \in \mathbb{R}_{\geq 0}$ is a cost associated with the edge $k$. The term that we are minimizing is the total edge error that is analagous to the potential energy in a flexible structure which we want to *relax* into a minimum energy state.

■ Fig. 6.16 shows two vertices from the pose graph, specifically vertex $i$ and vertex $j$, which represent the estimated poses $\hat{\boldsymbol{\xi}}_i$ to $\hat{\boldsymbol{\xi}}_j$, and edge $k$, which connects them. We have two values for the relative pose between the vertices. One is the explicit sensor measurement $\boldsymbol{z}_{i,j}$ and the other, shown dashed, is implicit and based on the current estimates from the pose graph. Any difference between the two indicates that one or both vertices need to be moved.

The first step is to express the error associated with the graph edge in terms of the sensor measurement and our best estimates of the vertex poses with respect to the world. For edge $k$ the relative pose, based on the current estimates in the pose graph, is

$${}^i\hat{\boldsymbol{\xi}}_j = \ominus\hat{\boldsymbol{\xi}}_i \oplus \hat{\boldsymbol{\xi}}_j$$

which we can write as an **SE**(2) matrix ${}^i\mathbf{T}_j$. We can also estimate the relative pose from the observation, the odometry $\boldsymbol{z}_k = \boldsymbol{z}_{i,j} = (\delta_d, \delta_\theta)$ which leads to

$$\mathbf{T}_z(\boldsymbol{z}_k) = \begin{pmatrix} \cos\delta_\theta & -\sin\delta_\theta & \delta_d\cos\delta_\theta \\ \sin\delta_\theta & \cos\delta_\theta & \delta_d\sin\delta_\theta \\ 0 & 0 & 1 \end{pmatrix}.$$

The *difference* between these two relative poses is $\mathbf{T}_e = \mathbf{T}_z^{-1}{}^i\mathbf{T}_j$ that can be written as

$$f_k(x, z) = \left[\mathbf{T}_z^{-1}(z_{i,j})\mathbf{T}^{-1}(x_i)\mathbf{T}(x_j)\right]_{xy\theta} \in \mathbb{R}^3$$

which is a configuration $(x_e, y_e, \theta_e) \in \mathbb{R}^2 \times \mathbf{S}^1$ that will be zero if the observation and the relative pose of the vertices are equal. To obtain the scalar cost required by (6.30) from the error vector $e$ we use a quadratic expression

$$F_k(x, z_k) = f_k^\top(x, z_k)\mathbf{\Omega}_k f_k(x, z_k) \tag{6.31}$$

where $\mathbf{\Omega}_k \in \mathbb{R}^{3\times3}$ is a positive-definite information matrix used as a weighting term. ◀ Although (6.31) is written as a function of all poses $x$, it in fact depends only on elements $i$ and $j$ of $x$ and the measurement $z_k$. Solving (6.30) is a complex optimization problem that does not have a closed-form solution, but this kind of nonlinear least squares problem can be solved numerically if we have a good initial estimate of $\hat{x}$. Specifically, this is a sparse nonlinear least squares problem which is discussed in ▶ App. F.2.4.

> In practice this matrix is diagonal reflecting confidence in the $x$-, $y$- and $\theta$-directions. The *bigger* (in a matrix sense) $\mathbf{\Omega}$ is, the more the edge *matters* in the optimization procedure. Different sensors have different accuracy and this must be taken into account. Information from a high-quality sensor should be given more weight than information from a low-quality sensor.

The edge error $f_k(x)$ can be linearized (see ▶ App. E) about the current state $x_0$ of the pose graph

$$f'_k(\mathbf{\Delta}) \approx f_{0,k} + \mathbf{J}_k\mathbf{\Delta}$$

where $\mathbf{\Delta}$ is a displacement relative to $x_0$, $f_{0,k} = f_k(x_0)$ and

$$\mathbf{J}_k = \frac{\partial f_k(x)}{\partial x} \in \mathbb{R}^{3\times3N}$$

is a Jacobian matrix that depends only on the pose of its two vertices $\boldsymbol{\xi}_i$ and $\boldsymbol{\xi}_j$ and is therefore mostly zeros

$$\mathbf{J}_k = \begin{pmatrix} \mathbf{0} & \cdots & \mathbf{A}_i & \cdots & \mathbf{B}_j & \cdots & \mathbf{0} \end{pmatrix}, \quad \text{where} \quad \mathbf{A}_i = \frac{\partial f_k(x)}{\partial \boldsymbol{\xi}_i} \in \mathbb{R}^{3\times3},$$
$$\mathbf{B}_j = \frac{\partial f_k(x)}{\partial \boldsymbol{\xi}_j} \in \mathbb{R}^{3\times3}$$

There are many ways to compute the Jacobians but here we will demonstrate the use of SymPy

```
>>> import sympy
>>> xi, yi, ti, xj, yj, tj = sympy.symbols("xi yi ti xj yj tj")
>>> xm, ym, tm = sympy.symbols("xm ym tm")
>>> xi_e = SE2(xm, ym, tm).inv() * SE2(xi, yi, ti).inv() \
...       * SE2(xj, yj, tj);
>>> fk = sympy.Matrix(sympy.simplify(xi_e.xyt()));
```

and the Jacobian that describes how the function $f_k$ varies with respect to $\boldsymbol{\xi}_i$ is

```
>>> Ai = sympy.simplify(fk.jacobian([xi, yi, ti]))
Matrix([
  [-cos(ti + tm), -sin(ti + tm), xi*sin(ti + tm) - xj*sin(ti + tm)
   - yi*cos(ti + tm) + yj*cos(ti + tm)],
  [sin(ti + tm), -cos(ti + tm), xi*cos(ti + tm) - xj*cos(ti + tm)
   + yi*sin(ti + tm) - yj*sin(ti + tm)],
  [0, 0, -1]])
>>> Ai.shape
(3, 3)
```

and we follow a similar procedure for $\mathbf{B}_j$. SymPy can automatically generate code, in a variety of programming languages, to compute the Jacobians.

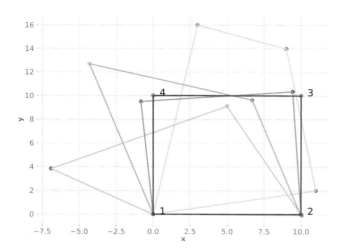

**◻ Fig. 6.17** Pose graph optimization showing the result at consecutive iterations with increasing darkness

It is quite straightforward to solve this type of pose-graph problem with the Toolbox. We load a simple pose graph, similar to ◻ Fig. 6.15, from a data file ▶

```
>>> pg = PoseGraph("data/pg1.g2o");
loaded g2o format file: 4 vertices, 4 edges
```

which returns a Toolbox `PoseGraph` object that describes the pose graph. We can visualize this by ▶

```
>>> pg.plot();
```

The optimization reduces the error in the network while animating the changing pose of the vertices

```
>>> pg.optimize(animate=True)
done in 4.60 msec.   Total cost 316.88
done in 2.66 msec.   Total cost 47.2186
...
done in 2.09 msec.   Total cost 3.14139e-11
```

The displayed text indicates that the total cost is decreasing while the graphics show the vertices moving into a configuration that minimizes the overall error in the network. The pose graph configurations are overlaid and shown in ◻ Fig. 6.17.

Now let's look a much larger example based on real robot data ▶

```
>>> pg = PoseGraph("data/killian-small.toro");
loaded TORO/LAGO format file: 1941 vertices, 3995 edges
```

which we can plot ▶

```
>>> pg.plot();
```

and this is shown in ◻ Fig. 6.18a. Note the mass of edges in the center of the graph, and if you zoom in you can see these in detail. We optimize the pose graph by

```
>>> pg.optimize()
solving....done in 0.91 sec. Total cost 1.78135e+06
  .

  .
solving....done in 1.1 sec. Total cost 5.44567
```

and the final configuration is shown in ◻ Fig. 6.18b. The original pose graph had severe pose errors from accumulated odometry error that meant the two trips along the corridor were initially very poorly aligned.

This simple text-based file format is used by the popular posegraph optimization package g²o which can be found at ▶ https://openslam-org.github.io/g2o.html.

The vertices have an orientation which is in the $z$-direction, rotate the graph to see this.

This simple text-based file format is used by the popular posegraph optimization package TORO which can be found at ▶ https://openslam-org.github.io/toro.html.

There are a lot of vertices and this takes a few seconds.

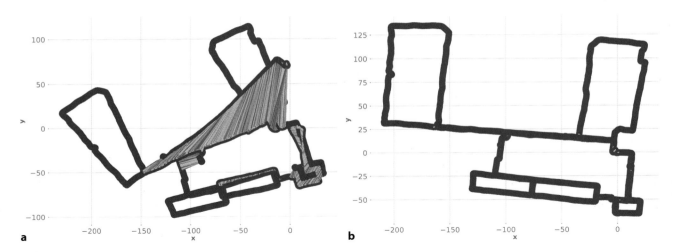

**a**   **b**

■ **Fig. 6.18** Pose graph with 1941 vertices in blue and 3995 edges in various colors, from the MIT Killian Court dataset. **a** Initial configuration; **b** final configuration after optimization

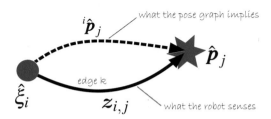

■ **Fig. 6.19** A single edge of the pose graph showing the estimated pose $\hat{\boldsymbol{\xi}}_i$ and estimated robot landmark position $\hat{\boldsymbol{p}}_j$, the relative position based on the sensor measurement $\boldsymbol{z}_{i,j}$, and the implicit relative position shown as a dashed line

The pose graph can also include landmarks as shown in ■ Fig. 6.19. Landmarks are described by a coordinate vector $\boldsymbol{p}_j \in \mathbb{R}^2$ not a pose, and therefore differ from the vertices discussed so far. To accommodate this, we redefine the state vector to be

$$\boldsymbol{x} = \{\boldsymbol{\xi}_0, \boldsymbol{\xi}_1, \ldots, \boldsymbol{\xi}_{N-1} | \boldsymbol{p}_0, \boldsymbol{p}_1, \ldots, \boldsymbol{p}_{M-1}\} \tag{6.32}$$

which includes $N$ robot poses and $M$ landmark positions. The robot at pose $i$ observes landmark $j$ at range and bearing $\boldsymbol{z}_{i,j} = (r^\#, \beta^\#)$, where $\cdot^\#$ denotes a measured value, and is converted to Cartesian form in frame $\{i\}$

$${}^i\boldsymbol{p}_j^\# = \left(r^\# \cos \beta^\#, \ r^\# \sin \beta^\#\right) \in \mathbb{R}^2 \ .$$

The estimated position of the landmark in frame $\{i\}$ can also be determined from the vertices of the pose graph

$${}^i\hat{\boldsymbol{p}}_j = \left(\ominus {}^0\hat{\boldsymbol{\xi}}_i\right) \cdot \hat{\boldsymbol{p}}_j \in \mathbb{R}^2$$

and the error vector is

$$\boldsymbol{f}_k(\boldsymbol{x}, \boldsymbol{z}_k) = {}^i\hat{\boldsymbol{p}}_j - {}^i\boldsymbol{p}_j^\# \in \mathbb{R}^2 \ .$$

Edge cost is given by (6.31) and $\boldsymbol{\Omega} \in \mathbb{R}^{2\times 2}$.

We follow a similar approach as earlier but the Jacobian matrix is now

$$\mathbf{J}_k = \frac{\partial \boldsymbol{f}_k(\boldsymbol{x})}{\partial \boldsymbol{x}} \in \mathbb{R}^{2\times(3N+2M)}$$

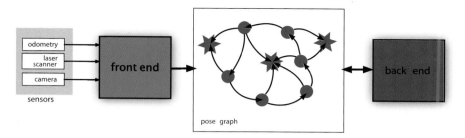

**▣ Fig. 6.20** Pose-graph SLAM system. The front end creates vertices as the robot travels, and creates edges based on sensor data. The back end adjusts the vertex positions to minimize total error

which again is mostly zero

$$\mathbf{J}_k = \begin{pmatrix} \mathbf{0} & \cdots & \mathbf{A}_i & \cdots & \mathbf{B}_j & \cdots & \mathbf{0} \end{pmatrix}$$

but the two nonzero blocks now have different widths

$$\mathbf{A}_i = \frac{\partial \boldsymbol{f}_k(\boldsymbol{x})}{\partial \boldsymbol{\xi}_i} \in \mathbb{R}^{2\times3}, \quad \mathbf{B}_j = \frac{\partial \boldsymbol{f}_k(\boldsymbol{x})}{\partial p_j} \in \mathbb{R}^{2\times2}$$

and the solution can be achieved as before, see ▶ App. F.2.4 for more details.

Pose graph optimization results in a graph that has optimal *relative* poses and positions but the absolute poses and positions are not necessarily correct. ▶ We can remedy this by fixing or *anchoring* one or more vertices (poses or landmarks) and not update them during the optimization, and this is discussed in ▶ App. F.2.4.

In practice a pose-graph SLAM system comprises two asynchronous subsystems as shown in ▣ Fig. 6.20: a *front end* and a *back end*, connected by a pose graph. The front end adds new vertices as the robot travels ▶ as well as edges that define constraints between vertices. For example, when moving from one place to another wheel odometry gives an estimate of distance and change in orientation which is a constraint. In addition, the robot's exteroceptive sensors may observe the relative position of a landmark and this also adds a constraint. Every measurement adds a constraint – an edge in the graph. There is no limit to the number of edges entering or leaving a vertex. The back end runs periodically to optimize the pose graph, and adjusts the poses of the vertices so that the constraints are satisfied as well as possible, that is, that the sensor observations are best explained. Since the graph is only ever extended in a local region it is possible to optimize just a local subset of the pose graph and less frequently optimize the entire graph. If vertices are found to be equivalent after optimization they can be merged. An incorrect landmark association will distort the entire graph and raise $\sum_k F_k(\hat{\boldsymbol{x}})$. One strategy to resolve this is progressively remove the highest cost edges until the graph is able to *relax* into a low energy state.

This is similar to the estimates in the map frame that we saw with the EKF SLAM system in ▶ Sect. 6.4.

Typically a new vertex is added by the front end every meter or so of travel, or after a sharp turn.

## 6.6 Sequential Monte-Carlo Localization

The estimation examples so far have assumed that the error in sensors such as odometry and landmark range and bearing have a Gaussian probability density function. In practice we might find that a sensor has a one-sided distribution (like a Poisson distribution) or a multi-modal distribution with several peaks. The functions we used in the Kalman filter, such as (6.2) and (6.8), are strongly nonlinear, which means that sensor noise with a Gaussian distribution will not result in a Gaussian error distribution on the value of the function – this is discussed further in ▶ App. H.2. The probability density function associated with a robot's configuration may have multiple peaks to reflect several hypotheses that equally well explain the data from the sensors as shown for example in ▣ Fig. 6.3c.

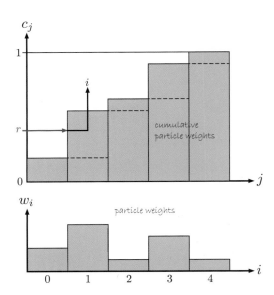

**□ Fig. 6.21** Random resampling. The cumulative histogram (top) is formed from the particle weights (bottom). A random number $r$ is used to select a particle $i$. The probability of selecting a highly weighted particle like 1 or 3 is clearly higher than for a low-weighted particle like 2 or 4

The Monte-Carlo estimator that we discuss in this section makes no assumptions about the distribution of errors. It can also handle multiple hypotheses for the state of the system. The basic idea is disarmingly simple. We maintain many *different* values of the robot's configuration or state vector. When a new measurement is available we score how well each particular value of the state explains what the sensor just observed. We keep the best fitting states and randomly sample from the prediction distribution to form a new generation of states. Collectively, these many possible states, and their scores, form a discrete approximation of the probability density function of the state we are trying to estimate. There is never any assumption about Gaussian distributions nor any need to linearize the system. While computationally expensive, it is quite feasible to use this technique with today's standard computers and it is well suited to parallel computation. If we plot these state vectors as points in the state space we have a cloud of particles each representing a plausible robot state. Hence this type of estimator is often referred to as a particle filter.

---

**Excurse 6.9: Monte Carlo Methods**

These are a class of computational algorithms that rely on repeated random sampling to compute their results. An early example of this idea is Buffon's needle problem posed in the eighteenth century by Georges-Louis Leclerc (1707–1788), Comte de Buffon: *Suppose we have a floor made of parallel strips of wood of equal width t, and a needle of length l is dropped onto the floor. What is the probability that the needle will lie across a line between the strips?* If $n$ needles are dropped and $h$ cross the lines, the probability can be shown to be $h/n = 2l/\pi t$ and in 1901 an Italian mathematician Mario Lazzarini performed the experiment, tossing a needle 3408 times, and obtained the estimate $\pi \approx 355/113$ (3.14159292).

Monte Carlo methods are often used when simulating systems with a large number of coupled degrees of freedom with significant uncertainty in inputs. Monte Carlo methods tend to be used when it is infeasible or impossible to compute an exact result with a deterministic algorithm. Their reliance on repeated computation and random or pseudo-random numbers make them well suited to calculation by a computer. The method was developed at Los Alamos as part of the Manhattan project during WW II by the mathematicians John von Neumann, Stanislaw Ulam and Nicholas Metropolis. The name Monte Carlo alludes to games of chance and was the code name for the secret project.

We will apply Monte-Carlo estimation to the problem of localization using odometry and a map. Estimating only three states $x = (x, y, \theta)$ is computationally tractable to solve with straightforward Python code. The estimator is initialized by creating $N$ particles $x_i$, $i = 0, \ldots, N-1$ distributed randomly over the configuration space of the robot. All particles have the same initial weight, importance or likelihood $w_i = 1/N$. The steps in the main iteration of the algorithm are:

1. Apply the state update to each particle

$$x_i^+\langle k+1 \rangle = f(x_i\langle k \rangle,\ u\langle k \rangle + r\langle k \rangle)$$

where $\hat{u}\langle k \rangle$ is the input to the system or the measured odometry $\hat{u}\langle k \rangle = \delta\langle k \rangle$ and $r\langle k \rangle \in \mathbb{R}^2$ is a random vector that represents uncertainty in the odometry. Often $r$ is drawn from a Gaussian random variable with covariance $\mathbf{R}$ but any physically meaningful distribution can be used. The state update is often simplified to

$$x_i^+\langle k+1 \rangle = f(x_i\langle k \rangle,\ u\langle k \rangle) + q\langle k \rangle$$

where $q\langle k \rangle \in \mathbb{R}^2 \times \mathbf{S}^1$ represents uncertainty in the pose of the robot.

2. We make an observation $z_j$ of landmark $j$ that has, according to the map, coordinate $p_j$. For each particle we compute the innovation

$$v_i = h(x_i^+,\ p_j) - z_j$$

which is the error between the predicted and actual landmark observation. A likelihood function provides a scalar measure of how well the particular particle explains this observation. In this example we choose a likelihood function

$$w_i = \mathrm{e}^{-v_i^\top \mathbf{L}^{-1} v_i} + w_0$$

to determine the weight of the particle, where $\mathbf{L}$ is a covariance-like matrix, and $w_0 > 0$ ensures that there is a finite probability of a particle being retained despite sensor error. We use a quadratic exponential function only for convenience, the function does not need to be smooth or invertible but only to adequately describe the likelihood of an observation. ▶

This particular type of estimator is known as a bootstrap type filter. The weight is computed at each step, with no dependence on previous values.

3. Select the particles that best explain the observation, a process known as resampling ▶ or importance sampling. A common scheme is to randomly select particles according to their weight. Given $N$ particles $x_i$ with corresponding weights $w_i$ we first normalize the weights

$$w_i' = \frac{w_i}{\sum_{i=0}^{N-1} w_i} \tag{6.33}$$

and construct a cumulative histogram

$$c_j = \Sigma_{i=0}^{j} w_i' . \tag{6.34}$$

As shown in ◼ Fig. 6.21 we then draw a uniform random number $r \in [0, 1]$ and find

$$i = \arg\min_j r < c_j$$

There are many resampling strategies for particle filters, both the resampling algorithm and the resampling frequency. Here we use the simplest strategy known variously as multinomial resampling, simple random resampling or select with replacement, at every time step. This is sometimes referred to as bootstrap particle filtering or condensation.

and select particle $i$ for the next generation. The process is repeated $N$ times. The end result will be a new set of $N$ particles with multiple copies of highly weighted particles while some lower weighted particles may not be copied at all. Step 1 of the next iteration will *spread out* the identical particles through the addition of the random vector $r\langle k \rangle$. Resampling is a critical component of the particle filter without which the filter would quickly produce a degenerate set of particles where a few have high weights and the bulk have almost zero weight which poorly represents the true distribution.

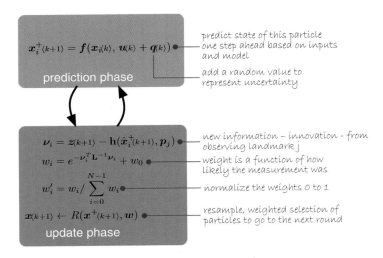

**Fig. 6.22** The particle filter estimator showing the prediction and update phases

These steps are summarized in ◧ Fig. 6.22. The Toolbox implementation is broadly similar to the previous examples. We create a map

```
>>> map = LandmarkMap(20, workspace=10);
```

and a robot with noisy odometry and an initial condition

```
>>> V = np.diag([0.02, np.deg2rad(0.5)]) ** 2;
>>> robot = Bicycle(covar=V, animation="car", workspace=map);
>>> robot.control = RandomPath(workspace=map)
```

and then a sensor with noisy readings

```
>>> W = np.diag([0.1, np.deg2rad(1)]) ** 2;
>>> sensor = RangeBearingSensor(robot, map, covar=W, plot=True);
```

For the particle filter we need to define two covariance matrices. The first is the covariance of the random noise added to the particle states at each iteration to represent uncertainty in configuration. We choose the covariance values to be comparable with those of **W**

```
>>> Q = np.diag([0.1, 0.1, np.deg2rad(1)]) ** 2;
```

and the covariance of the likelihood function applied to innovation

```
>>> L = np.diag([0.1, 0.1]);
```

Finally we construct a `ParticleFilter` estimator

```
>>> pf = ParticleFilter(robot, sensor=sensor, R=R, L=L,
...                      nparticles=1000);
```

which is configured with 1000 particles. The particles are initially uniformly distributed over the 3-dimensional configuration space.

We run the simulation for 10 seconds

```
>>> pf.run(T=10);
```

and watch the animation, snapshots of which are shown in ◧ Fig. 6.23. We see the particles moving as their states are updated by odometry, random perturbation and resampling. The initially randomly distributed particles begin to aggregate around those regions of the configuration space that best *explain* the sensor observations that are made. In Darwinian fashion these particles become more highly weighted and survive the resampling step, while the lower-weighted particles are extinguished.

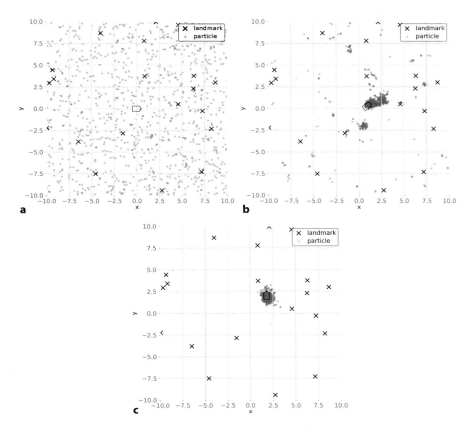

Particle filter results showing the evolution of the particle's configuration on the $xy$-plane over time. **a** initial particle distribution at $t = 0$; **b** particle distribution at $t = 1$; **a** particle distribution at $t = 3$. The initial robot configuration is shown by a gray vehicle icon

The particles approximate the probability density function of the robot's configuration. The most likely configuration is the expected value or mean of all the particles. A measure of uncertainty of the estimate is the spread of the particle cloud or its standard deviation. The `ParticleFilter` object keeps the history of the mean and standard deviation of the particle state at each time step, taking into account the particle weighting. ▶ As usual we plot the results of the simulation

```
>>> map.plot();
>>> robot.plot_xy();
```

and overlay the mean of the particle cloud

```
>>> pf.plot_xy();
```

which is shown in ▪ Fig. 6.24a. The initial part of the estimated path has quite high error since the particles have not converged on the true configuration. We can plot the standard deviation against time

```
>>> plt.plot(pf.get_std()[:100,:]);
```

and this is shown in ▪ Fig. 6.24b. We can see a sudden drop before $t = 2$ s as the particles that are distant from the true solution are rapidly eliminated. As mentioned at the outset, the particles are a sampled approximation to the PDF and we can display this as

```
>>> pf.plot_pdf()
```

The problem we have just solved is known in robotics as the kidnapped robot problem – a robot is placed in the world with no estimate of its initial pose, and needs

Here we have taken statistics over all particles. Other strategies are to estimate the kernel density at every particle – the sum of the weights of all neighbors within a fixed radius – and take the particle with the largest value.

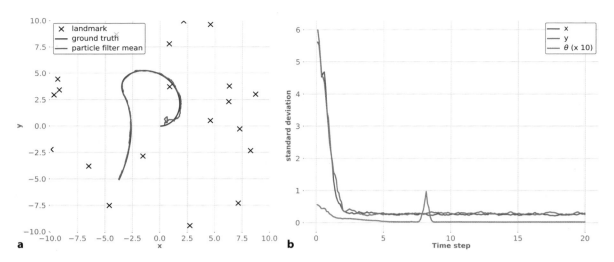

■ **Fig. 6.24** Particle filter results. **a** Robot path; **b** standard deviation of the particles versus time

to localize itself as quickly as possible. To represent this large uncertainty we uniformly distribute the particles over the 3-dimensional configuration space and their sparsity can cause the particle filter to take a long time to converge unless a very large number of particles is used. It is debatable whether this is a realistic problem. Typically we have some approximate initial pose of the robot and the particles would be initialized to that part of the configuration space. For example, if we know the robot is in a corridor then the particles would be placed in those areas of the map that are corridors, or if we know the robot is pointing north then set all particles to have that orientation.

Setting the parameters of the particle filter requires a little experience and the best way to learn is to experiment. For the kidnapped robot problem we set **Q** and the number of particles high so that the particles explore the configuration space but once the filter has converged lower values could be used. There are many variations of the particle filter concerned with the shape of the likelihood function and the resampling strategy.

## 6.7  Rao-Blackwellized SLAM

We will briefly and informally introduce the underlying principle of Rao-Blackwellized SLAM of which FastSLAM is a popular and well known instance. The approach is motivated by the fact that the size of the covariance matrix for EKF SLAM is quadratic in the number of landmarks, and for large-scale environments becomes computationally intractable.

If we compare the covariance matrices shown in ■ Figs. 6.11c and 6.14b we notice a stark contrast. In both cases we were creating a map of unknown landmarks but ■ Fig. 6.11c is mostly zero with a finite block diagonal structure whereas ■ Fig. 6.14b has no zero values at all. The difference is that for ■ Fig. 6.11c we assumed the robot trajectory was known exactly and that makes the landmark estimates *independent* – observing one landmark provides information about *only* that landmark. The landmarks are *uncorrelated*, hence all the zeros in the covariance matrix. If the robot trajectory is not known, the case for ■ Fig. 6.14b, then the landmark estimates are correlated – error in one landmark position is related to errors in robot pose and other landmark positions. The Kalman filter uses the correlation information so that a measurement of any one landmark provides information to improve the estimate of all the other landmarks and the robot's pose.

In practice we don't know the true pose of the robot but imagine a multi-hypothesis estimator ▶ where every hypothesis represents a robot trajectory that we *assume* is correct. This means that the covariance matrix will be block diagonal like ■ Fig. 6.11b – rather than a filter with a $2N \times 2N$ covariance matrix we can have $N$ simple filters that are each *independently* estimating the position of a single landmark and have a $2 \times 2$ covariance matrix. Independent estimation leads to a considerable saving in both memory and computation. Importantly though, we are only able to do this because we *assumed* that the robot's estimated trajectory is correct.

Each hypothesis also holds an estimate of the robot's trajectory to date. Those hypotheses that best explain the landmark measurements are retained and propagated while those that don't are removed and recycled. If there are $M$ hypotheses the overall computational burden falls from $O(N^2)$ for the EKF SLAM case to $O(M \log N)$ and in practice works well for $M$ in the order of tens to hundreds but can work for a value as low as $M = 1$.

Such as the particle filter that we discussed in ▶ Sect. 6.6.

## 6.8 Application: Lidar

To implement robot localization and SLAM we require a sensor that can make measurements of range and bearing to landmarks. Many such sensors are available, and are based on a variety of physical principles including ultrasonic ranging (■ Fig. 6.25a), computer vision or radar. One of the most common sensors used for robot navigation is lidar (■ Fig. 6.25b–d). A lidar emits short pulses of infrared laser light and measures the time for the reflected pulses to return. Operating range can be up to 100 m with an accuracy of the order of centimeters.

Common lidar configurations are shown in ■ Fig. 6.26. A 1D-lidar sensor, as shown in ■ Fig. 6.25b, measures the distance $r \in \mathbb{R}$, along the beam, to a surface – these are also called laser distance measuring devices or laser rangefinders. A 2D-lidar, as shown in ■ Fig. 6.25c, contains a 1D-lidar that rotates around a fixed axis. It returns a planar cross-section of the world in polar coordinate form $\mathbb{R} \times \mathbf{S}^1$, and emits a fixed number of pulses per revolution. ▶ For mobile robotic applications the lidar is typically configured to scan in a plane parallel to, and slightly above, the ground plane. A 3D-lidar, as shown in ■ Fig. 6.25d, contains multiple 1D-lidars that produce a fan of beams in a vertical plane, and that plane rotates around a fixed axis. It returns spherical coordinates $\mathbb{R} \times \mathbf{S}^2$ or 3D point coordinates $\mathbb{R}^3$ which can be visualized as a point cloud such as shown in ■ Fig. 6.27. Some lidars also

Typically every quarter, half or one degree. Rangefinders may have a reduced angular range of 180 or 270°, rather than a full revolution.

a      b      c      d

■ **Fig. 6.25** Robot range measuring sensors. **a** a low-cost ultrasonic rangefinder with maximum range of 6.5 m at 20 measurements per second (LV-MaxSonar-EZ1, image courtesy of SparkFun Electronics®); **b** a low-cost time-of-flight 1D-lidar with maximum range of 20 cm at 10 measurements per second (VL6180, image courtesy of SparkFun Electronics); **c** A 2D-lidar with a maximum range of 30 m, an angular range of 270° in 0.25° intervals at 40 scans per second (UTM-30LX, image courtesy of Hokuyo Automatic Co. Ltd.); **d** A 3D lidar with 16 beams, a $30 \times 360°$ field of view, 100 m range and 5-20 revolutions per second (VLP-16 Puck, image courtesy of Velodyne Lidar®)

**6**

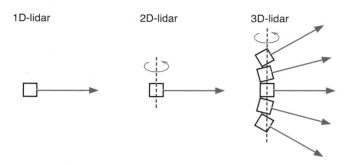

**Fig. 6.26** Common lidar configurations, the red arrow represents the laser beam

Velodyne Lidar

**Fig. 6.27** Point cloud created from a single scan of a Velodyne Alpha Prime 3D-lidar mounted on the vehicle shown closest (this vehicle is an inserted model and not part of the sensed point cloud). The lidar has a fan of 128 beams arranged in a vertical plane which trace out concentric circles on the ground plane, and the beam angles are not uniformly spaced. The trees on the right have cast *range shadows* and only their front surfaces are seen – this is sometimes called a "2.5D" representation. Points are colored according to the intensity (remission) of the returned laser pulse (image courtesy of Velodyne Lidar)

measure the return signal strength or remission which is a function of the infrared reflectivity of the surface.

Lidars have advantages and disadvantages compared to cameras and computer vision that we discuss in Parts IV and V of this book. On the positive side, lidars provide metric data, that is, they return actual range to points in the world in units of meters, and they can work in the dark. However lidars work less well than cameras outdoors since the returning laser pulse is overwhelmed by infrared light from the sun. Other disadvantages include the finite time to scan that is problematic if the sensor is moving; inability to discern fine texture or color; having moving parts; ◀ as well as being bulky, power hungry and expensive compared to cameras.

Solid-state lidars are an exciting emerging technology where optical beam steering replaces mechanical scanning.

### 6.8.1 Lidar-Based Odometry

A common application of lidars is lidar odometry, estimating the change in robot pose using lidar scan data rather than wheel encoder data. We will illustrate this with lidar scan data from a real robot

```
>>> pg = PoseGraph("data/killian.g2o.zip", lidar=True);
loaded g2o format file: 3873 vertices, 4987 edges
  3873 lidar scans: 180 beams, fov -90.0° to 90.0°, max range 50.0
```

and each scan is associated with a vertex of this already optimized pose graph. The range and bearing data for the scan at vertex 100 is

```
>>> [r, theta] = pg.scan(100);
>>> r.shape
(180,)
>>> theta.shape
(180,)
```

represented by two vectors each of 180 elements. We can plot these in polar form

```
>>> plt.clf()
>>> plt.polar(theta, r);
```

We will use lidar scans from two consecutive vertices

```
>>> p100 = pg.scanxy(100);
>>> p101 = pg.scanxy(100);
>>> p100.shape
(2, 180)
```

which are arrays whose columns are Cartesian point coordinates and these are overlaid in ◻ Fig. 6.28a. ▶ The robot is moving to the right, so the points on the wall toward the right of the figure are closer in the second scan.

To determine the change in pose of the robot between the two scans we need to align these two sets of points and this can be achieved with a scan-matching algorithm

Note that the points close to the lidar, at coordinate $(0, 0)$ in this sensor reference frame are much more tightly clustered and this is a characteristic of lidars where the points are equally spaced in angle not over an area.

```
>>> T = pg.scanmatch(100, 101);
>>> T.printline()
t = 0.506, -0.00872; 2.93°
```

which returns a relative pose $^{100}\mathbf{T}_{101} \in \mathbf{SE}(2)$ which is the motion of the robot – around 0.5 m in the $x$-direction and a slight rotation. We can also use $^{100}\mathbf{T}_{101}$ to transform points from frame $\{101\}$ to $\{100\}$ and as shown in ◻ Fig. 6.28b the points are now well aligned. The vertices of the graph also hold time stamp information and these two scans were captured

```
>>> pg.time(101) - pg.time(100)
1.76
```

seconds apart which indicates that the robot is moving quite slowly – a bit under $0.3 \, \text{ms}^{-1}$.

This scan matcher is based on the iterated closest point (ICP) algorithm which is covered in more detail in ▶ Sect. 14.7.2. At each iteration ICP assigns each point

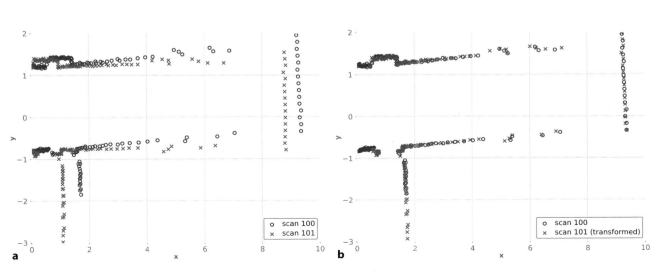

◻ **Fig. 6.28** Laser scan matching. **a** Laser scans from location 100 and 101; **b** location 100 points and transformed points from location 101

in the second set to the closest point in the first set, and then computes a transform that minimizes the sum of distances between all corresponding points. Some points may not actually be corresponding but as long as enough are, the algorithm will converge.

In practice there are additional challenges. Some lidar pulses will not return to the sensor if they fall on a surface with low reflectivity or on an oblique polished surface that specularly reflects the pulse away from the sensor – in these cases the sensor typically reports its maximum value. People moving through the environment change the shape of the world and temporarily cause a shorter range to be reported. In very large spaces all the walls may be beyond the maximum range of the sensor. Outdoors the beams can be reflected from rain drops, absorbed by fog or smoke and the return pulse can be overwhelmed by ambient sunlight. In long corridors, where the lidar scan pattern at consecutive poses are just two parallel lines, there is no way to gauge distance along the corridor – we would have to rely on wheel odometry only. Finally, the lidar, like all sensors, has measurement noise.

## 6.8.2 Lidar-Based Map Building

If the robot pose is sufficiently well known, through some localization process, then we can transform all the lidar scans to a global coordinate frame and build a map. Various map representations are possible but here we will outline how to build an occupancy grid as discussed in ▶ Chap. 5.

For a robot at a given pose, each beam in the scan is a directed ray that tells us several things. Firstly, the range measurement is the distance along the ray of the closest obstacle and we can determine the coordinates of the cell that contains an obstacle. However, we can tell nothing about cells further along the ray – the lidar only senses the surface of the object. Secondly, it is implicit that all the cells between the sensor and the first obstacle are obstacle free.

As the robot moves around the environment each cell is intersected by many rays and we use a simple voting scheme is used to determine whether cells are free or occupied. Based on the robot's pose and the beam we compute the equation of the ray and visit each cell in turn using the Bresenham line-drawing algorithm. All cells up to the returned range have their vote decreased (less occupied). The cell at the returned range has its vote increased (more occupied) and no further cells along the ray are visited. If the sensor receives no return laser pulse – the surface is not reflective or simply too distant – it sets the range to some maximum value, 50 m in this case, and we skip this ray.

```
>>> og = OccupancyGrid(workspace=[-100, 250, -100, 250],
                       cellsize=0.1, value=np.int32(0));
>>> pg.scanmap(og, maxrange=40)
>>> og.plot(cmap="gray")
```

and the result is shown in ◗ Fig. 6.29.

## 6.8.3 Lidar-Based Localization

We have mentioned landmarks a number of times in this chapter but avoided concrete examples of what they are. They could be distinctive visual features as discussed in ▶ Sect. 12.3 or artificial markers as discussed in ▶ Sect. 13.6.1. If we consider a lidar scan such as shown in ◗ Fig. 6.28a or ◗ Fig. 6.29b we see a fairly distinctive arrangement of points – a geometric signature – which we can use as a landmark. In many cases the signature will be ambiguous and of little value, for example a long corridor where all the points are collinear, but some signatures will

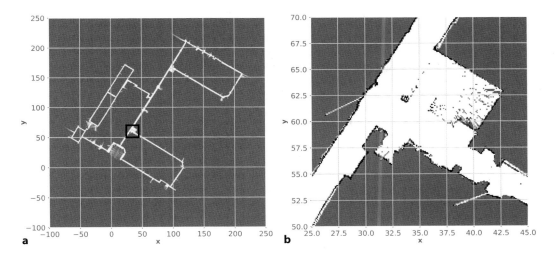

**◻ Fig. 6.29** **a** Laser scans rendered into an occupancy grid. An enlarged view of the area enclosed in the central square is dispayed in **b**. White cells are free space, black cells are occupied and gray cells are unknown. Grid cell size is 10 cm

be highly unique and can serve as a useful landmark. Naively, we could match the current lidar scan against all others and if the fit is good (the ICP error is low) we could add another constraint to the pose graph. However this strategy would be expensive with a large number of scans, so typically only scans in the vicinity of the robot's estimated position are checked, and this once again raises the data association problem.

## 6.9  Wrapping Up

In this chapter we learned about two very different ways of estimating a robot's position: by dead reckoning, and by observing landmarks whose true position is known from a map. Dead reckoning is based on the integration of odometry information, the distance traveled and the change in heading angle. Over time, errors accumulate leading to increased uncertainty about the pose of the robot.

We modeled the error in odometry by adding noise to the sensor outputs. The noise values are drawn from some distribution that describes the errors of that particular sensor. For our simulations we used zero-mean Gaussian noise with a specified covariance, but only because we had no other information about the specific sensor. The most realistic noise model available should be used. We then introduced the Kalman filter that provides an optimal estimate, in the least-squares sense, of the true configuration of the robot based on noisy measurements. The Kalman filter is however only optimal for the case of zero–mean Gaussian noise and a linear model. The model that describes how the robot's configuration evolves with time can be nonlinear in which case we approximate it with a linear model that included some partial derivatives expressed as Jacobian matrices – an approach known as extended Kalman filtering.

The Kalman filter also estimates uncertainty associated with the pose estimate and we see that the magnitude can never decrease and typically grows without bound. Only additional sources of information can reduce this growth and we looked at how observations of landmarks, with known positions, can be used. Once again we use the Kalman filter, but in this case we use both the prediction and the update phases of the filter. We see that in this case the uncertainty can be decreased by a landmark observation, and that over the longer term the uncertainty does not grow. We then applied the Kalman filter to the problem of estimating the positions of the landmarks given that we knew the precise position of the robot. In this case, the state vector of the filter comprised the coordinates of the landmarks.

Next we brought all this together and estimated the robot's configuration, the position of the landmarks and their uncertainties – simultaneous localization and mapping (SLAM). The state vector in this case contained the configuration of the robot and the coordinates of the landmarks.

An important problem when using landmarks is data association – determining which landmark has been observed by the sensor so that its position can be looked up in a map of known or estimated landmark positions. If the wrong landmark is looked up then an error will be introduced in the robot's position that may lead to future data associations errors and catastrophic divergence between the estimated and true state.

The Kalman filter scales poorly with an increasing number of landmarks and we introduced some alternative approaches. Firstly, pose-graph SLAM involves solving a large but sparse nonlinear least squares problem, turning the problem from one of (Kalman) filtering to one of optimization.

Secondly, we looked at Monte-Carlo estimation and introduced the particle filter. This technique is computationally intensive but makes no assumptions about the distribution of errors from the sensor or the linearity of the robot motion model, and supports multiple hypotheses. Particles filters can be considered as providing an approximate solution to the true system model, whereas a Kalman filter provides an exact solution to an approximate system model. Rao-Blackwellized SLAM combines Monte-Carlo and Kalman filter techniques.

We finished with a discussion of lidars which are a very common sensor for mobile robots, and showed how they can be applied to robot navigation, odometry and creating detailed floor plan maps.

### 6.9.1  Further Reading

■ ■ Localization and SLAM

The tutorials by Bailey and Durrant-Whyte (2006) and Durrant-Whyte and Bailey (2006) are a good introduction to this topic, while the textbook *Probabilistic Robotics* (Thrun et al. 2005) is a readable and comprehensive coverage of all the material touched on in this chapter. The book by Barfoot (2017) covers all the concepts of this chapter and more, albeit with different notation. The book by Siegwart et al. (2011) also has a good treatment of robot localization.

Particle filters are described by Thrun et al. (2005), Stachniss and Burgard (2014) and the tutorial introduction by Rekleitis (2004). There are many variations such as fixed or adaptive number of particles and when and how to resample – Li et al. (2015) provide a comprehensive review of resampling strategies. Determining the most likely pose was demonstrated by taking the weighted mean of the particles but many more approaches have been used. The kernel density approach takes the particle with the highest weight of neighboring particles within a fixed-size surrounding hypersphere.

Pose graph optimization, also known as GraphSLAM, has a long history starting with Lu and Milios (1997). More recently there has been significant progress with many publications (Grisetti et al. 2010) and open-source tools including g$^2$o (Kümmerle et al. 2011), $\sqrt{\text{SAM}}$ (Dellaert and Kaess 2006), iSAM (Kaess et al. 2007) and factor graphs leading to GTSAM ► https://gtsam.org/tutorials/intro.html. Agarwal et al. (2014) provides a gentle introduction to pose-graph SLAM and discusses the connection to land-based geodetic survey which is centuries old.

Rao-Blackwellized SLAM implementations include FastSLAM (Montemerlo et al. 2003; Montemerlo and Thrun 2007) as well as gmapping (► https://openslam-org.github.io/gmapping.html).

Popular implementations include Hector SLAM (▶ https://wiki.ros.org/hector_slam) and Cartographer (▶ https://github.com/cartographer-project/cartographer), both of which, along with gmapping, are available as part of the ROS ecosystem.

There are many online resources related to SLAM. A collection of open-source SLAM implementations available from OpenSLAM at ▶ http://www.openslam.org. An implementation of smoothing and mapping using factor graphs is available at ▶ https://gtsam.org and has C++, Python and MATLAB bindings. MATLAB implementations of SLAM algorithms can be found at ▶ https://www.iri.upc.edu/people/jsola/JoanSola/eng/toolbox.html and ▶ https://www-personal.acfr.usyd.edu.au/tbailey. Great teaching resources available online include Paul Newman's free ebook *C4B Mobile Robots and Estimation Resources* ebook at ▶ https://www.free-ebooks.net/ebook/C4B-Mobile-Robotics.

V-SLAM or visual SLAM is able to estimate 3-dimensional camera motion from images using natural images features (covered in ▶ Chap. 12) without the need for artificial landmarks. VI-SLAM is a V-SLAM system that also incorporates inertial sensor information from gyroscopes and accelerometers. VIO-SLAM is a VI-SLAM system that also incorporates odometry information. One of the first V-SLAM system was PTAM (parallel tracking and mapping) system by Klein and Murray (2007). PTAM has two parallel computational threads. One is the map builder that performs the front- and back-end tasks, adding landmarks to the pose graph based on estimated camera (robot) pose and performing graph optimization. The other thread is the localizer that matches observed landmarks to the estimated map to estimate the camera pose. ORB-SLAM is a powerful and mature family of V-SLAM implementations and the current version, ORB-SLAM3 (Campos 2021), supports V- and VI-SLAM with monocular, stereo and RGB-D cameras with regular and fisheye lenses. PRO-SLAM (▶ https://gitlab.com/srrg-software/srrg_proslam) is a simple stereo V-SLAM system designed for teaching.

#### ▪▪ Scan Matching and Map Making

Many versions and variants of the ICP algorithm exist and it is discussed further in ▶ Sect. 14.9.1. Improved convergence and accuracy can be obtained using the normal distribution transform (NDT), originally proposed for 2D by Biber and Straßer (2003), extended to 3D by Magnusson et al. (2007) and implementations are available at pointclouds.org. A comparison of ICP and NDT for a field robotic application is described by Magnusson et al. (2009). A fast and popular approach to lidar scan matching is that of Censi (2008) with additional resources at ▶ https://censi.science/research/robot-perception/plicp and this forms the basis of the canonical scan matcher in ROS.

When attempting to match a local geometric signature in a large point cloud (2- or 3D) to determine loop closure, we often wish to limit our search to a local spatial region. An efficient way to achieve this is to organize the data using a kd-tree which is provided in the `scipy.spatial` package. FLANN (Muja and Lowe 2009) is a fast approximation with a Python binding and available at ▶ https://github.com/flann-lib/flann.

For creating a map from robotic lidar scan data in ▶ Sect. 6.8 we used a naive approach – a more sophisticated technique is the beam model or likelihood field as described in Thrun et al. (2005).

#### ▪▪ Kalman Filtering

There are many published and online resources for Kalman filtering. Kálmán's original paper, Kálmán (1960), over 50 years old, is quite readable. The first use for aerospace navigation in the 1960s is described by McGee and Schmidt (1985). The book by Zarchan and Musoff (2005) is a very clear and readable introduction to Kalman filtering. I have always found the classic book, recently republished, Jazwinski (2007) to be very readable. Bar-Shalom et al. (2001) provide comprehensive coverage of estimation theory and also the use of GPS. Groves (2013) also covers Kalman filtering. Welch and Bishop's online resources at ▶ https://www. cs.unc.edu/~welch/kalman have pointers to papers, courses, software and links to other relevant web sites.

A significant limitation of the EKF is its first-order linearization, particularly for processes with strong nonlinearity. Alternatives include the iterated EKF described by Jazwinski (2007) or the Unscented Kalman Filter (UKF) (Julier and Uhlmann 2004) which uses discrete sample points (sigma points) to approximate the PDF. Some of these topics are covered in the Handbook (Siciliano and Khatib 2016, § 5 and § 35). The information filter is an equivalent filter that maintains an inverse covariance matrix which has some useful properties, and is discussed in Thrun et al. (2005) as the sparse extended information filter.

The insertion Jacobian (6.22) is important but details of its formulation are difficult to find, a derivation can be found at ▶ https://petercorke.com/robotics/ekf-covariance-matrix-update-for-a-new-landmark

### ▪▪ Data Association

SLAM techniques are critically dependent on accurate data association between observations and mapped landmarks, and a review of data association techniques is given by Neira and Tardós (2001). FastSLAM (Montemerlo and Thrun 2007) is capable of estimating data association as well as landmark position. Fiducials, such as April tags and ArUco markers, discussed in ▶ Sect. 13.6.1, can be used as artificial landmarks that eliminate the data association problem since the identity is encoded in the fiducial. Mobile robots can uniquely identify places based on their visual appearance using tools such as OpenFABMAP (Glover et al. 2012).

Data association for Kalman filtering is covered in the Robotics Handbook (Siciliano and Khatib 2016). Data association in the tracking context is covered in considerable detail in, the now very old, book by Bar-Shalom and Fortmann (1988).

### ▪▪ Sensors

The book by Kelly (2013) has a good coverage of sensors particularly lidar range finders. For aerial and underwater robots, odometry cannot be determined from wheel motion and an alternative, also suitable for wheeled robots, is visual odometry (VO). This is introduced in the tutorials by Fraundorfer and Scaramuzza (2012) and Scaramuzza and Fraundorfer (2011) and will be covered in ▶ Chap. 14. The Robotics Handbook (Siciliano and Khatib 2016) has good coverage of a wide range of robotic sensors. The principles of GPS and other radio-based localization systems are covered in some detail in the book by Groves (2013), and a number of links to GPS technical data are provided from this book's web site. The SLAM problem can be formulated in terms of bearing-only or range-only measurements. A camera is effectively a bearing-only sensor, giving the direction to a feature in the world. A VSLAM system is one that performs SLAM using bearing-only visual information, just a camera, and an introduction to the topic is given by Neira et al. (2008) and the associated special issue. Interestingly, the robotic VSLAM problem is the same as the bundle adjustment problem known to the computer vision community and which will be discussed in ▶ Chap. 14.

The book by Borenstein et al. (1996), although very dated, has an excellent discussion of robotic sensors in general and odometry in particular. It is out of print but can be found online. The book by Everett (1995) covers odometry, range and bearing sensors, as well as radio, ultrasonic and optical localization systems. Unfortunately the discussion of range and bearing sensors is now quite dated since this technology ha evolved rapidly over the last decade.

### ▪▪ General Interest

Bray (2014) gives a very readable account of the history of techniques to determine our location on the planet. If you ever wondered how to navigate by the stars or use a sextant Blewitt (2011) is a slim book that provides an easy to understand introduction.

The book *Longitude* (Sobel 1996) is a very readable account of the longitude problem and John Harrison's quest to build a marine chronometer.

## 6.9.2 Exercises

1. What is the value of the Longitude Prize in today's currency?
2. Implement a driver object (▶ Sect. 6.1.1) that drives the robot around inside a circle with specified center and radius.
3. Derive an equation for heading change in terms of the rotational rate of the left and right wheels for the car-like and differential-steer robot models.
4. Dead-reckoning (▶ Sect. 6.1)
    a) Experiment with different values of $\mathbf{P}_0$, $\mathbf{V}$ and $\hat{\mathbf{V}}$.
    b) ◼ Fig. 6.5 compares the actual and estimated position. Plot the actual and estimated heading angle.
    c) Compare the variance associated with heading to the variance associated with position. How do these change with increasing levels of range and bearing angle variance in the sensor?
    d) Derive the Jacobians in (6.6) and (6.7) for the case of a differential steer robot.
5. Using a map (▶ Sect. 6.2)
    a) Vary the characteristics of the sensor (covariance, sample rate, range limits and bearing angle limits) and investigate the effect on performance
    b) Vary $\mathbf{W}$ and $\hat{\mathbf{W}}$ and investigate what happens to estimation error and final covariance.
    c) Modify the `RangeBearingSensor` to create a bearing-only sensor, that is, as a sensor that returns angle but not range. The implementation includes all the Jacobians. Investigate performance.
    d) Modify the sensor model to return occasional errors (specify the error rate) such as incorrect range or beacon identity. What happens?
    e) Modify the EKF to perform data association instead of using the landmark identity returned by the sensor.
    f) ◼ Fig. 6.9 compares the actual and estimated position. Plot the actual and estimated heading angle.
    g) Compare the variance associated with heading to the variance associated with position. How do these change with increasing levels of range and bearing angle variance in the sensor?
6. Making a map (▶ Sect. 6.3)
    a) Vary the characteristics of the sensor (covariance, sample rate, range limits and bearing angle limits) and investigate the effect on performance.
    b) Use the bearing-only sensor from above and investigate performance relative to using a range and bearing sensor.
    c) Modify the EKF to perform data association instead of using identity returned by the sensor.
7. Simultaneous localization and mapping (▶ Sect. 6.4)
    a) Vary the characteristics of the sensor (covariance, sample rate, range limits and bearing angle limits) and investigate the effect on performance.
    b) Use the bearing-only sensor from above and investigate performance relative to using a range and bearing sensor.
    c) Modify the EKF to perform data association instead of using the landmark identity returned by the sensor.
    d) ◼ Fig. 6.13 compares the actual and estimated position. Plot the actual and estimated heading angle.

e) Compare the variance associated with heading to the variance associated with position. How do these change with increasing levels of range and bearing angle variance in the sensor?

8. Modify the pose-graph optimizer and test using the simple graph `pg1.g2o`
   a) anchor one vertex at a particular pose.
   b) add one or more landmarks. You will need to derive the relevant Jacobians first and add the landmark positions, constraints and information matrix to the data file.

9. Create a simulator for Buffon's needle problem, and estimate π for 10, 100, 1000 and 10,000 needle throws. How does convergence change with needle length?

10. Particle filter (▶ Sect. 6.6)
    a) Run the filter numerous times. Does it always converge?
    b) Vary the parameters $\mathbf{Q}$, $\mathbf{L}$, $w_0$ and $N$ and understand their effect on convergence speed and final standard deviation.
    c) Investigate variations to the kidnapped robot problem. Place the initial particles around the initial pose. Place the particles uniformly over the $xy$-plane but set their orientation to its actual value.
    d) Use a different type of likelihood function, perhaps inverse distance, and compare performance.

11. Experiment with ArUCo markers or April tags. Print some tags and extract them from images using the Toolbox functions described in ▶ Sect. 13.6.1.

12. Implement a lidar odometer and test it over the entire path saved in `killian.g2o`. Compare your odometer with the relative pose changes in the file.

13. In order to measure distance using lidar what timing accuracy is required to achieve 1 cm depth resolution?

14. Reformulate the localization, mapping and SLAM problems using a bearing-only landmark sensor.

15. Implement a localization or SLAM system using an external simulator such as CoppeliaSim. Obtain range measurements from the simulated robot, do lidar odometry and landmark recognition, and send motion commands to the robot. You can communicate with CoppeliaSim usings its Python API. Alternatively use the Gazebo simulators and communicate using the ROS protocol.

# Robot Manipulators

## Contents

Robot manipulator arms are a common and familiar type of robot. They are most commonly found in factories doing jobs such as assembly, welding and handling tasks. Other applications include picking items from shelves for e-commerce, packing fruit into trays, performing surgical procedures under human supervision, or even grappling cargo at the international space station. The first of these robots started work over 60 years ago and they have been enormously successful in practice – many millions of them are working in the world today, and they have assembled, packed or handled many products that we buy. Unlike the mobile robots we discussed in the previous part, robot manipulators do not move through the world. They have a static base and therefore operate within a limited workspace.

The chapters in this part cover the fundamentals of serial-link manipulators, a subset of all robot manipulator arms. ▶ Chap. 7 introduces the topic and is concerned with kinematics – the geometric relationship between the robot's joint angles or positions, and the pose of its end effector. We discuss the creation of smooth paths that the robot can follow and present an example of a robot drawing a letter on a plane and a 4-legged walking robot. ▶ Chap. 8 introduces the relationship between the rate of change of joint coordinates and the end-effector velocity which is described by the manipulator Jacobian matrix. We also cover alternative methods of generating paths in 3D space, and introduce the relationship between forces on the end effector, and torques or forces at the joints. ▶ Chap. 9 discusses independent joint control and some performance limiting factors such as weight and variable inertia. This leads to a discussion of the full nonlinear dynamics of serial-link manipulators – including inertia, gyroscopic forces, friction and gravity – and more sophisticated model-based and task-space control approaches.

# Robot Arm Kinematics

» *Take to kinematics. It will repay you.*
*It is more fecund than geometry; it adds a fourth dimension to space.*
– Chebyshev to Sylvester 1873

## Contents

© The Author(s), under exclusive license to Springer Nature Switzerland AG 2023
P. Corke, *Robotics, Vision and Control*, Springer Tracts in Advanced Robotics 146,
https://doi.org/10.1007/978-3-031-06469-2_7

**7**

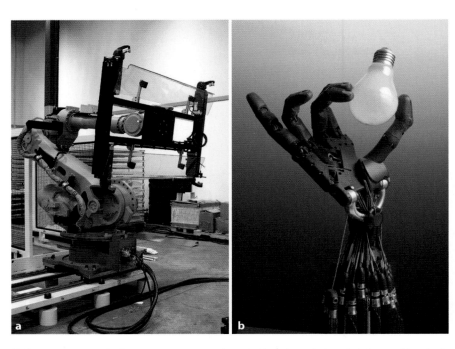

chap7.ipynb

► go.sn.pub/6RxahB

Degrees of freedom were covered in ► Sect. 2.4.9.

Kinematics is derived from the Greek word for motion.

A robot manipulator arm comprises mechanical links and computer-controllable joints. The base of the robot is generally fixed and motion of the joints changes the pose of the tool or end effector in space in order to perform useful work. End-effectors range in complexity from simple 2-finger or parallel-jaw grippers to complex human-like hands with multiple actuated finger joints and an opposable thumb – some examples are shown in ■ Fig. 7.1.

There are many different types of robot manipulators and ■ Fig. 7.2 shows some of that diversity. Most common is the six degree of freedom (DoF) ◄ robot manipulator arm comprising a series of rigid links and six actuated rotary joints. The SCARA (Selective Compliance Assembly Robot Arm) is rigid in the vertical direction and compliant in the horizontal plane which is advantageous for planar tasks such as electronic circuit board assembly. A gantry robot has two DoF of motion along overhead rails which gives it a very large working volume. In these robots the joints are connected in series – they are serial-link manipulators – and each joint has to support and move all the joints between itself and the end effector. In contrast, a parallel-link manipulator has all the motors at the base and connected directly to the end effector by mechanical linkages. Low moving mass, combined with the inherent stiffness of the parallel link structure, makes this class of robot capable of very high acceleration.

This chapter is concerned with *kinematics* – the branch of mechanics that studies the motion of a body, or a system of bodies, without considering mass or force. ◄ The pose of an end effector is a function of the state of each of the robot's joints and this leads to two important kinematic problems. Forward kinematics, covered in ► Sect. 7.1, is about determining the pose of the end effector from the joint configuration. Inverse kinematics, covered in ► Sect. 7.2, is about determining the joint configuration given the end-effector pose. ► Sect. 7.3 describes methods for generating smooth paths for the end effector. ► Sect. 7.4 works two complex applications: writing on a planar surface and a four- legged walking robot whose legs are simple robotic arms. ► Sect. 7.5 concludes with a discussion of some advanced topics.

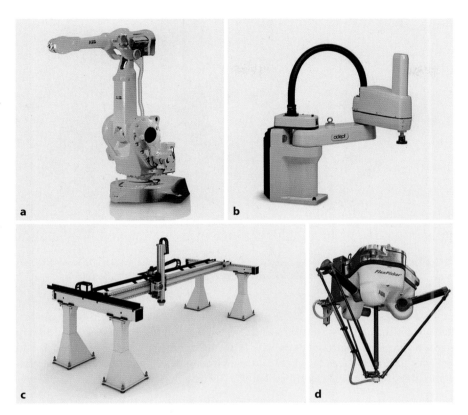

**◘ Fig. 7.2　a** A 6 DoF general-purpose industrial manipulator (source: ABB). **b** SCARA robot which has 4 DoF, typically used for electronic assembly (image of Adept Cobra s600 SCARA robot courtesy of Adept Technology, Inc.). **c** A gantry robot; the arm moves along an overhead rail (image courtesy of Güdel AG Switzerland, Mario Rothenbühler, ▶ https://www.gudel.com). **d** A parallel-link manipulator, the end effector is driven by 6 parallel links (source: ABB)

## 7.1　Forward Kinematics

The robots shown in ◘ Fig. 7.2a–c are serial-link manipulators and comprise a series of rigid links connected by actuated joints. Each joint controls the relative pose of the two links it connects, and the pose of the end effector is a function of all the joints. A series of links and joints is a kinematic chain or a rigid-body chain. Most industrial robots like ◘ Fig. 7.2a have six joints but increasingly we are seeing robots with seven joints which enables greater dexterity and a larger usable workspace – we will discuss that topic in ▶ Chap. 8. The YuMi® robot shown in ◘ Fig. 7.3a has two kinematic chains, each with seven joints. The Atlas™ robot in ◘ Fig. 7.3b has four chains: two of the chains are arms, and two of the chains are legs, with the end effectors being hands and feet respectively.

In robotics the most common type of joint is revolute, where one link rotates relative to another, about an axis fixed to both. Less common is a prismatic joint, also known as a sliding or telescopic joint, where one link slides relative to another along an axis fixed to both. The robots in ◘ Fig. 7.2b,c have one and three prismatic joints respectively. Many other types of joint are possible such as screw joints (1 DoF), cylindrical joints (2 DoF, one revolute and one prismatic) or spherical joints (3 DoF) but they are rare in robotics and will not be discussed here. ▶

We are interested in understanding the relationship between the joints, which we control, and the pose of the end effector which does the work. This is the forward kinematics, a mapping from joint coordinates, or robot configuration, to end-effector pose. We start in ▶ Sect. 7.1.1 by describing the problem in terms

In mechanism theory, all these joints are collectively known a  lower pairs and involve contact between two surfaces. Higher pairs involve contact between a point and a surface or a line and a surface.

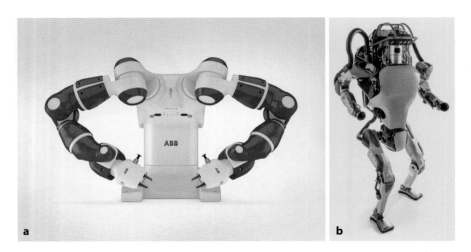

**◻ Fig. 7.3** **a** ABB Yumi® has two kinematic chains and 14 joints (source: ABB); **b** Boston Dynamics Atlas™ humanoid robot has four kinematic chains and 28 joints (source: Boston Dynamics)

of pose graphs, in ► Sect. 7.1.2 we introduce the notion of rigid-body chains, and in ► Sect. 7.1.3 we introduce chains with multiple branches or rigid-body trees. ► Sect. 7.1.4 introduces the URDF format for exchange of robot models and ► Sect. 7.1.5 introduces the classical Denavit-Hartenberg notation. In each section we treat the simpler 2D case first, then the 3D case.

### 7.1.1  Forward Kinematics from a Pose Graph

See ► Sect. 2.1.3 for an introduction to pose graphs.

In this section we apply our knowledge of pose graphs ◄ to solve the forward kinematics problem for robots that move in 2D and 3D.

#### 7.1.1.1  2-Dimensional (Planar) Robotic Arms

We will start simply with a single kinematic chain operating in the plane. Consider the very simple example shown in ◻ Fig. 7.4a which comprises one revolute joint and one link and is somewhat like the hand of a clock. The pose graph depicts the composition of two relative poses. The difference to the pose graphs we discussed in ► Sect. 2.1.3 is that one arrow is a function of a joint variable and is shown here in red. The arrows represent, in order, a rotation by the joint angle $q$ and then a fixed translation in the $x$-direction by a distance $a_1$. The pose of the end effector is simply

$$^0\xi_E = \xi^r(q) \oplus \xi^{t_x}(a_1)$$

where the relative pose $\xi$ is written as a function with a parameter which is either a variable, for example $q$, or a constant, for example $a_1$. The superscript indicates the specific function: $\xi^r$ is a relative pose that is a planar rotation and $\xi^{t_x}$ is a relative pose that is a planar translation in the $x$-direction.

So called because the rigid-body transformations are the simplest, or most elemental, transformations possible: a rotation, a translation in the $x$-direction, or a translation in the $y$-direction.

Using the Toolbox we represent this by an elementary transformation sequence (ETS) ◄ which we create by composing two elementary transformations which, for the 2-dimensional case, are created by methods of the ET2 class

```
>>> a1 = 1;
>>> e = ET2.R() * ET2.tx(a1);
```

and we use ⋆ to denote composition since ⊕ is not a Python operator. The result is a 2D elementary transformation sequence encapsulated in an ETS2 instance, a

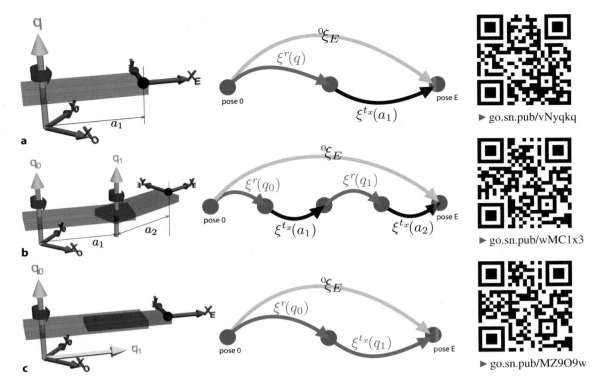

▶ go.sn.pub/vNyqkq

▶ go.sn.pub/wMC1x3

▶ go.sn.pub/MZ9O9w

◻ **Fig. 7.4** Some simple planar robotic arms and their associated pose graphs. The end effector is shown as a black sphere. The generalized joint variables are $q_i$. In the pose graph, edges can be variable due to joint variables (red) or constant (black). Frame {0} is the fixed world-reference frame

list-like object

```
>>> len(e)
2
```

that contains two rigid-body transforms: one variable and one constant. The variable transformation is created by the method with no arguments, while the constant corresponds to the method with one argument – a constant numeric robot dimension, in this case, equal to one. When the object is displayed ▶

```
>>> e
R(q) ⊕ tx(1)
```

This is the "pretty printed" version shown by ipython. In the regular Python REPL use `print(e)`.

we see that the empty parentheses have been replaced with q to make it clear that it represents a joint variable rather than a constant. The ⊕ symbol reminds us that this is a relative pose resulting from compounding two elementary motions.

For a particular value of the joint variable, say $q = \frac{\pi}{6}$ radians, the end-effector pose is

```
>>> e.fkine(pi / 6)
   0.866    -0.5      0.866
   0.5       0.866    0.5
   0         0         1
```

which is an SE2 object computed by composing the two relative poses in the ETS. The argument to `fkine` is substituted for the joint variable in the expression and is equivalent to

```
>>> SE2.Rot(pi / 6) * SE2.Tx(a1)
   0.866    -0.5      0.866
   0.5       0.866    0.5
   0         0         1
```

which is the forward kinematics for our very simple robot.

**7**

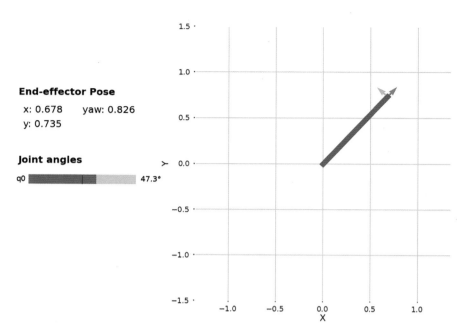

**End-effector Pose**

x: 0.678    yaw: 0.826
y: 0.735

**Joint angles**

q0                47.3°

■ **Fig. 7.5** Toolbox depiction of the 1-joint planar robot from ■ Fig. 7.4a using the `teach` method. The left-hand panel displays the position and orientation (yaw angle) of the end effector (in radians) and contains the joint angle slider (in degrees). The right-hand panel shows the robot and its end-effector coordinate frame

An easy and intuitive way to understand how this simple robot behaves is by

```
>>> e.teach()
```

which generates an interactive graphical representation of the robot arm as shown in ■ Fig. 7.5. As we adjust the joint angle using the slider, the arm's shape changes and the end-effector position and orientation are updated. This is not a particularly useful robot arm since its end effector can only reach points that lie on a circle.

Consider now the robot shown in ■ Fig. 7.4b which has two revolute joints $q_0$ and $q_1$. From the pose graph, the pose of the end effector is

$$^0\xi_E = \xi^r(q_0) \oplus \xi^{t_x}(a_1) \oplus \xi^r(q_1) \oplus \xi^{t_x}(a_2)$$

and using the Toolbox we write

```
>>> a1 = 1; a2 = 1;
>>> e = ET2.R() * ET2.tx(a1) * ET2.R() * ET2.tx(a2);
```

When the object is displayed

```
>>> e
R(q0) ⊕ tx(1) ⊕ R(q1) ⊕ tx(1)
```

we see that the empty parentheses have been replaced with sequential joint variables q0 and q1. We can evaluate this expression for specific joint angles, in this case given in degrees ◄

*Any iterable type can be passed, for example a Python list or tuple, or a NumPy array.*

```
>>> e.fkine(np.deg2rad([30, 40])).printline()
t = 1.21, 1.44; 70°
```

and the result is the end-effector pose when $q_0 = 30°$ and $q_1 = 40°$. This is equivalent to

```
>>> T = SE2.Rot(np.deg2rad(30)) * SE2.Tx(a1)
...      * SE2.Rot(np.deg2rad(40)) * SE2.Tx(a2);
>>> T.printline()
t = 1.21, 1.44; 70°
```

The `ETS2` object has many methods. We can find the number of joints, and the ET elements in the sequence that are joints

```
>>> e.n
2
>>> e.joints()
[R(q0), R(q1)]
```

The joint structure of a robot is often given in a shorthand notation comprising the letters R (for revolute) or P (for prismatic) to indicate the number and types of its joints. For this kinematic chain

```
>>> e.structure
'RR'
```

indicates a revolute-revolute sequence of joints. We could display the robot interactively as in the previous example, or noninteractively by

```
>>> e.plot(np.deg2rad([30, 40]));
```

The `ETS2` object also acts like a list. We can slice it

```
>>> e[1]
tx(1)
```

and this single element, an `ET2` object, has a number of methods

```
>>> e[1].eta
1
>>> e[1].A()
array([[1, 0, 1],
       [0, 1, 0],
       [0, 0, 1]])
```

which return respectively, the transformation constant $\eta$ passed to the constructor, `a1` in this case, and the corresponding $\mathbf{SE}(2)$ matrix which in this case represents a translation of 1 in the $x$-direction.

This simple planar robot arm has some interesting characteristics. Firstly, most end-effector *positions* can be reached with two different sets of joint angles. Secondly, the robot can position the end effector at any point within its reach but cannot achieve an arbitrary orientation as well.

We describe the configuration of a robot manipulator with $N$ joints by a vector of generalized coordinates where $\boldsymbol{q}_j \in \mathbf{S}^1$ is an angle for a revolute joint or $\boldsymbol{q}_j \in \mathbb{R}$ is a length ▶ for a prismatic joint. We refer to $\boldsymbol{q}$ as the *joint configuration* or *joint coordinates* and for an all-revolute robot as the *joint angles*.

This is a real interval since the joint length will have a minimum and maximum possible value.

🛑 This is sometimes, confusingly, referred to as the pose of the manipulator, but in this book we use the term pose to mean $\boldsymbol{\xi}$ the position and orientation of a body in space.

Recalling our discussion of configuration and task space from ▶ Sect. 2.4.9 this robot has 2 degrees of freedom and its configuration space is $C \in \mathbf{S}^1 \times \mathbf{S}^1$ and $\boldsymbol{q} \in C$. This is sufficient to reach points in the task space $\mathcal{T} \subset \mathbb{R}^2$ since dim $C =$ dim $\mathcal{T}$. However if our task space includes orientation $\mathcal{T} \subset \mathbb{R}^2 \times \mathbf{S}^1$ then it is underactuated since dim $C <$ dim $\mathcal{T}$ and the robot can access only a subset of the task space, for example, positions but not orientation.

We will finish up with the robot shown in ◻ Fig. 7.4c which includes a prismatic joint and is commonly called a polar-coordinate robot arm. The end-effector pose is

$$^0\boldsymbol{\xi}_E = \boldsymbol{\xi}^r(q_0) \oplus \boldsymbol{\xi}^{t_x}(q_1)$$

and using the Toolbox this is ▶

```
>>> e = ET2.R() * ET2.tx(qlim=[1,2])
```

For a prismatic joint we need to specify the range of motion `qlim` so that the `teach` method can determine the dimensions of the plot surface and the scale of the slider. Unless otherwise specified, revolute joints are assumed to have a range $[-\pi, \pi)$.

**7**

```
R(q0)  ⊕  tx(q1)
>>> e.structure
'RP'
```

and like the previous objects it has `fkine`, `teach`, and `plot` methods.

We could easily add more joints and use the now familiar Toolbox functionality to explore the capability of robots we create. A robot with three or more revolute joints is able to access all points in the task space $\mathcal{T} \subset \mathbb{R}^2 \times \mathbf{S}^1$, that is, achieve any pose in the plane (limited by reach).

A serial-link manipulator with $N$ joints numbered from 0 to $N-1$, has $N+1$ links, numbered from 0 to $N$. Joint $j$ connects link $j$ to link $j+1$ and moves them relative to each other. Link $\ell$ connects joint $\ell - 1$ to joint $\ell$. Link 0 is the base of the robot, typically fixed, and link $N$, is the last link of the robot and carries the end effector or tool.

### 7.1.1.2  3-Dimensional Robotic Arms

Most real-world robot manipulators have a task space $\mathcal{T} \subset \mathbb{R}^3 \times (\mathbf{S}^1)^3$ which allows arbitrary position and orientation of the end effector within its 3D working envelope or workspace. This requires a robot with a configuration space $\dim C \geq \dim \mathcal{T}$ which can be achieved by a robot with six or more joints. Using the 3D version of the code example above, we can define a robot that is a simplified human arm

```
>>> a1 = 1; a2 = 1;
>>> e = ET.Rz() * ET.Ry() \
...       * ET.tz(a1) * ET.Ry() * ET.tz(a2) \
...       * ET.Rz() * ET.Ry() * ET.Rz();
```

which has a 2 DoF spherical shoulder joint (line 2); an upper-arm, elbow joint, and lower-arm (line 3); and a 3 DoF spherical wrist joint which is a ZYZ Euler angle sequence (line 4). `e` is an `ETS` object, the 3-dimensional version of `ETS2`, which supports the same methods as `ETS2` used in the previous section

```
>>> e.n
6
>>> e.structure
'RRRRRR'
```

and the end-effector pose for all zero joint angles is

```
>>> e.fkine(np.zeros((6,)))
   1        0        0        0
   0        1        0        0
```

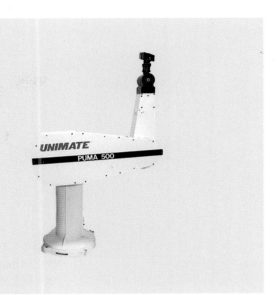

| 0 | 0 | 1 | 2 |
| 0 | 0 | 0 | 1 |

which is an SE3 instance.

## 7.1.2 Forward Kinematics as a Chain of Robot Links

The bulk of a robot's physical structure is due to its links and these dictate its shape and appearance, as well as having important properties such as mass, inertia, and collision potential. ◘ Fig. 7.6 shows the relationship between joints and link frames. The frame for link $\ell$ is indicated as $\{\ell\}$ and its pose is a function of the joint coordinates $\{q_j, j \in [0, \ell)\}$. We say that link $\ell$ is the parent of link $\ell + 1$ and the child of link $\ell - 1$. The fixed base is considered to be link 0 and has no parent, while the end effector has no child. The pose of the end effector, the forward kinematics is

$$
{}^{0}\boldsymbol{\xi}_N = {}^{0}\boldsymbol{\xi}_1(q_0) \oplus {}^{1}\boldsymbol{\xi}_2(q_1) \oplus \cdots \oplus {}^{N-1}\boldsymbol{\xi}_N(q_{N-1}) \tag{7.1}
$$

where ${}^{j}\boldsymbol{\xi}_{j+1}(q_j)$ is the pose of link frame $\{j+1\}$ with respect to link frame $\{j\}$. We can write this in functional form as

$$
{}^{0}\boldsymbol{\xi}_N = \mathcal{K}(\boldsymbol{q}) \tag{7.2}
$$

where $\mathcal{K}(\cdot)$ is specific to the robot and incorporates the joint and link parameters.

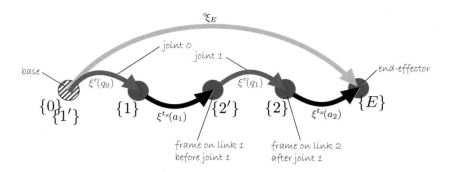

**Fig. 7.6** Pose graph showing link coordinate frames. Each relative pose in this graph is a function of a joint variable. Frame {0} is the base which is fixed and frame {E} holds the end effector

### 7.1.2.1 2-Dimensional (Planar) Case

We consider again the robot of ◻ Fig. 7.4b and rigidly attach a coordinate frame to each link as shown in ◻ Fig. 7.7. The pose of the link 1 frame depends only on $q_0$, while the pose of the link 2 frame depends on $q_0$ and $q_1$.

◻ Fig. 7.8 shows a more detailed pose graph for this robot with the joints and links highlighted. Frame $\{\ell'\}$ is the parent side of the joint, that is the joint is a transformation from $\{\ell'\}$ on the parent link to frame $\{\ell\}$ on the child. When $q_{\ell-1} = 0$ these two frames are identical.

We can describe the robot in a link-centric way by

```
>>> a1 = 1; a2 = 1;
>>> link1 = Link2(ET2.R(), name="link1");
>>> link2 = Link2(ET2.tx(a) * ET2.R(), name="link2", parent=link1);
>>> link3 = Link2(ET2.tx(a2), name="link3", parent=link2);
```

▶ go.sn.pub/uXIHHx

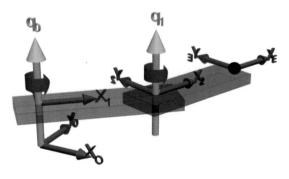

**Fig. 7.7** Coordinate frames are shown attached to the links of the robot from ◻ Fig. 7.4b. Link 1 is shown in red, and link 2 is in blue

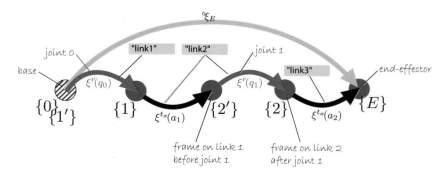

**Fig. 7.8** Pose graph for the robot from ◻ Fig. 7.4b highlighting the joints and link frames. Grey text boxes connect elements of the pose graph to the named `Link2` objects

which creates instances of `Link2` objects. Each constructor is passed a transformation from the parent's link frame to its own frame, a reference to its parent link, and an optional name. The transformation is described using an ETS and can include at most one joint variable, which if present, must be last in the sequence. In the example above we have also created a link frame for the end effector which has a constant transformation with respect to its parent.

We pass a list of these links objects to the robot object constructor and its display shows the rich structure we have created

```
>>> robot = ERobot2([link1, link2, link3], name="my robot")
ERobot2: my robot, 2 joints (RR)
```

| link | link | joint | parent | ETS: parent to link |
|------|------|-------|--------|---------------------|
| 0 | link1 | 0 | BASE | R(q0) |
| 1 | link2 | 1 | link1 | tx(1) $\oplus$ R(q1) |
| 2 | @link3 | | link2 | tx(1) |

At the top of the table we see the class type, the robot's name, the number of joints, and the joint structure. The @ symbol indicates a link frame that is also an end effector – it has no child links. ▶ The right-hand column shows the elementary transforms associated with this link. Sequential joint numbers have been assigned, starting at zero. This describes a kinematic chain, a sequence of rigid bodies and joints.

This robot subclass object has many methods. Forward kinematics returns an `SE2` instance

```
>>> robot.fkine(np.deg2rad([30, 40])).printline()
t = 1.21, 1.44; 70°
```

representing the pose of the `link3` frame. We can plot the robot at this configuration

```
>>> robot.plot(np.deg2rad([30, 40]));
```

but we can also plot a sequence of joint configurations as an animation. The sequence is provided as an array with the rows being consecutive configurations

```
>>> q = np.array([np.linspace(0, pi, 100),
...               np.linspace(0, -2 * pi, 100)]).T;
>>> q.shape
(100, 2)
>>> robot.plot(q);
```

The robot object behaves like a Python list and can be sliced

```
>>> robot[1]
Link2("link2", tx(1) ⊕ R(q), parent="link1")
```

which returns link 1, or it can be used as an iterator over the robot's links. It also behaves like a dictionary

```
>>> robot["link2"]
Link2("link2", tx(1) ⊕ R(q), parent="link1")
```

mapping a link name to a link reference. The robot's end effectors are

```
>>> robot.ee_links
[Link2("link3", tx(1), parent="link2")]
```

and in this case there is only one end effector in the returned list.

The link objects also have many methods and properties. We can obtain a reference to its parent link

```
>>> link2.parent
Link2("link1", R(q))
```

In the Python or IPython console that line is colored blue to indicate that it has a fixed transformation with respect to its parent.

as well as to its children

```
>>> link2.children
[Link2("link3", tx(1), parent="link2")]
```

which is a list of links. In this case there is only one child, but for branched robots there can be more.

The joint variable for this link has been assigned to $q_1$

```
>>> link2.jindex
1
```

and we can test the type of joint

```
>>> link2.isrevolute
True
>>> link2.isprismatic
False
```

`link3` would be neither revolute or prismatic – it is a fixed joint. Joints can have upper and lower limits, but in this case none are set

```
>>> print(link2.qlim)
None
```

For forward kinematics an important method is

```
>>> link2.A(pi / 6)
   0.866    -0.5      1
   0.5       0.866    0
   0         0        1
```

which is the relative pose of this link's frame with respect to its parent's frame or $^1\xi_2$. It evaluates the link's ETS

```
>>> link2.ets
tx(1)  ⊕  R(q)
```

for the particular value of $q$ passed to the A method. Forward kinematics is simply the product of the link transforms from base to tool.

### 7.1.2.2  3-Dimensional Case

For the 3D case we can create a robot model following a similar pattern but using the `Link`, `ERobot` and `ETS` classes instead. However, we can take some shortcuts. Firstly, we can create a robot directly from an ETS expression, in this case the example from ▶ Sect. 7.1.1.2

```
>>> a1 = 1; a2 = 1;
>>> robot6 = ERobot(ET.Rz() * ET.Ry() * ET.tz(a1) * ET.Ry() \
...                   * ET.tz(a2) * ET.Rz() * ET.Ry() * ET.Rz())
ERobot: noname, 6 joints (RRRRRR)
```

| link | link | joint | parent | ETS: parent to link |
|------|------|-------|--------|---------------------|
| 0 | link0 | 0 | BASE | Rz(q0) |
| 1 | link1 | 1 | link0 | Ry(q1) |
| 2 | link2 | 2 | link1 | tz(1) ⊕ Ry(q2) |
| 3 | link3 | 3 | link2 | tz(1) ⊕ Rz(q3) |
| 4 | link4 | 4 | link3 | Ry(q4) |
| 5 | @link5 | 5 | link4 | Rz(q5) |

The `ERobot` constructor has automatically partitioned the ETS expression and created link objects. We can now apply familiar methods like `fkine`, `plot` and `teach`.

The second short cut is to use one of the many models included in the Toolbox and these can be listed by

```
>>> models.list(type="ETS")
```

| class | name | manufacturer | type | DoF | di |
|-------|------|--------------|------|-----|-----|
| Panda | Panda | Franka Emika | ETS | 7 | 3d |
| Frankie | Frankie | Franka Emika, Omron | ETS | 9 | 3d |
| Puma560 | Puma560 | Unimation | ETS | 6 | 3d |
| Planar_Y | Planar-Y | | ETS | 6 | 3d |
| Planar2 | Planar2 | | ETS | 2 | 2d |
| GenericSeven | Generic Seven | Jesse's Imagination | ETS | 7 | 3d |
| XYPanda | XYPanda | Franka Emika | ETS | 9 | 3d |

*The result of this method is a wide table which has been cropped for inclusion in this book.

Each model is a subclass of the ERobot class. For example, to instantiate a model of the Panda robot is simply

```
>>> panda = models.ETS.Panda()
ERobot: Panda (by Franka Emika), 7 joints (RRRRRRR)
```

| link | link | joint | parent | ETS: parent to link |
|------|------|-------|--------|---------------------|
| 0 | link0 | 0 | BASE | tz(0.333) ⊕ Rz(q0) |
| 1 | link1 | 1 | link0 | Rx(-90°) ⊕ Rz(q1) |
| 2 | link2 | 2 | link1 | Rx(90°) ⊕ tz(0.316) ⊕ Rz(q2) |
| 3 | link3 | 3 | link2 | tx(0.0825) ⊕ Rx(90°) ⊕ Rz(q3) |
| 4 | link4 | 4 | link3 | tx(-0.0825) ⊕ Rx(-90°) ⊕ tz(0.3 |
| 5 | link5 | 5 | link4 | Rx(90°) ⊕ Rz(q5) |
| 6 | link6 | 6 | link5 | tx(0.088) ⊕ Rx(90°) ⊕ tz(0.107) |
| 7 | @ee | | link6 | tz(0.103) ⊕ Rz(-45°) |

| name | q0 | q1 | q2 | q3 | q4 | q5 | q6 |
|------|----|----|----|----|----|----|----|
| qr | 0° | -17.2° | 0° | -126° | 0° | 115° | 45° |
| qz | 0° | 0° | 0° | 0° | 0° | 0° | 0° |

*The result of this method is a wide table which has been cropped for inclusion in this book.

and the format of the first displayed table should be familiar by now. The second table shows some predefined joint configurations named qr and qz which appear as properties of the robot object. They are simply named joint configurations

```
>>> panda.qr
array([   0,   -0.3,    0,   -2.2,    0,    2,    0.7854])
```

We can add a configuration to this robot instance by ▶

```
>>> panda.addconfiguration("foo", [1, 2, 3, 4, 5, 6, 7])
```

and all configurations, predefined or user defined, can be accessed in the configurations dictionary

```
>>> panda.configs["foo"];
>>> panda.configs["qz"];
```

We can compute the forward kinematics

```
>>> panda.fkine(panda.qr).printline()
t = 0.484, 0, 0.413; rpy/zyx = 180°, -5.73°, 0°
```

or plot the robot at a particular named configuration

```
>>> panda.plot(panda.qr);
```

which creates the simple 3D *noodle* graphics plot shown in ◘ Fig. 7.9.

◘ Fig. 7.10a shows a more realistic visualization of a Panda robot which we will properly introduce in ▶ Sect. 7.1.4. Each link is a 3D mesh, as shown in ◘ Fig. 7.10b, whose vertices are defined with respect to its link frame. Computing

To make a configuration appear as an attribute of the robot object use addconfiguration_attr instead. Note that dynamically adding attributes to objects may confuse the Python type checking system in your IDE.

t = 1.25

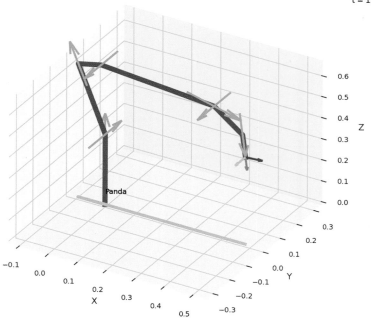

**Fig. 7.9** 3D plot of the Panda robot. This *noodle* plot uses the default PyPlot (Matplotlib) backend. The line segments connect the link frames rather than depicting the physical links of the robot

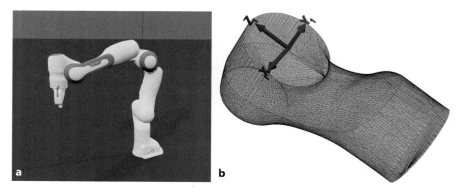

**Fig. 7.10** **a** 3D rendered view of a Panda robot; **b** wireframe mesh describing the shape of link 1, the points are defined with respect to the link coordinate frame. The pose of the link frame is computed by forward kinematics and the points that define the link's shape can be transformed to their proper position in the world frame

the forward kinematics is a sequential process that involves computing the pose of all the intermediate link frames. We can obtain all the link frames with a single method call

```
>>> T = panda.fkine_all(panda.qr);
>>> len(T)
9
```

The pose of the link 1 frame is

```
>>> T[1].printline()
t = 0, 0, 0.333; rpy/zyx = 0°, 0°, 0°
```

and `T[0]` is the base frame and `T[8]` is the tool frame. Given the pose of a link frame, we can transform all the points in that link's mesh and render it to the screen.

The center of mass (CoM) of the link is defined by a coordinate vector with respect to the link frame, and the inertia tensor is defined with respect to a coordinate

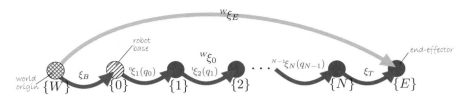

**Fig. 7.11** The kinematic chain from ◘ Fig. 7.6 extended to include a base and tool transformation

frame parallel to the link frame, but with its origin at the CoM. These parameters are important for modeling the dynamics of the robot which we will cover in ► Chap. 9.

### 7.1.2.3 Tools and Bases

It is useful to extend the forward kinematic expression of (7.1) by adding two extra transforms

$$
{}^{W}\boldsymbol{\xi}_E = \underbrace{{}^{W}\boldsymbol{\xi}_0}_{\boldsymbol{\xi}_B} \oplus {}^{0}\boldsymbol{\xi}_1 \oplus {}^{1}\boldsymbol{\xi}_2 \oplus \cdots \oplus {}^{N-1}\boldsymbol{\xi}_N \oplus \underbrace{{}^{N}\boldsymbol{\xi}_E}_{\boldsymbol{\xi}_T} \tag{7.3}
$$

which is shown in ◘ Fig. 7.11. We have used W to denote the world frame since here 0 designates link 0, the base link.

Conventionally, the start of the kinematic chain is the base of the robot, but the base transformation $\boldsymbol{\xi}_B$ allows us to place the base of the robot at an arbitrary pose within the world coordinate frame. In a manufacturing cell, the robot's base is fixed and defined relative to the cell. If the arm was mounted on a mobile robot the base transformation would be time varying.

There is no standard for kinematic models that dictates where the tool frame {N} is physically located on the robot. For a URDF robot model it is is likely to be the tool-mounting flange on the physical end of the robot as shown in ◘ Fig. 7.12b. For a Denavit-Hartenberg model, see ► Sect. 7.1.5, it is frequently the center of the spherical wrist mechanism which is physically inside the robot. The tool transformation $\boldsymbol{\xi}_T$ describes the pose of the tool tip – the bit that actually does the work and sometimes called the tool center point – with respect to frame {N}. A strong convention is that the robot's tool points in the $z$-direction as shown in ◘ Fig. 2.20. In practice $\boldsymbol{\xi}_T$ might vary from application to application, for instance a gripper, a screwdriver or a welding torch, or it might vary within an application if it involves tool changing. It might also consist of several components, perhaps a relative pose to a tool holder and then a relative pose specific to the selected tool.

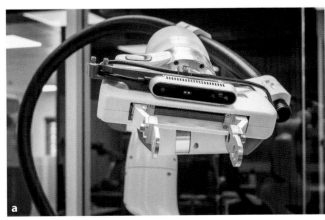

**Fig. 7.12** Panda robot **a** with gripper assembly that includes a camera; **b** without gripper and showing the DIN ISO 9409-1-A50 mounting flange to which tools can be attached (images by Dorian Tsai)

We can set the default base or tool of any Toolbox robot object

```
>>> panda.base = SE3.Tz(3) * SE3.Rx(pi) # robot 3m up, hanging down
>>> panda.tool = SE3.Tz(0.15); # tool is 150mm long in z-direction
```

To support the possibility of changing tools, the tool transformation can be overridden for a particular case

```
>>> panda.fkine(panda.qr, tool=SE3.Trans(0.05, 0.02, 0.20)
...             * SE3.Rz(np.deg2rad(45)));
```

The argument in all cases is an SE2 or SE3 object depending on the robot class.

### 7.1.3  Branched Robots

The simple robots discussed so far have a single end effector, but increasingly we see robots with multiple end effectors and some examples are shown in ◘ Fig. 7.3. The robot in ◘ Fig. 7.3a has two arms in a human-like configuration, and the humanoid robot in ◘ Fig. 7.3b has a torso with four manipulator arms (two of which we call legs).

#### 7.1.3.1  2D (Planar) Branched Robots

We will illustrate the key points using the planar robot shown in ◘ Fig. 7.13a which has two end effectors. Its pose graph is shown in ◘ Fig. 7.13b and using the Toolbox we express this as a list of link objects

```
>>> robot = ERobot2([
...     Link2(ET2.R(), name="link1"),
...     Link2(ET2.tx(1) * ET2.tx(1.2) * ET2.ty(-0.5) * ET2.R(),
...         name="link2", parent="link1"),
...     Link2(ET2.tx(1), name="ee_1", parent="link2"),
...     Link2(ET2.tx(1) * ET2.tx(0.6) * ET2.ty(0.5) * ET2.R(),
...         name="link3", parent="link1"),
...     Link2(ET2.tx(1), name="ee_2", parent="link3")],
...     name="branched");
```

We see that links named "link2" and "link3" share "link1" as their parent. Conversely, the link named "link1" has two children

```
>>> robot["link1"].children
[Link2("link2", tx(1) ⊕ tx(1.2) ⊕ ty(-0.5) ⊕ R(q), parent="link1"),
Link2("link3", tx(1) ⊕ tx(0.6) ⊕ ty(0.5) ⊕ R(q), parent="link1")]
```

Displaying the robot object

```
>>> robot
ERobot2: branched, 3 joints (RRR), 2 branches
```

| link | link | joint | parent | ETS: parent to link |
|------|------|-------|--------|---------------------|
| 0 | link1 | 0 | BASE | R(q0) |
| 1 | link2 | 1 | link1 | tx(1) ⊕ tx(1.2) ⊕ ty(-0.5) ⊕ R(q1) |
| 2 | @ee_1 |   | link2 | tx(1) |
| 3 | link3 | 2 | link1 | tx(1) ⊕ tx(0.6) ⊕ ty(0.5) ⊕ R(q2) |
| 4 | @ee_2 |   | link3 | tx(1) |

*The result of this method is a wide table which has been cropped for inclusion in this book.

shows two end effectors – the links designated with the @ symbol. Joint numbers have been sequentially assigned, starting at zero, in depth-first order, from the base to the end of the first branch, and then along the second branch. Joint coordinates are represented by a vector so it is important to understand how elements of that vector map to the joints of a branched robot.

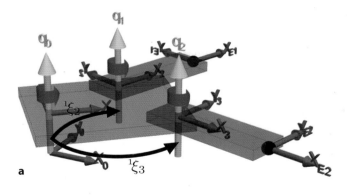

▶ go.sn.pub/eLbuEp

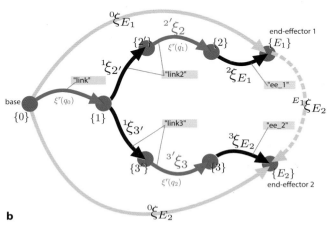

■ **Fig. 7.13** A branched robot. **a** The link frames; **b** the pose graph. Grey text boxes connect elements of the pose graph to the named `Link2` objects

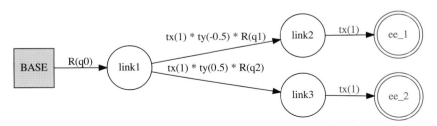

■ **Fig. 7.14** The rigid-body tree depicted as a graph, this is automatically generated by the `showgraph` method using the GraphViz package

A robot with this structure forms a rigid-body tree and we can display the tree as a graph

```
>>> robot.showgraph()
```

which is shown in ■ Fig. 7.14. As with the previous examples we can interactively teach this robot as shown in ■ Fig. 7.15 or plot it

```
>>> robot.teach()
>>> robot.plot([0.3, 0.4, -0.6]);
```

For forward kinematics we now need to specify which end effector we are interested in, for example ▶

The end effector can be specified by a reference to the link object or its name as a string.

```
>>> robot.fkine([0.3, 0.4, -0.6], end="ee_2")
   0.9553    0.2955    2.336
  -0.2955    0.9553    0.655
   0         0         1
```

**7**

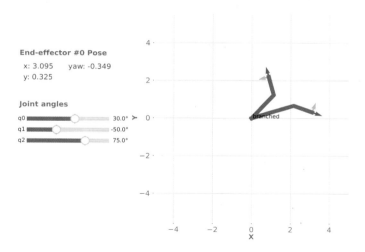

**☐ Fig. 7.15** The branched robot in teach mode. Both branches can be controlled by the sliders, but only the pose of the first end effector is displayed

and the relative pose of $\{EE_2\}$ with respect to $\{EE_1\}$, the dashed-gray arrow in ☐ Fig. 7.13b is simply

```
>>> robot.fkine([0.3, 0.4, -0.6], end="ee_2", start="ee_1")
    0.5403      0.8415    -0.6229
   -0.8415      0.5403     0.3132
    0           0          1
```

## 7.1.4 Unified Robot Description Format (URDF)

Many URDF files are written using Xacro, with a `.xacro` extension, which is an XML macro language that makes the files shorter and more readable. The Toolbox URDF importer includes an Xacro preprocessor to handle such files.

If a relative path is given it is first looked for relative to the current folder, and if not found it is looked for relative to the Robotics Toolbox data package, which is installed as part of the Toolbox.

So far we have used code to create rigid-body trees, but that is very specific to the programming language and software tools being used. To enable exchange of models we use URDF which is a portable XML-based file format. ◄ The format is widely used and models of many robots can be found online. The Toolbox can parse a URDF file ◄

```
>>> urdf, *_ = ERobot.URDF_read("ur_description/urdf/ur5_joint_\
                          limited_robot.urdf.xacro")
>>> urdf
[Link("base_link", SE3(), parent="world", m=4, r=[0, 0, 0],
    I=[0.00443, 0.00443, 0.0072, 0, 0, 0], Jm=0, B=0,
    Tc=[0, 0], G=0),
 Link("shoulder_link", SE3(0, 0, 0.08916) ⊕ Rz(q),
    parent="base_link", qlim=[-3.14, 3.14], m=3.7,
    r=[0, 0, 0], I=[0.0103, 0.0103, 0.00666, 0, 0, 0],
    Jm=0, B=0, Tc=[0, 0], G=0),
 Link("upper_arm_link",
    SE3(0, 0.1358, 0; 0°, 90°, -0°) ⊕ Ry(q),
    parent="shoulder_link", qlim=[-3.14, 3.14], m=8.39,
    r=[0, 0, 0.28], I=[0.227, 0.227, 0.0151, 0, 0, 0],
    Jm=0, B=0, Tc=[0, 0], G=0),
 Link("forearm_link", SE3(0, -0.1197, 0.425) ⊕ Ry(q),
    parent="upper_arm_link", qlim=[-3.14, 3.14],
    m=2.27, r=[0, 0, 0.196], I=[0.0312, 0.0312, 0.00409, 0, 0, 0],
    Jm=0, B=0, Tc=[0, 0], G=0),
 ...
```

Such as mass (`m`), center of mass (`r`), inertia tensor (`I`), motor inertia (`Jm`), motor friction (`B`) and gear ratio (`G`).

which returns a list of link objects initialized with parameters extracted from the URDF file, and the robot's name. Inertial parameters, if present in the file, will be included in the link objects. ◄

URDF models often represent the gripper fingers as links with their own joint variable – essentially creating a branched robot. In practice, gripper fingers are treated as part of an independent subsystem and not part of the kinematic chain. At most, we would represent the gripper by a fixed distance to the middle of the finger tips – the tool center point – and then open or close the fingers about that point.

The toolbox provides classes that read URDF files and return models as `Robot` instance subclasses

```
>>> ur5 = models.URDF.UR5()
ERobot: UR5 (by Universal Robotics), 6 joints (RRRRRR), 1 gripper...
```

| link | link | joint | parent | ETS: pare |
|------|------|-------|--------|-----------|
| 0 | world | | BASE | |
| 1 | base_link | | world | SE3() |
| 2 | shoulder_link | 0 | base_link | SE3(0, 0, 0.08916) |
| 3 | upper_arm_link | 1 | shoulder_link | SE3(0, 0.1358, 0; 0 |
| 4 | forearm_link | 2 | upper_arm_link | SE3(0, -0.1197, 0.4 |
| 5 | wrist_1_link | 3 | forearm_link | SE3(0, 0, 0.3922; 0 |
| 6 | wrist_2_link | 4 | wrist_1_link | SE3(0, 0.093, 0) ⊕ |
| 7 | @wrist_3_link | 5 | wrist_2_link | SE3(0, 0, 0.09465) |
| 8 | tool0 | | wrist_3_link | SE3(0, 0.0823, 0; - |
| 9 | base | | base_link | SE3(0°, -0°, -180°) |

| name | q0 | q1 | q2 | q3 | q4 | q5 |
|------|-----|------|-------|-----|--------|-----|
| qr | 180° | 0° | 0° | 0° | 90° | 0° |
| qz | 0° | 0° | 0° | 0° | 0° | 0° |
| qn | -40.4° | 20.7° | -85.6° | 65° | -40.4° | 0° |
| q1 | 0° | -90° | 90° | 0° | 90° | 0° |

*The result of this method is a wide table which has been cropped for inclusion in this book.

The first line indicates some key characteristics of the robot including the robot's name, number of joints and joint structure. The robot is described as having branches, even though it is a single kinematic chain. This is a consequence of the way the URDF file was written, and the "branches" can be seen clearly by

```
>>> ur5.showgraph()
```

The model also has one gripper

```
>>> ur5.grippers
[Gripper("ee_link", connected to wrist_3_link, 0 joints, 1 links)]
```

The tags at the end of the line above the first displayed table indicate that the model also includes inertial parameters (`dynamics`), 3D geometry of the links (`geometry`), and a collision model (`collision`). The geometry data allows for realistic 3D rendering of the robot as shown in ◘ Fig. 7.16. Each link is described by a meshfile ▶ which includes the 3D shape and optionally the surface color and texture. If the robot model includes geometry data then

```
>>> ur5.plot(ur5.qr);
```

will display a fully rendered view of the robot inside a new browser tab, rather than a noodle plot. Collision models are simple geometric approximations of the link shape that can be tested for intersection with other objects during simulation, and this is discussed in ▶ Sect. 7.5.5.

The mesh files are typically COLLADA (.dae) or STL format (.stl). COLLADA files also contain color and surface texture data which leads to more realistic rendering than for STL files which are geometry only.

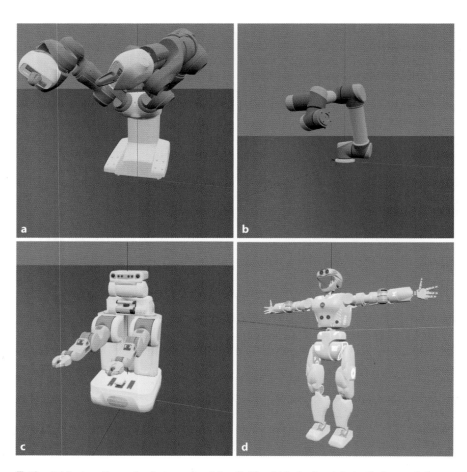

■ **Fig. 7.16** A gallery of robots rendered by Swift which is the default display tool for an `ERobot` instance with geometry data. **a** ABB YuMi (`models.URDF.YuMi`); **b** Universal Robotics UR5 (`models.URDF.UR5`); **c** Willow Garage PR2 (`models.URDF.PR2`); **d** NASA Valkyrie (`models.URDF.Valkyrie`). The YuMi is modeled by STL files that contain only geometry data, while the others are modeled by COLLADA files (.dae files) that contain geometry, color and texture

The inertial parameters can be listed in tabular form

```
>>> ur5.dynamics()
```

| j | m | r | I |
|---|---|---|---|
| world | 0 | 0,  0,  0 | 0,  0,  0,  0,  0,  0 |
| base_link | 4 | 0,  0,  0 | 0.00443,  0.00443,  0.0 |
| shoulder_link | 3.7 | 0,  0,  0 | 0.0103,  0.0103,  0.006 |
| upper_arm_link | 8.39 | 0,  0,  0.28 | 0.227,  0.227,  0.0151, |
| forearm_link | 2.27 | 0,  0,  0.196 | 0.0312,  0.0312,  0.004 |
| wrist_1_link | 1.22 | 0,  0.093,  0 | 0.00256,  0.00256,  0.0 |
| wrist_2_link | 1.22 | 0,  0,  0.0946 | 0.00256,  0.00256,  0.0 |
| wrist_3_link | 0.188 | 0,  0.065,  0 | 8.47e-05,  8.47e-05,  0 |
| tool0 | 0 | 0,  0,  0 | 0,  0,  0,  0,  0,  0 |
| base | 0 | 0,  0,  0 | 0,  0,  0,  0,  0,  0 |

*The result of this method is a wide table which has been cropped for inclusion in this book.

The six unique elements of the symmetric inertia tensor about the link's center of mass: $I_{xx}, I_{yy}, I_{zz}, I_{xy}, I_{yz}, I_{xz}$.

where `m` is the link's mass, `r` is the link's center of mass with respect to the link frame, `I` is the unique elements of the symmetric link inertia tensor, ◀ `Jm` is the motor inertia, `B` and `Tc` are the viscous and Coulomb friction, and `G` is the gear ratio. Robot dynamics is the topic of ▶ Chap. 9.

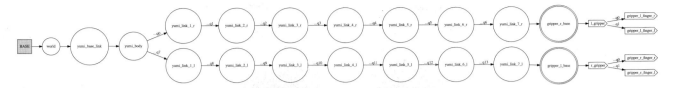

**Fig. 7.17** Kinematic structure of the YuMi robot as described by its URDF file. The two main branches represent the two arms of the robot, and each arm ends in a pair of branches which represent the fingers of the gripper. Joints are numbered in depth first order, but the finger joints are numbered with respect to the gripper

**!** **Inertia tensor reference frame**

The inertia tensor is defined with respect to the center of mass of the link which is convenient for many dynamics algorithms. URDF files describe inertia with respect to the inertia frame whose origin is at the center of mass. However this is not a standard and will vary across software packages, for instance the software used in the MATLAB version of this book stores the inertia with respect to the link frame.

We conclude this section with an example that builds on many of the concepts we have just covered. We load a model of a 3D branched robot from a URDF file

```
>>> yumi = models.URDF.YuMi()
ERobot: yumi (by ABB), 14 joints (RRRRRRRRRRRRRR), 2 grippers,
        2 branches, dynamics, geometry, collision
...
```

The displayed table indicates that the robot has 14 joints, all revolute, and two end effectors and the link structure

```
>>> yumi.showgraph(ets="brief")
```

is shown in **◻** Fig. 7.17. The robot has two grippers

```
>>> yumi.grippers
[Gripper("r_gripper", connected to gripper_l_base, 2 joints,
        3 links),
 Gripper("l_gripper", connected to gripper_r_base, 2 joints,
        3 links)]
```

each of which has two joints and three links – the links in the rigid-body tree that they attach to are displayed. Joint numbers, discussed earlier, have been assigned in a depth-first order. The robot can be visualised in a browser tab by

```
>>> yumi.plot(yumi.q1);
```

and this is shown in **◻** Fig. 7.16a.

The toolbox provides classes that define many robots based on URDF models shipped with the Toolbox. These classes can be listed by

```
>>> models.list(type="URDF")
```

| class | name | manufacturer | type | DoF | dim |
| --- | --- | --- | --- | --- | --- |
| Panda | panda | Franka Emika | URDF | 7 | 3d |
| Frankie | frankie | Franka Emika | URDF | 9 | 3d |
| FrankieOmni | FrankieOmni | Custom | URDF | 10 | 3d |
| UR3 | UR3 | Universal Robotics | URDF | 6 | 3d |
| UR5 | UR5 | Universal Robotics | URDF | 6 | 3d |
| UR10 | UR10 | Universal Robotics | URDF | 6 | 3d |
| Puma560 | Puma560 | Unimation | URDF | 6 | 3d |
| px100 | px100 | Interbotix | URDF | 7 | 3d |
| px150 | px150 | Interbotix | URDF | 8 | 3d |
| rx150 | rx150 | Interbotix | URDF | 8 | 3d |

*The result of this method is a wide table which has been cropped for inclusion in this book.

...

◘ **Table 7.1**   Standard Denavit-Hartenberg parameters for the PUMA 560 robot

| $\theta_j$ | $d_j$ | $a_j$ | $\alpha_j$ | $\sigma_j$ |
|---|---|---|---|---|
| $q_0$ | 0.6718 | 0 | 90° | 0 |
| $q_1$ | 0 | 0.4318 | 0° | 0 |
| $q_2$ | 0.1505 | 0.0203 | -90° | 0 |
| $q_3$ | 0.4318 | 0 | 90° | 0 |
| $q_4$ | 0 | 0 | -90° | 0 |
| $q_5$ | 0 | 0 | 0° | 0 |

One of the most complex models in the Toolbox is the PR2 robot shown in ◘ Fig. 7.16c. This has a mobile base, two arms each with a gripper, and a multitude of actuated sensors

```
>>> pr2 = models.URDF.PR2()
ERobot: pr2 (by Willow Garage), 31 joints
        (RRRRRRRRRRRPRRRRRRRRRRRRRRRRRRR), 27 branches,
        dynamics, geometry, collision
...
```

and comprises 31 joints, 66 links and 27 branches. We can display its kinematic structure in tabular, graph or visual form by

```
>>> pr2
>>> pr2.showgraph()
>>> pr2.plot(pr2.qz)
```

### 7.1.5  Denavit-Hartenberg Parameters

From the 1960s, well before the advent of URDF, the way to share a robot model was as a simple table of numbers like that shown in Table 7.1. There is one row per link-joint pair to describe the kinematics of the robot. If known, the inertial parameters of the link can be added as additional columns.

This compact notation is known as Denavit-Hartenberg notation. Each row in the table defines the spatial relationship between two consecutive link frames as shown in ◘ Fig. 7.18. A manipulator with $N$ joints numbered from 0 to $N-1$ has $N+1$ links, numbered from 0 to $N$. Joint $j$ connects link $j$ to link $j+1$. It follows that link $\ell$ connects joint $\ell-1$ to joint $\ell$. Link 0 is the base of the robot, typically fixed and link $N$, the last link of the robot, carries the end effector or tool.

**❶ Denavit-Hartenberg notation with zero-based indexing**

This description of Denavit-Hartenberg notation differs from standard textbooks which index the joints and parameters starting from one. This book adopts Pythonic base-zero numbering for joint indexing.

Each link is described by four parameters. The relationship between two link coordinate frames would ordinarily entail six parameters, three each for translation and rotation. Denavit-Hartenberg notation uses only four parameters but there are also two constraints: axis $x_j$ intersects $z_{j-1}$ and axis $x_j$ is perpendicular to $z_{j-1}$. One consequence of these constraints is that sometimes the link coordinate frames are not actually located on the physical links of the robot. Another consequence is that the robot must be placed into a particular configuration – the zero-angle configuration – in order to assign the link frames and this is discussed further in ▶ Sect. 7.5.1. The Denavit-Hartenberg parameters are summarized in ◘ Table 7.2.

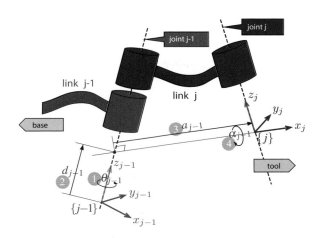

**Fig. 7.18** Definition of standard Denavit and Hartenberg link parameters. The colors red and blue denote all things associated with links $j-1$ and $j$ respectively. The numbers in circles represent the order in which the elementary transforms are applied. $\boldsymbol{x}_j$ is parallel to $\boldsymbol{z}_{j-1} \times \boldsymbol{z}_j$ and if those two axes are parallel then $d_j$ can be arbitrarily chosen

**Table 7.2** Denavit-Hartenberg parameters: their physical meaning, symbol and formal definition

| | | | |
|---|---|---|---|
| Joint angle | $\theta_j$ | The angle between the $\boldsymbol{x}_j$ and $\boldsymbol{x}_{j+1}$ axes about the $\boldsymbol{z}_{j-1}$ axis | Revolute joint variable or constant |
| Link offset | $d_j$ | The distance from the origin of frame {j} to the $\boldsymbol{x}_{j+1}$ axis along the $\boldsymbol{z}_j$ axis | Constant or prismatic joint variable |
| Link length | $a_j$ | The distance between the $\boldsymbol{z}_j$ and $\boldsymbol{z}_{j+1}$ axes along the $\boldsymbol{x}_{j+1}$ axis; for intersecting axes is parallel to $\boldsymbol{z}_j \times \boldsymbol{z}_{j+1}$ | Constant |
| Link twist | $\alpha_j$ | The angle from the $\boldsymbol{z}_j$ axis to the $\boldsymbol{z}_{j+1}$ axis about the $\boldsymbol{x}_{j+1}$ axis | Constant |
| Joint type | $\sigma_j$ | $\sigma = R$ for a revolute joint, $\sigma = P$ for a prismatic joint. By convention $R=0$ and $P=1$ | Constant |

The coordinate frame {j} is attached to the far (distal) end of link $j$. The $z$-axis of frame {j} is aligned with the axis of joint $j$. The transformation from link coordinate frame {j} to frame {j+1} is defined in terms of elementary transformations as

$$^{j}\boldsymbol{\xi}_{j+1} = \boldsymbol{\xi}^{r_z}(\theta_j) \oplus \boldsymbol{\xi}^{t_z}(d_j) \oplus \boldsymbol{\xi}^{t_x}(a_j) \oplus \boldsymbol{\xi}^{r_x}(\alpha_j) \qquad (7.4)$$

which can be expanded as an **SE**(3) matrix

$$^{j}\mathbf{A}_{j+1} = \begin{pmatrix} \cos\theta_j & -\sin\theta_j \cos\alpha_j & \sin\theta_j \sin\alpha_j & a_j \cos\theta_j \\ \sin\theta_j & \cos\theta_j \cos\alpha_j & -\cos\theta_j \sin\alpha_j & a_j \sin\theta_j \\ 0 & \sin\alpha_j & \cos\alpha_j & d_j \\ 0 & 0 & 0 & 1 \end{pmatrix} \qquad (7.5)$$

The parameters $\alpha_j$ and $a_j$ are always constant. For a revolute joint, $\theta_j$ is the joint variable and $d_j$ is constant, while for a prismatic joint, $d_j$ is variable and $\theta_j$ is constant. The generalized joint coordinates are

$$q_j = \begin{cases} \theta_j & \text{if } \sigma_j = R \\ d_j & \text{if } \sigma_j = P \end{cases}$$

**7**

### Excurse 7.4: Denavit and Hartenberg

Jacques Denavit and Richard Hartenberg introduced many of the key concepts of kinematics for serial-link manipulators in a 1955 paper (Denavit and Hartenberg 1955) and their later classic text *Kinematic Synthesis of Linkages* (Hartenberg and Denavit 1964).

Jacques Denavit (1930–2012) was born in Paris where he studied for his Bachelor degree before pursuing his masters and doctoral degrees in mechanical engineering at Northwestern University, Illinois. In 1958 he joined the Department of Mechanical Engineering and Astronautical Science at Northwestern where the collaboration with Hartenberg was formed. In addition to his interest in dynamics and kinematics Denavit was also interested in plasma physics and kinetics. After the publication of the book he moved to Lawrence Livermore National Lab, Livermore, California, where he undertook research on computer analysis of plasma physics problems.

Richard Hartenberg (1907–1997) was born in Chicago and studied for his degrees at the University of Wisconsin, Madison. He served in the merchant marine and studied aeronautics for two years at the University of Göttingen with space-flight pioneer Theodore von Kármán. He was Professor of mechanical engineering at Northwestern University where he taught for 56 years. His research in kinematics led to a revival of interest in this field in the 1960s, and his efforts helped put kinematics on a scientific basis for use in computer applications in the analysis and design of complex mechanisms. He also wrote extensively on the history of mechanical engineering.

A revolute robot joint and link can be created by

```
>>> link = RevoluteDH(a=1)
RevoluteDH(d=0, a=1, α=0)
```

which is a `RevoluteDH` instance and a subclass of the generic `DHLink` object. The displayed value of the object shows the kinematic and dynamic parameters (most of which have defaulted to zero), and the name implies the joint type (Revolute) and that standard Denavit-Hartenberg (DH) convention is used. ◄

A variant form, *modifed* Denavit-Hartenberg notation, is discussed in ▶ Sect. 7.5.3 and would be represented by a `RevoluteMDH` object.

A `DHLink` object shares many attributes and methods with the `Link` object we used earlier. For example, the link transformation (7.5) for $q=0.5$ rad is an SE3 object

```
>>> link.A(0.5)
   0.8776   -0.4794    0        0.8776
   0.4794    0.8776    0        0.4794
   0         0         1        0
   0         0         0        1
```

The forward kinematics is a function of the joint coordinates and is simply the composition of the relative pose due to each link. In the Denavit-Hartenberg representation link 0 is the base of the robot and commonly for the first link

$d_0 = 0$, but we could set $d_0 > 0$ to represent the height of the first joint above the world coordinate frame. For the final link, link $N$, the parameters $d_{N-1}$, $a_{N-1}$ and $\alpha_{N-1}$ provide a limited means to describe the tool-tip pose with respect to the $\{N\}$ frame. It is common to add a more general tool transformation as described in ▶ Sect. 7.1.2.3.

Denavit-Hartenberg kinematic descriptions of many robots can be found in manufacturer's data sheets and in the literature. They are very compact compared to a URDF model, and a table of Denavit-Hartenberg parameters is sufficient to compute the forward and inverse kinematics and to create a stick figure (*noodle plot*) or animation such as shown in ◘ Fig. 7.9. They can also be used to compute Jacobians, which will be covered in ▶ Chap. 8, and when combined with inertial parameters can be used to compute the rigid-body dynamics which will be covered in ▶ Chap. 9.

Determining the Denavit-Hartenberg parameters for a particular robot is challenging, but the toolbox includes many robot models defined in this way and you can list them by

```
>>> models.list(type="DH")
```

| class | name | manufacturer | type | DoF | dims | st |
|-------|------|--------------|------|-----|------|-----|
| Panda | Panda | Franka Emika | DH | 7 | 3d | RRR |
| Puma560 | Puma 560 | Unimation | DH | 6 | 3d | RRR |
| Stanford | Stanford arm | Victor Scheinman | DH | 6 | 3d | RRP |
| Ball | ball | | DH | 10 | 3d | RRR |
| Hyper | Hyper10 | | DH | 10 | 3d | RRR |
| Coil | Coil10 | | DH | 10 | 3d | RRR |
| Cobra600 | Cobra600 | Omron | DH | 4 | 3d | RRP |
| IRB140 | IRB 140 | ABB | DH | 6 | 3d | RRR |
| KR5 | KR5 | KUKA | DH | 6 | 3d | RRR |
| Orion5 | Orion 5 | RAWR Robotics | DH | 4 | 3d | RRR |
| Planar3 | Planar 3 link | | DH | 3 | 3d | RRR |
| Planar2 | Planar 2 link | | DH | 2 | 3d | RR |
| LWR4 | LWR-IV | Kuka | DH | 7 | 3d | RRR |
| Sawyer | Sawyer | Rethink Robotics | DH | 6 | 3d | RRR |

*The result of this method is a wide table which has been cropped for inclusion in this book.

...

We can instantiate a Denavit-Hartenberg model of a popular industrial robot by

```
>>> irb140 = models.DH.IRB140();
```

and displaying the object

```
>>> irb140
DHRobot: IRB 140 (by ABB), 6 joints (RRRRRR), dynamics, geometry, ...
```

| $\theta_j$ | $d_j$ | $a_j$ | $\alpha_j$ | $q^-$ | $q^+$ |
|------------|-------|-------|-----------|-------|-------|
| q1 | 0.352 | 0.07 | $-90.0°$ | $-180.0°$ | $180.0°$ |
| q2 | 0 | 0.36 | $0.0°$ | $-100.0°$ | $100.0°$ |
| q3 | 0 | 0 | $-90.0°$ | $-220.0°$ | $60.0°$ |
| q4 | 0.38 | 0 | $90.0°$ | $-200.0°$ | $200.0°$ |
| q5 | 0 | 0 | $-90.0°$ | $-120.0°$ | $120.0°$ |
| q6 | 0.065 | 0 | $0.0°$ | $-400.0°$ | $400.0°$ |

| name | q0 | q1 | q2 | q3 | q4 | q5 |
|------|-----|------|------|-----|------|------|
| qr | 0° | $-90°$ | 90° | 0° | 90° | $-90°$ |
| qz | 0° | 0° | 0° | 0° | 0° | 0° |
| qd | 0° | $-90°$ | 180° | 0° | 0° | $-90°$ |

shows a table of its Denavit-Hartenberg parameters and joint-angle limits. We can apply all the methods introduced previously such as forward kinematics

```
>>> irb140.fkine(irb140.qr).printline("rpy/xyz")
t = 0.005, 0, 0.332; rpy/xyz = 0°, -90°, -90°
```

creating a *noodle plot*

```
>>> irb140.plot(irb140.qr);
```

or interactive control using sliders

```
>>> irb140.teach()
```

The robot, defined using Denavit-Hartenberg notation, can also be converted to ETS format

```
>>> irb140.ets()
Rz(q0) ⊕ tz(0.352) ⊕ tx(0.07) ⊕ Rx(-90°) ⊕ Rz(q1)
  ⊕ tx(0.36) ⊕ Rz(q2) ⊕ Rx(-90°) ⊕ Rz(q3) ⊕ tz(0.38)
  ⊕ Rx(90°) ⊕ Rz(q4) ⊕ Rx(-90°) ⊕ Rz(q5) ⊕ tz(0.065)
```

## 7.2    Inverse Kinematics

A problem of real practical interest is the inverse of that just discussed: given the desired pose of the end effector $\xi_E$ what are the required joint coordinates? For example, if we know the Cartesian pose of an object, what should the robot's joint coordinates be in order to reach it? This is the inverse kinematics problem which is written in functional form as

$$q = \mathcal{K}^{-1}(\xi_E) \tag{7.6}$$

and in general is not unique, that is, a particular end-effector pose can be achieved by more than one joint configuration.

Two approaches can be used to determine the inverse kinematics. Firstly, a closed-form or analytic solution can be found using geometric or algebraic techniques. However, this becomes increasingly challenging as the number of robot joints increases and for some serial-link manipulators no closed-form solution exists. Secondly, an iterative numerical solution can be used. In ► Sect. 7.2.1 we again use the simple 2-dimensional case to illustrate the principles and then in ► Sect. 7.2.2 extend these to robot arms that move in 3-dimensions.

### 7.2.1    2-Dimensional (Planar) Robotic Arms

We will solve the inverse kinematics for the 2-joint robot of ◻ Fig. 7.4b in two ways: algebraic closed-form and numerical.

#### 7.2.1.1    Closed-Form Solution

We start by computing the forward kinematics symbolically and will use SymPy to help us with the algebra. We define some symbolic constants for the robot's lengths and then create an ETS for the robot of ◻ Fig. 7.4b in the now familiar way

```
>>> import sympy
>>> a1, a2 = sympy.symbols("a1 a2")
>>> e = ET2.R() * ET2.tx(a1) * ET2.R() * ET2.tx(a2);
```

Next, we define some symbolic variables to represent the joint angles

```
>>> q0, q1 = sympy.symbols("q0 q1")
```

and then compute the forward kinematics as an **SE**(2) matrix

```
>>> TE = e.fkine([q0, q1])
cos(q0 + q1)   -sin(q0 + q1)    a1*cos(q0) + a2*cos(q0 + q1)
sin(q0 + q1)    cos(q0 + q1)    a1*sin(q0) + a2*sin(q0 + q1)
    0               0               1
```

which is an algebraic representation of the robot's forward kinematics – the end-effector pose as a function of the joint variables. We will work with just the end-effector position

```
>>> x_fk, y_fk = TE.t;
```

Finally, we define two more symbolic variables to represent the desired end-effector position $(x, y)$

```
>>> x, y = sympy.symbols("x y")
```

Now we have two equations

$$x_{fk} = x, \quad y_{fk} = y$$

in two unknowns ($q_0$ and $q_1$) but SymPy is not able to solve such trigonometric equations directly – it needs a little guidance. For this type of problem it is often helpful to square both equations and add them $x_{fk}^2 + y_{fk}^2 = x^2 + y^2$ which we rewrite as

$$x_{fk}^2 + y_{fk}^2 - x^2 - y^2 = 0$$

and express in code as ▶

```
>>> eq1 = (x_fk**2 + y_fk**2 - x**2 - y**2).trigsimp()
a1**2 + 2*a1*a2*cos(q1) + a2**2 - x**2 - y**2
```

SymPy represents an equation as an expression which is implicitly equal to zero.

and now we have an equation with only one unknown, $q_1$, and the solution for `eq1` = 0 is found by

```
>>> q1_sol = sympy.solve(eq1, q1)
[-acos(-(a1**2 + a2**2 - x**2 - y**2)/(2*a1*a2)) + 2*pi,
  acos((-a1**2 - a2**2 + x**2 + y**2)/(2*a1*a2))]
```

which is a list. We observed in ▶ Sect. 7.1.1 that two different joint configurations can give the same end-effector position, and here we have solutions corresponding to a positive or negative value of $q_1$. ▶

The range of $\cos^{-1}$ is limited to quadrants 1 and 2.

To solve for $q_0$ we first expand the two equations mentioned above

```
>>> eq0 = tuple(map(sympy.expand_trig, [x_fk - x, y_fk - y]))
(a1*cos(q0) + a2*(-sin(q0)*sin(q1) + cos(q0)*cos(q1)) - x,
 a1*sin(q0) + a2*(sin(q0)*cos(q1) + sin(q1)*cos(q0)) - y)
```

and solve them for `sin(q0)` and `cos(q0)`

```
>>> q0_sol = sympy.solve(eq0, [sympy.sin(q0), sympy.cos(q0)]);
```

which is a Python dictionary containing the sine and cosine of `q0`. The ratio of these is `tan(q0)`

```
>>> sympy.atan2(q0_sol[sympy.sin(q0)],
...             q0_sol[sympy.cos(q0)]).simplify()
atan2((a1*y - a2*x*sin(q1) + a2*y*cos(q1))/(a1**2 + 2*a1*a2*cos(q1)
      + a2**2),
      (a1*x + a2*x*cos(q1) + a2*y*sin(q1))/(a1**2 + 2*a1*a2*cos(q1)
      + a2**2))
```

**7**

**Excurse 7.5: Spherical Wrist**

A spherical wrist is a key component of many modern robot manipulators. The wrist has three axes of rotation that are orthogonal and intersect at a common point. This is a gimbal-like mechanism, and as discussed in ▶ Sect. 2.3.1.2 will have a singularity.

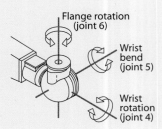

Flange rotation (joint 6)

Wrist bend (joint 5)

Wrist rotation (joint 4)

The robot end-effector pose, its position and an orientation, is defined at the center of the wrist for mathematical convenience. Since the wrist axes intersect at a common point, wrist motion causes zero translational motion, therefore end-effector position is a function only of the first three joints. This is a critical simplification that makes it possible to find closed-form inverse kinematic solutions for 6-axis industrial robots. An arbitrary end-effector orientation is achieved independently of position by means of the three wrist joints.

---

SymPy is the standard tool in the Python universe but is comparatively weak at complex trigonometric expressions.

and with some simplification we have a solution for $q_0$. It is a function of the particular solution for $q_1$ that we chose, as well as the robot's length constants. More powerful symbolic tools ◀ could do this example with less hand holding, but the complexity of algebraic solution increases dramatically with the number of joints and this is hard to completely automate.

### 7.2.1.2 Numerical Solution

We can think of the inverse kinematics problem as that of adjusting the joint coordinates until the forward kinematics matches the desired pose. More formally this is an optimization problem – to minimize the error between the forward kinematic solution and the desired end-effector pose $\boldsymbol{\xi}_E^*$

$$\boldsymbol{q}^* = \arg \min_{\boldsymbol{q}} \| \mathcal{K}(\boldsymbol{q}) \ominus \boldsymbol{\xi}_E^* \|$$

Once again we use the robot from ◻ Fig. 7.4b

```
>>> e = ET2.R() * ET2.tx(1) * ET2.R() * ET2.tx(1);
```

and define an error function based on the end-effector position error, not its orientation

$$E(q) = \| [\mathcal{K}(\boldsymbol{q})]_t - (x^*, \ y^*)^\top \|$$

which we can define as a lambda function

```
>>> pstar = np.array([0.6, 0.7]);   # desired position
>>> E = lambda q: np.linalg.norm(e.fkine(q).t - pstar);
```

We minimize this using the SciPy multi-variable minimization function

```
>>> sol = optimize.minimize(E, [0, 0]);
```

where the arguments are the cost function and the initial estimate of the joint coordinates. The joint angles that minimize the cost are

```
>>> sol.x
array([ -0.2295,    2.183])
```

and computing the forward kinematics

```
>>> e.fkine(sol.x).printline()
t = 0.6, 0.7; 112°
```

we can verify that this solution is correct. As already discussed there are two solutions for $q$ but the solution that is found using this approach depends on the initial choice of $q$ passed to the minimizer.

### 7.2.2 3-Dimensional Robotic Arms

#### 7.2.2.1 Closed-Form Solution

Closed-form solutions have been developed for most common types of 6-axis industrial robots, typically those defined by Denavit-Hartenberg models. A necessary condition for a closed-form solution of a 6-axis robot is a spherical wrist mechanism. Some robot models included with the Toolbox have closed-form inverse kinematics and we will illustrate this for the classical PUMA 560 robot

```
>>> puma = models.DH.Puma560();
```

At the *nominal* joint coordinates shown in ◻ Fig. 7.19a

```
>>> puma.qn
array([  0, 0.7854, 3.142, 0, 0.7854, 0])
```

the end-effector pose is

```
>>> T = puma.fkine(puma.qn);
>>> T.printline()
t = 0.596, -0.15, 0.657; rpy/zyx = 0°, 90°, 0°
```

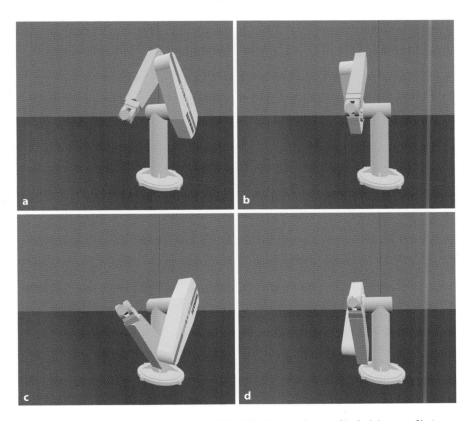

◻ **Fig. 7.19** Different configurations of the PUMA 560 robot. **a** Left-up-noflip; **b** right-up-noflip (nominal configuration qn); **c** left-down-noflip; **d** right-down-noflip

**7**

Modern robots typically have no shoulder lateral offset but these two solutions still exist, they are just less apparent. For this case the robot rotates 180° at the waist, *facing backwards*, then rotates the shoulder over the top to be in front again.

More precisely the elbow is above or below the shoulder.

For a two-fingered gripper the flip makes no functional difference to finger placement.

The PUMA's wrist is a ZXZ Euler angle sequence as discussed in ▶ Sect. 2.3.1.2.

Since the PUMA 560 is a 6-axis robot arm with a spherical wrist we can compute the inverse kinematics using an analytic closed-form solution.

```
>>> sol = puma.ikine_a(T)
IKSolution: q=[2.649, 2.356, 0.09396, -0.609, -0.9743, -2.768],
success=True
```

which contains status information indicating that a solution was found, and the joint configuration as a NumPy array. Surprisingly, the solution

```
>>> sol.q
array([ 2.649, 2.356, 0.09396, -0.609, -0.9743, -2.768])
```

is quite different to the joint coordinates we started with, but we can verify that this solution is correct

```
>>> puma.fkine(sol.q).printline()
t = 0.596, -0.15, 0.657; rpy/zyx = 0°, 90°, 0°
```

These two different joint configurations are shown in ◘ Fig. 7.19a, b.

The PUMA 560 robot is uncommon in having its shoulder joint horizontally offset from the waist, so in one solution the arm is to the left of the waist (like our left arm), in the other solution it is to the right (like our right arm). These are referred to as the left- and right-handed kinematic configurations respectively. In general, there are eight different joint coordinates that give the same end-effector pose. ◀ We can force a right-handed solution

```
>>> sol = puma.ikine_a(T, "r");
>>> sol.q
array([     0,   0.7854,  -3.142,   -3.142,  -0.7854,    3.142])
```

which gives the original joint configuration, noting that for $q_2$ the values of $\pi$ and $-\pi$ are equivalent.

In addition to the left- and right-handed solutions, there are solutions with the elbow either up or down, ◀ and with the wrist flipped or not flipped. The first wrist joint, $q_3$, of the PUMA 560 robot has a large rotational range and can adopt one of two angles that differ by $\pi$ radians. ◀

Four different kinematic solutions are shown in ◘ Fig. 7.19. The joint configuration returned by `ikine_a` is controlled by one or more of the flag characters:

left or right handed          `"l"`, `"r"`
elbow up or down           `"u"`, `"d"`
wrist flipped or not flipped  `"f"`, `"n"`

provided as an optional string.

Due to mechanical limits on joint angles and possible collisions between links not all eight solutions are physically achievable. It is also possible that no solution can be achieved. For example

```
>>> puma.ikine_a(SE3.Tx(3))
IKSolution: q=None, success=False, reason=Out of reach
```

has failed because the arm is simply not long enough to achieve the requested pose.

A pose may also be unachievable due to singularity where the alignment of axes reduces the effective degrees of freedom (the gimbal lock problem again). The PUMA 560 has a wrist singularity when $q_4$ is equal to zero and the axes of joints 3 and 5 become aligned. ◀ In this case the best that `ikine_a` can do is to constrain $q_3 + q_5$ but their individual values are arbitrary. For example consider the configuration

```
>>> q = [0, pi/4, pi, 0.1, 0, 0.2];
```

The inverse kinematic solution is

```
>>> puma.ikine_a(puma.fkine(q), "ru").q
array([       0,    0.7854,    -3.142,         0,         0,      0.3])
```

which has quite different values for $q_3$ and $q_5$ but the sum $q_3 + q_5 = 0.3$ for both cases.

#### 7.2.2.2 Numerical Solution

For the case of robots which do not have six joints and a spherical wrist or for which no analytical solution is available, we need to use an iterative numerical solution. Continuing with the example of the previous section, we use the method `ikine_LM` to compute the inverse kinematic solution numerically ▶

```
>>> T = puma.fkine(puma.qn);
>>> T.printline("rpy/xyz")
t = 0.596, -0.15, 0.657; rpy/xyz = 0°, 90°, 0°
>>> sol = puma.ikine_LM(T, q0=[0, 0, 0, 0, 0, 0])
IKSolution: q=[0, -0.8335, 0.09396, 0, -0.8312, 0],
success=True, iterations=5, searches=1,
residual=4.15e-09
```

which indicates success after just 5 iterations and with a very small solution error or residual. The joint configuration however is different to the original value

```
>>> puma.qn
array([       0,    0.7854,     3.142,         0,    0.7854,        0])
```

but does result in the correct end-effector pose

```
>>> puma.fkine(sol.q).printline("rpy/xyz")
t = 0.596, -0.15, 0.657; rpy/xyz = 0°, 90°, 0°
```

Plotting the pose

```
>>> puma.plot(sol.q);
```

shows clearly that `ikine_LM` has found the elbow-down configuration.

A limitation of this general numeric approach is that it does not provide explicit control over which solution is found as the analytic approach did – the only control is implicit via the initial estimate of joint coordinates (which we set to zero). ▶ If we specify different initial joint coordinates

```
>>> puma.ikine_LM(T, q0=[0, 0, 3, 0, 0, 0])
IKsolution(q=array([ 0, 0.7854, -3.142,    0, 0.7854, 0]),
         success=True, reason=None, iterations=10,
         residual=2.450e-11)
```

we have nudged the minimizer to converge on the elbow-up configuration. As would be expected, the general numerical approach of `ikine_LM` is considerably slower than the analytic approach of `ikine_a`. However, it has the great advantage of being able to work with manipulators at singularities and for manipulators with less than, or more than, six joints. The principles behind `ikine_LM` are discussed in ▶ Sect. 8.5.

#### 7.2.3 Underactuated Manipulator

An underactuated manipulator is one that has fewer than six joints, and SCARA robots such as shown in ◼ Fig. 7.2b are a common example. They have an RRPR joint structure which is optimized for planar-assembly tasks which require control of end-effector position in 3D and orientation in the $xy$-plane. The task space is $\mathcal{T} \subset \mathbb{R}^3 \times \mathbf{S}^1$ and the configuration space is $C \subset (\mathbf{S}^1)^2 \times \mathbb{R} \times \mathbf{S}^1$. Since $\dim C < 6$ the robot is limited in the end-effector poses that it can achieve. We will load a model of the Omron Cobra 600 SCARA robot

The Toolbox provides several numerical inverse kinematic solvers. Many can find solutions that respect joint limits by using the `joint_limit=True` option. However, they will be slower to execute.

Zero is not necessarily a good choice. If `q0` is not specified, then the solver will choose random initial values and keep retrying until it finds a solution. The argument `seed` can be used to initialize the random generator to allow for repeatable results.

```
>>> cobra = models.DH.Cobra600()
DHRobot: Cobra600 (by Omron), 4 joints (RRPR), dynamics, standard ...
```

| θⱼ | dⱼ | aⱼ | αⱼ | q⁻ | q⁺ |
|---|---|---|---|---|---|
| q1 | 0.387 | 0.325 | 0.0° | −50.0° | 50.0° |
| q2 | 0 | 0.275 | 180.0° | −88.0° | 88.0° |
| 0.0° | q3 | 0 | 0.0° | 0.0 | 0.21 |
| q4 | 0 | 0 | 0.0° | −180.0° | 180.0° |

| name | q0 | q1 | q2 | q3 |
|---|---|---|---|---|
| qz | 0° | 0° | 0 | 0° |
| qr | 0° | 0° | 0 | 0° |

and then define a desired end-effector pose

```
>>> TE = SE3.Trans(0.4, -0.3, 0.2)
...     * SE3.RPY(np.deg2rad([30, 0, 170]), order="xyz");
```

where the end-effector approach vector is pointing downward but not parallel to the vertical axis – it is 10° off vertical. This pose is *over-constrained* for the 4-joint SCARA robot, the robot physically cannot meet the orientation requirement for an approach vector that is not vertical and the inverse kinematic solution

```
>>> sol = cobra.ikine_LM(TE, seed=0)
IKSolution: q=[-0.111, -1.176, 0.187, -0.7634], success=False,
reason=iteration and search limit reached, iterations=3000,
searches=100, residual=0.0152
```

has failed to converge. We need to relax the requirement and ignore error in rotation about the $x$- and $y$-axes, and we achieve that by specifying a mask vector

```
>>> sol = cobra.ikine_LM(TE, mask=[1, 1, 1, 0, 0, 1], seed=0)
IKSolution: q=[-0.111, -1.176, 0.187, -0.7634],
success=True, iterations=8, searches=1,
residual=7.82e-10
```

The elements of the mask vector correspond respectively to the three translations and three orientations: $t_x$, $t_y$, $t_z$, $r_x$, $r_y$, $r_z$ in the end-effector coordinate frame. In this example we specified that error in rotation about the $x$- and $y$-axes is to be ignored (the zero elements). The resulting joint angles correspond to an achievable end-effector pose

```
>>> cobra.fkine(sol.q).printline("rpy/xyz")
t = 0.4, -0.3, 0.2; rpy/xyz = 30°, 0°, 180°
```

which has the desired translation and roll angle, but the yaw angle is incorrect, as we *allowed* it to be. They are what the robot mechanism actually permits. We can also compare the desired and achievable poses graphically

```
>>> TE.plot(color="r");
>>> cobra.fkine(sol.q).plot(color="b");
```

### 7.2.4    Overactuated (Redundant) Manipulator

A redundant manipulator is a robot with more than six joints. As mentioned previously, six joints are theoretically sufficient to achieve any desired pose in a Cartesian taskspace $\mathcal{T} \subset \mathbb{R}^3 \times \mathbf{S}^3$. In practice, issues such as joint limits, self collision, and singularities mean that not all poses within the robot's reachable space can be achieved. Adding additional joints is one way to overcome this problem but results in an infinite number of joint-coordinate solutions. To find a single solution we need to introduce constraints – a common one is the minimum-norm constraint which returns a solution where the norm of the joint-coordinate vector has the smallest magnitude.

We will illustrate this with the Panda robot shown in ◼ Fig. 7.3a which has 7 joints. We load the model

```
>>> panda = models.ETS.Panda();
```

then define the desired end-effector pose

```
>>> TE = SE3.Trans(0.7, 0.2, 0.1) * SE3.OA((0, 1, 0), (0, 0, -1));
```

which has its approach vector pointing downward. The numerical inverse kinematics solution is

```
>>> sol = panda.ikine_LM(TE, seed=0)
IKSolution: q=[0.4436, 1.388, 2.532, 0.4795, -2.528, 1.783,
-2.186], success=True, iterations=37, searches=2,
residual=9.34e-07
```

which is the joint configuration vector with the smallest norm that results in the desired end-effector pose, as we can verify

```
>>> panda.fkine(sol.q).printline("angvec")
t = 0.7, 0.2, 0.1; angvec = (180° | 0, 1, 0)
```

## 7.3 Trajectories

A very common requirement in robotics is to move the end effector smoothly from one pose to another. Building on what we learned in ▶ Sect. 3.3 we will discuss two approaches to generating such trajectories: straight lines in configuration space and straight lines in task space. These are known respectively as joint-space and Cartesian motion.

### 7.3.1 Joint-Space Motion

Consider the end effector moving between two poses

```
>>> TE1 = SE3.Trans(0.4, -0.2, 0) * SE3.Rx(3);
>>> TE2 = SE3.Trans(0.4, 0.2, 0) * SE3.Rx(1);
```

which describe two points in the $xy$-plane with different end-effector orientations. The joint configurations for these poses are ▶

```
>>> sol1 = puma.ikine_a(TE1, "ru");
>>> sol2 = puma.ikine_a(TE2, "ru");
```

We could also have used numerical inverse kinematics here.

and we have specified the right-handed elbow-up configuration. We require the motion to occur over a time period of 2 seconds in 20 ms time steps, and we create an array of time values

```
>>> t = np.arange(0, 2, 0.02);
```

A joint-space trajectory is formed by smoothly interpolating between the two joint configurations. The scalar-interpolation functions `quintic` or `trapezoidal` from ▶ Sect. 3.3.1 can be used in conjunction with the multi-axis *driver* function `mtraj`

```
>>> traj = mtraj(quintic, sol1.q, sol2.q, t);
```

or

```
>>> traj = mtraj(trapezoidal, sol1.q, sol2.q, t);
```

These result in a `Trajectory` instance and the joint coordinates

```
>>> traj.q.shape
(100, 6)
```

are a 100×6 array with one row per time step and one column per joint. A `Trajectory` object also holds the joint velocity and acceleration vectors, as a function of time, in the properties `qd` and `qdd`.

This is equivalent to `mtraj` with `quinitic` interpolation but optimized for the multi-axis case and also allowing initial and final velocity to be set using additional arguments.

From here on we will use the equivalent `jtraj` convenience function ◄

```
>>> traj = jtraj(sol1.q, sol2.q, t)
Trajectory created by jtraj: 100 time steps x 6 axes
```

For `mtraj` and `jtraj` the final argument can be a time vector, as here, or an integer specifying the number of time steps.

The trajectory is best viewed as an animation

```
>>> puma.plot(traj.q);
```

but we can also plot the joint angles versus time

```
>>> xplot(t, traj.q);
```

as shown in ◻ Fig. 7.20a. The joint-space trajectory is smooth, which was one of our criteria, but we do not know how the robot's end effector will move in Cartesian space. We can easily determine this by applying forward kinematics to the joint

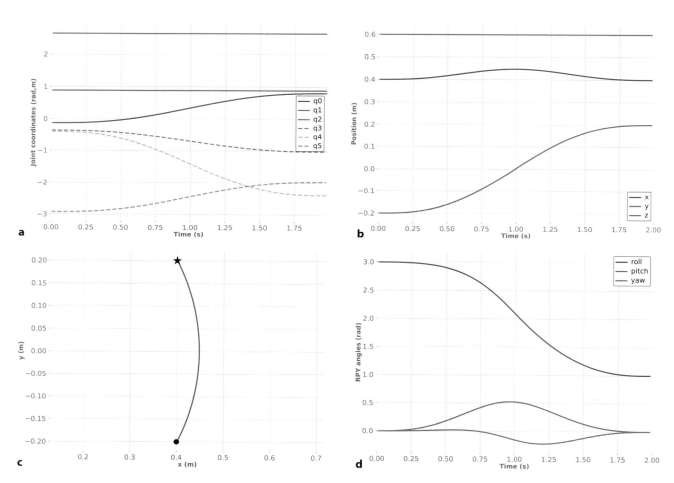

◻ **Fig. 7.20** Joint-space motion. **a** Joint coordinates versus time; **b** Cartesian position versus time; **c** Cartesian position locus in the *xy*-plane **d** roll-pitch-yaw angles versus time

coordinate trajectory

```
>>> T = puma.fkine(traj.q);
>>> len(T)
100
```

which results in an SE3 instance with 100 values. The translational part of this trajectory

```
>>> p = T.t;
>>> p.shape
(100, 3)
```

is an array with one row per time step and one column per coordinate. This is plotted against time

```
>>> xplot(t, T.t, labels="x y z");
```

in ◻ Fig. 7.20b. The path of the end effector in the $xy$-plane is shown in ◻ Fig. 7.20c and it is clear that the path is not a straight line. This is to be expected since we only specified the Cartesian coordinates of the end-points. As the robot rotates about its waist joint during the motion the end effector will naturally follow a circular arc. In practice this could lead to collisions between the robot and nearby objects, even if they do not lie on the direct path between the start and end poses. The orientation of the end effector, in XYZ roll-pitch-yaw angle form, can also be plotted against time

```
>>> xplot(t, T.rpy("xyz"), labels="roll pitch yaw");
```

as shown in ◻ Fig. 7.20d. Note that the roll angle ► varies from 3 to 1 radians as we specified. The roll and pitch angles have met their boundary conditions but have deviated along the path.

Rotation about the $x$-axis for a robot end effector using the XYZ angle sequence, see ► Sect. 2.3.1.2.

### 7.3.2 Cartesian Motion

For many applications we require straight-line motion in Cartesian space which is known as Cartesian motion. This is implemented using the Toolbox function `ctraj` which was introduced in ► Sect. 3.3.5. Its usage is very similar to `jtraj`

```
>>> Ts = ctraj(TE1, TE2, t);
```

where the arguments are the initial and final pose and the time vector. It returns the trajectory as a multi-valued SE3 instance. As for the previous joint-space example, we extract and plot the translation

```
>>> xplot(t, Ts.t, labels="x y z");
```

and orientation components

```
>>> xplot(t, Ts.rpy("xyz"), labels="roll pitch yaw");
```

which are shown in ◻ Fig. 7.21b and d.

◻ Fig. 7.21c shows that the task-space path is now a straight line. The corresponding joint-space trajectory is obtained by applying the inverse kinematics

```
>>> qc = puma.ikine_a(Ts);
```

and is shown in ◻ Fig. 7.21a. ► While broadly similar to ◻ Fig. 7.20a the minor differences are what make the difference between a curved and straight line in the task space.

When solving for a trajectory using numerical inverse kinematics, the solution for one point is used to initialize the solution for the next point on the trajectory.

**7**

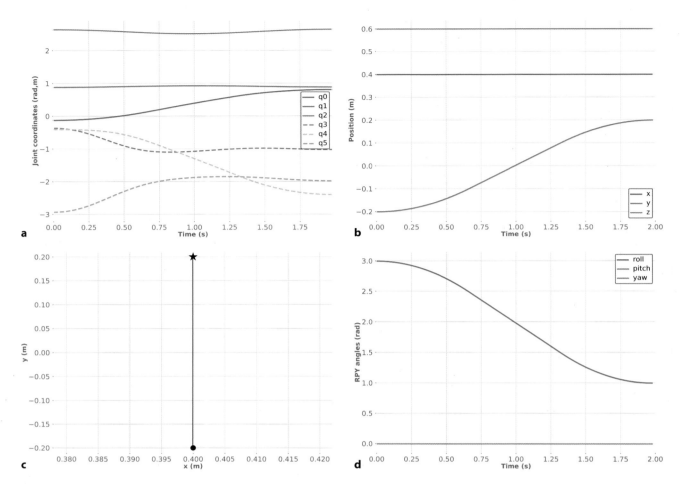

**◘ Fig. 7.21** Cartesian motion. **a** Joint coordinates versus time; **b** Cartesian position versus time; **c** Cartesian position locus in the *xy*-plane; **d** roll-pitch-yaw angles versus time

### 7.3.3 Kinematics in a Block Diagram

We can also implement this example in the bdsim block-diagram environment

```
>>> %run -m jointspace -H
```

and the block diagram model is shown in ◘ Fig. 7.22. The parameters of the Jtraj block are the initial and final values for the joint configuration. The smoothly varying joint angles are wired to an Armplot block which will animate a robot in a separate window, and to an Fkine block – both blocks have a parameter which is an instance of a Robot subclass, in this case an instance of the models.DH.Puma560 class. The Cartesian position of the end-effector pose is extracted by the translation block which is analogous to the function transl or the .t property of an SE3 object. The SCOPEXY block plots one input against the other.

### 7.3.4 Motion Through a Singularity

We briefly touched on the topic of singularities in ► Sect. 7.2.2.1 and we will revisit them again in ► Chap. 8. In this next example we deliberately choose a trajectory that moves through a robot wrist singularity. We repeat the previous example but choose different start and end poses

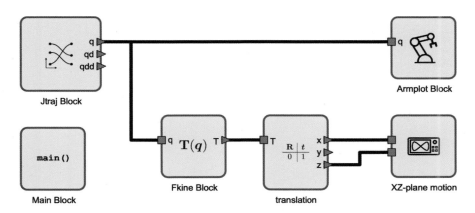

**Fig. 7.22** Block diagram model `models/jointspace` for joint-space motion

```
>>> TE1 = SE3.Trans(0.5, -0.3, 1.12) * SE3.OA((0, 1, 0), (1, 0, 0));
>>> TE2 = SE3.Trans(0.5, 0.3, 1.12) * SE3.OA((0, 1, 0), (1, 0, 0));
```

which results in motion in the $y$-direction with the end-effector $z$-axis pointing in the world $x$-direction. The Cartesian path is

```
>>> Ts = ctraj(TE1, TE2, t);
```

which we convert to joint coordinates

```
>>> sol = puma.ikine_a(Ts, "lu");
```

for a left-handed elbow-up configuration and plot against time

```
>>> xplot(t, sol.q, unwrap=True);
```

which is shown in ◪ Fig. 7.23a. We have used the `unwrap` option to remove the distracting jumps when angles pass through $\pm\pi$.

At time $t \approx 1.3$ s we observe that the wrist joint angles $q_3$ and $q_5$ change very rapidly. ▶ At this time $q_4$ is close to zero which means that the $q_3$ and $q_5$ rotational axes of the wrist are almost aligned – another gimbal lock situation or singularity. The robot has lost one degree of freedom and is now effectively a 5-axis robot.

As discussed in ▶ Sect. 7.2.2.1 this axis alignment means that we can only solve for the sum $q_3 + q_5$ meaning that there are an infinite number of solutions for $q_3$ and $q_5$ that have this sum. From ◪ Fig. 7.23b we observe that the numerical inverse kinematics method `ikine_LM` handles the singularity with far less unnecessary joint motion. This is a consequence of the minimum-norm solution which has returned $q_3$ and $q_5$ which have the correct sum but the smallest magnitude. The joint-space motion between the two poses, ◪ Fig. 7.23c, is immune to this problem since it is does not involve inverse kinematics. However, it will not maintain the orientation of the tool in the $x$-direction for the whole path – only at the two end points.

The dexterity of a manipulator is concerned with how *easily* its end effector can translate along, or rotate about, any axis. A common scalar measure of dexterity is manipulability which can be computed for each point along the trajectory

```
>>> m = puma.manipulability(sol.q);
```

and is plotted in ◪ Fig. 7.23d. At around $t = 1.3$ s the manipulability for the path computed using the analytic inverse kinematics (red curve) was almost zero, indicating a significant loss of dexterity. A consequence of this is the very rapid wrist joint motion, seen in ◪ Fig. 7.23a, which is required to follow the end-effector trajectory. We can see that the numerical inverse kinematics was able to keep manipulability high throughout the trajectory since it implicitly minimizes the velocity of all the joints. Manipulability and the numerical inverse kinematics function `ikine_LM` are based on the manipulator's Jacobian matrix which is the topic of ▶ Chap. 8.

$q_5$ has increased rapidly, while $q_3$ has decreased rapidly. This counter-rotational motion of the two joints means that the gripper does not actually rotate but the two motors are working hard.

7

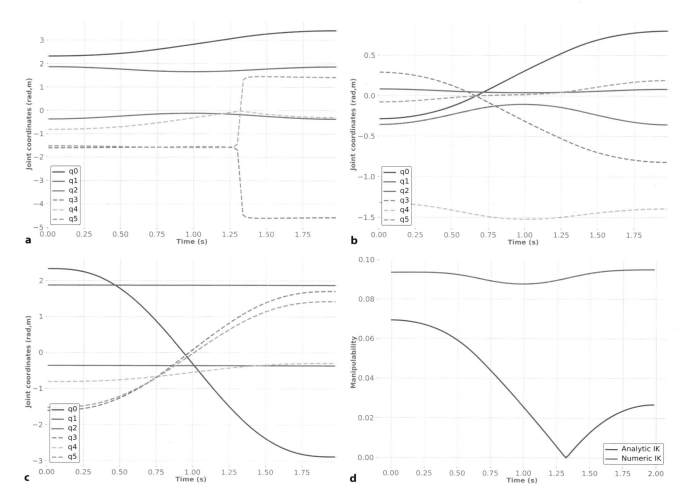

■ **Fig. 7.23** Joint angles for a path through a wrist singularity. **a** Cartesian trajectory computed using analytic inverse kinematics (`ikine_a`); **b** Cartesian trajectory computed using numerical inverse kinematics (`ikine_LM`); **c** joint-space motion (`jtraj`); **d** manipulability for the two Cartesian trajectories

### 7.3.5 Configuration Change

■ Fig. 7.19 shows the manipulator in four configurations that have the same end-effector pose, working in a left- or right-handed manner or with the elbow up or down. Consider the problem of a robot that is working for a while left-handed at one work station, then working right-handed at another. Movement from one configuration to another ultimately results in no change in the end-effector pose since both configurations have the *same* forward kinematic solution – therefore we *cannot* create a trajectory in Cartesian space. This is a case where we have to use joint-space motion.

For example, to move the robot arm from the right- to left-handed configuration we first define an end-effector pose

```
>>> TE = SE3.Trans(0.4, 0.2, 0.6) * SE3.Rx(pi);
```

and then determine the joint configuration for the right- and left-handed elbow-up cases

```
>>> sol_r = puma.ikine_a(TE, "ru");
>>> sol_l = puma.ikine_a(TE, "lu");
```

and then create a joint-space trajectory between these two joint coordinate vectors

```
>>> traj = jtraj(sol_r.q, sol_l.q, t);
```

Although the initial and final end-effector pose is the same, the robot makes some quite significant joint space motion which is best visualized by animation

```
>>> puma.plot(traj.q);
```

In a real-world robotics application you need to be careful that the robot does not collide with nearby objects.

## 7.4 Applications

### 7.4.1 Writing on a Surface

Our goal is to create a trajectory that will allow a robot to draw a letter. The Toolbox comes with a preprocessed version of the Hershey font ▶

```
>>> font = rtb_load_jsonfile("data/hershey.json");
```

as a dictionary of character descriptors. An upper-case 'B' is itself a dictionary

```
>>> letter = font["B"]
{'strokes': [[(0.16, 0.12), (0.16, -0.72)],
            [(0.16, 0.12), (0.52, 0.12), ... ],
            [(0.16, -0.28), (0.52, -0.28), ...
            (0.16, -0.72)]],
 'width': 0.84, 'top': 0.72, 'bottom': -0.12}
```

Developed by Dr. Allen V. Hershey at the Naval Weapons Laboratory in 1967, data from ▶ https://paulbourke. net/dataformats/hershey.

This is a variable-width font and all characters fit within a unit-grid. This particular character has a width of 0.84.

The "strokes" item is a list of lists of tuples that define the $x$- and $y$-coordinates of points in the plane for each stroke in the character. We perform some processing

```
>>> lift = 0.1; # height to raise the pen
>>> scale = 0.25;
>>> via = np.empty((0, 3));
>>> for stroke in letter["strokes"]:
...     xyz = np.array(stroke) * scale # convert stroke to nx2 array
...     xyz = np.pad(xyz, ((0, 0), (0, 1))) # add third column, z=0
...     via = np.vstack((via, xyz))   # append rows to via points
...     via = np.vstack((via, np.hstack([xyz[-1,:2], lift]))) # lift
```

to create an array of via points, one per row. The strokes are scaled by 0.25 so that the character is around 20 cm tall, and the third column is zero when writing or lift to raise the pen between strokes. Next we convert this to a continuous trajectory

```
>>> xyz_traj = mstraj(via, qdmax=[0.5, 0.5, 0.5], q0=[0, 0, lift],
...                    dt=0.02, tacc=0.2).q;
```

where the arguments are the via point array, the maximum speed in the $x$-, $y$- and $z$-directions, the initial point above the character, the sample interval and the acceleration time. The number of steps in the interpolated path is

```
>>> len(xyz_traj)
500
```

and will take

```
>>> len(xyz_traj) * 0.02
10
```

**7**

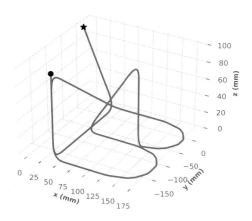

■ **Fig. 7.24**   The path of the end effector drawing the letter 'B'

seconds to execute at the 20 ms sample interval. The trajectory can be plotted

```
>>> fig = plt.figure(); ax = fig.add_subplot(111, projection="3d");
>>> plt.plot(xyz_traj[:,0], xyz_traj[:,1], xyz_traj[:,2]);
```

as shown in ■ Fig. 7.24.

The orientation of the pen around its longitudinal axis is not important in this application.

We now have a sequence of 3-dimensional points but for inverse kinematics we require a sequence of end-effector poses. We will create a coordinate frame at every point and assume that the robot is writing on a horizontal surface so these frames must have their approach vector pointing downward, that is, $a = (0, 0, -1)$. For this application the gripper can be arbitrarily ◄ oriented and we will choose alignment with the $y$-axis, that is, $o = [0, 1, 0]$. The character will be placed at $(0.6, 0, 0)$ in the workspace, and all this is achieved by

```
>>> T_pen = SE3.Trans(0.6, 0, 0.7) * SE3.Trans(xyz_traj)
...              * SE3.OA( [0, 1, 0], [0, 0, -1]);
```

Now we can apply inverse kinematics

```
>>> puma = models.DH.Puma560();
>>> sol = puma.ikine_a(T_pen, "lu");
```

to determine the joint coordinates and then animate it

```
>>> puma.plot(sol.q);
```

To run the complete example is simply

```
>>> %run -m writing
```

We have not considered the force that the robot-held pen exerts on the paper, we cover force control in ► Chap. 9. A very simple solution is to use a spring to push the pen against the paper with sufficient force to allow it to write.

and we see that the robot is drawing the letter 'B', and lifting its pen in between strokes. The approach is quite general and we could easily change the size of the letter, write whole words and sentences, write on an arbitrary plane or use a robot with quite different kinematics. ◄

## 7.4.2   A 4-Legged Walking Robot

■ Fig. 7.3b shows a sophisticated bipedal walking robot. In this example we will tackle something simpler, a four-legged (quadruped) walking robot with a gait that ensures it is statically stable at all times.

Kinematically, a robot leg is the same as a robot arm. For this application a three-joint serial-link manipulator is sufficient since the foot has point contact with the ground and orientation is not important. As always we start by defining our coordinate frames. This is shown in ■ Fig. 7.25 along with the robot leg in its

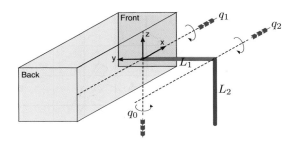

■ **Fig. 7.25** The coordinate frame and axis rotations for the simple leg. The leg is shown in its zero angle pose

zero-angle configuration. We have chosen a coordinate convention with the $x$-axis forward and the $z$-axis upward, constraining the $y$-axis to point to the left-hand side. The dimensions of the leg are

```
>>> mm = 0.001;  # millimeters
>>> L1 = -100 * mm;
>>> L2 = -100 * mm;
```

The robot has a 2 degree-of-freedom spherical hip. The first joint creates forward and backward motion, which is rotation about the $z$-axis. The second joint is hip up and down motion, which is rotation about the $x$-axis. The knee is translated by $L1$ in the $-y$-direction, and the third joint is knee motion, toward and away from the body, which is rotation about the $x$-axis. Finally, the foot is translated by $L2$ in the $-z$-direction. In ETS format this is

```
>>> leg = ERobot(ET.Rz() * ET.Rx() * ET.ty(L1) * ET.Rx() * ET.tz(L2))
ERobot: noname, 3 joints (RRR)
```

| link | link | joint | parent | ETS: parent to link |
|------|------|-------|--------|---------------------|
| 0 | link0 | 0 | BASE | Rz(q0) |
| 1 | link1 | 1 | link0 | Rx(q1) |
| 2 | link2 | 2 | link1 | ty(-0.1) $\oplus$ Rx(q2) |
| 3 | @link3 |  | link2 | tz(-0.1) |

A quick sanity check shows that for zero joint angles the foot is at

```
>>> leg.fkine([0,0,0]).t
array([      0,     -0.1,      -0.1])
```

as we designed it. We could also visualize the zero-angle pose by

```
>>> leg.plot([0, 0, 0]);
```

### 7.4.2.1 Motion of One Leg

Next we define the path that the end effector of the leg, its foot, will follow. The first consideration is that the end effector of all feet move backwards at the same speed in the ground plane – propelling the robot's body forward without its feet slipping. Each leg has a limited range of movement so it cannot move backward for very long. At some point we must reset the leg – lift the foot, move it forward and place it on the ground again. The second consideration comes from static stability – the robot must have at least three feet on the ground at all times so each leg must take its turn to reset. This requires that any leg is in contact with the ground for 75% of the cycle and is resetting for 25% of the cycle. A consequence of this is that the leg has to move much faster during reset since it has a longer path and less time to do it in. This pattern of coordinated leg motion is known as a gait – in this case a crawl gait or a wave gait.

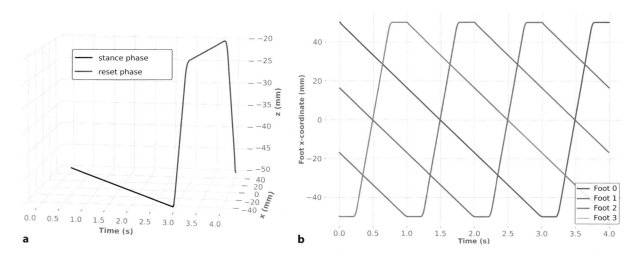

**Fig. 7.26 a** Trajectory taken by a single foot. Recall from ☐ Fig. 7.25 that the $z$-axis is downward. **b** The $x$-direction motion of each leg (offset vertically) to show the gait. The leg reset is the period of high $x$-direction velocity

During the stance phase, the foot is 50 mm below the hip and in contact with the ground. It moves from 50 mm forward of the hip to 50 mm behind, pushing the robot's body forward. During the reset phase the foot is raised to 20 mm below the hip, swung forward, and placed on the ground. The complete cycle – the start of the stance phase, the end of stance, top of the leg lift, top of the leg return and the start of stance – is defined by the via points

```
>>> xf = 50; xb = -xf;  y = -50; zu = -20; zd = -50;
>>> via = np.array([
...    [xf, y, zd],
...    [xb, y, zd],
...    [xb, y, zu],
...    [xf, y, zu],
...    [xf, y, zd]]) * mm;
```

where `xf` and `xb` are the forward and backward limits of leg motion in the $x$-direction (in units of mm), `y` is the distance of the foot from the body in the $y$-direction, and `zu` and `zd` are respectively the height of the foot motion in the $z$-direction for foot up and foot down.

Next we sample the multi-segment path at 100 Hz

```
>>> x = mstraj(via, tsegment=[3, 0.25, 0.5, 0.25], dt=0.01,
...            tacc=0.1).q
```

and we have specified a vector of desired segment times rather than maximum joint velocities to ensure that the reset takes exactly one quarter of the cycle. The final three arguments are the initial leg configuration, the sample interval and the acceleration time. This trajectory is shown in ☐ Fig. 7.26a and has a total time of 4 s and therefore comprises 400 points.

We apply inverse kinematics to determine the joint angle trajectories required for the foot to follow the computed Cartesian trajectory. This robot is underactuated so we use numerical inverse kinematics and set the mask so as to solve only for end-effector translation

```
>>> sol = leg.ikine_LM(SE3.Trans(x), mask=[1, 1, 1, 0, 0, 0]);
```

We can view the motion of the leg in animation

```
>>> leg.plot(sol.q);
```

to verify that it does what we expect: slow motion along the ground, then a rapid lift, forward motion and foot placement.

### 7.4.2.2 Motion of Four Legs

Our robot has width and length

```
>>> W = 100 * mm; L = 200 * mm;
```

We create multiple instances of the leg by cloning the leg object we created earlier, and providing different base transformations so as to attach the legs to different points on the body

```
>>> Tf = SE3.Rz(pi);
>>> legs = [
...    ERobot(leg, name="leg0", base=SE3.Trans( L/2,  W/2, 0)),
...    ERobot(leg, name="leg1", base=SE3.Trans(-L/2,  W/2, 0)),
...    ERobot(leg, name="leg2", base=SE3.Trans( L/2, -W/2, 0) * Tf),
...    ERobot(leg, name="leg3", base=SE3.Trans(-L/2, -W/2, 0) * Tf)];
```

The result is a list of ERobot objects. Note that legs 2 and 3, on the left-hand side of the body have been rotated about the $z$-axis so that they point away from the body.

As mentioned earlier, each leg must take its turn to reset. Since the trajectory is a cycle, we achieve this by having each leg run the trajectory with a phase shift equal to one quarter of the total cycle time. Since the total cycle has 400 points, each leg's trajectory is offset by 100, and we use modulo arithmetic to index into the cyclic gait for each leg. The result is the gait pattern shown in ◻ Fig. 7.26b. The core of the walking program is

```
>>> for i in range(4000):
...    legs[0].q = gait(qcycle, i, 0, False)
...    legs[1].q = gait(qcycle, i, 100, False)
...    legs[2].q = gait(qcycle, i, 200, True)
...    legs[3].q = gait(qcycle, i, 300, True)
>>> env.step(dt=0.02)  # render the graphics
```

where the function

```
>>> def gait(cycle, k, phi, flip):
...    k = (k + phi) % cycle.shape[0]  # modulo addition
...    q = cycle[k, :]
...    if flip:
...       q[0] = -q[0]  # for right-side legs
...    return q
```

returns the $(k+phi)^{th}$ element of q with modulo arithmetic that considers q as a cycle. The argument flip reverses the sign of the joint 0 motion for legs on the left-hand side of the robot.

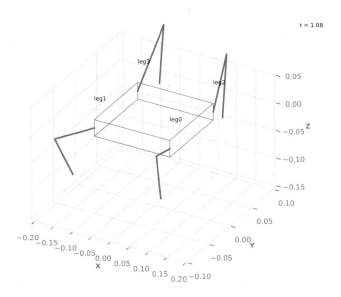

◻ **Fig. 7.27** The walking robot

To run the complete example is simply

```
>>> %run -m walking
```

and we see, after an initial delay to compute the trajectories, an animation of the robot walking. A snapshot from the simulation is shown in ◼ Fig. 7.27.

## 7.5    Advanced Topics

### 7.5.1    Zero-Angle Configuration

The shape of the robot when the joint coooordinates are all zero is completely arbitrary – there is no standard or convention. It is a function of the robot's motor controller hardware and the kinematic model. ◄

The configuration may not even be mechanically achievable.

❶ A Denavit-Hartenberg model and a URDF model of the *same* robot may have a different zero-joint shape, and so too might two URDF models of the same robot.

For a model defined:
- using ETS notation, the zero-joint shape is defined by evaluating the ETS for each link with the joint variable set to zero. To achieve a particular shape for zero joint coordinates, use the procedure in ► Sect. 7.5.2 to determine the ETS that will achieve that.
- by a URDF file, the zero-joint shape is defined by the `<origin>` element of the joints which define the rigid-body transformations of the links they connect. For joints with axes parallel to the *x*-, *y*- or *z*-axes, adjusting the roll, pitch or yaw parameters of the `<origin>` element is a straightforward way to adjust the zero-joint shape.
- using Denavit-Hartenburg parameters, the zero-joint shape is a consequence of the way the link frames are assigned. Due to the many constraints that apply to that assignment, the zero-joint shape is often quite unintuitive. The `DHRobot` class provides a joint coordinate offset to allow an arbitrary shape for the zero-joint coordinate case. The offset vector, $q_0$, is added to the user-specified joint angles before any kinematic or dynamic function is invoked, for example ◄

It is actually implemented within the `DHLink` object.

$$\xi_E = \mathcal{K}(q + q_0) \tag{7.7}$$

The offset is set by assigning the `offset` attribute of a `DHLink` subclass object, or passing the `offset` option to a `DHLink` subclass constructor.

### 7.5.2    Creating the Kinematic Model for a Robot

The classical method of determining Denavit-Hartenberg parameters is to systematically assign a coordinate frame to each link. The link frames for the PUMA 560 robot using the standard Denavit-Hartenberg formalism are shown in ◼ Fig. 7.28a. There are constraints on placing each frame since joint rotation must be about the *z*-axis and the link displacement must be in the *x*-direction. This in turn imposes constraints on the placement of the coordinate frames for the base and the end effector, and ultimately dictates the zero-angle shape discussed in ► Sect. 7.5.1. Determining the Denavit-Hartenberg parameters and link coordinate frames for a completely new mechanism is therefore more difficult than it should be – even for an experienced roboticist. Here we will introduce an alternative approach.

We will use the classic PUMA 560 as an exemplar of the class of all-revolute six-axis robot manipulators with $C = (\mathbf{S}^1)^6$. ◼ Fig. 7.28b shows a side view of the

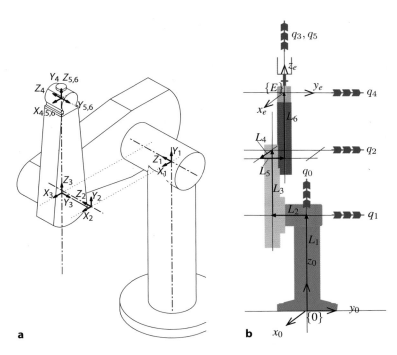

**■ Fig. 7.28** PUMA 560 robot coordinate frames. **a** Standard Denavit-Hartenberg link coordinate frames and zero-angle pose (Corke 1996b). Note that base-1 link frame indexing is used; **b** Zero-joint-angle configuration for ETS kinematic modeling, also showing dimensions and joint axes (indicated by blue triple arrows) (after Corke 2007)

robot in the desired zero-joint configuration shape. Its key dimensions are shown and have the values

```
>>> L1 = 0.672; L2 = -0.2337; L3 = 0.4318;
>>> L4 = 0.0203; L5 = 0.0837; L6 = 0.4318;
```

We will create a kinematic model of this robot by stepping along the kinematic chain from base to tip and recording the elementary transformations as we go. Starting with the world frame 0 we move up a distance of $L_1$, perform the waist rotation about the $z$-axis by $q_0$, move to the left ($-y$-direction) by $L_2$, perform the shoulder rotation about the $y$-axis by $q_1$, move up ($z$-direction) by $L_3$, move forward ($x$-direction) by $L_4$, move sideways by $L_5$, perform the elbow rotation about the $y$-axis by $q_2$, move up by $L_6$, then rotate by $q_3$ about the $z$-axis, $q_4$ about the $y$-axis and $q_5$ about the $z$-axis. The result of our journey is

```
>>> e = ET.tz(L1) * ET.Rz() * ET.ty(L2) * ET.Ry() \
...     * ET.tz(L3) * ET.tx(L4) * ET.ty(L5) * ET.Ry() \
...     * ET.tz(L6) * ET.Rz() * ET.Ry() * ET.Rz();
```

and we turn this into a fully-fledged robot object

```
>>> robot = ERobot(e)
ERobot: noname, 6 joints (RRRRRR)
```

| link | link | joint | parent | ETS: parent to link |
|------|------|-------|--------|---------------------|
| 0 | link0 | 0 | BASE | tz(0.672) ⊕ Rz(q0) |
| 1 | link1 | 1 | link0 | ty(-0.2337) ⊕ Ry(q1) |
| 2 | link2 | 2 | link1 | tz(0.4318) ⊕ tx(0.0203) ⊕ ty(0.083 |
| 3 | link3 | 3 | link2 | tz(0.4318) ⊕ Rz(q3) |
| 4 | link4 | 4 | link3 | Ry(q4) |
| 5 | @link5 | 5 | link4 | Rz(q5) |

*The result of this method is a wide table which has been cropped for inclusion in this book.

which we can use for kinematics and plotting. This approach avoids the complexities associated with the Denavit-Hartenberg convention, particularly in choosing the robot configuration and link frames.

### 7.5.3 Modified Denavit-Hartenberg Parameters

It first appeared in Craig (1986).

The Denavit-Hartenberg notation introduced in ▶ Sect. 7.1.5 dates back to the 1960s and is covered in many robotics textbooks. The modified Denavit-Hartenberg notation was introduced later ◄ and differs by placing the link coordinate frames at the near (proximal), rather than the far (distal), end of each link as shown in ◘ Fig. 7.29. This modified notation is in some ways clearer and tidier and is now quite commonly used.

> ❗ **Standard and modified Denavit-Hartenberg confusion**
> Having two Denavit-Hartenberg notations is confusing, particularly for those who are new to robot kinematics. Algorithms for kinematics, Jacobians and dynamics depend on the kinematic conventions used. Knowing the "Denavit-Hartenberg parameters" for a robot is useless unless you know whether they are standard or modified parameters.

In modified Denavit-Hartenberg notation the link transformation matrix is

$$^{j}\boldsymbol{\xi}_{j+1} = \boldsymbol{\xi}^{t_x}(a_{j-1}) \oplus \boldsymbol{\xi}^{r_x}(\alpha_{j-1}) \oplus \boldsymbol{\xi}^{r_z}(\theta_j) \oplus \boldsymbol{\xi}^{t_z}(d_j) \tag{7.8}$$

which has the same terms as (7.4) but in a different order – remember rotations are not commutative. In both conventions $a_j$ is the link length but in the standard form it is the displacement between the origins of frame {j} and frame {j + 1} while in the modified form it is from frame {j + 1} to frame {j + 2}.

> ❗ This description of Denavit-Hartenberg notation differs from standard textbooks which index the joints and parameters starting from one. This book adopts Pythonic base-zero numbering for all indexing.

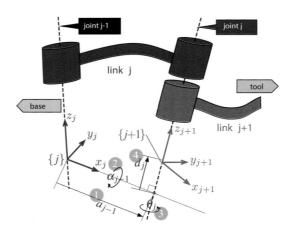

◘ **Fig. 7.29** Definition of modified Denavit and Hartenberg link parameters. The colors red and blue denote all things associated with links $j - 1$ and $j$ respectively. The numbers in circles represent the order in which the elementary transformations are applied

The Toolbox can handle either form, it only needs to be specified, and this is achieved by choosing the appropriate class when creating a link object, for example

```
>>> L1 = RevoluteMDH(d=1)
RevoluteMDH(d=1, a=0, α=0)
```

Everything else from here on, creating the robot object, kinematic and dynamic functions works as previously described.

The two forms can be interchanged by considering the link transformation as a string of elementary rotations and translations as in (7.4) or (7.8). Consider the transformation chain for standard Denavit-Hartenberg notation

$$\underbrace{\boldsymbol{\xi}^{r_z}(\theta_0) \oplus \boldsymbol{\xi}^{t_z}(d_0) \oplus \boldsymbol{\xi}^{t_x}(a_0) \oplus \boldsymbol{\xi}^{r_x}(\alpha_0)}_{\text{DH}_0 = {}^0\boldsymbol{\xi}_1} \oplus \underbrace{\boldsymbol{\xi}^{r_z}(\theta_1) \oplus \boldsymbol{\xi}^{t_z}(d_1) \oplus \boldsymbol{\xi}^{t_x}(a_1) \oplus \boldsymbol{\xi}^{r_x}(\alpha_1)}_{\text{DH}_1 = {}^1\boldsymbol{\xi}_2} \cdots$$

which we can regroup as

$$\underbrace{\boldsymbol{\xi}^{r_z}(\theta_0) \oplus \boldsymbol{\xi}^{t_z}(d_0)}_{\text{MDH}_0 = {}^0\boldsymbol{\xi}_1} \oplus \underbrace{\boldsymbol{\xi}^{t_x}(a_0) \oplus \boldsymbol{\xi}^{r_x}(\alpha_0) \oplus \boldsymbol{\xi}^{r_z}(\theta_1) \oplus \boldsymbol{\xi}^{t_z}(d_1)}_{\text{MDH}_1 = {}^1\boldsymbol{\xi}_2} \oplus \underbrace{\boldsymbol{\xi}^{t_x}(a_1) \oplus \boldsymbol{\xi}^{r_x}(\alpha_1)}_{\text{MDH}_2 = {}^2\boldsymbol{\xi}_3} \cdots$$

where the terms marked as $\text{MDH}_j$ have the form of (7.8).

### 7.5.4 Products of Exponentials

We introduced 2D and 3D twists in ▶ Chap. 2 and recall that a twist is a rigid-body transformation defined by a screw-axis direction, a point on the screw axis, and a screw pitch. A 2D twist is a vector $\boldsymbol{S} \in \mathbb{R}^3$ and a 3D twist is a vector $\boldsymbol{S} \in \mathbb{R}^6$.

For the case of the single-joint robot of ◨ Fig. 7.4a we can place a screw along the joint axis that rotates the end effector frame about the screw axis

$$\mathbf{T}_E(q) = e^{q[\hat{S}]} \mathbf{T}_E(0)$$

where $\hat{S}$ is a unit twist representing the screw axis and $\mathbf{T}_E(0)$ is the pose of the end effector when $q = 0$.

For the 2-joint robot of ◨ Fig. 7.4b we would write

$$\mathbf{T}_E(q) = e^{\mathring{q}_0[\hat{S}_0]} \left( e^{q_1[\hat{S}_1]} \mathbf{T}_E(0) \right)$$

where $\hat{S}_0$ and $\hat{S}_1$ are unit twists representing the screws aligned with the joint axes at the zero-joint configuration $q_1 = q_2 = 0$, and $\mathbf{T}_E(0)$ is the end-effector pose at that configuration. The term in parentheses is similar to the single-joint robot above, and the first twist rotates that joint and link about $\hat{\boldsymbol{S}}_0$. Using the Toolbox we define the link lengths and compute $\mathbf{T}_E(0)$

```
>>> a1 = 1; a2 = 1;
>>> TE0 = SE2(a1 + a2, 0, 0);
```

define the two 2-dimensional twists, in $\mathbf{SE}(2)$, for this example

```
>>> S0 = Twist2.UnitRevolute([0, 0]);
>>> S1 = Twist2.UnitRevolute([a1, 0]);
```

and apply them to $\boldsymbol{T}_E(0)$

```
>>> TE = S0.exp(np.deg2rad(30)) * S1.exp(np.deg2rad(40)) * TE0
    0.342    -0.9397    1.208
    0.9397    0.342     1.44
    0         0         1
```

For a general robot we can write the forward kinematics in product of exponential (PoE) form as

$$\mathbf{T}_E = \left( e^{q_0[\hat{S}_0]} \cdots e^{q_{N-1}[\hat{S}_{N-1}]} \right) \mathbf{T}_E(0)$$

The tool and base transformation are effectively included in $\mathbf{T}_E(0)$, but an explicit base transformation could be added if the screw axes are defined with respect to the robot's base rather than the world coordinate frame, or use the adjoint matrix to transform the screw axes from base to world coordinates.

where $^0\mathbf{T}_E(0)$ is the end-effector pose when the joint coordinates are all zero and $\hat{\boldsymbol{S}}_j$ is the unit twist for joint $j$ expressed in the world frame at that configuration. ◄ This can also be written as

$$\mathbf{T}_E = \mathbf{T}_E(0) \left( e^{q_0[^E\hat{S}_0]} \cdots e^{q_{N-1}[^E\hat{S}_{N-1}]} \right)$$

and $e^{q_j[^E s_j]}$ is the twist for joint $j$ expressed in the end-effector frame which is related to the twists above by $^E\hat{\boldsymbol{S}}_j = \text{Ad}(^E\boldsymbol{\xi}_0)\hat{\boldsymbol{S}}_j$.

For a particular robot we can compute the joint twists

```
>>> irb140 = models.DH.IRB140();
>>> S, TE0 = irb140.twists()
>>> S
  0: (0 0 0; 0 0 1)
  1: (-0.352 0 0.07; 0 1 0)
  2: (-0.352 0 0.43; 0 1 0)
  3: (0 0.43 0; 0 0 -1)
  4: (0.028 0 0.43; 0 1 0)
  5: (0 0.43 0; 0 0 -1)
```

where S is a `Twist` instance with six elements and `TE0` is an SE3 instance. For a particular joint configuration the forward kinematics is given by the composition of the exponentiated twists and `TE0`

```
>>> T = S.exp(irb140.qr).prod() * TE0
    0         0        -1        0.005
    1         0         0        0
    0        -1         0        0.332
    0         0         0        1
```

In this case S and `irb140.qr` both have six elements so an element-wise exponentiation is performed and then the product of those results is computed. To visualize the lines of action of the twists we first display the robot in its zero-angle configuration

```
>>> irb140.plot(irb140.qz);
```

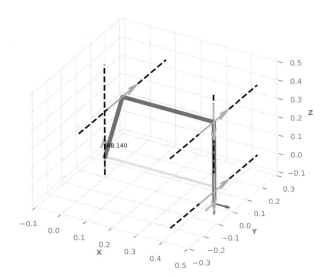

**◻ Fig. 7.30**  ABB IRB-140 robot showing the line of action of each joint twist

and overlay the lines of action

```
>>> lines = S.line()
  0: { 0 0 0; 0 0 1}
  1: { 0.352 0 -0.07; 0 1 0}
  2: { 0.352 0 -0.43; 0 1 0}
  3: { 0 -0.43 0; 0 0 -1}
  4: { -0.028 0 -0.43; 0 1 0}
  5: { 0 -0.43 0; 0 0 -1}
>>> lines.plot("k:")
```

where `lines` is a `Line3` instance that represents six lines using Plücker coordinates and these are shown in ◻ Fig. 7.30.

The Toolbox also supports robot models defined directly in terms of twists using the product of exponentials formalism. In a similar fashion to other approaches we create some `PoELink` instances and $\mathbf{T}_0$ as an SE3 instance. We can directly create a model of the robot from ◻ Fig. 7.4b, but embedded in 3D, using twists

```
>>> link1 = PoERevolute([0, 0, 1], [0, 0, 0]);  # Rz through (0,0,0)
>>> link2 = PoERevolute([0, 0, 1], [1, 0, 0]);  # Rz through (1,0,0)
>>> TE0 = SE3.Tx(2);  # end-effector pose when q=[0,0]
```

where the arguments to `PoELink.Revolute` are the same as those to `Twist.UnitRevolute`. The link objects are then combined into a `Robot` subclass object in a familiar way

```
>>> robot = PoERobot([link1, link2], TE0);
```

and the forward kinematics is

```
>>> robot.fkine([0, 0]).printline()
t = 2, 0, 0; rpy/zyx = 0°, 0°, 0°
```

## 7.5.5  Collision Detection

For many robotic applications we require the ability to check for potential collisions between the robot and objects that are fixed or moving in the world. This is typically achieved by modeling every object with one or more collision shapes as shown in ◻ Fig. 7.31. These shapes can be simple geometric primitives such as rectangular prisms, spheres or cylinders whose intersections can be computed cheaply, or more general meshes. ▶

As defined by an STL or COLLADA file.

**7**

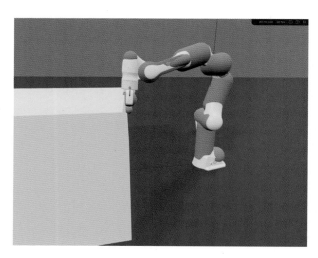

**◻ Fig. 7.31**   A Panda robot with its collision model objects shown in orange and a box-shaped obstacle

To illustrate this, we will instantiate a model of the Panda robot

```
>>> panda = models.URDF.Panda();
```

which includes collision models and a box with 1 m sides centered at $(1.1, 0, 0)$ in the workspace

```
>>> from spatialgeometry import Cuboid
>>> box = Cuboid([1, 1, 1], pose=SE3.Tx(1.1));
```

We test for a collision between the robot in a specified configuration and the box by

```
>>> panda.iscollided(panda.qr, box)
False
```

which in this case is false – none of the robot's links will intersect the box. If we set the center of the box to $(1.0, 0, 0)$

```
>>> box.T = SE3.Tx(1)
>>> panda.iscollided(panda.qr, box)
True
```

then there is a collision.

To visualize this we can extend the example above to instantiate the robot and the box in the Swift simulation environment

```
>>> # plot robot and get reference to graphics environment
>>> env = panda.plot(panda.qr, backend="swift");
>>> env.add(box);  # add box to graphics
>>> env.step()     # update the graphics
```

which will display in a browser tab. Looking closely we see that the gripper is not quite touching the box but the collision model, shown in ◻ Fig. 7.31, is very conservative. You can change the configuration of the robot by setting its q property, or the pose of the box by setting its T property, and then updating the simulator's display by

```
>>> env.step()
```

## 7.6 Wrapping Up

In this chapter we have learned how to determine the forward and inverse kinematics of a serial-link manipulator arm. Forward kinematics involves compounding the relative poses due to each joint and link, giving the pose of the robot's end effector relative to its base. We have considered this from the perspective of pose graphs and rigid-body trees in both 2D and 3D, and we have discussed other ways of representing models such as URDF files, Denavit-Hartenberg parameters, and the product of exponentials using twists.

Inverse kinematics is the problem of determining the joint coordinates given the end-effector pose. This inverse is not unique and the robot may have several joint configurations that result in the same end-effector pose. For simple robots, or those with six joints and a spherical wrist we can compute the inverse kinematics using an analytic solution which provides explicit control over which solution is found.

For robots which do not have six joints and a spherical wrist we can use an iterative numerical approach to solve the inverse kinematic problem. We showed how this could be applied to an underactuated 4-joint SCARA robot and a redundant 7-link robot. We also touched briefly on the topic of singularities which are due to the alignment of joint axes.

We also learned about creating paths to move the end effector smoothly between poses. Joint-space paths are simple to compute but in general do not result in straight-line paths in Cartesian space which may be problematic for some applications. Straight-line paths in Cartesian space can be generated but singularities in the workspace may lead to very high joint rates.

### 7.6.1 Further Reading

Serial-link manipulator kinematics are covered in all the standard robotics textbooks such as the Robotics Handbook (2016), Siciliano et al. (2009), Spong et al. (2006) and Paul (1981). Craig's text (2005) is also an excellent introduction to robot kinematics and uses the modified Denavit-Hartenberg notation, and the examples in the third edition are based on an older version of the Robotics Toolbox for MATLAB®. Lynch and Park (2017) and Murray et al. (1994) cover the product of exponential approach. An emerging alternative to Denavit-Hartenberg notation is URDF (unified robot description format) which is described at ► https://wiki.ros.org/urdf.

Siciliano et al. (2009) provide a very clear description of the process of assigning Denavit-Hartenberg parameters to an arbitrary robot. An alternative approach based on symbolic factorization of elementary transformation sequences was described in detail by Corke (2007). The definitive values for the parameters of the PUMA 560 robot are described in the paper by Corke and Armstrong-Hélouvry (1995).

Robotic walking is a huge field in its own right and the example given here is very simplistic. Machines have been demonstrated with complex gaits such as running and galloping that rely on dynamic rather than static balance. A good introduction to legged robots is given in the Robotics Handbook (Siciliano and Khatib 2016, § 17). Robotic hands, grasping and manipulation is another large topic which we have not covered – there is a good introduction in the Robotics Handbook (Siciliano and Khatib 2016, § 37, 38).

Parallel-link manipulators have not been covered in this book. They have advantages such as increased actuation force and stiffness (since the actuators form a truss-like structure) and are capable of very high-speed motion. For this class of mechanism, the inverse kinematics is usually closed-form and it is the forward kinematics that requires numerical solution. Useful starting points for this class of robots are the handbook (Siciliano and Khatib 2016, 2016, § 18), a brief section in Siciliano et al. (2009) and Merlet (2006).

Closed-form inverse kinematic solutions can be derived algebraically by writing down a number of kinematic relationships and solving for the joint angles, as described in Paul (1981). Software packages to automatically generate the forward and inverse kinematics for a given robot have been developed and these include SYMORO (Khalil and Creusot 1997) which is now available as open-source, and OpenRAVE (▶ https://openrave.org/), which contains the IKFast kinematic solver (Diankov 2010).

■■ **Historical**

The original work by Denavit and Hartenberg was their 1955 paper (1955) and their textbook (Hartenberg and Denavit 1964). The book has an introduction to the field of kinematics and its history but is currently out of print, although a version can be found online. The first full description of the kinematics of a six-link arm with a spherical wrist was by Paul and Zhang (1986). Lipkin (2008) provides a discussion about the various forms of Denavit-Hartenberg notation beyond the two variants covered in this chapter.

### 7.6.2 **Exercises**

1. Forward kinematics for planar robot from ◘ Fig. 7.4.
   a) For the 2-joint robot use the `teach` method to determine the two sets of joint angles that will position the end effector at (0.5, 0.5).
   b) Experiment with the three different models in ◘ Fig. 7.4 using the `fkine` and `teach` methods.
   c) Vary the models: adjust the link lengths, create links with a translation in the $y$-direction, or create links with a translation in the $x$- and $y$-direction.
2. Experiment with the `teach` method for the PUMA 560 robot.
3. Inverse kinematics for the 2-link robot (▶ Sect. 7.2.1).
   a) What happens to the solution when a point is out of reach?
   b) Most end-effector positions can be reached by two different sets of joint angles. What points can be reached by only one set?
   c) Find a solution given $x$ and $\theta$ where $y$ is unconstrained.
4. Compare the solutions generated by `ikine_a` and `ikine_LM` for the PUMA 560 robot at different poses. Is there any difference in accuracy? How much slower is `ikine_LM`?
5. For the PUMA 560 at configuration `qn` demonstrate a configuration change from elbow up to elbow down.
6. For a PUMA 560 robot investigate the errors in end-effector pose due to manufacturing errors.
   a) Make link 2 longer by 0.5 mm. For 100 random joint configurations what is the mean and maximum error in the components of end-effector pose?
   b) Introduce an error of 0.1° in the joint 2 angle and repeat the analysis above.
7. Investigate the redundant robot models `DH.Hyper` and `DH.Coil`. Manually control them using the `teach` method, compute forward kinematics and numerical inverse kinematics.
8. Experiment with the `ikine_LM` method to solve inverse kinematics for the case where the joint coordinates have limits (modeling mechanical end stops). Joint limits are set with the `qlim` property of the `Link` class.
9. Drawing a 'B' (▶ Sect. 7.4.1)
   a) Change the size of the letter.
   b) Write a word or sentence.
   c) Write on a vertical plane.

    d) Write on an inclined plane.

    e) Change the robot from a PUMA 560 to a IRB140.

    f) Write on a sphere. Hint: Write on a tangent plane, then project points onto the sphere's surface.

    g) This writing task does not require 6DoF since the rotation of the pen about its axis is not important. Remove the final link from the PUMA 560 robot model and repeat the exercise.

10. Walking robot (▶ Sect. 7.4.2)

    a) Shorten the reset trajectory by reducing the leg lift during reset.

    b) Increase the stride of the legs.

    c) Determine the position of the center of the robot as a function of time.

    d) Figure out how to steer the robot by changing the stride length on one side of the body.

    e) Change the gait so the robot moves sideways like a crab.

    f) Add another pair of legs. Change the gait to reset two legs or three legs at a time.

    g) Currently in the simulation the legs move but the body does not move forward. Modify the simulation so the body moves.

11. A robot hand comprises a number of fingers, each of which is a small serial-link manipulator. Create a model of a hand with 2, 3 or 5 fingers and animate the finger motion.

12. Create a simulation with two robot arms next to each other, whose end effectors are holding a basketball at diametrically opposite points in the horizontal plane. Write code to move the robots so as to rotate the ball about the vertical axis.

# Manipulator Velocity

## Contents

The velocity of a robot manipulator's end effector – the rate of change of its pose – is known as spatial velocity and was introduced in ▶ Sect. 3.1. This task-space velocity has a translational and rotational component and for 3-dimensional motion is represented by a 6-vector. That velocity is a consequence of the rate of change of the joint coordinates, the joint-space velocity, or simply joint velocity.

In this chapter we introduce the manipulator Jacobian matrix which describes the relationship between the joint velocity and the spatial velocity of the end effector. ▶ Sect. 8.1 uses a simple planar robot to introduce the manipulator Jacobian matrix and then extends the principle to more general robots. ▶ Sect. 8.2 uses the inverse Jacobian to generate straight-line Cartesian paths without requiring inverse kinematics. ▶ Sect. 8.3 discusses the numerical properties of the Jacobian matrix which provide insight into the dexterity of the manipulator – the directions in which it can move easily and those in which it cannot. It also introduces the concepts of over- and underactuated robots. ▶ Sect. 8.4 demonstrates how the Jacobian transpose is used to transform forces and moments on the end-effector to forces or moments at the joints. ▶ Sect. 8.5 revisits the problem of inverse kinematics with a more detailed discussion of the numerical solution that was introduced in the previous chapter, and its dependence on the Jacobian matrix is fully described. We finish in ▶ Sect. 8.6 with some advanced topics.

## 8.1 Manipulator Jacobian

In the last chapter we discussed the relationship between joint coordinates and end-effector pose – the manipulator kinematics. Now we investigate the relationship between the rate of change of these quantities – between joint velocity and the end-effector spatial velocity. This is the differential kinematics of the manipulator.

### 8.1.1 Jacobian in the World Coordinate Frame

We illustrate the basics with our now familiar 2-dimensional example, see ◼ Fig. 8.1, this time defined with symbolic rather than numeric parameters

```
>>> import sympy
>>> a1, a2 = sympy.symbols("a1, a2")
>>> e = ERobot2(ET2.R() * ET2.tx(a1) * ET2.R() * ET2.tx(a2))
```

and define a symbolic 2-vector to represent the joint angles

```
>>> q = sympy.symbols("q:2")
(q0, q1)
```

The forward kinematics are

```
>>> TE = e.fkine(q);
```

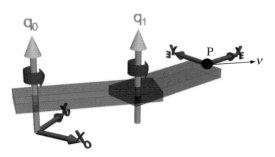

◼ **Fig. 8.1** Two-link robot showing the end-effector position P, represented by the coordinate vector $p = (x, y)$, and the task-space velocity vector $v = dp/dt$. This is the same as the robot in ◼ Fig. 7.7

---

**Excurse 8.1: Jacobian Matrix**

A Jacobian is the multidimensional form of the derivative – the derivative of a vector-valued function of a vector with respect to a vector. If $y = f(x)$ and $x \in \mathbb{R}^n$ and $y \in \mathbb{R}^m$ then the Jacobian is the $m \times n$ matrix

$$J = \frac{\partial f}{\partial x} = \begin{pmatrix} \frac{\partial y_1}{\partial x_1} & \cdots & \frac{\partial y_1}{\partial x_n} \\ \vdots & \ddots & \vdots \\ \frac{\partial y_m}{\partial x_1} & \cdots & \frac{\partial y_m}{\partial x_n} \end{pmatrix}$$

The Jacobian is named after Carl Jacobi, and more details are given in ▶ App. E.

---

and the position of the end effector $p = (x, y) \in \mathbb{R}^2$ is

```
>>> p = TE.t
array([a1*cos(q0) + a2*cos(q0 + q1),
       a1*sin(q0) + a2*sin(q0 + q1)], dtype=object)
```

and now we compute the derivative of $p$ with respect to the joint configuration $q$. Since $p$ and $q$ are both vectors, the derivative

$$\frac{\partial p}{\partial q} = \mathbf{J}(q) \tag{8.1}$$

will be a matrix – a Jacobian (see ▶ App. E) matrix

```
>>> J = sympy.Matrix(p).jacobian(q)
Matrix([
[-a1*sin(q0) - a2*sin(q0 + q1), -a2*sin(q0 + q1)],
[ a1*cos(q0) + a2*cos(q0 + q1),  a2*cos(q0 + q1)]])
>>> J.shape
(2, 2)
```

which is typically denoted by the symbol $\mathbf{J}$ and in this case is $2 \times 2$.

To determine the relationship between *joint* velocity and *end-effector* velocity we rearrange (8.1) as

$$\partial p = \mathbf{J}(q)\, \partial q$$

and dividing through by d$t$ we obtain

$$\dot{p} = \mathbf{J}(q)\dot{q}$$

where the Jacobian matrix maps velocity from the joint configuration space to the end-effector velocity in task space, and is itself a function of the joint configuration.

To obtain the general form we write the forward kinematics as a function, repeating (7.2)

$$^0\xi = \mathcal{K}(q)$$

where $q \in \mathbb{R}^N$ and $N = \dim C$ is the dimension of the configuration space – the number of robot joints. Taking the derivative we write the spatial velocity of the end effector

$$^0v = {}^0\mathbf{J}(q)\,\dot{q} \in \mathbb{R}^M \tag{8.2}$$

**8**

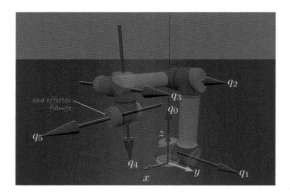

◻ **Fig. 8.2**   UR5 robot in joint configuration q1. The end-effector $z$-axis points in the world $x$-direction

where $M = \dim \mathcal{T}$ is the dimension of the task space. For the common case of operation in 3D space where $\boldsymbol{\xi} \in \mathbb{R}^3 \times \mathbf{S}^3$ then, as discussed in ▶ Sect. 3.1.1, $^0\boldsymbol{v} = (v_x, v_y, v_z, \omega_x, \omega_y, \omega_z) \in \mathbb{R}^6$ and comprises translational and rotational velocity components. The matrix $^0\mathbf{J}(\boldsymbol{q}) \in \mathbb{R}^{6 \times N}$ is the manipulator Jacobian or the geometric Jacobian. This relationship is sometimes referred to as the instantaneous forward kinematics.

The Jacobian matrix can be computed by the `jacob0` method of any Toolbox robot object. For a UR5 robot in the configuration shown in ◻ Fig. 8.2 the Jacobian

```
>>> ur5 = models.URDF.UR5();
>>> J = ur5.jacob0(ur5.q1)
array([[-0.02685,    0.3304,  -0.09465,  -0.09465,   -0.0823,        0],
       [  0.4745,         0,         0,         0,   -0.0823,        0],
       [       0,   -0.4745,   -0.4745,   -0.0823,         0,  -0.0823],
       [       0,         0,         0,         0,         0,        1],
       [       0,         1,         1,         1,         0,        0],
       [       1,         0,         0,         0,        -1,        0]])
```

is a $6 \times 6$ matrix. Each row corresponds to a task-space degree of freedom in the order $v_x, v_y, v_z, \omega_x, \omega_y,$ and $\omega_z$.

🛈 Some references, including the MATLAB® version of this book, and some software tools place the rotational components first, that is, the rows correspond to $\omega_x, \omega_y, \omega_z, v_x, v_y,$ and $v_z$.

Each column corresponds to a joint-space degree of freedom – it is the end-effector spatial velocity created by unit velocity of the corresponding joint. The total end-effector velocity is the linear sum of the columns weighted by the joint velocities. In this configuration, motion of joint 0, the first column, causes translation in the world $y$- and $x$-directions and rotation about the $z$-axis. Motion of joints 1 and 2 cause motion in the world $x$- and $z$-directions and negative rotation about the $y$-axis. The $3 \times 3$ block in the top right of the Jacobian relates wrist velocity ($\dot{q}_3, \dot{q}_4, \dot{q}_5$) to end-effector translational velocity. These values are small due to the

For a robot with a spherical wrist and no tool, this block will be zero.

small link lengths in the wrist mechanism. ◀

◻ Fig. 8.2 shows the joint axes in space and you can imagine how the end effector moves in space as each joint is slightly rotated. You could also use the `teach` method

```
>>> ur5.teach(ur5.q1)
```

to adjust individual joint angles and observe the corresponding change in end-effector pose.

### 8.1.2 Jacobian in the End-Effector Coordinate Frame

The Jacobian computed by the method `jacob0` maps joint velocity to the end-effector spatial velocity expressed in the world coordinate frame. We can use the velocity transformation (3.5) from the world frame to the end-effector frame, which is a function of the end-effector pose, to obtain the spatial velocity in the end-effector coordinate frame

$$
{}^E\nu = {}^E\mathbf{J}_0({}^E\boldsymbol{\xi}_0)\,{}^0\mathbf{J}(\boldsymbol{q})\,\dot{\boldsymbol{q}} = \begin{pmatrix} {}^E\mathbf{R}_0 & \mathbf{0}_{3\times3} \\ \mathbf{0}_{3\times3} & {}^E\mathbf{R}_0 \end{pmatrix} {}^0\mathbf{J}(\boldsymbol{q})\,\dot{\boldsymbol{q}} = {}^E\mathbf{J}(\boldsymbol{q})\,\dot{\boldsymbol{q}} \tag{8.3}
$$

which results in a new Jacobian for end-effector velocity. ▶

In the Toolbox this Jacobian is computed by the method `jacobe` and for the UR5 robot at the pose used above is

```
>>> ur5.jacobe(ur5.q1)
array([[ -0.4745,        0,       0,        0,  0.0823,       0],
       [ 0.02685,  -0.3303, 0.09465, 0.09465,  0.0823,       0],
       [       0,   0.4746,  0.4745,  0.0823,       0,  0.0823],
       [       0,       -1,      -1,      -1,       0,       0],
       [       0,        0,       0,        0,       0,      -1],
       [      -1,        0,       0,        0,       1,       0]])
```

For robots defined using Denavit-Hartenberg notation the Toolbox computes the end-effector Jacobian directly, and applies the inverse velocity transform to obtain the world frame Jacobian.

### 8.1.3 Analytical Jacobian

In (8.2) and (8.3) the spatial velocity is expressed in terms of translational and angular velocity vectors, however angular velocity is not a very intuitive concept. For some applications it can be more useful to consider the rotational velocity in terms of rates of change of roll-pitch-yaw angles, Euler angles, or exponential coordinates. Analytical Jacobians are those where the rotational velocity is expressed in a representation other than angular velocity.

**8**

Consider the case of XYZ roll-pitch-yaw angles $\boldsymbol{\Gamma} = (\alpha, \beta, \gamma) \in \mathbb{R}^3$ for which the rotation matrix is

$$\mathbf{R} = \mathbf{R}_x(\gamma)\,\mathbf{R}_y(\beta)\,\mathbf{R}_z(\alpha)$$

$$= \begin{pmatrix} c\beta c\alpha & -c\beta s\alpha & s\beta \\ c\gamma s\alpha + c\alpha s\beta s\gamma & -s\beta s\gamma s\alpha + c\gamma c\alpha & -c\beta s\gamma \\ s\gamma s\alpha - c\gamma c\alpha s\beta & c\gamma s\beta s\alpha + c\alpha s\gamma & c\beta c\gamma \end{pmatrix} \in \mathbf{SO}(3)$$

where we use the shorthand $c\theta$ and $s\theta$ to mean $\cos\theta$ and $\sin\theta$ respectively. Using the chain rule and a bit of effort we can write the derivative $\dot{\mathbf{R}}$, and recalling (3.1)

$$\dot{\mathbf{R}} = [\boldsymbol{\omega}]_\times \mathbf{R}$$

we can solve for the elements of $\boldsymbol{\omega}$ in terms of roll-pitch-yaw angles and their rates to obtain

$$\begin{pmatrix} \omega_x \\ \omega_y \\ \omega_z \end{pmatrix} = \begin{pmatrix} s\beta\dot{\alpha} + \dot{\gamma} \\ -c\beta s\gamma\dot{\alpha} + c\gamma\dot{\beta} \\ c\beta c\gamma\dot{\alpha} + s\gamma\dot{\beta} \end{pmatrix}$$

which can be factored as

$$\boldsymbol{\omega} = \begin{pmatrix} s\beta & 0 & 1 \\ -c\beta s\gamma & c\gamma & 0 \\ c\beta c\gamma & s\gamma & 0 \end{pmatrix} \begin{pmatrix} \dot{\alpha} \\ \dot{\beta} \\ \dot{\gamma} \end{pmatrix}$$

and written concisely as

$$\boldsymbol{\omega} = \mathbf{A}(\boldsymbol{\Gamma})\dot{\boldsymbol{\Gamma}}$$

The matrix $\mathbf{A}$ is itself a Jacobian that maps XYZ roll-pitch-yaw angle velocity to angular velocity. It can be computed by the Toolbox function

```
>>> rotvelxform((0.1, 0.2, 0.3), representation="rpy/xyz")
array([[  0.1987,        0,        1],
       [ -0.2896,   0.9553,        0],
       [  0.9363,   0.2955,        0]])
```

where the arguments are the roll, pitch and yaw angles. Provided that $\mathbf{A}$ is not singular, the analytical Jacobian is

$$\mathbf{J}_a(\boldsymbol{q}) = \begin{pmatrix} \mathbf{1}_{3\times3} & \mathbf{0}_{3\times3} \\ \mathbf{0}_{3\times3} & \mathbf{A}^{-1}(\boldsymbol{\Gamma}) \end{pmatrix} \mathbf{J}(\boldsymbol{q}) \ .$$

$\mathbf{A}$ is singular when $\cos\beta = 0$ (pitch angle $\beta = \pm\frac{\pi}{2}$) and this condition is referred to as a representational singularity. The analytical Jacobian is computed by

```
>>> ur5.jacob0_analytical(ur5.q1, "rpy/xyz");
```

and the associated spatial velocity vector is with respect to the world frame but the rotational velocity is expressed as the rate of change of XYZ roll-pitch-yaw angles. ◄ A similar approach can be taken for Euler angles using the argument `"eul"` to `rotvelxform` and `jacob0_analytical`.

Another useful analytical Jacobian relates angular velocity to the rate of change of exponential coordinates $\boldsymbol{s} = \hat{\boldsymbol{v}}\theta \in \mathbb{R}^3$ by

$$\boldsymbol{\omega} = \mathbf{A}(\boldsymbol{s})\dot{\boldsymbol{s}}$$

For `rotvelxform`, the option `inverse=True` will return the inverse computed in closed-form and `full=True` will return the block diagonal 6 × 6 matrix suitable for premultiplying the geometric Jacobian.

where

$$\mathbf{A}(s) = \mathbf{1}_{3\times3} - \frac{1-\cos\theta}{\theta}[\hat{\boldsymbol{v}}]_\times + \frac{\theta - \sin\theta}{\theta}[\hat{\boldsymbol{v}}]_\times^2$$

and $\hat{\boldsymbol{v}}$ and $\theta$ can be determined from the end-effector rotation matrix via the matrix logarithm, and is computed by passing the argument `"exp"` to `rotvelxform` and `jacob0_analytical`. This is singular when $\theta = 0$ which corresponds to a null rotation.

## 8.2 Application: Resolved-Rate Motion Control

We have discussed how the Jacobian matrix maps joint velocity to end-effector task-space velocity but the inverse problem has strong practical use – what joint velocities are needed to achieve a particular end-effector task-space velocity $\boldsymbol{v}$? Provided that the Jacobian matrix is square and nonsingular we can invert (8.2) and write

$$\dot{\boldsymbol{q}} = \mathbf{J}^{-1}(\boldsymbol{q})\boldsymbol{v} \tag{8.4}$$

which is the essence of resolved-rate motion control – a simple and elegant algorithm that generates constant velocity end-effector motion without requiring inverse kinematics. It uses the inverse Jacobian matrix to map or *resolve* the desired task-space velocity to joint velocity. In this example we will assume that the Jacobian is square ($6 \times 6$) and nonsingular but we will relax these constraints later.

The motion control scheme is typically implemented in discrete-time form as

$$\dot{\boldsymbol{q}}^*_{\langle k \rangle} = \mathbf{J}^{-1}(\boldsymbol{q}_{\langle k \rangle})\boldsymbol{v}^*$$
$$\boldsymbol{q}^*_{\langle k+1 \rangle} \leftarrow \boldsymbol{q}^*_{\langle k \rangle} + \delta_t \dot{\boldsymbol{q}}^*_{\langle k \rangle} \tag{8.5}$$

where $\delta_t$ is the sample interval and $\langle \cdot \rangle$ indicates the time step. The first equation computes the required joint velocity as a function of the current joint configuration and the desired end-effector spatial velocity $\boldsymbol{v}^*$. The second computes the desired joint coordinates at the next time step, using forward-rectangular integration.

An example of the algorithm in block-diagram form is shown in ◘ Fig. 8.3. The desired task-space velocity is $0.05\,\mathrm{m\,s}^{-1}$ in the world $y$-direction. The inputs to the `Jacobian` block are the current manipulator joint coordinates and the output is a $6 \times 6$ Jacobian matrix. This is inverted and multiplied by the desired task-space velocity to form the required joint velocity, and the robot is modeled by an integrator. ▶

This models an ideal robot whose actual joint velocity precisely matches the demanded joint velocity. A real robot exhibits tracking error and this is discussed in ▶ Sect. 9.1.7.

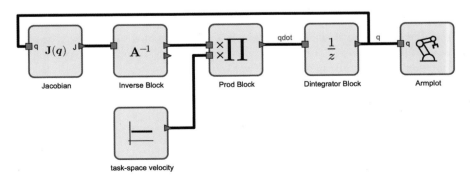

◘ **Fig. 8.3** Block-diagram model RRMC for resolved-rate motion control for constant end-effector velocity

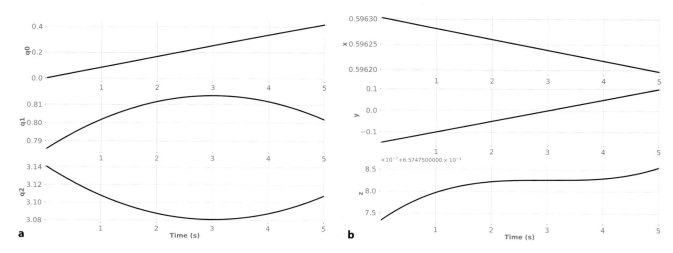

**◻ Fig. 8.4** Robot motion using resolved-rate motion control of ◻ Fig. 8.3 **a** joint-space motion **b** task-space end-effector motion. Note the small, but unwanted motion in the *x*- and *z*-directions;

Running the simulation

```
>>> %run -m RRMC -H
```

creates an animation of a Puma 560 manipulator with its end effector moving at constant velocity in task space. Simulation results are returned in the variable `out` from which we extract the discrete-time time and joint coordinates

```
>>> t = out.clock0.t;
>>> q = out.clock0.x;
```

The motion of the first three joints

```
>>> xplot(t, q[:, :3], stack=True);
```

is shown in ◻ Fig. 8.4a and are not linear with time – reflecting the changing kinematic configuration of the arm.

We apply forward kinematics to determine the end-effector position

```
>>> Tfk = puma.fkine(q);
```

The function `xplot` is a Toolbox utility that plots columns of a matrix in separate subgraphs.

which we plot ◂ as a function of time

```
>>> xplot(out.clock0.t, Tfk.t, stack=True);
```

and this is shown in ◻ Fig. 8.4b. The task-space motion is $0.05\,\text{m s}^{-1}$ in the *y*-direction as specified.

The approach just described, based purely on integration, suffers from an accumulation of error which we observe as small but unwanted motion in the *x*- and *z*-directions in ◻ Fig. 8.4b. The Jacobian is the gradient of a very nonlinear function – the forward kinematics – which is a function of configuration. The root cause of the error is that, with a finite time step, as the robot moves the previously computed Jacobian diverges from the true gradient.

To follow a more complex path we can change the algorithm to a *closed-loop* form based on the difference between the actual pose and the time-varying desired pose

$$\dot{\boldsymbol{q}}^*\langle k\rangle \leftarrow K_p \mathbf{J}^{-1}(\boldsymbol{q}\langle k\rangle)\,\Delta(\mathcal{K}(\boldsymbol{q}\langle k\rangle), \boldsymbol{\xi}^*\langle k\rangle) \tag{8.6}$$

where $K_p$ is a proportional gain, $\Delta(\cdot) \in \mathbb{R}^6$ is the spatial displacement from (3.11) and the desired pose $\boldsymbol{\xi}^*\langle k\rangle$ is a function of time.

A block-diagram model to demonstrate this for a circular path is shown in ◻ Fig. 8.5, where the tool of the robot traces out a circle of radius 50 mm in the horizontal plane. The desired pose is a frame whose origin follows a circular path

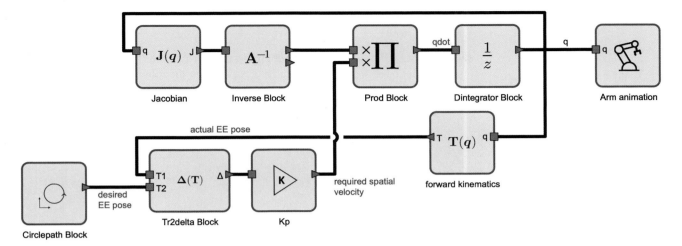

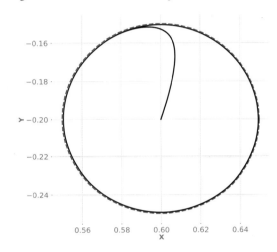

**◘ Fig. 8.5** Block-diagram model RRMC2 for closed-loop resolved-rate motion control with circular end-effector motion

**◘ Fig. 8.6** End-effector trajectory using closed-loop resolved-rate motion control of ◘ Fig. 8.5 with circular end-effector motion. The desired circular motion is shown as a red-dashed circle

as a function of time. The *difference*, using the $\Delta(\cdot)$ operator, between the desired pose and the current pose from forward kinematics is computed by the `tr2delta` block. The result is a spatial displacement, a translation and a rotation described by a 6-vector which is scaled by a proportional gain, to become the demanded spatial velocity. This is transformed by the inverse manipulator Jacobian matrix to become the demanded joint velocity that will drive the end effector toward the time varying pose. We run this by

```
>>> %run -m RRMC2 -H
```

and the results are shown in ◘ Fig. 8.6. This closed-loop form will also prevent the unbounded growth of integration error as shown in ◘ Fig. 8.4b.

## 8.3 Jacobian Condition and Manipulability

We have assumed, so far, that the Jacobian is square and non-singular but in practice this is not always the case. The Jacobian is a $\dim\mathcal{T} \times \dim\mathcal{C}$ matrix so for a robot operating in the task space $\mathcal{T} \in \mathbb{R}^3 \times \mathbf{S}^3$ where $\dim\mathcal{T} = 6$, a square Jacobian requires a robot with 6 joints.

### 8.3.1 Jacobian Singularities

If the Jacobian is square, a robot configuration $q$ where $\det(\mathbf{J}(q)) = 0$ is described as singular or degenerate. Singularities occur when the robot is at maximum reach, or when the axes of one or more joints become aligned. This results in the loss of degrees of freedom of motion, and this is the gimbal lock problem again.

For example, the UR5 robot at its *zero* configuration has a Jacobian matrix

```
>>> J = ur5.jacob0(ur5.qz)
array([[ -0.1915, -0.09465, -0.09465, -0.09465,    0.0823,        0],
       [  0.8996,        0,        0,        0,   -0.0823,        0],
       [       0,  -0.8996,  -0.4746,  -0.0823,        0,  -0.0823],
       [       0,        0,        0,        0,        0,        0],
       [       0,        1,        1,        1,        0,        1],
       [       1,        0,        0,        0,       -1,        0]])
```

which is singular

```
>>> np.linalg.det(J)
0
```

and digging a little deeper we see that the Jacobian rank has become

```
>>> np.linalg.matrix_rank(J)
5
```

compared to a maximum of six for a $6 \times 6$ matrix. The rank deficiency of $6 - 5 = 1$ means that one column is equal to some linear combination of the other columns. Looking at the Jacobian we see that columns 1, 2 and 3 are similar in that they all cause task space motion in the $x$- and $z$-directions and rotation about the $y$-axis. The function jsingu performs the column dependency analysis

```
>>> jsingu(J)
column 3 = - 0.923 column_1 + 1.92 column_2
```

indicating that any task space velocity induced by $\dot{q}_3$ can also be induced by some combination of $\dot{q}_1$ and $\dot{q}_2$. This means that joint 3 is not needed, and the robot has only 5 effective joints – making motion in one task-space direction inaccessible.

If the robot is close to, but not actually at, a singularity we still encounter problems. Consider a configuration close to the singularity at qz where all joints have a nonzero value of 5°

```
>>> qns = np.full((6,), np.deg2rad(5))
array([ 0.08727,  0.08727,  0.08727,  0.08727,  0.08727,  0.08727])
```

and the Jacobian is now

```
>>> J = ur5.jacob0(qns);
```

If we use resolved-rate motion control to generate a modest end-effector velocity of $0.1 \, \text{rad s}^{-1}$ about the $x$-axis

```
>>> qd = np.linalg.inv(J) @ [0, 0, 0, 0.1, 0, 0]
array([-0.003797,   -2.773,    5.764,   -4.088,  -0.02945,    1.093])
```

the result is very high-speed motion of many joints – the velocity of joint 2 is nearly 1 revolution per second and is over 50 times greater than the required task space rotation rate. ◄ Although the robot is no longer at a singularity, the determinant of the Jacobian is still very small

We observed a similar effect in ◘ Fig. 7.23.

```
>>> np.linalg.det(J)
-0.0009943
```

The closer we get to the singularity the higher the joint rates will be – at the singularity those rates will go to infinity. Alternatively, we can say that the condition number of the Jacobian is very high

```
>>> np.linalg.cond(J)
232.1
```

and the Jacobian is *poorly conditioned.*

For some motions, such as translation in this case, the poor condition of the Jacobian is not problematic. If we wished to translate the tool in the *y*-direction

```
>>> qd = np.linalg.inv(J) @ [0, 0.1, 0, 0, 0, 0]
array([  0.1269,  -0.3555,   0.7214,  -0.7412,   0.1226,   0.3768])
```

the required joint rates are modest and achievable. This particular joint configuration is therefore good for certain end-effector motions but poor for others.

## 8.3.2  Velocity Ellipsoid and Manipulability

It would be useful to have some measure of how well the robot is able to achieve a particular task-space velocity. We start by considering the set of generalized joint velocities $\dot{q} \in \mathbb{R}^N$ with a unit norm $\|\dot{q}\| = 1$, that is

$$\dot{q}^\top \dot{q} = 1$$

Geometrically, if we consider $\dot{q}$ to be a point in the N-dimensional joint-velocity space, this equation describes all points which lie on the surface of a hypersphere. Substituting (8.4) we can write

$$v^\top \left( \mathbf{J}(q)\mathbf{J}^\top(q) \right)^{-1} v = 1 \tag{8.7}$$

which is the equation of points $v$ on the surface of a hyper-ellipsoid within the dim $\mathcal{T}$-dimensional task-velocity space. Each point represents a possible velocity of the end effector and ideally these velocities are large in all possible task-space directions – an isotropic end-effector velocity space. That implies that the ellipsoid is close to spherical and its radii are of the same order of magnitude. However, if one or more radii are very small this indicates that the end effector cannot achieve velocity in the directions corresponding to those small radii.

For the planar robot arm of ◻ Fig. 8.1

```
>>> planar2 = models.ETS.Planar2();
```

we can interactively explore how the shape of the velocity ellipse changes with configuration using the teach method

```
>>> planar2.teach(np.deg2rad([30, 40]), vellipse=True);
```

and the initial view is shown in ◻ Fig. 8.7.

For a robot operating in 3-dimensional task space, (8.7) describes a 6-dimensional ellipsoid which we cannot visualize. However, we can extract that part of the Jacobian relating to *translational* velocity in the world frame. For the UR5 robot at the configuration shown in ◻ Fig. 8.2 the Jacobian is

```
>>> J = ur5.jacob0(ur5.q1);
```

and the top three rows, which correspond to translational velocity in task space, are

```
>>> Jt = J[:3, :];  # first 3 rows
```

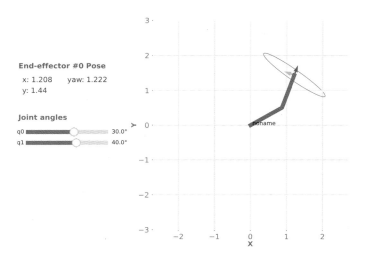

**Fig. 8.7** Two-link robot with overlaid velocity ellipse

from which we can compute and plot the velocity ellipsoid

```
>>> E = np.linalg.inv(Jt @ Jt.T)
array([[  9.005,    0.2317,      2.02],
       [  0.2317,    4.317,   0.05197],
       [   2.02,   0.05197,    2.608]])
>>> plot_ellipsoid(E);
```

Plotting an ellipsoid involves inverting the passed matrix, which we have computed using the matrix inverse. We can save ourselves two unnecessary matrix inversions by `plot_ellipsoid(Jt @ Jt.T, inverted=True)`. See ▶ App. C.1.4 for a refresher on ellipses and ellipsoids.

which is shown in ◘ Fig. 8.8a. ◀ Ideally this would be a sphere, indicating an ability to generate equal velocity in any direction – isotropic velocity capability. In this case the lenticular shape indicates that the end effector cannot achieve equal velocity in all directions. The ellipsoid radii, in the directions of the principal axes, are given by

```
>>> e, _ = np.linalg.eig(E);
>>> radii = 1 / np.sqrt(e)
array([  0.3227,    0.7029,    0.4819])
```

and we see that one radius is greater than the other two.

The rotational velocity ellipsoid close to, within 1° of, the singular configuration

```
>>> J = ur5.jacob0(np.full((6,), np.deg2rad(1)));
>>> Jr = J[3:, :];  # last 3 rows
>>> E = np.linalg.inv(Jr @ Jr.T);
>>> plot_ellipsoid(E);
```

This is much easier to see if you change the viewpoint interactively.

is shown in ◘ Fig. 8.8b. It is an elliptical plate with a very small thickness. ◀ The radii are

```
>>> e, x = np.linalg.eig(E);
>>> radii = 1 / np.sqrt(e)
array([  0.03998,     1.414,         2])
```

and the small radius corresponds to the direction

```
>>> x[:, 0]
array([  -0.9996,  -0.01308,   0.02617])
```

which is close to parallel with the $x$-axis. This indicates the robot's inability to rotate about the $x$-axis which is the degree of freedom that has been lost. Both joints 3 and 5 provide rotation about the world $y$-axis, joint 4 provides provides rotation about the world $z$-axis, but no joint generates rotation about the world $x$-axis. At the singularity the ellipsoid will have zero thickness.

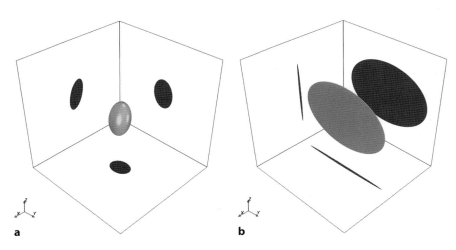

► go.sn.pub/HoVmEO

► go.sn.pub/gpcPbC

**◻ Fig. 8.8** End-effector velocity ellipsoids for the UR5 robot shown in blue. **a** Translational velocity ellipsoid for the pose shown in ◻ Fig. 8.2 (m s$^{-1}$); **b** rotational velocity ellipsoid for a near singular pose (rad s$^{-1}$), the ellipsoid is an elliptical plate. The black ellipses are shadows and show the shape of the ellipse as seen by looking along the $x$-, $y$- and $z$-directions. These plots were generated using PyVista rather than Matplotlib

The Toolbox provides a shorthand for displaying the velocity ellipsoid, which for the example above is

```
>>> ur5.vellipse(qns, "rot");
```

The size and shape of the ellipsoid describes how *well-conditioned* the manipulator is for making certain motions and  manipulability is a succinct scalar measure of that. The `manipulability` method computes Yoshikawa's manipulability measure

$$m(\boldsymbol{q}) = \sqrt{\det\big(\mathbf{J}(\boldsymbol{q})\mathbf{J}^\top(\boldsymbol{q})\big)} \in \mathbb{R}) \tag{8.8}$$

which is proportional to the volume of the ellipsoid. ► At the working configuration manipulability is

```
>>> ur5.manipulability(ur5.q1)
0.06539
```

compared to

```
>>> ur5.manipulability(ur5.qz)
0
```

at the singular configuration.

In these examples manipulability is the volume of the 6D task-space velocity ellipsoid but it is frequently useful to look at translational and rotational manipulability separately. This can be done by passing the `axes` parameter with the options `"trans"`, `"rot"` or `"both"`. For the singular configuration we see that

```
>>> ur5.manipulability(ur5.qz, axes="both")
(0.105, 2.449e-16)
```

the translational manipulability is finite and that it is the rotational manipulability which is zero.

In practice we find that the seemingly large workspace of a robot is greatly reduced by joint limits, self collision, singularities and regions of low manipulability. The manipulability measure discussed here is based only on the kinematics parameters of the robot. The fact that it is easier to move a small wrist joint than the larger waist joint suggests that mass and inertia should be considered, and such manipulability measures are discussed in ► Sect. 9.2.7.

The determinant is the product of the eigenvalues, and the ellipsoid radii are the square roots of the eigenvalues of $\mathbf{J}(\boldsymbol{q})\mathbf{J}^\top(\boldsymbol{q})$. ► App. C.1.4 is a refresher on these topics.

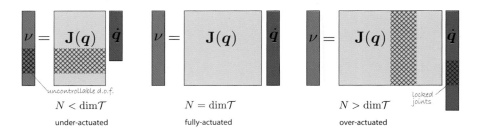

**Fig. 8.9** Schematic of Jacobian, $v$ and $\dot{q}$ for different numbers of robot joints $N$. The *hatched* areas represent matrix regions that could be deleted in order to create a square subsystem capable of solution

### 8.3.3 Dealing with Jacobian Singularity

For the case of a square Jacobian where $\det(\mathbf{J}(q)) = 0$ we cannot solve (8.2) directly for $\dot{q}$. One strategy to deal with singularity is to replace the inverse with the damped inverse

$$\dot{q} = \left(\mathbf{J}(q) + \lambda\mathbf{1}\right)^{-1} v$$

where $\lambda$ is a small constant added to the diagonal which places a *floor* under the determinant. This will introduces some error in $\dot{q}$ which, integrated over time, could lead to a significant discrepancy in tool position but a closed-loop resolved-rate motion scheme like (8.6) would minimize that.

An alternative is to use the pseudo-, generalized- or Moore-Penrose-pseudoinverse of the Jacobian

$$\dot{q} = \mathbf{J}(q)^+ v \tag{8.9}$$

> A matrix expression like $b = \mathbf{A}x$ is a system of scalar equations which we can solve for the unknowns $x$. At singularity, some of the equations become equivalent, so there are fewer unique equations than there are unknowns – we have an underdetermined system which may an infinite number of solutions or no solution.

which provides a solution that minimizes $\|\mathbf{J}(q)\dot{q} - v\|$, ◀ which is the error between the desired task-space velocity $v$ and the actual velocity $\mathbf{J}(q)\dot{q}$ in a least-squares sense. It is readily computed using the NumPy function `np.linalg.pinv`.

Yet another approach is to delete from the Jacobian all those columns that are linearly dependent on other columns. This is effectively the same as locking the joints corresponding to the deleted columns and we now have an underactuated system which we discuss in the next section.

### 8.3.4 Dealing with a Non-Square Jacobian

The Jacobian is a $\dim\mathcal{T} \times \dim\mathcal{C}$ matrix and, so far we have assumed that the Jacobian is square. For the nonsquare cases it is helpful to consider the velocity relationship

$$v = \mathbf{J}(q)\dot{q} \tag{8.10}$$

in the diagrammatic form shown in ◧ Fig. 8.9 for the common 3D case where $\dim\mathcal{T} = 6$. The Jacobian will be a $6 \times N$ matrix, the joint velocity an $N$-vector, and $v$ a 6-vector.

The case of $N < 6$ is referred to as an underactuated robot and there are too few joints to control the number of degrees of freedom in task space. Eq. (8.10)

> An overdetermined set of equations has more equations than unknowns and for which there is potentially no solution.

is an overdetermined system of equations ◀ so we can only approximate a solution, and the pseudoinverse will yield the solution which minimizes the error in a least-squares sense. Alternatively, we could *square up* the Jacobian by deleting some rows of $v$ and $\mathbf{J}$ – accepting that some task-space degrees of freedom are not controllable given the insufficient number of joints.

The case of $N > 6$ is referred to as overactuated or redundant and there are more joints than we actually need to control the number of degrees of freedom in task space. Eq. (8.10) is an underdetermined system of equations ▶ so we typically choose the solution with the smallest norm. Alternatively, we can *square up* the Jacobian by deleting some columns of $\mathbf{J}$ and rows of $\dot{q}$ – effectively locking the corresponding joints.

> An underdetermined set of equations has more unknowns than equations and for which there is an infinite number of solutions or no solution.

### 8.3.4.1 Jacobian for Under-Actuated Robot

An underactuated robot has $N < 6$, and a Jacobian that is taller than it is wide. For example a 2-joint manipulator at a nominal pose

```
>>> planar2 = models.ETS.Planar2();
>>> qn = [1, 1];
```

with $\mathcal{T} = \mathbb{R}^2 \times \mathbf{S}^1$ has the Jacobian

```
>>> J = planar2.jacob0(qn)
array([[ -1.751,   -0.9093],
       [  0.1242,  -0.4161],
       [      1,        1]])
```

If the Jacobian has full column rank ▶ then we can find a solution using the pseudoinverse which has the property

> A matrix with full column rank is one where the columns are all linearly independent, and the rank of the matrix equals the number of columns.

$$\mathbf{J}^+ \mathbf{J} = 1$$

just as the inverse does, and is defined as

$$\mathbf{J}^+ = \left(\mathbf{J}^\top \mathbf{J}\right)^{-1} \mathbf{J}^\top$$

and is implemented by the function `np.linalg.pinv`. ▶ For example, if the desired end-effector velocity is

> This is the left generalized- or pseudoinverse, since it is applied to the left of $\mathbf{J}$.

```
>>> xd_desired = [0.1, 0.2, 0];
```

and the pseudoinverse solution for joint velocity is

```
>>> qd = np.linalg.pinv(J) @ xd_desired
array([ 0.08309,   -0.1926])
```

The end-effector velocity

```
>>> J @ qd
array([ 0.02967,   0.09047,   -0.1095])
```

has *approximated* the desired velocity in the $x$- and $y$-directions, and has an undesired $z$-axis rotational velocity. Given that the robot is underactuated, there is no exact solution and this one is best in the least-squares sense – it is a compromise. The error in this case is

```
>>> np.linalg.norm(xd_desired - J @ qd)
0.1701
```

We have to confront the reality that we have only two inputs and we want to use them to control $v_x$ and $v_y$. We rewrite (8.2) in partitioned form as

$$\begin{pmatrix} v_x \\ v_y \\ \omega \end{pmatrix} = \begin{pmatrix} \mathbf{J}_{xy} \\ \mathbf{J}_\omega \end{pmatrix} \begin{pmatrix} \dot{q}_1 \\ \dot{q}_2 \end{pmatrix}$$

and taking the top partition, the first two rows, we write

$$
\begin{pmatrix} v_x \\ v_y \end{pmatrix} = \mathbf{J}_{xy} \begin{pmatrix} \dot{q}_1 \\ \dot{q}_2 \end{pmatrix}
$$

where $\mathbf{J}_{xy}$ is a $2 \times 2$ matrix which we can invert if $\det(\mathbf{J}_{xy}) \neq 0$

$$
\begin{pmatrix} \dot{q}_1 \\ \dot{q}_2 \end{pmatrix} = \mathbf{J}_{xy}^{-1} \begin{pmatrix} v_x \\ v_y \end{pmatrix}
$$

The joint velocity is now

```
>>> Jxy = J[:2, :];
>>> qd = np.linalg.inv(Jxy) @ xd_desired[:2]
array([  0.1667,  -0.4309])
```

which results in end-effector velocity

```
>>> xd = J @ qd
array([     0.1,      0.2,  -0.2642])
```

that has exactly the desired $x$- and $y$-direction velocity. The $z$-axis rotation is unavoidable since we have explicitly used the two degrees of freedom to control $x$- and $y$-translation, not $z$-rotation. The solution error

```
>>> np.linalg.norm(xd_desired - J @ qd)
0.2642
```

is $\approx 50\%$ higher than for the pseudoinverse approach, but now we have chosen the task-space constraints we wish to satisfy rather than have a compromise made for us. This same approach can be used for a robot operating in a higher-dimensional task space.

### 8.3.4.2 Jacobian for Overactuated Robot

An overactuated or redundant robot has $N > 6$, and a Jacobian that is wider than it is tall. In this case we can again apply the pseudoinverse but now the solution has an extra term

$$
\dot{q} = \mathbf{J}(q)^{+} v + \mathbf{P}(q)\, \dot{q}_0 \tag{8.11}
$$

The first term is, of the infinite number of solutions possible, the minimum-norm solution – the one with the smallest $\|\dot{q}\|$. The second term allows an infinite number of solutions since an arbitrary joint-space velocity $\dot{q}_0$ is added. The matrix

$$
\mathbf{P}(q) = \mathbf{1}_{N \times N} - \mathbf{J}(q)^{+}\, \mathbf{J}^{\top}(q) \in \mathbb{R}^{N \times N} \tag{8.12}
$$

projects any joint velocity $\dot{q}_0$ into the null space of $\mathbf{J}$ so that it will cause zero end-effector motion.

We will demonstrate this for the 7-axis Panda robot from ▶ Sect. 7.2.4 at a nominal pose

```
>>> panda = models.ETS.Panda();
>>> TE = SE3.Trans(0.5, 0.2, -0.2) * SE3.Ry(pi);
>>> sol = panda.ikine_LMS(TE);
```

and its Jacobian

```
>>> J = panda.jacob0(sol.q);
>>> J.shape
(6, 7)
```

is a $6 \times 7$ matrix. Now consider that we want the end effector to have the spatial velocity

```
>>> xd_desired = [0.1, 0.2, 0.3, 0, 0, 0];
```

then using the first term of (8.12) we can compute the required joint rates

```
>>> qd = np.linalg.pinv(J) @ xd_desired
array([  0.4045,  -0.5582,  0.04829,    0.167,   0.2591,   -0.6515,
          0.3502])
```

We see that all joints have nonzero velocity and contribute to the desired end-effector motion ▶ as specified

```
>>> J @ qd
array([      0.1,       0.2,       0.3,        0,        0,        0])
```

The Jacobian in this case has seven columns and a rank of six

```
>>> np.linalg.matrix_rank(J)
6
```

and therefore has a null space ▶ with just one basis vector, given as a column vector

```
>>> N = linalg.null_space(J);
>>> N.shape
(7, 1)
>>> N.T
array([[ -0.4231,  -0.1561,  -0.7747, -0.007397,   0.4094,  0.06011,
          0.1591]])
```

Any joint velocity that is a linear combination of its null-space basis vectors will result in zero end-effector motion. For this robot there is only one basis vector and we can easily show that this *null-space joint motion* causes zero end-effector motion

```
>>> np.linalg.norm( J @ N[:,0])
3.132e-16
```

If the basis vectors of the null space are arranged as columns in the matrix $\mathbf{N}(q) \in \mathbb{R}^{N \times R}$ where $R = N - \dim \mathcal{T}$ then an alternative expression for the projection matrix is

$$\mathbf{P}(q) = \mathbf{N}(q)\mathbf{N}(q)^{+}$$

Null-space motion can be used for redundant robots to avoid collisions between the links and obstacles (including other links), or to keep joint coordinates away from their mechanical limit stops. Consider that in addition to the desired task-space velocity, we wish to simultaneously increase $q_4$ in order to move the arm away from some obstacle. We set a desired joint velocity

```
>>> qd_0 = [0, 0, 0, 0, 1, 0, 0];
```

and project it into the null space

```
>>> qd = N @ np.linalg.pinv(N) @ qd_0
array([ -0.1732, -0.06391,  -0.3171, -0.003028,   0.1676,  0.02461,
          0.06513])
```

A scaling has been introduced but this joint velocity, or a scaled version of this, will increase $q_4$ without changing the end-effector pose. Other joints move as well – they provide the required compensating motion in order that the end-effector pose is not disturbed as shown by

```
>>> np.linalg.norm(J @ qd)
1.129e-16
```

A highly redundant snake robot like that shown in ◻ Fig. 8.10 would have a null space with $20 - 6 = 14$ dimensions. This could be used to provide fine control over the shape of the arm independently of the end-effector pose. This is a critical capability when operating in confined spaces.

If we use the pseudoinverse for RRMC and the robot end effector follows a repetitive path the joint coordinates will not generally follow a repetitive path. They are under constrained and will drift over time and potentially hit joint limits. We can use null-space control to provide additional constraints to prevent this.

See ▶ App. B.

**8**

◨ **Fig. 8.10**  20-DoF snake-robot arm: 2.5 m reach, 90 mm diameter and payload capacity of 25 kg (image courtesy of OC Robotics)

## 8.4  Force Relationships

In ▶ Sect. 3.2.2 we introduced wrenches $\boldsymbol{w} = (f_x, f_y, f_z, m_x, m_y, m_z) \in \mathbb{R}^6$ which are a vector of forces and moments. A wrench at the end effector is related to the joint torques or forces by the manipulator Jacobian matrix.

### 8.4.1  Transforming Wrenches to Joint Space

A wrench $^0\boldsymbol{w}_E$ applied at the end effector, and defined in the world coordinate frame, is transformed to joint space by the transpose of the manipulator Jacobian

$$\boldsymbol{Q} = {}^0\mathbf{J}^\top(\boldsymbol{q})\,{}^0\boldsymbol{w}_E \tag{8.13}$$

to the generalized joint force vector $\boldsymbol{Q} \in \mathbb{R}^N$ whose elements are joint torque or force for revolute or prismatic joints respectively.

If the wrench is defined in the end-effector coordinate frame then we use instead

$$\boldsymbol{Q} = {}^E\mathbf{J}^\top(\boldsymbol{q})\,{}^E\boldsymbol{w}_E \tag{8.14}$$

For the UR5 robot in the configuration shown in ◨ Fig. 8.2, a force of 20 N in the world $y$-direction is pushing the end effector *sideways* resulting in joint torques of

```
>>> tau = ur5.jacob0(ur5.q1).T @ [0, 20, 0, 0, 0, 0]
array([   9.491,       0,        0,        0,   -1.646,        0])
```

which is a significant torque on the waist joint due to a lever arm effect – this will cause the robot's waist joint to rotate. A force of 20 N applied in the world $x$-direction is pulling the end effector away from the base and results in joint torques of

```
>>> tau = ur5.jacob0(ur5.q1).T @ [20, 0,  0, 0, 0, 0]
array([  -0.537,    6.607,   -1.893,   -1.893,   -1.646,         0])
```

which results in a significant torque being applied to the shoulder joint.

The mapping between a wrench applied to the end effector and generalized joint force involves the transpose of the Jacobian, and this can never be singular. This is in contrast to the velocity relationship which involves the Jacobian inverse that can be singular. We exploit this property of the Jacobian transpose in ▶ Sect. 8.5 to solve the inverse-kinematic problem numerically.

## 8.4.2 Force Ellipsoids

In ▶ Sect. 8.3.2 we introduced the velocity ellipse and ellipsoid which describe the directions in which the end effector is best able to move. We can perform a similar analysis for the wrench at the end effector – the end-effector wrench. We start with a set of generalized joint forces with a unit norm $\| \boldsymbol{Q} \| = 1$

$$\boldsymbol{Q}^{\top} \boldsymbol{Q} = 1$$

and substituting (8.13) we can write

$$\boldsymbol{w}^{\top} \Big( \mathbf{J}(q) \mathbf{J}^{\top}(q) \Big) \boldsymbol{w} = 1$$

which is the equation of points $\boldsymbol{w}$ on the surface of a 6-dimensional hyper-ellipsoid in the end-effector wrench space. For the planar robot arm of ◘ Fig. 8.1 we can interactively explore how the shape of the force ellipse changes with configuration using the teach method

```
>>> planar2.teach(np.deg2rad([30, 40]), fellipse=True);
```

If this ellipse is close to circular, that is, its radii are of the same order of magnitude then the end effector can achieve an arbitrary wrench – an isotropic end-effector wrench space. However, if one or more radii are very small this indicates that the end effector cannot exert a force along, or a moment about, the axes corresponding to those small radii.

The force and velocity ellipsoids provide complementary information about how well suited the configuration of the arm is to a particular task. We know from personal experience that to throw an object quickly we have our arm outstretched and orthogonal to the throwing direction, whereas to lift something heavy we hold our arms close into our body.

## 8.5 Numerical Inverse Kinematics

In ▶ Sect. 7.2.2.1 we introduced the numerical approach to inverse kinematics for robots that move in three dimensions. This section reveals how that is implemented, and shows the important role that the Jacobian plays in the solution.

The principle is shown in ◘ Fig. 8.11 where the robot in its current joint configuration is shown in orange and the desired joint configuration is shown in blue.

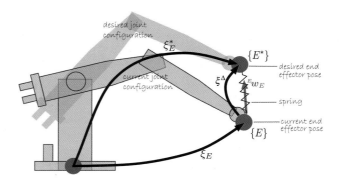

■ **Fig. 8.11** Schematic of the numerical inverse kinematic approach, showing the current and desired manipulator configuration. A pose graph is overlaid showing the current end effector pose $\boldsymbol{\xi}_E$, the desired end effector pose $\boldsymbol{\xi}_E^*$, and the required change in pose $\boldsymbol{\xi}^\Delta$

The overlaid pose graph shows the error between actual $\boldsymbol{\xi}_E$ and desired pose $\boldsymbol{\xi}_E^*$ is $\boldsymbol{\xi}^\Delta$ which can be described by a spatial displacement as discussed in ▶ Sect. 3.1.5

$$^E\boldsymbol{\Delta} = \Delta\big(\boldsymbol{\xi}_E, \boldsymbol{\xi}_E^*\big) = (\boldsymbol{t}, \hat{\boldsymbol{v}}\theta) \in \mathbb{R}^6$$

where the current pose is computed using forward kinematics $\boldsymbol{\xi}_E = \mathcal{K}(\boldsymbol{q})$.

Imagine a *special* spring between the end effector at the two poses which is pulling (and twisting) the robot's end effector toward the desired pose with a wrench proportional to the spatial displacement

$$^E\boldsymbol{w} = \gamma\,^E\boldsymbol{\Delta} \tag{8.15}$$

which is *resolved* to generalized joint forces

$$\boldsymbol{Q} = \,^E\mathbf{J}^\top(\boldsymbol{q})\,^E\boldsymbol{w}$$

using the Jacobian transpose (8.14). We assume that this virtual robot has no joint motors, only viscous dampers, so the joint velocity will be proportional to those joint forces

$$\dot{\boldsymbol{q}} = \frac{\boldsymbol{Q}}{B}$$

where $B$ is the joint damping coefficient (assuming all dampers are the same). Putting all this together, and combining constants $\alpha = \gamma/B$, we can write

$$\dot{\boldsymbol{q}} = \alpha\mathbf{J}^\top(\boldsymbol{q})\Delta\big(\mathcal{K}(\boldsymbol{q}),\ \boldsymbol{\xi}_E^*\big)$$

which is the joint velocity that will drive the forward kinematic solution toward the desired end-effector pose. This can be solved iteratively by

$$\begin{aligned} \boldsymbol{\delta}_{q\langle k\rangle} &= \alpha\mathbf{J}^\top(\boldsymbol{q}\langle k\rangle)\,\Delta\big(\mathcal{K}(\boldsymbol{q}),\ \boldsymbol{\xi}_E^*\big) \\ \boldsymbol{q}\langle k+1\rangle &\leftarrow \boldsymbol{q}\langle k\rangle + \boldsymbol{\delta}_{q\langle k\rangle} \end{aligned} \tag{8.16}$$

until the norm of the update $\|\boldsymbol{\delta}_{q\langle k\rangle}\|$ is sufficiently small and where $\alpha > 0$ is a well-chosen constant. Since the solution is based on the Jacobian transpose rather than the inverse, the algorithm works when the Jacobian is nonsquare or singular. In practice however, this algorithm is slow to converge and very sensitive to the choice of $\alpha$.

More practically we can formulate this as a least-squares problem in the world coordinate frame and minimize the scalar cost

$$e = \,^E\boldsymbol{\Delta}^\top\mathbf{M}\,^E\boldsymbol{\Delta}$$

where $\mathbf{M} = \mathrm{diag}(\boldsymbol{m}) \in \mathbb{R}^{6 \times 6}$ and $\boldsymbol{m}$ is the mask vector introduced in ▶ Sect. 7.2.3. The update becomes

$$\boldsymbol{\delta}_{q \langle k \rangle} = \big(\mathbf{J}^{\top}(\boldsymbol{q}_{\langle k \rangle})\, \mathbf{M}\, \mathbf{J}(\boldsymbol{q}_{\langle k \rangle})\big)^{-1} \mathbf{J}^{\top}(\boldsymbol{q}_{\langle k \rangle})\, \mathbf{M}\, \Delta\big(\mathcal{K}(\boldsymbol{q}_{\langle k \rangle}),\, \boldsymbol{\xi}_E^*\big)$$

which is much faster to converge but can behave poorly near singularities. We remedy this by introducing a damping constant $\lambda$

$$\boldsymbol{\delta}_{q \langle k \rangle} = \big(\mathbf{J}^{\top}(\boldsymbol{q}_{\langle k \rangle})\, \mathbf{M}\, \mathbf{J}(\boldsymbol{q}_{\langle k \rangle}) + \lambda \mathbf{1}_{N \times N}\big)^{-1} \mathbf{J}^{\top}(\boldsymbol{q}_{\langle k \rangle})\, \mathbf{M}\, \Delta\big(\mathcal{K}(\boldsymbol{q}_{\langle k \rangle}),\, \boldsymbol{\xi}_E^*\big)$$

which ensures that the term being inverted can never be singular.

An effective way to choose $\lambda$ is to test whether or not an iteration reduces the error, that is if $\|\boldsymbol{\delta}_{q \langle k \rangle}\| < \|\boldsymbol{\delta}_{q \langle k-1 \rangle}\|$. If the error is reduced, we can decrease $\lambda$ in order to speed convergence. If the error has increased, we revert to our previous estimate of $\boldsymbol{q}_{\langle k \rangle}$ and increase $\lambda$. This adaptive damping factor scheme is the basis of the well-known Levenberg-Marquardt optimization algorithm.

This algorithm is implemented by the `ikine_LM` method and works well in practice. As with all optimization algorithms it requires a reasonable initial estimate of $\boldsymbol{q}$ and this can be explicitly given using the argument `q0`. If not given, then random values of joint configuration are chosen until the optimization converges. This is controlled by the `slimit` option.

An important take away is that to compute the inverse kinematics, all we need is an iterative solver as discussed above, the forward kinematics and the Jacobian. These are straightforward to compute for a serial link manipulator with an arbitrary number of joints.

## 8.6 Advanced Topics

### 8.6.1 Manipulability Jacobian

It can be useful to know how manipulability changes with joint configuration and the manipulability Jacobian has elements $\frac{dm}{dq_i}$. For example

```
>>> panda = models.ETS.Panda()
>>> panda.jacobm(panda.qr).T
array([[  0, -0.002627,     0,  0.04064,     0, -0.02734,     0]])
```

indicates a strong relationship between $q_3$ and manipulability – increasing $q_3$ would increase manipulability, whereas increasing $q_5$ would decrease it. This vector could be interpreted as a joint-space velocity and be projected into the null space as $\dot{q}_0$ in (8.11). That would adjust the joint configuration so as to increase the volume of the velocity ellipsoid while leaving the end-effector pose unchanged.

### 8.6.2 Computing the Manipulator Jacobian Using Twists

In ▶ Sect. 7.5.4 we computed the forward kinematics as a product of exponentials based on the screws representing the joint axes in a zero-joint angle configuration. It is easy to differentiate the product of exponentials with respect to motion about each screw axis which leads to the Jacobian matrix

$$^0\bar{\mathbf{J}} = \Big(\hat{\boldsymbol{S}}_0 \quad \mathrm{Ad}\big(\mathrm{e}^{[\hat{S}_0]q_0}\big)\hat{\boldsymbol{S}}_1 \quad \cdots \quad \mathrm{Ad}\big(\mathrm{e}^{[\hat{S}_0]q_0} \cdots \mathrm{e}^{[\hat{S}_{N-2}]q_{N-2}}\big)\hat{\boldsymbol{S}}_{N-1}\Big)$$

for velocity in the world coordinate frame. The Jacobian is very elegantly expressed and can be easily built up column by column. Velocity in the end-effector coordinate frame is related to joint velocity by the Jacobian matrix

$$^E\bar{\mathbf{J}} = \mathrm{Ad}\big(^E\xi_0\big)\ ^0\bar{\mathbf{J}}$$

where Ad $(\cdot)$ is the adjoint matrix introduced in ▶ Sect. 3.1.3.

> ❗ The Jacobians $\bar{\mathbf{J}}$, distinguished by a bar, gives the velocity of the end effector as a velocity twist, not a spatial velocity. The difference is described in ▶ Sect. 3.1.3.

To obtain the Jacobian that gives spatial velocity as described in ▶ Sect. 8.1 we must apply a velocity transformation

$$^0\mathbf{J} = \begin{pmatrix} \mathbf{1}_{3\times3} & -\big[^0t_E\big]_\times \\ \mathbf{0}_{3\times3} & \mathbf{1}_{3\times3} \end{pmatrix} {}^0\bar{\mathbf{J}}$$

This is implemented by the `jacob0` method of the `PoERobot` class.

### 8.6.3  Manipulability, Scaling, and Units

The elements of the Jacobian matrix have values that depend on the choice of physical units used to describe link lengths and joint coordinates. Therefore, so too does the Jacobian determinant and the manipulability measure. A smaller robot will have lower manipulability than a larger robot, even if it has the same joint structure.

Yoshikawa's manipulability measure (8.8) is the volume of the velocity ellipsoid but not does measure velocity isotropy. It would give the same value for a large flat ellipsoid or a smaller more spherical ellipsoid. Many other measures have been proposed that better describe isotropy, for example, the ratio of maximum to minimum singular values.

For the redundant robot case we used the pseudoinverse to find a solution that minimizes the joint velocity norm. However, for robots with a mixture of revolute and prismatic joints the joint velocity norm is problematic since it involves quantities with different units. If the units are such that the translational and rotational velocities have very different magnitudes, then one set of velocities will dominate the other. A rotational joint velocity of $1\ \mathrm{rads}^{-1}$ is comparable in magnitude to a translational joint velocity of $1\ \mathrm{ms}^{-1}$. However, for a robot one-tenth the size the translational joint velocity would scale to $0.1\ \mathrm{ms}^{-1}$ while rotational velocities would remain the same and dominate.

### 8.7  Wrapping Up

Jacobians are an important concept in robotics, relating velocity in one space to velocity in another. We previously encountered Jacobians for estimation in ▶ Chap. 6 and will use them later for computer vision and control.

In this chapter we have learned about the manipulator Jacobian which describes the relationship between the rate of change of joint coordinates and the spatial velocity of the end effector expressed in either the world frame or the end-effector frame. We showed how the inverse Jacobian can be used to resolve desired Cartesian velocity into joint velocity as an alternative means of generating task-space paths for under- and overactuated robots. For overactuated robots we showed how null-space motions can be used to move the robot's joints without affecting the end-effector pose. The numerical properties of the Jacobian tell us about manipulability, that is how well the manipulator is able to move in different task-space directions, which we visualized as the velocity ellipsoid. At a singularity, indicated by linear dependence between columns of the Jacobian, the robot is unable to move in certain directions. We also created Jacobians to map angular velocity to roll-pitch-yaw or Euler angle rates, and these were used to form the analytical Jacobian matrix.

The force ellipsoid describes the forces that the end effector can exert in different directions. The Jacobian transpose is used to map wrenches applied at the end effector to joint torques, and also to map wrenches between coordinate frames. It is also the basis of numerical inverse kinematics for arbitrary robots and singular poses.

## 8.7.1 Further Reading

The manipulator Jacobian is covered by almost all standard robotics texts such as the robotics handbook (Siciliano and Khatib 2016), Lynch and Park (2017), Siciliano et al. (2009), Spong et al. (2006), Craig (2005), and Paul (1981). An excellent discussion of manipulability and velocity ellipsoids is provided by Siciliano et al. (2009), and the most common manipulability measure is that proposed by Yoshikawa (1984). Patel et al. (2015) provide a comprehensive survey of robot performance measures including manipulability and velocity isotropy measures. Computing the manipulator Jacobian based on Denavit-Hartenberg parameters, as used in this Toolbox, was first described by Paul and Shimano (1978).

The resolved-rate motion control scheme was proposed by Whitney (1969). Extensions such as pseudoinverse Jacobian-based control are reviewed by Klein and Huang (1983) and damped least-squares methods are reviewed by Deo and Walker (1995).

## 8.7.2 Exercises

1. For the simple 2-link example (▶ Sect. 8.1.1) compute the determinant symbolically and determine when it is equal to zero. What does this mean physically?
2. Add a tool to the UR5 robot that is 10 cm long in the end-effector $z$-direction. What happens to the top-right $3 \times 3$ block in the Jacobian? What if the tool was 30 cm long?
3. Derive the analytical Jacobian for Euler angles.
4. Resolved-rate motion control (▶ Sect. 8.2)
   a) Experiment with different Cartesian translational and rotational velocity demands, and combinations.
   b) Extend the block diagram system of ◨ Fig. 8.4 to also record the determinant of the Jacobian matrix.
   c) In ◨ Fig. 8.4 the robot's motion is simulated for 5 s. Extend the simulation time to 10 s and explain what happens.
   d) Set the initial pose and direction of motion to mimic that of ▶ Sect. 7.3.4. What happens when the robot reaches the singularity?
   e) Replace the Jacobian inverse block in ◨ Fig. 8.3 with the Numpy function `np.linalg.pinv`.
   f) Replace the Jacobian inverse block in ◨ Fig. 8.3 with a damped least squares function, and investigate the effect of different values of the damping factor.
   g) Replace the Jacobian inverse block in ◨ Fig. 8.3 with a block based on the NumPy function `np.linalg.lstsq`.

5. Use the interactive graphical teach tool for the PUMA 560 robot at its singular configuration. Adjust joints 3 and 5 and see how they result in the same end-effector motion.

6. Velocity and force ellipsoids for the two link manipulator (▶ Sects. 8.3.2 and 8.4.2). Determine, perhaps using the interactive `teach` method:
   a) What configuration gives the best manipulability?
   b) What configuration is best for throwing a ball in the positive $x$-direction?
   c) What configuration is best for carrying a heavy weight if gravity applies a force in the $-y$-direction?
   d) Plot the velocity ellipse ($x$- and $y$-velocity) for the two-link manipulator at a grid of end-effector positions in its workspace. Each ellipsoid should be centered on the end-effector position.

7. Velocity and force ellipsoids for the UR5 manipulator
   a) Find a configuration where manipulability is greater than at `q1`.
   b) As above, but find a configuration that maximizes manipulability.
   c) Use the `teach` method with the `vellipse` and `fellipse` options to interactively animate the ellipsoids.
   d) Repeat for a redundant robot such as the Panda.

8. The model `models.ETS.XYPanda()` describes a 9-joint robot (PPRRRRRRR) comprising an $xy$-base (PP) carrying a Panda arm (RRRRRRR).
   a) Compute the null space and show that combinations of its basis vectors cause zero end-effector motion. Can you physically interpret these vectors?
   b) Compute a tasks-space end-effector path and use numerical inverse kinematics to solve for the joint coordinates. Analyze how the motion is split between the base and the robot arm.
   c) With the end effector at a constant pose, explore null-space control. Set a velocity for the mobile base and see how the arm configuration accommodates that.
   d) Develop a null-space controller that keeps the last robot arm joints in the middle of their working range by using the first two joints to position the base of the arm. Modify this so as to maximize the manipulability of the XYPanda robot.
   e) Consider now that the Panda robot arm is mounted on a nonholonomic robot, create a controller that generates appropriate steering and velocity inputs to the mobile robot (challenging).
   f) For an arbitrary pose and end-point spatial velocity we will move six joints and lock two joints. Write an algorithm to determine which two joints should be locked.

9. Prove, or demonstrate numerically, that the two equations for null-space projection $\mathbf{1} - \mathbf{J}(\boldsymbol{q})^{+}\mathbf{J}^{\top}(\boldsymbol{q})$ and $\mathbf{N}(\boldsymbol{q})\mathbf{N}(\boldsymbol{q})^{+}$ are equivalent.

10. The model `models.DH.Hyper3d(20)` is a 20-joint robot that moves in 3-dimensional space.
   a) Explore the capabilities of this robot.
   b) Compute a task-space end-effector trajectory that traces a circle on the ground, and use numerical inverse kinematics to solve for the joint coordinates.
   c) Add a null-space control strategy that keeps all joint angles close to zero while it is moving.
   d) Define an end-effector target pose on the ground that the robot must reach after passing through a hole in a wall. Can you determine the joint configuration that allows this? Can you do this for holes in two walls?

11. Write code to compute the Jacobian of a robot object using twists as described in ▶ Sect. 8.6.2 and compare results.

12. Consider a vertical plane in the workspace of a UR5 robot. Divide the plane into 2-cm grid cells and for each cell determine if it is reachable by the robot, and if it is then determine the manipulability for the first three joints of the

robot arm and place that value in the corresponding grid cell. Display a heat map of the robot's manipulability in the plane. Move the plane with respect to the robot and see how the heatmap changes.

13. a) Create a version of the PUMA 560 Denavit-Hartenberg model with all the linear dimensions in inches instead of meters. For the same joint configuration as the metric version, how does the manipulability differ?

    b) Create a version of the PUMA 560 Denavit-Hartenberg model robot with all linear dimensions reduced by a factor of ten. For the same joint configuration, how does the manipulability differ between the two sized robots?

    c) As above, but make the robot ten times bigger.

# Dynamics and Control

## Contents

This chapter introduces the topics of dynamics and control for a serial-link manipulator arm. We start with control of the individual joints of a robot arm, and this is closely related to the problem of controlling the speed of wheels and propellers in mobile robots. ▶ Sect. 9.1 describes the key elements of a robot joint control system that enables a single joint to follow a desired trajectory, and introduces the challenges found in practice such as friction, gravity load and varying inertia.

The motion of the end effector is the composition of the motion of each link, and the links are ultimately moved by forces and torques exerted on them by the joint actuators. Each link in the serial-link manipulator is supported by a reaction force and torque from the preceding link, and is subject to its own weight as well as forces and torques from the links that it supports. ▶ Sect. 9.2 introduces the *rigid-body* equations of motion, a set of coupled nonlinear differential equations, that describe the joint torques necessary to achieve a particular manipulator state. These equations can be factored into terms describing inertia, gravity load and gyroscopic coupling which provide insight into how the motion of one joint exerts a disturbance force on other joints, and how those terms vary with configuration and payload. ▶ Sect. 9.3 introduces the forward dynamics which describe how the manipulator moves, that is, how its configuration evolves over time in response to the applied forces and torques that are due to the joint actuators, end effector contract and gravity.

▶ Sect. 9.4 introduces control systems that explicitly account for the rigid-body dynamics which leads to improved position control. Many applications require force and position control and ▶ Sect. 9.5 is an introduction to task- or operational-space dynamics where the robot's dynamics are abstracted to a point mass, and the dynamics and control are expressed in task space. ▶ Sect. 9.6 provides two application examples: operational-space control which allows simultaneous motion and force control, and series-elastic actuators for human-safe robots.

## 9.1 Independent Joint Control

A robot drivetrain comprises an actuator or motor, and a transmission to connect it to the robot link. We start by considering a common approach to robot joint control where each joint or axis is an independent control system that attempts to accurately follow its trajectory. However, as we shall see, this is complicated by various *disturbance* torques due to gravity and friction that act on the joint, as well as coupling torques due to the motion of other joints. A very common control structure is the nested control loop. The outer loop is responsible for maintaining position and determines the velocity of the joint that will minimize position error. The inner loop is responsible for maintaining the joint velocity as demanded by the outer loop.

### 9.1.1 Actuators

Most robots today are driven by rotary electric motors. Large industrial robots typically use brushless servo motors, while small laboratory or hobby robots use brushed DC motors or stepper motors. Pneumatic actuation, using cylinders or bladders, is possible though uncommon but has the advantage of being able to generate motion while being naturally compliant. For applications that require very high forces it is common to use electro-hydraulic actuation – hydraulic actuation with electrically operated hydraulic valves. Applications include manipulators for very large payloads as used in mining, forestry or construction, or highly dynamic humanoids such as the Atlas robot shown in ◗ Fig. 7.3b.

A servo mechanism, or servo is an automatic device that uses feedback of error between the desired and actual position of a mechanism to drive the device to the desired position. Perhaps the first such device was the ship steering engine (image courtesy WikiCommons) designed by John McFarlane Gray for Isambard Kingdom Brunel's Great Eastern in 1866. The word servo is derived from the Latin root *servus* meaning slave and the first usage was by J. J. L. Farcot in 1868 – "Le Servomoteur" – to describe the hydraulic and steam engines used for steering ships.

Error in position is measured by a sensor, then amplified to drive a motor that generates a force to move the device to reduce the error. Servo system development was spurred by WW II with the development of electrical servo systems for fire-control applications that used electric motors and electro-mechanical *amplidyne* power amplifiers. Later servo amplifiers used vacuum tubes and more recently solid state power amplifiers (electronic speed controllers). Today servo mechanisms are ubiquitous and are used to position the read/write heads in optical and magnetic disk drives, the lenses in autofocus cameras, remote control toys, satellite-tracking antennas, automatic machine tools and robot joints.

"Servo" is properly a noun or adjective but has become a verb "to servo". In the context of vision-based control we use the compound verb "visual servoing".

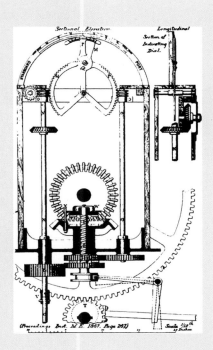

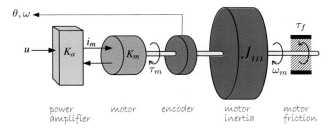

**◻ Fig. 9.1** Key components of a robot-joint actuator. A demand voltage $u$ controls the current $i_m$ flowing into the motor which generates a torque $\tau_m$ that accelerates the rotational inertia $J_m$ and is opposed by friction $\tau_f$. The encoder measures rotational speed and angle

The key elements of an electric drive system are shown in ◻ Fig. 9.1. Electric motors can be either current or voltage controlled. ▸ Here we assume current control where a motor driver or amplifier provides current

$$i_m = K_a u$$

that is proportional to the applied control voltage $u$, and where $K_a$ is the transconductance of the amplifier with units of A V$^{-1}$. The torque generated by the motor is proportional to the current

$$\tau_m = K_m i_m \tag{9.1}$$

where $K_m$ is the motor torque constant ▸ with units of N m A$^{-1}$. This torque accelerates the rotating part of the motor – the armature or rotor – which has rotational

Current control is implemented by an electronic constant current source, a variable voltage source with feedback of actual motor current. A variable voltage source is most commonly implemented by a pulse-width modulated (PWM) switching circuit. The alternative is voltage control, but this requires that the electrical dynamics of the motor due to its resistance and inductance, as well as back EMF (the voltage produced by a spinning motor that opposes the applied voltage), be taken into account when designing the control system.

Which is a function of the strength of the permanent magnets and the number of turns of wire in the motor's coils.

inertia $J_m$ and a rotational velocity of $\omega_m$. Motion is opposed by the frictional torque $\tau_f$ which has many components.

### 9.1.2  Friction

Any rotating machinery, motor or gearbox, will be affected by friction – a force or torque that *opposes* motion. The net torque from the motor is

$$\tau' = \tau_m - \tau_f$$

where $\tau_f$ is the friction torque which is function of velocity

$$\tau_f = B\omega_m + \tau_C \tag{9.2}$$

where the slope $B > 0$ is the viscous friction coefficient, and the offset is Coulomb friction which is generally modeled by

$$\tau_C = \begin{cases} \tau_C^+, & \omega_m > 0 \\ 0, & \omega_m = 0 \\ \tau_C^-, & \omega_m < 0 \end{cases} \tag{9.3}$$

In general, the friction coefficients depend on the direction of rotation and this asymmetry is more pronounced for Coulomb than for viscous friction.

The total friction torque as a function of rotational velocity is shown in ◖ Fig. 9.2. At very low speeds, highlighted in gray, Stribeck friction (pronounced streebeck) is dominant. The applied torque must exceed the stiction torque before rotation can occur – a process known as *breaking stiction*. Once the joint is moving, the Stribeck friction force rapidly decreases and viscous and Coulomb friction dominate. ◀

There are several sources of friction *experienced* by the motor. The first component is due to the motor itself: its bearings and, for a brushed motor, the brushes rubbing on the commutator. The friction parameters are often provided in the motor manufacturer's data sheet. Other sources of friction are the gearbox and the bearings that support the link.

Stribeck friction is difficult to parameterize and in robotics is generally ignored. Even viscous and Coulomb friction data for robot joints is rare, despite it generally being a significant dynamic effect.

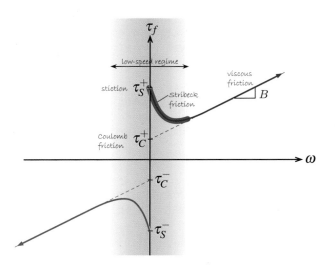

◖ **Fig. 9.2** Typical friction versus speed characteristic. The dashed lines depict a simple piecewise-linear friction model characterized by slope (viscous friction) and intercept (Coulomb friction). The low-speed regime is shaded and shown in exaggerated fashion

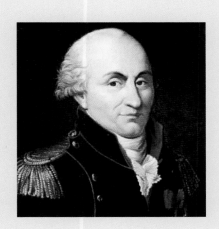

### 9.1.3  Link Mass

A motor in a robot arm does not exist in isolation, it is connected to a link as shown schematically in ◻ Fig. 9.3. The link has two significant effects on the motor – it adds extra inertia and it adds a torque due to the weight of the arm. Both vary with the configuration of the joint.

◻ Fig. 9.4 shows a simple robot with two revolute joints that operates in the vertical plane. Joint 0 rotates the red link with respect to the fixed base frame. We assume the mass of the red link is concentrated at its center of mass (CoM), so the extra inertia of the link will be $m_1 r_1^2$. The motor will also experience the inertia of the blue link and this will depend on the value of $q_1$ – the inertia of the arm is greatest when the arm is straight, and decreases as the arm is folded.

Gravity acts on the center of mass ▶ of the red link to create a torque on the joint 0 motor which will be proportional to $\cos q_0$. Gravity acting on the center of mass of the blue link also creates a torque on the joint 0 motor, and this is more pronounced since it is acting at a greater distance from the motor – the *lever arm effect* is greater.

These effects are clear from even a cursory examination of ◻ Fig. 9.4 but the reality is even more complex. Jumping ahead to material we will cover in the next section, we can use the Toolbox ▶ to determine the torque acting on each of the

Also referred to as the center of gravity.

This requires SymPy.

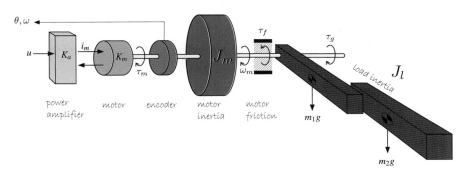

◻ **Fig. 9.3**  Robot joint actuator with attached links. The center of mass of each link is indicated by ⊕

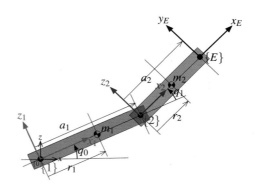

**◻ Fig. 9.4** A two-link arm, similar to ◻ Fig. 7.7, that moves in the vertical plane, and gravity acts in the downwards direction. Link frames and relevant dimensions are shown. The center of mass (CoM) of each link is indicated by 🌀 and is a distance of $r_i$ in the $x$-direction of the frame {i}

joints as a function of the position, velocity and acceleration of the joints. We first define a robot model in a similar way as in ► Chap. 7 but this time the link lengths are symbolic values. The inertial parameters of each link, mass and position of the center of mass, and gravity are also defined symbolically

```
>>> import sympy
>>> a1, a2, r1, r2, m1, m2, g = sympy.symbols("a1 a2 r1 r2 m1 m2 g")
>>> link1 = Link(ET.Ry(flip=True), m=m1, r=[r1, 0, 0], name="link0")
Link("link0", Ry(-q), isflip=1, m=m1, r=[r1, 0, 0],
    I=[0, 0, 0, 0, 0, 0], Jm=0, B=0, Tc=[0, 0], G=0)
>>> link2 = Link(ET.tx(a1) * ET.Ry(flip=True), m=m2, r=[r2, 0, 0],
        name="link1")
Link("link1", tx(a1) ⊕ Ry(-q), isflip=1, m=m2, r=[r2, 0, 0],
    I=[0, 0, 0, 0, 0, 0], Jm=0, B=0, Tc=[0, 0], G=0)
>>> robot = ERobot([link1, link2])
ERobot: noname, 2 joints (RR), dynamics
```

| link | link | joint | parent | ETS: parent to link |
|------|-------|-------|--------|---------------------|
| 0 | link0 | 0 | BASE | Ry(-q0) |
| 1 | @link1 | 1 | link0 | tx(a1) ⊕ Ry(-q1) |

In ◻ Fig. 9.4, the $y$-axis is into the page which means that $q_0$ and $q_1$ are rotations about the $-y$-axis. We specify this negative joint rotation by the `flip` option to `ETS.ry()`.

which shows a summary of the kinematic model. ◄ The dynamic parameters of the robot are

```
>>> robot.dynamics()
```

| j | m | r | I | | | | | | Jm | B | Tc | |
|-------|----|---------|----|----|----|----|----|----|----|---|----|----|
| link0 | m1 | r1, 0, 0 | 0, | 0, | 0, | 0, | 0, | 0 | 0 | 0 | 0, | 0 |
| link1 | m2 | r2, 0, 0 | 0, | 0, | 0, | 0, | 0, | 0 | 0 | 0 | 0, | 0 |

*The result of this method is a wide table which has been cropped for inclusion in this book.

Next, we define some symbolic tuples for joint angle, velocity and acceleration

```
>>> q = sympy.symbols("q:2")
(q0, q1)
>>> qd = sympy.symbols("qd:2")
(qd0, qd1)
>>> qdd = sympy.symbols("qdd:2")
(qdd0, qdd1)
```

The joint torques as a function of $q$, $\dot{q}$ and $\ddot{q}$ are

```
>>> tau = robot.rne(q, qd, qdd, gravity=[0, 0, g], symbolic=True);
```

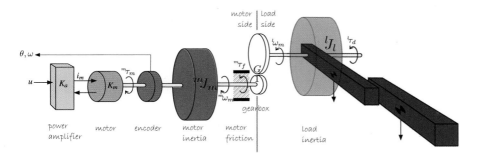

**◻ Fig. 9.5** Schematic of complete robot joint including gearbox. The reference frame for all quantities is explicitly indicated by an $m$ or $l$ pre-superscript for motor or load side respectively. The effective inertia of the links is shown as ${}^l J_l$ and the disturbance torque due to the link motion is ${}^l \tau_d$

which is a very long symbolic 2-vector, one element per joint, with surprisingly many terms which we can summarize as

$$\tau_0 = m_{00}(q_1)\ddot{q}_0 + \underbrace{m_{01}(q_1)\ddot{q}_1 + c_1(q_1)\dot{q}_0\dot{q}_1 + c_2(q_1)\dot{q}_1^2 + g(q_0, q_1)}_{\text{disturbance}}$$

$$m_{00}(q_1) = m_1 r_1^2 + m_2(a_1^2 + r_2^2 + 2a_1 r_2 \cos q_1),$$
$$m_{01}(q_1) = m_2 r_2(r_2 + a_1 \cos q_1),$$
$$c_1(q_1) = 2a_1 m_2 r_2 \sin q_1,$$
$$c_2(q_1) = a_1 m_2 r_2 \sin q_1,$$
$$g(q_0, q_1) = (m_1 r_1 + a_1 m_2)\text{g} \cos q_0 + m_2 r_2 \cos(q_0 + q_1)\text{g}$$

$$(9.4)$$

where g is gravitational acceleration which acts downward. This is Newton's second law where the inertia is dependent on $q_1$, plus a number of disturbance terms. As already discussed, the gravity torque $g(\cdot)$ is dependent on $q_0$ and $q_1$. What is perhaps most surprising is that the torque applied to joint 0 depends on the velocity and the acceleration of joint 1 and this will be discussed further in ▶ Sect. 9.2.

In summary, the effect of joint motion in a series of mechanical links is nontrivial. Each joint experiences a torque related to the position, velocity and acceleration of *all* the other joints and for a robot with many joints this becomes quite complex.

### 9.1.4 Gearbox

Electric motors are compact and efficient and can rotate at very high speed, but produce very low torque. Therefore it is common to use a reduction gearbox to tradeoff speed for increased torque. For a prismatic joint, the gearbox might also convert rotary motion to linear. The disadvantage of a gearbox is increased cost, weight, friction, backlash, mechanical noise and, for harmonic gears, torque ripple. Very high-performance robots, such as those used in high-speed electronic assembly, use expensive high-torque motors with a direct drive or a very low gear ratio achieved using cables or thin metal bands rather than gears.

◻ Fig. 9.5 shows the complete drivetrain of a typical robot joint. For a $G{:}1$ reduction drive, the torque at the link is $G$ times the torque at the motor. The quantities measured at the link, reference frame $l$, are related to the motor referenced quantities, reference frame $m$, as shown in ◻ Table 9.1. The inertia of the load is reduced by a factor of $G^2$ and the disturbance torque by a factor of $G$. ▶

There are two components of inertia *felt* by the motor. The first is due to the rotating part of the motor itself, its rotor, which is denoted by ${}^m J_m$. It is a constant

If you turned the motor shaft by hand, you would *feel* the inertia of the load through the gearbox but it would be reduced by $G^2$.

**◘ Table 9.1**  Relationship between load ($^l$·) and motor ($^m$·) referenced quantities for reduction gear ratio $G$

| $^lJ = G^2\,{}^mJ$ | $^l\tau = G\,{}^m\tau$ | $^l\omega = {}^m\omega/G$ |
|---|---|---|
| $^lB = G^2\,{}^mB$ | $^l\tau_C = G\,{}^m\tau_C$ | $^l\dot\omega = {}^m\dot\omega/G$ |

intrinsic characteristic of the motor and the value is provided in the motor manufacturer's data sheet. The second component is the variable load inertia $^lJ_l$ which is the inertia of the driven link and all the other links that are attached to it. For joint $j$, this is element $m_{jj}$ of the configuration dependent inertia matrix which will be introduced in ▶ Sect. 9.2.2.

### 9.1.5  Modeling the Robot Joint

The complete motor drive comprises the motor to generate torque, the gearbox to amplify the torque and reduce the effects of the load, and an encoder to provide feedback of position and velocity. A schematic of a typical integrated motor unit is shown in ◘ Fig. 9.6.

Collecting the various equations above, we can write the torque balance on the motor shaft, referenced to the motor, as

$$K_m K_a u - B'\omega - \tau'_C(\omega) - \frac{\tau_d(\boldsymbol{q})}{G} = J'\dot\omega \tag{9.5}$$

where $B'$, $\tau'_C$ and $J'$ are the effective total viscous friction, Coulomb friction and inertia due to the motor, gearbox, bearings and the load

$$B' = B_m + \frac{B_l}{G^2}, \quad \tau'_C = \tau_{C,m} + \frac{\tau'_{C,l}}{G}, \quad J' = J_m + \frac{J_l}{G^2} \,. \tag{9.6}$$

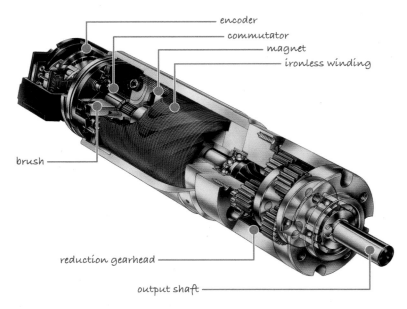

**◘ Fig. 9.6**  Schematic of an integrated motor-encoder-gearbox assembly (courtesy of maxon precision motors, inc.)

| ▣ **Table 9.2** Motor and drive parameters for PUMA 560 joint one (shoulder) with respect to the motor side of the gearbox (Corke 1996b) | | | |
|---|---|---|---|
| Motor torque constant | $K_m$ | 0.228 | $N\,m\,A^{-1}$ |
| Motor inertia | $J_m$ | $200 \times 10^{-6}$ | $kg\,m^2$ |
| Motor + joint viscous friction | $B_m$ | $817 \times 10^{-6}$ | $N\,m\,s\,rad^{-1}$ |
| Motor + joint Coulomb friction | $\tau_C^+$ | 0.126 | $N\,m$ |
| | $\tau_C^-$ | $-0.0709$ | $N\,m$ |
| Gear ratio | G | 107.815 | |
| Maximum torque | $\tau_{max}$ | 0.900 | $N\,m$ |
| Maximum speed | $\dot{q}_{max}$ | 165 | $rad\,s^{-1}$ |

In order to analyze the dynamics of (9.5) we must first linearize it, and this can be done simply by setting all additive constants to zero ▶

$$J'\dot{\omega} + B'\omega = K_m K_a u$$

and then applying the Laplace transformation

$$sJ'\Omega(s) + B'\Omega(s) = K_m K_a U(s)$$

where $\Omega(s)$ and $U(s)$ are the Laplace transform of the time domain signals $\omega(t)$ and $u(t)$ respectively. This can be rearranged as a linear transfer function

$$\frac{\Omega(s)}{U(s)} = \frac{K_m K_a}{J's + B'} \qquad (9.7)$$

relating motor speed to control input. It has a single pole, the mechanical pole, at $s = -B'/J'$.

We will use data for joint one of the PUMA 560 robot since its parameters are well known and are listed in ▣ Table 9.2. We have no data about link friction, so we will assume $B' = B_m$. The link inertia $\mathbf{M}_{11}$ experienced by the joint 1 motor as a function of configuration is shown in ▣ Fig. 9.16c and we see that it varies significantly – from 3.66 to 5.21 $kg\,m^2$. Using the mean value of the extreme inertia values, which is 4.43 $kg\,m^2$, the effective inertia is

$$J' = J_m + \frac{1}{G^2}\mathbf{M}_{11}$$
$$= 200 \times 10^{-6} + \frac{4.43}{(107.815)^2}$$
$$= 200 \times 10^{-6} + 380 \times 10^{-6} = 580 \times 10^{-6} kg\,m^2$$

and we see that the inertia of the link referred to the motor side of the gearbox is comparable to the inertia of the motor itself.

The Toolbox can generate a linearized dynamic model

```
>>> puma = models.DH.Puma560(); # load model with dynamic parameters
>>> tf = puma.jointdynamics(puma.qn);
```

which is a list of tuples containing the numerator and denominator polynomials of continuous-time linear-time-invariant (LTI) models, computed for the particular configuration. For joint one that we are considering here, the transfer function is

```
>>> tf[1]
((1,), (0.0005797, 0.000817))
```

Rather than ignoring Coulomb friction, which is generally quite significant, it can be replaced by additional equivalent viscous friction $B' = B_m + \frac{B_l}{G^2} + \frac{\tau_C'}{\omega}$ at the operating point.

For example, as implemented by the package `python-control`.

which is similar to (9.7) except that it does not account for $K_m$ and $K_a$ since these electrical parameters are not stored in the `Robot` object. Such a model can be used for a range of standard control system analysis and design tasks. ◀

### 9.1.6 Velocity Control Loop

A common approach to controlling the position output of a motor is the nested control loop. The outer loop is responsible for maintaining position and determines the velocity of the joint that will minimize position error. The inner loop – the velocity loop – is responsible for maintaining the velocity of the joint as demanded by the outer loop. Motor speed control is important for all types of robots, not just manipulators, for example, controlling the wheel speeds for a mobile robot, the rotor speeds of a quadrotor as discussed in ▶ Chap. 4, or the propeller speeds of an underwater robot. We will start with the inner velocity loop and work our way outwards.

The motor velocity is typically computed by taking the difference in motor position at each sample time, and the position is measured by a shaft encoder. This can be problematic at very low speeds where the encoder tick rate is lower than the sample rate. For slow-speed motion, a better strategy is to measure the time between encoder ticks.

A block diagram of a velocity control loop is shown in ◻ Fig. 9.7. The input to the motor power amplifier is based on the error between the demanded and actual velocity. ◀ A saturator models the finite maximum torque that the motor can deliver. To test this velocity controller, we use the test harness shown in ◻ Fig. 9.8 which provides a trapezoidal velocity demand.

We first consider the case of proportional control where $K_i = 0$ and

$$u^* = K_v(\dot{q}^* - \dot{q}) \, . \tag{9.8}$$

The `-H` option stops the function from blocking until all the figures are dismissed.

Simulating the velocity controller and its test harness ◀

```
>>> %run -m vloop_test -H
```

allows us to experiment, and we find that a gain of $K_v = 0.6$ gives satisfactory performance as shown in ◻ Fig. 9.9. There is some minor overshoot at the discontinuity but less gain leads to increased velocity error while more gain leads to oscillation – control engineering is all about tradeoffs.

We also observe a very slight steady-state error – the actual velocity is less than the demand at all times. From a classical control system perspective, this velocity loop is a Type 0 system since it contains no integrator. A characteristic of such systems is that they exhibit a finite error for a constant input. Intuitively, for the motor to move at constant speed, it must generate a finite torque to overcome friction, and since motor torque is proportional to velocity error there must be a finite velocity error.

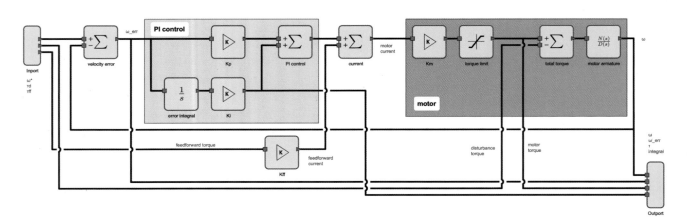

◻ **Fig. 9.7**   Block-diagram model `models/vloop` of a robot joint-velocity control loop

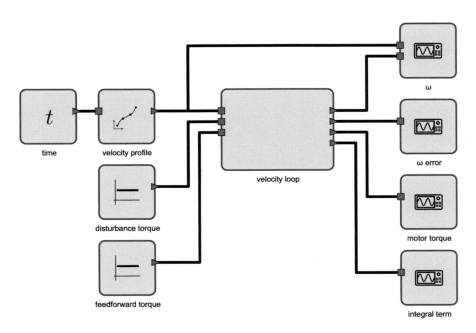

**Fig. 9.8** Block-diagram model `models/vloop-test` of the test harness for the velocity-control loop

Now, we will investigate the effect of inertia variation on the closed-loop response. Using (9.6) and the data from ◘ Fig. 9.16c, we find that the minimum and maximum joint inertia at the motor are $515 \times 10^{-6}$ and $648 \times 10^{-6}$ kg m² respectively. ◘ Fig. 9.10 shows the velocity tracking error using the control gains chosen above for various values of link inertia. We can see that the tracking error decays more slowly for larger inertia. For a case where the inertia variation is more extreme, the gain should be chosen to achieve satisfactory closed-loop performance at both extremes.

◘ Fig. 9.15a shows that the gravity torque on this joint varies from approximately $-40$ to $40$ N m. We now add a disturbance torque equal to just half that maximum amount, $20$ N m applied on the load side of the gearbox. We do this by setting a nonzero value in the torque disturbance block of ◘ Fig. 9.8 and rerunning the simulation. The results shown in ◘ Fig. 9.11 indicate that the control performance has been badly degraded – the tracking error has increased to more

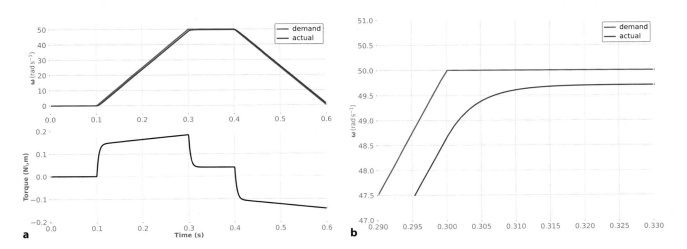

**Fig. 9.9** Velocity-loop response for a trapezoidal demand. **a** Velocity tracking and motor torque; **b** close-up view of velocity tracking

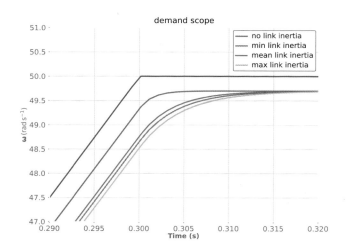

■ **Fig. 9.10**   Velocity-loop response with a trapezoidal demand for varying inertia $M_{22}$

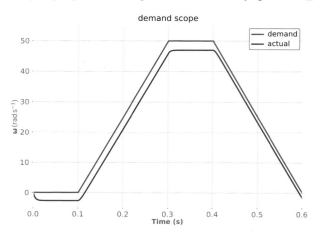

■ **Fig. 9.11**   Velocity-loop response to a trapezoidal demand with a gravity disturbance of $20\,\mathrm{N\,m}$ and mean link inertia

---

**Excurse 9.3: Motor Limits**

Electric motors are limited in both torque and speed. The maximum torque is defined by the maximum current the power amplifier can provide. A motor also has a maximum rated current beyond which the motor can be damaged by overheating or demagnetization of its permanent magnets which irreversibly reduces its torque constant $K_m$. As speed increases so too does friction, and the maximum speed is $\omega_{\mathrm{max}} = \tau_{\mathrm{max}}/B$.

The product of motor torque and speed is the mechanical output power, and this also has an upper bound. Motors can tolerate some overloading, peak power and peak torque, for short periods of time but the sustained rating is significantly lower than the peak.

---

than $2\,\mathrm{rad\,s^{-1}}$. This has the same root cause as the very small error we saw in
■ Fig. 9.9 – a Type 0 system exhibits a finite error for a constant input or a constant disturbance.

There are three common approaches to counter this error. The first, and simplest, is to increase the gain. This will reduce the tracking error but will also push the system toward instability and increase the overshoot.

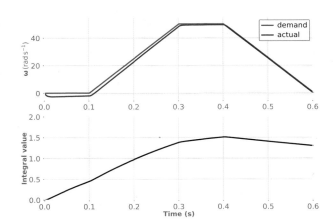

☐ **Fig. 9.12** Velocity loop response to a trapezoidal demand with a gravity disturbance of 20 N m and proportional-integral control

The second approach, commonly used in industrial motor drives, is to add integral action – adding an integrator changes the system to Type 1 which has zero error for a constant input or constant disturbance. We change (9.8) to a proportional-integral (PI) controller

$$u^* = K_v(\dot{q}^* - \dot{q}) + K_i \int_0^t (\dot{q}^* - \dot{q})dt, \quad K_i > 0 \; .$$

In the block diagram of ☐ Fig. 9.7 this is achieved by setting `Ki` to a value greater than zero. With some experimentation, we find the gains $K_v = 1$ and $K_i = 10$ work well and the performance is shown in ☐ Fig. 9.12. The integrator state evolves over time to cancel out the disturbance term and we can see the error decaying to zero. In practice, the disturbance varies over time, and the integrator's ability to track it depends on the value of the integral gain $K_i$. In reality, other disturbances affect the joint, for instance, Coulomb friction and torques due to velocity and acceleration coupling. The controller needs to be well tuned so that these have minimal effect on the tracking performance.

As always in engineering there are tradeoffs. The integral term can lead to increased overshoot, so increasing $K_i$ usually requires some compensating reduction of $K_v$. If the joint actuator is pushed to its performance limit, for instance, the torque limit is reached, then the tracking error will grow with time since the motor acceleration will be lower than required. The integral of this increasing error will grow and can lead to a condition known as integral windup. When the joint finally reaches its destination, the accumulated integral keeps driving the motor until the integral decays – leading to overshoot. Various strategies are employed to combat this, such as limiting the maximum value of the integrator, or only allowing integral action when the motor is close to its setpoint. The approaches just mentioned are collectively referred to as disturbance rejection since they are concerned with reducing the effect of an unknown disturbance.

In the robotics context, the gravity disturbance is actually known. In ▶ Sect. 9.1.3, we showed that the torque due to gravity acting on the joints is a function of the joint angles, and, in a robot, these angles are always known. If we know this torque, and the motor torque constant, we can *add* a compensating control to the output of the PI controller. ▶ Predicting the disturbance and canceling it out – a strategy known as torque feedforward control – is an alternative to disturbance rejection. We can experiment with this control approach by setting the feedforward torque block of ☐ Fig. 9.8 to the same, or approximately the same, value as the disturbance.

Even if the gravity load is known imprecisely, this trick will reduce the magnitude of the disturbance.

**9**

> **Excurse 9.4: Back EMF**
>
> A spinning motor acts like a generator and produces a voltage $V_b$ called the back EMF which opposes the current flowing into the motor. Back EMF is proportional to motor speed $V_b = K_m \omega$, where $K_m$ is the motor torque constant whose units can also be interpreted as $\text{V s rad}^{-1}$. When this voltage equals the maximum possible voltage from the power amplifier, then no more current can flow into the motor and torque falls to zero – this sets an upper bound on motor speed. As speed increases, the reduced current flow reduces the torque available, and this looks like an additional source of damping.

### 9.1.7 Position Control Loop

The outer loop is responsible for maintaining position and we use a proportional controller ◄ based on the error between demanded $q^*(t)$ and actual $q^\#$ position to compute the desired speed of the motor

Another common approach is to use a proportional-integral-derivative (PID) controller for position but it can be shown that the D gain of this controller is related to the P gain of the inner velocity loop.

$$\dot{q}^* = K_p\big(q^*(t) - q^\#\big) \tag{9.9}$$

which is the input to the velocity loop. In many cases, particularly when following a trajectory, we also know the desired velocity $\dot{q}^*(t)$ and we can use this as a feedforward signal and the position control becomes

$$\dot{q}^* = K_p\big(q^*(t) - q^\#\big) + K_{\text{ff}}\,\dot{q}^*(t) \tag{9.10}$$

where $K_{\text{ff}}$ would have a value ranging from zero (no feedforward) to one (maximum feedforward).

A block diagram model of the position loop is shown in ◻ Fig. 9.13a and the position and velocity demand, $q^*(t)$ and $\dot{q}^*(t)$, are smooth functions of time computed by a trajectory generator ◄ that moves from 0 to 0.5 rad in 1 s. Joint position is obtained by integrating joint velocity, obtained from the motor velocity loop, and scaled by the inverse gearbox ratio. The error between the motor and desired position provides the velocity demand for the inner loop.

We use a trapezoidal trajectory as discussed in ► Sect. 3.3.1.

We run the test harness shown in ◻ Fig. 9.13b by

```
>>> %run -m ploop_test -H
```

and its performance is tuned by adjusting the four gains: $K_p$, $K_v$, $K_i$, $K_{\text{ff}}$ in order to achieve good tracking performance along the trajectory. For $K_p = 40$, $K_v = 0.6$, $K_i = 0$ and $K_{\text{ff}} = 0$, the tracking and error responses are shown in ◻ Fig. 9.14a. We see that the final error is zero but there is some tracking error along the path where the motor position lags behind the demand. The error between the demand and actual curves is due to the cumulative velocity error of the inner loop which has units of angle.

The position loop, like the velocity loop, is based on classical negative feedback. Having zero position error while tracking a ramp would mean zero demanded velocity to the inner loop which is actually contradictory. This model contains an integrator, after the velocity loop, which makes it a Type 1 system that will exhibit a constant error to a *ramp* input. There are two common approaches to reducing this tracking error. Firstly, we can add an integrator to the position loop – making it a proportional-integral controller – which adds an extra integrator and creates a Type 2 system. However, this presents yet another parameter to tune. Secondly, we can introduce feedforward control – by setting $K_{\text{ff}} = 1$ we add the desired velocity to the output of the proportional controller, which is the input to the velocity loop. Conveniently, the trapezoidal trajectory function computes velocity as a function

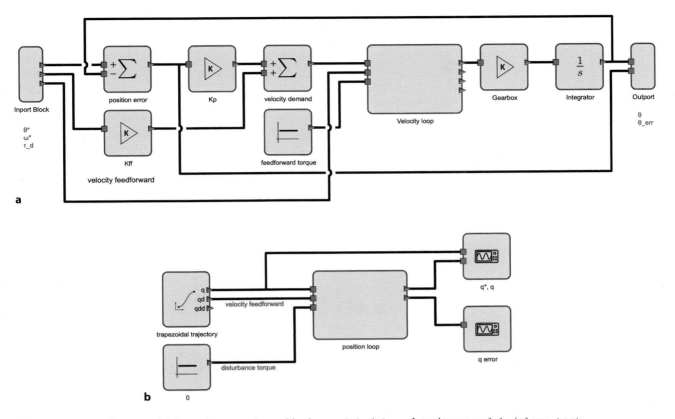

a

b    0

■ **Fig. 9.13** Block-diagram models for position control. **a** position loop `models/ploop`; **b** test harness `models/ploop_test`

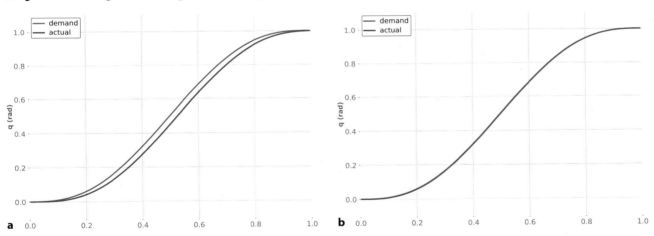

a

b

■ **Fig. 9.14** Position loop following a trapezoidal trajectory. **a** Proportional control only **b** proportional control plus velocity demand feedforward

of time as well as position. The time response with velocity feedforward is shown in ■ Fig. 9.14b and we see that the tracking error is greatly reduced.

## 9.1.8 Summary

A common structure for robot joint control is the nested control loop. The inner loop uses a proportional or proportional-integral control law to generate a torque so that the actual velocity closely follows the velocity demand. The outer loop uses a proportional-control law to generate the velocity demand so that the actual position closely follows the position demand. Disturbance torques due to gravity and other dynamic coupling effects impact the performance of the velocity loop as do variation in the parameters of the plant being controlled, and this in turn leads to errors in position tracking. Gearing reduces the magnitude of disturbance torques by $1/G$ and the variation in inertia and friction by $1/G^2$ but at the expense of cost, weight, increased friction and mechanical noise.

The velocity-loop performance can be improved by adding an integral control term, or by feedforward of the disturbance torque which is largely predictable. The position-loop performance can also be improved by feedforward of the desired joint velocity. In practice, control systems use both feedforward and feedback control. Feedforward is used to inject signals that we can compute, in this case the joint velocity, and in the earlier case the gravity torque. Feedback control compensates for all remaining sources of error including variation in inertia due to manipulator configuration and payload, changes in friction with time and temperature, and all the disturbance torques due to velocity and acceleration coupling. In general, the use of feedforward allows the feedback gain to be reduced since a large part of the control signal now comes from the feedforward.

## 9.2 Rigid-Body Equations of Motion

Consider the motor which actuates joint $j \in \{0, \ldots, N-1\}$ of a serial-link manipulator. From ◻ Fig. 7.18, we recall that joint $j$ connects link $j$ to link $j + 1$. The motor exerts a torque that causes the outward link (link $j + 1$) to rotationally accelerate but it also exerts a reaction torque on the inward link (link $j$). The outward links $j + 1$ to $N$ exert a weight force due to gravity, and rotating links also exert gyroscopic forces on each other. The inertia that the motor *experiences* is a function of the configuration of the outward links.

The situation at the individual link is quite complex but, for the *series* of links, the result can be written elegantly and concisely as a set of coupled differential equations in matrix form

$$Q = \mathbf{M}(q)\ddot{q} + \mathbf{C}(q, \dot{q})\dot{q} + f(\dot{q}) + g(q) + \mathbf{J}^\top(q)w \tag{9.11}$$

where $q, \dot{q}, \ddot{q} \in \mathbb{R}^N$ are respectively the vector of generalized joint coordinates, velocities and accelerations, $\mathbf{M} \in \mathbb{R}^{N \times N}$ is the joint-space inertia matrix, $\mathbf{C} \in \mathbb{R}^{N \times N}$ is the Coriolis and centripetal coupling matrix, $f \in \mathbb{R}^N$ is the friction force, $g \in \mathbb{R}^N$ is the gravity loading, $w \in \mathbb{R}^6$ is a wrench applied at the end effector, $\mathbf{J}(q) \in \mathbb{R}^{6 \times N}$ is the manipulator Jacobian, and $Q \in \mathbb{R}^N$ is the vector of generalized actuator forces. This equation describes the manipulator rigid-body dynamics and is known as the inverse dynamics – given the configuration, velocity and acceleration, it computes the required joint forces or torques.

These equations can be derived using any classical dynamics method such as Newton's second law and Euler's equation of motion, as discussed in ▶ Sect. 3.2.1, or a Lagrangian energy-based approach. A very efficient way of computing (9.11) is the recursive Newton-Euler algorithm which starts at the base, and working outward using the velocity and acceleration of each joint, computes the velocity and acceleration of each link. Then, working from the tool back to the base, it computes the forces and moments acting on each link and thus the joint torques. ◀ The recursive Newton-Euler algorithm has $O(N)$ complexity and can be written in functional form as

$$Q = \mathcal{D}^{-1}(q, \dot{q}, \ddot{q}) . \tag{9.12}$$

The recursive form of the inverse dynamics does not explicitly calculate the matrices $\mathbf{M}$, $\mathbf{C}$ and $g$ of (9.11). However, we can use the recursive Newton-Euler algorithm to calculate these matrices, and the Toolbox functions `inertia` and `coriolis` use Walker and Orin's (1982) '*Method 1*'. While the recursive forms are computationally efficient for the inverse dynamics, computing the coefficients of the individual dynamic terms ($\mathbf{M}$, $\mathbf{C}$ and $g$) in (9.11) is quite costly – $O(N^3)$ for an $N$-axis manipulator.

In the Toolbox, this is implemented by the `rne` method of `Robot` subclass objects. ▶
Consider the PUMA 560 robot

```
>>> puma = models.DH.Puma560();
```

at the nominal configuration, and with zero joint velocity and acceleration. To
achieve this state, the required generalized joint forces, or joint torques in this case,
must be

```
>>> zero = np.zeros((6,));
>>> Q = puma.rne(puma.qn, zero, zero)
array([      0,   31.64,   6.035,       0, 0.02825,       0])
```

Since the robot is not moving (we specified $\dot{q} = \ddot{q} = 0$), these torques must be those
required to *hold the robot up* against gravity. We can confirm this by computing the
torques required in the absence of gravity

```
>>> Q = puma.rne(puma.qn, zero, zero, gravity=[0, 0, 0])
array([      0,       0,       0,       0,       0,       0])
```

by overriding the object's default gravity vector.

Like most Toolbox methods, `rne` can operate on a trajectory

```
>>> traj = jtraj(puma.qz, puma.qr, 10);
>>> Q = puma.rne(traj.q, traj.qd, traj.qdd);
```

which has returned

```
>>> Q.shape
(10, 6)
```

a $10 \times 6$ matrix with each row representing the generalized force required for the
corresponding row of `traj.q`. The joint torques corresponding to the fifth time step
are

```
>>> Q[5, :]
array([ -1.681,   52.44,  -14.38,       0, 0.003379,       0])
```

Consider now a case where the robot is moving. It is *instantaneously* at the nominal
configuration but joint 0 is moving at 1 rad s$^{-1}$ and the acceleration of all joints is
zero. Then, in the absence of gravity, the required joint torques

```
>>> puma.rne(puma.qn, [1, 0, 0, 0, 0, 0], zero, gravity=[0, 0, 0])
array([30.53, 0.628, -0.3607, -0.0003056, 0, 0])
```

are nonzero. The torque on joint 0 is that needed to overcome friction which always
opposes the motion. More interesting is that nonzero torque needs to be exerted on
joints 1, 2 and 3 to oppose the gyroscopic torques that joint 0 motion is exerting on
those joints.

The elements of **M**, **C**, *f* and *g* are complex functions of the links' kinematic
and inertial parameters. Each link has ten independent inertial parameters: the link
mass $m_i \in \mathbb{R}$; the center of mass (CoM) $r_i \in \mathbb{R}^3$ with respect to the link coordinate
frame; and six second moments which represent the inertia tensor of the link about
the CoM, but with respect to axes aligned with the link frame {j}. We can view the
dynamic parameters of a robot's link by

```
>>> print(puma[1].dyn())
m    =       17
r    =    -0.36    0.006     0.23
     |     0.13        0        0 |
I    = |       0     0.52        0 |
     |       0        0     0.54 |
Jm   =   0.0002
B    =   0.00082
Tc   =     0.13(+)   -0.071(-)
G    =   1.1e+02
qlim =    -1.9  to     1.9
```

Not all robot arm models in the
Toolbox have dynamic parameters, see
the "dynamics" column in the output of
the `models.list()` command, or use
`models.list(keywords=("dynamics",))`
to list models with dynamic
parameters. The PUMA 560 robot is
used for the examples in this chapter
since it has a full dynamic model that
includes inertia and friction.

In this case, the gear ratio is negative which means that the motor controller's sense of positive rotation is the opposite to the kinematic convention for the robot.

which in order are: the link mass, CoM position, link inertia tensor, motor inertia, motor friction, motor Coulomb friction for positive and negative velocities, and reduction gear ratio ◄.

The remainder of this section examines the various matrix components of (9.11) in order of decreasing significance for most robotics applications.

### 9.2.1 Gravity Term

$$Q = \mathbf{M}(q)\ddot{q} + \mathbf{C}(q,\dot{q})\dot{q} + f(\dot{q}) + \underline{g(q)} + \mathbf{J}^{\top}(q)w$$

We start our detailed discussion with the gravity term because it is generally the dominant term in (9.11) and is present even when the robot is stationary. Some robots use counterbalance weights ◄ or even springs to reduce the gravity torque that needs to be provided by the motors – this allows the motors to be smaller and thus lower in cost.

Counterbalancing will however increase the inertia associated with a joint since it adds additional mass at the end of a lever arm, and will also increase the total mass of the robot.

In the previous section, we used the `rne` method to compute the gravity load by setting the joint velocity and acceleration to zero. A more convenient approach is to use the `gravload` method

```
>>> Q = puma.gravload(puma.qn)
array([      0,    31.64,    6.035,        0,  0.02825,        0])
```

The `Robot` object contains a default gravitational acceleration vector which is initialized to the nominal value for Earth ◄

This is the Newtonian gravitational acceleration, as discussed in ► Sect. 3.4.2.1, which is vertically downward. The `gravity` argument for the `Robot` constructor can change this.

```
>>> puma.gravity
array([      0,        0,    -9.81])
```

We could change gravity to the lunar value

```
>>> puma.gravity /= 6
```

resulting in reduced joint torques

```
>>> puma.gravload(puma.qn)
array([      0,    5.273,    1.006,        0,  0.004709,        0])
```

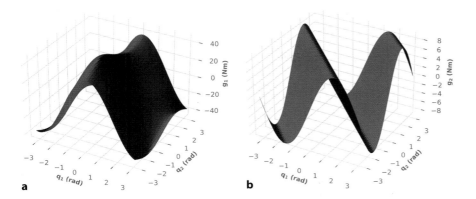

**Fig. 9.15** Gravity load variation with manipulator configuration. **a** Shoulder gravity load, $g_2(q_1, q_2)$; **b** elbow gravity load $g_3(q_1, q_2)$

or we could turn our lunar robot upside down

```
>>> puma.base = SE3.Rx(pi);
>>> puma.gravload(puma.qn)
array([       0,    -5.273,    -1.006,          0,   -0.004709,          0])
```

and see that the torques have changed sign. Before proceeding, we bring our robot back to Earth and right-side up

```
>>> puma = models.DH.Puma560();
```

The torque exerted on a joint, due to gravity acting on the robot, depends very strongly on the robot's configuration. Intuitively, the torque on the shoulder joint is much greater when the arm is stretched out horizontally

```
>>> Q = puma.gravload(puma.qs)
array([       0,     46.01,     8.772,          0,    0.02825,          0])
```

than when the arm is pointing straight up

```
>>> Q = puma.gravload(puma.qr)
array([       0,   -0.7752,     0.2489,         0,          0,          0])
```

The gravity torque on the elbow is also very high in the first configuration since it has to support the lower arm and the wrist. We can investigate how the gravity load on joints 2 and 3 varies with joint configuration by

```
>>> N = 100;
>>> Q1, Q2 = np.meshgrid(np.linspace(-pi, pi, N),
...                      np.linspace(-pi, pi, N));
>>> G1, G2 = np.zeros((N,N)), np.zeros((N,N));
>>> for i in range(N):
...     for j in range(N):
...         g = puma.gravload(np.array([0, Q1[i,j], Q2[i,j], 0, 0, 0]))
...         G1[i, j] = g[1]   # shoulder gravity load
...         G2[i, j] = g[2]   # elbow gravity load
>>> plt.axes(projection="3d").plot_surface(Q1, Q2, G1);
```

and the results are shown in ◻ Fig. 9.15. The gravity torque on joint 1 varies between $\pm 40\,\mathrm{N\,m}$ and for joint 2 varies between $\pm 10\,\mathrm{N\,m}$. This type of analysis is very important in robot design to determine the required torque capacity for the motors.

### 9.2.2  Inertia Matrix

$$Q = \underline{\mathbf{M}(q)\ddot{q}} + \mathbf{C}(q, \dot{q})\dot{q} + f(\dot{q}) + g(q) + \mathbf{J}^{\top}(q)w$$

The diagonal elements of this matrix include the motor armature inertias, multiplied by $G^2$.

The joint-space inertia matrix, sometimes called the mass matrix, is symmetric and positive definite. It is a function of the manipulator configuration ◄

```
>>> M = puma.inertia(puma.qn)
array([[   3.659,   -0.4044,    0.1006,  -0.002517,         0,          0],
       [  -0.4044,    4.414,    0.3509,          0,   0.00236,          0],
       [   0.1006,    0.3509,    0.9378,          0,   0.00148,          0],
       [-0.002517,         0,         0,     0.1925,         0,  2.828e-05],
       [        0,   0.00236,   0.00148,          0,    0.1713,          0],
       [        0,         0,         0,  2.828e-05,         0,    0.1941]])
```

and the diagonal elements $m_{jj}$ describe the inertia *experienced* by joint $j$, that is, $Q_j = m_{jj}\ddot{q}_j$. Note that the first two diagonal elements, corresponding to the robot's waist and shoulder joints, are large since motion of these joints involves rotation of the heavy upper- and lower-arm links. The off-diagonal terms $m_{ij} = m_{ji}, i \neq j$ are the products of inertia which couple the acceleration of joint $j$ to the generalized force on joint $i$.

We can investigate some of the elements of the inertia matrix and how they vary with robot configuration quite simply

```
>>> N = 100;
>>> Q1, Q2 = np.meshgrid(np.linspace(-pi, pi, N),
...                      np.linspace(-pi, pi, N));
```

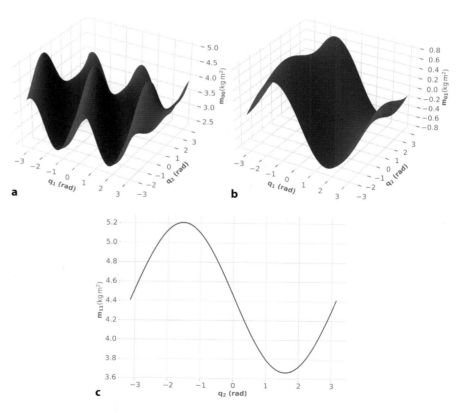

☐ **Fig. 9.16**  Variation of inertia matrix elements as a function of manipulator configuration. **a** Joint 0 inertia as a function of joint 1 and 2 angles $m_{00}(q_1, q_2)$; **b** product of inertia $m_{01}(q_1, q_2)$; **c** joint 1 inertia as a function of joint 2 angle $m_{11}(q_2)$

```
>>> M00, M01 = np.zeros((N,N)), np.zeros((N,N))
>>> M11 = np.zeros((N,N))
>>> for i in range(N):
...    for j in range(N):
...        M = puma.inertia(np.array([0, Q1[i,j], Q2[i,j], 0, 0, 0]))
...        M00[i, j] = M[0, 0]
...        M01[i, j] = M[0, 1]
...        M11[i, j] = M[1, 1]
>>> plt.axes(projection="3d").plot_surface(Q1, Q2, M00);
```

The results are shown in ◻ Fig. 9.16 and we see significant variation in the value of $m_{11}$ which changes by a factor of

```
>>> M00.max() / M00.min()
2.156
```

This is important for robot design since, for a fixed maximum motor torque, inertia sets the upper bound on acceleration which in turn affects motion time and path following accuracy.

The off-diagonal term $m_{01}$ represents coupling between the angular acceleration of joint 1 and the torque on joint 0. That is, if joint 1 accelerates then a torque will be exerted on joint 0 and vice versa.

### 9.2.3 Friction

$$Q = \mathbf{M}(q)\ddot{q} + \mathbf{C}(q,\dot{q})\dot{q} + \underline{f(\dot{q})} + g(q) + \mathbf{J}(q)^{\top}w$$

Friction is generally the next most dominant joint force after gravity. ▶ If the robot is moving with a joint 0 velocity of $1\,\mathrm{rad\,s^{-1}}$ then the torque due to viscous and Coulomb friction is

Joint friction for the PUMA 560 robot varies from 10 to 47% of the maximum motor torque, for the first three joints (Corke 1996b).

```
>>> puma.friction([1, 0, 0, 0, 0, 0])
array([ -30.53,    0,    0,    0,    0,    0])
```

which has a negative sign, it acts in the opposite direction to the velocity, and affects joint 0 only. The friction values are lumped, and motor referenced, that is, they apply to the motor side of the gearbox.

### 9.2.4 Coriolis and Centripetal Matrix

$$Q = \mathbf{M}(q)\ddot{q} + \underline{\mathbf{C}(q,\dot{q})\dot{q}} + f(\dot{q}) + g(q) + \mathbf{J}(q)^{\top}w$$

The matrix $\mathbf{C}$ is a function of joint coordinates and joint velocity, and element $c_{ij}$ couples the velocity of joint $j$ to a generalized force acting on joint $i$. The coupling is due to gyroscopic effects: the centripetal torques are proportional to $\dot{q}_j^2$, while the Coriolis torques are proportional to $\dot{q}_i\dot{q}_j$.

For example, at the nominal configuration with the elbow joint moving at $1\,\mathrm{rad\,s^{-1}}$

```
>>> qd = [0, 0, 1, 0, 0, 0];
```

the Coriolis matrix is

```
>>> C = puma.coriolis(puma.qn, qd)
array([[  0.3607, -0.09566, -0.09566, 0.0005445, -0.0004321,   2e-05],
       [       0,   0.3858,   0.3858,         0, -4.134e-05,       0],
       [       0,        0,        0,         0, -0.0009207,       0],
       [-3.584e-05,      0,        0,         0,          0, -1.414e-05],
       [       0, 0.0009207, 0.0009207,        0,          0,       0],
       [   2e-05,        0,        0, 1.414e-05,          0,       0]])
```

Element $c_{12}$ represents significant coupling from joint 2 velocity to torque on joint 1 – rotational velocity of the elbow exerting a torque on the shoulder. The elements of this matrix represent a coupling from velocity to joint force and have the same units as viscous friction or damping, however the sign can be positive or negative. The joint disturbance torques, due to the motion of just this one joint, are

```
>>> C @ qd
array([-0.09566,    0.3858,    0,   0, 0.0009207,    0])
```

### 9.2.5 **Effect of Payload**

Any real robot has a specified maximum payload which is dictated by two dynamic effects. The first is that a mass at the end of the robot increases the inertia *experienced* by all the joint motors, and this reduces acceleration and dynamic performance. The second is that the mass generates a weight force which all the joints need to support. In the worst case, the increased gravity torque component might exceed the rating of one or more motors. However, even if the rating is not exceeded, there is less torque available for acceleration which again reduces dynamic performance.

To quantify these effects, we will take a snapshot of the robot's gravity torque and inertia matrix

```
>>> G = puma.gravload(puma.qn);
>>> M = puma.inertia(puma.qn);
```

and then add a 2.5 kg point mass to the PUMA 560 which is its rated maximum payload

```
>>> puma.payload(2.5, [0, 0, 0.1]);
```

The center of mass of the payload cannot be at the center of the wrist coordinate frame, that is inside the wrist mechanism, so we offset it 100 mm in the $z$-direction of the wrist frame.

The inertia at the nominal configuration is now

```
>>> M_loaded = puma.inertia(puma.qn);
```

and the ratio with respect to the unloaded case is ◄

Some elements of **M** are very small which would result in anomalous ratios, so we set those small values to nan to make this explicit.

```
>>> M_loaded / np.where(abs(M) < 1e-6, np.nan, M)
array([[   1.336,      nan,    2.149,      nan,      nan,      nan],
       [     nan,    1.267,    2.919,      nan,    74.01,      nan],
       [   2.149,    2.919,     1.66,      nan,    66.41,      nan],
       [     nan,      nan,      nan,    1.065,      nan,       1],
       [     nan,    74.01,    66.41,      nan,    1.145,      nan],
       [     nan,      nan,      nan,       1,      nan,       1]])
```

We see that the diagonal elements have increased, for instance the elbow joint inertia has increased by 66% which reduces its maximum acceleration by nearly 40%. The inertia of joint 5 is unaffected since this added mass lies on the axis of this joint's rotation. Some of the off-diagonal terms have increased significantly, particularly in row and column 4, which indicates that motion of that wrist joint swinging the offset mass creates large reaction forces that are *felt* by all the other joints.

The gravity load has also increased by some significant factors

```
>>> puma.gravload(puma.qn) / np.where(abs(G) < 1e-6, np.nan, G)
array([     nan,    1.522,    2.542,      nan,    86.81,      nan])
```

at the elbow and wrist. We set the payload of the robot back to zero before proceeding

```
>>> puma.payload(0)
```

### 9.2.6  Base Wrench

A moving robot exerts a wrench on its base – its weight as well as reaction forces and torques as the arm moves around. This wrench is returned as an optional output argument of the `rne` method, for example

```
>>> Q, wb = puma.rne(puma.qn, zero, zero, base_wrench=True);
```

The wrench

```
>>> wb
array([        0,        0,    229.2,   -48.27,   -31.09,        0])
```

needs to be applied to the base to keep it in equilibrium. The vertical force of 230 N opposes the downward weight force of the robot

```
>>> sum([link.m for link in puma]) * puma.gravity[2]
-229.2
```

There is also a moment about the $x$- and $y$-axes since the center of mass of the robot in this configuration is not over the origin of the base coordinate frame.

The base wrench is important in situations where the robot does not have a rigid base such as on a satellite in space, on a boat, an underwater vehicle or even on a ground vehicle with soft suspension.

### 9.2.7  Dynamic Manipulability

In ▶ Sect. 8.3.2, we discussed a kinematic measure of manipulability that describes how well configured the robot is to achieve velocity in any task-space direction. The force ellipsoid of ▶ Sect. 8.4.2 describes how well the manipulator is able to accelerate in different task-space directions but is based on the kinematic, not dynamic, parameters of the robot arm. Following a similar approach, we consider the set of generalized joint forces with unit norm

$$\boldsymbol{Q}^\top \boldsymbol{Q} = 1$$

From (9.11), ignoring gravity and assuming $\dot{\boldsymbol{q}} = 0$, we can write

$$\boldsymbol{Q} = \mathbf{M}(\boldsymbol{q})\,\ddot{\boldsymbol{q}}$$

Differentiating (8.2) and assuming $\dot{\boldsymbol{q}} = 0$, we write

$$\dot{\boldsymbol{\nu}} = \mathbf{J}(\boldsymbol{q})\,\ddot{\boldsymbol{q}}$$

where $\dot{\boldsymbol{\nu}}$ is spatial acceleration – a vector comprising translational and angular acceleration. Combining these, we write ▶

If the matrix is not square, we use the pseudoinverse.

$$\dot{\boldsymbol{\nu}}^\top \left( \mathbf{J}(\boldsymbol{q})\,\mathbf{M}^{-1}(\boldsymbol{q})\,\mathbf{M}^{-\top}(\boldsymbol{q})\,\mathbf{J}^\top(\boldsymbol{q}) \right)^{-1} \dot{\boldsymbol{\nu}} = 1$$

which is the equation of points $\dot{\boldsymbol{\nu}}$ on the surface of a hyperellipsoid in the task acceleration space.

If we consider just the translational acceleration, we take the translational part of the Jacobian

```
>>> Jt = puma.jacob0(puma.qn, half="trans");  # first 3 rows
```

from which we can compute and plot the translational-acceleration ellipsoid

```
>>> M = puma.inertia(puma.qn);
>>> E = (Jt @ np.linalg.inv(M) @ np.linalg.inv(M).T @ Jt.T);
>>> plot_ellipsoid(E);
```

▶ go.sn.pub/IfXBCw

**9**

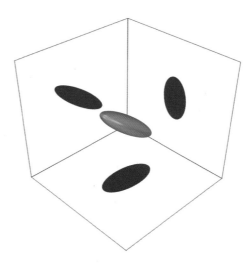

■ **Fig. 9.17** Translational acceleration ellipsoid for PUMA 560 robot in its nominal configuration. The black ellipses are shadows and show the shape of the ellipse as seen by looking along the $x$-, $y$- and $z$-directions. This plot was generated using PyVista rather than Matplotlib

which is shown in ■ Fig. 9.17. The major axis of this ellipsoid is the direction in which the manipulator has maximum acceleration at this configuration. The radii of the ellipsoid are the square roots of the eigenvalues

```
>>> e, _ = np.linalg.eig(E)
>>> radii = 1 / np.sqrt(e)
array([    2.219,     9.582,     5.905])
```

and the direction of maximum acceleration is given by the corresponding eigenvector. The ratio of the minimum to maximum radii

```
>>> radii.min() / radii.max()
0.2315
```

The 6-dimensional ellipsoid has dimensions with different units: $ms^{-2}$ and $rad\, s^{-2}$. This makes comparison of all 6 radii problematic.

is a measure of the nonuniformity of end-effector acceleration. ◀ It would be unity for isotropic acceleration capability.

Another common scalar manipulability measure takes the inertia into account, and considers the ratio of the minimum and maximum eigenvalues of the generalized inertia matrix

$$\mathbf{M}_x = \mathbf{J}^{-\top}(\boldsymbol{q})\,\mathbf{M}(\boldsymbol{q})\,\mathbf{J}(\boldsymbol{q})^{-1}$$

which is a measure $m \in [0, 1]$ where 1 indicates uniformity of acceleration in all directions. For this example

```
>>> puma.manipulability(puma.qn, method="asada")
0.004389
```

## 9.3 Forward Dynamics

To determine the motion of the manipulator in response to the forces and torques applied to its joints, we require the forward or integral dynamics. Rearranging the equations of motion (9.11), we obtain the joint acceleration

$$\ddot{\boldsymbol{q}} = \mathbf{M}^{-1}(\boldsymbol{q})\Big(\boldsymbol{Q} - \mathbf{C}(\boldsymbol{q},\dot{\boldsymbol{q}})\dot{\boldsymbol{q}} - \boldsymbol{f}(\dot{\boldsymbol{q}}) - \boldsymbol{g}(\boldsymbol{q}) - \mathbf{J}^{\top}(\boldsymbol{q})\boldsymbol{w}\Big) \qquad (9.13)$$

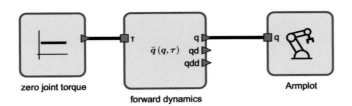

**Fig. 9.18** Block diagram model `models/zerotorque` for the PUMA 560 manipulator with zero applied joint torques. This model removes Coulomb friction in order to simplify the numerical integration

and $\mathbf{M}(q)$ is positive-definite and therefore always invertible. This function is computed by the `accel` method of the `Robot` class

```
>>> qdd = puma.accel(puma.q, puma.qd, Q)
```

given the joint coordinates, joint velocity and applied joint torques.

A simple demonstration of forward dynamics is the block diagram model shown in ◨ Fig. 9.18 where the joint torques are set to zero. The simulation is run by

```
>>> %run -m zerotorque -H
```

which integrates (9.13) over time. Since there are no torques to counter gravity and hold it upright, we see that the robot collapses under its own weight. The robot's upper arm falls and swings back and forth as does the lower arm, while the waist joint rotates because of Coriolis coupling. The motion will slowly decay as the energy is dissipated by viscous friction.

To see this in more detail we can plot the first three joint angles, returned by the simulation in the object `out`, as a function of time

```
>>> xplot(out.t, out.q[:3, :])
```

which is shown in ◨ Fig. 9.19.

We can achieve the same result using the Toolbox directly

```
>>> torque_func = lambda t, q, qd: np.zeros((6,))
>>> traj = puma.nofriction().fdyn(T=5, q0=puma.qr, Q=torque_func)
>>> xplot(traj.t, traj.q)
```

which again integrates (9.13) over time. The joint torques at each integration time step are returned by the function object passed as the option `Q` – in this case returning zero.

This example is simplistic and typically the joint torques would be computed by some control law as a function of the actual and desired robot joint coordinates and rates. This is the topic of the next section.

### ⊟ Coulomb friction and forward dynamics integration

Coulomb friction is a strong nonlinearity and can cause difficulty when using numerical integration routines to solve the forward dynamics. This is usually manifested by very long integration times. Fixed-step solvers tend to be more tolerant but are not yet available in this implementation.

A work around is to remove the problematic Coulomb friction from the robot model, but this reduces the fidelity of the model and the results. The default PUMA 560 model, `models.DH.Puma560`, has nonzero viscous and Coulomb friction parameters for each joint. Coulomb friction can be removed by

```
>>> puma_nf = puma.nofriction();
```

which returns a copy of the robot object that is similar in all respects except that the Coulomb friction is zero. The method has options to remove Coulomb and/or viscous friction, the default is to only remove Coulomb friction.

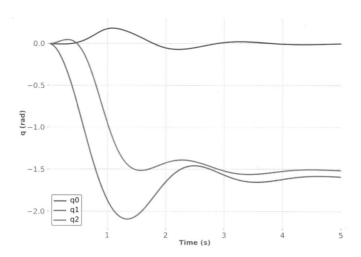

**▣ Fig. 9.19** Joint angle trajectory for PUMA 560 robot, with zero Coulomb friction, collapsing under gravity from an initial zero joint-angle configuration

## 9.4  Rigid-Body Dynamics Compensation

In ▶ Sect. 9.1, we discussed some of the challenges for independent joint control and introduced the concept of feedforward to compensate for the gravity disturbance torque. Inertia variation and other dynamic coupling forces were not explicitly dealt with and were left for the feedback controller to handle as a disturbance. However, inertia and coupling torques can be computed according to (9.11) given knowledge of joint angles, joint velocities and accelerations, and the inertial parameters of the links. We can incorporate these torques into the control law using one of two *model-based* approaches: feedforward control, and computed torque control. The structural differences are contrasted in ▣ Figs. 9.20 and 9.21.

### 9.4.1  Feedforward Control

The torque feedforward controller shown in ▣ Fig. 9.20 is

$$
\boldsymbol{Q}^* = \underbrace{\hat{\mathcal{D}}^{-1}(\boldsymbol{q}^*, \dot{\boldsymbol{q}}^*, \ddot{\boldsymbol{q}}^*)}_{\text{feedforward}} + \underbrace{\{\mathbf{K}_v(\dot{\boldsymbol{q}}^* - \dot{\boldsymbol{q}}) + \mathbf{K}_p(\boldsymbol{q}^* - \boldsymbol{q})\}}_{\text{feedback}}
$$

$$
= \hat{\mathbf{M}}(\boldsymbol{q}^*)\ddot{\boldsymbol{q}}^* + \hat{\mathbf{C}}(\boldsymbol{q}^*, \dot{\boldsymbol{q}}^*)\dot{\boldsymbol{q}}^* + \hat{\boldsymbol{f}}(\dot{\boldsymbol{q}}^*) + \hat{\boldsymbol{g}}(\boldsymbol{q}^*)
$$
$$
+ \{\mathbf{K}_v(\dot{\boldsymbol{q}}^* - \dot{\boldsymbol{q}}) + \mathbf{K}_p(\boldsymbol{q}^* - \boldsymbol{q})\}
\tag{9.14}
$$

The feedforward term is the estimated joint force required to achieve the desired manipulator motion state of position $\boldsymbol{q}^*$, velocity $\dot{\boldsymbol{q}}^*$ and acceleration $\ddot{\boldsymbol{q}}^*$. The feedback term generates joint forces related to tracking error due to errors such as uncertainty in the inertial parameters, unmodeled forces or external disturbances. $\mathbf{K}_p$ and $\mathbf{K}_v$ are respectively the position and velocity gain (or damping) matrices and are typically diagonal. $\hat{\mathbf{M}}$, $\hat{\mathbf{C}}$, $\hat{\boldsymbol{f}}$ and $\hat{\boldsymbol{g}}$ are respectively estimates of the mass matrix, Coriolis coupling matrix, friction and gravity. ◀

This assumes that we have accurate knowledge of the robot's inertial parameters: link mass, link center of mass and link inertia tensor, as well as friction.

In the block diagram, the feedforward torque is computed by the inverse dynamics block and added to the feedback torque computed from position and velocity error. The desired joint angles, velocity and acceleration are generated using quintic polynomials by the joint trajectory block whose parameters are the initial and final joint angles. Since the robot configuration changes relatively slowly, the

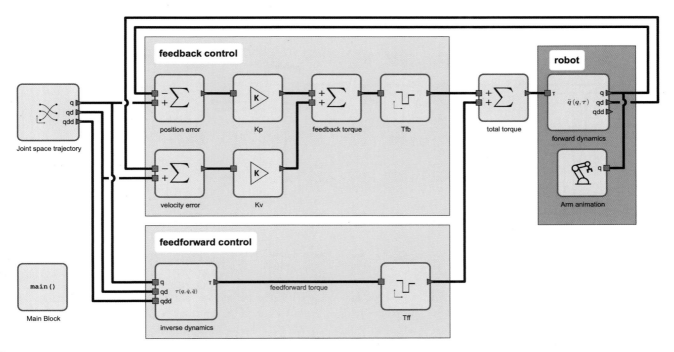

feedforward torque can be evaluated at a greater interval, $T_{\text{ff}}$, than the error feedback loops, $T_{\text{fb}}$.

We run the simulation by

```
>>> %run -m feedforward -H
```

and the time history results are saved in the variable `out`.

The feedforward term linearizes the nonlinear dynamics about the operating point $(\boldsymbol{q}^*, \dot{\boldsymbol{q}}^*, \ddot{\boldsymbol{q}}^*)$. If the linearization is ideal – that is $\hat{\mathbf{M}} = \mathbf{M}$, $\hat{\mathbf{C}} = \mathbf{C}$, $\hat{\boldsymbol{f}} = \boldsymbol{f}$ and $\hat{\boldsymbol{g}} = \boldsymbol{g}$ – then the dynamics of the error $\boldsymbol{e} = \boldsymbol{q}^* - \boldsymbol{q}$ can be obtained by combining (9.11) and (9.14)

$$\mathbf{M}(\boldsymbol{q}^*)\ddot{\boldsymbol{e}} + \mathbf{K}_v\dot{\boldsymbol{e}} + \mathbf{K}_p\boldsymbol{e} = 0 \tag{9.15}$$

For well chosen $\mathbf{K}_p$ and $\mathbf{K}_v$, the error will decay to zero but the joint errors are coupled due to the nondiagonal matrix $\mathbf{M}$, and their dynamics are dependent on the manipulator configuration.

### 9.4.2  Computed-Torque Control

The computed-torque controller is shown in □ Fig. 9.21. It belongs to a class of controllers known as inverse-dynamic control in which a nonlinear system is cascaded with its inverse so that the overall system has unity gain. The computed torque control is

$$\begin{aligned}
\boldsymbol{Q} &= \hat{\mathcal{D}}^{-1}\Big(\boldsymbol{q}^*, \dot{\boldsymbol{q}}^*, \ddot{\boldsymbol{q}}^* + \mathbf{K}_v(\dot{\boldsymbol{q}}^* - \dot{\boldsymbol{q}}) + \mathbf{K}_p(\boldsymbol{q}^* - \boldsymbol{q})\Big) \\
&= \hat{\mathbf{M}}(\boldsymbol{q})\{\ddot{\boldsymbol{q}}^* + \mathbf{K}_v(\dot{\boldsymbol{q}}^* - \dot{\boldsymbol{q}}) + \mathbf{K}_p(\boldsymbol{q}^* - \boldsymbol{q})\} + \hat{\mathbf{C}}(\boldsymbol{q}^*, \dot{\boldsymbol{q}}^*)\dot{\boldsymbol{q}}^* \\
&\quad + \hat{\boldsymbol{f}}(\dot{\boldsymbol{q}}^*) + \hat{\boldsymbol{g}}(\boldsymbol{q}^*)
\end{aligned} \tag{9.16}$$

where $\hat{\mathcal{D}}^{-1}(\cdot)$ is the estimated joint force required to achieve the desired manipulator motion state. In practice, the inverse is not perfect and there will be tracking

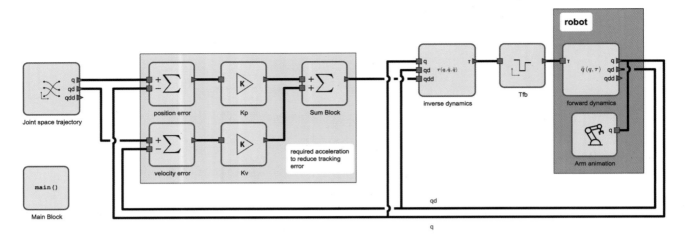

◘ **Fig. 9.21** Block-diagram model `models/computed-torque` for a PUMA 560 robot with computed-torque control

This assumes that we have accurate knowledge of the robot's inertial parameters: link mass, link center of mass and link inertia tensor, as well as friction.

errors in position and velocity. These contribute to a correctional acceleration via the position and velocity gain (or damping) matrices $\mathbf{K}_p$ and $\mathbf{K}_v$ respectively, and these are typically diagonal. $\hat{\mathbf{M}}$, $\hat{\mathbf{C}}$, $\hat{f}$ and $\hat{g}$ are respectively estimates of the mass, Coriolis, friction and gravity. ◄

The desired joint angles and velocity are generated using quintic polynomials by the joint trajectory block whose parameters are the initial and final joint angles. In this case, the inverse dynamics must be evaluated at each servo interval, although the coefficient matrices $\hat{\mathbf{M}}$, $\hat{\mathbf{C}}$, and $\hat{g}$ could be evaluated at a lower rate since the robot configuration changes relatively slowly.

We run the simulation by

```
>>> %run -m computed-torque -H
```

and the time history results are saved in the variable `out`.

If the linearization is ideal – that is $\hat{\mathbf{M}} = \mathbf{M}$, $\hat{\mathbf{C}} = \mathbf{C}$, $\hat{f} = f$ and $\hat{g} = g$ – then the dynamics of the error $e = q^* - q$ are obtained by combining (9.11) and (9.16)

$$\ddot{e} + \mathbf{K}_v \dot{e} + \mathbf{K}_p e = 0 \tag{9.17}$$

Unlike (9.15), the joint errors here are uncoupled and their dynamics are independent of manipulator configuration. In the case of model error, there will be some coupling between axes, and the right-hand side of (9.17) will be a nonzero forcing function.

## 9.5 Task-Space Dynamics and Control

Our focus so far has been on accurate control of the robot's joints which is important for industrial applications where speed and precision in positioning are critical. Today, however, we expect robots to safely perform everyday tasks like opening doors or drawers, cleaning a surface or exchanging objects with people. These are tasks that humans perform with ease but they involve a complex interplay of forces. When we open a door, our hand exerts forces on the door to accelerate it, but the door exerts reaction forces that guide our hand along a circular arc. Exerting forces and responding to reaction forces is critical to many real-world applications.

To consider this class of robot problems, it is useful to take an end-effector-centric view, as shown in ◘ Fig. 9.22. We have abstracted away the robot and consider the end effector as a point mass in 3D space. The environment can exert a wrench on the mass – that could be from a human it is interacting with, or it could

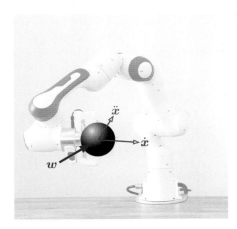

■ **Fig. 9.22** We consider the end effector as a mass depicted by the black sphere, and we are interested in how it moves in response to applied forces and torques. $w$ is the total wrench applied by the environment and the joint actuators, and $\dot{x}$ and $\ddot{x}$ are the resulting task-space velocity and acceleration (robot image courtesy of Franka Emika)

be a reaction wrench from some contact task. The robot's actuators also exert a wrench on the mass. In response to the net wrench, the mass moves with some velocity and acceleration.

To understand these dynamics, we start with the relationship between joint-space and task-space velocity given in (8.2)

$$\boldsymbol{v} = \mathbf{J}(\boldsymbol{q})\dot{\boldsymbol{q}} \in \mathbb{R}^M \tag{9.18}$$

where $\mathbf{J}(\boldsymbol{q}) \in \mathbb{R}^{M \times N}$ is the manipulator Jacobian, $N$ is the number of joints, and $M = \dim \mathcal{T}$ is the dimension of the task space. The temporal derivative is

$$\dot{\boldsymbol{v}} = \mathbf{J}(\boldsymbol{q})\ddot{\boldsymbol{q}} + \dot{\mathbf{J}}(\boldsymbol{q})\dot{\boldsymbol{q}} \tag{9.19}$$

where $\dot{\boldsymbol{v}}$ is spatial acceleration which is a vector comprising translational and angular acceleration, and $\ddot{\boldsymbol{q}}$ is the joint acceleration.

$\dot{\mathbf{J}}(\boldsymbol{q})$ is the time derivative of the Jacobian, or the rate of change of the elements of the Jacobian matrix, which is computed by the method `jacob0_dot`. ▶ We can expand this as

This could be derived directly by symbolic differentiation of the elements of the Jacobian matrix expressed as functions of joint coordinates.

$$\dot{\mathbf{J}}(\boldsymbol{q}) = \frac{\mathrm{d}\mathbf{J}(\boldsymbol{q})}{\mathrm{d}t} = \frac{\partial \mathbf{J}(\boldsymbol{q})}{\partial \boldsymbol{q}}\frac{\mathrm{d}\boldsymbol{q}}{\mathrm{d}t} = \mathbf{H}(\boldsymbol{q})\frac{\mathrm{d}\boldsymbol{q}}{\mathrm{d}t} \in \mathbb{R}^{M \times N}$$

where $\mathbf{H}(\boldsymbol{q}) \in \mathbb{R}^{N \times M \times N}$ is the manipulator Hessian which is a rank-3 tensor that is computed by the `hessian0` method. ▶

The tensor-vector product is a tensor mode-1 product, or contraction, $\mathbf{H}(\boldsymbol{q}) \times_1 \dot{\boldsymbol{q}}$ that is computed by $\sum_{j,k} h_{i,j,k}\dot{q}_i$, $i = 0, \ldots, N-1$ or `np.tensordot(H, qd, (0,0))`.

If $\mathbf{J}(\boldsymbol{q})$ is square we can rearrange (9.19) as

$$\ddot{\boldsymbol{q}} = \mathbf{J}^{-1}(\boldsymbol{q})\left(\dot{\boldsymbol{v}} - \dot{\mathbf{J}}(\boldsymbol{q})\dot{\boldsymbol{q}}\right) . \tag{9.20}$$

Substituting (9.20) and the inverse of (9.18) into (9.11), we can rewrite the rigid-body dynamics in task space as

$$\mathbf{M}_x(\boldsymbol{q})\ddot{\boldsymbol{x}} + \mathbf{C}_x(\boldsymbol{q},\dot{\boldsymbol{q}})\dot{\boldsymbol{x}} + \boldsymbol{f}_x(\dot{\boldsymbol{q}}) + \boldsymbol{g}_x(\boldsymbol{q}) = \boldsymbol{w} \tag{9.21}$$

where $\boldsymbol{x} \in \mathbb{R}^M$ is the task-space pose as a vector and $\boldsymbol{w} \in \mathbb{R}^M$ is the total wrench applied to the end effector. This formulation is also called the operational-space dynamics. The dynamics of the arm are abstracted into the mass depicted by the black sphere in ◻ Fig. 9.22. However, because the mass is connected to the robot arm, it *inherits* the dynamics of the robot. For example, if we let go of the sphere it would fall, but not straight down.

The terms $\mathbf{M}_x$, $\mathbf{C}_x$, $\boldsymbol{f}_x$, $\boldsymbol{g}_x$ are respectively the task-space inertia matrix, Coriolis and centripetal coupling, friction, and gravity load which, for a non-redundant robot, are related to the corresponding terms in (9.11) by

$$
\begin{aligned}
\mathbf{M}_x(\boldsymbol{q}) &= \mathbf{J}_a^{-\top}(\boldsymbol{q})\,\mathbf{M}(\boldsymbol{q})\,\mathbf{J}_a^{-1}(\boldsymbol{q}) \in \mathbb{R}^{M\times M} \\
\mathbf{C}_x(\boldsymbol{q},\dot{\boldsymbol{q}}) &= \Big(\mathbf{J}_a^{-\top}(\boldsymbol{q})\mathbf{C}(\boldsymbol{q},\dot{\boldsymbol{q}}) - \mathbf{M}_x(\boldsymbol{q})\,\dot{\mathbf{J}}_a(\boldsymbol{q})\Big)\mathbf{J}_a^{-1}(\boldsymbol{q}) \in \mathbb{R}^{M\times M} \\
\boldsymbol{f}_x(\dot{\boldsymbol{q}}) &= \mathbf{J}_a^{-\top}(\boldsymbol{q})\,\boldsymbol{f}(\dot{\boldsymbol{q}}) \in \mathbb{R}^{M} \\
\boldsymbol{g}_x(\boldsymbol{q}) &= \mathbf{J}_a^{-\top}(\boldsymbol{q})\,\boldsymbol{g}(\boldsymbol{q}) \in \mathbb{R}^{M}
\end{aligned}
\tag{9.22}
$$

where $\mathbf{J}_a(\boldsymbol{q})$ is the analytical Jacobian introduced in ▸ Sect. 8.1.3. A common vector representation of pose in task space is $\boldsymbol{x} = (\boldsymbol{p}, \boldsymbol{\Gamma}) \in \mathbb{R}^6$ where $\boldsymbol{p} \in \mathbb{R}^3$ is the position of the end effector and $\boldsymbol{\Gamma} \in \mathbb{R}^3$ is its orientation represented by Euler angles, roll-pitch-yaw angles or exponential coordinates. The derivatives are simply $\dot{\boldsymbol{x}} = (\dot{\boldsymbol{p}}, \dot{\boldsymbol{\Gamma}})$ and $\ddot{\boldsymbol{x}} = (\ddot{\boldsymbol{p}}, \ddot{\boldsymbol{\Gamma}})$. Using this representation means that $\dot{\boldsymbol{x}}$ is *not* spatial velocity $\boldsymbol{v}$ as used in ▸ Chaps. 3 and 8, ◂ and therefore we must use the appropriate analytical Jacobian $\mathbf{J}_a \in \mathbb{R}^{M\times N}$.

For a redundant robot where $M \neq N$, we use instead

$$
\begin{aligned}
\mathbf{M}_x(\boldsymbol{q}) &= \big(\mathbf{J}_a(\boldsymbol{q})\,\mathbf{M}^{-1}(\boldsymbol{q})\,\mathbf{J}_a^{\top}(\boldsymbol{q})\big)^{-1} \in \mathbb{R}^{M\times M} \\
\mathbf{C}_x(\boldsymbol{q},\dot{\boldsymbol{q}}) &= \Big(\bar{\mathbf{J}}_a^{\top}(\boldsymbol{q})\mathbf{C}(\boldsymbol{q},\dot{\boldsymbol{q}}) - \mathbf{M}_x(\boldsymbol{q})\dot{\mathbf{J}}_a(\boldsymbol{q})\Big)\bar{\mathbf{J}}_a(\boldsymbol{q}) \in \mathbb{R}^{M\times M} \\
\boldsymbol{f}_x(\dot{\boldsymbol{q}}) &= \bar{\mathbf{J}}_a^{\top}(\boldsymbol{q})\,\boldsymbol{f}(\dot{\boldsymbol{q}}) \in \mathbb{R}^{M} \\
\boldsymbol{g}_x(\boldsymbol{q}) &= \bar{\mathbf{J}}_a^{\top}(\boldsymbol{q})\,\boldsymbol{g}(\boldsymbol{q}) \in \mathbb{R}^{M}
\end{aligned}
\tag{9.23}
$$

where $\bar{\mathbf{J}}_a(\boldsymbol{q}) = \mathbf{M}^{-1}(\boldsymbol{q})\mathbf{J}_a^{\top}(\boldsymbol{q})\mathbf{M}_x(\boldsymbol{q}) \in \mathbb{R}^{N\times M}$ is the dynamically consistent inverse of $\mathbf{J}_a(\boldsymbol{q})$ – a generalized inverse that minimizes the manipulator's instantaneous kinetic energy. These are computed by the methods `inertia_x`, `coriolis_x`, `friction_x`, `gravload_x` which are analogous to their joint-space counterparts `inertia`, `coriolis`, `friction`, `gravload`.

The wrench applied to the mass in ◻ Fig. 9.22 has two components

$$
\boldsymbol{w} = \boldsymbol{w}_e + \boldsymbol{w}_c
\tag{9.24}
$$

where $\boldsymbol{w}_e$ is the wrench applied by the environment, for example, a person pushing on the end effector or a reaction force arising from some contact task. $\boldsymbol{w}_c$ is the control, the wrench that we apply to the end effector by controlling the torques applied at the joints.

We start our discussion of task-space control by applying the simple controller

$$
\boldsymbol{w}_c = \hat{\mathbf{g}}_x(\boldsymbol{q})
\tag{9.25}
$$

where $\hat{\mathbf{g}}_x$ is our best estimate of the operational-space gravity wrench. ◂ Substituting this into (9.24) and (9.21), the dynamics of the end effector become

$$
\mathbf{M}_x(\boldsymbol{q})\ddot{\boldsymbol{x}} + \mathbf{C}_x(\boldsymbol{q},\dot{\boldsymbol{q}})\dot{\boldsymbol{x}} + \boldsymbol{f}_x(\dot{\boldsymbol{q}}) = \boldsymbol{w}_e
$$

which has no gravity term – the end effector is weightless. With zero applied wrench, that is $\boldsymbol{w}_e = 0$, and if the robot is not moving, $\dot{\boldsymbol{q}} = 0$, the end effector will not accelerate. If we pushed on the end effector, that is, we applied a nonzero $\boldsymbol{w}_e$, it would accelerate, and we would feel inertia. We would also feel some effect like damping due to the $\dot{\boldsymbol{x}}$ term.

---

*The rotational component of spatial velocity is angular velocity $\boldsymbol{\omega}$ as introduced in ▸ Sect. 3.1.1, whereas in this case it is $\dot{\boldsymbol{\Gamma}}$.*

*This assumes that we have accurate knowledge of the robot's inertial parameters: link mass and link center of mass. Accurate dynamic models for robots are rare and only a small subset of Toolbox robot models have dynamic parameters. The most accurate dynamic model is* `models.DH.Puma560`.

To implement this controller, the control wrench from (9.25) is mapped to the generalized joint forces using the analytical Jacobian ▸

$$\boldsymbol{Q}_c = \mathbf{J}_a^\top(\boldsymbol{q})\,\boldsymbol{w}_c \tag{9.26}$$

to become the torque or force demand for the joint actuators.

Damping, or viscous friction, is a force proportional to velocity that opposes motion. Superficially, the term $\mathbf{C}_x(\boldsymbol{q}, \dot{\boldsymbol{q}})$ looks like a viscous friction coefficient, but it is a complex non-linear function of joint configuration and the products of joint-space velocities. For the robot at its nominal configuration, and with the end effector moving at a constant $0.1\ \mathrm{ms}^{-1}$ in the $y$-direction

```
>>> xd = [0, 0.1, 0, 0, 0, 0];
>>> qd = np.linalg.inv(puma.jacob0_analytical(puma.qn, "eul")) @ xd;
>>> Cx = puma.coriolis_x(puma.qn, qd, representation="eul");
```

the end effector would experience a wrench of

```
>>> Cx @ xd
array([  0.226, -0.02948,  -0.1431, 0.006606, -0.0001601, 0.003991])
```

which opposes motion in the $y$-directions, as viscous friction does, but it also pushes the end effector in the $x$- and $-z$-directions. ▸ The velocity term is highly non-linear and there is coupling between the task-space axes, and as a consequence this will not *feel* like proper viscous friction.

As we did with gravity, we can effectively remove this velocity coupling, as well as friction, using the control

$$\boldsymbol{w}_c = \hat{\mathbf{C}}_x(\boldsymbol{q}, \dot{\boldsymbol{q}})\dot{\boldsymbol{x}} + \hat{\mathbf{g}}_x(\boldsymbol{q}) + \hat{\boldsymbol{f}}_x(\dot{\boldsymbol{q}})$$

and the dynamics of the end effector becomes simply

$$\mathbf{M}_x(\boldsymbol{q})\ddot{\boldsymbol{x}} = \boldsymbol{w}_e\ .$$

However, in this multi-dimensional case, the inertia matrix is not diagonal

```
>>> Mx = puma.inertia_x(puma.qn, representation="eul")
array([[   17.21,   -2.754,   -9.623,        0,   0.2795,        0],
       [  -2.754,     12.1,    1.246,   -0.968, -0.07034,  -0.3254],
       [  -9.623,    1.246,    13.24,        0,   0.2795,        0],
       [       0,   -0.968,        0,    0.579,        0,   0.1941],
       [  0.2795, -0.07034,   0.2795,        0,   0.1713,        0],
       [       0,  -0.3254,        0,   0.1941,        0,   0.1941]])
```

so we will still experience cross coupling, but this time acceleration cross coupling. For example, a force of $10\,\mathrm{N}$ in the $x$-direction

```
>>> np.linalg.inv(Mx) @ [10, 0, 0, 0, 0, 0]
array([  1.167,   0.1774,   0.9012,   0.2962,   -3.304, 0.001352])
```

will result in acceleration in the $x$- and $z$-directions as well as considerable rotational acceleration about the $y$-axis.

To follow a trajectory $\boldsymbol{x}^*(t)$ we choose the control

$$\boldsymbol{w}_c = \hat{\mathbf{M}}_x(\boldsymbol{q})\Big(\ddot{\boldsymbol{x}}^* + \mathbf{K}_p(\boldsymbol{x}^* - \boldsymbol{x}) + \mathbf{K}_v(\dot{\boldsymbol{x}}^* - \dot{\boldsymbol{x}})\Big) + \hat{\mathbf{C}}_x(\boldsymbol{q}, \dot{\boldsymbol{q}})\dot{\boldsymbol{x}}$$
$$+ \hat{\boldsymbol{f}}_x(\dot{\boldsymbol{q}}) + \hat{\mathbf{g}}_x(\boldsymbol{q}) \tag{9.27}$$

which is driven by the error between desired and actual task space pose and its derivatives. If the estimates of the task-space inertial terms are exact, then the dynamics of the task-space error $\boldsymbol{e} = \boldsymbol{x}^* - \boldsymbol{x}$ are

$$\ddot{\boldsymbol{e}} + \mathbf{K}_v\dot{\boldsymbol{e}} + \mathbf{K}_p\boldsymbol{e} = -\boldsymbol{w}_e$$

The wrench here is different to that described in ▸ Sects. 3.2.2 and 8.4.1. The gravity wrench computed by (9.22) is a function of an analytical Jacobian which in turn depends on the representation chosen for $\boldsymbol{x}$. Therefore to transform the wrench to joint space, we must use the analytical Jacobian rather than the geometric Jacobian of ▸ Sect. 8.4.1.

We chose Euler angles in this example to represent task-space orientation since the orientation is not singular using this representation. It is singular for the case of roll-pitch-yaw angles.

and will converge to zero for well-chosen values of $\mathbf{K}_p > 0$ and $\mathbf{K}_v > 0$ if the external wrench $\boldsymbol{w}_e = 0$. The non-linear rigid-body dynamics have been linearized, and this control approach is called feedback linearization. ◄ This controller maps a displacement $\boldsymbol{x}^* - \boldsymbol{x}$ to a force $\boldsymbol{w}_c$ and therefore acts as a mechanical impedance – this type of robot control is called impedance control.

This is the operational-space equivalent of the computed-torque control from ► Sect. 9.4.2.

We can go one step further and actually specify the desired inertia $\mathbf{M}^*$ of the closed-loop dynamics using the control

$$
\boldsymbol{w}_c = \hat{\mathbf{M}}_x(\boldsymbol{q})\mathbf{M}_x^{*-1}\Big(\mathbf{M}_x^*\ddot{\boldsymbol{x}}^* + \mathbf{K}_p(\boldsymbol{x}^* - \boldsymbol{x}) + \mathbf{K}_v(\dot{\boldsymbol{x}}^* - \dot{\boldsymbol{x}})\Big)
$$
$$
+ \hat{\mathbf{C}}_x(\boldsymbol{q},\dot{\boldsymbol{q}})\dot{\boldsymbol{x}} + \hat{\boldsymbol{f}}_x(\dot{\boldsymbol{q}}) + \hat{\boldsymbol{g}}_x(\boldsymbol{q}) \tag{9.28}
$$

and, assuming exact linearization, the closed-loop error dynamics become

$$
\mathbf{M}_x^*\ddot{\boldsymbol{e}} + \mathbf{K}_v\dot{\boldsymbol{e}} + \mathbf{K}_p\boldsymbol{e} = -\underline{\mathbf{M}_x^*\mathbf{M}_x^{-1}(\boldsymbol{q})}\boldsymbol{w}_e \ .
$$

This technique is sometimes referred to as inertia shaping. We can now specify the dynamics of the end effector in task space in terms of apparent inertia, damping and stiffness. However, the external wrench is coupled by the underlined term which is not guaranteed to be diagonal, and this will lead to undesirable motion in response to the wrench.

To remedy this we need to measure the external wrench and several approaches are commonly used. Joint motor current can be measured and is related to joint torque by (9.1), but in practice, friction and other effects mask the signal we are interested in. More sophisticated robots have torque sensors built into their joint actuators, for example as shown in ◘ Fig. 9.23. Joint torques are related to the end-effector wrench by the Jacobian transpose according to (8.13). An alternative approach is to use a force/torque sensing wrist such as shown in ◘ Fig. 9.24. If we choose a control that also includes the measured external wrench $\boldsymbol{w}_e^{\#}$

$$
\boldsymbol{w}_c = \hat{\mathbf{M}}_x(\boldsymbol{q})\mathbf{M}_x^{*-1}\Big(\mathbf{K}_p(\boldsymbol{x}^* - \boldsymbol{x}) - \mathbf{K}_v(\dot{\boldsymbol{x}}^* - \dot{\boldsymbol{x}})\Big) + \hat{\mathbf{C}}_x(\boldsymbol{q},\dot{\boldsymbol{q}})\dot{\boldsymbol{x}} + \hat{\boldsymbol{f}}_x(\dot{\boldsymbol{q}})
$$
$$
+ \hat{\boldsymbol{g}}_x(\boldsymbol{q}) + (\mathbf{1} - \hat{\mathbf{M}}_x(\boldsymbol{q})\mathbf{M}_x^{*-1})\boldsymbol{w}_e^{\#}
$$
$$
\tag{9.29}
$$

then the closed-loop error dynamics become

$$
\mathbf{M}_x^*\ddot{\boldsymbol{e}} + \mathbf{K}_v\dot{\boldsymbol{e}} + \mathbf{K}_p\boldsymbol{e} = -\boldsymbol{w}_e \ .
$$

The robot now acts like a 6-dimensional spring and if we pushed on the end effector that is, we applied a nonzero $\boldsymbol{w}_e$, we would experience the dynamics of a physical mass-spring-damper system. The translational and torsional damping and stiffness are specified by $\mathbf{K}_v$ and $\mathbf{K}_p$ respectively. ◄

$\mathbf{K}_v$ and $\mathbf{K}_p$ are typically diagonal matrices so that the robot behaves like a physically realizable spring-mass-damper system. The diagonal elements of these matrices have units of $\mathrm{N\,s\,m}^{-1}$ or $\mathrm{N\,s\,m\,rad}^{-1}$, and $\mathrm{N\,m}^{-1}$ or $\mathrm{N\,m\,rad}^{-1}$ respectively.

So far, we have assumed that the robot's joints can be commanded to exert a specified force or torque as computed by (9.26) and the particular control law. While some robots are capable of this, the majority of robots today are position controlled which reflects the heritage of robot arms for performing accurate position-controlled tasks in factories. Torque control of robot arms is non-trivial because it needs to account for friction, as well as complex phenomena such as motor and gearbox torque variation with angle – torque ripple. High-quality joint-torque control requires sophisticated joint actuators, such as shown in ◘ Fig. 9.23, that can measure torque in order to compensate for those nonidealities.

A standard position-controlled robot is very unforgiving when it comes into contact with the environment and interaction forces will rise very quickly, potentially damaging the robot or its environment. To perform compliant tasks with such a robot, we need to measure the environmental wrench $\boldsymbol{w}_e^{\#}$ and use that to

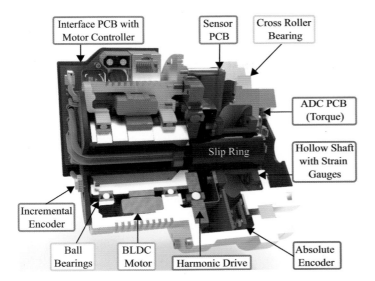

**Fig. 9.23** An integrated sensor-actuator-controller unit that can measure and control joint position and torque (image courtesy Tamim Asfour and Karlsruhe Institute of Technology)

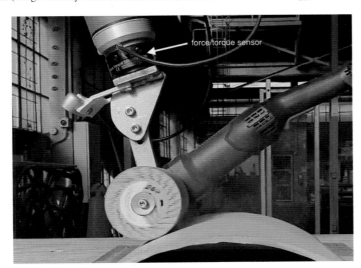

**Fig. 9.24** A robot performing a linishing operation with a wrist force-torque sensor (image courtesy Amelia Luu)

modify the end-effector position. Using a force/torque sensor, such as shown in Fig. 9.24, we can choose a control

$$\Delta x = \mathbf{K}_s^{-1}(w_e^* - w_e^{\#}) - \mathbf{K}_v \dot{x}$$

to determine the change in the robot's position to achieve the desired force. The desired robot stiffness matrix is $\mathbf{K}_s$, and the damping is $\mathbf{K}_v > 0$. This controller maps a wrench $w_e^{\#}$ to a displacement $\Delta x$ and therefore acts as a mechanical admittance – this type of robot control is called admittance or compliance control. In practice, it requires a very high-sample rate and low-latency between the force-torque measurements and the corrective position commands being sent to the joint actuators.

## 9.6 Applications

### 9.6.1 Operational Space Control

Imagine a robot tasked with wiping a table when the table's height is unknown and its surface is only approximately horizontal. The robot's end effector is pointing downward. To achieve the task, we need to move the end effector along a path in the $xy$-plane to achieve coverage and hold the wiper at a constant orientation about the $z$-axis. Simultaneously, we want to maintain a constant force in the $z$-direction to hold the wiper against the table and a constant torque about the $x$- and $y$-axes in order to keep the wiper parallel to the table top. The first group of axes need to be position controlled, while the second group need to be force controlled.

The operational-space controller allows control of position or force for each task-space degree of freedom – defined by the axes of the task coordinate frame. The control wrench is

$$
\begin{aligned}
\boldsymbol{w}_c = {} & \hat{\mathbf{M}}_x(\boldsymbol{q})\Big(\boldsymbol{\Omega}_p\big(\ddot{\boldsymbol{x}} + \mathbf{K}_p(\boldsymbol{x}^* - \boldsymbol{x}) + \mathbf{K}_v(\dot{\boldsymbol{x}}^* - \dot{\boldsymbol{x}})\big) - \boldsymbol{\Omega}_f \mathbf{K}_{vf}\dot{\boldsymbol{x}}\Big) \\
& + \boldsymbol{\Omega}_p\big(\mathbf{K}_f(\boldsymbol{w}^* - \boldsymbol{w}) + \boldsymbol{w}^*\big) \\
& + \hat{\mathbf{C}}_x(\boldsymbol{q}, \dot{\boldsymbol{q}})\dot{\boldsymbol{x}} + \hat{\boldsymbol{g}}_x(\boldsymbol{q})
\end{aligned}
$$

where $\mathbf{K}_i$ are gains. The task specification matrices are

$$
\boldsymbol{\Omega}_p = \begin{pmatrix} \mathbf{R}^\top \boldsymbol{\Sigma}_t \mathbf{R} & \mathbf{0} \\ \mathbf{0} & \mathbf{R}^\top \boldsymbol{\Sigma}_r \mathbf{R} \end{pmatrix}, \ \boldsymbol{\Omega}_f = \begin{pmatrix} \mathbf{R}^\top \bar{\boldsymbol{\Sigma}}_t \mathbf{R} & \mathbf{0} \\ \mathbf{0} & \mathbf{R}^\top \bar{\boldsymbol{\Sigma}}_r \mathbf{R} \end{pmatrix}
$$

where $\bar{\boldsymbol{\Sigma}} = \mathbf{1} - \boldsymbol{\Sigma}$, $\mathbf{R}$ is the orientation of the task frame, and the position and rotation specification matrices are respectively

$$
\boldsymbol{\Sigma}_t = \begin{pmatrix} \sigma_x & 0 & 0 \\ 0 & \sigma_y & 0 \\ 0 & 0 & \sigma_z \end{pmatrix}, \ \boldsymbol{\Sigma}_r = \begin{pmatrix} \rho_x & 0 & 0 \\ 0 & \rho_y & 0 \\ 0 & 0 & \rho_z \end{pmatrix}
$$

where the elements $\sigma_i$ or $\rho_i$ are 1 if that task-space degree of freedom is position or angle controlled, or 0 if it is force or torque controlled. The task specification – the task coordinate frame, and the task specification matrices – can be changed at any time during the motion.

A block-diagram model of the controller for a simplified table cleaning scenario is shown in ◨ Fig. 9.25 and can be run by

```
>>> %run -m opspace -H
```

In this simulation, the task-space coordinate frame is parallel to the end-effector co-ordinate frame. Motion is position controlled in the $x$- and $y$-directions and about the $x$-, $y$- and $z$-axes of this frame – the robot moves from its initial pose to a nearby pose using 5 out of the 6 task-space degrees of freedom.

Motion is force controlled in the $z$-direction with a setpoint of $-5\,\text{N}$ in the world frame. To achieve this, the controller drives the end effector downward until it touches the surface at $z = z_t$ which is modeled as a stiffness of $100\,\text{N}\,\text{m}^{-1}$. Results in ◨ Fig. 9.26 show the $x$- and $y$-positions converging on the goal $(0.8, 0.2)$ while the $z$-position decreases until the end effector makes contact with the table surface. The controller is able to simultaneously satisfy position and force constraints. The end effector is around $0.05\,\text{m}$ below the table surface which is due to the table being modeled with a relatively low stiffness – the robot has pressed into the soft table top.

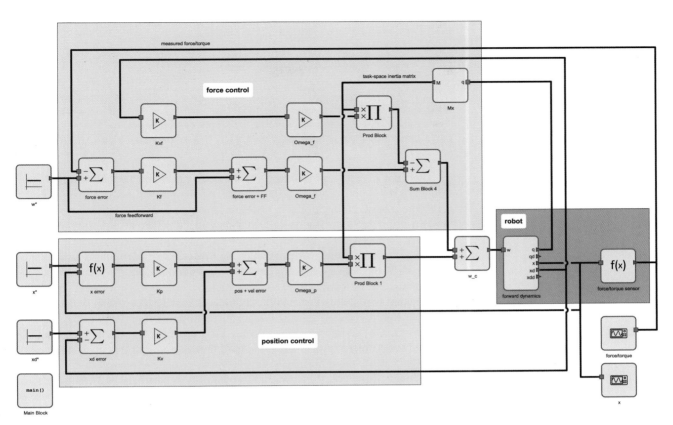

**◼ Fig. 9.25**  Block-diagram model `models/opspace` of an operational-space control system for a PUMA 560 robot as described by (Khatib 1987)

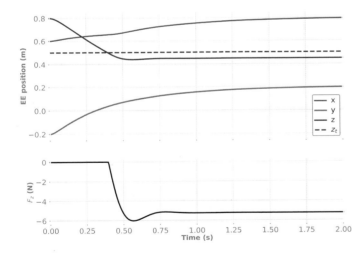

**◼ Fig. 9.26**  Operational space controller results for translational degrees of freedom. The end effector moves to a desired $x$- and $y$-position while also moving in the $-z$-direction until it contacts the work piece at height $z_t$ where it is able to exert the specified force of $-5$ N

## 9.6.2  Series-Elastic Actuator (SEA)

For high-speed robots, the elasticity of the links and the joints becomes a significant dynamic effect which will affect the path-following accuracy. Joint elasticity is typically caused by elements of the transmission such as: a harmonic gearbox which is inherently elastic, torsional elasticity of a motor shaft or spiral couplings, or longitudinal elasticity of a toothed belt or cable drive.

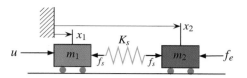

■ **Fig. 9.27** Schematic of a series-elastic actuator. The two masses represent the motor and the load, and they are connected by an elastic element or spring. We use a control force applied to $m_1$ to indirectly control the position of $m_2$. $f_s$ is the spring force, and $f_e$ is the environmental force

Or the robot is not very accurate.

There are advantages in having some flexibility between the motor and the load. Imagine a robot performing a task that involves the gripper picking an object off a table whose height is uncertain. ◄ A simple strategy to achieve this is to move down until the gripper touches the table, close the gripper and then lift up. However, at the instant of contact a large and discontinuous force will be exerted on the robot which has the potential to damage the object or the robot. This is particularly problematic for robots with large inertia that are moving quickly – the kinetic energy must be instantaneously dissipated. An elastic element – a spring – between the motor and the joint would help here. At the moment of contact, the spring would start to compress and the kinetic energy is transferred to potential energy in the spring – the robot control system has time to react and stop or reverse the motors. We have changed the problem from a damaging hard impact to a soft impact. In addition to shock absorption, the deformation of the spring provides a means of determining the force that the robot is exerting. This capability is particularly useful for robots that interact closely with people since it makes the robot less dangerous in case of collision, and a spring is simple technology that cannot fail. For robots that must exert a force as part of their task, this is a simpler approach than the operational-space controller introduced in ► Sect. 9.6.1. Unfortunately, position control is now more challenging because there is an elastic element between the motor and the load.

In a real robot, this is most commonly a rotary system with a torsional spring.

Consider the 1-dimensional case shown in ■ Fig. 9.27 where the motor is represented by a mass $m_1$ to which a controllable force $u$ is applied. ◄ It is connected via a linear elastic element or spring to the load mass $m_2$. If we apply a positive force to $m_1$, it will move to the right and compress the spring, and this will exert a positive force on $m_2$ which will also move to the right. Controlling the position of $m_2$ is not trivial since this system has no friction and is marginally stable. It can be stabilized by feedback of position and velocity of the motor *and* of the load – all of which are potentially measurable.

In robotics, such a system built into a robot joint, is known as a series-elastic actuator or SEA. A block-diagram model of an SEA system is shown in ■ Fig. 9.28 which attempts to drive the load $m_2$ to a position $x_2^* = 1$. The model computes the control force using a state-feedback controller based on motor and load position and velocity. We can simulate this system by

```
>>> %run -m SEA -H
```

and results are shown in ■ Fig. 9.29. In the first case, there is no obstacle and the controller achieves its goal with minimal overshoot, but note the complex force profile applied to $m_1$. In the second case, the load mass is stopped at $x_2 = 0.8$ and the elastic force changes to accommodate this.

Knowing $x_1$ and $x_2$ and the spring stiffness $K_s$ allows us to compute the environmental force $f_e$ applied to $m_2$. The actuator is also a sensor, and some of these principles are employed by the torque-controlled actuator shown in ■ Fig. 9.23.

A more complex problem to analyze is elasticity of the links. Most robots have very stiff links to avoid this problem, but for robots operating with extreme acceleration it may need to be taken into account.

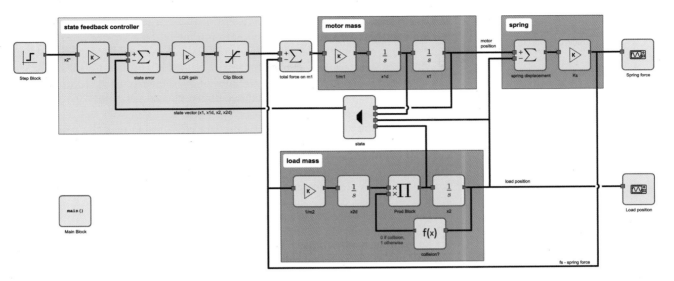

**Fig. 9.28** Block-diagram model `models/SEA` of a series-elastic actuator colliding with an obstacle

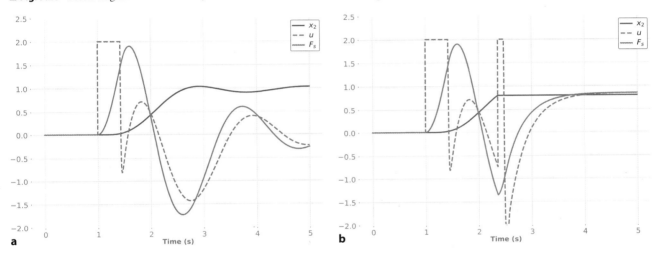

**Fig. 9.29** Response of the series-elastic actuator to a unit-step demand at $t = 1$ s, showing load position (m), motor force (N) and spring force (N). **a** Moving to $x_2^* = 1$ with no collision; **b** moving to $x_2^* = 1$ with an obstacle at $x_2 = 0.8$ which is reached at $t \approx 2.3$

## 9.7  Wrapping Up

This chapter has introduced several different approaches to robot manipulator control. We started with the simplest case of independent joint control, explored the effect of disturbance torques and variation in inertia, and showed how feedforward of disturbances such as gravity could provide significant improvement in performance. We then learned how to model the forces and torques acting on the individual links of a serial-link manipulator. The equations of motion, or inverse dynamics, allow us to determine the joint forces required to achieve particular joint velocity and acceleration. The equations have terms corresponding to inertia, gravity, velocity coupling, friction and externally applied forces. We looked at the significance of these terms and how they vary with manipulator configuration and payload. The equations of motion provide insight into important issues such as how the velocity or acceleration of one joint exerts a disturbance force on other joints which is important for control design. We then discussed the forward dynamics which describe how the configuration evolves with time in response to forces and torques applied at the joints by the actuators and by external forces such as gravity. We extended the feedforward notion to full model-based control using torque feedforward and computed torque controllers. We then reformulated the rigid-body dynamics in task space which allowed us to design a pose controller where we can also specify the dynamic characteristics of the end effector as a spring-mass-damper system. Finally, we discussed two applications: operational-space control where we can control position or force of different task-space degrees of freedom, and series-elastic actuators where a compliant element between the robot motor and the link enables force control and people-safe operation.

### 9.7.1 Further Reading

The engineering design of motor control systems is covered in mechatronics textbooks such as Bolton (2015). The dynamics of serial-link manipulators is well covered by all the standard robotics textbooks such as Paul (1981), Spong et al. (2006), Siciliano et al. (2009) and the Robotics Handbook (Siciliano and Khatib 2016). The efficient recursive Newton-Euler method we use today is the culmination of much research in the early 1980s and described in Hollerbach (1982). The equations of motion can be derived via a number of techniques, including Lagrangian (energy based), Newton-Euler, d'Alembert (Fu et al. 1987, Lee et al. 1983) or Kane's method (Kane and Levinson 1983). The computational cost of Lagrangian methods (Uicker 1965; Kahn 1969) is enormous, $O(N^4)$, which made it infeasible for real-time use on computers of that era and many simplifications and approximation had to be made. Orin et al. (1979) proposed an alternative approach based on the Newton-Euler (NE) equations of rigid-body motion applied to each link. Armstrong (1979) then showed how recursion could be applied resulting in $O(N)$ complexity. Luh et al. (1980) provided a recursive formulation of the Newton-Euler equations with linear and angular velocities referred to link coordinate frames which resulted in a thousand-fold improvement in execution time making it practical to implement in real-time. Hollerbach (1980) showed how recursion could be applied to the Lagrangian form, and reduced the computation to within a factor of 3 of the recursive NE form, and Silver (1982) showed the equivalence of the recursive Lagrangian and Newton-Euler forms, and that the difference in efficiency was due to the representation of angular velocity.

The recursive Newton-Euler algorithm is typically computed using vector representation of translational and angular velocities and accelerations, but this is cumbersome. Featherstone's spatial vector notation (Featherstone 1987, 2010a, 2010b) is elegant and compact, and is supported by the Spatial Maths Toolbox classes `SpatialVelocity`, `SpatialAcceleration`, `SpatialForce`, `SpatialMomentum` and `SpatialInertia`.

The forward dynamics, ▶ Sect. 9.3, is computationally more expensive. An $O(N^3)$ method was proposed by Walker and Orin (1982) and is used in the Toolbox. Featherstone's (1987) articulated-body method has $O(N)$ complexity but for $N < 9$ is more expensive than Walker and Orin's method.

Critical to any consideration of robot dynamics is knowledge of the inertial parameters, ten per link, as well as the motor's parameters. Corke and Armstrong-Hélouvry (1994, 1995) published a meta-study of PUMA 560 parameters and provided a consensus estimate of inertial and motor parameters for the PUMA 560 robot. Some of this data was obtained by painstaking disassembly of the robot and determining the mass and dimensions of the components. Inertia of components can be estimated from mass and dimensions by assuming mass distribution, or it can be measured using a bifilar pendulum as discussed in Armstrong et al. (1986).

Alternatively, the parameters can be estimated by measuring the joint torques or the base reaction force and moment as the robot moves. A number of early works in this area include Mayeda et al. (1990), Izaguirre and Paul (1985), Khalil and Dombre (2002) and a more recent summary is Siciliano and Khatib (2016, § 6). Key to successful identification is that the robot moves in a way that is sufficiently exciting (Gautier and Khalil 1992; Armstrong 1989). Friction is an important dynamic characteristic and is well described in Armstrong's (1988) thesis. The survey by Armstrong-Hélouvry et al. (1994) is a very readable and thorough treatment of friction modeling and control. Motor parameters can be obtained directly from the manufacturer's data sheet or determined experimentally, without having to remove the motor from the robot, as described by Corke (1996a). The parameters used in the Toolbox PUMA 560 model are the best estimates from Corke and Armstrong-Hélouvry (1995) and Corke (1996a).

The discussion on control has been quite brief and has strongly emphasized the advantages of feedforward control. Robot joint control techniques are well covered by Spong et al. (2006), Craig (2005) and Siciliano et al. (2009) and summarized in Siciliano and Khatib (2016, § 8). Siciliano et al. (2009) have a good discussion of actuators and sensors as does the, now quite old, book by Klafter et al. (1989). The control of flexible joint robots is discussed in Spong et al. (2006). Adaptive control can be used to accommodate the time-varying inertial parameters and there is a large literature on this topic but some good early references include the book by Craig (1987) and key papers include Craig et al. (1987), Spong (1989), Middleton and Goodwin (1988) and Ortega and Spong (1989). The operational-space control structure was proposed in Khatib (1987) and the example in ▶ Sect. 9.6.1 is based on that paper with some minor adjustment to the notation used. There has been considerable recent interest in series-elastic as well as variable stiffness actuators (VSA) whose position and stiffness can be independently controlled much like our own muscles – a good collection of articles on this technology can be found in the special issue by Vanderborght et al. (2008).

Dynamic manipulability is discussed in Spong et al. (2006) and Siciliano et al. (2009). The Asada measure used in the Toolbox is described in Asada (1983).

■ ■ **Historical and General**

Newton's second law is described in his 1687 master work *Principia Naturalis* (mathematical principles of natural philosophy), written in Latin but an English translation is available online at ▶ https://www.archive.org/details/newtonspmathema00newtrich. His writing on other subjects, including transcripts of his notebooks, can be found online at ▶ http://www.newtonproject.sussex.ac.uk.

## 9.7.2 Exercises

1. Independent joint control (▶ Sect. 9.1.8)
   a) Investigate different values of `Kv` and `Ki` as well as demand signal shape and amplitude.
   b) Perform a root-locus analysis of `vloop` to determine the maximum permissible gain for the proportional case. Repeat this for the PI case.
   c) Consider that the motor is controlled by a voltage source instead of a current source, and that the motor's impedance is 1 mH and 1.6 $\Omega$. Modify `vloop` accordingly. Extend the model to include the effect of back EMF.
   d) Increase the required speed of motion so that the motor torque becomes saturated. With integral action, you will observe a phenomenon known as integral windup – examine what happens to the state of the integrator during the motion. Various strategies are employed to combat this, such as limiting the maximum value of the integrator, or only allowing integral action when the motor is close to its setpoint. Experiment with some of these.

e) Create a model of the PUMA robot with each joint controlled by `ploop`. Parameters for the different motors in the PUMA are described in Corke and Armstrong-Hélouvry (1995).

2. The motor torque constant has units of $N\,m\,A^{-1}$ and is equal to the back EMF constant which has units of $V\,s\,rad^{-1}$. Show that these units are equivalent.

3. Simple two-link robot arm of ◘ Fig. 9.4
   a) Plot the gravity load as a function of both joint angles. Assume $m_1 = 0.45\,kg$, $m_2 = 0.35\,kg$, $r_1 = 8\,cm$ and $r_2 = 8\,cm$.
   b) Plot the inertia for joint 1 as a function of $q_2$. To compute link inertia, assume that we can model the link as a point mass located at the center of mass.

4. Run the code that generates ◘ Fig. 9.15 to compute gravity loading on joints 2 and 3 as a function of configuration. Add a payload and repeat.

5. Run the code that generates ◘ Fig. 9.16 to show how the inertia of joints 1 and 2 vary with payload.

6. Generate the curve of ◘ Fig. 9.16c. Add a payload and compare the results. By what factor does this inertia vary over the joint angle range?

7. Why is the manipulator inertia matrix symmetric?

8. The robot exerts a wrench on the base as it moves (▶ Sect. 9.2.6). Consider that the robot is sitting on a frictionless horizontal table (say on a large air puck). Create a simulation model that includes the robot arm dynamics and the sliding dynamics on the table. Show that moving the arm causes the robot to translate and spin. Can you devise an arm motion that moves the robot base from one position to another and stops?

9. Overlay the dynamic manipulability ellipsoid on the display of the robot. Compare this with the force ellipsoid from ▶ Sect. 8.4.2.

10. Model-based control (▶ Sect. 9.4)
    a) Compute and display the joint tracking error for the torque feedforward and computed torque cases. Experiment with different motions, control parameters and sample rate $T_{fb}$.
    b) Reduce the rate at which the feedforward torque is computed and observe its effect on tracking error.
    c) In practice, the dynamic model of the robot is not exactly known, we can only invert our best estimate of the rigid-body dynamics. In simulation, we can model this by using the `perturb` method, see the online documentation, which returns a robot object with inertial parameters varied by plus and minus the specified percentage. Modify the block diagram models so that the inverse dynamics block is using a robot model with parameters perturbed by 10%. This means that the inverse dynamics are computed for a slightly different dynamic model to the robot under control and shows the effect of model error on control performance. Investigate the effects on error for both the torque feedforward and computed torque cases.
    d) Expand the operational-space control example to include a sensor that measures all the forces and torques exerted by the robot on an inclined table surface. Move the robot end effector along a circular path in the $xy$-plane while exerting a constant downward force – the end effector should move up and down as it traces out the circle. Show how the controller allows the robot tool to conform to a surface with unknown height and surface orientation.

11. Operational-space control (▶ Sect. 9.6.1)
    a) Increase the stiffness of the table surface and see the effect of steady-state end effector height.
    b) Investigate the effect of changing the various gain terms.
    c) Add a more sophisticated trajectory for the $xy$-plane motion that performs a coverage pattern.

d) Modify the force-torque sensor function to model a table that slopes in the world $x$-direction.

e) As above but the table slopes in the world $x$- and $y$-direction.

f) Experiment with a different task-space pose representation, for example, roll-pitch-yaw angles or exponential coordinates.

g) Add compliance about the $x$- and $y$-axes to allow a flat table cleaning pad to conform to the table's normal vector.

h) Use a redundant robot such as the Panda and repeat the questions above.

12. Series-elastic actuator (▶ Sect. 9.6.2)

a) Experiment with different values of stiffness for the elastic element and control parameters. Try to reduce the settling time.

b) Modify the simulation so that the robot arm moves to touch an object at unknown distance and applies a force of 5 N to it.

c) Plot the frequency response function $X_2(s)/X_1(s)$ for different values of $K_s$, $m_1$ and $m_2$.

d) Simulate the effect of a collision between the load and an obstacle by adding a step to the spring force.

# Light and Color

*In nature, light creates the color.*
*In the picture, color creates the light.*
– Hans Hoffman

## Contents

© The Author(s), under exclusive license to Springer Nature Switzerland AG 2023
P. Corke, *Robotics, Vision and Control*, Springer Tracts in Advanced Robotics 146,
https://doi.org/10.1007/978-3-031-06469-2_10

**10**

In ancient times it was believed that the eye radiated a cone of visual flux which mixed with visible objects in the world to create a sensation in the observer – like the sense of touch, but at a distance – which we call the extromission theory of vision. Today, the intromission theory considers that light from an illuminant falls on the scene, some of which is reflected into the eye of the observer to create a perception about that scene. The light that reaches the eye, or the camera, is a function of the illumination impinging on the scene and a material property of the scene known as reflectivity.

This chapter is about light itself and our perception of light in terms of brightness and color. ▶ Sect. 10.1 describes light in terms of electromagnetic radiation, and mixtures of light as continuous spectra. ▶ Sect. 10.2 provides a brief introduction to colorimetry, the science of color perception, human trichromatic color perception and how colors can be represented in various color spaces. ▶ Sect. 10.3 covers a number of advanced topics such as color constancy, gamma correction and white balancing. ▶ Sect. 10.4 has two application examples concerned with distinguishing different colored objects in an image, and the removal of shadows in an image.

## 10.1 Spectral Representation of Light

Around 1670, Sir Isaac Newton discovered that white light was a mixture of different colors. We now know that each of these colors is a single frequency or wavelength of electromagnetic radiation. We perceive the wavelengths between 400 and 700 nm as different colors as shown in ▢ Fig. 10.1.

In general, the light that we observe is a mixture of many wavelengths and can be represented as a function $E(\lambda)$ that describes intensity as a function of wavelength $\lambda$. Monochromatic light, such as emitted by an LED or laser, comprises a single wavelength, in which case $E$ is an impulse or Dirac function.

The most common source of light is incandescence, the emission of light from a hot body such as the Sun or the filament of an old-fashioned incandescent light

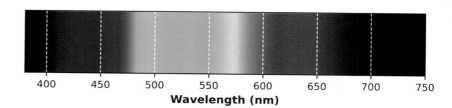

▢ **Fig. 10.1**    The spectrum of visible colors as a function of wavelength in nanometers. The visible range depends on viewing conditions and the individual but is generally accepted as being 400–700 nm. Wavelengths greater than 700 nm are termed infrared and those below 400 nm are ultraviolet

---

**Excurse 10.1: Spectrum of Light**
During the plague years of 1665–1666 Isaac Newton developed his theory of light and color. He demonstrated that a prism could decompose white light into a spectrum of colors, and that a lens and a second prism could recompose the multi-colored spectrum into white light. Importantly, he showed that the color of the light did not change when it was reflected from different objects, from which he concluded that color is an intrinsic property of light, not the object. (The sketch is from Newton's notebook)

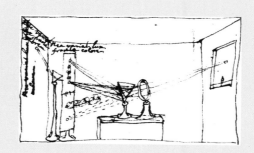

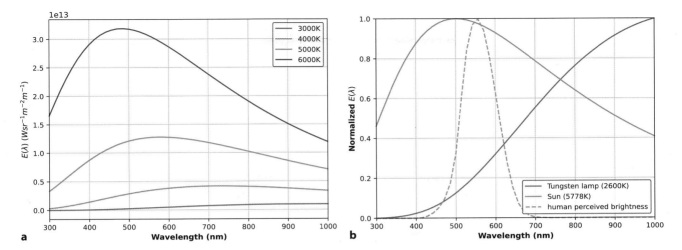

🔲 **Fig. 10.2** Blackbody spectra. **a** Blackbody emission spectra for temperatures from 3000–6000 K. **b** Blackbody emissions for the Sun (5778 K), a tungsten lamp (2600 K) and perceived brightness to the human eye – all normalized to unity for readability

bulb. In physics, this is modeled as a blackbody radiator or Planckian source. The emitted power as a function of wavelength $\lambda$ is given by Planck's radiation formula

$$E(\lambda) = \frac{2hc^2}{\lambda^5 \left(e^{hc/k\lambda T} - 1\right)} \; W\,sr^{-1}\,m^{-2}\,m^{-1} \qquad (10.1)$$

where $T$ is the absolute temperature (K) of the source, h is Planck's constant, k is Boltzmann's constant, and c the speed of light. ▶ This is the power emitted per steradian ▶ per unit area per unit wavelength.

We can plot the emission spectra for a blackbody at different temperatures. First we define a range of wavelengths ▶

```
>>> nm = 1e-9;
>>> lmbda = np.linspace(300, 1_000, 100) * nm;
```

in this case from 300 to 1000 nm, and then compute the blackbody spectra

```
>>> for T in np.arange(3_000, 6_001, 1_000):
...     plt.plot(lmbda / nm, blackbody(lmbda, T))
```

as shown in 🔲 Fig. 10.2a. We see that as temperature increases the maximum amount of power increases, while the wavelength at which the peak occurs decreases. The total amount of power radiated (per unit area) is the area under the blackbody curve and is given by the Stefan-Boltzman law

$$P(\lambda) = \frac{2\pi^5 k^4}{15c^2 h^3} T^4 \; W\,m^{-2}$$

and the wavelength corresponding to the peak of the blackbody curve is given by Wien's displacement law

$$\lambda_{\max} = \frac{2.8978 \times 10^{-3}}{T} \; m \; .$$

The inverse relationship between peak wavelength and temperature is familiar to us in terms of what we observe when heating an object. As shown in 🔲 Table 10.1, an object starts to glow faintly red at around 800 K, and moves through orange and yellow toward white as temperature increases.

$c = 2.998 \times 10^8 \; ms^{-1}$

$h = 6.626 \times 10^{-34} \; Js$

$k = 1.381 \times 10^{-23} \; JK^{-1}$

Solid angle is measured in steradians (sr). A full sphere is $4\pi$ sr.

We use the variable name `lmbda` since `lambda` is a Python keyword.

**10**

### Excurse 10.2: Infrared Radiation

Infrared radiation was discovered in 1800 by William Herschel (1738–1822), the German-born British astronomer. He was Court Astronomer to George III; built a series of large telescopes; with his sister Caroline performed the first sky survey discovering double stars, nebulae and the planet Uranus; and studied the spectra of stars. Using a prism and thermometers to measure the amount of heat in the various colors of sunlight he observed that temperature increased from blue to red, and increased even more beyond red where there was no visible light (Image from Herschel 1800)

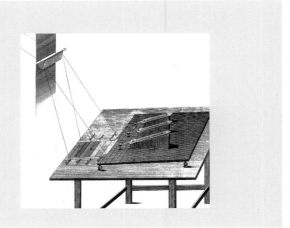

### Excurse 10.3: The Incandescent Lamp

Sir Humphry Davy demonstrated the first electrical incandescent lamp using a platinum filament in 1802. Sir Joseph Swan demonstrated his first light bulbs in 1850 using carbonized paper filaments. However, it was not until advances in vacuum pumps in 1865 that such lamps could achieve a useful lifetime. Swan patented a carbonized cotton filament in 1878 and a carbonized cellulose filament in 1881. His lamps came into use after 1880 and the Savoy Theatre in London was completely lit by electricity in 1881. In the USA Thomas Edison did not start research into incandescent lamps until 1878, but he patented a long-lasting carbonized bamboo filament the next year and was able to mass produce them. The Swan and Edison companies merged in 1883.

The light bulb subsequently became the dominant source of light on the planet, but is now being phased out due to its poor energy efficiency. (image by Douglas Brackett, Inv., Edisonian.com)

| ▣ **Table 10.1**  Color of a heated object versus temperature | |
|---|---|
| Incipient red heat | 770–820 K |
| Dark red heat | 920–1020 K |
| Bright red heat | 1120–1220 K |
| Yellowish red heat | 1320–1420 K |
| Incipient white heat | 1520–1620 K |
| White heat | 1720–1820 K |

The filament of a tungsten lamp has a temperature of 2600 K and glows *white hot*. The Sun has a surface temperature of 5778 K. The spectra of these sources

```
>>> lamp = blackbody(lmbda, 2_600);
>>> sun = blackbody(lmbda, 5_778);
>>> plt.plot(lmbda / nm, np.c_[lamp / np.max(lamp),
...                           sun / np.max(sun)]);
```

are compared in ◙ Fig. 10.2b. The tungsten lamp curve is much lower in magnitude, but has been scaled up (by 56) for readability. The peak of the Sun's emission is around 500 nm and it emits a significant amount of power in the visible part of the spectrum. The peak for the tungsten lamp is around 1100 nm, well outside the visible band, and perversely most of its power falls in the infrared band that we perceive as heat, not light.

### 10.1.1 Absorption

The Sun's spectrum at ground level on the Earth has been measured and tabulated

```
>>> sun_ground = loadspectrum(lmbda, "solar");
>>> plt.plot(lmbda / nm, sun_ground);
```

and is shown in ◙ Fig. 10.3a. It differs markedly from that of a blackbody since some wavelengths have been absorbed more than others by the atmosphere. Our eye's peak sensitivity has evolved to be closely aligned to the peak of the spectrum of atmospherically filtered sunlight.

Transmittance $T$ is the inverse of absorptance, and is the fraction of light passed as a function of wavelength and distance traveled. It is described by Beer's law

$$T = 10^{-Ad} \tag{10.2}$$

where $A$ is the absorption coefficient in units of $m^{-1}$, that is a function of wavelength, and $d$ is the optical path length. The absorption spectrum $A(\lambda)$ for water is loaded from tabulated data

```
>>> lmbda = np.linspace(300, 1_000, 100) * nm;
>>> A = loadspectrum(lmbda, "water");
```

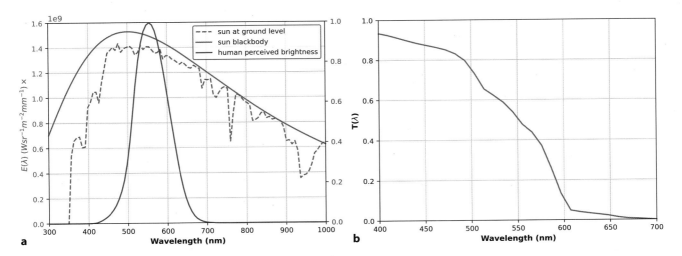

◙ **Fig. 10.3** **a** Modified solar spectrum at ground level (*blue*). The dips in the solar spectrum correspond to various water absorption bands. $CO_2$ absorbs radiation in the infrared region, and ozone $O_3$ absorbs strongly in the ultraviolet region. **b** Transmission through 5 m of water. The longer wavelengths, reds, have been strongly attenuated

and the transmission through 5 m of water is

```
>>> d = 5;
>>> T = 10 ** (-A * d);
>>> plt.plot(lmbda / nm, T);
```

which is plotted in ◨ Fig. 10.3b. We see that the red light is strongly attenuated, which makes the scene appear more blue. Differential absorption of wavelengths is a significant concern when imaging underwater, and we revisit this topic in ▶ Sect. 10.3.4.

### 10.1.2 Reflectance

Surfaces reflect incoming light. The reflection might be specular (as from a mirror-like surface, see ▶ Exc. 13.10), or Lambertian (diffuse reflection from a matte surface, see ▶ Exc. 10.15). The fraction of light that is reflected $R(\lambda) \in [0, 1]$ is the reflectivity, reflectance or albedo of the surface, and is a function of wavelength. White paper for example has a reflectance of around 70 %. The reflectance spectra of many materials have been measured and tabulated. ◀ Consider, for example, the reflectivity of a red house brick over a wide spectral range

From ▶ https://speclib.jpl.nasa.gov, weathered red brick (0412UUUBRK).

```
>>> lmbda = np.linspace(100, 10_000, 100) * nm;
>>> R = loadspectrum(lmbda, "redbrick");
>>> plt.plot(lmbda / (1_000 * nm), R);
```

which is plotted in ◨ Fig. 10.4a. In the visible part of the spectrum, ◨ Fig. 10.4b, it reflects red light more than blue.

### 10.1.3 Luminance

The light reflected from a surface, its luminance, has a spectrum given by

$$L(\lambda) = E(\lambda)R(\lambda) \, \mathrm{W\,m^{-2}} \tag{10.3}$$

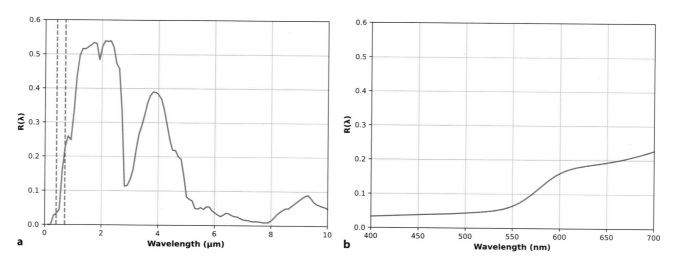

**a**          **b**

◨ **Fig. 10.4** Reflectance of a weathered red house brick (data from ASTER, Baldridge et al. 2009). **a** Full range measured from 300 nm visible to 10,000 nm (infrared) with the visible band delimited by dashed lines; **b** close-up view of the visible region

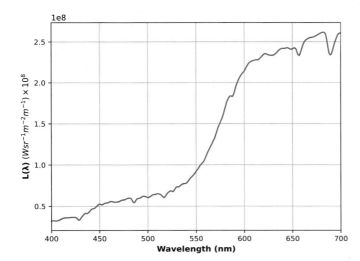

**Fig. 10.5** Luminance of the weathered red house brick under illumination from the Sun at ground level, based on data from Figs. 10.3a and 10.4b

where $E$ is the incident illumination and $R$ is the reflectance. The illuminance of the Sun in the visible region is

```
>>> lmbda = np.arange(400, 701) * nm;
>>> E = loadspectrum(lmbda, "solar");
```

at ground level. The reflectivity of the brick is

```
>>> R = loadspectrum(lmbda, "redbrick");
```

and the light reflected from the brick is

```
>>> L = E * R;
>>> plt.plot(lmbda / nm, L);
```

which is shown in Fig. 10.5. It is this spectrum that is interpreted by our eyes as the color red.

## 10.2 Color

» *Color is the general name for all sensations arising from the activity of the retina of the eye and its attached nervous mechanisms, this activity being, in nearly every case in the normal individual, a specific response to radiant energy of certain wavelengths and intensities.*
– T.L. Troland, Report of Optical Society of America
Committee on Colorimetry 1920–1921

We have described the spectra of light in terms of power as a function of wavelength, but our own perception of light is in terms of subjective quantities such as brightness and color.

### 10.2.1 The Human Eye

Our eyes contain two types of light-sensitive photoreceptors as shown in Fig. 10.6: rod cells and cone cells. Rod cells are much more sensitive than cone cells, respond to intensity only, and are used at night. In normal daylight conditions our cone cells are active and these are color sensitive.

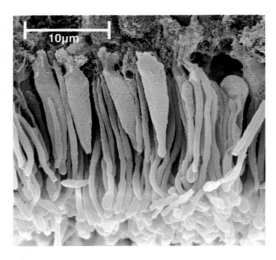

■ **Fig. 10.6** A colored scanning electron micrograph of rod cells (white) and cone cells (yellow) in the human eye. The cells diameters are in the range 0.5–4 μm. The cells contain different types of light-sensitive opsin proteins. Surprisingly, the rods and cones are not on the surface of the retina, they are behind that surface which is a network of nerves and blood vessels

**10**

Humans, like most primates, are trichromats and have three types of cones that respond to different parts of the spectrum. They are referred to as long (L), medium (M) and short (S) cones according to the wavelength of their peak response, or more commonly as red, green and blue cones. Most other mammals are dichromats, they have just two types of cone cells. Many birds and fish are tetrachromats and have a type of cone cell that is sensitive to ultra violet light. Dragonflies have up to thirty different types of cone cells.

Human rod and cone cells have been extensively studied and the spectral response of cone cells can be loaded

```
>>> cones = loadspectrum(lmbda, "cones");
>>> plt.plot(lmbda / nm, cones);
```

The different spectral characteristics are due to the different photopsins in the cone cell.

where `cones` has three columns corresponding to the L, M and S cone responses and each row corresponds to the wavelength in `lmbda`. The spectral responses of the cones $L(\lambda)$, $M(\lambda)$ and $S(\lambda)$ are shown in ■ Fig. 10.7a. ◄

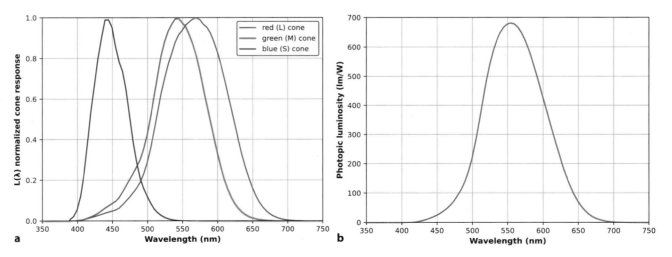

■ **Fig. 10.7** **a** Spectral response of human cones (normalized). **b** Luminosity curve for the standard human observer. The peak response is 683 lmW$^{-1}$ at 555 nm (*green*)

### Excurse 10.4: Opsins

Opsins are the photoreceptor molecules used in the visual systems of all animals. They belong to the class of G protein-coupled receptors (GPCRs) and comprise seven helices that pass through the cell's membrane. They change shape in response to particular molecules outside the cell and initiate a cascade of chemical signaling events inside the cell that results in a change in cell function. Opsins contain a chromophore (indicated by the arrow), a light-sensitive molecule called retinal derived from vitamin A, that stretches across the opsin. When retinal absorbs a photon its shape changes, deforming the opsin and activating the cell's signalling pathway. The basis of all vision is a fortuitous genetic mutation 700 million years ago that made a chemical sensing receptor light sensitive. There are many opsin variants across the animal kingdom – our rod cells contain rhodopsin and our cone cells contain photopsins. The American biochemist George Wald (1906–1997) received the 1967 Nobel Prize in Medicine for his discovery of retinal and characterizing the spectral absorbance of photopsins. (image by Dpryan from Wikipedia)

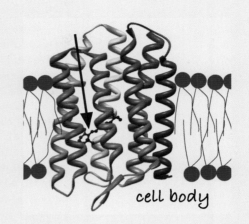

cell body

### Excurse 10.5: Trichromatic Theory

The trichromatic theory of color vision suggests that our eyes have three discrete types of receptors that when stimulated produce the sensations of red, green and blue, and that all color sensations are "psychological mixes" of these fundamental colors. It was first proposed by the English scientist Thomas Young (1773–1829) in 1802 but made little impact. It was later championed by Hermann von Helmholtz and James Clerk Maxwell. The figure shows how beams of red, green and blue light mix. Helmholtz (1821–1894) was a prolific German physician and physicist. He invented the opthalmascope for examing the retina in 1851, and in 1856 he published the *Handbuch der physiologischen Optik* (Handbook of Physiological Optics) which contained theories and experimental data relating to depth perception, color vision, and motion perception. Maxwell (1831–1879) was a Scottish scientist best known for his electromagnetic equations, but who also extensively studied color perception, color-blindness, and color theory. His 1860 paper *On the Theory of Colour Vision* won a Rumford medal, and in 1861 he demonstrated color photography in a Royal Institution lecture.

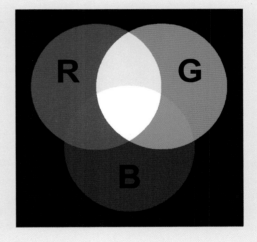

The retina of the human eye has a central or foveal region which is only 0.6 mm in diameter, has a 5 degree field of view and contains most of the 6 million cone cells: 65 % sense red, 33 % sense green and only 2 % sense blue. We unconsciously scan our high-resolution fovea over the world to build a large-scale mental image of our surrounds. In addition there are 120 million rod cells, distributed over the entire retina, and strongly linked to our ability to detect motion.

**10**

This is the photopic response for a light-adapted eye using the cone photoreceptor cells. The dark adapted, or scotopic , using the eye's monochromatic rod photoreceptor cells is different, and peaks at around 510 nm.

The LED on an infrared remote control can be seen as a bright light in most digital cameras – try this with your mobile phone camera and TV remote. Some security cameras provide infrared scene illumination for covert night time monitoring. Note that some cameras are fitted with infrared filters to prevent the sensor becoming saturated by ambient infrared radiation.

### 10.2.1.1 Perceived Brightness

The brightness we associate with a particular wavelength is given by the luminosity function with units of lumens per watt. For our daylight (photopic) vision the luminosity as a function of wavelength has been experimentally determined, tabulated and forms the basis of the 1931 CIE standard that represents the average human observer. ◄ The photopic luminosity function is provided by the Toolbox

```
>>> human = luminos(lmbda);
>>> plt.plot(lmbda / nm,  human);
```

and is shown in ◘ Fig. 10.7b. Consider two light sources emitting the same power (in watts) but one has a wavelength of 550 nm (green) and the other has a wavelength of 450 nm (blue). Our perception of the brightness of these two lights is quite different, in fact the blue light appears only

```
>>> luminos(450 * nm) / luminos(550 * nm)
0.0382
```

or 3.8 % as bright as the green one. The silicon sensors used in digital cameras have strong sensitivity in the red and infrared part of the spectrum. ◄

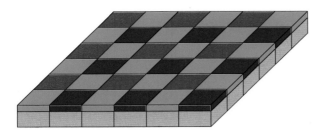

**Fig. 10.8** Bayer filtering. The gray blocks represent the array of light-sensitive silicon photosites over which is an array of red, green and blue filters. Invented by Bryce E. Bayer of Eastman Kodak, U.S. Patent 3,971,065

### 10.2.2 Camera sensor

The sensor in a digital camera is analogous to the retina, but instead of rod and cone cells there is a regular array of light-sensitive photosites on a silicon chip. ▶ Each photosite is of the order $1–10\,\mu$m square and outputs a signal proportional to the intensity of the light falling over its area. ▶

Most common color cameras have an array of tiny color filters over the photosites as shown in ◻ Fig. 10.8. These filters pass either red, green or blue light to the photosites and the spectral response of these filters is chosen to be similar to that of human cones $M(\lambda)$ shown in ◻ Fig. 10.7a. A very common arrangement of color filters is the Bayer pattern – a regular $2\times 2$ pattern comprising two green filters, one red and one blue.

Each photosite is the analog of a cone cell, sensitive to just one color. A consequence of this arrangement is that we cannot make independent measurements ▶ of red, green and blue at every pixel, but it can be estimated. For example, the amount of red at a blue sensitive pixel is obtained by interpolation from its red filtered neighbors. Digital camera *raw image* files contain the actual outputs of the Bayer-filtered photosites.

Larger filter patterns are possible and $3\times 3$ or $4\times 4$ arrays of filters are known as assorted pixel arrays. Using more than 3 different color filters leads to a multispectral camera with better color resolution, a range of neutral density (gray) filters leads to a high-dynamic-range camera, or these various filters can be mixed to give a camera with better dynamic range and color resolution.

The photosites generally correspond directly with pixels in the resulting image.

The output is proportional to the total number of photons captured by the photosite since the last time it was read. See ▶ Exc. 11.3.

More expensive "3 CCD" (charge coupled device) cameras make independent measurements at each pixel since the light is split by a set of prisms, filtered and presented to one CCD image sensor for each primary color.

### 10.2.3 Measuring Color

The path taken by the light entering the eye shown in ◻ Fig. 10.9a. The spectrum of the luminance $L(\lambda)$ is a function of the light source and the reflectance of the object as given by (10.3). The response from each of the three cones is

$$
\rho = \int_\lambda L(\lambda) M_r(\lambda)\,d\lambda
$$

$$
\gamma = \int_\lambda L(\lambda) M_g(\lambda)\,d\lambda
$$

$$
\beta = \int_\lambda L(\lambda) M_b(\lambda)\,d\lambda \tag{10.4}
$$

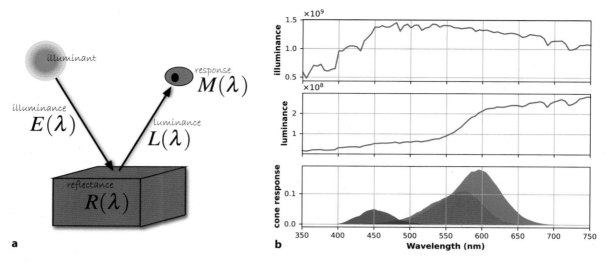

**Fig. 10.9** The tristimulus response of the human eye. **a** Path of light from illuminant to the eye. **b** Within the eye three filters are applied and the total output of these filters, the areas shown in solid color, are the tristimulus value

where $M_r(\lambda)$, $M_g(\lambda)$ and $M_b(\lambda)$ are the spectral response of the red, green and blue cones respectively as shown in ◘ Fig. 10.7a. The response is a 3-vector $(\rho, \gamma, \beta)$ which is known as a tristimulus value.

For the case of the red brick the integrals correspond to the areas of the solid color regions in ◘ Fig. 10.9b. We can compute the tristimulus values by approximating the integrals of (10.4) as a summation with $d\lambda = 1\,\text{nm}$

```
>>> np.sum(np.c_[L, L, L] * cones * nm, axis=0)
array([  16.36,     10.07,     2.822])
```

The dominant response is from the L cone, which is unsurprising since we know that the brick is red.

An arbitrary continuous spectrum is an infinite-dimensional vector and cannot be uniquely represented by just 3 parameters. However, it is clearly *sufficient* for

---

**Excurse 10.8: Lightmeters, Illuminance and Luminance**

A photographic lightmeter measures luminous flux which has units of $\text{lm}\,\text{m}^{-2}$ or lux (lx). The luminous intensity $I$ of a point light source (emitting light in all directions) is the luminous flux per unit solid angle measured in $\text{lm}\,\text{sr}^{-1}$ or candelas (cd). The illuminance $E$ falling normally onto a surface is

$$E = \frac{I}{d^2}\,\text{lx}$$

where $d$ is the distance between source and the surface. Outdoor illuminance on a bright sunny day is approximately 10,000 lx. Office lighting levels are typically around 1000 lx and moonlight is 0.1 lx.

The luminance or *brightness* of a surface is

$$L_s = E_i \cos\theta\,\text{nt}$$

which has units of $\text{cd}\,\text{m}^{-2}$ or nit (nt), and where $E_i$ is the incident illuminance at an angle $\theta$ to the surface normal.

**Excurse 10.9: Color Blindness**

Color blindness, or color deficiency, is the inability to perceive differences between some of the colors that others can distinguish. Protanopia, deuteranopia, and tritanopia refer to the absence of the L, M and S cones respectively. More common conditions are protanomaly, deuteranomaly and tritanomaly where the cone pigments are mutated and the peak response frequency changed. It is most commonly a genetic condition since the red and green photopsins are coded in the X chromosome. The most common form (occurring in 6 % of males including the author) is deuteranomaly where the M-cone's response is shifted toward the red end of the spectrum resulting in reduced sensitivity to greens and poor discrimination of hues in the red, orange, yellow and green region of the spectrum.

The English scientist John Dalton (1766–1844) confused scarlet with green, and pink with blue. He hypothesized that the vitreous humor in his eyes was tinted blue and instructed that his eyes be examined after his death. This revealed that the humors were perfectly clear but DNA recently extracted from his preserved eye showed that he was a deuteranope. Color blindness is still referred to as Daltonism.

☐ **Table 10.2** The CIE 1931 primaries (Commission Internationale de L'Éclairage 1987) are spectral colors corresponding to the emission lines in a mercury vapor lamp

|  | red | green | blue |
|---|---|---|---|
| λ (nm) | 700.0 | 546.1 | 435.8 |

our species and has allowed us to thrive in diverse environments. A consequence of using this tristimulus value representation is that many *different* spectra will produce the *same* visual stimulus and these are referred to as metamers. More important is the corollary – many, but not all, color stimuli can be generated by a mixture of just three monochromatic stimuli. These are the three primary colors we learned about as children. ▶ There is no unique set of primaries – any three will do so long as none of them can be matched by a combination of the others. The CIE has defined a set of monochromatic primaries, and their wavelengths are given in ☐ Table 10.2.

Primary colors are not a fundamental property of light – they are a fundamental property of the observer. There are three primary colors only because we, as trichromats, have three types of cones. Birds would have four primary colors and dogs would have two.

### 10.2.4 Reproducing Colors

A computer or television monitor is able to produce a variable amount of each of three primaries at every pixel. For a liquid crystal display (LCD) the colors are obtained by color filtering and attenuating white light emitted by the backlight, and an OLED display comprises a stack of red, green, and blue LEDs at each pixel. The important problem is to determine how much of each primary is required to match a given color stimulus.

We start by considering a monochromatic stimulus of wavelength $\lambda_S$ which is defined as

$$L(\lambda) = \begin{cases} L_\lambda & \text{if } \lambda = \lambda_S \\ 0 & \text{otherwise} . \end{cases}$$

> **Excurse 10.10: What are Primary Colors?**
>
> The notion of primary colors is very old, but their number (anything from two to six) and their color was the subject of much debate. Much of the confusion was due to there being additive primaries (red, green and blue) that are used when mixing lights, and subtractive primaries (cyan, magenta, yellow) used when mixing paints or inks. Whether or not black and white were primary colors was also debated.

The response of the cones to this stimulus is given by (10.4) but because $L(\lambda)$ is an impulse we can drop the integral to obtain the tristimulus values

$$
\begin{aligned}
\rho &= L_\lambda M_r(\lambda_S) \\
\gamma &= L_\lambda M_g(\lambda_S) \\
\beta &= L_\lambda M_b(\lambda_S) \, .
\end{aligned}
\tag{10.5}
$$

The units are chosen such that equal quantities of the primaries appear to be white.

Consider next three monochromatic primary light sources denoted **R**, **G** and **B**; with wavelengths $\lambda_r, \lambda_g$ and $\lambda_b$; and intensities $R$, $G$ and $B$ respectively. ◄ The tristimulus values from these light sources is

$$
\begin{aligned}
\rho &= R M_r(\lambda_r) + G M_r(\lambda_g) + B M_r(\lambda_b) \\
\gamma &= R M_g(\lambda_r) + G M_g(\lambda_g) + B M_g(\lambda_b) \\
\beta &= R M_b(\lambda_r) + G M_b(\lambda_g) + B M_b(\lambda_b) \, .
\end{aligned}
\tag{10.6}
$$

For the perceived color of these three light sources combined to match that of the monochromatic stimulus, the two tristimuli must be equal. We equate (10.5) and (10.6) and write compactly in matrix form as

$$
L_\lambda \begin{pmatrix} M_r(\lambda_S) \\ M_g(\lambda_S) \\ M_b(\lambda_S) \end{pmatrix} = \begin{pmatrix} M_r(\lambda_r) & M_r(\lambda_g) & M_r(\lambda_b) \\ M_g(\lambda_r) & M_g(\lambda_g) & M_g(\lambda_b) \\ M_{br}(\lambda_r) & M_b(\lambda_g) & M_b(\lambda_b) \end{pmatrix} \begin{pmatrix} R \\ G \\ B \end{pmatrix}
$$

and then solve for the required amounts of primary colors

$$
\begin{pmatrix} R \\ G \\ B \end{pmatrix} = L_\lambda \begin{pmatrix} M_r(\lambda_r) & M_r(\lambda_g) & M_r(\lambda_b) \\ M_g(\lambda_r) & M_g(\lambda_g) & M_g(\lambda_b) \\ M_{br}(\lambda_r) & M_b(\lambda_g) & M_b(\lambda_b) \end{pmatrix}^{-1} \begin{pmatrix} M_r(\lambda_S) \\ M_g(\lambda_S) \\ M_b(\lambda_S) \end{pmatrix} \, .
\tag{10.7}
$$

This visual stimulus has a spectrum comprising three impulses (one per primary), yet has the same visual appearance as the original continuous spectrum – this is the basis of trichromatic matching. The $3 \times 3$ matrix is constant, but depends upon the spectral response of the cones to the chosen primaries ($\lambda_r, \lambda_g, \lambda_b$).

The right-hand side of (10.7) is simply a function of $\lambda_S$ which we can write in an even more compact form

$$
\begin{pmatrix} R \\ G \\ B \end{pmatrix} = \begin{pmatrix} \bar{r}(\lambda_S) \\ \bar{g}(\lambda_S) \\ \bar{b}(\lambda_S) \end{pmatrix}
\tag{10.8}
$$

where $\bar{r}(\lambda), \bar{g}(\lambda), \bar{b}(\lambda)$ are known as color matching functions. These functions have been empirically determined from human test subjects and tabulated. They

> **Excurse 10.11: Color Matching Experiments**
>
> Color matching experiments are performed using a light source comprising three adjustable lamps that correspond to the primary colors and whose intensity can be individually adjusted. The lights are mixed and diffused and compared to some test color. In color matching notation the primaries, the lamps, are denoted by **R**, **G** and **B**, and their intensities are $R$, $G$ and $B$ respectively. The three lamp intensities are adjusted by a human subject until they appear to match the test color. This is denoted
>
> $$\mathbf{C} \equiv R\mathbf{R} + G\mathbf{G} + B\mathbf{B}$$
>
> which is read as the visual stimulus **C** (the test color) is matched by, or looks the same as, a mixture of the three primaries with brightness $R$, $G$ and $B$. The notation $R\mathbf{R}$ can be considered as the lamp **R** at intensity $R$.
>
> Experiments show that color matching obeys the algebraic rules of additivity and linearity which are known as Grassmann's laws. For example two light stimuli $\mathbf{C}_1$ and $\mathbf{C}_2$
>
> $$\mathbf{C}_1 \equiv R_1\mathbf{R} + G_1\mathbf{G} + B_1\mathbf{B}$$
> $$\mathbf{C}_2 \equiv R_2\mathbf{R} + G_2\mathbf{G} + B_2\mathbf{B}$$
>
> when mixed will match
>
> $$\mathbf{C}_1 + \mathbf{C}_2 \equiv (R_1 + R_2)\mathbf{R} + (G_1 + G_2)\mathbf{G} + (B_1 + B_2)\mathbf{B}\ .$$

can be loaded using the function `cmfrgb`

```
>>> cmf = cmfrgb(lmbda);
>>> plt.plot(lmbda / nm, cmf);
```

and are shown graphically in ◼ Fig. 10.10a. Each curve indicates how much of the corresponding primary is required to match the monochromatic light of wavelength $\lambda$. ◼ Fig. 10.10b shows the same data as a curve in the three-dimensional tristimulus value space.

For example to create the sensation of light at 500 nm (green) we would need

```
>>> green = cmfrgb(500 * nm)
array([-0.07137,  0.08536,  0.04776])
```

Surprisingly, this requires a significant *negative* amount of the red primary and this is problematic since a light source cannot have a negative luminance.

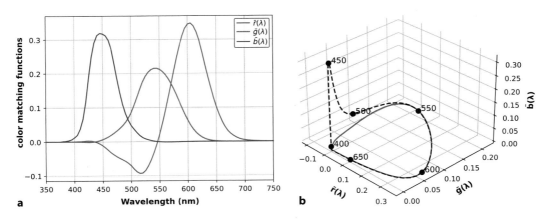

◼ **Fig. 10.10** **a** The CIE 1931 color matching functions for the standard observer, based on 2° field of view. **b** The color matching locus as a function of wavelength, with its projection onto the $\bar{r}\bar{g}$-plane shown in red

We reconcile this by adding some white light ($R = G = B = w$, see
► Sect. 10.2.8) so that the tristimulus values are all positive. For instance

```
>>> w = -np.min(green)
0.07137
>>> feasible_green = green + w
array([       0,    0.1567,    0.1191])
```

If we looked at this color side-by-side with the desired 500 nm green we would say
that the generated color had the correct hue but was not as *saturated*. Saturation
refers to the purity of the color. Spectral colors are *fully saturated* but become less
saturated (more pastel) as increasing amounts of white is added. In this case we
have mixed in a stimulus of light gray (7 % gray).

This leads to a very important point about color reproduction – it is *not* possible
to reproduce every possible color using just three primaries. This makes intuitive
sense since a color is properly represented as an infinite-dimensional spectral function and a 3-vector can only approximate it.

The Toolbox function `cmfrgb` can also compute the CIE tristimulus values for
an arbitrary spectrum. The luminance spectrum of the redbrick illuminated by sunlight at ground level was computed in ► Sect. 10.1.3, and its tristimulus values are

```
>>> RGB_brick = cmfrgb(lmbda, L)
array([ 0.01554, 0.006651, 0.003096])
```

These are the respective amounts of the three CIE primaries that are perceived –
by the average human – as having the same color as the original brick under those
lighting conditions.

## 10.2.5 Chromaticity Coordinates

The tristimulus values describe color as well as brightness. For example, if we
double R, G and B we would perceive approximately the same color, but describe
it as brighter.

Relative tristimulus values are obtained by normalizing the tristimulus values

$$r = \frac{R}{R + B + G}, \quad g = \frac{G}{R + B + G}, \quad b = \frac{B}{R + B + G} \tag{10.9}$$

which results in chromaticity coordinates $r$, $g$ and $b$ that are invariant to overall
brightness. By definition $r + g + b = 1$ so one coordinate is redundant and typically
only $r$ and $g$ are considered. Since the effect of intensity has been eliminated, the
2-dimensional quantity $(r, g)$ represents color, or more precisely, *chromaticity*.

We can plot the locus of spectral colors, the colors of the rainbow, on the chromaticity diagram using a variant of the color matching functions

```
>>> rg = lambda2rg(np.linspace(400, 700, 100) * nm);
>>> plt.plot(rg[:, 0], rg[:, 1]);
```

which results in the horseshoe-shaped curve shown in ◘ Fig. 10.11. This is a scaled
version of the red curve in ◘ Fig. 10.10b. The red and blue primaries are joined
by a straight line segment called the purple boundary and is the locus of saturated
purples.

The Toolbox function `lambda2rg` computes the color matching function
◘ Fig. 10.10 for the specified wavelength and then converts the tristimulus values
to chromaticity coordinates using (10.9). The plot shown in ◘ Fig. 10.11, with
wavelength annotations, is produced by

```
>>> plot_spectral_locus("rg")
```

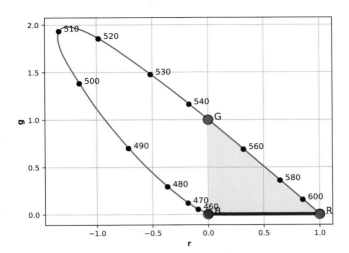

☐ **Fig. 10.11** The spectral locus on the *rg* chromaticity plane. Monochromatic stimuli lie on the locus and the wavelengths (in nm) are marked. All possible colors lie on, or within, this locus. The 1931 CIE standard primary colors are marked and the yellow triangle indicates the gamut of colors that can be represented by these primaries

The chromaticity of the CIE primaries listed in ☐ Table 10.2 are arranged in columns

```
>>> primaries = lambda2rg(cie_primaries()).T
array([[      1, 0.001084, -0.0003924],
       [      0,   0.9987, 0.0002112]])
```

and can be added to the plot as circles

```
>>> plot_point(primaries, "o");
```

which are shown in ☐ Fig. 10.11.

Grassmann's center of gravity law states that a mixture of two colors lies along a line between those two colors on the chromaticity plane. A mixture of *N* colors lies within a region bounded by those colors. Considered with respect to ☐ Fig. 10.11, this has significant implications. Firstly, since all color stimuli are combinations of spectral stimuli, all real color stimuli must lie on, or inside the spectral locus. Secondly, any colors we create from mixing the primaries can only lie *within* the triangle bounded by the primaries – the color gamut. It is clear from

**Excurse 10.12: Colorimetric Standards**
Colorimetry is a complex topic and standards are very important. Two organizations, CIE and ITU, play a leading role in this area.

The Commission Internationale de l'Eclairage (CIE) or International Commission on Illumination was founded in 1913 and is an independent nonprofit organization that is devoted to worldwide cooperation and the exchange of information on all matters relating to the science and art of light and lighting, color and vision, and image technology. The CIE's eighth session was held at Cambridge, UK, in 1931 and established international agreement on colorimetric specifications and formalized the XYZ color space. The CIE is recognized by ISO as an international standardization body.

See ▶ https://www.cie.co.at for more information and CIE datasets.

The International Telecommunication Union (ITU) is an agency of the United Nations and was established to standardize and regulate international radio and telecommunications. It was founded as the International Telegraph Union in Paris on 17 May 1865. The International Radio Consultative Committee or CCIR (Comité Consultatif International des Radiocommunications) became, in 1992, the Radiocommunication Bureau of ITU or ITU-R. It publishes standards and recommendations relevant to colorimetry in its BT series (broadcasting service television). See ▶ https://www.itu.int for more detail.

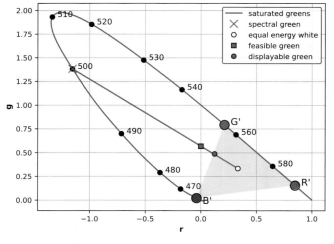

**☐ Fig. 10.12** Chromaticity diagram showing the color gamut for nonstandard primaries at 600, 555 and 450 nm

☐ Fig. 10.11 that the CIE primaries define only a small subset of all possible colors – those within the yellow triangle. Very many real colors *cannot* be created using these primaries, in particular the colors of the rainbow which lie on the spectral locus from 460–545 nm. In fact no matter where the primaries are located, not all possible colors can be produced. ◄ In geometric terms, there are no three points within the gamut that form a triangle that includes the entire gamut. Thirdly, we observe that much of the locus requires a negative amount of the red primary and cannot be represented.

We spend much of our life looking at screens, and this theoretical inability to display all real colors is surprising and perplexing. To illustrate how this is resolved we revisit the problem from ► Sect. 10.2.4 which is concerned with displaying 500 nm green – the green we see in a rainbow. ☐ Fig. 10.12 shows the chromaticity of spectral green

```
>>> green_cc = lambda2rg(500 * nm)
array([  -1.156,    1.382])
>>> plot_point(green_cc, "x");
```

We could increase the gamut by choosing different primaries, perhaps using a different green primary would make the gamut larger, but there is the practical constraint of finding a light source (LED or phosphor) that can efficiently produce that color.

as a cross-shaped marker. White is by definition $R = G = B = 1$ and its chromaticity

```
>>> white_cc = tristim2cc([1, 1, 1])
array([  0.3333,    0.3333])
>>> plot_point(white_cc, "o");
```

is shown as a circular marker. According to Grassmann's law, the mixture of our desired green and white must lie along the indicated green line. The chromaticity of the feasible green computed earlier is indicated by a square, but is outside the *displayable* gamut of the primaries used in this example. The least-saturated displayable green lies at the intersection of the green line and the gamut boundary and is indicated by the round-green marker.

Luminance here has different meaning to that defined in ► Sect. 10.1.3 and can be considered synonymous to brightness here.

Earlier we said that there are no three points within the gamut that form a triangle that includes the entire gamut. The CIE proposed, in 1931, a system of *imaginary nonphysical primaries* known as **X**, **Y** and **Z** that define a minimally enclosing triangle of the entire spectral locus. **X** and **Z** have zero luminance – the luminance is contributed entirely by **Y** ◄. All real colors can thus be matched by positive amounts of these three primaries. ◄ The corresponding tristimulus values are denoted $(x, y, z)$.

The units are chosen such that equal quantities of the primaries are required to match the equal-energy white stimulus.

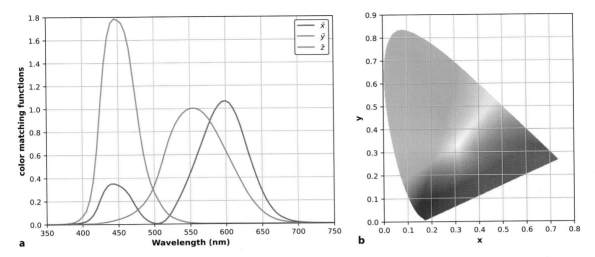

**Fig. 10.13** **a** The color matching functions for the standard observer, based on the imaginary primaries **X**, **Y** (intensity) and **Z** are tabulated by the CIE. **b** Colors on the *xy*-chromaticity plane

The XYZ color matching functions defined by the CIE

```
>>> cmf = cmfxyz(lmbda);
>>> plt.plot(lmbda / nm, cmf);
```

are shown graphically in ◻ Fig. 10.13a. This shows the amount of each CIE XYZ primary that is required to match a spectral color and we note that points on these curves always have positive coordinates. The corresponding chromaticity coordinates are

$$x = \frac{X}{X+Y+Z}, \quad y = \frac{Y}{X+Y+Z}, \quad z = \frac{Z}{X+Y+Z} \qquad (10.10)$$

and once again $x + y + z = 1$ so only two parameters are required – by convention $y$ is plotted against $x$ in a chromaticity diagram. The spectral locus can be plotted in a similar way as before

```
>>> xy = lambda2xy(lmbda);
>>> plt.plot(xy[0], xy[1], "ko");
```

A more sophisticated plot, showing the colors within the spectral locus, can be created

```
>>> plot_chromaticity_diagram("xy")
```

and is shown in ◻ Fig. 10.13b. ▶ These coordinates are a *standard* way to represent color for graphics, printing and other purposes. For example the chromaticity coordinates of peak green (550 nm) is

```
>>> lambda2xy(550 * nm)
array([ 0.3016,   0.6923])
```

and the chromaticity coordinates of a standard tungsten illuminant at 2600 K is

```
>>> lamp = blackbody(lmbda, 2_600);
>>> lambda2xy(lmbda, lamp)
array([ 0.4677,   0.4127])
```

The colors depicted in figures such as ◻ Figs. 10.1 and 10.13b can only approximate the true color due to the gamut limitation of the technology you use to view this book: the inks used to print the page or your computer's display. No display technology has a gamut large enough to present an accurate representation of the chromaticity at every point.

### 10.2.6 Color Names

Chromaticity coordinates provide a quantitative way to describe and compare colors, however humans refer to colors by name. Many computer operating systems

**10**

### Excurse 10.13: Colors

Colors are important to human beings and there are over 4000 color-related words in the English language (Steinvall 2002). All languages have words for black and white, and red is the next most likely color word to appear followed by yellow, green, blue and so on. The ancient Greeks only had words for black, white, red and yellowish-green. We also associate colors with emotions, for example red is angry and blue is sad, but this varies across cultures. In Asia, orange is generally a positive color whereas in the west it is the color of road hazards and bulldozers. Technologies, including chemistry, have made a huge number of colors available to us in the last 700 years, yet with this choice comes confusion about color naming – people may not necessarily agree on the linguistic tag to assign to a particular color.

In 2010 XKCD conducted a survey ► https://xkcd.com/color/rgb/ to find the names that people associated with colors. These 954 colors are in the database used by `name2color` and the names are all prefixed by `xkcd:`.

(Word cloud by tagxedo.com using color frequency data from Steinvall 2002)

With the X11 package the file is named `rgb.txt`. The Toolbox uses Matplotlib to perform this function.

contain a database or file ◄ that maps human understood names of colors to their corresponding $(R, G, B)$ tristimulus values. The Toolbox uses the color database provided by Matplotlib which defines over one thousand colors. For example, the RGB tristimulus values for the color orange are

```
>>> name2color("orange")
array([ 1,  0.6471,   0])
```

with the values normalized to the interval [0,1]. We could also request *xy*-chromaticity coordinates

```
>>> bs = name2color("orange", "xy")
array([ 0.4548,   0.4771])
```

With reference to ◻ Fig. 10.13b, we see that this point lies in the yellow-red part of the colorspace. The color can also be specified with a Python regular expression

```
>>> name2color(".*coral.*")
['xkcd:coral pink', 'xkcd:dark coral', 'xkcd:coral', 'coral',
 'lightcoral']
```

which returns matching color names.

We can also solve the inverse problem, mapping tristimulus values

```
>>> color2name([0.45, 0.48], "xy")
'xkcd:orange yellow'
```

to the name of the closest color – in terms of Euclidean distance in the chromaticity plane.

### 10.2.7  Other Color and Chromaticity Spaces

A color space is a 3-dimensional space that contains all possible tristimulus values – all colors and all levels of brightness. There are an infinite number of choices of Cartesian coordinate frame with which to define colors and we have already

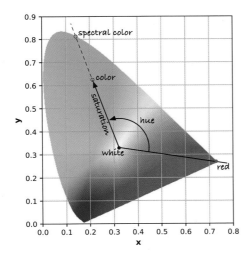

**◘ Fig. 10.14**  Hue and saturation. A line is extended from the white point through the chromaticity in question to the spectral locus. The angle of this line is hue, and saturation is the length of the vector normalized with respect to distance to the locus

discussed two of these: RGB and XYZ. However, we could also use polar, spherical or hybrid coordinate systems.

The 2-dimensional *rg*- or *xy*-chromaticity spaces do not account for brightness – we normalized it out in (10.9) and (10.10). Brightness, frequently referred to as luminance in this context, is denoted by *Y* and the definition from ITU Recommendation 709

$$Y^{709} = 0.2126R + 0.7152G + 0.0722B \qquad (10.11)$$

is a weighted sum of the RGB-tristimulus values, and reflects the eye's high sensitivity to green and low sensitivity to blue. Chromaticity plus luminance leads to 3-dimensional color spaces such as *rgY* or *xyY*.

Humans seem to more naturally consider chromaticity in terms of two characteristics: hue and saturation. Hue is the dominant color, the closest spectral color, and saturation refers to the purity, or absence of mixed white. Stimuli on the spectral locus are completely saturated while those closer to its centroid are less saturated. The concepts of hue and saturation are illustrated in geometric terms in ◘ Fig. 10.14.

The color spaces that we have discussed lack easy interpretation in terms of hue and saturation, so alternative color spaces have been proposed. The two most commonly known are HSV and CIE L*C*h. In color-space notation H is hue, and S is saturation which is also known as C or chroma. The intensity dimension is named either V for value or L for lightness but they are computed quite differently. ▶

The function `colorspace_convert` can be used to convert between different color spaces. For example the hue, saturation and intensity for each of pure red, green and blue RGB-tristimulus values ▶ are

```
>>> colorspace_convert([1, 0, 0], "RGB", "HSV")
array([  0,   1,    1])
>>> colorspace_convert([0, 1, 0], "RGB", "HSV")
array([ 120,   1,    1])
>>> colorspace_convert([0, 0, 1], "RGB", "HSV")
array([ 240,   1,    1])
```

In each case the saturation is 1, the colors are pure, and the intensity is 1. As shown in ◘ Fig. 10.14 hue is represented as an angle in the range [0, 360)° with red at

$L^*$ is a nonlinear function of relative luminance and approximates the nonlinear response of the human eye. Value is given by $V = \frac{1}{2}(\min(R, G, B) + \max(R, G, B))$.

This function assumes that *RGB* values are linear, that is, not gamma encoded, see ▶ Sect. 10.3.6. The particular numerical values chosen here are invariant under gamma encoding.

**◻ Fig. 10.15** Flower scene. **a** Original color image; **b** hue image; **c** saturation image. Note that the white flowers have low saturation (they appear dark); **d** intensity or monochrome image; **e** a* image (green to red); **f** b* image (blue to yellow)

**10**

0° increasing through the spectral colors associated with decreasing wavelength (orange, yellow, green, blue, violet).

If we reduce the amount of the green primary

```
>>> colorspace_convert([0, 0.5, 0], "RGB", "HSV")
array([ 120,    1, 0.5])
```

For very dark colors numerical problems lead to imprecise hue and saturation coordinates.

we see that intensity drops but hue and saturation are unchanged. ◄ For a medium gray

```
>>> colorspace_convert([0.4, 0.4, 0.4], "RGB", "HSV")
array([  0,    0, 0.4])
```

the saturation is zero, it is only a mixture of white, and the hue has no meaning since there is no color. If we add the green to the gray

```
>>> colorspace_convert(np.array([0, 0.5, 0])
...                           + np.array([0.4, 0.4, 0.4]), "RGB", "HSV")
array([ 120, 0.5556, 0.9])
```

we have the green hue and a medium saturation value.

The `colorspace` function can also be applied to a color image, a topic we will cover fully in the next chapter

```
>>> flowers = Image.Read("flowers4.png")
Image: 640 x 426 (uint8), R:G:B [.../images/flowers4.png]
```

which is shown in ◻ Fig. 10.15a and comprises several different colored flowers and background greenery. The image `flowers` has 3 dimensions and the third is the color plane that selects the red, green or blue pixels. ◄

This is covered in greater detail in the next chapter.

To convert the image to hue, saturation and value is simply

```
>>> hsv = flowers.colorspace("HSV")
Image: 640 x 426 (uint8), H:S:V
```

and the result is another 3-dimensional matrix but this time the color planes represent hue, saturation and value. We can display these planes

```
>>> hsv.plane("H").disp();
>>> hsv.plane("S").disp();
>>> hsv.plane("V").disp();
```

as images which are shown in ◘ Fig. 10.15b–d respectively. In the hue image, black represents red and white represents violet. The red flowers appear as both a very small hue angle (dark) and a very large angle (white) close to 360°. The yellow flowers and the green background can be seen as distinct hue values. The saturation image shows that the red and yellow flowers are highly saturated, while the green leaves and stems are less saturated. The white flowers have very low saturation since, by definition, the color white contains a lot of white.

A limitation of many color spaces is that the *perceived* color difference between two points is not directly related to their Euclidean distance. In some parts of the chromaticity space two distant points might appear quite similar, whereas in another region two close points might appear quite different. This has led to the development of perceptually uniform color spaces such as the CIE L\*u\*v\*(CIELUV) and L\*a\*b\*(CIELAB) spaces. ▶

> L\*u\*v\* is a 1976 update to the 1960 CIE Luv uniform color space (UCS). L\*a\*b\* is a 1976 CIE standardization of Hunter's Lab color space defined in 1948.

The `colorspace` function can convert between thirteen different color spaces including L\*a\*b\*, L\*u\*v\*, YUV and $YC_BC_R$. To convert this image to L\*a\*b\* color space follows the same pattern

```
>>> Lab = flowers.colorspace("L*a*b*")
Image: 640 x 426 (uint8), L*:a*:b*
```

which again results in an image with 3 dimensions. The chromaticity ▶ is encoded in the a\* and b\* planes.

> Relative to a white illuminant, which this function assumes as CIE $D_{65}$ with $Y = 1$. a\*b\* are not invariant to overall luminance.

```
>>> Lab.plane("a*").disp();
>>> Lab.plane("b*").disp();
```

and these are shown in ◘ Fig. 10.15e, f respectively. L\*a\*b\* is an opponent color space where a\* spans colors from green (black) to red (white) while b\* spans blue (black) to yellow (white), with white at the origin where $a^* = b^* = 0$.

### 10.2.8 Transforming Between Different Primaries

The CIE standards were defined in 1931, which was well before the introduction of color television in the 1950s. The CIE primaries in ◘ Table 10.2 are based on the emission lines of a mercury lamp which are highly repeatable and suitable for laboratory use. Early television receivers used CRT monitors where the primary colors were generated by phosphors that emit light when bombarded by electrons. The phosphors used, and their colors, have varied over the years in pursuit of brighter and more efficient displays. An international agreement, ITU recommendation 709, defines the primaries for high definition television (HDTV) and these are listed in ◘ Table 10.3.

This raises the problem of converting tristimulus values from one set of primaries to another. Using the notation we introduced earlier we define two sets of primaries: $\mathbf{P}_1, \mathbf{P}_2, \mathbf{P}_3$ with tristimulus values $(S_1, S_2, S_3)$, and $\mathbf{P}'_1, \mathbf{P}'_2, \mathbf{P}'_3$ with tristimulus values $(S'_1, S'_2, S'_3)$. We can always express one set of primaries as a linear

◘ **Table 10.3** *xyz*-chromaticity of standard primaries and whites. The CIE primaries of ◘ Table 10.2 and the more recent ITU recommendation 709 primaries defined for HDTV. $D_{65}$ is the white of a blackbody radiator at 6500 K, and *E* is equal-energy white

|   | $R_{CIE}$ | $G_{CIE}$ | $B_{CIE}$ | $R_{709}$ | $G_{709}$ | $B_{709}$ | $D_{65}$ | $E$ |
|---|---|---|---|---|---|---|---|---|
| *x* | 0.7347 | 0.2738 | 0.1666 | 0.640 | 0.300 | 0.150 | 0.3127 | 0.3333 |
| *y* | 0.2653 | 0.7174 | 0.0089 | 0.330 | 0.600 | 0.060 | 0.3290 | 0.3333 |
| *z* | 0.0000 | 0.0088 | 0.8245 | 0.030 | 0.100 | 0.790 | 0.3582 | 0.3333 |

The coefficients can be negative so the new primaries do not have to lie within the gamut of the old primaries.

combination ◄ of the other

$$
\begin{pmatrix} \mathbf{P}_1 \\ \mathbf{P}_2 \\ \mathbf{P}_3 \end{pmatrix} = \begin{pmatrix} a_{11} & a_{21} & a_{31} \\ a_{12} & a_{22} & a_{32} \\ a_{13} & a_{23} & a_{33} \end{pmatrix} \begin{pmatrix} \mathbf{P}'_1 \\ \mathbf{P}'_2 \\ \mathbf{P}'_3 \end{pmatrix}
\tag{10.12}
$$

and since the two tristimuli match then

$$
\begin{pmatrix} S'_1 & S'_2 & S'_3 \end{pmatrix} \begin{pmatrix} \mathbf{P}'_1 \\ \mathbf{P}'_2 \\ \mathbf{P}'_3 \end{pmatrix} = \begin{pmatrix} S_1 & S_2 & S_3 \end{pmatrix} \begin{pmatrix} \mathbf{P}_1 \\ \mathbf{P}_2 \\ \mathbf{P}_3 \end{pmatrix}
\tag{10.13}
$$

Substituting (10.12), equating tristimulus values and then transposing we obtain

$$
\begin{pmatrix} S'_1 \\ S'_2 \\ S'_3 \end{pmatrix} = \begin{pmatrix} a_{11} & a_{21} & a_{31} \\ a_{12} & a_{22} & a_{32} \\ a_{13} & a_{23} & a_{33} \end{pmatrix}^\mathsf{T} \begin{pmatrix} S_1 \\ S_2 \\ S_3 \end{pmatrix} = \mathbf{C} \begin{pmatrix} S_1 \\ S_2 \\ S_3 \end{pmatrix}
\tag{10.14}
$$

which is simply a linear transformation of the tristimulus values.

Consider the concrete problem of transforming from CIE primaries to XYZ-tristimulus values. We know from ◨ Table 10.3 the CIE primaries in terms of XYZ primaries

```
>>> C = np.array([[0.7347,  0.2738, 0.1666],
...               [0.2653,  0.7174, 0.0088],
...               [0,       0.0089, 0.8245]]);
```

which is exactly the first three columns of ◨ Table 10.3. The transform is therefore

$$
\begin{pmatrix} X \\ Y \\ Z \end{pmatrix} = \mathbf{C} \begin{pmatrix} R \\ G \\ B \end{pmatrix}
$$

Recall from ► Sect. 10.2.5 that luminance is contributed entirely by the **Y** primary. It is common to apply the constraint that unity $R$, $G$, $B$ values result in unity luminance $Y$ and a white with a specified chromaticity. We will choose $D_{65}$ white, whose chromaticity is given in ◨ Table 10.3 and which we will denote $(x^\mathrm{w}, y^\mathrm{w}, z^\mathrm{w})$. We can now write

$$
\frac{1}{y_\mathrm{w}} \begin{pmatrix} x_\mathrm{w} \\ y_\mathrm{w} \\ z_\mathrm{w} \end{pmatrix} = \mathbf{C} \begin{pmatrix} J_R & 0 & 0 \\ 0 & J_G & 0 \\ 0 & 0 & J_B \end{pmatrix} \begin{pmatrix} 1 \\ 1 \\ 1 \end{pmatrix}
$$

where the left-hand side has $Y = 1$ and we have introduced a diagonal matrix **J** which scales the luminance of the primaries. We can solve for the elements of **J**

$$
\begin{pmatrix} J_R \\ J_G \\ J_B \end{pmatrix} = \mathbf{C}^{-1} \begin{pmatrix} x_\mathrm{w} \\ y_\mathrm{w} \\ z_\mathrm{w} \end{pmatrix} \frac{1}{y_\mathrm{w}}
$$

Substituting real values we obtain

```
>>> white = np.array([0.3127, 0.3290, 0.3582]);
>>> J = np.linalg.inv(C) @ white / white[1]
array([ 0.5609,    1.17,    1.308])
>>> C @ np.diag(J)
array([[ 0.4121,   0.3205,   0.2179],
       [ 0.1488,   0.8397,   0.01151],
       [      0,   0.01042,   1.078]])
```

The middle row of this matrix leads to the luminance relationship

$$
Y = 0.1488R + 0.8395G + 0.0116B
$$

which is similar to (10.11). The small variation is due to the different primaries used – CIE in this case versus Rec. 709 for (10.11).

The RGB-tristimulus values of the redbrick was computed earlier and we can determine its XYZ-tristimulus values

```
>>> XYZ_brick = (C @ np.diag(J) @ RGB_brick).T
array([0.009212, 0.007933, 0.003408])
```

which we convert to chromaticity coordinates by (10.10)

```
>>> tri = tristim2cc(XYZ_brick)
array([ 0.4482,    0.386])
```

Referring to ◘ Fig. 10.13b we see that this $xy$-chromaticity lies in the red region and is named

```
>>> color2name(tri, "xy")
'xkcd:red brown'
```

which is plausible for a "weathered red brick".

### 10.2.9 What Is White?

In the previous section we touched on the subject of white. White is both the absence of color and also the sum of all colors. One definition of white is *standard daylight* which is taken as the mid-day Sun in Western/Northern Europe which has been tabulated by the CIE as illuminant $D_{65}$. It can be closely approximated by a blackbody radiator at 6500 K

```
>>> d65 = blackbody(lmbda, 6_500);
>>> lambda2xy(lmbda, d65)
array([ 0.3136,    0.3243])
```

which we see is close to the $D_{65}$ chromaticity given in ◘ Table 10.3.

Another definition is based on white light being an equal mixture of all spectral colors. This is represented by a uniform spectrum

```
>>> ee = np.ones(lmbda.shape);
```

which is also known as the equal-energy stimulus and has chromaticity

```
>>> lambda2xy(lmbda, ee)
array([ 0.3334,    0.334])
```

which is close to the defined value of $(\frac{1}{3}, \frac{1}{3})$.

## 10.3 Advanced Topics

Color is a large and complex subject, and in this section we will briefly introduce a few important remaining topics. Color temperature is a common way to describe the spectrum of an illuminant. The effect of illumination color on the apparent color of an object is the color constancy problem, a very real concern for robots using color cues in an environment with natural lighting. White balancing is one way to overcome this. Another source of color change, in media such as water, is the absorption of certain wavelengths. Most cameras actually implement a nonlinear relationship, between actual scene luminance and the output tristimulus values – gamma correction – which must be considered in colormetric operations. Finally we look at a more realistic model of surface reflection which has both specular and diffuse components, each with different spectral characteristics.

### 10.3.1 Color Temperature

Photographers often refer to the color temperature of a light source – the temperature of a black body whose spectrum according to (10.1) is most similar to that of

**10**

■ **Table 10.4** Color temperatures of some common light sources

| Light source | Color temperature (K) |
|---|---|
| Candle light | 1900 |
| Dawn/dusk sky | 2000 |
| 40 W tungsten lamp | 2600 |
| 100 W tungsten lamp | 2850 |
| Tungsten halogen lamp | 3200 |
| Direct sunlight | 5800 |
| Overcast sky | 6000–7000 |
| Standard daylight (sun + blue sky) | 6500 |
| Hazy sky | 8000 |
| Clear blue sky | 10,000–30,000 |

the light source. The color temperatures of a number of common lighting sources are listed in ■ Table 10.4. We describe low-color-temperature illumination as warm – it appears reddy orange to us. High-color-temperature is more harsh – it appears as brilliant white, perhaps with a tinge of blue.

### 10.3.2 Color Constancy

We adapt our perception of color so that the integral, or average, over the entire scene is gray. This works well over a color temperature range 5000–6500 K.

Studies show that human perception of what is white is adaptive and has a remarkable ability to *tune out* the effect of scene illumination so that white objects always appear to be white. ◀ For example at night, under a yellowish tungsten lamp, the pages of a book still appear white to us, but a photograph of that scene viewed later under different lighting conditions will look yellow. All of this poses real problems for a robot that is using color to understand the scene, because the observed chromaticity varies with lighting. Outdoors, a robot has to contend with an illumination spectrum that depends on the time of day and cloud cover, as well as colored re-

---

**Excurse 10.14: Scene Luminance**
Scene luminance is the product of illuminance and reflectance but reflectance, is key to scene understanding since it can be used as a proxy for the type of material. Illuminance can vary in intensity and color across the scene and this complicates image understanding. Unfortunately separating luminance into illuminance and reflectance is an ill-posed problem, yet humans are able to do this very well as this illusion illustrates – the squares labeled A and B actually have the same gray level.

The American inventor and founder of Polaroid Corporation Edwin Land (1909–1991) proposed the retinex theory (retinex=retina + cortex) to explain how the human visual system factorizes reflectance from luminance. (Checker shadow illusion courtesy of Edward H. Adelson, ▶ http://persci.mit.edu/gallery)

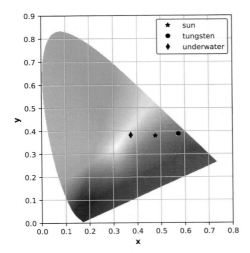

**◻ Fig. 10.16** Chromaticity of the red-brick under different illumination conditions

flections from buildings and trees. This affects the luminance and apparent color of the object. To illustrate this problem we revisit the red brick

```
>>> lmbda = np.arange(400, 701) * nm;
>>> R = loadspectrum(lmbda, "redbrick");
```

under two different illumination conditions, the Sun at ground level

```
>>> sun = loadspectrum(lmbda, "solar");
```

and a tungsten lamp

```
>>> lamp = blackbody(lmbda, 2_600);
```

and compute the *xy*-chromaticity for each case

```
>>> xy_sun = lambda2xy(lmbda, sun * R)
array([ 0.4764,   0.3787])
>>> xy_lamp = lambda2xy(lmbda, lamp * R)
array([ 0.5721,   0.3879])
```

and we can see that the chromaticity, or apparent color, has changed significantly. These values are plotted on the chromaticity diagram in ◻ Fig. 10.16.

### 10.3.3 White Balancing

Photographers need to be aware of the illumination color temperature. An incandescent lamp appears more yellow than daylight, so a photographer would place a blue filter on the camera to attenuate the red part of the spectrum to compensate. We can achieve a similar function by choosing the matrix **J**

$$
\begin{pmatrix} R' \\ G' \\ B' \end{pmatrix} = \begin{pmatrix} J_R & 0 & 0 \\ 0 & J_G & 0 \\ 0 & 0 & J_B \end{pmatrix} \begin{pmatrix} R \\ G \\ B \end{pmatrix}
$$

to adjust the gains of the color channels. ▶ For example, increasing $J_B$ would compensate for the lack of blue under tungsten illumination. This is the process of white balancing – ensuring the appropriate chromaticity of objects that we know are white (or gray).

Some cameras allow the user to set the color temperature of the illumination through a menu, typically with options for tungsten, fluorescent, daylight and flash, which select different preset values of **J**. In manual white balancing the camera

Typically $J_G = 1$ and $J_R$ and $J_B$ are adjusted.

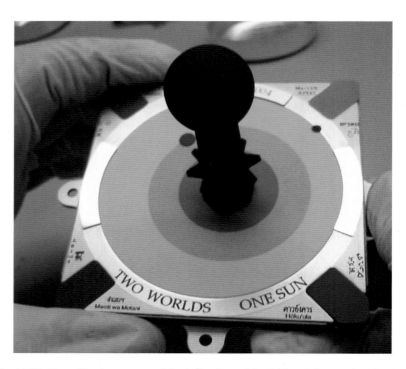

**◨ Fig. 10.17** The calibration target used for the PanCam of the Spirit and Opportunity Mars Rovers which landed in 2004. Regions of known reflectance and chromaticity (red, yellow, green, blue and shades of gray) are used to set the white balance of the camera. The central stalk has a very low reflectance and also serves as a sundial. In the best traditions of sundials it bears a motto (image courtesy NASA/JPL/Cornell/Jim Bell)

is pointed at a gray or white object and a button is pressed. The camera adjusts its channel gains $\mathbf{J}$ so that equal tristimulus values are produced $R' = G' = B'$, which as we recall results in the desired white chromaticity. For colors other than white, this correction introduces some color error but nevertheless has a satisfactory appearance to the eye. Automatic white balancing is commonly used and involves heuristics to estimate the color temperature of the light source, but it can be fooled by scenes with a predominance of a particular color.

The most practical solution is to use the tristimulus values of three objects with known chromaticity in the scene. This allows the matrix $\mathbf{C}$ in (10.14) to be estimated directly, mapping the tristimulus values from the sensor to XYZ coordinates which are an absolute lighting-independent representation of surface reflectance. From this, the chromaticity of the illumination can also be estimated. This approach has been used for cameras on various Mars rovers, using a calibration target like that shown in ◨ Fig. 10.17. The target can be imaged periodically to update the white balance under changing Martian illumination.

### 10.3.4 **Color Change Due to Absorption**

A final, and extreme, example of problems with color occurs underwater. Consider a robot trying to find a docking station identified by colored targets. As discussed earlier in ▶ Sect. 10.1.1, water acts as a filter that absorbs more red light than blue light. For an object underwater, this filtering affects both the illumination falling on the object and the reflected light, the luminance, on its way to the camera. Consider again the red brick

```
>>> lmbda = np.arange(400, 701) * nm;
>>> R = loadspectrum(lmbda, "redbrick");
```

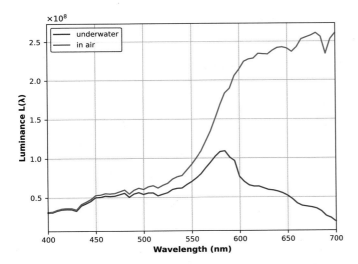

■ **Fig. 10.18** Comparison of the luminance spectrum for the red brick when viewed in air and underwater

which is now 1 m underwater and with a camera a further 1 m from the brick. The illumination on the water's surface is that of sunlight at ground level

```
>>> sun = loadspectrum(lmbda, "solar");
```

The absorption spectrum of water is

```
>>> A = loadspectrum(lmbda, "water");
```

and the total optical path length through the water is

```
>>> d = 2;
```

The transmission $T$ is given by Beer's law (10.2).

```
>>> T = 10.0 ** (-d * A);
```

and the resulting luminance of the brick is

```
>>> L = sun * R * T;
```

which is shown in ■ Fig. 10.18. We see that the longer wavelengths, the reds, have been strongly attenuated. The apparent color of the brick is

```
>>> xy_water = lambda2xy(lmbda, L)
array([ 0.3742,   0.3823])
```

**Excurse 10.15: Lambertian Reflection**
A non-mirror-like or matte surface is a diffuse reflector and the amount of light reflected at a particular angle from the surface normal is proportional to the cosine of the reflection angle $\theta_r$. This is known as Lambertian reflection after the Swiss mathematician and physicist Johann Heinrich Lambert (1728–1777). A consequence is that the object has the same apparent brightness at all viewing angles. See also specular reflection in ▶ Exc. 13.10.

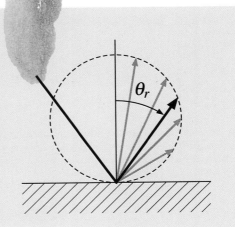

which is also plotted in the chromaticity diagram of ◻ Fig. 10.16. The brick appears much more blue than it did before. In reality, underwater imaging is more complex than this due to the scattering of light by tiny suspended particles which reflect ambient light into the camera that has not been reflected from the target.

### 10.3.5 Dichromatic Reflection

The simple reflection model introduced in ▸ Sect. 10.1.3 is suitable for objects with matte surfaces (e.g. paper, unfinished wood) but if the surface is somewhat shiny the light reflected from the object will have two components – the dichromatic reflection model – as shown in ◻ Fig. 10.19a. One component is the illuminant specularly reflected from the surface without spectral change – the interface or Fresnel reflection. The other is light that interacts with the surface: penetrating, scattering, undergoing selective spectral absorbance and being re-emitted in all directions as modeled by Lambertian reflection. The relative amounts of these two components depends on the material and the geometry of the light source, observer and surface normal.

A good example of this can be seen in ◻ Fig. 10.19b. Both tomatoes appear red which is due to the scattering lightpath where the light has interacted with the surface of the fruit. However each fruit has an area of specular reflection that appears to be white, the color of the light source, not the surface of the fruit.

The real world is more complex still, due to inter-reflections. For example green light reflected from the leaves will fall on the red fruit and be scattered. Some of that light will be reflected off the green leaves again, and so on – nearby objects influence each other's color in complex ways. To achieve photorealistic results in computer graphics all these effects need to be modeled based on detailed knowledge of surface reflection properties and the geometry of all surfaces. In robotics we rarely have this information, so we need to develop algorithms that are robust to these effects.

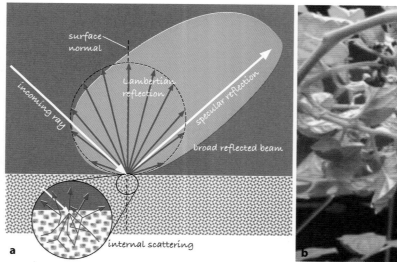

◻ **Fig. 10.19** Dichromatic reflection. **a** Some incoming light undergoes specular reflection from the surface (white arrow), while light that penetrates the surface is scattered, filtered and re-emitted in all directions according to the Lambertian reflection model (red arrows). **b** Specular surface reflection can be seen clearly in the nonred *highlight* areas on the two tomatoes, these are reflections of the ceiling lights (image courtesy of Distributed Robot Garden project, MIT)

### 10.3.6 Gamma

CRT monitors were once ubiquitous and the luminance produced at the face of the display was nonlinearly related to the control voltage $V$ according to

$$L = V^\gamma \tag{10.15}$$

where $\gamma \approx 2.2$. To correct for this, early video cameras applied the inverse nonlinearity $V = L^{1/\gamma}$ to their output signal which resulted in a system that was linear from end to end. ▶ Both transformations are commonly referred to as gamma correction though more properly the camera-end operation is gamma encoding and the display-end operation is gamma decoding. ▶

LCD displays have a stronger nonlinearity than CRTs, but correction tables are applied within the display to make it follow the *standard* $\gamma = 2.2$ behavior of the obsolete CRT. ▶

To show the effect of display gamma we create a simple test pattern

```
>>> wedge = np.linspace(0, 1, 11).reshape(1,-1);
>>> Image(wedge).disp();
```

that is shown in ◼ Fig. 10.20 and is like a photographer's *grayscale step wedge*. If we display this on our computer screen it will appear differently to the one printed in the book. We will most likely observe a large change in brightness between the second and third block – the effect of the gamma-decoding nonlinearity (10.15) in the display of your computer.

If we apply gamma encoding

```
>>> Image(wedge ** (1 / 2.2)).disp();
```

we observe that the intensity changes appear to be more linear and closer to the one printed in the book.

🛈 **Color Spaces and Gamma**
The chromaticity coordinates of (10.9) and (10.10) are computed as ratios of tristimulus values which are linearly related to luminance in the scene. The nonlinearity applied to the camera output must be corrected, i.e. gamma decoded *before* any colometric operations. The Toolbox function `gamma_decode` performs this operation. Gamma decoding can also be performed when an image is loaded using the `gamma` argument to the class method `Image.Read()`.

Today most digital cameras ▶ encode images in sRGB format (IEC 61966-2-1 standard) which uses the ITU Rec. 709 primaries and the gamma-encoding function

$$E' = \begin{cases} 12.92L, & L \leq 0.0031308 \\ 1.055L^{1/2.4} - 0.055, & L > 0.0031308 \end{cases} \tag{10.16}$$

which comprise a linear function for small values and a power law for larger values. The overall gamma is approximately $1/2.2$.

Some cameras have an option to choose gamma as either 1 or 0.45 ($= 1/2.2$).

Gamma encoding and decoding are often referred to as gamma compression and gamma decompression respectively, since the encoding operation compresses the range of the signal, while decoding decompresses it.

Macintosh computers are an exception and prior to macOS 10.6 used $\gamma = 1.8$ which made colors appear brighter and more vivid.

The EXIF metadata of a JPEG file (JFIF file format) has a tag `Color Space` which is set to sRGB, or `Uncalibrated` if the gamma or color model is not known. See ▶ Sect. 11.1.1.

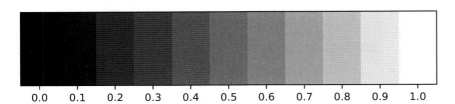

◼ **Fig. 10.20**  The linear intensity wedge

**⚠ Video Color Spaces: YUV and YC$_B$C$_R$**

For most colorspaces such as *HSV* or *xyY* the chromaticity coordinates are invariant to changes in intensity. Many digital video devices provide output in YUV or YC$_B$C$_R$ format which has a luminance component Y and two chromaticity components which are color difference signals such that U, C$_B$ ∝ B′ − Y′ and V, C$_R$ ∝ R′ − Y′ where R′, B′ are gamma-*encoded* tristimulus values, and Y′ is gamma-*encoded* intensity. The gamma nonlinearity means that UV or C$_B$C$_R$ will not be a constant as overall lighting level changes.

The tristimulus values from the camera must be first converted to linear tristimulus values, by applying the appropriate gamma decoding, and then computing chromaticity.

## 10.4 Application: Color Images

### 10.4.1 Comparing Color Spaces

In this section we bring together many of the concepts and tools introduced in this chapter. We will compare the chromaticity coordinates of the colored squares (top three rows) of the GretagMacbeth ColorChecker® chart shown in ◻ Fig. 10.21 using the *xy*- and L*a*b*-color spaces. We compute chromaticity from first principles using the spectral reflectance information for each square which is provided with the Toolbox

```
>>> lmbda = np.linspace(400, 701, 100) * nm;
>>> macbeth = loadspectrum(lmbda, "macbeth");
```

which has 24 columns, one per square of the test chart. We load the relative power spectrum of the D$_{65}$ standard white illuminant

```
>>> d65 = loadspectrum(lmbda, "D65") * 3e9;
```

and scale it to a brightness comparable to sunlight as shown in ◻ Fig. 10.3a. Then for each nongray square, squares 0 to 17,

```
>>> XYZ = np.empty((18, 3));
>>> Lab = np.empty((18, 3));
>>> for i in range(18):
...    L = macbeth[:,i] * d65;
```

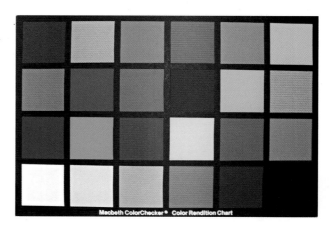

**◻ Fig. 10.21** The GretagMacbeth ColorChecker® is an array of 24 printed color squares (numbered left to right, top to bottom), which includes different grays and colors as well as spectral simulations of skin, sky, foliage etc. Spectral data for the squares is provided with the Toolbox

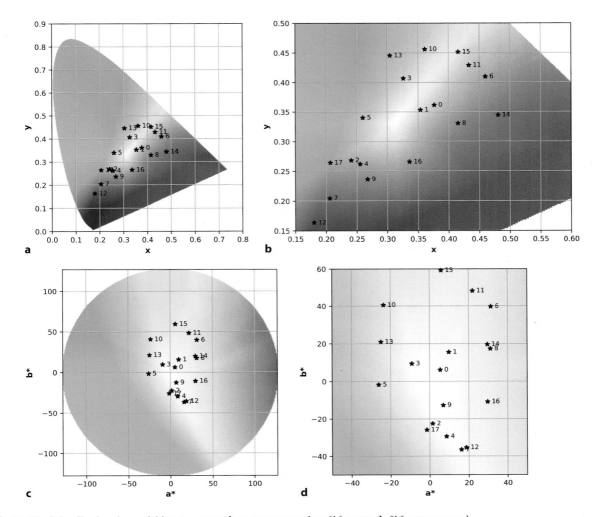

■ **Fig. 10.22** Color Checker chromaticities. **a** *xy*-space; **b** *xy*-space zoomed; **c** a*b*-space; **d** a*b*-space zoomed

```
...     RGB = np.maximum(cmfrgb(lmbda, L), 0);
...     XYZ[i,:] = colorspace_convert(RGB, "rgb", "xyz");
...     Lab[i,:] = colorspace_convert(RGB, "rgb", "L*a*b*");
```

we compute the luminance spectrum (line 4). Then we use the CIE color matching functions to determine the eye's tristimulus response and impose the gamut limits (line 5). This is converted to the XYZ color space (line 6), and the L*a*b* color space (line 7). Next we convert xyz to a chromaticity coordinate, and extract the a*b* values of Lab

```
>>> xy = tristim2cc(XYZ);
>>> ab = Lab[:, 1:];
>>> xy.shape, ab.shape
((18, 2), (18, 2))
```

giving two matrices, each 18 × 2, with one row per colored square. Finally, we plot these points on their respective color planes and the results are displayed in ■ Fig. 10.22. We see, for example, that square 15 is closer to 9 and further from 7 in the a*b*-plane. The L*a*b* color space was designed so that the Euclidean distance between points is approximately proportional to the color difference perceived by humans. If we are using algorithms to distinguish objects by color then L*a*b* would be preferred over RGB or XYZ.

### 10.4.2  **Shadow Removal**

For a robot vision system that operates outdoors, shadows are a significant problem as we can see in ◨ Fig. 10.23a. Shadows cause surfaces of the same type to appear quite different and this is problematic for a robot trying to use vision to understand the scene and plan where to drive. Even more problematic is that this effect is not constant – it varies with the time of day and cloud condition. The image in ◨ Fig. 10.23b has had the effects of shadowing removed, and we can now see very clearly the different types of terrain – grass and gravel.

The key to removing shadows comes from the observation that the bright parts of the scene are illuminated directly by the sun, while the darker shadowed regions are illuminated by the sky. Both the sun and the sky can be modeled as blackbody radiators with color temperatures as listed in ◨ Table 10.4. Shadows therefore have two defining characteristics: they are dark and they have a slight blue tint.

We model the camera using (10.4), but assume that the spectral response of the camera's color sensors are Dirac functions $M(\lambda) = \delta(\lambda - \lambda_x)$ which allows us to eliminate the integrals

$$R = E(\lambda_R)R(\lambda_R)M_R(\lambda_R)$$
$$G = E(\lambda_G)R(\lambda_G)M_G(\lambda_G)$$
$$B = E(\lambda_B)R(\lambda_B)M_B(\lambda_B)$$

For each pixel we compute chromaticity coordinates $r = R/G$ and $b = B/G$ which are invariant to change in illumination magnitude.

$$r = \frac{E(\lambda_R)R(\lambda_R)M_R(\lambda_R)}{E(\lambda_G)R(\lambda_G)M_R(\lambda_G)} = \frac{\frac{2hc^2}{\lambda^5\left(e^{hc/k\lambda_R T}-1\right)}R(\lambda_R)M_R(\lambda_R)}{\frac{2hc^2}{\lambda^5\left(e^{hc/k\lambda_G T}-1\right)}R(\lambda_G)M_R(\lambda_G)}$$

To simplify further we apply the Wien approximation, eliminating the $-1$ term, which is a reasonable approximation for color temperatures in the range under consideration, and now we can write

$$r \approx \frac{e^{hc/k\lambda_G T}R(\lambda_R)M_R(\lambda_R)}{e^{hc/k\lambda_R T}R(\lambda_G)M_G(\lambda_G)} = e^{hc(1/\lambda_G - 1/\lambda_R)/kT}\frac{M_R(\lambda_R)}{M_G(\lambda_G)}\frac{R(\lambda_R)}{R(\lambda_G)}$$

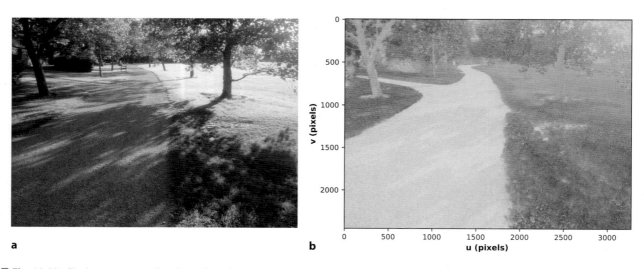

**a**    **b**

◨ **Fig. 10.23** Shadows create confounding effects in images. **a** View of a park with strong shadows; **b** the shadow invariant image in which the variation lighting has been almost entirely removed (Corke et al. 2013) which is a function of color temperature $T$ and various constants: physical constants c, h and k; sensor response wavelength $\lambda_x$ and magnitude $M_x(\lambda_x)$, and material properties $R(\lambda_x)$. Taking the logarithm we obtain the very simple form

$$\log r = c_1 - \frac{c_2}{T} \tag{10.17}$$

and repeating the process for blue chromaticity we can write

$$\log b = c_1' - \frac{c_2'}{T} \tag{10.18}$$

Every color pixel $(R, G, B) \in \mathbb{R}^3$ can be mapped to a point $(\log r, \log b) \in \mathbb{R}^2$ and as the color temperature changes the points will all move along lines with a slope of $c_2'/c_2$. Therefore a projection onto the orthogonal direction, a line with slope $c_2/c_2'$, results in a 1-dimensional quantity

$$s = -c_2 \log r + c_2' \log b$$

that is invariant to the color temperature of the illuminant. We can compute this for every pixel in an image

```
>>> im = Image.Read("parks.png", gamma="sRGB", dtype="float")
Image: 3264 x 2448 (float32), R:G:B [.../images/parks.png]
>>> s = shadow_invariant(im.image, 0.7);
>>> Image(s).disp(interpolation="none", badcolor="red");
```

where the second argument to `shadow_invariant` is the slope of the line in radians, and the function accepts and returns an image as a NumPy array. The result is shown in ◘ Fig. 10.23b, and pixels have a grayscale value that is a complex function of material reflectance and camera sensor properties. The options to the `disp` method help to highlight fine detail in the image and flag, in red, the pixels that had a divide-by-zero error in computing chromaticity.

To achieve this result we have made some approximations and a number of rather strong assumptions: the camera has a linear response from scene luminance to RGB-tristimulus values, the color channels of the camera have non-overlapping spectral response, and the scene is illuminated by blackbody light sources. The first assumption means that we need to use a camera with $\gamma = 1$ or apply gamma decoding to the image before we proceed. The second is far from true, especially for the red and green channels of a color camera, yet the method works well in practice. The biggest effect is that the points move along a line with a slope different to $c_2'/c_2$ but we can estimate the slope empirically by looking at a set of shadowed and nonshadowed pixels corresponding to the same material in the scene

```
>>> theta = esttheta(im)
```

which will prompt you to select a region and returns an angle which can be passed to `shadow_invariant`. The final assumption means that the technique will not work for non-incandescent light sources, or where the scene is partly illuminated by reflections from colored surfaces. More details are provided in the function source code.

## 10.5 Wrapping Up

We have learned that the light we see is electromagnetic radiation with a mixture of wavelengths, a continuous spectrum, which is modified by reflectance and absorption. The spectrum elicits a response from the eye which we interpret as color – for humans the response is a tristimulus value, a 3-vector that represents the outputs

10

**Excurse 10.16: Beyond Human Vision**

Consumer cameras are functionally equivalent to the human eye and are sensitive to the visible spectrum and return tristimulus data. More sophisticated cameras do not have these limitations.

**■■ Infrared Cameras**

are sensitive to infrared radiation and a number of infrared bands are defined by CIE:

- IR-A (700–1400 nm), near infrared (NIR)
- IR-B (1400–3000 nm), short-wavelength infrared (SWIR)
- IR-C (3000 nm–1000 μm) which has subbands:
  - medium-wavelength (MWIR, 3000–8000 nm)
  - long-wavelength (LWIR, 8000–15,000 nm), Cameras in this band are also called thermal or thermographic.

**■■ Ultraviolet Cameras**

are sensitive to ultraviolet region (NUV, 200–380 nm) and are used in industrial applications such as detecting corona discharge from high-voltage electrical systems.

**■■ Hyperspectral Cameras**

have more than three classes of photoreceptors. They sample the incoming spectrum, typically from infrared to ultraviolet, at tens or even hundreds of wavelengths. Hyperspectral cameras are used for applications including aerial survey classification of land-use and identification of the mineral composition of rocks.

of the three different types of cones in our eye. A digital color camera is functionally equivalent. The tristimulus can be considered as a 1-dimensional brightness coordinate and a 2-dimensional chromaticity coordinate which allows colors to be plotted on a plane. The spectral colors form a locus on this plane and all real colors lie within this locus. Any three primary colors form a triangle on this plane which is the gamut of those primaries. Any color within the triangle can be matched by an appropriate mixture of those primaries. No set of primaries can define a gamut that contains all colors. An alternative set of imaginary primaries, the CIE XYZ system, does contain all real colors and is the standard way to describe colors. Tristimulus values can be transformed using linear transformations to account for different sets of primaries. Nonlinear transformations can be used to describe tristimulus values in terms of human-centric qualities such as hue and saturation. We also discussed the definition of white, color temperature, color constancy, the problem of white balancing, the nonlinear response of display devices and how this effects the common representation of images and video.

We learned that the colors and brightness we perceive is a function of the light source and the surface properties of the object. While humans are quite able to "factor out" illumination change this remains a significant challenge for robotic vision systems. We finished up by showing how to remove shadows in an outdoor color image.

### 10.5.1 Further Reading

At face value color is a simple concept that we learn in kindergarten, but as we delve in we find it is a fascinating and complex topic with a massive literature. In this chapter we have only begun to scrape the surface of photometry and colorimetry. Photometry is the part of the science of radiometry concerned with measurement of visible light. It is challenging for engineers and computer scientists since it makes use of uncommon units such as lumen, steradian, nit, candela and lux. One source of complexity is that words like intensity and brightness are synonyms in everyday speech but have very specific meanings in photometry. Colorimetry is the science of color perception, and is also a large and complex area since human perception of color depends on the individual observer, ambient illumination and even the field of view. Colorimetry is however critically important in the design of cameras, computer displays, video equipment and printers. Comprehensive online information about computer vision is available through CVonline at ► https://homepages.inf.ed.ac.uk/rbf/CVonline, and the material in this chapter is covered by the section *Image Physics*.

The computer vision textbooks by Gonzalez and Woods (2018) and Forsyth and Ponce (2012) each have a discussion on color and color spaces. The latter also has a discussion on the effects of shading and inter-reflections. The book by Gevers et al. (2012) is a solid introduction to color vision theory and covers the dichromatic reflectance model in detail. It also covers computer vision algorithms that deal with the challenges of color constancy. The Retinex theory is described in Land and McCann (1971). Other resources related to color constancy can be found at ► http://colorconstancy.com.

Readable and comprehensive books on color science include Berns (2019), Koenderink (2010), Hunt (1987) and from a television or engineering perspective Benson (1986). A more conversational approach is given by Hunter and Harold (1987), which also covers other aspects of appearance such as gloss and luster. Charles Poynton has for a long time maintained excellent online tutorials about color spaces and gamma at ► http://www.poynton.com. His book (Poynton 2012) is an excellent and readable introduction to these topics while also discussing digital video systems in great depth. Comprehensive Python code for color science can be found at ► https://www.colour-science.org.

The CIE standard (Commission Internationale de l'Éclairage 1987) is definitive but hard reading. The work of the CIE is ongoing and its standards are periodically updated at ► https://www.cie.co.at. The color matching functions were first tabulated in 1931 and revised in 1964.

■■ **General Interest**

Crone (1999) covers the history of theories of human vision and color. How the human visual system works, from the eye to perception, is described in two very readable books Stone (2012) and Gregory (1997). Land and Nilsson (2002) describe the design principles behind animal eyes and how characteristics such as acuity, field of view and low light capability are optimized for different species.

### 10.5.2 Data Sources

The Toolbox contains a number of data files describing various spectra which are summarized in ◘ Table 10.5. Each file has as its first column the wavelength in meters. The files have different wavelength ranges and intervals but the helper function `loadspectrum` interpolates the data to the user specified wavelengths.

Several internet sites contain spectral data in tabular format and this is linked from the book's web site. This includes reflectivity data for over 2000 materials provided by NASA's online ASTER spectral library 2.0 (Baldridge et al. 2009) at ► https://speclib.jpl.nasa.gov and the Spectral Database from the University of Eastern Finland Color Research Laboratory at ► https://uef.fi/en/spectral. Data on cone responses and CIE color matching functions is available from the Colour & Vision Research Laboratory at University College London at ► http://www.cvrl.org. CIE data is also available online at ► https://cie.co.at.

### 10.5.3 Exercises

1. You are a blackbody radiator! Plot your own blackbody emission spectrum. What is your peak emission frequency? What is the name of that region of the electromagnetic spectrum? What sort of sensor would you use to detect this?

■ **Table 10.5** Various spectra are shipped with the Toolbox. Relative luminosity values lie in the interval [0,1], and relative spectral power distribution (SPD) are normalized to a value of 1.0 at 550 nm. These files can be loaded using the Toolbox `loadspectrum` function

| Filename | Units | Description |
|----------|-------|-------------|
| cones | Rel. luminosity | Spectral response of human cones |
| bb2 | Rel. luminosity | Spectral response of Sony ICX 204AK sensor used in Point Grey BumbleBee2 camera |
| photopic | Rel. luminosity | CIE 1924 photopic response |
| scotopic | Rel. luminosity | CIE 1951 scoptic response |
| redbrick | Reflectivity | Reflectivity spectrum of a weathered red brick |
| macbeth | Reflectivity | Reflectivity of the Gretag-Macbeth Color Checker array (24 squares), see ■ Fig. 10.21 |
| solar | $W\,m^{-2}\,m^{-1}$ | Solar spectrum at ground level |
| water | $1\,m^{-1}$ | Light absorption spectrum of water |
| D65 | Rel. SPD | CIE Standard $D_{65}$ illuminant |

2. Consider a sensor that measures the amount of radiated power $P_1$ and $P_2$ at wavelengths $\lambda_1$ and $\lambda_2$ respectively. Write an equation to give the temperature $T$ of the blackbody in terms of these quantities.

3. Using the Stefan-Boltzman law compute the power emitted per square meter of the Sun's surface. Compute the total power output of the Sun.

4. Use numerical integration to compute the power emitted in the visible band 400–700 nm per square meter of the Sun's surface.

5. Why is the peak luminosity defined as 683 lmW$^{-1}$?

6. Given typical outdoor illuminance as per ► Sect. 10.2.3 determine the luminous intensity of the Sun.

7. Sunlight at ground level. Of the incoming radiant power determine, in percentage terms, the fraction of infrared, visible and ultraviolet light.

8. Use numerical integration to compute the power emitted in the visible band 400–700 nm per square meter for a tungsten lamp at 2600 K. What fraction is this of the total power emitted?

9. The fovea of the human eye has a 5° field of view. For a vertical surface 2 m away, what diameter circle does this correspond to?

10. Plot and compare the human photopic and scotopic spectral response.
    a) Compare the response curves of human cones and the RGB channels of a color camera. Use `cones.dat` and `bb2.dat`.

11. Can you create a metamer for the red brick?

12. Prove Grassmann's center of gravity law mentioned in ► Sect. 10.2.4.

13. On the $xy$-chromaticity plane plot the locus of a blackbody radiator with temperatures in the range 1000–10,000 K.

14. Plot the XYZ primaries on the $rg$-plane.

15. For ■ Fig. 10.12 determine the chromaticity of the feasible green.

16. Determine the tristimulus values for the red brick using the Rec. 709 primaries.

17. Take a picture of a white object under incandescent illumination (turn off auto white balancing). Determine the average RGB tristimulus value and compute the $xy$-chromaticity. How far off white is it? Determine the color balance matrix $J$ to correct the chromaticity. What is the chromaticity of the illumination?

18. What is the name of the color of the red brick, from ► Sect. 10.2.3 when viewed underwater.

19. Imagine a target like ■ Fig. 10.17 that has three colored patches of known chromaticity. From their observed chromaticity determine the transform from

observed tristimulus values to Rec. 709 primaries. What is the chromaticity of the illumination?

20. Consider an underwater application where a target $d$ meters below the surface is observed through $m$ meters of water, and the water surface is illuminated by sunlight. From the observed chromaticity can you determine the true chromaticity of the target? How sensitive is this estimate to incorrect estimates of $m$ and $d$? If you knew the true chromaticity of the target could you determine its distance?

21. Is it possible that two different colors look the same under a particular lighting condition? Create an example of colors and lighting that would cause this.

22. Use one of your own pictures and the approach of ▶ Sect. 10.4.1. Can you distinguish different objects in the picture?

23. Show analytically or numerically that scaling a tristimulus value has no effect on the chromaticity. What happens if the chromaticity is computed on gamma encoded tristimulus values?

24. Create an interactive tool with sliders for R, G and B that vary the color of a displayed patch. Now modify this for sliders $X$, $Y$ and $Z$ or $x$, $y$ and $Y$.

25. Take a color image and determine how it would appear through 1, 5 and 10 m of water.

26. Determine the names of the colors in the Gretag-Macbeth color checker chart.

27. Plot the color-matching function components shown in ◼ Fig. 10.10 as a 3D curve. Rotate it to see the locus as shown in ◼ Fig. 10.11.

28. What temperature does a blackbody need to be at so that its peak emission matches the peak sensitivity of the human blue cone? Is there any metal still solid at that temperature?

29. Research the retinex algorithm, implement and experiment with it.

# Images and Image Processing

## Contents

© The Author(s), under exclusive license to Springer Nature Switzerland AG 2023
P. Corke, *Robotics, Vision and Control*, Springer Tracts in Advanced Robotics 146,
https://doi.org/10.1007/978-3-031-06469-2_11

`chap11.ipynb`

▶ go.sn.pub/XuZbLo

Image processing is a computational process that transforms one or more input images into an output image. Image processing is frequently used to enhance an image for *human* viewing or interpretation, for example to improve contrast. Alternatively, and of more interest to robotics, it is the foundation for the process of feature extraction which will be discussed in much more detail in the next chapter.

An image is a rectangular array of picture elements (pixels) so we will use a NumPy array to represent an image within a Python session. This allows us to use NumPy's fast and powerful armory of functions and operators for array computing.

We start in ▶ Sect. 11.1 by describing how to load images into a Python session from sources such as files (images and videos), cameras, and the internet. Next, in ▶ Sect. 11.2, we introduce image histograms which provide information about the distribution of pixel values in an image. That sets the scene for discussion about various classes of image processing algorithms. These algorithms operate pixel-wise on a single image, a pair of images, or on local groups of pixels within an image, and we refer to these as monadic, dyadic, and spatial operations respectively. Monadic and dyadic operations are covered in ▶ Sects. 11.3 and 11.4. Spatial operators are described in ▶ Sect. 11.5 and include operations such as smoothing, edge detection, and template matching. A closely related technique is shape-specific filtering, or mathematical morphology, and this is described in ▶ Sect. 11.6. Finally, in ▶ Sect. 11.7 we discuss shape changing operations such as cropping, shrinking, expanding, as well as more complex operations such as scaling, rotation and generalized image warping.

Robots will always gather imperfect images of the world due to noise, shadows, reflections and uneven illumination. In this chapter we discuss some fundamental tools and "tricks of the trade" that can be applied to real-world images.

## 11.1    Obtaining an Image

Today, digital images are ubiquitous since cameras are built into our digital devices and images cost almost nothing to create and share. We each have ever growing personal collections, as well as access to massive online collections of digital images such as Flickr, Instagram and Google Images. We also have access to live image streams from other people's cameras – there are tens of thousands of webcams around the world broadcasting images to the internet, as well images of Earth from space, the Moon and Mars.

### 11.1.1    Images from Files

The images and other data are part of the package `mvtb-images` which is a dependency of the Machine Vision Toolbox. The `iread` function attempts first to read the image in the current folder, and then in the collection of images shipped with the Toolbox.

We start with images stored in files that are provided with the Toolbox, ◄ but you can easily substitute your own images. We import an image into a Python session using the Toolbox function `iread`

```
>>> street, _ = iread("street.png");
```

which returns a NumPy array

```
>>> street
array([[193, 193, 191, ...,    8,    7,    6],
       [189, 190, 191, ...,    7,    7,    8],
       [190, 189, 190, ...,    8,    8,    7],
       ...,
       [ 50,  51,  52, ...,   30,   28,   29],
       [ 53,  54,  53, ...,   28,   28,   29],
       [ 53,  56,  54, ...,   27,   27,   28]], dtype=uint8)
```

that contains a lot of small integers of type `uint8` – unsigned 8-bit integers in the interval $[0, 255]$. These elements are the pixel values, or gray values and represent the brightness of points in the original scene. ▶ For this 8-bit image the pixel values vary from 0 (black) to 255 (white).

The array is two dimensional

```
>>> street.shape
(851, 1280)
```

and has 851 rows and 1280 columns. We normally describe the dimensions of an image in terms of its width × height, so this would be a 1280 × 851 pixel image. The pixel at image coordinate (200, 400)

```
>>> street[400, 200]
24
```

has a low value since it belongs to a point in a dark doorway.

> ❶ **Pixel coordinates and array indices are not the same**
> We write the coordinates of a pixel as $(u, v)$ which, in the Cartesian convention, are the horizontal and vertical coordinates respectively. In NumPy, arrays are indexed by row and column, which is the array element $(v, u)$ – note the reversal of coordinates.

We can easily display this array as an image using Matplotlib

```
>>> idisp(street);
```

and this is shown in ◘ Fig. 11.1. We denote the horizontal axis of an image as $u$ and the coordinate increases left to right. The vertical axis is $v$ and the coordinate increases vertically downward. The top-left pixel is $(0, 0)$.

The toolbar at the bottom of the window allows us to zoom into the image, and then pan and scroll. The coordinates and value of the pixel underneath the cursor is displayed in the right of the toolbar.

Actually, they are the gamma-encoded brightness values as discussed in ▶ Sect. 10.3.6. The values we loaded from the file are approximately the square root of the scene brightness values. Use the gamma argument to iread to perform gamma decoding which results in pixel values proportional to scene luminance.

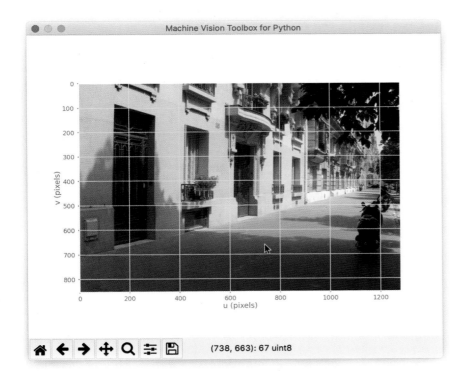

◘ **Fig. 11.1** Grayscale image. The Toolbox image browsing window displayed by the idisp function or the Image class disp method. The toolbar at the bottom allows zooming, panning and scrolling within the image. The gray value of the pixel at the cursor tip, and its coordinate, is displayed in the right of the toolbar

**11**

A very large number of image file formats have been developed and are comprehensively catalogued at ▶ https://en.wikipedia.org/wiki/Image_file_formats. The most popular is JPEG which is used by smart phones, digital cameras and webcams. PNG and GIF are widely used on the web, while TIFF is common in desktop publishing. The internal format of these files is complex, but there are many open-source packages which can read and write them. The Toolbox leverages some of those tools in order to read many common image file formats.

A much simpler set of formats, widely used on Linux systems, are PBM, PGM and PPM (generically PNM) which represent images without compression, and optionally as readable ASCII text. A host of open-source tools such as ImageMagick provide format conversions and image manipulation under Linux, MacOS X and Windows. (word map by ▶ https://tagxedo.com)

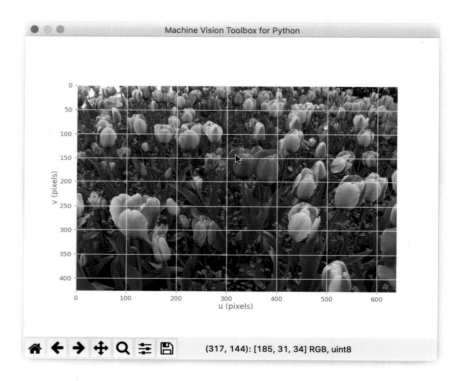

☐ **Fig. 11.2** Color image. The Toolbox image browsing window displayed by the `idisp` function or the `Image` class `disp` method. The toolbar at the bottom allows zooming and scrolling within the image. The tristimulus value of the pixel at the cursor tip (in red, green, blue order), and its coordinate, is displayed in the right of the toolbar. The pixel is on a red petal and we see that the tristimulus has a large red component

Lossless means that the compressed image, when uncompressed, will be exactly the same as the original image.

This image was read from a file called `street.png` which is encoded in the portable network graphics (PNG) format – a lossless compression format ◄ widely used on the internet.

This particular image has no color, it is a grayscale or monochromatic image, but we can just as easily load a color image

```
>>> flowers, _ = iread("flowers8.png")
```

**Excurse 11.2: JPEG**

The JPEG standard defines a *lossy* compression algorithm to reduce the size of an image file. Unlike normal file compression (eg. zip, rar, etc.) and decompression, the decompressed image isn't the same as the original image and this allows much greater levels of compression. JPEG compression exploits limitations of the human eye, discarding information that won't be noticed such as very small color changes (which are perceived less accurately than small changes in brightness) and fine texture. It is very important to remember that JPEG is intended for compressing images that will be *viewed by humans*. The loss of color detail and fine texture may be problematic for computer algorithms that analyze images.

JPEG was designed to work well for natural scenes but it does not do so well on lettering and line drawings with high spatial-frequency content. The degree of loss can be varied by adjusting the "quality factor" which allows a tradeoff between image quality and file size. JPEG can be used for grayscale or color images.

What is commonly referred to as a JPEG file, often with an extension of .jpg or .jpeg, is more correctly a JPEG EXIF (Exchangeable Image File Format) file. EXIF is the format of the file that holds a JPEG-compressed image as well as metadata such as focus, exposure time, aperture, flash and so on. This can be accessed using the metadata method of the Image object, or by command-line utilities such as exiftool (▶ https://exiftool.org). See the Independent JPEG group web site ▶ https://jpegclub.org for more details.

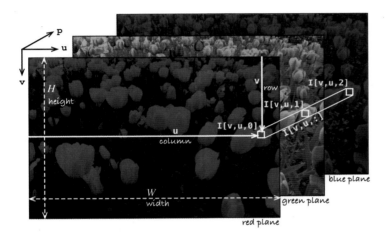

☐ **Fig. 11.3** Color image shown as a 3-dimensional structure with dimensions: row, column, and color plane

Just as for the previous example, the elements of the array are

```
>>> flowers.dtype
dtype('uint8')
```

uint8 values and we can display the image using Matplotlib

```
>>> idisp(flowers);
```

as shown in ☐ Fig. 11.2. Once again there is a toolbar which allows us to zoom into the image, and then pan and scroll. The coordinates and the tristimulus value of the pixel underneath the cursor is shown in the right of the toolbar.

This array is three dimensional

```
>>> flowers.shape
(426, 640, 3)
```

and has 426 rows and 640 columns. The third dimension of the array, shown in ☐ Fig. 11.3, is known as the color plane index. For example

```
>>> idisp(flowers[:, :, 0]);
```

displays the red color plane as a grayscale image – showing the red stimulus at each pixel. The index 1 or 2 would select the green or blue plane respectively.

We can also think of this array as a two dimensional array of one dimensional RGB tristimulus values, each of which is a 3-element vector. For example the pixel at $(318, 276)$

```
>>> pix = flowers[276, 318, :]
array([ 77, 109, 218], dtype=uint8)
```

has a tristimulus value $(77, 109, 218)$. This pixel has a large blue (third) component and corresponds to one of the small blue flowers.

The `iread` function has many options. The argument `mono=True` forces a grayscale image to be returned, by default computed using ITU Rec. 709 which is defined by (10.11). The argument `gamma="sRGB"` performs gamma decoding and returns a linear image where the graylevels, or tristimulus values, are proportional to the luminance of the original scene. The function can also accept a URL instead of a filename, allowing it to load an image from anywhere on the internet.

To simplify working with images, for this and future chapters, we will introduce a class to represent an image. The class holds additional information about the image such as the filename and color plane names, has methods to process images, supports overloaded operators, and hides a lot of low-level detail that gets in the way of understanding concepts. We will illustrate the use of the class by repeating the examples above. For the grayscale image

```
>>> street = Image.Read("street.png")
Image: 1280 x 851 (uint8) [.../images/street.png]
```

which creates an instance of an `Image` object. The class method `Read` takes the same optional arguments as `iread`. The object behaves in many ways just like a NumPy array

```
>>> street.shape
(851, 1280)
```

and we can display the image using its `disp` method

```
>>> street.disp();
```

which behaves just the same as the `idisp` function, and takes the same optional arguments.

The minimum and maximum pixel values are

```
>>> street.min()
0
>>> street.max()
255
```

and a summary of the pixel value statistics can be displayed by

```
>>> street.stats()
range=0 - 255, mean=92.431, sdev=70.340
```

The NumPy array can be accessed by

```
>>> img = street.image;
>>> type(img)
numpy.ndarray
```

and the value of the pixel at $(200, 400)$ is

```
>>> street.image[400, 200]
24
```

We could also slice the NumPy array

```
>>> subimage = street.image[100:200, 200:300];
>>> type(subimage)
numpy.ndarray
```

which is a $100 \times 100$ chunk of the image. Similarly, we could write

```
>>> subimage = street[100:200, 200:300];
>>> type(subimage)
machinevisiontoolbox.classes.Image
```

but now the slice is applied to the `Image` object and it returns another `Image` object representing a $100 \times 100$ chunk of the image.

For the color image

```
>>> flowers = Image.Read("flowers8.png")
Image: 640 x 426 (uint8), R:G:B [.../images/flowers8.png]
```

creates an `Image` instance and we see additional information about the color planes. The range of pixel values, in each plane, can be displayed by

```
>>> flowers.stats()
R: range=0 - 255, mean=83.346, sdev=74.197
G: range=0 - 255, mean=77.923, sdev=42.822
B: range=0 - 255, mean=72.085, sdev=61.307
```

The NumPy array can again be accessed by the `image` property and the tristimulus value of the pixel at $(318, 276)$ is

```
>>> flowers.image[276, 318, :]
array([ 77, 109, 218], dtype=uint8)
```

and the red color plane is

```
>>> flowers.image[:, :, 0];
```

as a 2D NumPy array. Alternatively, we could write

```
>>> flowers[:, :, 0].disp();
```

where the `Image` slice returns an `Image` object rather than a NumPy array. We could also write this as

```
>>> flowers.plane(0).disp();
```

or specify the plane by name

```
>>> flowers.plane("R").disp();
```

or

```
>>> flowers.red().disp();
```

The methods `red`, `green` and `blue` are guaranteed to return the specified color plane, if it is present, irrespective of the color plane order. If multiple color plane names are given, a new multi-plane image will be created, for example

```
>>> flowers.plane("BRG")
Image: 640 x 426 (uint8), B:R:G
```

would return an image with three planes in the order blue, red and green.

The Toolbox can handle arbitrary color plane order, and names. For example, the results of color space conversions will result in multiplane image where the planes are named after the color space components, for example, a hue-saturation-value image has planes labeled "H", "S", and "V". Color plane names can also be separated by colons, for example

```
>>> flowers.plane("B:G:R").disp();
```

which is particularly useful for longer names like "a*" or "b*".

**❗ OpenCV uses BGR color ordering**

The red, green, blue color plane order is almost ubiquitous today, but OpenCV uses a blue, green, red ordering. As much as possible the Toolbox keeps to the RGB ordering.

Many image file formats also contain rich metadata – data about the data in the file. JPEG files, generated by most digital cameras, are particularly comprehensive and the metadata can be retrieved as a dictionary

```
>>> church = Image.Read("church.jpg");
>>> church.metadata()
{'Make': 'Panasonic',
 'Model': 'DMC-FZ30',
 'Software': 'Ver.1.0  ',
 'Orientation': 1,
 'DateTime': '2009:08:22 15:50:06',
 'CompressedBitsPerPixel': 4.0,
 'MaxApertureValue': 3.0,
 'LightSource': 0,
 'Flash': 16,
 'FocalLength': 7.4,
 ...}
```

EXIF tags are standardized, so for any camera the focal length will always be given by the `FocalLength` key.

### 11.1.2  Images from File Sequences

To access a sequence of files we can create an image iterator

```
>>> images = ImageCollection("seq/*.png");
>>> len(images)
9
```

which loads nine images in PNG format from the folder `seq`. We can obtain a particular image by slicing

```
>>> images[3]
Image: 512 x 512 (uint8), id=3 [.../seq/im.004.png]
```

or we can iterate over the images

```
>>> for image in images:
...   image.disp()  # do some operation
```

For a sequence of images in a zip file, we can access the individual image files without having to first unzip the file

```
>>> images = ZipArchive("bridge-l.zip", "*.pgm");
>>> len(images)
251
```

### 11.1.3  Images from an Attached Camera

Most laptop computers have a builtin camera, and additional cameras can be easily attached via a USB port. Accessing a camera is operating system specific, but the Toolbox provides a general interface ◀ to cameras which are identified by consecutive integers starting from zero. We can connect to camera zero by

Under the hood, this is done by OpenCV.

```
>>> camera = VideoCamera(0)
VideoCamera(0) 1920 x 1080 @ 29fps
```

which returns an instance of a `VideoCamera` object and the camera's recording light will turn on. Image size and frame rate are properties of this object, and the constructor accepts the same additional arguments as `iread` and `Image.Read`. The camera is turned off by

```
>>> camera.release()
```

**Excurse 11.3: Photons to Pixel Values**

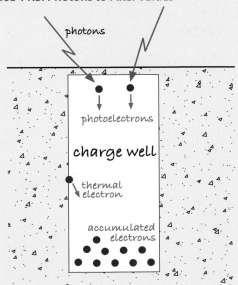

A lot goes on inside a camera. Photosites are the light-sensitive elements of the sensor chip, and are arranged in a dense array. Each photosite is typically square with a side length in the range 1–10 μm. Over a fixed time interval, the number of photons falling on a photosite follows a Poisson distribution where the mean *and* the variance are proportional to the luminance – this variance appears as *shot noise* on the pixel value. A fraction of these photons are converted to electrons – this is the quantum efficiency of the sensor – and they accumulate in a charge well at the photosite. The number of photons captured is proportional to surface area, but not all of a photosite is light sensitive due to the presence of transistors and other devices. The fraction of the photosite's area that is sensitive is called the fill factor and for CMOS sensors can be less than 50%, but this can be improved by fabricating microlenses above each photosite.

The charge well also accumulates thermally generated electrons, the dark current, which is proportional to temperature and is a source of noise – extreme low-light cameras are cooled to reduce this. Another source of noise is pixel nonuniformity due to adjacent pixels having a different gain or offset – uniform illumination therefore leads to pixels with different values which appears as additive noise. The charge well has a maximum capacity and with excessive illumination surplus electrons can overflow into adjacent charge wells leading to artefacts such as flaring and streaking in the image.

At the end of the exposure interval the accumulated charge (thermal *plus* photo-electrons) is read. For low-cost CMOS sensors the charge wells are connected sequentially via a switching network to one or more on-chip analog to digital converters. This results in a rolling shutter and for high speed relative motion this leads to tearing or jello effect as shown in the image. More expensive CMOS and CCD sensors have a global shutter – they make a temporary snapshot copy of the charge in a buffer which is then digitized sequentially.

The exposure on the sensor is

$$H = qL\frac{T}{N^2} \text{ lx s}$$

where $L$ is scene luminance (in nit), $T$ is exposure time, $N$ is the $f$-number (inverse aperture diameter) and $q \approx 0.7$ is a function of focal length, lens quality and vignetting. Exposure time $T$ has an upper bound equal to the inverse frame rate. To avoid motion blur, a short exposure time is needed, but this leads to darker and noisier images.

The integer pixel value is

$$x = kH$$

where $k$ is a gain related to the ISO setting[a] of the camera. To obtain an sRGB (see ▶ Sect. 10.3.6) image with an average value of 118[b] the required exposure is

$$H = \frac{10}{S_{\text{SOS}}} \text{ lx s}$$

where $S_{\text{SOS}}$ is the ISO rating – standard output sensitivity (SOS) – of the digital camera. Higher ISO increases image brightness by greater amplification of the measured charge, but the various noise sources are also amplified leading to increased image noise which is manifested as graininess.

In photography, the camera settings that control image brightness can be combined into a single exposure value (EV)

$$\text{EV} = \log_2 \frac{N^2}{T}$$

and all combinations of $f$-number and shutter speed that have the same EV value yield the same exposure. This allows a tradeoff between aperture (depth of field) and exposure time (motion blur). For most low-end cameras the aperture is fixed and the camera controls exposure using $T$ instead of relying on an expensive, and slow, mechanical aperture. A difference of 1 EV is a factor of two change in exposure, which photographers refer to as a *stop*. Increasing EV – "stopping down" – results in a darker image. Most DSLR cameras allow you to manually adjust EV relative to what the camera's lightmeter has determined.

[a] Which is backward compatible with historical scales (ASA, DIN) devised to reflect the sensitivity of chemical films for cameras – a higher number reflected a more sensitive or "faster" film.

[b] 18% saturation, middle gray, of 8-bit pixels with gamma of 2.2.

### Excurse 11.4: Dynamic Range

The dynamic range of a sensor is the ratio of its largest value to its smallest value. For images, it is useful to express the $\log_2$ of this ratio which makes it equivalent to the photographic concepts of stops or exposure value. Each photosite contains a charge well in which photon-generated electrons are captured during the exposure period (see ▶ Exc. 11.3). The charge well has a finite capacity beyond which the photosite saturates and this defines the maximum value. The minimum number of electrons is not zero, it is a finite number of thermally generated electrons.

An 8-bit image has a dynamic range of around 8 stops, a high-end 10-bit camera has a range of 10 stops, and photographic film is perhaps in the range 10–12 stops but is quite nonlinear.

At a particular state of adaptation, the human eye has a range of 10 stops, but the total adaptation range is an impressive 20 stops. This is achieved by using the iris and slower (tens of minutes) chemical adaptation of the sensitivity of rod cells. Dark adaptation to low luminance is slow, whereas adaptation from dark to bright is faster but sometimes painful.

**11**

An image is obtained using the `grab` method

```
>>> image = camera.grab()
Image: 1920 x 1080 (uint8), R:G:B, id=0
```

The frames are generated at a fixed rate of $R$ frames per second, so the worst case wait is uniformly distributed in the interval $[0, 1/R]$. Ungrabbed images don't accumulate, they are discarded.

which waits until the next frame becomes available then returns an `Image` instance. ◄ The object is also an iterator, so successive frames can be obtained by

```
>>> for image in camera:
...    # process the image
```

### 11.1.4  Images from a Video File

A video file contains a sequence of compresseed images encoded in a video file format such as MPEG4 or AVI. The Toolbox can read individual frames from many popular video file formats, for example

```
>>> video = VideoFile("traffic_sequence.mp4")
VideoFile(traffic_sequence.mp4) 704 x 576, 350 frames @ 30fps
```

returns a `VideoFile` object which is polymorphic with the `VideoCamera` class just described. This video has 350 frames and was captured at 30 frames per second.

The size of each frame ◄ within the video is

This is the shape of the NumPy array containing the image which is `(height, width, ncolors)`. The dimensions given above are `width x height`.

```
>>> video.shape
(576, 704, 3)
```

which indicates that the frames are color images. Frames from the video are obtained by iterating over the `VideoFile` object and we can write a very simple video player

```
>>> for frame in video:
...    frame.disp(reuse=True) # display frame in the same axes
```

where the `reuse` argument prevents `disp` from creating a new set of axes on every iteration.

### 11.1.5 Images from the Web

The term "web camera" has come to mean any USB-connected local camera, but here we use it in the traditional sense as an *internet*-connected camera that runs a web server that can deliver images on request. There are tens of thousands of these web cameras around the world that are pointed at scenes from the mundane to the spectacular. Given the URL of a webcam ▶ we can acquire an image from a camera anywhere in the world.

For example we can connect to a camera at Dartmouth College in New Hampshire

```
>>> dartmouth = WebCam("https://webcam.dartmouth.edu/webcam/
                       image.jpg");
```

which returns a `WebCam` object which is polymorphic with the `VideoCamera` and `VideoFile` classes previously described.

Webcams support a variety of options that can be embedded in the URL but there is no clear standard for these. This example shows a common URL format that works with many web cams.

**◻ Fig. 11.4**  An image from the Dartmouth University webcam which looks out over the main college green

The next image is obtained by iteration, or using the `grab` method

```
>>> dartmouth.grab().disp();
```

and returns a color image such as the one shown in ◻ Fig. 11.4. Webcams are configured by their owner to take pictures periodically, anything from once per second to once per minute. Repeated access will return the same image until the camera takes its next picture.

### 11.1.6  Images from Space

To obtain an API key you need to register on the Google Maps Platform, provide credit card details, and agree to abide by Google's terms and conditions of usage. Google provide tens of thousands of free maploads per month.

We can access satellite views and road maps of anywhere on the planet from inside a Python session. We first create an instance of an `EarthView` object

```
>>> world = EarthView();
```

which requires a Google API key that is given either by passing it as an argument `key="YOUR KEY"` or by setting it as the environment variable `GOOGLE_KEY`. ◄ To

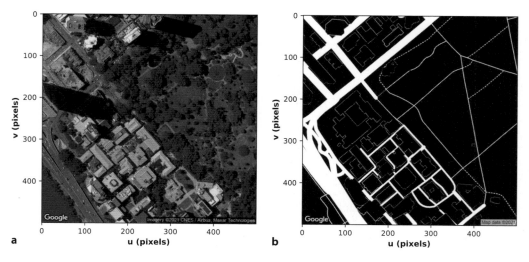

**◻ Fig. 11.5**  Aerial views of Brisbane, Australia. **a** Color aerial image; **b** binary image or occupancy grid where white pixels are driveable roads (images provided by Google, Airbus, Maxar Technologies)

display a satellite image of my university is simply

```
>>> world.grab(-27.475722, 153.0285, 17).disp();
```

which is shown in ◘ Fig. 11.5a. The arguments are latitude, longitude and zoom level ▶.

Instead of a satellite view we can select a road map view

```
>>> world.grab(-27.475722,153.0285, 15, type="map").disp();
```

which shows rich mapping information such as road and place names. A simpler representation is given by

```
>>> world.grab(-27.475722,153.0285, 15, type="roads").disp();
```

which is shown in ◘ Fig. 11.5b as a binary image where white pixels correspond to roads and everything else is black. We could use this as an occupancy grid for robot path planning as discussed in ▶ Sect. 5.4.

Zoom level is an integer such that zero corresponds to a view of the entire Earth. Every increase by one, halves the field of view in the latitude and longitude directions

### 11.1.7 Images from Code

When debugging an algorithm it can be helpful to start with a perfect and simple image, before moving on to more challenging real-world images. The image can be created using any drawing package, exported as an image file, and then imported to Python as already discussed.

Alternatively, we could create the image directly with Python code. The `Image` class has a number of methods to generate a variety of simple patterns including lines, grids of dots or squares, intensity ramps and intensity sinusoids. For example, these class methods create images containing specific patterns

```
>>> image = Image.Ramp(cycles=2, size=500, dir="x");
>>> image = Image.Sin(cycles=5, size=500, dir="y");
>>> image = Image.Squares(number=5, size=500);
>>> image = Image.Circles(number=2, size=500);
```

which are shown in ◘ Fig. 11.6a. The `size` argument specifies the size of the created image, in this case they are all $500 \times 500$ pixels, and the remaining arguments are specific to the type of pattern requested.

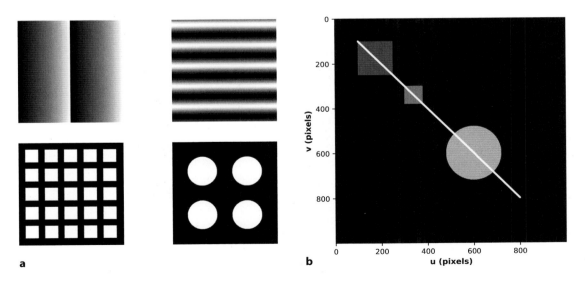

**a**    **b**

◘ **Fig. 11.6** Images from code. **a** Some Toolbox generated test patterns; **b** Simple image created from graphical primitives

We can also construct an image from simple graphical primitives. First, we create an image with `uint8` pixels all equal to zero (black)

```
>>> canvas = Image.Zeros(1000, 1000, dtype="uint8")
Image: 1000 x 1000 (uint8)
```

The `Image` object has methods `draw_xx` which are similar to the Spatial Math Toolbox functions `plot_xx`. The latter use Matplotlib to draw in an onscreen `Axes`, while the former draw into a NumPy array.

and then we draw two squares into it ◄

```
>>> canvas.draw_box(lt=(100, 100), wh=(150, 150), color=100,
...                  thickness=-1);
>>> canvas.draw_box(lt=(300, 300), wh=(80, 80), color=150,
...                  thickness=-1);
```

Top in this case means the lowest $v$ coordinate since $v$ increases downward in the image.

where the arguments are the coordinates of the left-top corner; ◄ the width and height; the value to set the pixels to; and the thickness of the border where `-1` means completely fill the box. We can draw a circle in a similar way

```
>>> canvas.draw_circle((600, 600), 120, color=200, thickness=-1)
```

centered at $(600, 600)$ with a radius of 120 pixels and filled with a gray value of 200. Finally, we draw a line segment onto our canvas

```
>>> canvas.draw_line((100, 100), (800, 800), color=250, thickness=8)
```

which extends from $(100, 100)$ to $(800, 800)$ with a thickness of 8 pixels, and those pixels are all set to 200. The result

```
>>> canvas.disp();
```

is shown in ◘ Fig. 11.6b. We can clearly see that the shapes have different brightness, and we note that the line and the circle show the effects of quantization which results in a *steppy* or jagged shape. ◄

In computer graphics it is common to apply anti-aliasing, use the `antialias=True` option, where edge pixels and edge-adjacent pixels are set to fractional gray values which give the impression of a smoother edge.

❗ Note that all these functions take coordinates expressed in $(u, v)$ notation not NumPy row column notation. The top-left pixel is $(0, 0)$.

## 11.2  Pixel Value Distribution

The distribution of pixel values provides useful information about the quality of the image and the composition of a scene. For the scene

```
>>> church = Image.Read("church.png", mono=True)
Image: 1280 x 851 (uint8) [.../images/church.png]
```

shown in ◘ Fig. 11.9a we can access some simple pixel statistics as `Image` methods

```
>>> church.min()
5
>>> church.max()
238
>>> church.mean()
132.9
>>> church.median()
143
>>> church.std()
57.44
```

or print a single-line summary

```
>>> church.stats()
range=5 - 238, mean=132.906, sdev=57.445
```

The full pixel-value distribution can be seen in the histogram

```
>>> h = church.hist()
histogram with 256 bins: xrange 0.0 - 255.0, yrange 0.0 - 35830.0
>>> h.plot();
```

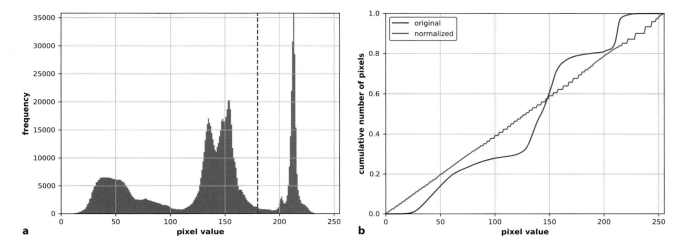

**◘ Fig. 11.7** Church scene. **a** Histogram, **b** normalized cumulative histogram before and after normalization

shown in ◘ Fig. 11.7a. The horizontal axis is the pixel value and the vertical axis is the pixel-value frequency – the number of times that pixel value occurs in the image.

The `hist` method of the `Image` object returns a `Histogram` object which has a number of methods including plotting. The histogram is shown in ◘ Fig. 11.7a and we see that the gray values (horizontal axis) span the range from 5 to 238 which is close to the full range of possible values. If the image was underexposed the histogram area would be shifted to the left. If the image was overexposed the histogram would be shifted to the right and many pixels would have the maximum value. A normalized cumulative histogram

```
>>> h.plot("ncdf", color="blue")
```

is shown as the blue line in ◘ Fig. 11.7b and its use will be discussed in the next section. Histograms can also be computed for color images, in which case the `Histogram` object contains three histograms – one for each image plane.

For this scene, the distribution of pixel values is far from uniform and we see that there are three significant peaks. However if we look more closely we see lots of very minor peaks – the notion of a peak depends on the scale at which we consider the data. The `peaks` method will automatically find the position of the peaks

```
>>> x = h.peaks();
>>> x.shape
(45,)
```

and in this case has found 45 peaks most of which are quite minor. Peaks that are *significant* are not only greater than their immediate neighbors, they are greater than all other values *nearby* – the problem now is to specify what we mean by nearby. For example, the peaks that are separated in the horizontal direction by more than ±25 pixel values are

```
>>> x = h.peaks(scale=25)
array([   213,      154,       40])
```

These are the positions of the three significant peaks that we observe by eye, in order of descending height. The critical part of finding the peaks is choosing the appropriate scale. Peak finding is a topic that we will encounter again in ► Sect. 11.5.2, and is also discussed in ► App. J.

The peaks in the histogram correspond to particular populations of pixels in the image. The lowest peak corresponds to the dark pixels which mostly belong to the ground and the roof. The middle peak mostly corresponds to the sky pixels, and the highest peak mostly corresponds to the white walls. The next section shows how relational operations can be applied to find areas of the image with a particular range of brightness. However, each of the scene elements has a distribution of gray values and for most real scenes we cannot simply map gray level to a scene element. For example, some sky pixels are brighter than some wall pixels, and a very small number of ground pixels are brighter than some sky and wall pixels.

## 11.3    Monadic Operations

Monadic image-processing operations are shown schematically in ■ Fig. 11.8. The result $\mathbf{Y}$ is an image of the same size $W \times H$ as the input image $\mathbf{X}$, and each output pixel is a function of the corresponding input pixel

$$y_{u,v} = f(x_{u,v}), \quad \forall (u, v) \in \mathbf{X} \ .$$

One useful class of monadic functions changes the type of the pixel data. For example to change from `uint8` (integer pixels in the range $[0, 255]$) to floating point values in the range $[0, 1]$ we use the `Image` object's `to` method

```
>>> church_float = church.to("float")
Image: 1280 x 851 (float64)
>>> church.max()
238
>>> church_float.max()
0.9333
```

or vice versa

```
>>> church_float.to("uint8")
Image: 1280 x 851 (uint8)
```

Floating point values can be either 4-byte `float32` type or 8-byte `float64` type. We can also specify this when an image is loaded

```
>>> street_float = Image.Read("street.png", dtype="float")
Image: 1280 x 851 (float32) [.../images/street.png]
```

A color image has three dimensions which we can also consider as a 2-dimensional image where each pixel value is a 3-vector. A monadic operation can convert a color image to a grayscale image, where each output pixel value is a scalar representing the luminance of the corresponding input pixel

```
>>> gray = flowers.mono()
Image: 640 x 426 (uint8)
```

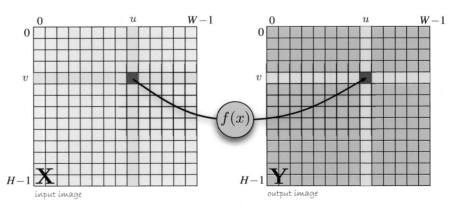

■ **Fig. 11.8**  Monadic image processing operations. Each output pixel is a function of the corresponding input pixel (shown in red)

The inverse operation

```
>>> color = gray.colorize()
Image: 640 x 426 (uint8), R:G:B
```

returns a color image where each color plane is equal to `gray` – when displayed it still appears as a monochrome image. We can create a color image where only the red plane is equal to the input image by

```
>>> color = gray.colorize((1, 0, 0))
Image: 640 x 426 (uint8), R:G:B
```

The result is a red-tinted version of the original image where each output color pixel is the tristimulus $(1, 0, 0)$ scaled by the corresponding grayscale value.

A very common monadic operation is thresholding. This is a logical monadic operation which separates the pixels into two classes according to their intensity

```
>>> bright = (church >= 180)
Image: 1280 x 851 (bool)
```

The result is a logical image where the pixels have the logical or boolean values `True` or `False`. When displayed

```
>>> bright.disp();
```

as shown in ◙ Fig. 11.9b the value `True` corresponds to white pixels – those image pixels in the interval $[180, 255]$ – while `False` pixels are shown as black. Such images, where the pixels have only two values, are known as binary images.

Looking at the image histogram in ◙ Fig. 11.7a, we see that the gray value of 180 lies midway between the second and third peak. This is a good approximation to the optimal strategy for separating pixels belonging to these two populations.

Many monadic operations are concerned with altering the distribution of gray levels within the image. Sometimes an image does not span the full range of available gray levels, for example the image is under- or over-exposed. We can apply a linear mapping to the gray-scale values which ensures that pixel values span the full range ► which might be $[0, 1]$ or $[0, 255]$ depending on the image pixel type, for example

The histogram of such an image will have gaps. If $M$ is the maximum possible pixel value, and $N < M$ is the maximum value in the image, then the stretched image will have at most $N$ unique pixel values, meaning that $M - N$ values cannot occur.

```
>>> church.stretch().stats()
range=0.0 - 1.0, mean=0.549, sdev=0.247
```

A more sophisticated version is histogram normalization or histogram equalization

```
>>> im = church.normhist();
```

and the result, shown in ◙ Fig. 11.9c, has accentuated textural details of wall and sky which previously had a very small gray-level variation. The image pixels have been mapped via the inverse cumulative histogram, the blue line in ◙ Fig. 11.7b. The input pixel values are scaled to the interval $[0, 1]$, and for that value on the vertical axis of the cumulative histogram, we read off the corresponding value on the horizontal axis. The cumulative distribution of the resulting image, the red line in ◙ Fig. 11.7b, is now linear – all gray values are now equally likely to occur.

**❗ Image enhancement does not add information to the image**

Operations such as `stretch` and `normhist` can *enhance* the image from the perspective of a human observer, but it is important to remember that no new information has been added to the image. Subsequent image processing steps will not be improved.

As discussed in ► Sect. 10.3.6 the output of a camera is generally gamma encoded so that the pixel value is a nonlinear function $L^\gamma$ of the luminance sensed at the photosite. Such images can be gamma decoded by a nonlinear monadic operation

```
>>> im = church.gamma_decode(2.2);
```

11

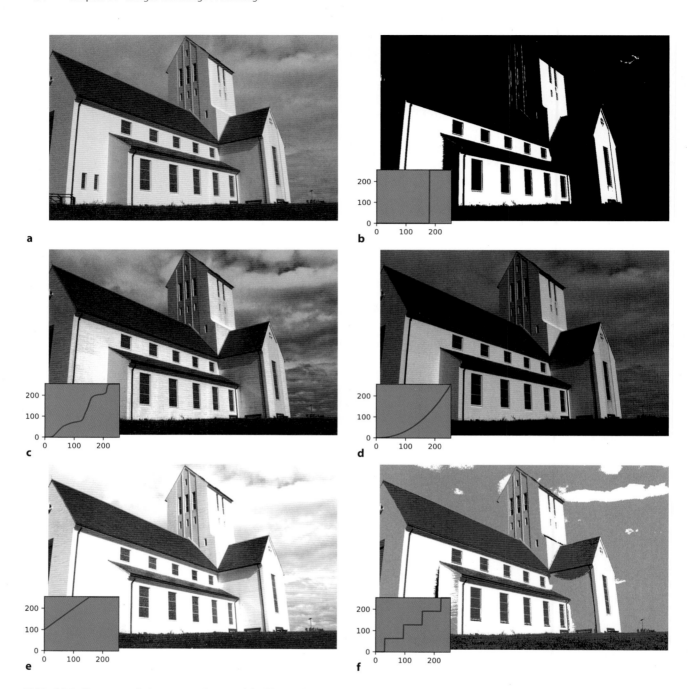

□ **Fig. 11.9** Some monadic image operations: **a** original image; **b** after thresholding; **c** after histogram normalization; **d** after gamma correction; **e** after brightness increase; **f** after posterization. Inset in each figure is a graph showing the mapping from image gray level on the horizontal axis to the output value on the vertical axis

The gamma correction has now been applied twice: once by the gamma_decode method and once in the display device. This makes the resulting image appear to have unnaturally high contrast.

that raises each pixel to the specified power as shown in □ Fig. 11.9d, or

```
>>> im = church.gamma_decode("sRGB");
```

for images encoded with the sRGB gamma standard. ◄

The Image class supports all the Python arithmetic operators like scalar multiplication and division, addition or unary negation. For example

```
>>> (church // 64).disp();
```

creates the pop-art posterization effect by reducing the number of gray levels as shown in ◻ Fig. 11.9f. Integer division results in an image with pixel values in the range [0, 3] and therefore just four different shades of gray.

An `Image` instance has methods to implement many common mathematical functions such as `abs` or `sqrt` which are applied to each pixel in the image. We can apply *any* function that takes, and returns, a scalar by

```
>>> church.apply(lambda x: x // 64, vectorize=True).disp()
```

and in this case the function will be called

```
>>> church.npixels
1089280
```

times which will be quite slow. A more efficient approach, provided that the function can accept and return a NumPy array, is

```
>>> church.apply(lambda x: x // 64).disp();
```

which passes the image as a NumPy array to the function and converts the resulting NumPy array to a new `Image` instance.

If the image is of `uint8` type we can pass every pixel through a 256-element lookup table (LUT). For example

```
>>> lut = [x // 64 for x in range(256)]; # create lookup table
>>> church.LUT(lut).disp();
```

which allows an arbitrary mapping of pixel values – if the pixel value is `i` the result will be `lut[i]`. This is performed very efficiently under the hood by OpenCV.

## 11.4  Dyadic Operations

Dyadic operations are shown schematically in ◻ Fig. 11.10. Two input images result in a single output image, and all three images are of the same size. Each output pixel is a function of the corresponding pixels in the two input images

$$y_{u,v} = f(x_{1:u,v}, x_{2:u,v}), \quad \forall (u, v) \in \mathbf{X}_1, \mathbf{X}_2 .$$

Examples of common dyadic operations include binary arithmetic operators such as addition, subtraction, element-wise multiplication, or NumPy dyadic array functions such as `max`, `min`, and `atan2`.

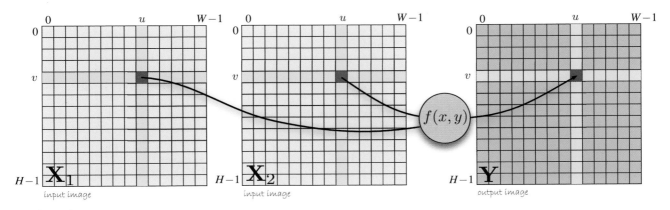

◻ **Fig. 11.10**  Dyadic image processing operations. Each output pixel is a function of the two corresponding input pixels (shown in red)

The Toolbox `Image` objects support Python operator overloading

```
>>> church / 2       # new Image, all pixel values halved
Image: 1280 x 851 (float32)
>>> church + 20      # new Image, all pixel values increased by 20
Image: 1280 x 851 (uint8)
>>> church - church # new Image, all pixel values equal to 0
Image: 1280 x 851 (uint8)
```

We can also apply arbitrary functions to corresponding pairs of pixels

```
>>> church.apply2(church, lambda x, y: x - y, vectorize=True).disp()
```

which is an inefficient way to compute the pixel-wise difference between the church image and itself, since the function will be called over one million times. A more efficient approach is to exploit the fact that the underlying images are stored as NumPy arrays, so

```
>>> church.apply2(church, lambda x, y: x - y).disp();
```

will pass the image array to the lambda function and convert the resulting NumPy array to a new `Image` instance.

There are some important subtleties when performing arithmetic with NumPy `uint8` values. We can demonstrate these by defining two `uint8` values

```
>>> a = np.uint8(100)
100
>>> b = np.uint8(200)
200
```

and performing some simple arithmetic

```
>>> a + b
RuntimeWarning: overflow encountered in ubyte_scalars
44
>>> a - b
RuntimeWarning: overflow encountered in ubyte_scalars
156
```

NumPy performs modular arithmetic and the results are all modulo 256, that is, they have wrapped around from 255 to 0 or vice versa. A stark example of this is

```
>>> -a
156
```

where negating a postive integer results in a positive integer, and no warning. A common remedy is to convert the images to signed integers of a larger size, or floating-point values using `.to("int16")` or `.to("float32")` respectively.

For division

```
>>> a / b
0.5
```

the integers have been automatically promoted to floats leading to a result that is a float.

### 11.4.1 Applications

We will illustrate dyadic operations with two examples.

#### 11.4.1.1 Application: Chroma Keying

Chroma keying is a technique commonly used in television to superimpose the image of a person over some background, for example a weather presenter superimposed over a weather map. The subject is filmed against a blue or green background which makes it quite easy, using just the pixel values, to distinguish between background and the subject. We load an image of a subject taken in front of a green screen

```
>>> foreground = Image.Read("greenscreen.png", dtype="float")
Image: 1024 x 768 (float32), R:G:B [.../images/greenscreen.png]
```

and this is shown in ◨ Fig. 11.11a. We compute the chromaticity coordinates using (10.9)

```
>>> cc = foreground.gamma_decode("sRGB").tristim2cc()
Image: 1024 x 768 (float32), r:g
```

after first converting the gamma-encoded color image to linear tristimulus values. The resulting image has two planes named $r$ and $g$ which are the red and green chromaticity values respectively. In this case, the $g$ image plane is sufficient to distinguish the background pixels. Its histogram

```
>>> cc.plane("g").hist().plot()
```

is shown in ◨ Fig. 11.11b and indicates a large population of pixels around 0.55 which is the green background, and a lower-valued population which belongs to the subject. We can safely say that the subject corresponds to any pixel for which $g < 0.45$ and create a *mask* image

```
>>> mask = cc.plane("g") < 0.45;
>>> mask.disp();
```

where a pixel is `True` (displayed as white) if it is part of the subject as shown in ◨ Fig. 11.11c.

The image of the subject without the background

```
>>> (foreground * mask).disp();
```

is shown in ◨ Fig. 11.11d. Multiplication by a logical image like `mask` performs a per-pixel switching function – the result is the `foreground` pixel value if multiplied by `True` otherwise zero. In this case, `foreground` has three planes so `mask` is applied to each plane, resulting in an `Image` with three planes.

Next we load the desired background image

```
>>> background = Image.Read("road.png", dtype="float")\
...                    .samesize(foreground)
Image: 1024 x 768 (float32), R:G:B
```

which is scaled and cropped to be the same size as our original image. Then we compute the background with a *cutout* for the subject

```
>>> (background * ~mask).disp();
```

which is shown in ◨ Fig. 11.11e. The ~ operator invert the logical image, swapping `True` and `False` values. Finally, we add the subject with no background, to the background with no subject to obtain the subject superimposed over the background

```
>>> composite = foreground * mask  + background * ~mask;
>>> composite.disp();
```

which is shown in ◨ Fig. 11.11f. The technique will of course fail if the subject contains any colors that match the color of the background. ▶ The last step could also be performed using the per-pixel switching method `choose`

```
>>> background.choose(foreground, mask).disp();
```

where each output pixel is selected from the corresponding pixel in `foreground` if the corresponding mask pixel is `True`, otherwise it selects the corresponding pixel from `background`.

Distinguishing foreground objects from the background is an important problem in robot vision, but the terms *foreground* and *background* are ill-defined and

In the early days of television a blue screen was used. Today a green background is more popular because of problems that occur with blue eyes and blue denim clothing.

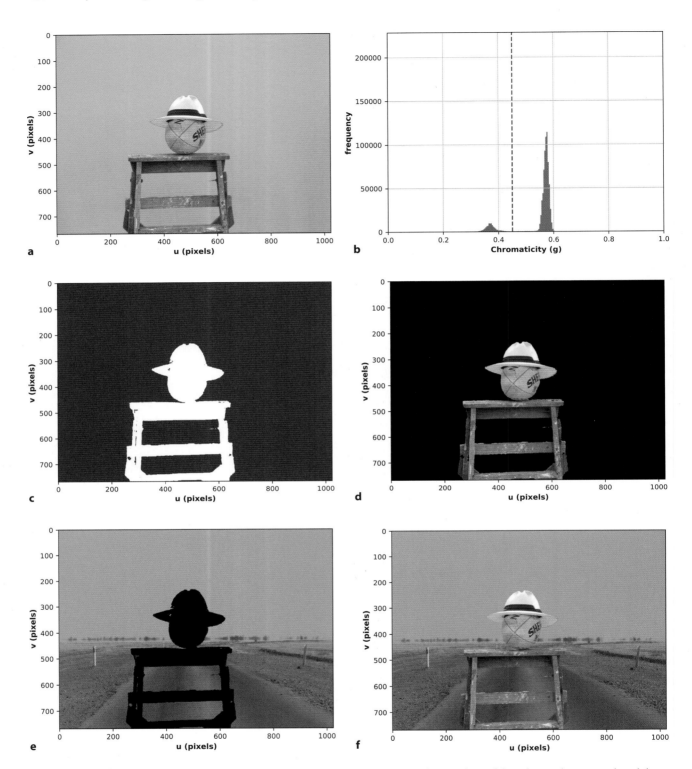

■ **Fig. 11.11** Chroma-keying. **a** The subject against a green background; **b** a histogram of green chromaticity values; **c** the computed mask image where true is white; **d** the subject masked; **e** the background with inverse mask applied; **f** the subject masked into a background scene (green screen image courtesy of Fiona Corke)

application specific. In robotics, we rarely have the luxury of a special background as we did for the chroma-key example. We could instead take a picture of the scene without a foreground object present and consider this to be the background, but that requires that we have special knowledge about when the foreground object is not present. It also assumes that the background does not vary over time. Variation is a significant problem in real-world scenes where ambient illumination and shadows change over quite short time intervals, and the scene may be structurally modified over very long time intervals.

### 11.4.1.2 Application: Motion Detection

In this example we process an image sequence and *estimate* the background, even though there are a number of objects moving in the scene. We will use a recursive algorithm that updates the estimated background image $\hat{\mathbf{B}}$ at each time step, based on the previous estimate and the current image

$$\hat{\mathbf{B}}_{\langle k+1 \rangle} \leftarrow \hat{\mathbf{B}}_{\langle k \rangle} + c(\mathbf{X}_{\langle k \rangle} - \hat{\mathbf{B}}_{\langle k \rangle})$$

where $k$ is the time step and $c(\cdot)$ is a monadic image saturation function

$$c(x) = \begin{cases} \sigma, & x > \sigma \\ x, & -\sigma \leq x \leq \sigma \\ -\sigma, & x < -\sigma \end{cases}$$

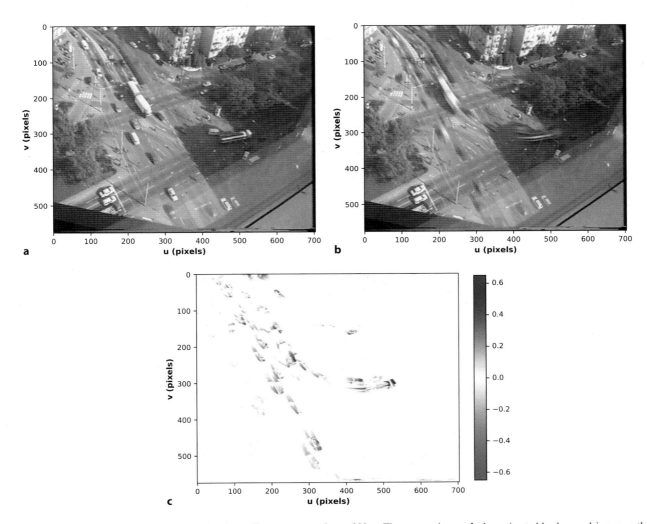

**a**    **b**    **c**

■ **Fig. 11.12** Example of motion detection for the traffic sequence at frame 200. **a** The current image; **b** the estimated background image; **c** the difference between the current and estimated background images where white is zero, red and blue are negative and positive values respectively and magnitude is indicated by color intensity

To demonstrate this we open a video file showing traffic moving through an intersection

```
>>> video = VideoFile("traffic_sequence.mp4", mono=True,
...                   dtype="float")
VideoFile(traffic_sequence.mp4) 704 x 576, 350 frames @ 30fps
```

and then iterate over the frames

```
>>> sigma = 0.02;
>>> background = None
>>> for im in video:
...    if background is None:
...       background = im  # first frame only
...    else:
...       d = im - background
...       background += d.clip(-sigma, sigma)
...    background.disp(reuse=True, fps=video.fps)
```

One frame from this sequence is shown in ◘ Fig. 11.12a. The estimated background image shown in ◘ Fig. 11.12b reveals the static elements of the scene and the moving vehicles have become a faint blur. Subtracting the scene from the estimated background creates an image where pixels are colored where they are different to the background as shown in ◘ Fig. 11.12c. Applying a threshold to the absolute value of this difference image shows the area of the image where there is motion. Of course, where the cars are stationary for long enough they will become part of the background.

## 11.5 Spatial Operations

Spatial operations are shown schematically in ◘ Fig. 11.13. Each pixel in the output image is a function of all pixels in a *region* surrounding the corresponding pixel in the input image

$$y_{u,v} = f(\mathcal{W}_{u,v}), \quad \forall(u,v) \in \mathbf{X}$$

where $\mathcal{W}_{u,v}$, the red shaded region in ◘ Fig. 11.13, is known as a window – a $w \times w$ square region centered on $(u,v)$. $w$ is generally odd and, to ensure this, relevant Toolbox methods accept the half width $h \in \mathbb{N}$ such that $w = 2h + 1$.

Spatial operations are powerful because of the variety of possible functions $f(\cdot)$, linear or nonlinear, that can be applied. The remainder of this section discusses linear spatial operators such as smoothing and edge detection, and some

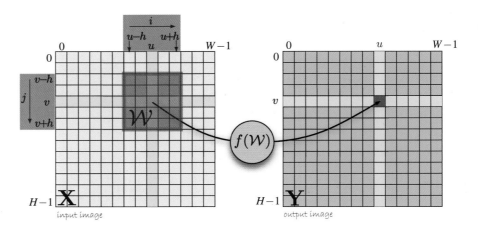

◘ **Fig. 11.13** Spatial image processing operations. The red shaded region shows the window $\mathcal{W}$ that is the set of pixels used to compute the output pixel (shown in red)

nonlinear functions such as template matching, rank filtering, census and rank transforms, and mathematical morphology.

### 11.5.1    Linear Spatial Filtering

A very important linear spatial operator is correlation

$$y_{u,v} = \sum_{(i,j) \in \mathcal{W}} x_{u+i,v+j} k_{i,j}, \quad \forall (u,v) \in \mathbf{X} \tag{11.1}$$

where $\mathbf{K}$ is the kernel, a $w \times w$ array, ▶ whose elements are referred to as the filter coefficients. For every output pixel, the corresponding window of pixels from the input image $\mathcal{W}$ is multiplied element-wise with the kernel $\mathbf{K}$. The center of the window and kernel is considered to be coordinate $(0, 0)$ and $i, j = -h, \ldots, h \subset \mathbb{Z} \times \mathbb{Z}$. This can be considered as the weighted sum of pixels within the window where the weights are defined by the kernel $\mathbf{K}$. Correlation is often written in operator form as

> For implementation purposes the elements of $\mathbf{K}$ might be signed or unsigned integers, or floating point values.

$$\mathbf{Y} = \mathbf{K} \otimes \mathbf{X}$$

A closely-related operation is convolution

$$y_{u,v} = \sum_{(i,j) \in \mathcal{W}} x_{u-i,v-j} k_{i,j}, \quad \forall (u,v) \in \mathbf{X} \tag{11.2}$$

where $\mathbf{K}$ is the convolution kernel, a $w \times w$ array. Note that the sign of the $i$ and $j$ indices has changed in the first term compared to (11.1). Convolution is often written in operator form as

$$\mathbf{Y} = \mathbf{K} * \mathbf{X}$$

As we will see, convolution is the workhorse of image processing and the kernel $\mathbf{K}$ can be chosen to perform functions such as smoothing, gradient calculation or edge detection. Using the Toolbox, convolution is performed using the `convolve` method

```
>>> O = I.convolve(K);
```

---

**Excurse 11.7: Correlation or Convolution?**

These two terms are often used loosely and they have similar, albeit distinct, definitions. Correlation is simply the sum over the element-wise product of the image window and the kernel – for the 1D-case it is the same as the dot product. Convolution is the spatial domain equivalent of frequency domain multiplication and the kernel is the impulse response of a frequency domain filter. Convolution also has many useful mathematical properties that are described in ▶ Exc. 11.8.

The difference in indexing between (11.1) and (11.2) is equivalent to reflecting the kernel – flipping it horizontally and vertically about its center point (as performed by the NumPy `flip` function). Many kernels are symmetric, in which case correlation and convolution yield the same result. However edge detection is always based on non-symmetric kernels so we must take care to apply convolution. We will only use correlation for template matching in ▶ Sect. 11.5.2.

> **Excurse 11.8: Properties of Convolution**
>
> Convolution obeys the familiar rules of algebra. It is commutative
>
> $$\mathbf{A} * \mathbf{B} = \mathbf{B} * \mathbf{A}$$
>
> associative
>
> $$\mathbf{A} * \mathbf{B} * \mathbf{C} = (\mathbf{A} * \mathbf{B}) * \mathbf{C} = \mathbf{A} * (\mathbf{B} * \mathbf{C})$$
>
> distributive (superposition applies)
>
> $$\mathbf{A} * (\mathbf{B} + \mathbf{C}) = \mathbf{A} * \mathbf{B} + \mathbf{A} * \mathbf{C}$$
>
> linear
>
> $$\mathbf{A} * (\alpha \mathbf{B}) = \alpha(\mathbf{A} * \mathbf{B})$$
>
> and shift invariant – the spatial equivalent of time invariance in 1D signal processing – the result of the operation is the same everywhere in the image.

**11**

If `I` has multiple color planes then so will the output image – each output color plane is the convolution of the corresponding input image plane with the kernel `K`.

Convolution and correlation are computationally expensive – an $N \times N$ input image with a $w \times w$ kernel requires $w^2 N^2$ multiplications and additions.

### 11.5.1.1 Image Smoothing

Consider a convolution kernel which is a square $21 \times 21$ array containing equal elements

```
>>> K = Image.Constant(21, 21, value=1/21**2);
>>> K.shape
(21, 21)
```

and of unit volume, that is, its values sum to one. The result of convolving an image with this kernel is an image where each output pixel is the mean of the pixels in a corresponding $21 \times 21$ neighborhood in the input image. As you might expect this averaging

```
>>> mona = Image.Read("monalisa.png", mono=True, dtype="float")
Image: 677 x 700 (float32) [.../images/monalisa.png]
>>> mona.convolve(K).disp();
```

Defocus is actually slightly different, it involves a kernel which is a 2-dimensional Airy pattern or sinc function. The Gaussian function is similar in shape, but is always positive whereas the Airy pattern has low amplitude negative going rings.

See ▶ App. G for more details.

leads to smoothing, blurring or *defocus* ◀ which we see in ◙ Fig. 11.14b.

The kernel can also be created using a classmethod of the `Kernel` class

```
>>> K = Kernel.Box(h=10);
```

which returns the kernel, commonly known as a box filter, as a NumPy array. We have specified the half width of the kernel $h = 10$ and the width will be $w = 2h + 1$ and guaranteed to be an odd number.

A more suitable kernel for smoothing is the 2-dimensional ◀ Gaussian function

$$\mathbf{G}(u, v) = \frac{1}{2\pi\sigma^2} e^{-\frac{u^2+v^2}{2\sigma^2}} \tag{11.3}$$

which is symmetric about the origin and the volume under the curve is unity. Pixels are weighted by their distance from the center. The spread of the Gaussian is

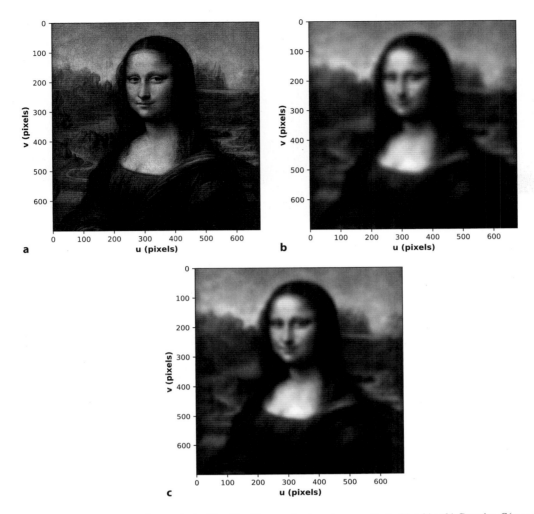

**�‣ Fig. 11.14** Smoothing. **a** Original image; **b** smoothed with a 21 × 21 averaging kernel; **c** smoothed with a 31 × 31 Gaussian $G(\sigma = 5)$ kernel

controlled by the standard deviation parameter $\sigma$. Applying this kernel to the image

```
>>> K = Kernel.Gauss(sigma=5);
>>> mona.convolve(K).disp();
```

produces the result shown in ◪ Fig. 11.14c. Here we have specified the standard deviation of the Gaussian to be 5 pixels. This discrete approximation to the Gaussian is

```
>>> K.shape
(31, 31)
```

a 31 × 31 NumPy array. Smoothing can be achieved conveniently using the `Image` class `smooth` method

```
>>> mona.smooth(sigma=5).disp();
```

Blurring is a counter-intuitive image processing operation since we typically go to a lot of effort to obtain a clear and crisp image. To deliberately *ruin it* seems, at face value, somewhat reckless. However as we will see later, Gaussian smoothing turns out to be extremely useful.

The kernel is itself a 2D array, and therefore we can display it as an image

```
>>> idisp(K);
```

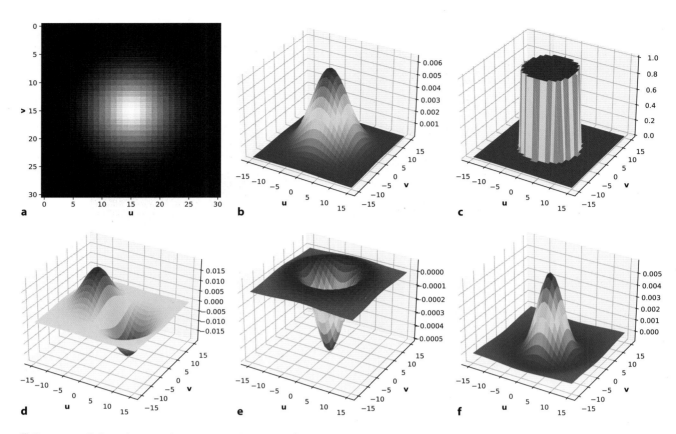

**◻ Fig. 11.15** Gallery of commonly used convolution kernels. $h = 15$, $\sigma = 5$. **a** +Gaussian shown as intensity image, **b** Gaussian, **c** Top hat ($r = 8$), **d** Derivate of Gaussian (DoG), **e** Laplacian of Gaussian (LoG), **f** Difference of Gaussian (DiffG)

which is shown in ◻ Fig. 11.15a. We clearly see the large value at the center of the kernel, and that it falls off smoothly and symmetrically in all directions. We can also display the kernel as a surface

```
>>> span = np.arange(-15, 15 + 1);
>>> X, Y = np.meshgrid(span, span);
>>> plt.subplot(projection="3d").plot_surface(X, Y, K);
```

as shown in ◻ Fig. 11.15b. A crude approximation to the Gaussian is the top-hat kernel which is cylinder with vertical sides rather than a smooth and gentle fall off

### Excurse 11.9: How Wide is my Gaussian?

When choosing a Gaussian kernel we need to consider the standard deviation and the dimensions of the kernel **K** that contains the discrete Gaussian function. Computation time is proportional to $w^2$ so ideally we want the kernel to be no bigger than it needs to be. The Gaussian decreases monotonically in all directions but never reaches zero. Therefore we choose the half-width $h$ of the kernel such that the smallest value of the Gaussian is less than some threshold outside the $w \times w$ convolution kernel.

At the edge of the kernel, a distance $h$ from the center, the value of the Gaussian will be $e^{-h^2/2\sigma^2}$. For $\sigma = 1$ and $h = 2$ the Gaussian will be $e^{-2} \approx 0.14$, for $h = 3$ it will be $e^{-4.5} \approx 0.01$, and for $h = 4$ it will be $e^{-8} \approx 3.4 \times 10^{-4}$. If $h$ is not specified, the Toolbox chooses $h = 3\sigma$. For $\sigma = 1$ that is a $7 \times 7$ kernel which contains all values of the Gaussian greater than 1% of the peak value.

### Excurse 11.10: Separable Filters

Some kernels are separable which means we can obtain the same result by convolving each row with a 1-dimensional kernel, and then each column with another 1-dimensional kernel. The total number of operations is reduced to $2wN^2$, better by a factor of $w/2$. A filter is separable if the kernel is a matrix of rank 1. For example, the simple box filter

```
>>> K=Kernel.Box(h=1)
array([[ 0.1111,    0.1111,    0.1111],
       [ 0.1111,    0.1111,    0.1111],
       [ 0.1111,    0.1111,    0.1111]])
>>> np.linalg.matrix_rank(K)
1
```

is separable. The 1-dimensional kernels are obtained by singular value decomposition

```
>>> U, s, Vh = np.linalg.svd(K, full_matrices=True)
>>> Kh = s[0] * U[:, 0]   # 1D horizontal kernel
array([ -0.1925,  -0.1925,  -0.1925])
>>> Kv = Vh[0, :]         # 1D vertical kernel
array([ -0.5774,  -0.5774,  -0.5774])
```

To cross check, we can convolve the two 1-dimensional kernels, using the outer product, to form the equivalent 2-dimensional kernel

```
>>> np.outer(Kh, Kv)
array([[ 0.1111,    0.1111,    0.1111],
       [ 0.1111,    0.1111,    0.1111],
       [ 0.1111,    0.1111,    0.1111]])
```

which was our original kernel.

The kernel we started with was exactly separable, but we can use this approach to create a separable approximation to a non-separable kernel.

### Excurse 11.11: Properties of the Gaussian

The Gaussian function $G(\cdot)$ has some special properties. The convolution of two Gaussians is another Gaussian

$$\mathbf{G}(\sigma_1) * \mathbf{G}(\sigma_2) = \mathbf{G}\left(\sqrt{\sigma_1^2 + \sigma_2^2}\right)$$

For the case where $\sigma_1 = \sigma_2 = \sigma$ then

$$\mathbf{G}(\sigma) * \mathbf{G}(\sigma) = \mathbf{G}\left(\sqrt{2}\sigma\right)$$

A Gaussian also has the same shape in the spatial and frequency domains.

The 2-dimensional Gaussian is separable – it can be written as the product of two 1-dimensional Gaussians which we can show by

$$\mathbf{G}(u, v) = \left(\frac{1}{\sigma\sqrt{2\pi}}e^{-\frac{u^2}{2\sigma^2}}\right)\left(\frac{1}{\sigma\sqrt{2\pi}}e^{-\frac{v^2}{2\sigma^2}}\right)$$

in amplitude

```
>>> K = Kernel.Circle(radius=8, h=15);
```

as shown in ◾ Fig. 11.15c. The arguments specify a radius of 8 pixels within a window of half width $h = 15$.

### 11.5.1.2 Border Extrapolation

For all spatial operations there is a problem when the window is close to the border of the input image as shown in ◘ Fig. 11.16. In this case the output pixel is a function of a window that contains pixels *beyond the border* of the input image – these pixels have no defined value.

Several techniques are used to extrapolate pixel values outside the image borders, and this is controlled by the optional argument `border` passed to `convolve`. Firstly, we can assume the pixels beyond the border have a particular value. Zero is a common choice, but will tend to darken the output pixels within a distance $h$ of the border. Secondly, we can assume that the border pixels are replicated outside the image. Thirdly, we can consider the border is like a mirror and the image is reflected across the border – this is the default behavior implemented by the `convolve` method.

Another option is to consider that the result is invalid when the window crosses the border of the image and the invalid output pixels are indicated with hatching in ◘ Fig. 11.16. The valid result is a slightly smaller output image of size $(W - 2h) \times (H - 2h)$. This option can be selected by passing the option `border="valid"` to `convolve`.

### 11.5.1.3 Edge Detection

Frequently we are interested in finding the edges of objects in a scene. Consider the image

```
>>> castle = Image.Read("castle.png", mono=True, dtype="float")
Image: 1280 x 960 (float32) [.../images/castle.png]
```

shown in ◘ Fig. 11.17a. It is informative to look at the pixel values along a 1-dimensional profile through the image. A horizontal profile of the image at $v = 360$ is

```
>>> profile = castle.image[360, :];
>>> profile.shape
(1280,)
```

which is a 1D array that we can plot

```
>>> plt.plot(profile);
```

against the horizontal coordinate $u$ in ◘ Fig. 11.17b. The tall spikes correspond to the white letters and other markings on the sign. Looking at one of the spikes more closely, ◘ Fig. 11.17c, we see the intensity profile across the vertical stem of the letter T. The background intensity is $\approx 0.3$ and the white paint intensity is $\approx 0.9$

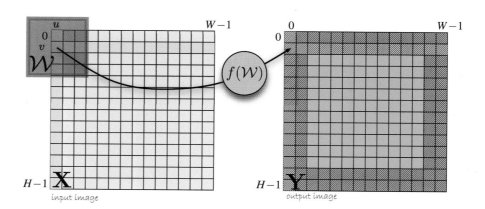

◘ **Fig. 11.16** The output pixel at $(u, v)$ is not defined when the window $\mathcal{W}$ extends beyond the border of the input image. The hatched pixels in the output image are all those for which the output value is not defined

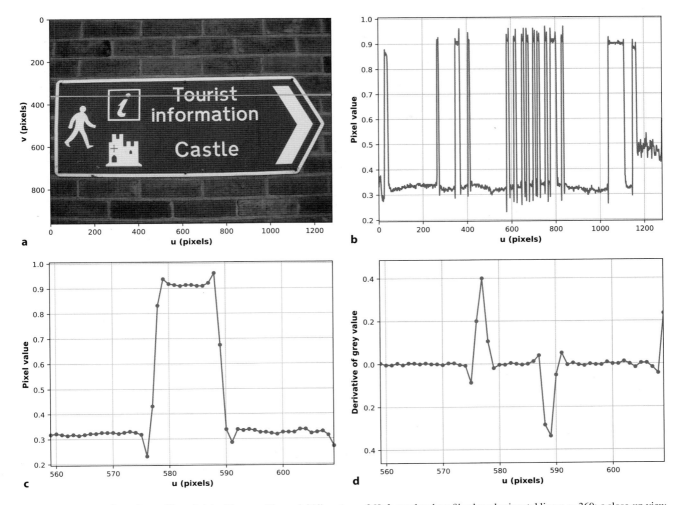

**Fig. 11.17** Edge intensity profile. **a** Original image with overlaid line at $v = 360$; **b** graylevel profile along horizontal line $v = 360$; **c** close-up view of the spike at $u \approx 580$; **d** derivative of **c** (image from the ICDAR 2005 OCR dataset; Lucas 2005)

but both will vary with lighting conditions. The very rapid increase in brightness, over the space of a few pixels, is quite distinctive and a more reliable indication of an edge than any decision based on the actual gray levels.

The discrete first-order derivative along this cross-section is

$$\frac{dp}{du}\bigg|_u = p_u - p_{u-1}$$

which can be computed using the NumPy function `diff`

```
>>> plt.plot(np.diff(profile));
```

and is shown in ◼ Fig. 11.17d. The signal is nominally zero with clear nonzero responses at the edges of an object, in this case the edges of the stem of the letter T.

The derivative at point $u$ can also be written as a *symmetrical* first-order difference

$$\frac{dp}{du}\bigg|_u = \frac{1}{2}(-p_{u-1} + p_{u+1})$$

where the terms have been ordered to match the left-to-right increase in the pixel's $u$-coordinate. This is equivalent to a dot product or *correlation* with the

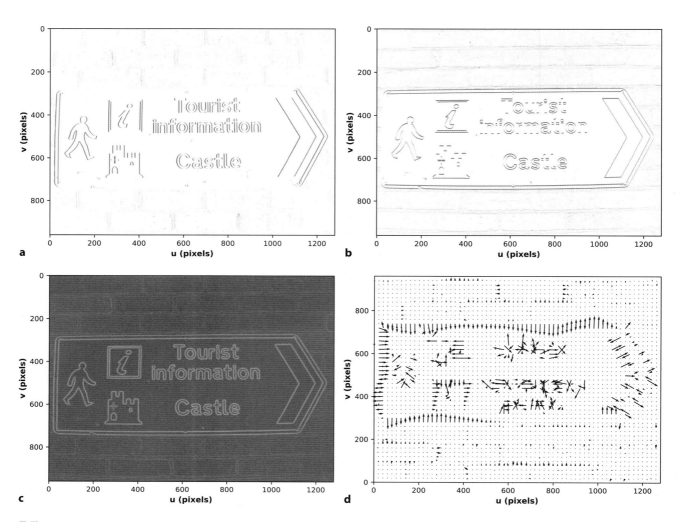

**Fig. 11.18** Edge gradient. **a** *u*-direction gradient computed using 1D kernel; **b** *v*-direction gradient computed using Sobel kernel; **c** derivative of Gaussian magnitude; **d** derivative of Gaussian direction. Gradients shown with blue as positive, red as negative, and white as zero

**11**

1-dimensional kernel of odd length

$$\mathbf{K}_{\mathrm{corr}} = \left( \begin{array}{ccc} -\frac{1}{2} & 0 & \frac{1}{2} \end{array} \right)$$

or *convolution* with the kernel

$$\mathbf{K} = \left( \begin{array}{ccc} \frac{1}{2} & 0 & -\frac{1}{2} \end{array} \right)$$

due to the difference in the indexing between (11.1) and (11.2). Convolving the image with this kernel

```
>>> K = [0.5, 0, -0.5];
>>> castle.convolve(K).disp(colormap="signed");
```

produces the result shown in ■ Fig. 11.18a in which vertical edges, high horizontal gradients, are clearly seen. Since this kernel has signed values the result of the convolution will also be signed, that is, the gradient at a pixel can be positive or negative.

This kernel is often written with the signs reversed which is correct if it is used for correlation. For convolution, the kernel must be written as shown here.

Many convolution kernels have been proposed for computing horizontal gradient. An old, but popular, choice is the Sobel kernel ◀

```
>>> Du = Kernel.Sobel()
array([[  0.125,        0,  -0.125],
```

```
[      0.25,          0,     -0.25],
[      0.125,         0,     -0.125]])
```

and we see that each row is a scaled version of the 1-dimensional kernel κ defined above. The overall result is a weighted sum of the horizontal gradient for the current row, and the rows above and below. Convolving our image with this kernel

```
>>> castle.convolve(Du).disp(colormap="signed");
```

generates a horizontal gradient image very similar to that shown in ◘ Fig. 11.18a for the 1D kernel. Vertical gradient is computed using the transpose of the kernel

```
>>> castle.convolve(Du.T).disp(colormap="signed");
```

and highlights horizontal edges ▸ as shown in ◘ Fig. 11.18b. The notation used for gradients varies considerably across the literature. Most commonly the horizontal and vertical gradient are denoted respectively as $\partial \mathbf{X}/\partial u, \partial \mathbf{X}/\partial v$; $\nabla_u \mathbf{X}, \nabla_v \mathbf{X}$ or $\mathbf{X}_u, \mathbf{X}_v$. In operator form this is written

$$\mathbf{X}_u = \mathbf{D}_u * \mathbf{X}$$
$$\mathbf{X}_v = \mathbf{D}_u^\top * \mathbf{X}$$

where **D** is a horizontal gradient kernel such as Sobel.

Taking the derivative of a signal accentuates high-frequency noise, and all images have noise as discussed in ▸ Exc. 11.3. At the pixel level, noise is a stationary random process – the values are not correlated between pixels. By contrast, the features that we are interested in, such as edges, have correlated changes in pixel value over a larger spatial scale as shown in ◘ Fig. 11.17c. We can reduce the effect of noise by smoothing the image before taking the derivative

$$\mathbf{X}_u = \mathbf{D}_u * (\mathbf{G}(\sigma) * \mathbf{X})$$

Instead of convolving the image with the Gaussian and *then* the derivative, we exploit the associative property of convolution to write

$$\mathbf{X}_u = \mathbf{D}_u * (\mathbf{G}(\sigma) * \mathbf{X}) = (\mathbf{D}_u * \mathbf{G}(\sigma)) * \mathbf{X}$$

We obtain a smooth gradient by convolving the image with the *derivative of the Gaussian*. We can evaluate the kernel numerically by convolving the Sobel and Gaussian kernels

```
>>> from scipy.signal import convolve2d
>>> Gu = convolve2d(Du, Kernel.Gauss(sigma=1));
>>> Gu.shape
(9, 9)
```

or analytically by taking the derivative, in the *u*-direction, of the Gaussian (11.3)

$$\mathbf{G}_u(u, v) = -\frac{u}{2\pi\sigma^4} e^{-\frac{u^2+v^2}{2\sigma^2}} \tag{11.4}$$

which is computed by the class method `Kernel.DGauss` and is shown in ◘ Fig. 11.15d.

The standard deviation $\sigma$ controls the *scale* of the edges that are detected. For large $\sigma$, which implies increased smoothing, edges due to fine texture will be attenuated, leaving only the edges of large features. This ability to find edges at different spatial scale is important, and underpins the concept of scale space that we will discuss in ▸ Sect. 12.3.2. Another interpretation of this operator is as a spatial *bandpass filter* – a cascade of a low-pass filter (smoothing) with a high-pass filter (differentiation).

Computing the horizontal and vertical components of the image gradient at each pixel

```
>>> Xu = castle.convolve(Kernel.DGauss(sigma=2));
>>> Xv = castle.convolve(Kernel.DGauss(sigma=2).T);
```

Filters can be designed to respond to edges at an arbitrary angle. The Sobel kernel itself can be considered as an image and rotated using `Image(Du).rotate()`. To obtain angular precision generally requires a larger kernel such as that generated by (11.4) and implemented by `kdgauss`.

**Excurse 11.12: Carl Friedrich Gauss**

Gauss (1777–1855) was a German mathematician who made major contributions to fields such as number theory, differential geometry, magnetism, astronomy and optics. He was a child prodigy, born in Brunswick, Germany, the only son of uneducated parents. At the age of three he corrected, in his head, a financial error his father had made, and made his first mathematical discoveries while in his teens. Gauss was a perfectionist and a hard worker, but not a prolific writer. He refused to publish anything he did not consider complete and above criticism. It has been suggested that mathematics could have been advanced by fifty years if he had published all of his discoveries. According to legend, Gauss was interrupted in the middle of a problem and told that his wife was dying – he responded "Tell her to wait a moment until I am through".

The normal distribution, or Gaussian function, was not one of his achievements. It was first discovered by de Moivre in 1733 and again by Laplace in 1778. The SI unit for magnetic flux density is named in his honor.

**11**

allows us to compute the magnitude of the gradient at each pixel

```
>>> m = (Xu ** 2 + Xv ** 2).sqrt()
Image: 1280 x 960 (float32)
```

This *edge-strength* image shown in ◨ Fig. 11.18c reveals the edges very distinctly. The direction of the gradient at each pixel is

```
>>> th = Xv.apply2(Xu, np.arctan2);  # arctan2(Xv, Xu)
```

but gradient is better viewed as a sparse quiver plot

```
>>> plt.quiver(castle.uspan(20), castle.vspan(20),
...    Iu.image[::20, ::20], Iv.image[::20, ::20], scale=10);
```

as shown in ◨ Fig. 11.18d. This is much noisier than the magnitude plot but where the edge gradient is strong, on the border of the sign or the edges of letters, the direction is normal to the edge. The fine-scale brick texture appears as almost random edge direction.

The gradient images can be computed conveniently using the method

```
>>> Xu, Xv = castle.gradients(Kernel.DGauss(sigma=2))
```

where the argument overrides the default Sobel kernel.

A well known and very effective edge detector is the Canny edge detector or operator. It uses the edge magnitude and direction that we have just computed, and performs two additional steps. The first is nonlocal-maxima suppression. Consider the gradient magnitude image of ◨ Fig. 11.18c as a 3-dimensional surface where height is proportional to brightness as shown in ◨ Fig. 11.19. We see a series of hills and ridges and we wish to find the pixels that lie along the *ridge lines*. By examining pixel values in a local neighborhood *normal* to the edge direction, that is in the direction of the edge gradient, we can find the maximum value and set all other pixels to zero. The result is a set of nonzero pixels corresponding to peaks and ridge lines. The second step is hysteresis thresholding. For each nonzero pixel that exceeds the upper threshold, a chain is created of adjacent pixels that exceed the lower threshold. Any other pixels are set to zero.

**Excurse 11.13: Pierre-Simon Laplace**

Laplace (1749–1827) was a French mathematician and astronomer who consolidated the theories of mathematical astronomy in his five volume *Mécanique Céleste* (Celestial Mechanics). While a teenager his mathematical ability impressed d'Alembert who helped to procure him a professorship. When asked by Napoleon why he hadn't mentioned God in his book on astronomy he is reported to have said "Je n'avais pas besoin de cette hypothèse-là" ("I have no need of that hypothesis"). He became a count of the Empire in 1806 and later a marquis.

The Laplacian operator, a second-order differential operator, and the Laplace transform are named after him.

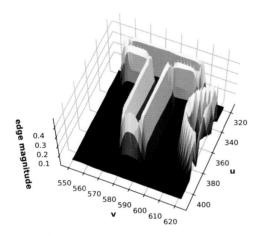

**◼ Fig. 11.19** Closeup of gradient magnitude around the letter T shown as a 3-dimensional surface

To apply the Canny detector to our example image is straightforward

```
>>> edges = castle.canny()
Image: 1280 x 960 (uint8)
```

and returns the image shown in ◼ Fig. 11.20a where the edges are marked by intensity values of 255. We observe that the edges are much thinner than those for the magnitude of derivative of Gaussian operator which is shown in ◼ Fig. 11.20b. The hysteresis threshold parameters can be set with optional arguments.

So far we have considered an edge as a point of high gradient, and nonlocal-maxima suppression has been used to *search* for the maximum value in local neighborhoods. An alternative means to find the point of maximum gradient is to compute the second derivative and determine where it is equal to zero. The Laplacian operator

$$\nabla^2 \mathbf{X} = \frac{\partial^2 \mathbf{X}}{\partial u^2} + \frac{\partial^2 \mathbf{X}}{\partial v^2} = \mathbf{X}_{uu} + \mathbf{X}_{vv} \tag{11.5}$$

is the sum of the second spatial derivative in the horizontal and vertical directions. This is an isotropic filter that responds equally to edges in any direction.

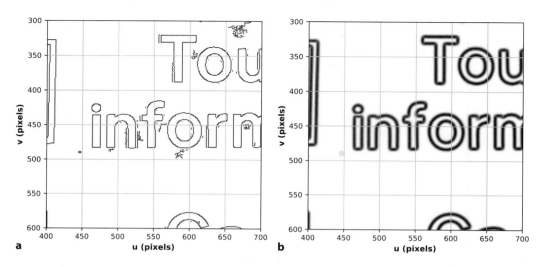

**◻ Fig. 11.20** Comparison of two edge operators: **a** Canny operator with default parameters; **b** Magnitude of derivative of Gaussian kernel ($\sigma = 2$). The Gaussian derivative magnitude operator requires less computation than Canny, but generates thicker edges. For both cases results are shown inverted, white is zero

For a discrete image this can be computed by convolution with the Laplacian kernel

```
>>> L = Kernel.Laplace()
array([[ 0,  1,   0],
       [ 1,  -4,  1],
       [ 0,  1,   0]])
```

The second derivative is even more sensitive to noise than the first derivative and is again commonly used in conjunction with a Gaussian smoothed image

$$\nabla^2 \mathbf{X} = \mathbf{L} * (\mathbf{G}(\sigma) * \mathbf{X}) = \underbrace{(\mathbf{L} * \mathbf{G}(\sigma))}_{\text{LoG}} * \mathbf{X} \tag{11.6}$$

which we combine into the Laplacian of Gaussian (LoG) kernel, and $\mathbf{L}$ is the Laplacian kernel given above. This can be written analytically as

$$\mathbf{LoG}(u, v) = \frac{\partial^2 \mathbf{G}}{\partial u^2} + \frac{\partial^2 \mathbf{G}}{\partial v^2} \tag{11.7}$$

$$= \frac{1}{\pi \sigma^4}\left(\frac{u^2 + v^2}{2\sigma^2} - 1\right)e^{-\frac{u^2+v^2}{2\sigma^2}} \tag{11.8}$$

which is known as the Marr-Hildreth operator or the *Mexican hat* kernel and is shown in ◻ Fig. 11.15e.

We apply this kernel to our image by

```
>>> lap = castle.convolve(Kernel.LoG(sigma=2));
```

and the result is shown in ◻ Fig. 11.21a and b. The maximum gradient occurs where the second derivative is zero, but a significant edge is a zero crossing from a strong positive value (blue) to a strong negative value (red). Consider the close-up view of the Laplacian of the letter T shown in ◻ Fig. 11.21b. We generate a horizontal cross-section of the stem of the letter T at $v = 360$

```
>>> profile = lap.image[360, 570:601];
>>> plt.plot(np.arange(570, 601), profile, "-o");
```

which is shown in ◻ Fig. 11.21c. We see that the zero values of the second derivative lies *between* the pixels. A zero crossing detector selects pixels adjacent to the zero crossing points

```
>>> zc = lap.zerocross();
```

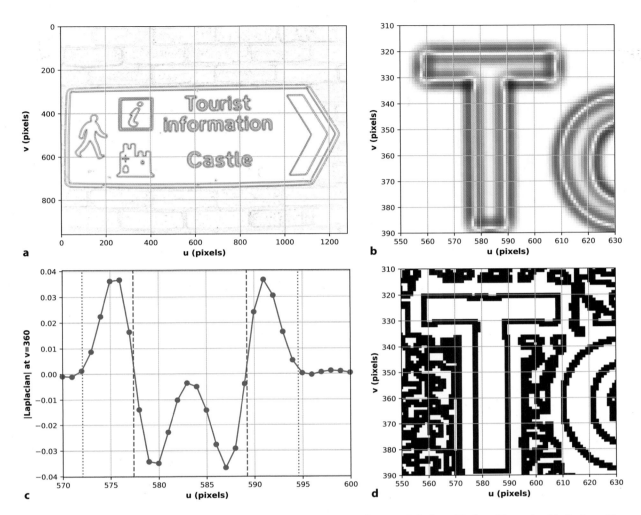

**Fig. 11.21** Laplacian of Gaussian. **a** Laplacian of Gaussian; **b** close-up view of **a** around the letter T where blue and red indicate positive and negative values respectively; **c** a horizontal cross-section of the LoG through the stem of the T; **d** close-up view of the zero-crossing detector output at the letter T

---

**Excurse 11.14: Difference of Gaussians**

The Laplacian of Gaussian (LoG) can be approximated by the difference of two Gaussian functions

$$\mathbf{DoG}(u, v; \sigma_1, \sigma_2) = \mathbf{G}(\sigma_1) - \mathbf{G}(\sigma_2) = \frac{1}{2\pi\sigma_1^2\sigma_2^2}\left(\sigma_2^2 e^{-\frac{u^2+v^2}{2\sigma_1^2}} - \sigma_1^2 e^{-\frac{u^2+v^2}{2\sigma_2^2}}\right)$$

where $\sigma_1 > \sigma_2$ and commonly $\sigma_1 = 1.6\sigma_2$. This is computed by the Toolbox class-method `Kernel.DoG` and shown in ◻ Fig. 11.14f.

This approximation is useful in scale-space sequences which will be discussed in ▶ Sect. 12.3.2. Consider an image sequence where $\mathbf{X}_{\langle k+1 \rangle} = \mathbf{G}(\sigma) \otimes \mathbf{X}_{\langle k \rangle}$, that is, the images are increasingly smoothed. The difference between any two images in the sequence is therefore equivalent to $\mathbf{DoG}(\sqrt{2}\sigma, \sigma)$ applied to the original image.

---

and this is shown in ◻ Fig. 11.21d. We see that each edge appears twice. Referring again to ◻ Fig. 11.21c, we observe weak zero crossings at the points indicated by the dotted lines, and much more definitive zero crossing at the points indicated by the dashed lines.

> ❗ **Images edges are not necessarily the same as object edges**
> A fundamental limitation of all edge detection approaches is that intensity edges do not necessarily delineate the boundaries of objects. The object may have poor contrast with the background which results in weak boundary edges. Conversely, the object may have a stripe on it which is not its edge. Shadows frequently have very sharp edges but are not real objects. Object texture will result in a strong output from an edge detector at points not just on its boundary, as for example with the bricks in ◨ Fig. 11.17b.

### 11.5.2  Template Matching

In our discussion so far we have used kernels that represent mathematical functions such as the Gaussian, its derivative, or its Laplacian. We have also considered the convolution kernel as a matrix, as an image and as a 3-dimensional surface as shown in ◨ Fig. 11.15. In this section we will consider that the kernel *is an image*, or a part of an image, which we refer to as a template. In template matching we wish to find which parts of the input image are most similar to the template.

Template matching is shown schematically in ◨ Fig. 11.22. Each pixel in the output image is given by

$$y_{u,v} = s(\mathcal{W}_{u,v}, \mathbf{T}), \quad \forall (u, v) \in \mathbf{X}$$

where $\mathcal{W}_{u,v}$ is the $w \times w$ window centered at $(u, v)$ in the input image as for convolution and correlation, and $\mathbf{T}$ is the $w \times w$ template, the pattern of pixels we are searching for. Once again $w$ is an odd number.

The function $s(\mathbf{A}, \mathbf{B})$ is a scalar measure that describes the *similarity* of two equally-sized images $\mathbf{A}$ and $\mathbf{B}$. A number of common similarity measures ◄ are given in ◨ Table 11.1. The most intuitive are computed from the pixel-wise difference $\mathbf{A} - \mathbf{B}$ as either the sum of the absolute differences (SAD) or the sum of the squared differences (SSD). These metrics are zero if the images are identical and increase with dissimilarity. It is not easy to say what value of the measure constitutes a poor match but a ranking of similarity measures can be used to determine the *best* match.

More complex measures such as normalized cross correlation yield a score in the interval $[-1, +1]$ with $+1$ for identical regions. In practice, a value greater than 0.8 is considered to be a good match. Normalized cross correlation is computationally expensive – requiring multiplication, division and square root operations. Note that it is possible for the result to be undefined if the denominator is zero,

These measures can be augmented with a Gaussian weighting to deemphasize the differences that occur at the edges of the two windows.

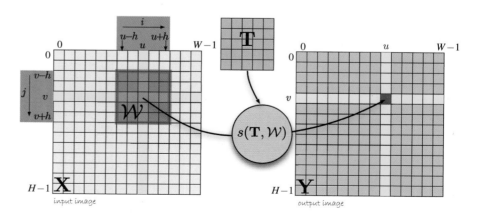

◨ **Fig. 11.22** Template matching. The similarity $s(\cdot)$ between the red shaded region $\mathcal{W}$ and the template $\mathbf{T}$ is computed as the output pixel (shown in red)

**Table 11.1** Similarity measures for two equal sized image regions **A** and **B**. The $Z$-prefix indicates that the measure accounts for the zero-offset or the difference in mean of the two images (Banks and Corke 2001). $\overline{\mathbf{A}}$ and $\overline{\mathbf{B}}$ are the mean of image regions **A** and **B** respectively. `Image` class method names are indicated in the last column

| **Sum of absolute differences** | | |
|---|---|---|
| SAD | $s = \sum_{(u,v)\in\mathbf{A},\mathbf{B}} \lvert a_{u,v} - b_{u,v} \rvert$ | `sad` |
| ZSAD | $s = \sum_{(u,v)\in\mathbf{A},\mathbf{B}} \lvert (a_{u,v} - \overline{\mathbf{A}}) - (b_{u,v} - \overline{\mathbf{B}}) \rvert$ | `zsad` |
| **Sum of squared differences** | | |
| SSD | $s = \sum_{(u,v)\in\mathbf{A},\mathbf{B}} (a_{u,v} - b_{u,v})^2$ | `ssd` |
| ZSSD | $s = \sum_{(u,v)\in\mathbf{A},\mathbf{B}} \left( (a_{u,v} - \overline{\mathbf{A}}) - (b_{u,v} - \overline{\mathbf{B}}) \right)^2$ | `zssd` |
| **Cross correlation** | | |
| NCC | $s = \dfrac{\sum_{(u,v)\in\mathbf{A},\mathbf{B}} a_{u,v} b_{u,v}}{\sqrt{\sum_{(u,v)\in\mathbf{A}X} a_{u,v}^2 \cdot \sum_{(u,v)\in\mathbf{B}} b_{u,v}^2}}$ | `ncc` |
| ZNCC | $s = \dfrac{\sum_{(u,v)\in\mathbf{A},\mathbf{B}} (a_{u,v}-\overline{\mathbf{A}})(b_{u,v}-\overline{\mathbf{B}})}{\sqrt{\sum_{(u,v)\in\mathbf{A}} (a_{u,v}-\overline{\mathbf{A}})^2 \cdot \sum_{(u,v)\in\mathbf{B}} (b_{u,v}-\overline{\mathbf{B}})^2}}$ | `zncc` |

which occurs if either **A** or **B** are all zero for NCC, or uniform (all pixels have the same value) for ZNCC.

To illustrate we will use the Mona Lisa's eye as a $51 \times 51$ template

```
>>> mona = Image.Read("monalisa.png", mono=True, dtype="float");
>>> A = mona.roi([170, 220, 245, 295]);
```

and evaluate the three common measures. If $\mathbf{B} \equiv \mathbf{A}$ then it is easily shown that SAD = SSD = 0 and NCC = 1 indicating a perfect match.

```
>>> B = A
>>> A.sad(B)
0.0
>>> A.ssd(B)
0.0
>>> A.ncc(B)
1.0
```

Now consider the case where the two images are of the same scene but one image is darker than the other – the illumination or the camera exposure has changed. In this case $\mathbf{B} = \alpha\mathbf{A}$ and now

```
>>> B = 0.9 * A
>>> A.sad(B)
24.428635
>>> A.ssd(B)
0.2672252
```

these measure indicate a degree of dissimilarity, while the normalized cross-correlation

```
>>> A.ncc(B)
1.0
```

is invariant to the scaling.

Next, consider that the pixel values have an offset ▶ so that $\mathbf{B} = \mathbf{A} + \beta$ and we find that

This could be due to an incorrect black level setting. A camera's black level is the value of a pixel corresponding to no light and is often $> 0$.

```
>>> B = A + 0.1
>>> A.sad(B)
260.1
>>> A.ssd(B)
26.010002
>>> A.ncc(B)
0.9817292
```

all measures now indicate a degree of dissimilarity. The problematic offset can be dealt with by first subtracting from each of **A** and **B** their mean value

```
>>> A.zsad(B)
2.5782734e-05
>>> A.zssd(B)
2.8106684e-13
>>> A.zncc(B)
1.0
```

and these measures now all indicate a perfect match. The $z$-prefix denotes variants of the similarity measures, described above, that are invariant to intensity offset. Only the ZNCC measure

```
>>> B = 0.9 * A + 0.1
>>> A.zncc(B)
1.0000001
```

Note the result here is $> 1$ while it should lie in the interval $[-1, 1]$. This is due to accumulated error in the thousands of arithmetic operations required to compute this value.

is invariant to both gain and offset variation. ◄ All these methods will fail if the images have even a small change in relative rotation or scale.

#### 11.5.2.1 Application: Finding Wally

Consider the problem from the well known children's book "Where's Wally" or "Where's Waldo" – the fun is trying to find Wally's face in a crowd ◄

For copyright reasons the images cannot be reproduced, but the images are included with the Toolbox.

```
>>> crowd = Image.Read("wheres-wally.png", mono=True, dtype="float")
Image: 640 x 640 (float32) [.../images/wheres-wally.png]
>>> crowd.disp();
```

Fortunately we know roughly what he looks like and the template

```
>>> T = Image.Read("wally.png", mono=True, dtype="float")
Image: 21 x 25 (float32) [.../images/wally.png]
>>> T.disp();
```

was extracted from a different image and scaled so that the head is approximately the same width as other heads in the crowd scene (25 pixels wide).

The similarity of our template T to every possible window location is computed by

```
>>> sim = crowd.similarity(T, "zncc")
Image: 620 x 616 (float32)
```

using the matching measure ZNCC. The result

```
>>> sim.disp(colormap="signed", colorbar=True);
```

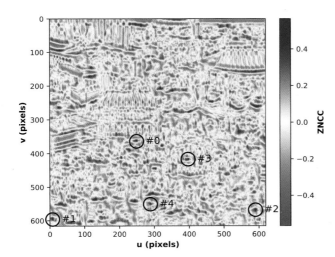

■ **Fig. 11.23** Similarity image S with the top five Wally candidates marked. The color bar indicate the similarity scale

is shown in ■ Fig. 11.23 and the pixel color indicates the ZNCC similarity. We can see a number of spots of high similarity (dark blue) which are candidate positions for Wally. The peak values, with respect to a local $3 \times 3$ window, are

```
>>> maxima, location = sim.peak2d(scale=2, npeaks=5)
>>> maxima
array([ 0.5258,     0.523,    0.5222,    0.5032,     0.5023],
      dtype=float32)
```

in descending order. The first argument controls nonlocal-maxima suppression, it specifies the half-width $h = 2$ of a window in which a valid peak must be greater than all its neigbours. The second argument specifies the number of peaks to return. The largest value 0.5258 is the similarity of the strongest match found.

These matches occur at the coordinates given by the columns of second return value

```
>>> location
array([[250,    8, 590, 397, 289],
       [364, 596, 568, 417, 551]])
```

and we highlight these points on the similarity image

```
>>> crowd.disp();
>>> plot_circle(centre=location, radius=20, color="k");
>>> plot_point(location, color="none", marker="none", text="  #{}");
```

using circles that are numbered sequentially. The best match at $(261, 377)$ is in fact the correct answer – we found Wally! It is interesting to look at the other highly ranked candidates. The two next-best matches, at the bottom of the image, are also characters wearing baseball caps and who therefore look similar.

There are some important points to note from this example. The images have quite low resolution and the template is only $21 \times 25$ – it is a very crude likeness to Wally. The match is not a strong one – only 0.5258 compared to the maximum possible value of 1.0 and there are several contributing factors. The matching measure is not invariant to scale, that is, as the relative scale (zoom) changes the similarity score falls quite quickly. In practice perhaps a 10–20% change in scale between $\mathbf{T}$ and $\mathcal{W}$ can be tolerated. For this example the template was only approximately scaled. Secondly, not all Wallys are the same. Wally in the template is facing forward but the Wally we found in the image is looking to our left. Another problem is that the template and window typically includes pixels from the background as well as the object of interest. As the object moves, the background pixels may change, leading to a lower similarity score. This is known as the mixed pixel problem and is discussed in the next section. Ideally the template should bound the object of interest as tightly as possible. In practice, with real images, another problem arises due to perspective distortion. A square pattern of pixels in the center of the image will appear keystone shaped at the edge of the image and thus will match less well with the square template.

A common problem with template matching is that false matches can occur. In the example above, the second-best candidate had a similarity score only 0.5% lower than the best, the fifth-best candidate was only than 5% lower. In practice a number of rules are applied before a match is accepted: the similarity must exceed some threshold and the first candidate must exceed the second candidate by some factor to ensure there is no ambiguity – this is called a ratio test.

Another approach is to bring more information to bear on the problem, for example, knowledge of camera or object motion. If we tracked Wally from frame to frame in an image sequence, then we would pick the best Wally closest to the his position in the previous frame. Alternatively, we could create a motion model, typically a constant velocity model which assumes he moves approximately the same distance and direction from frame to frame. In this way we could predict Wally's future position, and then we would pick the best match closest to that predicted position. To gain efficiency, we could limit our search to the vicinity of the predicted position. Tracking brings some extra complexity such as having to deal with Wally stopping, changing direction or being temporarily obscured.

### 11.5.2.2 Nonparameteric Local Transforms

Nonparametric similarity measures are more robust to the mixed pixel problem and we can apply a local transform to the image and template before matching. Two common transforms from this class are the census transform and the rank transform.

The census transform maps pixel values from a local region to an integer considered as a bit string – each bit corresponds to one pixel in the region as shown in ◘ Fig. 11.24. If a pixel is greater than the center pixel its corresponding bit is set to one, else it is zero. For a $w \times w$ window the string will be $w^2-1$ bits long. ◄ The two bit strings are compared using a Hamming distance which is the num-

For a 32-bit integer `uint32` this limits the window to $5 \times 5$ unless a sparse mapping is adopted (Humenberger et al. 2009). A 64-bit integer `uint64` supports a $7 \times 7$ window.

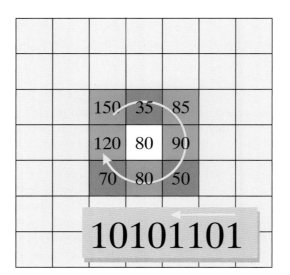

◘ **Fig. 11.24** Example of census and rank transform for a $3 \times 3$ window. Pixels are marked red or blue if they are less than or greater than or equal to the center pixel respectively. These boolean values are then packed into a binary word, in the direction shown, from least significant bit upwards. The census value is $10101101_2$ or decimal 173. The rank transform value is the total number of one bits and is 5

ber of bits that are different. This can be computed by counting the number of set bits in the exclusive-or of the two bit strings. Thus very few arithmetic operations are required compared to the more conventional methods – no square roots or division – and such algorithms are amenable to implementation in special purpose hardware or FPGAs. Another advantage is that intensities are considered relative to the center pixel of the window making it invariant to overall changes in intensity or gradual intensity gradients.

The rank transform maps the pixel values in a local region to a scalar which is the number of elements in the region that are greater than the center pixel. This measure captures the *essence* of the region surrounding the center pixel, and like the census transform it is invariant to overall changes in intensity since it is based on local relative gray-scale values.

These transforms are implemented by `Image` class methods `censusxform` and `rankxform`. They can be used as a pre-processing step applied to each of the images before using a simple classical similarity measure such as SAD.

### 11.5.3 Nonlinear Operations

Another class of spatial operations is based on nonlinear functions of pixels within the window. For example

```
>>> out = mona.window(np.var, h=3);
```

computes the variance of the pixels in *every* $7 \times 7$ window. The arguments specify a function, in this case the NumPy function to compute the variance of an array, and the window half width. The function is called with a $49 \times 1$ vector argument comprising the pixels in the window arranged as a 1D NumPy array, and the function's return value becomes the corresponding output pixel value. This operation acts as an edge detector since it has a low value for homogeneous regions irrespective of their brightness. It is however computationally expensive because the function is called over 470,000 times. Any function that accepts a vector input and returns a scalar can be used in this way.

Rank filters sort the pixels within the window by value, in descending order, and return the specified element from the sorted list. The maximum value over a $5 \times 5$ window about each pixel is the first ranked pixel in the window

```
>>> mx = mona.rank(rank=0, h=2);
```

where the arguments are the pixel rank and the window half-width. The median over a $5 \times 5$ window is the thirteenth in rank

```
>>> med = mona.rank(rank=12, h=2);
```

and is useful as a filter that removes local outlier values. To demonstrate this we will add significant noise to a copy of the Mona Lisa image

```
>>> spotty = mona.copy()
Image: 677 x 700 (float32) [.../images/monalisa.png]
>>> pixels = spotty.view1d();  # create a NumPy 1D view
>>> npix = mona.npixels    # total number of pixels
473900
>>> # choose 10,000 unique pixels
>>> k = np.random.choice(npix, 10_000, replace=True);
>>> pixels[k[:5_000]] = 0  # set half of them to zero
>>> pixels[k[5_000:]] = 1  # set half of them to one
>>> spotty.disp();
```

and this is shown in ◼ Fig. 11.25a. We have set 5000 random pixels to be zero, and another 5000 random pixels to the maximum value. This type of noise is often referred to as impulse noise or salt and pepper noise. We apply a $3 \times 3$ median filter

```
>>> spotty.rank(rank=4, h=1).disp();
```

**11**

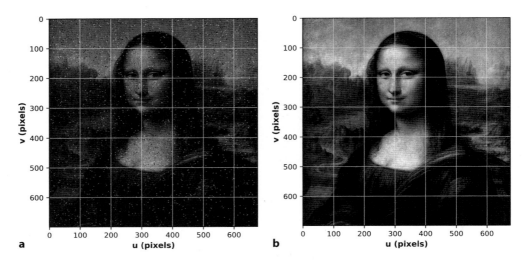

■ **Fig. 11.25**   Median filter cleanup of impulse noise. **a** Noise corrupted image; **b** median filtered result

and the result shown in ■ Fig. 11.25b is considerably improved. Smoothing would also attenuate the noise, but it would also blur the image – median filtering preserves edges in the scene and does not introduce blur.

Instead of specifying the window in terms of its half width, we can specify it by a logical 2D array which is sometimes called a *footprint*

```
>>> M = np.full((3, 3), True);
>>> M[1, 1] = False
>>> M
array([[ True,   True,   True],
       [ True,  False,   True],
       [ True,   True,   True]])
```

which is 3 × 3 in this case

```
>>> max_neighbors = mona.rank(rank=0, footprint=M);
```

The return value is the first in rank (maximum) over a *subset* of pixels from the window corresponding to the `True` elements of `M`. The result is the maximum of the eight neighbors of each pixel in the input image. We can use this

```
>>> (mona > max_neighbors).disp();
```

This how the `peak2d` method works.

to display all those points where the pixel value is greater than its local neighbors. This is an example of nonlocal-maxima suppression. ◄ The array we used here is very similar to a structuring element which we will meet in the next section.

## 11.6   **Mathematical Morphology**

Mathematical morphology is a class of nonlinear spatial operators shown schematically in ■ Fig. 11.26. Each pixel in the output image is a function of a *subset* of pixels in a region surrounding the corresponding pixel in the input image

$$y_{u,v} = f(\mathcal{W}_{u,v}, \mathbf{S}), \qquad \forall (u, v) \in \mathbf{X} \tag{11.9}$$

where $\mathbf{S}$ is the structuring element, a small binary or logical image with odd side lengths. The structuring element is similar to the convolution kernel discussed previously except that now it controls *which pixels* in the window $\mathcal{W}_{u,v}$ the function $f(\cdot)$ is applied to – it specifies a subset of pixels within the window. The selected pixels are those for which the corresponding values of the structuring element are nonzero or `True` – these are shown in red in ■ Fig. 11.26. Mathematical morphology, as its name implies, is concerned with the form or *shape* of objects in the image.

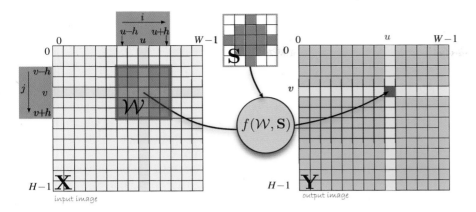

**Fig. 11.26** Morphological image processing operations. The operation is defined only for the elements of $\mathcal{W}$ selected by the corresponding elements of the structuring element **S** shown in red

The easiest way to explain the concept is with a simple example, in this case a synthetic binary image

```
>>> im = Image.Read("eg-morph1.png")
Image: 50 x 40 (uint8) [.../images/eg-morph1.png]
>>> im.disp();
```

which is shown, repeated, down the first column of ◻ Fig. 11.27. The structuring element is shown in red at the end of each row. If we consider the top-most row, the structuring element is a $5 \times 5$ square

```
>>> S1 = np.ones((5, 5));
```

and is applied to the original image using the morphological minimum operation

```
>>> e1 = im.morph(S1, op="min")
Image: 50 x 40 (uint8)
```

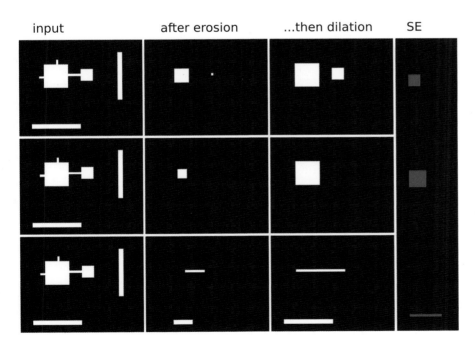

**Fig. 11.27** Mathematical morphology example. Pixels are either 0 (black) or 1 (white). Each column corresponds to processing using the structuring element, shown in the last column in red

and the result is shown in the second column. For each pixel in the input image we take the *minimum* of all pixels in the $5 \times 5$ window. If *any* of those pixels are zero the resulting pixel will be zero. We can see this in animation by

```
>>> morphdemo(im, S, op="min")
```

The result is dramatic – two objects have disappeared entirely and the two squares have become separated and smaller. The two objects that disappeared were not *consistent* with the shape of the structuring element. This is where the connection to morphology or shape comes in – only shapes that could *contain* the structuring element will be present in the output image.

The structuring element could define any shape: a circle, an annulus, a 5-pointed star, a line segment 20 pixels long at 30° to the horizontal, or the silhouette of a duck. Mathematical morphology allows very powerful shape-based filters to be created. The second row shows the results for a larger $7 \times 7$ structuring element which has resulted in the complete elimination of the small square and the further reduction of the large square. The third row shows the results for a structuring element which is a horizontal line segment 14 pixel wide, and the only remaining shapes are long horizontal lines.

The operation we just performed is known as *erosion*, since large objects are eroded and become smaller – in this case the $5 \times 5$ structuring element has caused two pixels ◄ to be *shaved off* all the way around the perimeter of each shape. The small square, originally $5 \times 5$, is now only $1 \times 1$. If we repeated the operation the small square would disappear entirely, and the large square would be reduced even further.

The inverse operation is dilation which makes objects larger. In ◘ Fig. 11.27 we apply dilation to the second column results

```
>>> d1 = e1.morph(S1, op="max")
Image: 50 x 40 (uint8)
```

and the results are shown in the third column. For each pixel in the input image we take the *maximum* of all pixels in the $5 \times 5$ window. If *any* of those neighbors is one, then the resulting pixel will be one. In this case we see that the two squares have returned to their original size, but the large square has lost its protrusions.

Morphological operations can be written in operator form. Erosion is

$$\mathbf{Y} = \mathbf{X} \ominus \mathbf{S}$$

where in (11.9) $f(\cdot) = \min(\cdot)$, and dilation is

$$\mathbf{Y} = \mathbf{X} \oplus \mathbf{S}$$

where in (11.9) $f(\cdot) = \max(\cdot)$. These operations are also known as Minkowski subtraction and addition respectively.

Erosion and dilation are related by

$$\mathbf{X} \oplus \mathbf{S} = \overline{\overline{\mathbf{X}} \ominus \mathbf{S}'}$$

where the bar denotes the logical complement of the pixel values, and the prime denotes reflection about the center pixel. Essentially this states that eroding the white pixels is the same as dilating the dark pixels and vice versa. For morphological operations

$$(\mathbf{X} \oplus \mathbf{S}_1) \oplus \mathbf{S}_2 = \mathbf{X} \oplus (\mathbf{S}_1 \oplus \mathbf{S}_2)$$
$$(\mathbf{X} \ominus \mathbf{S}_1) \ominus \mathbf{S}_2 = \mathbf{X} \ominus (\mathbf{S}_1 \oplus \mathbf{S}_2)$$

which means that successive erosion or dilation with a structuring element is equivalent to the application of a single larger structuring element, but the former is

The half width of the structuring element.

**11**

input        after dilation        ...then erosion        SE

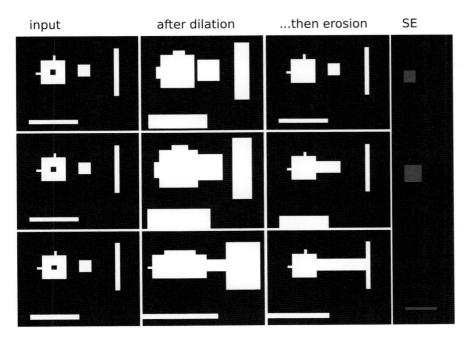

**◘ Fig. 11.28**  Mathematical morphology example. Pixels are either 0 (black) or 1 (white). Each row corresponds to processing using the structuring element, shown in the last column in red

computationally cheaper. ▸ The shorthand functions

```
>>> out = im.erode(S);
>>> out = im.dilate(S);
```

can be used instead of the low-level method `morph`.

The sequence of operations, erosion then dilation, is known as opening since it opens up gaps. In operator form it is written as

$$\mathbf{X} \circ \mathbf{S} = (\mathbf{X} \ominus \mathbf{S}) \oplus \mathbf{S}$$

Not only has the opening selected particular shapes, it has also *cleaned up* the image: the squares have been separated and the protrusions on the large square have been removed since they are not consistent with the shape of the structuring element.

In ◘ Fig. 11.28, using the image `eg-morph2.png`, we perform the operations in the opposite order, dilation then erosion. In the first row no shapes have been lost, they grew then shrank, and the large square still has its protrusions. The hole has been filled since it is not consistent with the shape of the structuring element. In the second row, the larger structuring element has caused the two squares to join together. This sequence of operations is referred to as closing since it closes gaps and is written in operator form as

$$\mathbf{X} \bullet \mathbf{S} = (\mathbf{X} \oplus \mathbf{S}) \ominus \mathbf{S}$$

Note that in the bottom row, the line segment has become attached to the left edge – this is due to the default behavior when the processing window extends beyond the image border. Just as for the case of convolution with `convolve`, these methods have options to control how to handle the situation when the processing window extends beyond the border of the image.

Opening and closing ▸ are implemented by the methods `open` and `close` respectively. Unlike erosion and dilation, repeated application of opening or closing is futile since those operations are idempotent

$$(\mathbf{X} \circ \mathbf{S}) \circ \mathbf{S} = \mathbf{X} \circ \mathbf{S}$$
$$(\mathbf{X} \bullet \mathbf{S}) \bullet \mathbf{S} = \mathbf{X} \bullet \mathbf{S}$$

For example a $3 \times 3$ square structuring element applied twice is equivalent to a $5 \times 5$ square structuring element. The former involves $2 \times (3 \times 3 \times N^2) = 18N^2$ operations whereas the later involves $5 \times 5 \times N^2 = 25N^2$ operations.

These names make sense when considering what happens to white objects against a black background. For black objects the operations perform the inverse function.

These operations can also be applied to grayscale images to emphasize particular-shaped objects in the scene prior to an operation like thresholding. Dilation is the maximum of the selected elements in the moving window, and erosion is the minimum.

### 11.6.1 Noise Removal

A common use of morphological opening is to remove noise in an image. The image

```
>>> objects = Image.Read("segmentation.png")
Image: 640 x 480 (uint8) [.../images/segmentation.png]
>>> objects.disp();
```

shown in ◘ Fig. 11.29a is a noisy binary image from the output of a rather poor thresholding operation. We wish to remove the white pixels that do not belong to the objects, and to fill in the holes in the four white rectangular objects.

We choose a symmetric *circular* structuring element of radius 3

```
>>> S_circle = Kernel.Circle(3)
array([[  0,    0,    0,    1,    0,    0,    0],
       [  0,    1,    1,    1,    1,    1,    0],
```

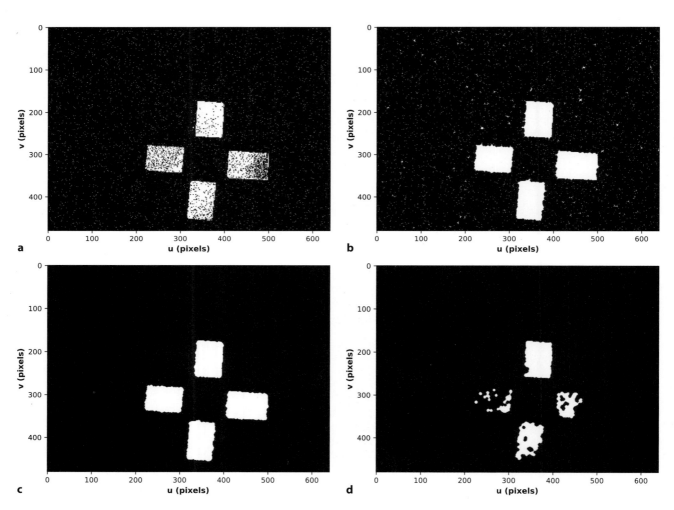

◘ **Fig. 11.29** Morphological cleanup. **a** Original binary image, **b** original after opening, **c** opening then closing, **d** closing then opening. Structuring element is a circle of radius 3

```
    [  0,   1,   1,   1,   1,   1,   0],
    [  1,   1,   1,   1,   1,   1,   1],
    [  0,   1,   1,   1,   1,   1,   0],
    [  0,   1,   1,   1,   1,   1,   0],
    [  0,   0,   0,   1,   0,   0,   0]])
```

and apply a closing operation to fill the holes in the objects

```
>>> closed = objects.close(S_circle);
```

and the result is shown in ◼ Fig. 11.29b. The holes have been filled, but the noise pixels have grown to be small circles and some have agglomerated. We eliminate these by an opening operation

```
>>> clean = closed.open(S_circle);
```

and the result shown in ◼ Fig. 11.29c is a considerably cleaned up image. These operations are not commutative, and if we apply them in the inverse order, opening then closing

```
>>> objects.open(S_circle).close(S_circle).disp();
```

the results as shown in ◼ Fig. 11.29d are much poorer. Although the opening has removed the isolated noise pixels it has removed large chunks of the targets which cannot be restored.

### 11.6.2 Boundary Detection

The top-hat transform uses morphological operations to detect the edges of objects. Continuing the example from above, and using the image `clean` shown in ◼ Fig. 11.29c, we compute its erosion using the circular structuring element

```
>>> eroded = clean.erode(S_circle)
Image: 640 x 480 (uint8)
```

The objects in this image are slightly smaller since the structuring element has caused three pixels to be *shaved off* the outside of each object. Subtracting the eroded image from the original

```
>>> edge = clean - eroded
Image: 640 x 480 (uint8)
>>> edge.disp();
```

results, as shown in ◼ Fig. 11.30, in a set of pixels around the edge of each object.

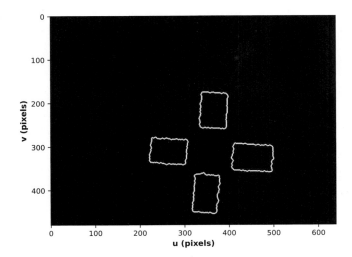

◼ **Fig. 11.30**  Boundary detection by morphological processing

### 11.6.3  Hit or Miss Transform

The hit or miss transform is a morphological operation defined by two structuring elements

$$\mathbf{Y} = (\mathbf{X} \ominus \mathbf{S}_1) \cap (\overline{\mathbf{X}} \ominus \mathbf{S}_2)$$

which is the logical-and of the binary image and its complement, eroded by two different structuring elements. This preserves pixels where ones in the window are consistent with $\mathbf{S}_1$ and zeros in the window are consistent with $\mathbf{S}_2$.

The two structuring elements can be combined as an *interval* $\mathbf{S}_1 - \mathbf{S}_2$ with values of 1 (the corresponding pixel must be a one), -1 (the corresponding pixel must be a zero), or 0 (don't care) as shown in ◘ Fig. 11.31a. The interval must match the underlying image pixels in order for the result to be a one, as shown in ◘ Fig. 11.31b. If there is any mismatch, as shown in ◘ Fig. 11.31c, then the result will be zero. The transform is implemented by the `hitormiss` method, for example

```
>>> out = image.hitormiss(S);
```

where S is the interval.

The hit or miss transform can be applied iteratively with a sequence of intervals to perform complex operations such as skeletonization and linear feature detection. The skeleton of the objects is computed by

```
>>> skeleton = clean.thin();
```

and is shown in ◘ Fig. 11.32a. The lines are a single pixel wide and are the edges of a generalized Voronoi diagram – they delineate sets of pixels according to the shape boundary they are closest to. We can then find the endpoints of the skeleton

```
>>> ends = skeleton.endpoint()
Image: 640 x 480 (bool)
```

and also the triplepoints

```
>>> joins = skeleton.triplepoint();
```

which are points at which three lines join. These are shown in ◘ Fig. 11.32b and c respectively.

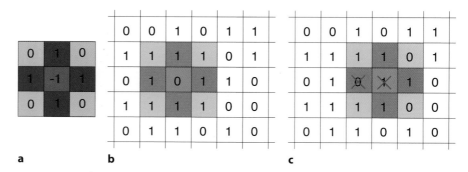

a          b                              c

◘ **Fig. 11.31**  Hit or miss transform. **a** The interval has values of 1 (must be one, blue), -1 (must be zero, red), or 0 (don't care, green); **b** an example of a hit, the underlying pixel pattern is consistent with the overlaid interval; **c** an example of a miss, the underlying pixel pattern is inconsistent with the overlaid interval, the inconsistent pixels are indicated by a cross

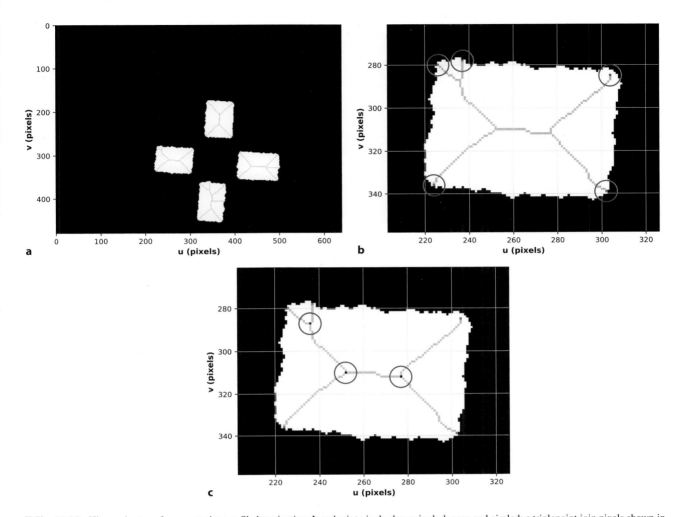

**◻ Fig. 11.32** Hit or miss transform operations. **a** Skeletonization; **b** endpoint pixels shown in dark gray and circled; **c** triplepoint join pixels shown in black and circled

### 11.6.4 Distance Transform

We introduced the distance transform for robot path planning in ▶ Sect. 5.4.1. Given an occupancy grid, it computed the distance of every free cell from the goal location. The distance transform we discuss here ▶ operates on a binary image and the output value, corresponding to every zero pixel in the input image, is the distance to the nearest nonzero pixel.

For path planning in ▶ Sect. 5.4.1 we used a slow iterative wavefront approach to compute the distance transform. Here we use the OpenCV function `distanceTransform`.

Consider the problem of fitting a model to a shape in an image. We create the outline of a rotated square

```
>>> im = Image.Squares(1, size=256).rotate(0.3).canny()
Image: 256 x 256 (uint8)
```

which is shown in ◻ Fig. 11.33a and then compute the distance transform

```
>>> dx = im.distance_transform(norm="L2")
Image: 256 x 256 (float32)
```

which is shown in ◻ Fig. 11.33b. The value of a pixel is its distance from the nearest non-zero pixel in the input image.

An initial estimate of the square is shown in red. The value of the distance transform at any point on this red square indicates how far away it is from the nearest point on the original square. If we summed the distance transform for every

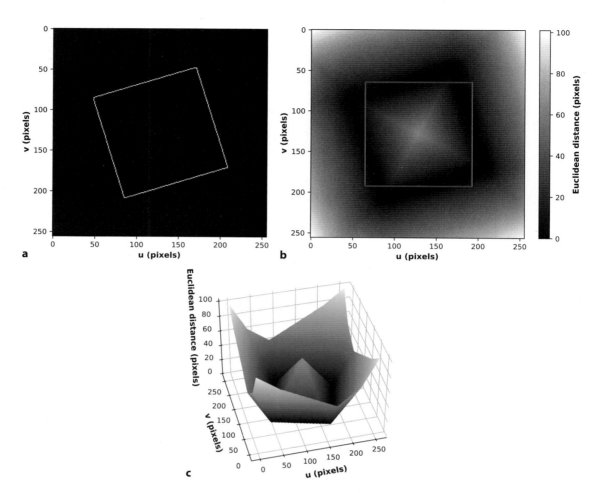

■ **Fig. 11.33** Distance transform. **a** Input binary image; **b** distance transformed input image with overlaid model square; **c** distance transform as a surface

### Excurse 11.16: Distance Transform

The **distance transform** of a binary image has a value at each pixel equal to the distance from that pixel to the nearest nonzero pixel in the input image. The distance metric is typically either Euclidean ($L_2$ norm) or Manhattan distance ($L_1$ norm). It is zero for pixels that are nonzero in the input image. This transform is closely related to the **signed distance function** whose value at any point is the distance of that point to the nearest boundary of a shape, and is positive inside the shape and negative outside the shape. The figure shows the signed distance function for a unit circle, and has a value of zero, indicated by the red plane, at the object boundary. If we consider a shape to be defined by its signed distance transform then its zero contour defines the shape boundary.

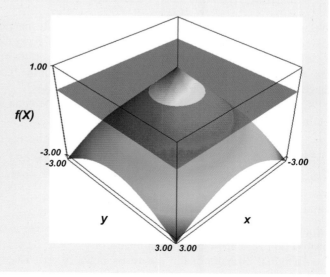

point on the red square, or even just the vertices, we obtain a total distance measure which will only be zero when our model square, in red, overlays the original square. The total distance is a cost function which we can minimize using an optimization routine that adjusts the position, orientation and size of the model. Considering the distance transform as a 3-dimensional surface in ◘ Fig. 11.33c, our problem is analogous to dropping an extensible square hoop into the valley of the distance transform. Note that the distance transform only needs to be computed once, and during model fitting the cost function is simply a lookup of the computed distance. This is an example of chamfer matching and a full example, with optimization, is given in `examples/chamfer_match.py`.

## 11.7 Shape Changing

The final class of image processing operations that we will discuss are those that change the shape or size of an image.

### 11.7.1 Cropping

The simplest shape change of all is selecting a rectangular region from an image which is the familiar cropping operation. Consider the image

```
>>> mona.disp();
```

shown in ◘ Fig. 11.34a from which we interactively specify a region of interest or ROI

```
>>> eyes, roi = mona.roi();
>>> eyes
Image: 137 x 41 (float32)
```

by clicking and dragging a selection box over the image. In this case we selected the eyes, and the corners of the selected region were

```
>>> roi
array([239, 375, 172, 212])
```

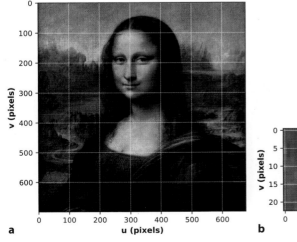
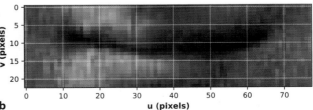

◘ **Fig. 11.34**  Example of region of interest or image cropping. **a** Original image, **b** selected region of interest

where the elements are the left, right, top and bottom coordinates of the box edges. The function can be used noninteractively by specifying a ROI

```
>>> smile = mona.roi([265, 342, 264, 286])
Image: 78 x 23 (float32)
>>> smile.disp();
```

which in this case selects the Mona Lisa's smile shown in ▫ Fig. 11.34b.

### 11.7.2  Image Resizing

Often we wish to reduce the dimensions of an image, perhaps because the large number of pixels results in long processing time, or requires too much memory. We demonstrate this with a high-resolution image

```
>>> roof = Image.Read("roof.png", mono=True)
Image: 2009 x 1668 (uint8) [.../images/roof.png]
```

which is shown in ▫ Fig. 11.35a. The simplest means to reduce image size is subsampling or decimation which selects every $m$-th pixel in the $u$- and $v$-directions, where $m \in \mathbb{N}$ is the subsampling factor. For example with $m = 2$, an $N \times N$ image

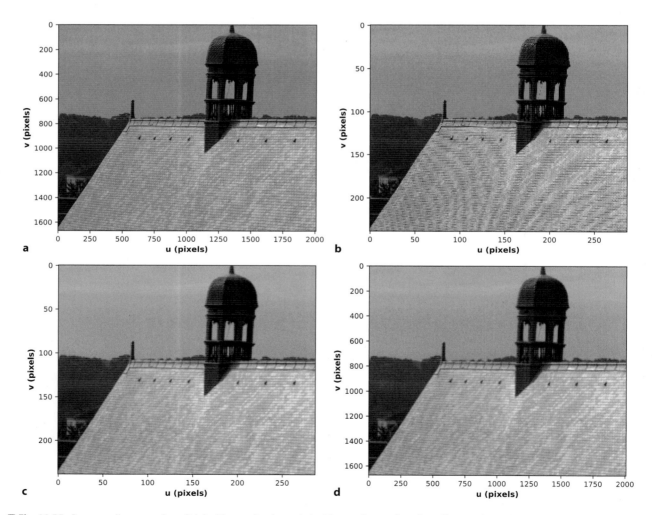

▫ **Fig. 11.35** Image scaling example. **a** Original image; **b** subsampled with $m = 7$, note the axis scaling; **c** subsampled with $m = 7$ after smoothing; **d** image **c** restored to original size by pixel replication

becomes an $N/2 \times N/2$ images which has one quarter the number of pixels of the original image.

For this example, we will reduce the image size by a factor of seven in each direction

```
>>> roof[::7, ::7].disp();
```

which selects every seventh pixel along the rows and columns. The result is shown is shown in ◘ Fig. 11.35b and we observe some pronounced curved lines on the roof ▶ which were not in the original image. These are artifacts of the sampling process. Subsampling reduces the spatial sampling rate of the image which can lead to spatial aliasing of high-frequency components due to texture or sharp edges. To ensure that the Shannon-Nyquist sampling theorem is satisfied, an anti-aliasing low-pass spatial filter must be applied to reduce the spatial bandwidth of the image before it is subsampled. ▶ This is another application for image blurring and the Gaussian kernel is a suitable low-pass filter for this purpose.

```
>>> roof.smooth(sigma=3)[::7, ::7].disp();
```

and the results for $m = 7$ are shown in ◘ Fig. 11.35c. We note that the curved line artifacts are no longer present. We can write this more concisely as

```
>>> smaller = roof.scale(1/7, sigma=3)
Image: 287 x 238 (uint8)
```

The inverse operation is pixel replication, where each input pixel is replicated as an $m \times m$ tile in the output image

```
>>> smaller.replicate(7).disp();
```

which is shown in ◘ Fig. 11.35d and appears a little *blocky* along the edge of the roof and along the skyline. The subsampling step removed 98% of the pixels, and restoring the image to its original size has not added any new information.

*These are an example of a Moiré pattern.*

*Any realizable low-pass filter has a finite response above its* cutoff *frequency. In practice the cutoff frequency is selected to be far enough below the theoretical cutoff that the filter's response at the Nyquist frequency is* sufficiently *small. As a rule of thumb for subsampling by $m$ a Gaussian with $\sigma = m/2$ is used.*

### 11.7.3 Image Pyramids

An important concept in computer vision, and one that we return to in the next chapter, is scale space. The `Image` method `pyramid` returns a pyramidal decomposition of the image

```
>>> pyramid = mona.pyramid()
[Image: 677 x 700 (float32),
 Image: 339 x 350 (float32),
 Image: 170 x 175 (float32),
 Image: 85 x 88 (float32),
 Image: 43 x 44 (float32),
 Image: 22 x 22 (float32),
 Image: 11 x 11 (float32),
 Image: 6 x 6 (float32),
 Image: 3 x 3 (float32),
 Image: 2 x 2 (float32),
 Image: 1 x 1 (float32)]
>>> len(pyramid)
11
```

as a list of images at successively lower resolution. Note that the last element is the $1 \times 1$ resolution version – a single dark gray pixel! These images are pasted into a composite image which is displayed in ◘ Fig. 11.36.

Image pyramids are the basis of many so-called coarse-to-fine strategies. Consider the problem of looking for a pattern of pixel values that represent some object of interest. The smallest image can be searched very quickly for the object since it comprises only a small number of pixels. The search is then refined using the next larger image, but now we know which area of that larger image to search. The process is repeated until the object is located in the highest resolution image.

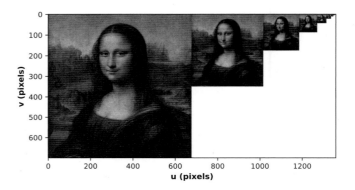

**◻ Fig. 11.36** Image pyramid, a succession of images each half (by side length) the resolution of the one to the left

### 11.7.4 Image Warping

Image warping is a transformation of the pixel *coordinates* rather than the pixel values. Warping can be used to scale an image up or down in size, rotate an image or apply quite arbitrary shape changes. The coordinates of a pixel in the original view $(u, v)$ are expressed as functions

$$u = f_u(u', v'), \quad v = f_v(u', v') \tag{11.10}$$

of the coordinates $(u', v')$ in the new view.

Consider the simple example, shown in ◻ Fig. 11.37a, where the image is reduced in size by a factor of 4 in both dimensions and offset so that its origin, its top-left corner, is shifted to the coordinate $(100, 200)$. We can express this concisely as

$$u' = \frac{u}{4} + 100, \quad v' = \frac{v}{4} + 200 \tag{11.11}$$

which we rearrange into the form of (11.10) as

$$u = 4(u' - 100), \quad v = 4(v' - 200) \tag{11.12}$$

We first create a pair of coordinate matrices

```
>>> Up, Vp = Image.meshgrid(width=500, height=500)
```

which define the coordinates $(u', v')$ of the pixels in a $500 \times 500$ output image. These matrices are such that `Up[v,u]` = u and `Vp[v,u]` = v. The corresponding coordinates in the input image are given by (11.12)

```
>>> U = 4 * (Up - 100);
>>> V = 4 * (Vp - 200);
```

These matrices are maps $(u', v') \mapsto (u, v)$ from output image coordinates to input image coordinates. For example, the output image pixel at $(200, 300)$ comes from

```
>>> p = (300, 200);  # (v, u)
>>> (U[p], V[p])
(400, 400)
```

in the input image, while the output image pixel at $(200, 100)$ comes from

```
>>> p = (100, 200);  # (v, u)
>>> (U[p], V[p])
(400, -400)
```

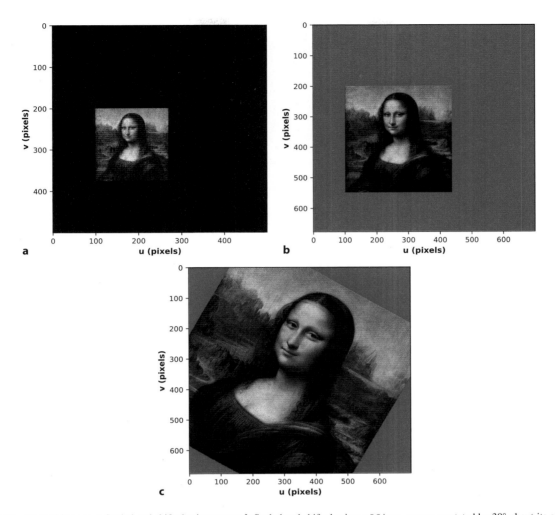

**◨ Fig. 11.37** Warped images. **a** Scaled and shifted using `warp`; **b** Scaled and shifted using `affine_warp`; **c** rotated by 30° about its center using `affine_warp`. Pixels displayed as red were set to a value of `nan` by `affine_warp` – they were not interpolated from any image pixels

which is outside the bounds of the input image. We can now warp the input image

```
>>> mona.warp(U, V).disp();
```

and the result is shown in ◨ Fig. 11.37a. Much of this image is black because those pixels have no corresponding input pixels – they map to points outside the bounds of the image.

To recap, the shape of U and V define the shape of the output image, while their elements describe the mapping $(u', v') \mapsto (u, v)$. This is a very general approach to warping and can be used for complex nonlinear maps, for example, to undistort images as shown in ◨ Fig. 13.9.

Frequently there is an affine relationship between the images – a planar translation, rotation and scaling. For the example above, we can write the affine transformation in homogeneous form as

$$
\begin{pmatrix} u' \\ v' \\ w' \end{pmatrix} = \begin{pmatrix} \frac{1}{4} & 0 & 100 \\ 0 & \frac{1}{4} & 200 \\ 0 & 0 & 1 \end{pmatrix} \begin{pmatrix} u \\ v \\ 1 \end{pmatrix}
$$

**11**

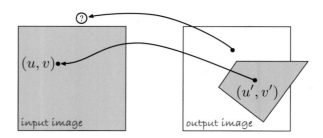

**Fig. 11.38** Coordinate notation for image warping. The pixel $(u', v')$ in the output image is sourced from the pixel at $(u, v)$ in the input image as indicated by the arrow. The warped image is not necessarily polygonal, nor entirely contained within the output image

which is the product of a diagonal scaling matrix and an $\mathbf{SE}(2)$ matrix describing planar rigid-body motion

$$
\begin{pmatrix} u' \\ v' \\ w' \end{pmatrix} = \begin{pmatrix} \frac{1}{4} & 0 & 0 \\ 0 & \frac{1}{4} & 0 \\ 0 & 0 & 1 \end{pmatrix} \underbrace{\begin{pmatrix} 1 & 0 & 100 \\ 0 & 1 & 200 \\ 0 & 0 & 1 \end{pmatrix}}_{\mathbf{SE}(2) \text{ matrix}} \begin{pmatrix} u \\ v \\ 1 \end{pmatrix}
$$

which we can evaluate by

```
>>> M = np.diag([0.25, 0.25, 1]) * SE2(100, 200)
array([[   0.25,        0,      100],
       [      0,     0.25,      200],
       [      0,        0,        1]])
```

A general $3 \times 3$ matrix multiplied by an $\mathbf{SE}(2)$ matrix is not in the group $\mathbf{SE}(2)$ so a general NumPy array is returned.

The affine transformation is a $2 \times 3$ matrix, and `affine_warp` uses only the first two rows of the passed $3 \times 3$ matrix.

Different interpolation modes can be selected. The default is bilinear interpolation where a pixel at coordinate $(u + \delta_u, v + \delta_v)$ is a linear combination of the pixels $(u, v)$, $(u + 1, v)$, $(u, v + 1)$ and $(u + 1, v + 1)$, and $u, v \in \mathbb{N}_0$ and $\delta_u, \delta_v \in [0, 1)$. The interpolation function acts as a weak anti-aliasing filter, but for very large reductions in scale the image should be smoothed first using a Gaussian kernel.

As an `SE2` object which `affine_warp` converts to a NumPy array.

The result is a $3 \times 3$ NumPy array ◄ which specifies the affine warp

```
>>> out = mona.affine_warp(M, bgcolor=np.nan);
>>> out.disp(badcolor="r");
```

and the result is shown in ☐ Fig. 11.37b. We also specified that output pixels with no corresponding input pixels are set to `nan`, and the option to `disp` displays those as red. By default the output image is the same size as the input. ◄

Some subtle things happen under the hood. Firstly, while $(u', v')$ are integer coordinates in the output image, the input image coordinates $(u, v)$ will not necessarily be integers. The pixel values must be interpolated ◄ from neighboring pixels in the input image. Secondly, not all pixels in the output image have corresponding pixels in the input image as illustrated in ☐ Fig. 11.38 – these are displayed as red. In case of mappings that are extremely distorted it may be that many adjacent output pixels map to the same input pixel and this leads to pixelation or *blockyness* in the output image.

Now let's try something a bit more ambitious and rotate the image by 30° about its center. We could use an $\mathbf{SE}(2)$ matrix again but, as discussed in ▶ Sect. 2.2.2.4, rotating about an arbitrary point is more concisely described by a unit revolute twist

```
>>> S = Twist2.UnitRevolute(mona.centre)
(350 -338.5; 1)
```

which we exponentiate to find the $\mathbf{SE}(2)$ transformation ◄

```
>>> M = S.exp(pi / 6)
   0.866     -0.5      220.4
   0.5        0.866   -122.4
   0          0          1
>>> out = mona.affine_warp(M, bgcolor=np.nan)
Image: 700 x 677 (float32)
```

and the result is shown in ▫ Fig. 11.37c. Note the direction of rotation – our definition of the $x$- and $y$-axes (parallel to the $u$- and $v$-axes respectively) is such that the $z$-axis is defined as being into the page making a clockwise rotation a positive angle. Also note that the corners of the original image have been lost, they fall outside the bounds of the output image.

Affine warping is used by the `scale` method to change image scale, and the `rotate` method to perform rotation. The example above could also be achieved by

```
>>> twisted_mona = mona.rotate(pi/6);
```

An affine transformation can perform any combination of translation, rotation, scaling, shearing and reflection of the input image, but it cannot model the change due to a different viewpoint. While some changes of viewpoint simply result in translation, scaling or rotation, more complex changes of viewpoint will induce perspective foreshortening that cannot be properly represented by an affine transformation. A *perspective* transformation performs a general change in viewpoint that is described by a $3 \times 3$ matrix – this is introduced in ▶ Sect. 14.8.1.

## 11.8  Wrapping Up

In this chapter we learned how to acquire images from a variety of sources such as image files, video files, video cameras and the internet, and load them into a Python session. Once loaded, we can treat them as 2D or 3D arrays and conveniently manipulate them using NumPy. The elements of the image matrices can be integer, floating-point or logical values. We then introduced the `Image` class which encapsulates the NumPy array to hide a lot of low-level detail which makes it easier to grasp important concepts with the minimum amount of code. There are many methods which provide a consistent interface to the underlying algorithms which are implemented using NumPy, SciPy or OpenCV. The source code can be read to reveal the details.

Next, we introduced many image processing operations and used the `Image` class to illustrate these. Operations on a single image include: unary arithmetic operations, type conversion, graylevel stretching and various color transformations; nonlinear operations such as histogram normalization and gamma encoding or decoding; and logical operations such as thresholding. We also discussed operations on pairs of images such as arithmetic and boolean operations which support applications such as chroma-keying, background estimation and moving object detection.

The largest and most diverse class of operations are spatial operators. Linear operations are defined by a kernel which can be chosen to perform functions such as image smoothing (to reduce the effect of image noise or as a low-pass anti-aliasing filter prior to decimation) or for edge detection. Nonlinear spatial operations were used for template matching, computing rank statistics (including the median filter which eliminates impulse noise) and mathematical morphology which filters an image based on shape and can be used to cleanup binary images. A variant form, the hit or miss transform, can be used iteratively to perform functions such as skeletonization.

Finally we discussed shape changing operations such as regions of interest, scale changing and the problems that can arise due to aliasing, and generalized image warping which can be used for scaling, translation, rotation or undistorting an image. All these image processing techniques are the foundations of feature extraction algorithms that we discuss in the next chapter.

### 11.8.1  Further Reading

Image processing is a large field and this chapter has provided an introduction to many of the most useful techniques from a robotics perspective. More comprehensive coverage of the topics introduced here, as well as other topics such as grayscale morphology, image restoration, wavelet and frequency domain methods, deformable shape analysis, and image compression can be found in Nixon and Aguado (2019), Gonzalez and Woods (2018), Davies (2017), Klette (2014), and Szeliski (2022). The book by Szeliski (2022) can be downloaded, for personal use, from ► https://szeliski.org/Book. Nixon and Aguado (2019) includes examples in Python and MATLAB, Gonzalez and Woods (2018) has examples in MATLAB, Davies has examples in Python and C++.

Online information about computer vision is available through CVonline at ► https://homepages.inf.ed.ac.uk/rbf/CVonline, and the material in this chapter is covered under the section *Image Transformations and Filters*.

Edge detection is a subset of image processing, but has a huge literature of its own. Forsyth and Ponce (2012) have a comprehensive introduction to edge detection and a useful discussion on the limitations of edge detection. Nixon and Aguado (2019) also cover phase congruency approaches to edge detection and compare various edge detectors. The Sobel kernel for edge detection was described in an unpublished 1968 publication from the Stanford AI lab by Irwin Sobel and Jerome Feldman: *A 3×3 Isotropic Gradient Operator for Image Processing*. The Canny edge detector was originally described in Canny (1983, 1987).

Nonparametric measures for image similarity became popular in the 1990s with a number of key papers such as Zabih and Woodfill (1994), Banks and Corke (2001), Bhat and Nayar (2002). The application to real-time image processing systems using high-speed logic such as FPGAs has been explored by several groups (Corke et al. 1999; Woodfill and Von Herzen 1997).

Mathematical morphology is another very large topic and we have only scraped the surface – important techniques such as grayscale morphology and watersheds have not been covered at all. The general image processing books mentioned above all have useful discussion on this topic. Most of the specialist books in this field are now quite old: Soille (2003) provides a thorough and theoretical treatment, Shih (2009) is a good general introduction, while Dougherty and Lotufo (2003) has a more hands-on tutorial approach.

The approach to computer vision covered in this book is often referred to as bottom-up processing. This chapter has been about *low-level* vision techniques which are operations on pixels. The next chapter is about *high-level* vision techniques where sets of pixels are grouped and then described so as to represent objects in the scene.

### 11.8.2  Sources of Image Data

All the images used in this part of the book are provided by the Python package `mvtb-data` which is installed as a dependency of the Machine Vision Toolbox for Python.

### 11.8.3  Software Tools

There is a great variety of open-source software for image and video manipulation, in almost every programming language ever created. The Toolbox used by this book makes strong use of OpenCV (► https://opencv.org) which is a mature open-source computer vision software project with thousands of algorithms, interfaces for C++, C, Python and Java and runs on Windows, Linux, Mac OS, iOS and Android. There are several books about OpenCV, and Kaehler and Bradski (2017) provides a good introduction to the software and to computer vision in general.

### 11.8.4 Exercises

1. Become familiar with the functions `iread` and `idisp` for grayscale and color images, as well as the equivalent class-based operations `Image.Read` and the `disp` method. Explore pixel values in the image as well as the zoom, pan and scroll buttons in the image toolbar.
2. Use the `roi` method to extract the Mona Lisa's smile.
3. Grab some frames from the camera on your computer, or from a video file, and display them.
4. Write a loop that grabs a frame from your camera and displays it. Add some effects to the image before display such as "negative image", thresholding, posterization, false color, edge filtering etc.
5. Look at the histogram of grayscale images that are under, well and over exposed. For a color image look at the histograms of the RGB color channels for scenes with different dominant colors. Combine real-time image capture with computation and display of the histogram.
6. Create two copies of a grayscale image into NumPy arrays A and B. Write code to time how long it takes to compute the difference of A and B using the NumPy shorthand A-B or using two nested `for` loops. Use the Python package `timeit` to perform the timing. Compare the times for images with `uint8`, `float32` and `float64` pixel types.
7. Given a scene with luminance of 800 nit and a camera with ISO of 1000, $q = 0.7$ and $f$-number of 2.2 what exposure time is needed so that the average gray level of the 8-bit image is 150?
8. Images from space, ► Sect. 11.1.6:
   a) Obtain a map of the roads in your neighborhood. Use this to find a path between two locations, using the robot motion planners discussed in ► Sect. 5.4.
   b) For the images returned by the `EarthView` class write a function to convert pixel coordinate to latitude and longitude. Use the `coordformat` option to `Image.disp` so that the pixel's latitude and longitude is shown in the window's toolbar.
   c) Upload GPS track data from your phone and overlay it on a satellite image.
9. Motion detection:
   a) Modify the Traffic example from ► Sect. 11.4 and highlight the moving vehicles. Explore the effect of changing the parameter $\sigma$.
   b) Write a loop that performs background estimation using frames from your camera. What happens as you move objects in the scene, or let them sit there for a while? Explore the effect of changing the parameter $\sigma$.
   c) Combine concepts from motion detection and chroma-keying to put pixels from the camera where there is motion into the desert scene.
10. Convolution:
    a) Compare the results of smoothing using a $21 \times 21$ uniform kernel and a Gaussian kernel. Can you observe the ringing artifact in the former?
    b) Why do we choose a smoothing kernel that sums to one?
    c) Compare the performance of the simple horizontal gradient kernel $\mathbf{K} = \begin{pmatrix} 0.5 & 0 & -0.5 \end{pmatrix}$ with the Sobel kernel.
    d) Investigate filtering with the Gaussian kernel for different values of $\sigma$ and kernel size.

e) Create a $31 \times 31$ kernel to detect lines at $60°$.

f) Derive analytically the derivative of the Gaussian in the $x$-direction (11.4).

g) Derive analytically the Laplacian of Gaussian (11.8).

h) Show the difference between difference of Gaussian and derivative of Gaussian.

11. Show analytically the effect of an intensity scale error on the SSD and NCC similarity measures.

12. Template matching using the Mona Lisa image; convert it first to grayscale:

a) Use the `roi` method to select one of Mona Lisa's eyes as a template. The template should have odd dimensions.

b) Use the `similarity` method to compute the similarity image. What is the best match and where does it occur? What is the similarity to the other eye? Where does the second best match occur and what is its similarity score?

c) Scale the intensity of the Mona Lisa image and investigate the effect on the peak similarity.

d) Add an offset to the intensity of the Mona Lisa image and investigate the effect on the peak similarity.

e) Repeat steps (c) and (d) for different similarity measures such as SAD, SSD. Investigate the effect of preprocessing with the rank and census transform.

f) Scale the template size by different factors (use the `scale` method) in the range 0.5 to 2.0 in steps of 0.05 and investigate the effect on the peak similarity. Plot peak similarity vs scale.

g) Repeat (f) for rotation of the template in the range $-0.2$ to $0.2$ rad in steps of 0.05.

13. Perform the sub-sampling example from ▶ Sect. 11.7.2 and examine aliasing artifacts around sharp edges and the regular texture of the roof tiles.

14. Write a function to create ◘ Fig. 11.36 from the output of `pyramid`.

15. Warp the image to polar coordinates $(r, \theta)$ with respect to the center of the image, where the horizontal axis is $r$ and the vertical axis is $\theta$.

16. Create a warp function that mimics your favorite funhouse mirror.

# Image Feature Extraction

**Contents**

© The Author(s), under exclusive license to Springer Nature Switzerland AG 2023
P. Corke, *Robotics, Vision and Control*, Springer Tracts in Advanced Robotics 146,
https://doi.org/10.1007/978-3-031-06469-2_12

`chap12.ipynb`

▶ go.sn.pub/HWpVq3

**12**

In the last chapter we discussed the acquisition and processing of images. We learned that images are simply large arrays of pixel values but for robotic applications images have too much *data* and not enough *information*. We need to be able to answer pithy questions such as what is the pose of the object? what type of object is it? how fast is it moving? how fast am I moving? and so on. The answers to such questions are *measurements* obtained from the image and which we call image features. Features are the *gist* of the scene and the raw material that we need for robot control.

The image processing operations from the last chapter operated on one or more input images and returned another image. In contrast, *feature extraction* operates on a single image and returns one or more *image features*. Features are typically scalars (for example, object area or aspect ratio) or short vectors (for example the coordinate of an object, the parameters of a line, the corners of a bounding box). Image feature extraction is an essential first step in using image data to control a robot. It is an *information concentration* step that reduces the data rate from $10^6$–$10^8$ bytes s$^{-1}$ at the output of a camera to something of the order of tens of features per frame that can be used as input to a robot's control system.

In this chapter we discuss features and how to extract them from images. Drawing on image processing techniques from the last chapter we will discuss three classes of features: regions, lines and interest points. ▶ Sect. 12.1 discusses region features which are contiguous groups of pixels that are homogeneous with respect to some pixel property. For example the set of pixels that represent a red object against a nonred background. We also briefly introduce deep-learning-based approaches to the problems of finding semantically homogeneous pixel regions and detecting objects. ▶ Sect. 12.2 discusses line features which describe straight lines in the world. Straight lines are distinct and very common in human-made environments – for example the edges of doorways, buildings or roads. The final class of features, discussed in ▶ Sect. 12.3, are point features. These are distinctive *points* in an image that can be reliably detected in different views of the same scene and are critical to the content of ▶ Chap. 14.

▶ Sect. 12.4 presents two applications: optical character recognition, and image retrieval. The latter is the problem of finding which image, from an existing set of images, is most similar to some new image. This can be used by a robot to determine whether it has visited a particular place, or seen the same object, before.

❗ It is important to always keep in mind that image features are a summary of the information present in the pixels that comprise the image. The information lost by summarization is countered by assumptions based on our knowledge of the scene, but our system will only ever be as good as the validity of our assumptions. For example, we might use image features to describe the position and shape of a group of red pixels that correspond to a red object. However the size feature, typically the number of pixels, does not say anything about the size of the red object in the world – we need extra information such as the distance between the camera and the object, and the camera's focal length. We also need to assume that the object is not partially occluded – that would make the observed size less than the true size. Further we need to assume that the illumination is such that the chromaticity of the light reflected from the object is considered to be red. We might also find features in an image that do not correspond to a physical object – decorative markings, shadows, or reflections from bright light sources.

☐ **Fig. 12.1** Examples of pixel classification. The left-hand column is the input image and the right-hand column is the classification. The classification is application specific and the pixels have been classified into $C$ categories represented as different colors. The objects of interest are **a** the individual letters on the sign ($C = 2$); **b** the red tomatoes ($C = 2$); **c** locally homogeneous regions of pixels ($C = 27$); **d** semantic classification by a deep network showing background, car, dog and person pixels ($C = 21$) ▶

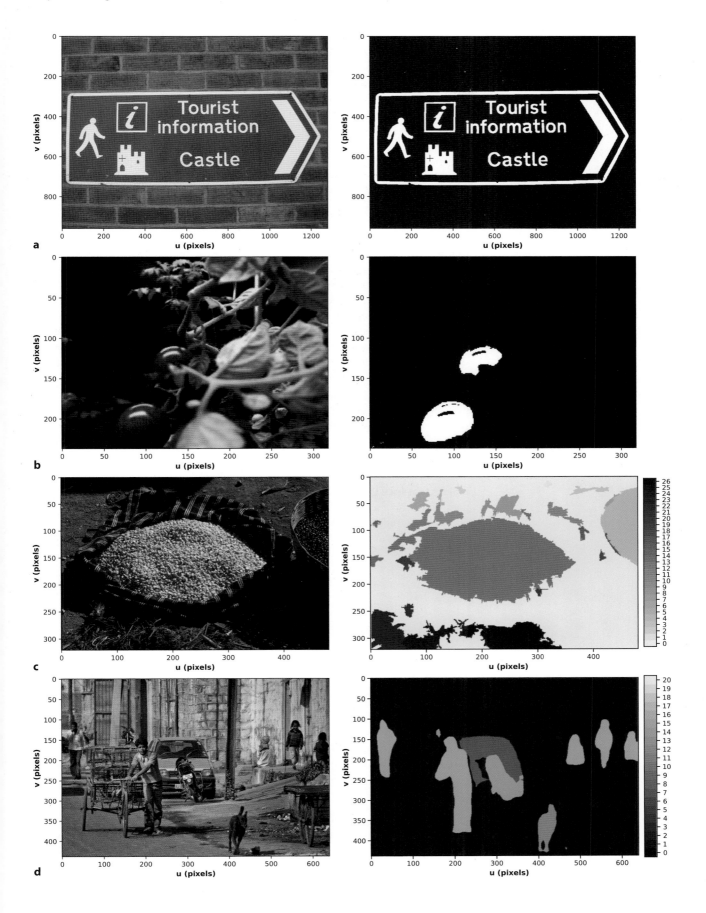

## 12.1 **Region Features**

Image segmentation is the process of partitioning an image into *application mean-ingful* regions as illustrated in ◾ Fig. 12.1. The aim is to segment, or separate, those pixels that represent objects of interest from all other pixels in the scene. This is one of the oldest approaches to scene understanding and while conceptually straight-forward it is very challenging in practice. A key requirement is *robustness* – how gracefully the method degrades as the underlying assumptions are violated, for ex-ample changing scene illumination or viewpoint. Segmentation of an image into regions comprises three subproblems:

**Pixel Classification** – is a decision process applied to each pixel that assigns the pixel to one of $C$ classes $c = 0, \ldots, C - 1$. Commonly we use $C = 2$ which is known as binary classification or binarization and some examples are shown in ◾ Fig. 12.1a–b. The pixels have been classified as object ($c = 1$) or not-object ($c = 0$) which are displayed as white or black pixels respectively. ◾ Fig. 12.1c–d are multi-level classifications where $C > 2$ and the pixel's class is reflected in its displayed color.

The underlying *assumption* in the examples of ◾ Fig. 12.1 is that regions are *ho-mogeneous* with respect to some characteristic such as brightness, color, texture or semantic meaning – the classification is *always application specific*. In practice we accept that this stage is imperfect and that pixels may be misclassified – subsequent processing steps will have to deal with this.

**Object Instance Representation** – where adjacent pixels of the same class are *connected* to form $m$ spatial sets $S_0, \ldots, S_{m-1}$. For example in ◾ Fig. 12.1b there are two sets, each representing an individual tomato – a tomato *instance*. The sets can be represented by assigning a set label to each pixel, by lists of the coordinates of all pixels in each set, or lists of pixels that define the perimeter of each set. This is sometimes refered to as instance segmentation.

**Object Instance Description** – where the sets $S_i$ are *described* in terms of com-pact scalar or vector-valued *features* such as size, position, and shape. In some applications there is an additional step – semantic classification – where we as-sign an application meaningful label to the pixel classes or sets, for example, "cat", "dog", "coffee cup" etc.

These topics are expanded on in the next three subsections.

### 12.1.1 **Pixel Classification**

Pixels, whether monochrome or color, are assigned to a class that is represented by an integer $c = 0, \ldots, C - 1$ where $C$ is the number of classes. In many of the examples we will use binary classification with just two classes corresponding to object and not-object, or foreground and background.

#### 12.1.1.1 **Monochrome Image Classification**

A common approach to binary classification of monochrome pixels is the monadic operator introduced in the last chapter

$$c_{u,v} = \begin{cases} 0, & \text{if} \quad x_{u,v} < t \\ 1, & \text{if} \quad x_{u,v} \geq t \end{cases} \quad \forall (u, v) \in \mathbf{X}$$

where the decision is based simply on the value of the pixel $x_{u,v}$. This approach is called thresholding and $t$ is the *threshold*.

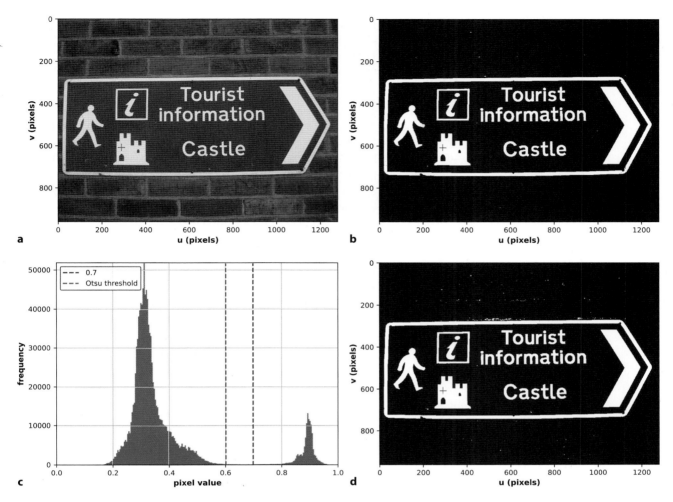

◻ **Fig. 12.2** Binary classification. **a** Original image (image sourced from the ICDAR collection; Lucas 2005); **b** binary classification with threshold of 0.7; **c** histogram of grayscale pixel values, threshold values indicated; **d** binary classification with Otsu threshold

Thresholding is very simple to implement. Consider the image

```
>>> castle = Image.Read("castle.png", dtype="float");
```

which is shown in ◻ Fig. 12.2a. The thresholded image

```
>>> (castle >= 0.7).disp();
```

is shown in ◻ Fig. 12.2b. The pixels have been quite accurately classified as corresponding to white paint or not. This classification is based on the seemingly reasonable *assumption* that the white paint objects are brighter than everything else in the image.

In the early days of computer vision, when computer power was limited, this approach was widely used – it was easier to contrive a world of white objects and dark backgrounds than to implement more sophisticated classification. Many modern industrial vision inspection systems still use this fast and simple approach since it allows the use of modest embedded computers – it works very well if the objects are on a conveyor belt of a suitable contrasting background or in silhouette at an inspection station. In a real-world robot environment, we generally have to work a little harder in order to achieve useful gray-level classification. An important question, and a hard one, is where did the threshold value of 0.7 come from? The most common approach is trial and error and the `Image` method

```
>>> castle.ithresh()
```

displays the image and a threshold slider that can be adjusted until a satisfactory classification is obtained. However, if the image was captured on a day with different lighting conditions a different threshold would be required.

A more principled approach than trial and error is to analyze the histogram of the image

```
>>> castle.hist().plot();
```

which is shown in ◘ Fig. 12.2c. The histogram has two clearly defined peaks, a bimodal distribution, which correspond to two *populations* of pixels. The smaller peak for gray values around 0.9 corresponds to the pixels that are bright and it has quite a small range of variation in value. The wider and taller peak for gray values around 0.3 corresponds to pixels in the darker background of the sign and the bricks, and this peak has a much larger variation in brightness.

To separate the two classes of pixels we choose the decision boundary, the threshold, to lie in the valley between the peaks. In this regard the choice of $t = 0.7$ is a good one. The *valley* in this case is very wide, so we have quite a range of choice for the threshold. For example $t = 0.75$ would also work well.

The optimal threshold can be computed using Otsu's method

```
>>> t = castle.otsu()
0.6039216
```

which separates an image into two classes of pixels, see ◘ Fig. 12.2d, in a way that minimizes the variance of values within each class and maximizes the variance of values between the classes – assuming that the histogram has just two peaks. Sadly, as we shall see, the real world is rarely this facilitating.

Consider a different image of the same scene which has a highlight

```
>>> castle2 = Image.Read("castle2.png", dtype="float");
```

which is shown in ◘ Fig. 12.3a. The result of applying the Otsu threshold

```
>>> t = castle2.otsu()
0.59607846
```

is shown ◘ Fig. 12.3b and the pixel classification is poor – the highlight overlaps and obscures several of the characters. The histogram shown in ◘ Fig. 12.3c is similar to the previous one – it is still bimodal – but we see that the peaks are wider and the valley is less deep. The pixel gray-level populations are now overlapping and, unfortunately for us, there is no single threshold that can separate them. The result of applying a threshold of 0.79 is shown in ◘ Fig. 12.3d and successfully classifies the characters under the highlight, but the other characters are lost.

Since no one threshold is appropriate for the whole image we can choose a threshold for every pixel, often called an adaptive threshold. Typically the threshold is of the form

$$t_{u,v} = \mu(\mathcal{W}_{u,v}) - k\sigma(\mathcal{W}_{u,v})$$

where $\mathcal{W}_{u,v}$ is a local window around the pixel at $(u, v)$ and $\mu(\cdot)$ and $\sigma(\cdot)$ are the mean and standard deviation respectively. For the image in ◘ Fig. 12.3a

```
>>> castle2.adaptive_threshold(h=15).disp();
```

the adaptive threshold classification is shown in ◘ Fig. 12.4. All the pixels belonging to the letters have been correctly classified but compared to ◘ Fig. 12.3c there are many false positives – nonobject pixels classified as objects. Later in this section we will discuss techniques to eliminate these false positives. Note that the classification process is no longer a function of just the input pixel, it is now a complex function of the pixel and its neighbors. In this case, the algorithm has done a good job using default parameters but for other scenes we might not be so lucky. We would need to do some trial and error to choose suitable window size and offset parameter $k$.

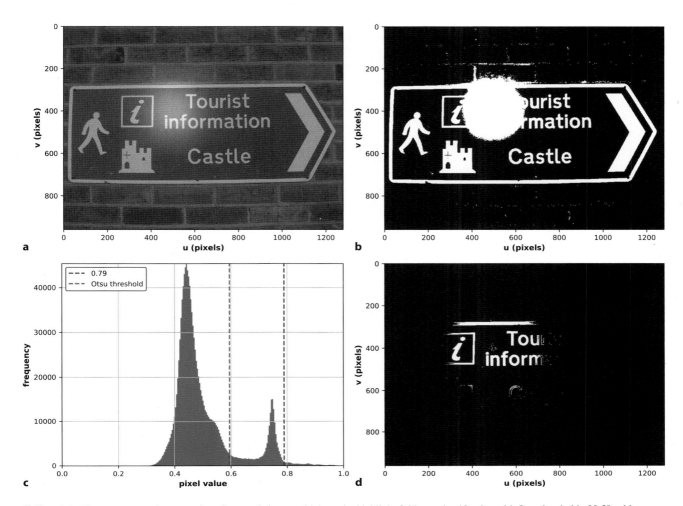

**Fig. 12.3** Binary segmentation example. **a** Gray-scale image with intensity highlight; **b** binary classification with Otsu threshold of 0.59; **c** histogram with thresholds indicated; **d** binary classification with threshold of 0.79

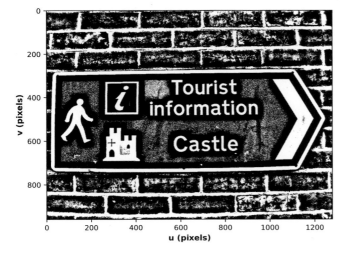

**Fig. 12.4** Result of adaptive thresholding

> **! Thresholding is generally not a robust approach**
> Segmentation approaches based on thresholding are notoriously brittle – a slight change in illumination of the scene means that the thresholds we chose would no longer be appropriate. In most real scenes there is no simple mapping from pixel values to particular objects – we cannot for example choose a threshold that would select a motorbike or a duck.

### 12.1.1.2 Color Image Classification

Color is a powerful cue for segmentation but roboticists tend to shy away from using it because of the problems with color constancy discussed in ▶ Sect. 10.3.2. In this section we consider two examples that use color images. The first is a fairly simple scene with four yellow markers that were used for an indoor drone landing experiment

```
>>> targets = Image.Read("yellowtargets.png", dtype="float",
...                       gamma="sRGB");
```

shown in ◻ Fig. 12.5a. The second is from the MIT Robotic Garden project

```
>>> garden = Image.Read("tomato_124.png", dtype="float",
...                      gamma="sRGB");
```

and is shown in ◻ Fig. 12.6a. Our objective is to determine the centroids of the yellow targets and the red tomatoes respectively. The initial stages of processing are the same for each image but we will illustrate the process in detail for the image of the yellow targets shown in ◻ Fig. 12.5.

The first step is to remove the influence of overall scene illumination level by converting the color image to chromaticity coordinates that were introduced in ▶ Sect. 10.2.5

```
>>> ab = targets.colorspace("L*a*b*").plane("a*:b*")
Image: 640 x 480 (float32), a*:b*
```

which is an image with just two planes – the a*b*-chromaticity values. We chose a*b*-chromaticity, from ▶ Sect. 10.2.7, since Euclidean distance in this space matches the human perception of difference between colors.

The b* color plane spans the range of colors from blue to yellow with white in the middle

```
>>> ab.plane("b*").disp();
```

and is shown in ◻ Fig. 12.5b. This is a grayscale image and applying a threshold of 70 will provide a good segmentation.

A threshold-free approach is to consider the pixels as a set of points with values $(a^*, b^*) \in \mathbb{R}^2$ which we can cluster into sets of similar chromaticity. We will use the $k$-means algorithm to find the clusters but we do have to specify the number of clusters. We will use our knowledge that this particular scene has essentially two differently colored elements: yellow targets and everything else which is essentially gray: floor, metal drain cover and shadows. The pixels are clustered into *two* chromaticity classes ($C = 2$) by

```
>>> targets_labels, targets_centroids, resid = \
... ab.kmeans_color(k=2, seed=0)
```

and we can display the pixel classification

```
>>> targets_labels.disp(colormap="jet", colorbar=True);
```

as a false color image as shown in ◻ Fig. 12.5c. The pixels in this image have values $c = \{0, 1\}$ indicating which class the corresponding input pixels has been assigned to. We see that the yellow targets have been cleanly assigned to class $c = 0$ which is displayed as blue.

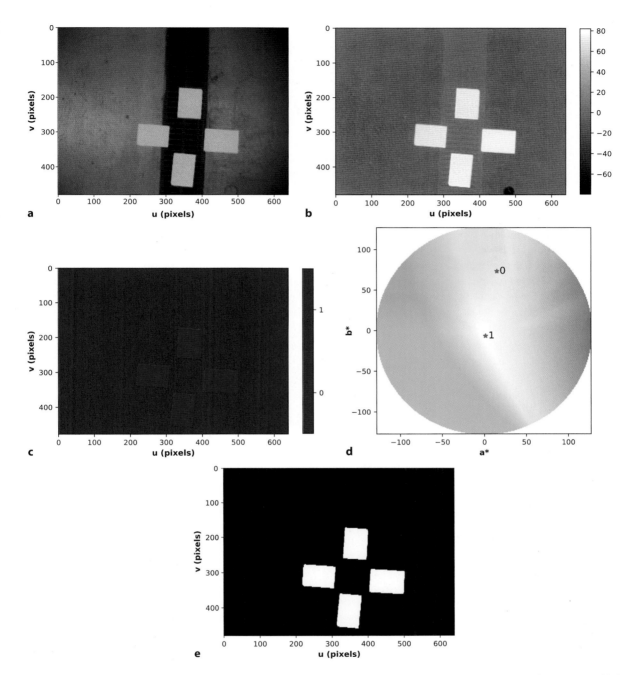

■ **Fig. 12.5** Target image example. **a** Original image; **b** b* color plane; **c** pixel classification ($C = 2$) shown in false color; **d** cluster centroids in the a*b*–chromaticity space; **e** all pixels of class $c = 0$

The function also returns the a*b*-chromaticity of the cluster centroids

```
>>> targets_centroids
array([[  14.71,    1.106],
       [  73.63,   -6.511]], dtype=float32)
```

as one column per cluster. We can plot these cluster centroids on the a*b*-chromaticity plane

```
>>> plot_chromaticity_diagram(colorspace="a*b*");
>>> plot_point(targets_centroids, marker="*", text="{}");
```

which is shown in ◩ Fig. 12.5d. We see that cluster 0 is the closest to yellow. The names of the colors associated with the cluster centroids are

```
>>> [color2name(c, "a*b*") for c in targets_centroids.T]
['xkcd:squash', 'ghostwhite']
```

which are consistent with yellow and gray.

The residual is the sum of the distance of every point from its assigned cluster centroid, and indicates the compactness of the clusters. We can normalize by the number of pixels

```
>>> resid / ab.npixels
59.11
```

Since the algorithm uses a random initialization we will obtain different clusters and classification on every run, and therefore different residuals. ◀

$k$-means clustering is computationally expensive and therefore not very well suited to real-time applications. However we can divide the process into a training phase and a classification phase. In the training phase a number of example images would be concatenated and passed to `kmeans_color` which would identify the centers of the clusters for each class. Subsequently, we can assign pixels to their closest cluster relatively cheaply

```
>>> labels = ab.kmeans_color(centroids=targets_centroids)
Image: 640 x 480 (uint8)
```

The pixels belonging to class 0 can be selected

```
>>> objects = (labels == 0)
Image: 640 x 480 (bool)
```

which is a *logical image* that can be displayed

```
>>> objects.disp();
```

as shown in ◩ Fig. 12.5e. All `True` pixels are displayed as white and correspond to the yellow targets in the original image.

For the garden image of ◩ Fig. 12.6a we follow a very similar procedure. The a* color plane is shown as a grayscale image in ◩ Fig. 12.6b and we could apply a threshold of 35 to achieve a good segmentation. We could also use clustering again, but this time with three clusters ($C = 3$) based on our knowledge that the scene contains: red tomatoes, green leaves, and dark background

```
>>> ab = garden.colorspace("L*a*b*").plane("a*:b*")
Image: 318 x 238 (float32), a*:b*
>>> garden_labels, garden_centroids, resid \
...    = ab.kmeans_color(k=3, seed=0);
>>> garden_centroids
array([[ -19.46,   -1.375,    39.86],
       [  31.04,    2.673,    24.09]], dtype=float32)
```

The pixel classes are shown in false color in ◩ Fig. 12.6c. Pixels corresponding to the tomato have been assigned to class $c = 2$ which are (coincidentally) displayed as red. The cluster centroids are marked on the a*b*-chromaticity plane in ◩ Fig. 12.6d. The names of the colors associated with the cluster centroids are

```
>>> [color2name(c, "a*b*") for c in garden_centroids.T]
['xkcd:military green', 'xkcd:grey', 'xkcd:dark salmon']
```

The red pixels can be selected

```
>>> tomatoes = (garden_labels == 2);
```

and the resulting logical image is shown in ◩ Fig. 12.6e.

This segmentation is far from perfect. Both tomatoes have holes due to specular reflection as discussed in ▶ Sect. 10.3.5. A few pixels at the bottom left have also been erroneously classified as a tomato. We can improve the result by applying a morphological closing operation with a large circular kernel which is consistent

One option is to run $k$-means a number of times, and accept the clustering for which the residual is lowest. If we don't know $k$ we could iterate over a range of values and choose $k$ that leads to the lowest residual. For repeatable results we can explicitly set the random number seed using the `seed` argument.

**12**

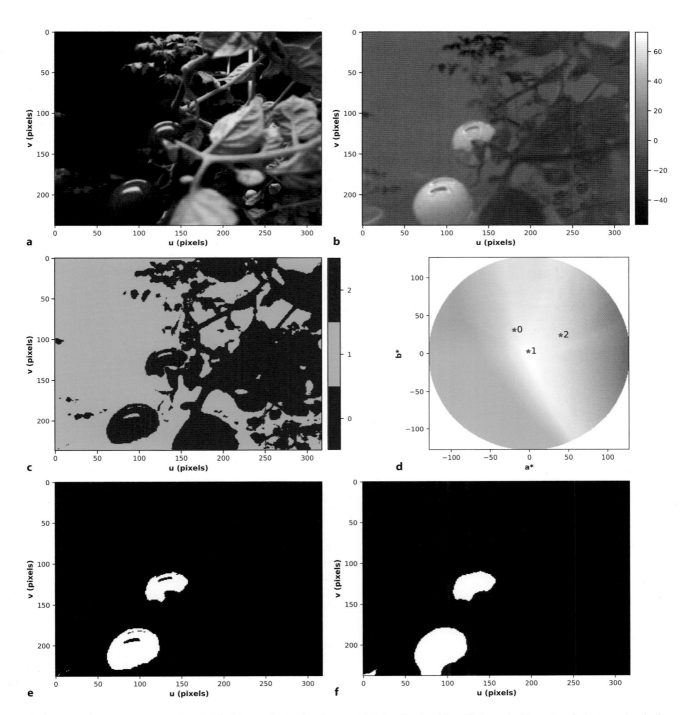

**◨ Fig. 12.6** Garden image example. **a** Original image; **b** a* color plane; **c** pixel classification ($C = 3$) shown in false color; **d** cluster centers in the a*b*–chromaticity space **e** all pixels of class $c = 2$; **f** after morphological closing with a circular structuring element of radius 15 (input image courtesy of Distributed Robot Garden project, MIT)

with the shape of the tomato

```
>>> tomatoes_binary = tomatoes.close(Kernel.Circle(radius=15));
>>> tomatoes_binary.disp();
```

and the result is shown in ◨ Fig. 12.6f. The closing operation has somewhat restored the shape of the fruit, but with the unwanted consequence that the group of misclassified pixels in the bottom-left corner have been enlarged. Nevertheless, this image contains a workable classification of pixels into two classes: tomato and not-tomato.

**Excurse 12.1: *k*-Means Clustering**

$k$-means is an iterative algorithm for grouping $n$-dimensional points into $k$ spatial clusters. Each cluster is defined by a center point $c_i \in \mathbb{R}^n, i = 0, \ldots, k-1$. At each iteration, all points are assigned to the *closest* cluster centroids based on Euclidean distance, and then each cluster centroid is updated to be the mean of all the points assigned to the cluster.

To demonstrate with 500 random 2-dimensional points

```
>>> data = np.random.rand(500, 2);
```

as a $500 \times 2$ matrix with one point per row. We will cluster this data into three sets

```
>>> from scipy.cluster.vq import kmeans2
>>> centroids, labels = kmeans2(data, k=3)
```

where `labels` is a 500-vector whose elements specify the class of the corresponding row of `data`. `centroids` is a $3 \times 2$ matrix whose rows specify the center of each 2-dimensional cluster.

We plot the points in each cluster with different colors

```
>>> for i in range(3):
...     plot_point(data[labels==i, :].T,
...         color="rgb"[i], marker=".",
...         markersize=10)
```

and it is clear that the points have been effectively partitioned. The centroids center have been superimposed as larger dots.

The $k$-means algorithm requires an initial estimate of the center of each cluster and this is generally achieved using random numbers, see the SciPy documentation for details. A consequence is that the algorithm may return different results at each invocation.

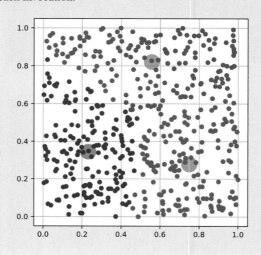

---

**12**

Observe that they have the same chromaticity as the dark background, class 1 pixels, which are situated close to the white point on the a\*b\*-plane.

The garden image illustrates two common real-world imaging artifacts: specular reflection and occlusion. The surface of the tomato is sufficiently shiny, and oriented in such a way, that the camera sees a reflection of the room light – those pixels are white rather than red. ◀ The top tomato is also partly obscured by leaves and branches. Depending on the application requirements, this may or may not be a problem. The occluded tomato might not be reachable from the direction the picture was taken, because of the occluding material, so it might in fact be appropriate to not classify this as a tomato.

These examples have achieved a workable classification of the image pixels into object and not-object. The resulting groups of white pixels are commonly known as blobs. It is interesting to note that we have not specified any threshold or any definition of the object color, but we did have to specify the number of classes and

---

**Excurse 12.2: Specular Highlights**

Specular highlights in images are reflections of bright light sources and can complicate segmentation as shown in ◼ Figs. 12.3 and 12.6e.

As discussed in ▶ Exc. 13.10 the light reflected by most real objects has two components: the specular surface reflectance which does not change the spectrum of the light; and the diffuse body reflectance which filters the reflected light.

There are several ways to reduce the problem of specular highlights. Firstly, move or remove the problematic light source, or move the camera. Secondly, use a diffuse light source near the camera, for instance a ring illuminator that fits around the lens of the camera. Thirdly, attenuate the specular reflection using a polarizing filter since light that is specularly reflected from a dielectric surface will be polarized.

determine which of those classes corresponded to the objects of interest. ▶ In the garden image case we had to choose and design some additional bespoke image processing steps in order to achieve a good classification. Pixel classification is a difficult problem but we can get quite good results by exploiting knowledge of the problem, having a good collection of image processing tricks, and experience.

The color of the object of interest is known, yellow and red respectively in these examples. We could use the `name2color` function to find the chromaticity of the named color, then choose the cluster with the closest centroid.

### 12.1.1.3 Semantic Classification

Over the last decade a new approach to segmenting complex scenes has emerged which is based on deep-learning technology. A deep network is trained using large amounts of labeled data, that is images where the outlines of objects have been drawn by humans and the object type determined by humans. ▶ The training process adjusts millions of weights in the neural network so that the predicted output for a set of training images best matches the human labeled data. Training may take from hours to days on Graphical Processing Unit (GPU) hardware, and the result is a neural network model, literally a set of millions of parameters.

This is the basis of a whole new industry in data labeling.

Using the trained neural network to map an input image to a set of outputs is referred to as *inference* and is relatively fast, typically a small fraction of a second, and does not require a GPU. Neural networks have the ability to generalize, that is, to recognize an object even if it has never encountered that exact image before. They are not perfect, but can give very impressive results.

The details of neural network architectures and training networks is beyond the scope of this book, but it is instructive to take a pre-trained network and apply it to an image. Semantic classification labels each pixel according to the *type* of object it belongs to, for example a person, car, motorbike etc. We will use the PyTorch package and the specific network we will use is FCN-ResNet, a fully convolutional network that has been pretrained using the PASCAL VOC dataset ▶ which has 21 classes that includes background, person, bicycle, car, cat, dog, and sheep.

Visual Object Classes, see ▶ http://host.robots.ox.ac.uk/pascal/VOC.

The first step is to load the image we wish to segment

```
>>> scene = Image.Read("image3.jpg")
Image: 640 x 438 (uint8), R:G:B [.../images/image3.jpg]
>>> scene.disp();
```

which is shown in ◻ Fig. 12.7a. This is a complex scene and all the approaches we have discussed so far would be unable to segment this scene in a meaningful way. Assuming that PyTorch is installed, we import the relevant packages

```
>>> import torch
>>> import torchvision as tv
```

and then normalize the image to have the same pixel value distribution as the training images, and convert it to a PyTorch tensor ▶

A PyTorch tensor is a multidimensional NumPy-like array with support for operations on a GPU, and methods to transfer it to and from a GPU.

```
>>> transform = tv.transforms.Compose([
...     tv.transforms.ToTensor(),
...     tv.transforms.Normalize(mean=[0.485, 0.456, 0.406],
...                             std=[0.229, 0.224, 0.225])]);
>>> in_tensor = transform(scene.image);
```

Next we load the FCN-ResNet network, configure it for evaluation (inference) and pass the image through it ▶

On the first execution the network model will be downloaded.

```
>>> model = tv.models.segmentation.fcn_resnet50(pretrained=True)\
...             .eval();
>>> outputs = model(torch.stack([in_tensor]));
```

and then convert the resulting tensor to a Toolbox `Image` instance

```
>>> labels = Image(torch.argmax(outputs["out"].squeeze(),
...                 dim=0).detach().cpu().numpy());
>>> labels.disp(colormap="viridis", ncolors=20, colorbar=True);
```

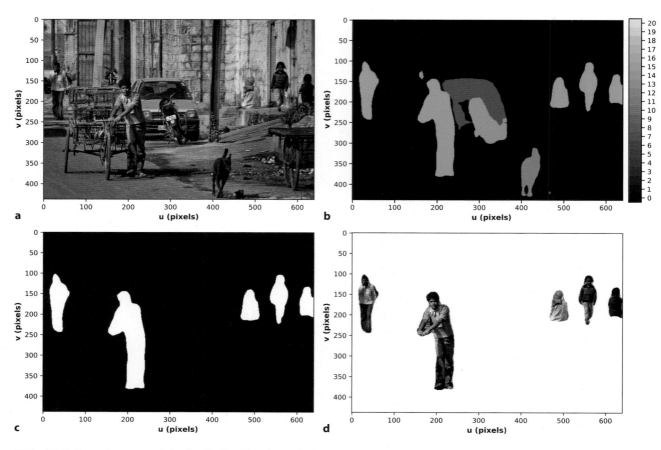

**■ Fig. 12.7** Semantic segmentation using ResNet-50. **a** the original scene; **b** pixel classification; **c** binary image showing pixels belonging to the "person" class; **c** input pixels belonging to the `person` class (image is included as part of ► https://github.com/OlafenwaMoses/ImageAI)

Pascal VOC 20 classes are enumerated starting from zero: background, aeroplane, bicycle, bird, boat, bottle, bus, car, cat, chair, cow, diningtable, dog, horse, motorbike, person, pottedplant, sheep, sofa, train, tvmonitor.

which is shown in ■ Fig. 12.7b. The PASCAL VOC ordinal value for the "person" class is 15 ◄ and we can select and mask out all the pixels corresponding to people

```
>>> (labels == 15).disp();
>>> scene.choose("white", labels != 15).disp();
```

which are shown in ■ Fig. 12.7c and d respectively.

### 12.1.2   Object Instance Representation

In the previous section we took grayscale or color images and processed them to produce binary images that contain one or more regions of white pixels. When we look at the binary image in ■ Fig. 12.8a we see four white objects but the image is just an array of black and white pixels – there is no notion of objects. What we mean when we talk about objects is a set of pixels, of the same class, that are *adjacent* to, or *connected* to, each other. More formally we can say that an object is a spatially contiguous region of pixels of the same class. Objects in a binary scene are commonly known as blobs, regions or connected components. In this section we consider the problem of allocating pixels to spatial sets $S_0, \ldots, S_{m-1}$ which correspond to the human notion of objects, and where $m$ is the number of objects in the scene.

If we think in terms of object classes, then ■ Fig. 12.8a contains only two types of object: "S" and shark. There are three *instances* of the shark object class

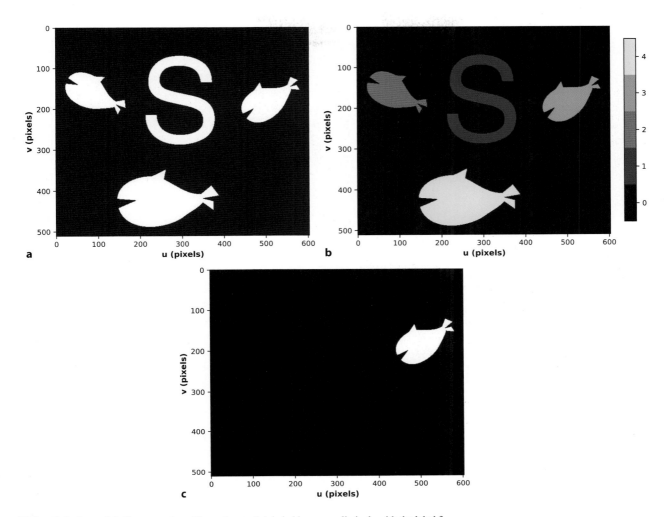

**Fig. 12.8** Image labeling example. **a** Binary image; **b** labeled image; **c** all pixels with the label 3

and one instance of an "S" object. The terms object and instance are often used interchangeably.

### 12.1.2.1 Creating Binary Blobs

Consider the binary image

```
>>> sharks = Image.Read("sharks.png");
>>> sharks.disp();
```

which is shown in ◻ Fig. 12.8a. We assign the pixels to spatial sets by

```
>>> labels, m = sharks.labels_binary()
>>> m
5
```

which has found five blobs. The pixel assignments are given by `labels` which is an image, the same size as the binary image

```
>>> labels.disp(colorbar=True);
```

which is shown as a false color image in ◻ Fig. 12.8b – each connected region has a unique label and hence unique color. This process of label assignment is often refered to as connectivity analysis, blob labeling or blob coloring. Each pixel in the label image has an integer value in the range $[0, 5)$ that is the label of the spatially contiguous set or blob to which the corresponding input pixel belongs. Looking at the label values in the figure, or by interactively exploring the pixel values in the displayed image, we see that the background has the label 0 and the right shark has the label 3. For the first time we have been able to say something quantitative about the structure of an image: there are five objects, one of which is the background.

To obtain an image containing a particular blob is now very easy. To select all pixels belonging to the right shark we create a logical image

```
>>> right_shark = (labels == 3);
>>> right_shark.disp();
```

which is shown in ◨ Fig. 12.8c.

In this example we have assumed 4-way connectivity, that is, pixels are connected to their north, south, east and west neighbors of the same class. The 8-way connectivity option allows connection via any of a pixel's eight neighbors of the same class. ◄

8-way connectivity can lead to surprising results. For example a black and white checkerboard would have just two regions; all white squares are one region and all the black squares another.

### 12.1.2.2 Maximally Stable Extremal Regions (MSER)

The results shown in ◨ Fig. 12.3 were disappointing because no single threshold was able classify all the character pixels. However all characters are correctly classified for some, but not all, thresholds. In fact, each character region is correctly segmented for some *range* of thresholds and we would like the union of regions classified over the range of all thresholds. The maximally stable extremal region or MSER algorithm does exactly this. It is able to label the regions in grayscale image in just a single step

```
>>> labels, m = castle2.labels_MSER()
```

and for this image

```
>>> m
394
```

Although no explicit threshold has been given labels_MSER has a number of parameters and in this case their default values have given satisfactory results. See the online documentation for details of the parameters.

sets were found. ◄ The label image

```
>>> labels.disp(colormap="viridis_r", ncolors=m);
```

is shown in ◨ Fig. 12.9 as a false color image and pixels with the same label correspond to a stable region – a blob that is consistent over some range of thresholds. The pixel classification step has been integrated into the representation step. All the character objects are correctly classified but the number of stable sets greatly exceeds the thirty or so characters and icons on the sign.

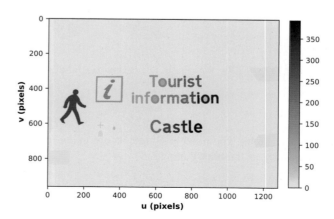

◨ **Fig. 12.9** Segmentation using maximally stable extremal regions (MSER). The identified regions are uniquely color coded. Some non-text regions can be seen faintly on the left and right edges

### 12.1.2.3 **Graph-Based Segmentation**

Consider the complex scene

```
>>> grain = Image.Read("58060.png")
Image: 481 x 321 (uint8), R:G:B [.../images/58060.png]
>>> grain.disp();
```

shown in ◻ Fig. 12.10a. The Gestalt principle of emergence says that we identify objects as a whole rather than as a collection of parts – we see a bowl of grain rather than deducing a bowl of grain by recognizing its individual components. However when it comes to a detailed pixel by pixel segmentation things become quite subjective – different people would perform the segmentation differently based on judgment calls about what is *important.* ► For example, should the colored stripes on the cloth be segmented? If segments represent real world objects, then the Gestalt view would be that the cloth should be just one segment. However the stripes are real, some effort was made to create them, so perhaps they should be segmented. This is why segmentation is a *hard* problem – humans cannot agree on what is correct. No computer algorithm could, or could be expected to, make this type of judgment.

The Berkeley segmentation site ► https://www.eecs.berkeley.edu/ Research/Projects/CS/vision/bsds hosts these images plus a number of different human-made segmentations.

The graph-based segmentation labels the regions in a color image in just a single step

```
>>> labels, m = grain.labels_graphseg()
>>> m
27
```

and in this case finds 27 regions. The label image

```
>>> labels.disp(colormap="viridis_r", ncolors=m);
```

is shown in ◻ Fig. 12.10b. The pixel classification step has been integrated into the representation step.

Graph-based segmentation considers the image as a graph (see ► App. I) where each pixel is a vertex and has 8 edges connecting it to its neighboring pixels. The weight of each edge is a nonnegative measure of the dissimilarity between the two pixels – the absolute value of the difference in color. The algorithm starts with every vertex assigned to its own set. At each iteration the edge weights are examined and if the vertices are in different sets but the edge weight is below a threshold the two vertex sets are merged. The threshold is a function of the size of the set and a parameter k which sets the scale of the segmentation – a larger value expresses a preference for larger connected components. Other parameters include sigma and min_size which are the scale of the Gaussian smoothing and the minimum region size that is returned.

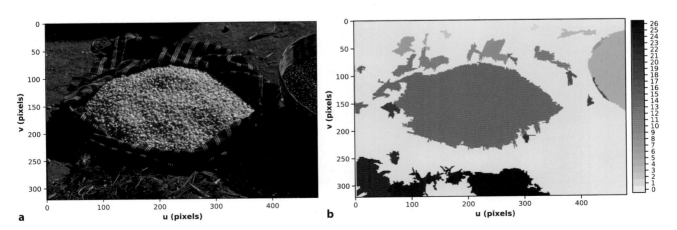

◻ **Fig. 12.10** Complex segmentation example. **a** Original color image (image from the Berkeley Segmentation Dataset; Martin et al. 2001); **b** graph-based segmentation

### 12.1.3 Object Instance Description

In the previous section we learned how to find different instances of objects in the scene using connected component analysis as shown in ◨ Fig. 12.8c. However this representation of the object instances is still just an image, with logical pixel values rather than a concise numeric description of object's size, position and shape.

#### 12.1.3.1 Area

The size, or area, of a blob is simply the number of pixels. For a binary image we can simply sum the pixels

```
>>> right_shark.sum()
7728
```

since the object pixels are `True` (with a value of 1) and the background pixels are `False` (with a value of 0).

#### 12.1.3.2 Bounding Boxes

A useful measure of the extent of a blob is the bounding box – the smallest rectangle with sides parallel to the $u$- and $v$-axes that encloses the region. The $(u, v)$ coordinates of all the nonzero (object) pixels are the corresponding elements of

```
>>> u, v = right_shark.nonzero()
```

where `u` and `v` are equal-length 1D arrays

```
>>> u.shape
(7728,)
```

The bounds of the region are

```
>>> umin = u.min()
442
>>> umax = u.max()
580
>>> vmin = v.min()
124
>>> vmax = v.max()
234
```

These bounds define a rectangle which we can superimpose on the image

```
>>> plot_box(lrbt=[umin, umax, vmin, vmax], color="g");
```

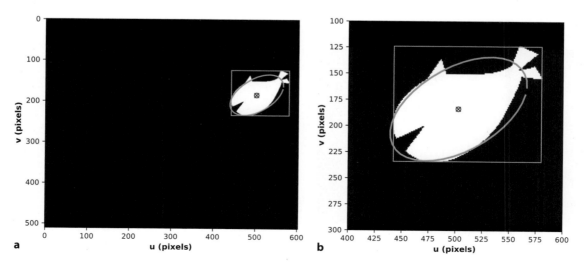

◨ **Fig. 12.11**  A single shark (blob 3) isolated from the binary image of ◨ Fig. 12.8a. **a** Bounding box (green), centroid (blue) and equivalent ellipse (cyan) for one shark; **b** zoomed view

as shown in ◨ Fig. 12.11. The bounding box fits snugly around the blob and its center could be considered as the center of the blob. The sides of the bounding box are horizontal and vertical which means that as the blob rotates the size and shape of the bounding box would change, even though the size and shape of the blob does not.

### 12.1.3.3 Moments

Moments are a rich and computationally cheap class of image features which can describe region size and location as well as orientation and shape. The moment of an image $\mathbf{X}$ is a scalar

$$m_{pq}(\mathbf{X}) = \sum_{(u,v) \in \mathbf{X}} u^p v^q \, \mathbf{X}_{u,v} \tag{12.1}$$

where $(p + q)$ is the *order* of the moment. The zero-order moment $p = q = 0$ is

$$m_{00}(\mathbf{X}) = \sum_{(u,v) \in \mathbf{X}} \mathbf{X}_{u,v} \tag{12.2}$$

and for a binary image, where the background pixels are zero, this is simply the number of nonzero (white) pixels – the area of the region.

Moments are calculated using the `Image` method `mpq` and for the single shark the zero-order moment is

```
>>> m00 = right_shark.mpq(0, 0)
7728
```

which is the area of the region in units of pixels.

Moments can be given a physical interpretation by regarding the image as a mass distribution. Consider that the shark is made of thin plate, where each pixel has one unit of area and one unit of mass. The total mass of the region is $m_{00}$, and the center of mass or centroid of the region is

$$u_c = \frac{m_{10}}{m_{00}}, \ v_c = \frac{m_{01}}{m_{00}} \tag{12.3}$$

where $m_{10}$ and $m_{01}$ are the first-order moments. For our example the centroid of the target region is

```
>>> uc = right_shark.mpq(1, 0) / m00
502.5
>>> vc = right_shark.mpq(0, 1) / m00
183.7
```

which we can overlay on the plot

```
>>> plot_point((uc, vc), ["bo", "bx"]);
```

as shown in ◨ Fig. 12.11.

The central moments $\mu_{pq}$ are computed with respect to the centroid

$$\mu_{pq}(\mathbf{X}) = \sum_{(u,v) \in \mathbf{X}} (u - u_c)^p (v - v_c)^q \, \mathbf{X}_{u,v} \tag{12.4}$$

and are invariant to the position of the region. They are related to the moments $m_{pq}$ by

$$
\begin{aligned}
&\mu_{10} = 0, && \mu_{01} = 0 \\
&\mu_{20} = m_{20} - \frac{m_{10}^2}{m_{00}}, && \mu_{02} = m_{02} - \frac{m_{01}^2}{m_{00}}, && \mu_{11} = m_{11} - \frac{m_{10}m_{01}}{m_{00}}
\end{aligned}
\tag{12.5}
$$

and are computed by the `Image` method `upq`.

Using the thin plate analogy again, the inertia of the region about axes parallel to the $u$- and $v$-axes and intersecting at the centroid of the region is given by the symmetric inertia tensor

$$\mathbf{J} = \begin{pmatrix} \mu_{20} & \mu_{11} \\ \mu_{11} & \mu_{02} \end{pmatrix}. \tag{12.6}$$

The central second moments $\mu_{20}$, $\mu_{02}$ are the moments of inertia of the shape. The product of inertia, $\mu_{11}$, is nonzero if the shape is asymmetric with respect to the $uv$-axes.

The *equivalent ellipse* is the ellipse that has the same inertia tensor as the region. For our example

```
>>> u20 = right_shark.upq(2, 0); u02 = right_shark.upq(0, 2);
>>> u11 = right_shark.upq(1, 1);
>>> J = np.array([[u20, u11], [u11, u02]])
array([[7.83e+06, -2.917e+06],
       [-2.917e+06, 4.733e+06]])
```

and we can superimpose the equivalent ellipse over the region

```
>>> plot_ellipse(4 * J  / m00, centre=(uc, vc), inverted=True,
...                color="blue");
```

and the result is shown in ◨ Fig. 12.11.

The eigenvalues and eigenvectors of $\mathbf{J}$ are related to the radii of the ellipse and the orientation of its major and minor axes (see ▶ App. C.1.4). For this example, the eigenvalues

```
>>> lmbda, x = np.linalg.eig(J)
>>> lmbda
array([9.584e+06, 2.979e+06])
```

are the principal moments of inertia of the region. The maximum and minimum radii of the equivalent ellipse are respectively

$$a = 2\sqrt{\frac{\lambda_1}{m_{00}}}, \; b = 2\sqrt{\frac{\lambda_0}{m_{00}}} \tag{12.7}$$

NumPy does not guarantee the ordering of eigenvalues. The function `np.linalg.eigh`, for a symmetric matrix, does sort the eigenvalues into ascending order.

where $\lambda_1 \geq \lambda_0$. In Python this is ◀

```
>>> a = 2 * np.sqrt(lmbda.max() / m00)
70.43
>>> b = 2 * np.sqrt(lmbda.min() / m00)
39.27
```

in units of pixels. These lengths are characteristic of this particular shape and are invariant to scale and rotation. The aspect ratio of the region

```
>>> b / a
0.5575
```

is a scalar that crudely characterizes the shape, and is invariant to position, scale and rotation of the region.

The eigenvectors of $\mathbf{J}$ are the principal axes of the ellipse – the directions of its major and minor axes. The major, or principal, axis is the eigenvector $v$ corresponding to the maximum eigenvalue. For this example the eigenvectors are the columns of

```
>>> x
array([[  0.857,    0.5153],
       [ -0.5153,    0.857]])
```

◻ **Table 12.1** Region features and their invariance to camera motion: translation, rotation about the object's centroid and scale factor

|  | Translation | Rotation | Scale |
|---|---|---|---|
| Area | ✓ | ✓ | ✗ |
| Centroid | ✗ | ✓ | ✓ |
| Orientation $\theta$ | ✓ | ✗ | ✓ |
| Aspect ratio | ✓ | ✓ | ✓ |
| Circularity | ✓ | ✓ | ✓ |
| Hu moments | ✓ | ✓ | ✓ |

and the corresponding eigenvector is

```
>>> i = np.argmax(lmbda)  # get index of largest eigenvalue
0
>>> v = x[:, i]
array([  0.857,  -0.5153])
```

The angle of this eigenvector, with respect to the horizontal axis, is

$$\theta = \tan^{-1}\frac{v_y}{v_x}$$

and for our example this is

```
>>> np.rad2deg(np.arctan2(v[1], v[0]))
-31.02
```

degrees which indicates that the major axis of the equivalent ellipse is approximately 30° *above* horizontal. ▶

To summarize, we have created an image containing a spatially contiguous set of pixels corresponding to one of the objects in the scene. We have determined its area, a box that entirely contains it, its position (the location of its centroid), and its orientation. Some parameters are a function only of the region's shape and are independent of, or invariant to, the object's position and orientation. Aspect ratio is a very crude shape measure and additional shape measures, or invariants, are summarized in ◻ Table 12.1 – these will be introduced later in this section.

With reference to ◻ Fig. C.2b the angle increases clockwise down from the horizontal since the *y*-axis of the image is downward so the *z*-axis is into the page.

### 12.1.3.4 **Blob Descriptors**

In the previous sections we went through a lengthy process to isolate a single blob in the scene and to compute some descriptive parameters for it. A more concise way to obtain this same information, for all blobs in the scene, is to use the `blobs` method of a binary image

```
>>> blobs = sharks.blobs();
```

which returns an instance of a `Blobs` object. This contains descriptive parameters for multiple blobs which can be displayed in tabular form

```
>>> blobs
```

| id | parent | centroid | area | touch | perim | circul |
|---|---|---|---|---|---|---|
| 0 | -1 | 245.6, 425.9 | 1.85e+04 | False | 811.2 | 0.392 |
| 1 | -1 | 502.3, 183.9 | 7.51e+03 | False | 514.8 | 0.396 |
| 2 | -1 | 82.9, 159.6 | 7.52e+03 | False | 513.4 | 0.399 |
| 3 | -1 | 297.0, 180.0 | 1.44e+04 | False | 1235.4 | 0.132 |

*The result of this method is a wide table which has been cropped for inclusion in this book.

where `id` is the region label. The object behaves like a Python list, and its length

```
>>> len(blobs)
4
```

indicate that it describes 4 blobs. We can slice it or iterate on it, for example

```
>>> blobs[3]
```

| id | parent | centroid | area | touch | perim | circu |
|----|--------|----------|------|-------|-------|-------|
| 3 | -1 | 297.0, 180.0 | 1.44e+04 | False | 1235.4 | 0.13 |

*The result of this method is a wide table which has been cropped for inclusion in this book.

Each blob in the object has many attributes which can be accessed as properties

```
>>> blobs[3].area
1.436e+04
>>> blobs[3].umin
214
>>> blobs[3].aspect
0.7013
>>> blobs[3].centroid
(297, 180)
```

The touch property is True if the blob touches the boundary of the image. The moments are a named tuple

```
>>> blobs[3].moments.m00    # moment p=q=0
1.436e+04
>>> blobs[3].moments.mu11   # central moment p=q=1
3.187e+06
>>> blobs[3].moments.nu03   # normalized central moment p=0, q=3
-0.004655
```

In this case blobs[3] is a Blobs instance containing just one blob, but if we access the properties of a Blobs instance containing multiple blobs the result is an array with the attribute for all blobs. For example

```
>>> blobs.area
array([1.847e+04, 7510, 7523, 1.436e+04])
```

is a 1D array of the area of all the blobs.

The object also has methods, for example

```
>>> blobs[3].plot_box(color="red")
```

will draw a red bounding box around blob[3], while

```
>>> blobs[:2].plot_box(color="red")
```

will draw a red box around the first two blobs. To annotate all blobs with centroids and bounding boxes is

```
>>> blobs.plot_centroid(marker="+", color="blue")
>>> blobs.plot_box(color="red")
```

which is shown in ◼ Fig. 12.12a.

We can extract a ROI containing a particular blob and rotate it so that its major axis is horizontal by

```
>>> sharks.roi(blobs[1].bbox).rotate(blobs[1].orientation).disp();
```

which is shown in ◼ Fig. 12.12b.

The Blobs object can also be sliced by a logical array

```
>>> blobs[blobs.area > 10_000]
```

| id | parent | centroid | area | touch | perim | circu |
|----|--------|----------|------|-------|-------|-------|
| 0 | -1 | 245.6, 425.9 | 1.85e+04 | False | 811.2 | 0.39 |
| 3 | -1 | 297.0, 180.0 | 1.44e+04 | False | 1235.4 | 0.13 |

*The result of this method is a wide table which has been cropped for inclusion in this book.

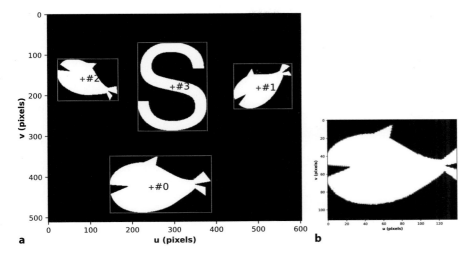

**Fig. 12.12** **a** Graphical annotations for the blobs shown in Fig. 12.8, showing centroids, bounding boxes and blob labels; **b** blob 1 extracted from the image and rotated so that its major axis is horizontal

to select blobs with more than 10,000 pixels. There is also a `filter` method, and for the tomato image

```
>>> tomato_blobs = tomatoes_binary.blobs()
```

| id | parent | centroid | area | touch | perim | circul |
|----|--------|----------|------|-------|-------|--------|
| 0 | -1 | 8.9, 235.1 | 39 | True | 37.9 | 0.380 |
| 1 | -1 | 90.7, 206.5 | 2.56e+03 | False | 201.5 | 0.880 |
| 2 | -1 | 130.2, 126.9 | 1.14e+03 | False | 144.5 | 0.763 |

*The result of this method is a wide table which has been cropped for inclusion in this book.

we could set bounds on the allowable area

```
>>> tomato_blobs.filter(area=(1_000, 5_000))
```

| id | parent | centroid | area | touch | perim | circul |
|----|--------|----------|------|-------|-------|--------|
| 1 | -1 | 90.7, 206.5 | 2.56e+03 | False | 201.5 | 0.880 |
| 2 | -1 | 130.2, 126.9 | 1.14e+03 | False | 144.5 | 0.763 |

*The result of this method is a wide table which has been cropped for inclusion in this book.

to exclude small blobs due to noise – selecting only blobs with an area between 1000 and 5000 pixels. Other filter parameters include aspect ratio and boundary touching, and more details are provided in the online documentation. To exclude blobs that touch the boundary

```
>>> tomato_blobs.filter(touch=False)
```

| id | parent | centroid | area | touch | perim | circul |
|----|--------|----------|------|-------|-------|--------|
| 1 | -1 | 90.7, 206.5 | 2.56e+03 | False | 201.5 | 0.880 |
| 2 | -1 | 130.2, 126.9 | 1.14e+03 | False | 144.5 | 0.763 |

*The result of this method is a wide table which has been cropped for inclusion in this book.

The filter rules can also be cascaded, for example

```
>>> tomato_blobs.filter(area=[1000, 5000], touch=False, color=1)
```

| id | parent | centroid | area | touch | perim | circul |
|----|--------|----------|------|-------|-------|--------|
| 1 | -1 | 90.7, 206.5 | 2.56e+03 | False | 201.5 | 0.880 |
| 2 | -1 | 130.2, 126.9 | 1.14e+03 | False | 144.5 | 0.763 |

*The result of this method is a wide table which has been cropped for inclusion in this book.

and a blob must pass all rules in order to be accepted. The `sort` method will sort the blobs on criteria including area, aspect ratio, perimeter length and circularity.

### 12.1.3.5  Blob Hieararchy

Consider the binary image

```
>>> multiblobs = Image.Read("multiblobs.png");
>>> multiblobs.disp();
```

which is shown in ◘ Fig. 12.13a and, as we have done previously, we label the regions

```
>>> labels, m = multiblobs.labels_binary()
>>> m
6
>>> multiblobs.disp();
```

which is shown in ◘ Fig. 12.13b. This has labeled the white blobs and the black background, but when we look at the image we perceive some of those black blobs as holes in the white blobs. Those holes in turn can contain white blobs which could also have holes – there is a hierarchy of black and white blobs. The blob analysis introduced in ▶ Sect. 12.1.3.4 includes hierarchy information

```
>>> blobs = multiblobs.blobs()
```

| id | parent | centroid | area | touch | perim | circ |
|----|--------|----------|------|-------|-------|------|
| 0 | -1 | 907.9, 735.1 | 1.94e+05 | False | 2219.0 | 0.5 |
| 1 | 0 | 1025.0, 813.7 | 1.06e+05 | False | 1386.9 | 0.7 |
| 2 | 1 | 938.1, 855.2 | 1.72e+04 | False | 489.7 | 1.0 |
| 3 | 1 | 988.1, 697.2 | 1.21e+04 | False | 411.5 | 0.9 |
| 4 | 0 | 846.0, 511.7 | 1.75e+04 | False | 495.9 | 0.9 |
| 5 | -1 | 291.7, 377.8 | 1.7e+05 | False | 1711.6 | 0.8 |
| 6 | 5 | 312.7, 472.1 | 1.75e+04 | False | 494.5 | 1.0 |
| 7 | 5 | 241.9, 245.0 | 1.75e+04 | False | 495.9 | 0.9 |
| 8 | -1 | 1228.0, 254.3 | 8.14e+04 | False | 1214.2 | 0.7 |
| 9 | 8 | 1225.2, 220.0 | 1.75e+04 | False | 495.9 | 0.9 |

*The result of this method is a wide table which has been cropped for inclusion in this book.

and the second column is the `id` of the blob's parent – the blob that encloses or contains it. Blob `-1` is the background that contains all white blobs and no parameters are computed for it. Additional properties of the blobs include

```
>>> blobs[1].children
[2, 3]
>>> blobs[1].parent
0
```

We can turn this hierarchical blob data into a label image that reflects the hierarchy

```
>>> blobs.label_image().disp();
```

which is shown in ◘ Fig. 12.13c and the enclosed black regions now have distinct labels. There are a total of 11 unique labels, the 10 listed in the table above plus the background blob with `id` of `-1`. The label values are the `id` listed above plus one.

We can also display the blob hierarchy as a graph

```
>>> blobs.dotfile(show=True);
```

which is shown in ◘ Fig. 12.14.

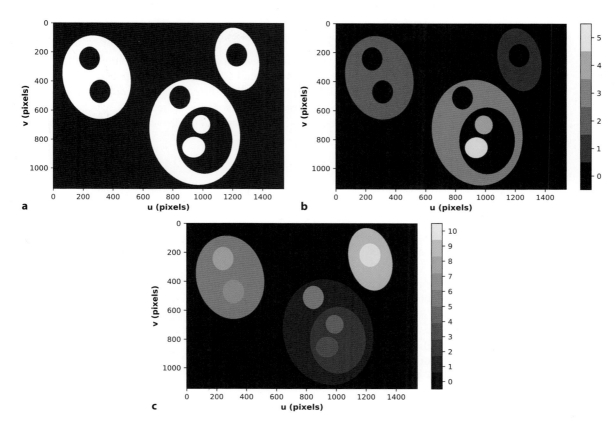

**◼ Fig. 12.13** Image labeling example. **a** Binary image; **b** labeled image; **c** labeled image with regions considered as a hierarchy

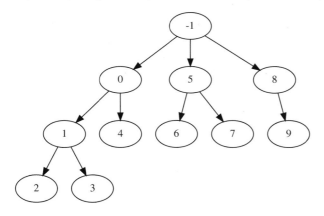

**◼ Fig. 12.14** Blob hierarchy as a tree. Nodes are labeled by the blob id and `-1` is the background

### 12.1.3.6 Shape from Moments

In order to recognize particular objects we need some measures of shape that are invariant to the rotation and scale of the image. Returning to the sharks image

```
>>> blobs = sharks.blobs()
```

| id | parent | centroid | area | touch | perim | circu |
|----|--------|----------|------|-------|-------|-------|
| 0 | -1 | 245.6, 425.9 | 1.85e+04 | False | 811.2 | 0.39 |
| 1 | -1 | 502.3, 183.9 | 7.51e+03 | False | 514.8 | 0.39 |
| 2 | -1 | 82.9, 159.6 | 7.52e+03 | False | 513.4 | 0.39 |
| 3 | -1 | 297.0, 180.0 | 1.44e+04 | False | 1235.4 | 0.13 |

*The result of this method is a wide table which has been cropped for inclusion in this book.

```
>>> blobs.aspect
array([  0.5609,     0.5619,      0.562,     0.7013])
```

we can see that the scene contains two different shapes: there are three blobs with an aspect ratio around 0.56 and one blob with a larger aspect ratio around 0.70.

The aspect ratio is not particularly discriminating and more powerful shape descriptions can be obtained from the ratios of moments which are invariant to position, orientation and scale. For this scene the seven Hu moment invariants of the blobs

```
>>> blobs.humoments
[array([   0.21,   0.01199, 0.001997, 0.0005748, 6.032e-07,
    6.149e-05,  1.245e-07]),
 array([   0.21,   0.01193, 0.002015, 0.0005745, 6.045e-07,
    6.115e-05,  1.291e-07]),
 array([ 0.2097,   0.01189, 0.001979, 0.0005618, 5.801e-07,
    5.989e-05,  1.199e-07]),
 array([ 0.4705,   0.02568, 0.000624, 1.115e-05, 9.217e-10,
    -1.041e-06, -1.232e-10])]
```

In practice the discrete nature of the pixel data means that the invariance will only be approximate.

can be considered as *fingerprints* of the blobs. They indicate the similarity of blobs 0, 1 and 2 despite their different position, orientation and scale. ◄ It also indicates the difference in shape of blob 3. This shape descriptor can be considered as a point in $\mathbb{R}^7$, and similarity to other shapes can be defined in terms of Euclidean distance in this descriptor space. If we knew that there were only two different shaped objects in the scene we could apply $k$-means clustering to this 7-dimensional data.

---

**Excurse 12.3: Moment Invariants**

The normalized image moments

$$v_{pq} = \frac{\mu_{pq}}{\mu_{00}^{\gamma}}, \ \gamma = \frac{1}{2}(p+q) + 1 \text{ for } p+q = 2, \ 3, \ \cdots \tag{12.8}$$

are invariant to translation and scale, and are computed from the central moments by the `Image` method.

Third-order moments allow for the creation of quantities that are invariant to translation, scale and orientation within a plane. One such set of moments, defined by Hu (1962), are

$$\varphi_1 = v_{20} + v_{02}$$
$$\varphi_2 = (v_{20} - v_{02})^2 + 4v_{11}^2$$
$$\varphi_3 = (v_{30} - 3v_{12})^2 + (3v_{21} - v_{03})^2$$
$$\varphi_4 = (v_{30} + v_{12})^2 + (v_{21} + v_{03})^2$$
$$\varphi_5 = (v_{30} - 3v_{12})(v_{30} + v_{12})\left[(v_{30} + v_{12})^2 - 3(v_{21} + v_{03})^2\right]$$
$$\qquad + (3v_{21} - v_{03})(v_{21} + v_{03})\left[3(v_{30} + v_{12})^2 - (v_{21} + v_{03})^2\right]$$
$$\varphi_6 = (v_{20} - v_{02})\left[(v_{30} + v_{12})^2 - (v_{21} + v_{03})^2\right]$$
$$\qquad + 4v_{11}(v_{30} + v_{12})(v_{21} + v_{03})$$
$$\varphi_7 = (3v_{21} - v_{03})(v_{30} + v_{12})\left[(v_{30} + v_{12})^2 - 3(v_{21} + v_{03})^2\right]$$
$$\qquad + (3v_{12} - v_{30})(v_{21} + v_{03})\left[3(v_{30} + v_{12})^2 - (v_{21} + v_{03})^2\right]$$

and computed by the `humoments` method of the `Image` class and the `Blobs` class.

### 12.1.3.7  Shape from Perimeter

The shape of a region is concisely described by its perimeter, boundary, contour or perimeter pixels – sometimes called edgels. ◘ Fig. 12.15 shows three common ways to represent the perimeter of a region – each will give a slightly different estimate of the perimeter length. A chain code is a list of the outermost pixels of the region whose center's are linked by short line segments. In the case of a 4-neighbor chain code, the successive pixels must be adjacent and the perimeter segments have an orientation of $k \times 90°$, where $k \in \{0, 1, 2, 3\}$. With an 8-neighbor chain code, or Freeman chain code, the perimeter segments have an orientation of $k \times 45°$, where $k \in \{0, \cdots, 7\}$. The crack code has its segments in the *cracks* between the pixels on the perimeter of the region and the pixels outside the region. These have orientations of $k \times 90°$, where $k \in \{0, 1, 2, 3\}$.

The perimeter can be encoded as a list of pixel coordinates $(u_i, v_i)$ or very compactly as a bit string using just 2 or 3 bits to represent $k$ for each segment. These various representations are equivalent and any representation can be transformed to another.

> ❗ Note that for chain codes, the perimeter follows a path that is on average half a pixel inside the true perimeter and therefore underestimates the perimeter length. The error is most significant for small regions.

The blob object contains perimeter information ▶ which includes a list of points on the perimeter. In this case there are

Obtained using the OpenCV `findContours` function which by default returns an 8-neighbor chain code.

```
>>> blobs[1].perimeter.shape
(2, 435)
```

435 perimeter points. The first five points on the perimeter of blob 1 are

```
>>> blobs[1].perimeter[:, :5]
array([[559, 558, 558, 558, 557],
       [124, 125, 126, 127, 128]], dtype=int32)
```

which is a 2D array with one column per perimeter point. The perimeter can be overlaid on the current plot by

```
>>> blobs[1].plot_perimeter(color="orange")
```

in this case as an orange line. The plotting methods can also be invoked for all blobs

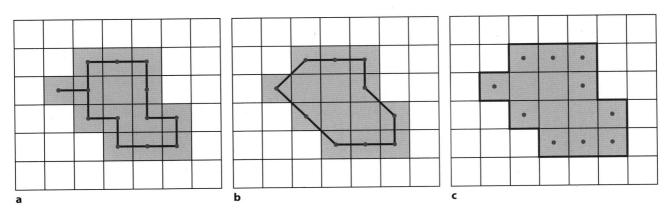

a　　b　　c

◘ **Fig. 12.15** Perimeter representations with region pixels shown in gray, perimeter segments shown in blue and the center of perimeter pixels marked by a red dot. **a** Chain code with 4 directions; **b** Freeman chain code with 8 directions; **c** crack code. The perimeter lengths for this example are respectively 14, 12.2 and 18 pixels

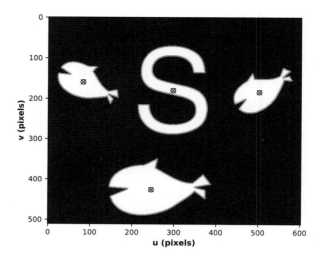

**◘ Fig. 12.16** Boundaries (orange) and centroids of four blobs

```
>>> sharks.disp();
>>> blobs.plot_perimeter(color="orange")
>>> blobs.plot_centroid()
```

which is shown in ◘ Fig. 12.16.

The perimeter length, taking into account the angle of each perimeter segment is

```
>>> p = blobs[1].perimeter_length
514.8
```

from which we can compute circularity – a commonly used and intuitive shape feature. It is defined as

$$\rho = \frac{4\pi m_{00}}{p^2} \tag{12.9}$$

where $p$ is the region's perimeter length. Circularity has a maximum value of $\rho = 1$ for a circle, is $\rho = \frac{\pi}{4}$ for a square and zero for an infinitely long line. Circularity is also invariant to translation, rotation and scale. For the sharks image

```
>>> blobs.circularity
array([ 0.3923,    0.3962,    0.399,    0.1315])
```

For small blobs, quantization effects can lead to significant errors in circularity.

shows again that blob 3 has a different shape to the others. Its low circularity is due to it being effectively a long line. ◄

Every object has one external perimeter, which may include a section of the image border if the object touches the border. An object with holes will have one or more internal perimeters. The Toolbox only returns the external perimeter length – the inner perimeter length is the sum of the external perimeter lengths of its child regions.

The external perimeter contains all the essential information about the shape of a region. It is possible, assuming that the region has no holes, to compute the moments from just the perimeter. We can demonstrate this by first converting the perimeter to a polygon object and then computing the zero-order moment

```
>>> p = Polygon2(blobs[1].perimeter).moment(0, 0)
-7510
```

The value is negative in this case because the perimeter points are given in a clockwise order. The method `area` always returns a positive value.

which gives the same result as counting all the pixels. ◄

One way to analyze the rich shape information encoded in the perimeter is to compute the distance and angle to every perimeter point with respect to the object's

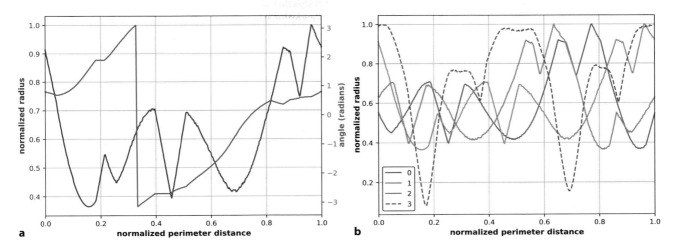

■ **Fig. 12.17** Radius signature matching. **a** Radius and angle signature for blob 1 (top right shark in ■ Fig. 12.16), **b** normalized radius signatures for all blobs

centroid. This is computed by

```
>>> r, th = blobs[1].polar();
>>> plt.plot(r, "r", th, "b");
```

and the radius and angle are shown in ■ Fig. 12.17a. These are computed for 400 points (default) evenly spaced along the entire perimeter of the object. Both the radius and angle signatures describe the shape of the object. The angle signature is invariant to the scale of the object while the amplitude of the radius signature scales with object size. The radius signatures of all four blobs can be compared by

```
>>> for blob in blobs:
...     r, theta = blob.polar()
...     plt.plot(r / r.max());
```

and are shown in ■ Fig. 12.17b. We have normalized by the maximum radius in order to remove the effect of object scale. The signatures are a function of normalized distance along the perimeter and they all start at the top left-most pixel on the object's perimeter. Different objects of the same shape have identical signatures but possibly shifted horizontally and wrapped around – the first and last points in the horizontal direction are adjacent.

To compare the shape profile of blob 1 with all the shape profiles, for all possible horizontal shifts, is

```
>>> similarity, _ = blobs.polarmatch(1)
>>> similarity
[0.9936, 1, 0.9856, 0.6434]
```

and indicates that the shape of blob 1 closely matches the shape of itself and blobs 0 and 2, but not the shape of blob 3. The `polarmatch` method computes the 1-dimensional normalized cross correlation, see ■ Table 12.1, for every possible rotation of one signature with respect to the other, and returns the highest value.

There are many variants to the approach described. The signature can be Fourier transformed and described more concisely in terms of the first few Fourier coefficients. The perimeter curvature can also be computed which highlights corners, allowing it to be approximated by straight line segments and arcs. The `perimeter_approx` method can approximate the perimeter by a series of line segments.

### 12.1.4  **Object Detection Using Deep Learning**

Deep networks, introduced in ▶ Sect. 12.1.1.3, are also able to detect instances of objects in a scene and directly compute their class and bounding box. These networks perform all the steps described above: pixel classification (semantic segmentation), distinguishing multiple objects (instance segmentation) and description using bounding boxes. We will illustrate this with Faster R-CNN, a region-based convolutional neural network, trained using the COCO dataset. ◄ This network has 80 object classes including person, bicycle, car, cat, dog, sheep, apple, pizza, kite, toothbrush, frisbee etc.

Common Objects in Context, see ▶ https://cocodataset.org.

We will load the same image used in ▶ Sect. 12.1.1.3

```
>>> scene = Image.Read("image3.jpg")
Image: 640 x 438 (uint8), R:G:B [.../images/image3.jpg]
>>> scene.disp();
```

which is shown in ◘ Fig. 12.18a. Assuming that PyTorch is installed, we first convert the image to a PyTorch tensor

```
>>> import torch
>>> import torchvision as tv
>>> transform = tv.transforms.ToTensor();
>>> in_tensor = transform(scene.image);
```

Next, we load the Faster R-CNN network, configure it for evaluation (inference) and pass the image through it

```
>>> model = tv.models.detection\
...              .fasterrcnn_resnet50_fpn(pretrained=True).eval();
>>> outputs = model(torch.stack([in_tensor]));
```

and then convert the results to Python lists

```
>>> # list of confidence scores
>>> scores = outputs[0]["scores"].detach().numpy();
>>> # list of class names as strings
>>> labels = outputs[0]["labels"].detach().numpy();
>>> # list of boxes as array([x1, y1, x2, y2])
>>> boxes = outputs[0]["boxes"].detach().numpy();
```

which describe

```
>>> len(scores)
42
```

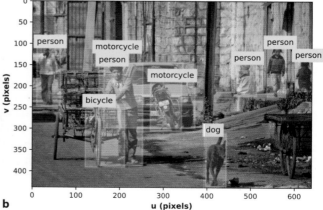

◘ **Fig. 12.18** Deep neural network for detecting objects in a complex real-world scene

objects, but some of these have very low confidence. To show the confident detections and their labels is

```
>>> classname_dict = {1: "person", 2: "bicycle", 3: "car",
...                   4: "motorcycle", 18: "dog"};
>>> for score, label, box in zip(scores, labels, boxes):
...    if score > 0.5:  # only confident detections
...       plot_labelbox(classname_dict[label], lbrt=box, filled=True,
...                     alpha=0.3, color="yellow", linewidth=2);
```

which is shown in ◻ Fig. 12.18b. We have defined a dictionary `classname_dict` to map a subset of the integer class labels to the corresponding COCO class names.

### 12.1.5 Summary

We have discussed the process of transforming an input image, gray scale or color, into concise descriptors of regions within the scene. The criteria for what constitutes a region is *application specific*. For a tomato picking robot it would be round red regions, for landing a drone it might be yellow targets on the ground.

The process outlined is the classical *bottom up* approach to machine vision applications and the key steps are:

- Classifying the pixels according to the application specific criterion, for example, "redness", "yellowness" or "person-ness". Each pixel is assigned to a class $c$.
- Grouping adjacent pixels of the same class into sets, and each pixel is assigned a label $S$ indicating to which spatial set it has been assigned.
- Describing the sets in terms of features derived from their spatial extent, moments, equivalent ellipse and perimeter.

These steps are a progression from *low-level* to *high-level* representation. The low-level operations consider pixels in isolation, whereas the high-level is concerned with more abstract concepts such as size and shape. The MSER and graphcuts algorithms are powerful because they combine steps 1 and 2 and consider regions of pixels and localized differences in order to create a segmentation.

Importantly none of these steps need to be perfect. Perhaps the first step has some false positives, isolated pixels misclassified as objects, that we can eliminate by morphological operations, or reject after connectivity analysis based on their small size. The first step may also have false negatives, for example specular reflection and occlusion may cause some object pixels to be classified incorrectly as non-object. In this case we need to develop some heuristics, perhaps morphological processing, to fill in the gaps in the blob. Another option is to oversegment the scene – increase the number of regions and use some application-specific knowledge to merge adjacent regions. For example a specular reflection colored region might be merged with surrounding regions to create a region corresponding to the whole fruit.

For some applications it might be possible to engineer the camera position and illumination to obtain a high quality image, but for a robot operating in the real world this luxury does not exist. A robot needs to glean as much useful information as it can from the image and move on.

Domain knowledge is always a powerful tool. Given that we know the scene contains tomatoes and plants, the observation of a large red region that is not circular allows us to infer that the tomato is occluded. The robot may choose to move to a location where the view of tomato is not occluded, or to seek a different tomato that is not occluded.

These techniques are suitable for relatively simple scenes and have the advantage of requiring only modest amounts of computation. However, in recent times these techniques have been eclipsed by those based on deep learning which is a powerful and practical solution to complex image segmentation problems. A network transforms a color image, in a single step, to a pixel classification or object detections using human-meaningful semantic labels like "person" or "bicycle". Many open-source tools exist and it is now relatively straightforward to train a network for the particular objects in your application. Training does however require a lot of labeled data and a lot of computation time.

## 12.2    Line Features

Lines are distinct visual features that are particularly common in human-made environments – for example the edges of roads, buildings and doorways. In ▸ Sect. 11.5.1.3 we discussed how image intensity gradients can be used to find edges within an image, and this section will be concerned with fitting line segments to such edges.

We will illustrate the principle using the very simple scene

```
>>> points5 = Image.Read("5points.png", dtype="float");
```

shown in ◘ Fig. 12.19a. Consider any one of these points – there are an infinite number of lines that pass through that point. If the point could vote for these lines, then each possible line passing through the point would receive one vote. Now consider another point that does the same thing, casting a vote for all the possible lines that pass through it. One line (the line that both points lie on) will receive a vote from each point – a total of two votes – while all the other possible lines receive either zero or one vote.

We want to describe each line in terms of a minimum number of parameters but the standard form $v = mu + c$ is problematic for the case of vertical lines where $m = \infty$. Instead it is common to represent lines using the $(\rho, \theta)$ parameterization shown in ◘ Fig. 12.20

$$u \cos \theta + v \sin \theta = \rho \tag{12.10}$$

where $\theta \in [0, \pi)$ is the angle from the horizontal axis to the line's normal, and $\rho \in [-\rho_{min}, \rho_{max}]$ is the perpendicular distance between the origin and the line as shown in ◘ Fig. 12.20. Line 2 could be described by $\theta_2 > \pi$ and $\rho_2 > 0$ but instead

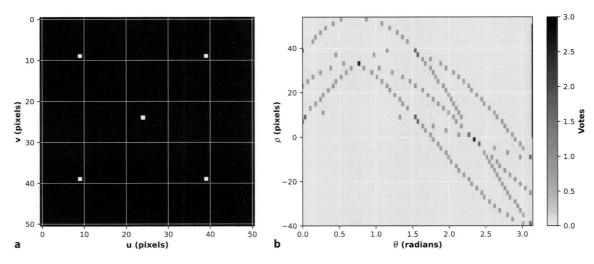

◘ **Fig. 12.19** Hough transform fundamentals. **a** Five points that define six lines; **b** the Hough accumulator array. The horizontal axis is an angle $\theta \in \mathbf{S}^1$ so we can imagine the graph wrapped around a cylinder and the left- and right-hand edges joined. The sign of $\rho$ also changes at the join so the curve intersections on the left- and right-hand edges are equivalent

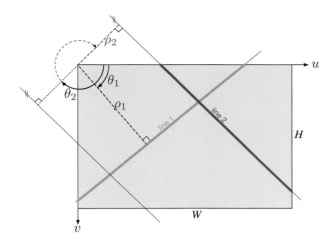

**■ Fig. 12.20** $(\theta, \rho)$ parameterization for two line segments. $\theta \in [0, \pi)$ is the angle from horizontal clockwise to the line's normal. Positive values of $\rho$ are shown in blue and negative values in red. Line 2 has a negative value of $\rho$

we keep $\theta_2 \in [0, \pi)$ and make $\rho_2 < 0$. ▶ A horizontal line has $\theta = \frac{\pi}{2}$ and a vertical line has $\theta = 0$. Any line can therefore be considered as a point $(\theta, \rho)$ in the 2-dimensional space of all possible lines.

> This is the convention adopted by OpenCV.

It is not practical to vote for one out of an infinite number of lines through each point, so we consider lines drawn from a finite set. The $\theta\rho$-space is quantized and a corresponding $N_\theta \times N_\rho$ array **A** is used to tally the votes – the *accumulator* array. For a $W \times H$ input image

$$\rho\text{max} = -\rho\text{min} = \max(W, H) \ .$$

The array **A** has $N_\rho$ elements spanning the interval $\rho \in [-\rho_{\max}, \rho_{\max}]$ and $N_\theta$ elements spanning the interval $\theta \in [0, \pi)$. The indices of the array are $(i, j) \subset \mathbb{N}_0{}^2$ such that

$$i = 0, \ldots, N_\theta - 1 \mapsto \theta \in [0, \pi)$$
$$j = 0, \ldots, N_\rho - 1 \mapsto \rho \in [-\rho\text{max}, \rho\text{max}] \ .$$

An edge point $(u, v)$ votes for *all* lines that satisfy (12.10). For every $i$ the corresponding value of $\theta$ is computed using the linear mapping above, then $\rho$ is computed according to

$$\rho = u \cos\theta + v \sin\theta$$

and mapped to a corresponding integer $j$, and finally $a_{i,j}$ is incremented to record the vote. Every edge point adds a vote to $N_\theta$ elements of **A** that lie along a sinuosidal curve.

At the end of the process, those elements of **A** with the largest number of votes correspond to dominant lines in the scene. For the example of ■ Fig. 12.19a, the resulting accumulator array is shown in ■ Fig. 12.19b. Most of the array contains zero votes (yellow) and the dark curves are trails of single votes corresponding to each of the five input points. These curves intersect and those points correspond to lines with more than one vote. We see four locations where two curves intersect, resulting in cells with two votes, and these correspond to the lines joining the four outside points of ■ Fig. 12.19a. The horizontal axis represents angle $\theta \in \mathbf{S}^1$ so the left- and right-hand ends are joined and $\rho$ changes sign – the curve intersection points on the left- and right-hand sides of the array are equivalent. We also see two locations where three curves intersect, resulting in cells with three votes, and these correspond to the diagonal lines that include the middle point of ■ Fig. 12.19a. This technique is known as the Hough (pronounced huff) transform.

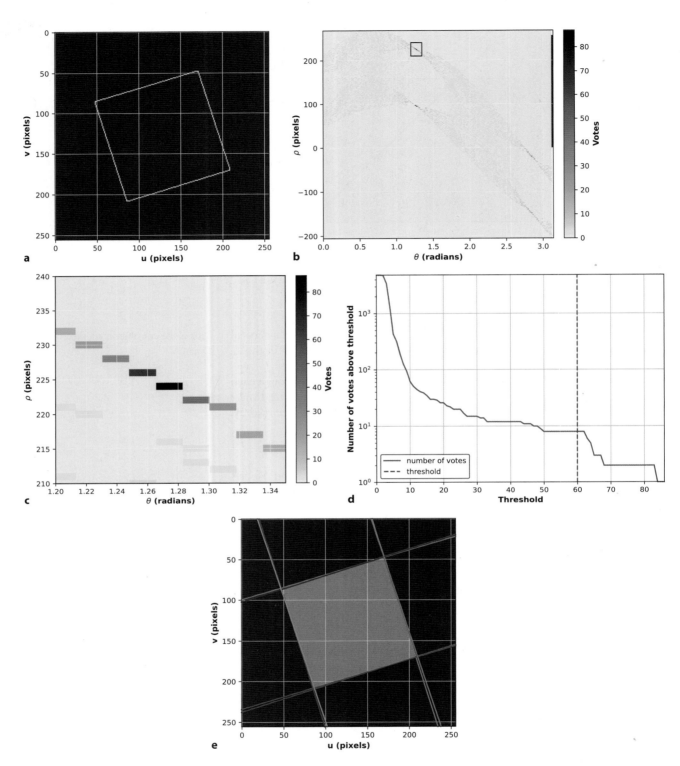

**◻ Fig. 12.21** Hough transform for a rotated square. **a** Edge image; **b** Hough accumulator; **c** close-up view of the Hough accumulator; **d** distribution of votes above threshold, versus threshold; **e** estimated lines overlaid on the original image

Consider the more complex example of a solid square rotated counter-clockwise by 0.3 rad

```
>>> square = Image.Squares(number=1, size=256, fg=128).rotate(0.3)
Image: 256 x 256 (uint8)
```

and the edge points

```
>>> edges = square.canny();
```

are shown in ◨ Fig. 12.21a. The Hough transform is computed by

```
>>> h = edges.Hough();
```

which returns an instance of the `Hough` class. The accumulator array can be visualized as an image ▶

```
>>> h.plot_accumulator()
```

which is shown in ◨ Figs. 12.21b, c. The four darkest spots correspond to dominant edges in the input image. We can see that many other possible lines have received a small number of votes as well. A histogram of the number of votes is available in the `votes` property

```
>>> plt.plot(h.votes);
>>> plt.yscale("log");
```

and is plotted in ◨ Fig. 12.21d. Although the object has only four sides there are many more than four peaks in the accumulator array, but most of them are very minor peaks.

Choosing lines with 60 or more votes we obtain a set of dominant lines

```
>>> lines = h.lines(60)
array([[   1.274,        224],
       [   1.274,         95],
       [   2.845,       -149],
       [   1.257,         97],
       [   1.257,        226],
       [   2.845,        -20],
       [   2.827,       -147],
       [   2.827,        -18]], dtype=float32)
```

which is a 2D array of line parameters where each row is $(\theta, \rho)$. We see that there are pairs of similar line parameters and this is due to quantization effects. ◨ Fig. 12.21c shows two accumulator cells with a value exceeding the threshold in the vicinity of what should be a single peak. ▶

The detected lines can be projected onto the original image

```
>>> square.disp();
>>> h.plot_lines(lines);
```

and the result is shown in ◨ Fig. 12.21e. We can clearly see multiple lines parallel to each edge of the square.

Choosing the threshold is critical to the quality of the result. Too high a threshold will lead to lines being missed, while too low a threshold will lead to many spurious lines.

The probabilistic Hough transform is a more efficient version that processes only a subset of points in the image ▶ and returns line segments rather than lines. A line segment has a minimum length and can contain gaps, provided that the length of the gap does not exceed a threshold. ▶ For the real image shown in ◨ Fig. 12.22

```
>>> church = Image.Read("church.png", mono=True)
Image: 1280 x 851 (uint8) [.../images/church.png]
>>> edges = church.canny()
Image: 1280 x 851 (uint8)
>>> h = edges.Hough();
>>> lines = h.lines_p(100, minlinelength=200, maxlinegap=5, seed=0);
```

we can plot the lines with 100 or more votes that meet the line length and gap criteria

```
>>> church.disp();
>>> h.plot_lines_p(lines, "r--")
```

The Toolbox uses the OpenCV function `HoughLines` to compute the Hough transform, but this does not return the accumulator array. The `plot_accumulator` method "reverse engineers" the accumulator array through a costly process of computing the Hough transform for all possible thresholds. This is helpful for pedagogy but very inefficient in practice.

Nonlocal-maxima suppression would be useful here but the underlying OpenCV function `HoughLines` does not support this.

The algorithm depends on random numbers and the results will differ from run to run unless OpenCV's random number seed is set.

If the gap exceeds the threshold, two line segments are returned provided their length meets the minimum length criteria.

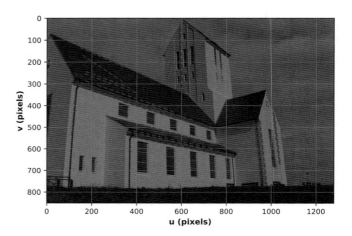

**Fig. 12.22**   Probabilistic Hough transform of a real image

and these are shown overlaid on the image. We can see that a number of lines are converging on a perspective vanishing point to the right of the image. However, many strong lines in the image have not been found, and the highly textured roof has generated some spurious line segments.

### 12.2.1  Summary

The Hough transform is elegant in principle and in practice it can work well or infuriatingly badly. It performs poorly when the scene contains a lot of texture or the edges are indistinct. Texture causes votes to be cast widely, but not uniformly, over the accumulator array which tends to mask the true peaks. In practice it can require lot of experimentation to find good parameters and these may not generalize to other scenes.

The Hough transform estimates the direction of the line by fitting lines to the edge pixels. It ignores rich information about the direction of the edge at each pixel which was discussed in ▶ Sect. 11.5.1.3. The consequence of not using all the information available is poorer estimation. There is little added expense in using the direction at each pixel since we have already computed the image gradients in order to evaluate edge magnitude.

### 12.3  Point Features

The final class of features that we will discuss are point features. These are visually distinct points in the image that are also known as interest points, salient points, keypoints or corner points. We will first introduce some classical techniques for finding point features and then discuss scale-invariant techniques.

### 12.3.1  Classical Corner Detectors

We recall from ▶ Sect. 11.5.1.3 that a point on a line has a strong gradient in a direction normal to the line. However gradient *along* the line is low which means that a pixel on the line will look very much like its neighbors along the line. In contrast, an *interest point* is a point that has a high image gradient in orthogonal directions. It might be single pixel that has a significantly different intensity to all of its neighbors, or it might literally be a pixel on the corner of an object. Since interest points

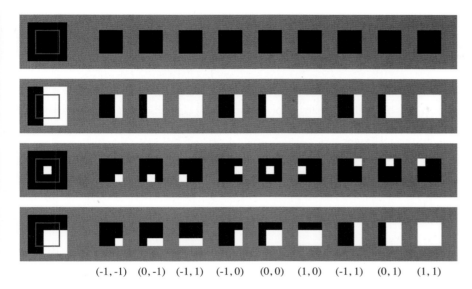

(-1,-1)  (0,-1)  (-1,1)  (-1,0)  (0,0)  (1,0)  (-1,1)  (0,1)  (1,1)

**▢ Fig. 12.23** Principle behind the Morovec detector. At the left of each row is an image region with with a $3 \times 3$ window indicated in red. To the right are the contents of that red window for various displacements as shown across the bottom of the figure as $(\delta_u, \delta_v)$

are quite distinct they have a much higher likelihood of being reliably detected in different views of the same scene. They are therefore key to multi-view techniques such as stereo and motion estimation which we will discuss in ▶ Chap. 14.

The earliest corner point detector was Moravec's *interest operator*, so called because it detected points in the scene that were *interesting* from a tracking perspective. It was based on the intuition that if a small image window $\mathcal{W}$ is to be unambiguously located in another image it must be quite different to the same-size window at any adjacent location. ▢ Fig. 12.23 shows, for each row, the contents of the red window as it is displaced horiontally and vertically. The first row is the case for a region of the image where all pixels have the same value, and we see that the displaced windows are all identical. The second row is the case for a vertical edge, and we see that the displaced windows have some variability but there are only three unique windows. The third and fourth rows show the case of an isolated point and a corner respectively – each of the displaced windows has a unique value. Clearly the isolated point and the corner point are the most distinctive.

Moravec defined the similarity between a region centered at $(u, v)$ and an adjacent region, displaced by $(\delta_u, \delta_v)$, as

$$s(u, v, \delta_u, \delta_v) = \sum_{(i,j) \in \mathcal{W}} \left( x_{u+\delta_u+i, v+\delta_v+j} - x_{u+i, v+j} \right)^2, \quad \forall u, v \in \mathbf{X} \qquad (12.11)$$

where $\mathcal{W}$ is some local region of the image $\mathbf{X}$ – typically a $w \times w$ square window where $w$ is odd. This is the sum of squared differences (SSD) similarity measure from ▢ Table 12.1 that we discussed previously – a low value means the regions are similar. Similarity is evaluated for displacements in eight cardinal ▶ directions $(\delta_u, \delta_v) \in \mathcal{D}$ and the minimum value is the interest measure or corner strength

$$C_M(u, v) = \min_{(\delta_u, \delta_v) \in \mathcal{D}} s(u, v, \delta_u, \delta_v) \qquad (12.12)$$

N, NE, E, ..., W, NW or $i, j \in \{-1, 0, 1\}$.

which has a large value only if all the displaced patches are different to the original patch. The function $C_M(\cdot)$ is evaluated for every pixel in the image and interest points are those where $C_M$ is *high*. The main limitation of the Moravec detector is that it is nonisotropic since it examines image change, essentially gradient, in a limited number of directions. Consequently the detector can give a strong output for a point on a line, which is not desirable.

We can generalize the approach by defining the similarity as the weighted sum of squared differences between the image region and the displaced region as

$$s(u, v, \delta_u, \delta_v) = \sum_{(i,j) \in \mathcal{W}} w_{i,j} \left( \underline{x_{u+\delta_u+i,v+\delta_v+j}} - x_{u+i,v+j} \right)^2$$

where **W** is a weighting matrix, for example a 2D Gaussian, that emphasizes points closer to the center of the window $\mathcal{W}$. The underlined term can be approximated by a truncated Taylor series ◄

See ► App. E.

$$x_{u+\delta_u,v+\delta_v} \approx x_{u,v} + \delta_u x_{u:u,v} + \delta_v x_{v:u,v}$$

where $\mathbf{X}_u$ and $\mathbf{X}_v$ are the horizontal and vertical image gradients respectively. We can now write

$$\begin{aligned}
s(u, v, \delta_u, \delta_v) = {} & \delta_u^2 \sum_{(i,j) \in \mathcal{W}} w_{i,j}\, x_{u:u+i,v+j}^2 \\
& + \delta_v^2 \sum_{(i,j) \in \mathcal{W}} w_{i,j}\, x_{v:u+i,v+j}^2 \\
& + \delta_u \delta_v \sum_{(i,j) \in \mathcal{W}} w_{i,j}\, x_{u:u+i,v+j}\, x_{v:u+i,v+j}
\end{aligned}$$

which can be written compactly in matrix quadratic form as

$$s(u, v, \delta_u, \delta_v) = (\delta_u\ \delta_v) \mathbf{A} \begin{pmatrix} \delta_u \\ \delta_v \end{pmatrix}$$

where

$$\mathbf{A} = \begin{pmatrix} \sum w_{i,j}\, x_{u:u+i,v+j}^2 & \sum w_{i,j}\, x_{u:u+i,v+j}\, x_{v:u+i,v+j} \\ \sum w_{i,j}\, x_{u:u+i,v+j}\, x_{v:u+i,v+j} & \sum w_{i,j}\, x_{v:u+i,v+j}^2 \end{pmatrix}.$$

where the summations are convolutions. Assuming the weighting matrix is a Gaussian kernel $\mathbf{W} = \mathbf{G}(\sigma_I)$, we can rewrite this in operator form as

$$\mathbf{A} = \begin{pmatrix} \mathbf{G}(\sigma_I) * \mathbf{X}_u^2 & \mathbf{G}(\sigma_I) * \mathbf{X}_u \mathbf{X}_v \\ \mathbf{G}(\sigma_I) * \mathbf{X}_u \mathbf{X}_v & \mathbf{G}(\sigma_I) * \mathbf{X}_v^2 \end{pmatrix} \tag{12.13}$$

The gradient images $\mathbf{X}_u$ and $\mathbf{X}_v$ are typically calculated using a derivative of Gaussian kernel method (► Sect. 11.5.1.3) with a smoothing parameter $\sigma_D$. $\mathbf{X}_u^2$, $\mathbf{X}_v^2$ and $\mathbf{X}_u \mathbf{X}_v$ are computed using element-wise squaring or multiplication.

Recall from ► Exc. 11.2 that lossy image compression such as JPEG removes high-frequency detail from the image, and this is exactly what defines a corner. Ideally, corner detectors should be applied to images that have not been compressed and decompressed.

which is a symmetric $2 \times 2$ matrix referred to variously as the structure tensor, autocorrelation matrix or second moment matrix. ◄ It captures the intensity structure of the local neighborhood and its eigenvalues provide a rotationally invariant description of the neighborhood. The elements of the **A** matrix are computed from the image gradients, squared or multiplied element-wise, and then smoothed using a Gaussian kernel. The latter reduces noise and improves the stability and reliability of the detector.

An interest point $(u, v)$ is one for which $s(u, v : \cdot)$ is high for *all* directions of the vector $(\delta_u, \delta_v)$. That is, in whatever direction we move the window it rapidly becomes dissimilar to the original window. If we consider the original image **X** as a surface, the principal curvatures of the surface at $(u, v)$ are $\lambda_1$ and $\lambda_2$ which are the eigenvalues of **A**. If both eigenvalues are small then the surface is flat, that is the image region has approximately constant local intensity. If one eigenvalue is high and the other low, then the surface is ridge shaped which indicates an edge. If both eigenvalues are high the surface is sharply peaked which we consider to be a corner. ◄

The Shi-Tomasi detector considers the strength of the corner, or *cornerness*, as the minimum eigenvalue

$$C_{\text{ST}}(u, v) = \min(\lambda_1, \lambda_2) \tag{12.14}$$

where $\lambda_i$ are the eigenvalues of $\mathbf{A}$. Points in the image for which this measure is high are referred to as "*good features to track*". The Harris detector ▶ is based on this same insight but defines corner strength as

Sometimes referred to in the literature as the Plessey corner detector.

$$C_H(u, v) = \det(\mathbf{A}) - k\,\text{tr}(\mathbf{A}) \tag{12.15}$$

and again a large value represents a strong, distinct, corner. Since $\det(\mathbf{A}) = \lambda_1\lambda_2$ and $\text{tr}(\mathbf{A}) = \lambda_1 + \lambda_2$ the Harris detector responds when both eigenvalues are large and elegantly avoids computing the eigenvalues of $\mathbf{A}$ which has a somewhat higher computational cost. ▶ A commonly used value for $k$ is 0.04. Another variant is the Noble detector

Evaluating eigenvalues for a $2 \times 2$ matrix involves solving a quadratic equation and therefore requires a square root operation.

$$C_N(u, v) = \frac{\det(\mathbf{A})}{\text{tr}(\mathbf{A})} \tag{12.16}$$

which is arithmetically simple but potentially singular.

The corner strength can be computed for every pixel and results in a corner-strength image. Then nonlocal maxima suppression is applied to only retain values that are greater than their immediate neighbors. A list of such points is created and sorted into descending corner strength. A threshold can be applied to only accept corners above a particular strength, or above a particular fraction of the strongest corner, or simply the $N$ strongest corners.

The Toolbox provides a Harris corner detector which we will demonstrate using a real image

```
>>> view1 = Image.Read("building2-1.png", mono=True);
>>> view1.disp();
```

The Harris features are computed by

```
>>> harris1 = view1.Harris(nfeat=500)
Harris features, 500 points
```

which returns an instance of a `HarrisFeature` object which is a subclass of `BaseFeature2D`. The object behaves like a Python list

```
>>> len(harris1)
500
```

and can be iterated or sliced

```
>>> harris1[0]
Harris feature: (597.0, 648.0), strength=1.64, id=-1
```

where the displayed values shows the essential properties of the feature. Each feature has properties such as position ▶ and strength

In this implementation the coordinates are integers. Other corner detectors provide the corner position to subpixel accuracy.

```
>>> harris1[0].p
array([[    597],
       [    648]])
>>> harris1[0].strength
1.636
```

The `id` property is inherited from the `id` of the image from which the features are derived. For the `VideoFile`, `ImageCollection` and `ZipArchive` image sources the image id takes on sequential integer values, while `Image.Read` sets it to $-1$.

We can also create expressions such as

```
>>> harris1[:5].p
array([[    597,      587,      576,      810,      149],
       [    648,      658,      636,      580,      490]])
>>> harris1[:5].strength
[1.636, 1.532, 1.514, 1.487, 1.348]
```

which are the positions and strengths of the first five features as 2D and 1D arrays respectively.

---

**Excurse 12.4: Determinant of the Hessian (DoH)**

Another approach to determining image curvature is to take the determinant of the Hessian (DoH), and this is used by the SURF corner detector. The Hessian **H** is the matrix of second-order gradients at a point

$$
h_{u,v} = \begin{pmatrix} x_{uu:u,v} & x_{uv:u,v} \\ x_{uv:u,v} & x_{vv:u,v} \end{pmatrix}
$$

where $\mathbf{X}_{uu} = \partial^2 \mathbf{X}/\partial u^2$, $\mathbf{X}_{vv} = \partial^2 \mathbf{X}/\partial v^2$ and $\mathbf{X}_{uv} = \partial^2 \mathbf{X}^2/\partial u \partial v$. The determinant det(**H**) has a large magnitude when there is gray-level variation in two directions. However, second derivatives accentuate image noise even more than first derivatives and the image must be smoothed first.

---

The corners can be overlaid on the image as yellow crosses

```
>>> view1.disp(darken=True);
>>> harris1.plot();
```

as shown in ◘ Fig. 12.24a. The `darken` option to `disp` reduces the brightness of the image to make the overlaid corner markers more visible. A close-up view is shown

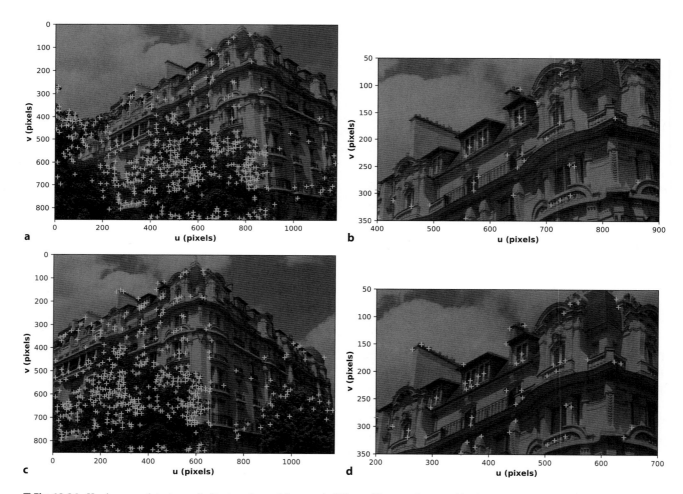

**a**    **b**    **c**    **d**

◘ **Fig. 12.24** Harris corner detector applied to two views of the same building. **a** View one; **b** zoomed in view one; **c** view two; **d** zoomed in view two. Notice that a number of the detected corners are attached to the same world features in the two views

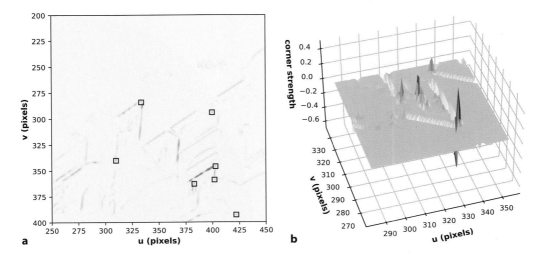

**Fig. 12.25** Harris corner strength. **a** Zoomed view of corner strength displayed as an image (blue is positive, red is negative); **b** zoomed view of corner strength image displayed as a surface

in Fig. 12.24b and we see the features are indeed often located on the corners of objects. To plot a subset of the corner features we could choose every fifth feature

```
>>> harris1[::5].plot()
```

or a subset of 20 evenly spaced corners

```
>>> harris1.subset(20).plot()
```

We see that the corners tend to cluster unevenly, with a greater density in regions of high texture such as the trees. For some applications this can be problematic, and to distribute them more evenly we can increase the distance used for nonlocal-maxima suppression

```
>>> harris1 = view1.Harris(nfeat=500, scale=15)
Harris features, 500 points
```

by specifying a minimum distance between corners of 15 pixels – the default is 7.

The Harris corner strength for all pixels can be computed and displayed

```
>>> view1.Harris_corner_strength().disp();
```

and is shown in Fig. 12.25a with detected corner features overlaid. We observe that the corner strength function is positive (blue) for corner features and negative (red) for linear features. A zoomed in view is shown in Fig. 12.25b which indicates that the detected corner is at the top of a peak of *cornerness* that is several pixels wide.

A cumulative histogram of the strength of the 500 detected corners is shown in Fig. 12.26. The strongest corner has $C_H \approx 1.64$ but most are much weaker than this, only 10% of corners exceed half this value.

Consider another image of the same building taken from a different location

```
>>> view2 = Image.Read("building2-2.png", mono=True);
```

and the detected corners

```
>>> harris2 = view2.Harris(nfeat=250);
>>> view2.disp(darken=True);
>>> harris2.plot();
```

are shown in Fig. 12.24c, d. For many applications in robotic vision – such as tracking, mosaicing and stereo vision that we will discuss in ▶ Chap. 14 – it is important that corner features are detected at the same world points irrespective of variation in illumination or changes in rotation and scale between the two views.

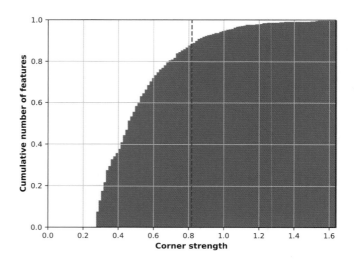

**◻ Fig. 12.26**   Normalized cumulative histogram of corner strengths

From ◻ Fig. 12.24 we see that many, but not all, of the features are indeed *attached* to the same world feature in both views.

The Harris detector is computed from image gradients and is therefore robust to offsets in illumination, and the eigenvalues of the structure tensor **A** are invariant to rotation. However the detector is not invariant to changes in scale. As we *zoom in* the gradients around the corner points become lower – the same change in intensity is spread over a larger number of pixels. This reduces the image curvature and hence the corner strength. The next section discusses a remedy for this using *scale-invariant* corner detectors.

### 12.3.2 Scale-Space Corner Detectors

The Harris corner detector introduced in the previous section works very well in practice but is poor at finding the same world points if the images have significantly different scale. Unfortunately change in scale, due to changing camera to scene distance or zoom, is common in many real applications. We also notice that the Harris detector responds strongly to fine texture, such as the leaves of the trees in ◻ Fig. 12.24 but we would like to be able to detect features that are associated with larger-scale scene structure such as windows and balconies.

◻ Fig. 12.27 illustrates the fundamental principle of scale-space feature detection. We first load a synthetic image

```
>>> foursquares = Image.Read("scale-space.png", dtype="float");
```

which is shown in ◻ Fig. 12.27a. The image contains four squares of different size: $5 \times 5$, $9 \times 9$, $17 \times 17$ and $33 \times 33$. The scale-space sequence is computed by applying a Gaussian kernel with increasing $\sigma$ that results in the regions becoming increasingly blurred and smaller regions progressively disappearing from view. At each step in the sequence, the Gaussian-smoothed image is convolved with the Laplacian kernel (12.5) which results in strong negative responses for these bright blobs. ◄

We actually compute the difference of Gaussian approximation to the Laplacian of Gaussian, as illustrated in ◻ Fig. 12.30.

With the Toolbox we compute the scale-space sequence by

```
>>> G, L, s = foursquares.scalespace(60, sigma=2);
```

where the input arguments are the number of scale steps to compute, and the $\sigma$ of the Gaussian kernel to be applied at each successive step. G is a list of images, the input image at increasing levels of smoothing, and L is a list of images which are the Laplacian of the corresponding smoothed images, and s is the corresponding

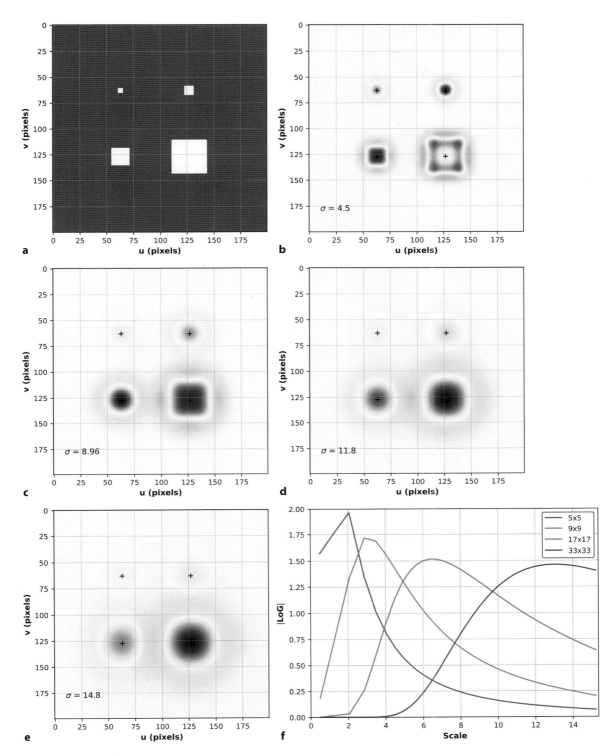

**☐ Fig. 12.27** Scale-space example. **a** Synthetic image **X** with blocks of sizes $5 \times 5$, $9 \times 9$, $17 \times 17$, and $33 \times 33$; **b–e** Normalized Laplacian of Gaussian $\sigma^2 \mathbf{L} * \mathbf{G}(\sigma) * \mathbf{X}$ for increasing values of scale, $\sigma$ value indicated in lower left. False color is used: blue is positive and red is negative; **f** magnitude of Laplacian of Gaussian at the center of each square (indicated by '+') versus $\sigma$

scale. For example the sixth image in the Laplacian of Gaussian sequence (LoG) is displayed by

```
>>> L[5].disp(colormap="signed");
```

and is shown in ◻ Fig. 12.27b and has a scale of

```
>>> s[5]
4.5
```

The Laplacian of Gaussian at three other points in the scale-space sequence are shown in ◻ Fig. 12.27c–e.

Consider now the value of the Laplacian of Gaussian at the center of the smallest square as a function of scale

```
>>> plt.plot(s[:-1], [-Ls.image[63, 63] for Ls in L]);
```

which is shown as the blue curve in ◻ Fig. 12.27f. In a similar way we can plot the response at the center of the other squares as a function of scale. Each curve has a well defined peak, and the scale associated with the peak is proportional to the size of the region – the characteristic scale of the region.

If we stack the 2D images from the sequence L into a 3D volume then a scale-space feature point is any pixel that is a local 3D maxima. That is, any element that is greater than its 26 neighbors in *all three* dimensions – its spatial neighbors at the current scale and at the scale above and below. Such points are detected by the function findpeaks3d

```
>>> features = findpeaks3d(np.stack([np.abs(Lk.image) for Lk in L],
...                        axis=2), npeaks=4)
array([[      63,       63,        1,    1.962],
       [     127,       63,        2,    1.716],
       [      63,      127,       11,    1.514],
       [     127,      127,       43,     1.46]])
```

which returns a 2D array with one row per maxima, and each row is $v, u$, the scale index $k$ and strength. We can superimpose the detected features on the original image

```
>>> foursquares.disp();
>>> for v, u, i, _ in features:
...    plt.plot(u, v, 'k+')
...    scale = s[int(i)]
...    plot_circle(radius=scale * np.sqrt(2), centre=(u, v),
...                color="y")
```

and the result is shown in ◻ Fig. 12.28. The scale associated with a feature can be visualized using circles of radius proportional to the feature scale, and the result is also shown in ◻ Fig. 12.28. We see that the identified features are located at the

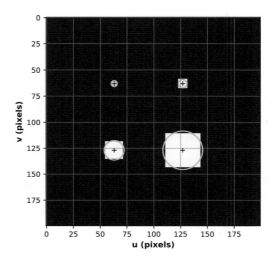

◻ **Fig. 12.28**  Synthetic image with overlaid feature center and scale indicator

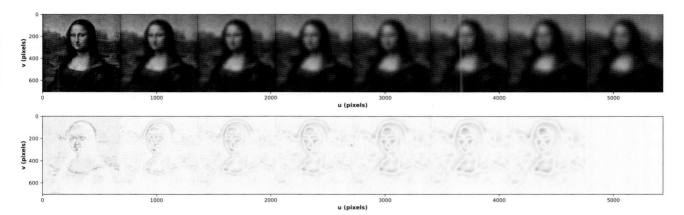

■ **Fig. 12.29**  Scale-space sequence for $\sigma = 2$, (top) Gaussian sequence, (bottom) Laplacian of Gaussian sequence

center of each object and that the scale of the feature is related to the size of the object. The region within the circle is known as the support region of the feature.

For a real image

```
>>> mona = Image.Read("monalisa.png", dtype="float");
```

we compute the scale-space in eight large steps with $\sigma = 8$

```
>>> G, L, _ = mona.scalespace(8, sigma=8);
```

and we can horizontally stack the images in the scale space sequence

```
>>> Image.Hstack(G).disp();
>>> Image.Hstack(L).disp();
```

as shown in ■ Fig. 12.29. From left to right we see the eight levels of scale. The Gaussian sequence of images becomes increasing blurry. In the Laplacian of Gaussian sequence the dark eyes are strongly positive (blue) blobs at low scale and her light colored forehead becomes a strongly negative (red) blob at high scale.

Convolving the original image with a Gaussian kernel of increasing $\sigma$ results in the kernel size, and therefore the amount of computation, growing at each scale step. Recalling the properties of a Gaussian from ▶ Exc. 11.11, a Gaussian convolved with a Gaussian is another wider Gaussian. Instead of convolving our original image with ever wider Gaussians, we can repeatedly apply the same Gaussian to the previous result. We also recall from ▶ Exc. 11.14 that the LoG kernel is approximated by the difference of two Gaussians. Using the properties of convolution we can write

$$\big(\mathbf{G}(\sigma_1) - \mathbf{G}(\sigma_2)\big) * \mathbf{X} = \mathbf{G}(\sigma_1) * \mathbf{X} - \mathbf{G}(\sigma_2) * \mathbf{X}$$

where $\sigma_1 > \sigma_2$. The difference of Gaussian operator applied to the image is equivalent to the difference of the image at two different levels of smoothing. If we perform the smoothing by successive application of a Gaussian we have a sequence of images at increased levels of smoothing. The difference between successive steps in the sequence is therefore an approximation to the Laplacian of Gaussian. ■ Fig. 12.30 shows this in diagrammatic form.

### 12.3.2.1  Scale-Space Point Feature

The scale-space concepts just discussed underpin a number of popular feature detectors which find salient points within an image and determines their scale and also their orientation. The Scale-Invariant Feature Transform (SIFT) is based on the maxima in a difference of Gaussian sequence. ▶

SIFT (Lowe 2004) uses image pyramids as shown in ■ Fig. 12.35 to speed the computation. The pyramid levels are referred to as *octaves* since the successive images are scaled by two in both dimensions. SIFT inserts an additional octave which is double the size of the original image. Within each octave a series of Gaussian blur operations, or *scales*, are performed. The number of octaves and scales are parameters of SIFT, but typically are set to 4 and 5 respectively.

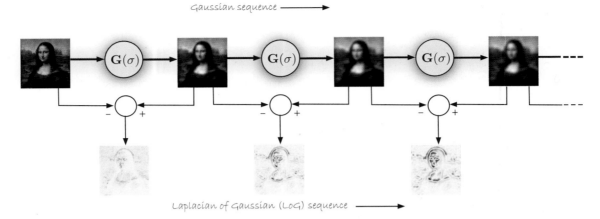

□ **Fig. 12.30**   Schematic for calculation of Gaussian and Laplacian of Gaussian scale-space sequence

To illustrate we will compute the SIFT features for the building image used previously

```
>>> sift1 = view1.SIFT(nfeat=200)
SIFT features, 200 points
```

which returns an instance of a `SIFTFeature` object, a subclass of `BaseFeature2D`, which behaves similarly to the `HarrisCorner` introduced earlier. For example the first feature is

```
>>> sift1[0]
SIFT feature: (2.3, 285.4), strength=0.05, scale=2.1, orient=148.5°,
id=-1
```

Each feature has a coordinate (estimated to subpixel precision), scale, and orientation, which is defined by the dominant edge direction within the support region.

This image contains over 10,000 SIFT features but, as we did earlier with the Harris corner features, we requested the 200 strongest which we plot

```
>>> view1.disp(darken=True);
>>> sift1.plot(filled=True, color="y", hand=True, alpha=0.3)
```

and the result is shown in □ Fig. 12.31. The `plot` method draws a circle around the feature's location with a radius that indicates its scale – the size of the support region. The option `hand` draws a radial line which indicates the orientation of the SIFT feature – the direction of the dominant gradient. ◄

If the support region contains multiple dominant directions, then multiple SIFT features are created. They have the same position, scale and descriptor, but differ in their orientation.

Feature scale varies widely and a histogram

```
>>> plt.hist(sift1.scale, bins=100);
```

shown in □ Fig. 12.32 indicates that there are many small features associated with fine image detail and texture. The bulk of the features have a scale less than 10 pixels but some have scales over 50 pixels.

The SIFT algorithm is more than just a scale-invariant feature detector, it also computes a very robust *descriptor*. The SIFT descriptor is a vector $d \in \mathbb{R}^{128}$ which encodes the image gradient in subregions of the support region in a way which is invariant to brightness, scale and rotation. We can think of it as a *fingerprint* that enables feature descriptors to be unambiguously matched to a descriptor of the same world point in another image even if their scale and orientation are quite different. The difference in position, scale and orientation of the matched features gives some indication of the relative camera motion between the two views. Matching features between scenes is crucial to the problems that we will address in ► Chap. 14.

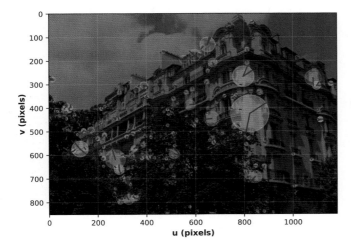

**◻ Fig. 12.31** SIFT features showing the support region (scale) and orientation as a radial blue line. Some features appear to have two orientations, but these are two features at the same location with different orientations

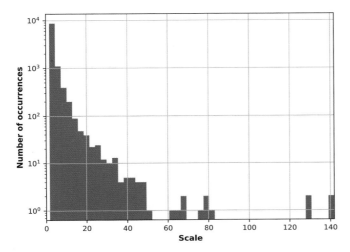

**◻ Fig. 12.32** Histogram of SIFT feature scales shown with logarithmic vertical scale

## 12.4 Applications

### 12.4.1 Character Recognition

An important class of visual objects are text characters. Our world is filled with informative text in the form of signs and labels that provide information about the names of places and directions to travel. We process much of this unconsciously, but this rich source of information is largely unavailable to robots.

For this example we will use the open-source Tesseract package, and you will first need to install the binary and the Python interface `pytessearct`. We start with the image of a sign shown in ◻ Fig. 12.33a and apply OCR to the entire image

```
>>> import pytesseract as tess
>>> penguins = Image.Read("penguins.png");
>>> ocr = tess.image_to_data(penguins.image < 100,
...                          output_type=tess.Output.DICT);
```

The image was first binarized such that the text is dark on a white background – this improves Tesseract's performance. The result was requested in dictionary format

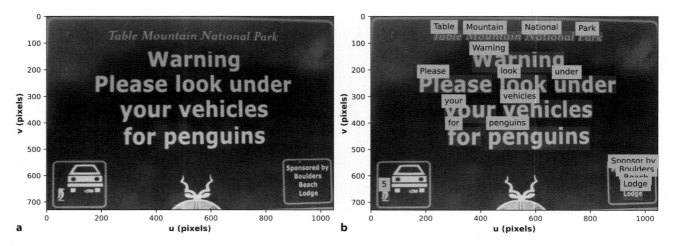

**◻ Fig. 12.33** Optical character recognition. The bounding boxes of detected words are shown in yellow

which is a dictionary of lists. We filter the results based on the confidence level, and whether the string contains any non-whitespace characters

```
>>> for confidence, text in zip(ocr["conf"], ocr["text"]):
...    if text.strip() != "" and float(confidence) > 0:
...        print(confidence, text)
96 Table
95 Mountain
95 National
96 Park
95 Warning
...
7 a
44 =
33 mae
14 --------
```

The words on the sign are correctly read, apart from some graphical icons which are incorrectly interpreted. We can highlight the confident detections on the original image

```
>>> for i, (text, confidence) in enumerate(zip(ocr["text"],
...                                            ocr["conf"])):
...    if text.strip() != "" and float(confidence) > 50:
...        plot_labelbox(text,
...            lb=(ocr["left"][i], ocr["top"][i]),
...            wh=(ocr["width"][i], ocr["height"][i]),
...            color="y", filled=True, alpha=0.2)
```

pytesseract returns the "top" coordinate in the image which is the minimum $v$-coordinate or "bottom" for a non-inverted $v$-axis.

with the result shown in ◻ Fig. 12.33. ◄

### 12.4.2 Image Retrieval

Given a set of images $\{\mathbf{X}_j, j = 0, \ldots, N-1\}$ and a new image $\mathbf{X}'$ the image retrieval problem is to determine $j$ such that $\mathbf{X}_j$ is the most similar to $\mathbf{X}'$. This is a difficult problem when we consider the effect of changes in viewpoint and exposure. Pixel-level similarity measures such as sum of absolute differences (SSD) or zero-mean normalized cross correlation (ZNCC) introduced in ► Sect. 11.5.2 are not suitable for this problem, since quite small changes in viewpoint will result in almost zero similarity.

Image retrieval is useful to a robot to determine if it has visited a particular place before, or seen the same object before. If the previous images have some associated semantic data, such as the name of an object or the name of a place, then by inference that semantic data applies to the new image. For example if a new image matches an existing image that has the semantic tag "kitchen" then it implies the robot is seeing the same scene and is therefore in or close to, the kitchen.

The particular technique that we will introduce is commonly referred to as "bag of words" and is useful for many robotic applications. It builds on techniques we have previously encountered such as SIFT point features and $k$-means clustering.

We start by loading a set of twenty images

```
>>> images = ImageCollection("campus/*.png", mono=True);
```

which is an image iterator. We use this to compute and concatenate the SIFT features for all images features

```
>>> features = [];
>>> for image in images:
...    features += image.SIFT()
>>> features.sort(by="scale", inplace=True);
```

which is an instance of a `SIFTFeature` object with

```
>>> len(features)
42194
```

features which are sorted in decreasing order of scale. The numerical parameters of the first ten features can be listed in tabular form

```
>>> features[:10].table()
```

| # | centroid | strength | scale | orient | id |
|---|----------|----------|-------|--------|-----|
| 0 | 350.0, 175.1 | 0.0309 | 140 | 204° | 4 |
| 1 | 462.5, 273.1 | 0.0367 | 104 | 268° | 7 |
| 2 | 499.7, 264.2 | 0.026 | 103 | 137° | 15 |
| 3 | 291.5, 90.7 | 0.0368 | 97.8 | 348° | 5 |
| 4 | 521.8, 286.6 | 0.0221 | 94.6 | 77.1° | 3 |
| 5 | 521.8, 286.6 | 0.0221 | 94.6 | 333° | 3 |
| 6 | 132.2, 111.8 | 0.0335 | 93.4 | 277° | 11 |
| 7 | 132.2, 111.8 | 0.0335 | 93.4 | 106° | 11 |
| 8 | 209.1, 220.2 | 0.0272 | 90.5 | 253° | 6 |
| 9 | 96.6, 274.9 | 0.0422 | 89.3 | 279° | 0 |

and we see the properties discussed earlier such as centroid, strength, scale and orientation for these SIFT features. The property `id` indicates the image in the collection that the feature was detected in.

To gain insight into what parts of the world these feature represent we can display their support regions. This is a square region of the image, centered at the feature's centroid and rotated by the feature's orientation with a side length proportional to the feature's scale. That region is *warped* out of the original image into a $50 \times 50$ pixel thumbnail image. We can display a grid of these support regions for the first 400 SIFT features

```
>>> supports = [];
>>> for feature in features[:400]:
...    supports.append(feature.support(images))
>>> Image.Tile(supports, columns=20).disp(plain=True);
```

which is shown in Fig. 12.34. We can see fragments of buildings and trees across a variety of scales – from massive porticos to smaller windows. We will investigate feature 108, shown circled in Fig. 12.34, which is a fairly typical type of window

```
>>> feature = features[108]
SIFT feature: (243.9, 200.7), strength=0.03, scale=41.8,
orient=207.9°, id=8
```

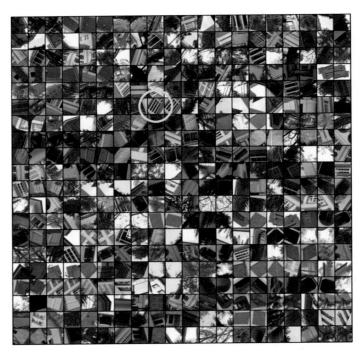

**Fig. 12.34** The support regions for the first 400 SIFT features sorted by decreasing scale. Feature 108 is highlighted

We can display that image and superimpose the feature

```
>>> images[feature.id].disp();
>>> feature.plot(filled=True, color="y", hand=True, alpha=0.5)
```

which is shown in ◘ Fig. 12.35a, along with a larger view of the support region in ◘ Fig. 12.35b.

The key insight behind the bag of words technique is that many of these features will describe visually similar scene elements such as shown in ◘ Fig. 12.34. If we consider each SIFT feature descriptor as a point in a 128-dimensional space

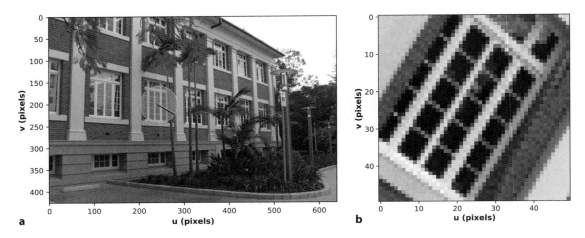

**Fig. 12.35** **a** Image 8 with visual word SIFT feature 108 indicated by green circle showing scale and a radial line showing orientation; **b** the square support region has the same area as the circle and the horizontal axis is parallel to the orientation direction, and has been rescaled to $50 \times 50$ pixels

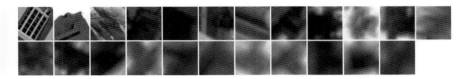

**◘ Fig. 12.36** Exemplars of visual word 965 from the various images in which it appears

then similar descriptors will form clusters, and this is a $k$-means problem. To find 2000 feature clusters

```
>>> bag = BagOfWords(features, 2_000, seed=0)
BagOfWords: 42194 features from 20 images, 2000 words, 0 stop words
```

returns a `BagOfWords` object that contains the original features, the center of each cluster, and various other information. Each cluster is referred to as a visual word and is described by a 128-element SIFT descriptor. The set of all visual words, 2000 in this case, is a visual vocabulary. Just as a document comprises a set of words drawn from some vocabulary, each image comprises a collection (or *bag*) of visual words drawn from the visual vocabulary.

The clustering step assigns a visual word index to every SIFT feature. For the particular feature shown above

```
>>> w = bag.word(108)
965
```

we find that the $k$-means clustering has assigned this image feature to word 965 in the vocabulary – it is an instance of visual word 965. That particular visual word appears

```
>>> bag.occurrence(w)
32
```

times across the set of images, and it appears at least once in each of the images

```
>>> bag.contains(w)
array([ 1,  2,  5,  6,  7,  8, 10, 11, 12, 13, 14, 16, 17, 18, 19])
```

We can display some of the different instances of word 965 by

```
>>> bag.exemplars(w, images).disp();
```

which is shown in ◘ Fig. 12.36. These exemplars do look quite different, but we need to keep in mind that we are viewing them as patterns of pixels whereas the similarity is in terms of the descriptor. ▶ The exemplars do however share some dominant horizontal and vertical structure.

The frequency of occurrence of the visual words is given by

```
>>> word, freq = bag.wordfreq();
```

where `word` is a vector containing all unique words and `freq` are their corresponding frequencies. The visual words occur with quite different frequencies and we observe

```
>>> np.max(freq)
704
>>> np.median(freq)
14
```

that the maximum value is significantly greater than the median value. This becomes clearer if we display the frequency values in descending order of frequency

```
>>> plt.bar(word, -np.sort(-freq), width=1);
```

as shown in ◘ Fig. 12.37. The very frequently occurring words have less meaning or power to discriminate between images, and are analogous to English stop words in text document retrieval. ▶

The descriptor comprises histograms of orientation gradients.

Search engines ignore stop words such as "a", "and", "the" etc.

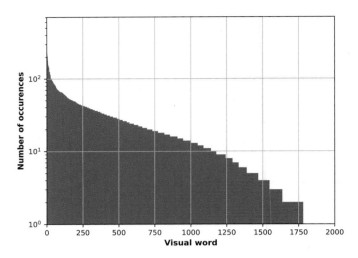

■ **Fig. 12.37** Histogram of the number of occurrences of each word (sorted). Note the small number of words that occur very frequently

The 50 most frequently occuring words will be considered as stop words and removed from the bag of words

```
>>> bag = BagOfWords(features, 2_000, nstopwords=50, seed=0)
Removing 7197 features (17.1%) associated with 50 most frequent words

BagOfWords: 34997 features from 20 images, 1950 words, 50 stop words
```

which leaves some 35,000 SIFT features behind. This method performs relabeling so that word labels are now in the interval $[0, 1950)$.

Our visual vocabulary comprises $k = 1950$ visual words. We apply a technique from text document retrieval and describe *each* image by a weighted word-frequency vector

$$\boldsymbol{v}_i = (t_0, \ldots, t_{K-1})^\top \in \mathbb{R}^k$$

which is a column vector whose elements describes the frequency of the corresponding visual words in an image.

$$t_j = \frac{n_{ij}}{n_i} \underbrace{\log \frac{N}{N_j}}_{\text{idf}} \tag{12.17}$$

where $j$ is the visual word label, $N$ is the total number of images in the database, $N_j$ is the number of images which contain word $j$, $n_i$ is the number of words in image $i$, and $n_{ij}$ is the number of times word $j$ appears in image $i$. The inverse document frequency (idf) term is a weighting that reduces the significance of words that are common across all images and which are thus less discriminatory. The weighted word frequency vectors are obtained by the wwvf method and for image 10 is

```
>>> v10 = bag.wwfv(10);
>>> v10.shape
(1950,1)
```

which is a column vector that concisely describes the image in terms of its constituent visual words. ◄

This is a very large vector but it contains less than 1% of the number of elements of the original image. It is a very concise summary.

The similarity between two images is the cosine of the angle between their corresponding word-frequency vectors

$$s(\boldsymbol{v}_1, \boldsymbol{v}_2) = \frac{\boldsymbol{v}_1^\top \boldsymbol{v}_2}{\|\boldsymbol{v}_1\| \, \|\boldsymbol{v}_2\|}$$

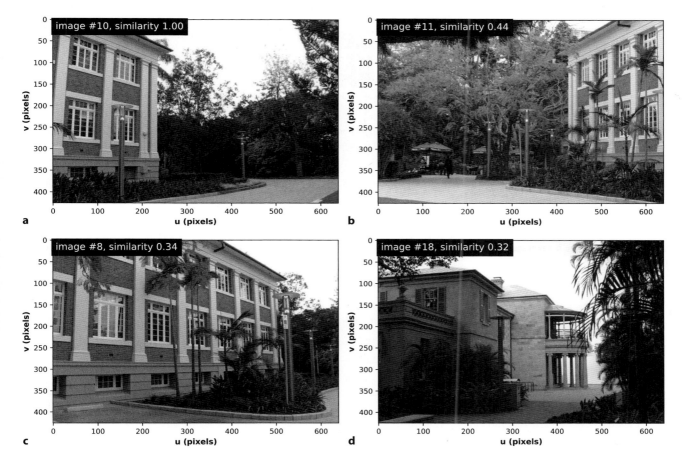

**Fig. 12.38** Image retrieval. Image 10 is the query, and in decreasing order of match quality we have retreived images 11, 8 and 18

and is implemented by the `similarity` method. A value of one indicates maximum similarity. The similarity of image 10 to all the images in the bag is simply

```
>>> sim_10 = bag.similarity(v10);
```

which returns a 20-element array of similarity values between the word frequency vector for image 10 and the others. We sort this into descending order of similarity

```
>>> k = np.argsort(-sim_10)
array([10, 11,  8, 18, 17,  4,  7, 15, 16,  9,  6, ...])
```

which shows that image 10 is most similar to image 10 as expected, and in decreasing order of similarity we have images 11, 8, 18 and so on. These are shown in ◖ Fig. 12.38 along with the similarity scores, and we see that the algorithm has mostly recalled different views of the same building.

Now consider that we have some new images – our robot is out in the world – and we wish to retrieve previous images of the places it is seeing. The steps are broadly similar to the previous case. We load five new images

```
>>> query = ImageCollection("campus/holdout/*.png", mono=True);
```

but we do not need to perform clustering. We simply assign the features in the new images to the already computed set of visual words, that is, to determine the closest visual word for each of the new image features This is achieved simply by

```
>>> S = bag.similarity(query);
```

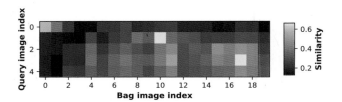

**◻ Fig. 12.39** Similarity matrix for 20 images where light colors indicate strong similarity. Element $(i, j)$ indicates the similarity between bag image $i$ and query image $j$

Which requires the image-word statistics from the existing bag of words to compute the idf weighting terms.

which computes the SIFT features for each input image, finds the closest visual word for each, removes the stop words, computes the word-frequency vector ◄ according to (12.17), and computes the distance to the word-frequency vectors for all the images in the bag. The result is a $5 \times 20$ array where the element $\mathtt{S[i, j]}$ indicates the similarity between the new image $i$ and existing image $j$. The result is best interpreted visually

```
>>> Image(S).disp(colorbar=True);
```

which is shown in ◻ Fig. 12.39. The best match for each new image is the location of the maxima in each row

```
>>> np.argmax(S, axis=1)
array([ 0, 10, 17, 17, 17])
```

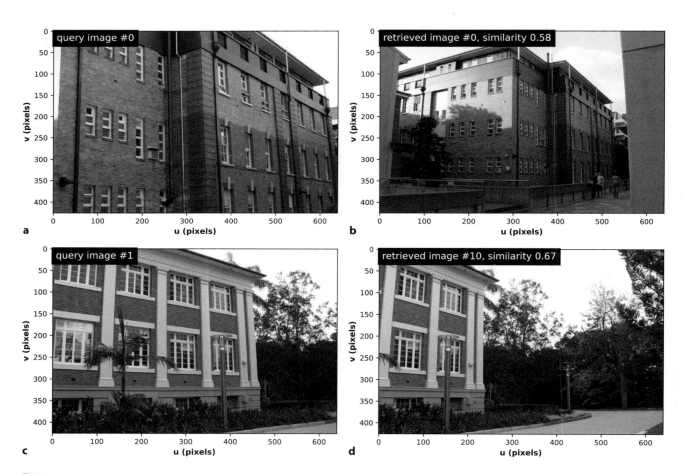

**◻ Fig. 12.40** Image retrieval for new images. The new query images **a** and **c** recall the database images **b** and **d** respectively

We could also use the `retrieve` method

```
>>> bag.retrieve(query[0])
(0, 0.5752)
>>> bag.retrieve(query[1])
(10, 0.6631)
```

which takes an image and returns a tuple containing the index of the best matching image in the bag and its similarity. New image 0 best matches image 0 in the original sequence, new image 1 matches image 10. Two of the new images and their closest existing images are shown in ◘ Fig. 12.40. The first recall has a low similarity score but is a reasonable result – the recall image includes the building from the test image at the right and another building that has many similarities.

**❗ Bag of words ignores the spatial arrangement of features**

Note that the match would be just as strong if the query image was upside down, or even if it was cut up into lots of pieces and rearranged. Bag of words looks for the presence of visual words, not their spatial arrangement.

## 12.5 Wrapping Up

In this chapter we have discussed the extraction of features from an image. Instead of considering the image as millions of independent pixel values we succinctly describe regions within the image that correspond to distinct objects in the world. For instance we can find regions that are homogeneous with respect to intensity or color or semantic meaning and describe them in terms of features such as a bounding box, centroid, equivalent ellipse, aspect ratio, circularity and perimeter shape. Features have invariance properties with respect to translation, rotation about the optical axis and scale which are important for object recognition. Straight lines are common visual features in human-made environments, and we showed how to find and describe distinct straight lines in an image using the Hough transform.

We also showed how to find interest points that can reliably *associate* to particular points in the world irrespective of the camera view. These are key to techniques such as camera motion estimation, stereo vision, image retrieval, tracking and mosaicing that we will discuss in the next chapter.

### 12.5.1 Further Reading

This chapter has presented a classical bottom up approach for feature extraction, starting with pixels and working our way up to higher level concepts such as regions and lines. Prince (2012) and Szeliski (2022) both provide a good introduction to high-level vision using probabilistic techniques that can be applied to problems such as object recognition, for example face recognition, and image retrieval. The book by Szeliski (2022) can be downloaded, for personal use, from ► https://szeliski.org/Book. In the last few years computer vision, particularly object recognition, has undergone a revolution using deep convolutional neural networks. These have demonstrated very high levels of accuracy in locating and recognizing objects against complex background despite changes in viewpoint and illumination. Resources to get started with PyTorch can be found at ► https://pytorch.org.

**■ ■ Region Features**

Region-based image segmentation and blob analysis are classical techniques covered in many books and papers. Gonzalez and Woods (2018) and Szeliski (2022) provide a thorough treatment of the methods introduced in this chapter, in particular thresholding and perimeter descriptors. Davies (2017) and Klette (2014) also covers thresholding, connectivity analysis, shape analysis and the Hough transform and its variants. Otsu's algorithm for threshold determination was introduced in Otsu (1975), and a popular algorithm for adaptive thresholding was introduced in Niblack (1985). Nixon and Aguado (2012) expands on material covered in this chapter and introduces techniques such as deformable templates and perimeter descriptors. The Freeman chain code was first described in Freeman (1974). Flusser (2000) has shown that the seven moments proposed by Hu (1962), and described in ▶ Sect. 12.1.3.3, are in fact not independent since $\phi_3 = (\phi_5^2 + \phi_7^2)/\phi_4^3$.

In addition to region homogeneity based on intensity and color it is also possible to describe the texture of regions – a spatial pattern of pixel intensities whose statistics can be described (Gonzalez and Woods 2018). Regions can then be segmented according to texture, for example a smooth road versus textured grass.

Clustering of data is an important topic in machine learning (Bishop 2006). In this chapter we have used a simple implementation of $k$-means, which is far from state-of-the-art in clustering, and requires the number of clusters to be known in advance. More advanced clustering algorithms are hierarchical and employ data structures such as kd-trees to speed the search for neighboring points. The initialization of the cluster centroids is also critical to performance. Szeliski (2022) introduces more general clustering methods as well as graph-based methods for computer vision. The graph-based segmentation (graphcuts) algorithm for segmentation was described by Felzenszwalb and Huttenlocher (2004). The maximally stable extremal region (MSER) algorithm is described by Matas et al. (2004). The Berkeley Segmentation Dataset at ▶ https://www.eecs.berkeley.edu/Research/Projects/CS/vision/bsds contains numerous complex real-world images each with several human-made segmentations.

Early work on using text recognition for robotics is described by Posner et al. (2010), while Lam et al. (2015) describe the application of OCR to parsing floor plans of buildings for robot navigation. A central challenge with OCR of real-world scenes is to determine which parts of the scene contain text and should be passed to the OCR engine. A powerful text detector is the stroke width transform described by Li et al. (2014). The Tesseract open-source OCR engine was developed at Hewlett Packard in the late 1980s, open sourced in 2005 and subsequently sponsored by Google. It is available at ▶ https://github.com/tesseract-ocr and is described in a paper by its original developer (2007).

#### ■ ■ Line Features

The Hough transform was first first described in U.S. Patent 3,069,654 "Method and Means for Recognizing Complex Patterns" by Paul Hough, and its history is discussed in Hart (2009). The original application was automating the analysis of bubble chamber photographs and it used the problematic slope-intercept parametrization for lines. The currently known form with the $(\theta, \rho)$ parameterization was first described in Duda and Hart (1972) as a "generalized Hough transform" and is available at ▶ https://www.ai.sri.com/pubs/files/tn036-duda71.pdf. The Hough transform is covered in textbooks such as Szeliski (2022), Gonzalez and Woods (2018), Davies (2017) and Klette (2014). The latter has a good discussion on shape fitting in general and estimators that are robust with respect to outlier data points. The basic Hough transform has been extended in many ways and there is a large literature. A useful review of the transform and its variants is presented in Leavers (1993). The transform can be generalized to other shapes (Ballard 1981) such as circles of a fixed size where votes are cast for the coordinates of the circle's center. For circles of unknown size a three-dimensional voting array is required for the circle's center and radius.

#### ■ ■ Point Features

The literature on interest operators dates back to the early work of Moravec (1980) and Förstner (Förstner and Gülch 1987; Förstner 1994). The Harris corner detector

(Harris and Stephens 1988) became very popular for robotic vision applications in the late 1980s since it was able to run in real-time on computers of the day, and the features were quite stable (Tissainayagam and Suter 2004) from image to image. The Noble detector is described in Noble (1988). The work of Shi, Tomasi, Lucas and Kanade (Shi and Tomasi 1994; Tomasi and Kanade 1991) led to the Shi-Tomasi detector and the Kanade-Lucas-Tomasi (KLT) tracker. Good surveys of the relative performance of many corner detectors include those by Deriche and Giraudon (1993) and Mikolajczyk and Schmid (2004).

Scale-space concepts have long been known in computer vision. Koenderink (1984), Lindeberg (1993) and ter Haar Romeny (1996) are a readable introduction to the topic. Scale-space was applied to classic corner detectors creating hybrid detectors such as scale-Harris (Mikolajczyk and Schmid 2004). An important development in scale-space feature detectors was the scale-invariant feature transform (SIFT) introduced in the early 2000s by Lowe (2004) and was a significant improvement for applications such as tracking and object recognition. Unusually, and perhaps unfortunately, it was patented which made researchers wary about using it. In response a very effective alternative called Speeded Up Robust Features (SURF) was developed (Bay et al. 2008) but it was also patented. The SIFT patent expired in March 2020. The SIFT and SURF detectors do give different results and they are compared in Bauer et al. (2007).

Many other interest point detectors and features have been, and continue to be proposed. FAST by Rosten et al. (2010) has very low computational requirements and high repeatability, and C and MATLAB software resources are available at ▶ https://www.edwardrosten.com/work/fast.html. CenSurE by Agrawal et al. (Agrawal et al. 2008) claims higher performance than SIFT, SURF and FAST at lower cost. BRIEF by Calonder et al. (2010) is not a feature detector but is a low cost and compact feature descriptor, requiring just 256 bits instead of 128 floating-point numbers per feature as required by SIFT. Other feature descriptors include histogram of oriented gradients (HOG), oriented FAST and rotated BRIEF (ORB), binary robust invariant scaleable keypoint (BRISK), fast retina keypoint (FREAK), aggregate channel features (ACF), vector of locally aggregated descriptors (VLAD), random ferns and many many more. These feature descriptors are often described as *engineered*, that is they are designed by humans with insight into the problem. An emerging alternative is to apply learning to this problem leading to descriptors such as LIFT (Yi et al. 2016) and GRIEF (Krajník et al. 2017).

Local features have many advantages and are quite stable from frame to frame, but for outdoor applications the feature locations and the descriptors vary considerably with changes in lighting conditions, see for example Valgren and Lilienthal (2010). Night and day are obvious examples but even over a period of a few hours the descriptors change considerably. Over seasons the appearance change can be drastic: trees with or without leaves; the ground covered by grass or snow, wet or dry. Enabling robots to recognize places despite their changing appearance is the research field of robust place recognition which is introduced in Lowry et al. (2015).

The "bag of words" technique for image retrieval was first proposed by Sivic and Zisserman (2003) and has been used by many other researchers since. A notable extension for robotic applications is FABMAP (Cummins and Newman 2008) which explicitly accounts for the joint probability of feature occurrence and associates a probability with the image match, and is available in OpenCV.

## 12.5.2 Exercises

1. Gray-level classification
   a) Experiment with the `ithresh` method on the images `castle.png` and `castle2.png`.
   b) Experiment with the adaptive threshold algorithm and vary its parameters.
   c) Develop an algorithm that finds only the letters in the MSER segmentation of ◻ Fig. 12.9.
   d) Explore the parameters of the `MSER` method.
   e) Apply `labels_graphseg` to the `castle2.png` image. Understand and adjust the parameters to improve performance.
   f) Load the image `adelson.png` and attempt to segment the letters A and B.
2. Color classification
   a) Change $k$, the number of clusters, in the color classification examples. Is there a best value?
   b) Run $k$-means several times and determine how different the final clusters are.
   c) Write a function that determines which of the clusters represents the targets, that is, the yellow cluster or the red cluster.
   d) Apply `labels_graphseg` to the targets and garden image. How does it perform? Understand and adjust the parameters to improve performance.
   e) Experiment with the parameters of the morphological "cleanup" used for the targets and garden images.
   f) Write code that loops over images captured from your computer's camera, applies a classification, and shows the result. The classification could be a grayscale threshold or color clustering to a pre-learned set of color clusters (see `kmeans_color, centroid=` ...).
3. Blobs. Create an image of an object with several holes in it. You could draw it and take a picture, export it from a drawing program, or write code to generate it.
   a) Determine the outer, *inner* and total boundaries of the object.
   b) Place small objects within the holes in the objects. Write code to display the topological hierarchy of the blobs in the scene.
   c) Create a blob whose centroid does not lie on the blob.
   d) For the same shape at different scales (use the `scale` method) plot how the circularity changes as a function of scale. Explain the shape of this curve?
   e) Create an image of a square object and plot the estimated and true perimeter as a function of the square's side length. What happens when the square is small?
   f) Create an image of a simple scene with a number of different shaped objects. Using the shape invariant features (aspect ratio, circularity) to create a simple shape classifier. How well does it perform? Repeat using the Hu moment features.
   g) Repeat the perimeter matching example with some objects that you create. Modify the code to create a plot of edge-segment angle ($k$) versus $\theta$ and repeat the perimeter matching example.
   h) Another commonly used feature, not supported by the Toolbox, is the aligned rectangle. This is the smallest rectangle whose sides are aligned with the axes of the equivalent ellipse and which entirely contains the blob. The aspect ratio of this rectangle and the ratio of the blob's area to the rectangle's area are each scale and rotation invariant features. Write code to compute this rectangle, overlay the rectangle on the image, and compute the two features.
   i) Write code to trace the perimeter of a blob.

4. Experiment with OCR.
   a) Capture your own image and attempt to read the text in it. How does accuracy vary with text size, contrast or orientation?
   b) How does performance vary between a scene with a lot of background and a ROI which is just a single word? How could you find bounding boxes for words in the image?
5. Hough transform
   a) Experiment with using the Sobel edge operator instead of Canny.
   b) Apply the Hough transform to one of your own images.
   c) Write code that loops over images captured from your computer's camera, finds the dominant lines, overlays them on the image and displays it.
   d) Write an implementation of the Hough transform from scratch.
      i. Experiment with varying the size of the Hough accumulator.
      ii. Write code to find the maxima using nonlocal-maxima suppression.
      iii. Write code to return the strongest $N$ peaks in the accumulator.
6. Corner detectors
   a) Experiment with the Harris detector by changing the parameters $k$, $\sigma_D$ and $\sigma_I$.
   b) Compare the performance of the Harris, Noble and Shi-Tomasi corner detectors.
   c) Implement the Moravec detector and compare to the Harris detector.
   d) Create a smoothed second derivative $\mathbf{X}_{uu}$, $\mathbf{X}_{vv}$ and $\mathbf{X}_{uv}$.
7. Experiment with the deep-learning approaches to pixel classification and object detection.
   a) Try other pretrained networks.
   b) Measure the inference time (note that the first inference may take much longer than subsequent ones).
   c) Create a real-time loop that takes input from your camera, applies an object detector and displays the result.
8. Experiment with the bag of words algorithm.
   a) Examine the different support regions of different visual words using the `exemplars` method.
   b) Investigate the effect of changing the number of stop words.
   c) Investigate the effect of changing the size of the vocabulary. Try 1000, 1500, 2500, 3000 and 5000.
   d) Try some different datasets. Take some images from around your house or neighborhood, or use one of the many large scale datasets available online, for example the Oxford Robot Car dataset (Maddern et al. 2016).
   e) Investigate the RootSIFT trick described by Arandjelović and Zisserman (2012).
   f) Investigate using SURF rather than SIFT features.
   g) Investigate using the SIFT corner detector with BRISK or FREAK features.

# Image Formation

» *Everything we see is a perspective,*
*not the truth.*
– Marcus Aurelius

## Contents

© The Author(s), under exclusive license to Springer Nature Switzerland AG 2023
P. Corke, *Robotics, Vision and Control*, Springer Tracts in Advanced Robotics 146,
https://doi.org/10.1007/978-3-031-06469-2_13

In this chapter we discuss how images, discussed in previous chapters, are formed and captured. It has long been known that a simple pinhole is able to create an inverted image on the wall of a darkened room. Some marine mollusks, for example the Nautilus, have pinhole camera eyes. The eyes of vertebrates use a lens to form an inverted image on the retina where the light-sensitive rod and cone cells, shown previously in ◻ Fig. 10.6, are arranged. A digital camera is similar in principle – a glass or plastic lens forms an image on the surface of a semiconductor chip where an array of light-sensitive devices converts the light to a digital image.

The process of image formation, in an eye or in a camera, involves a *projection* of the 3-dimensional world onto a 2-dimensional surface. The depth information is lost, and we can no longer tell from the image whether it is of a large object in the distance, or a smaller closer object. This transformation from 3 to 2 dimensions is known as perspective projection and is discussed in ▶ Sect. 13.1. ▶ Sect. 13.2 introduces the topic of camera calibration, the estimation of the parameters of the perspective transformation. ▶ Sects. 13.3 to 13.5 introduce alternative types of cameras capable of wide-angle, panoramic or light-field imaging. ▶ Sect. 13.6 introduces two applications: the use of artificial markers in scenes to determine the pose of objects, and the use of a planar homography for making image-based measurements of a planar surface. ▶ Sect. 13.7 introduces some advanced concepts such as projecting lines and conics, and nonperspective cameras.

## 13.1  Perspective Camera

### 13.1.1  Perspective Projection

A small hole in the wall of a darkened room will cast an inverted image of the outside world onto the opposite wall – a so-called pinhole camera. The pinhole camera produces a very dim image since, as shown in ◻ Fig. 13.1a, only a small fraction of the light leaving the object finds its way to the image. A pinhole camera has no focus adjustments – all objects are in focus irrespective of distance.

The key to brighter images is to use an objective lens, as shown in ◻ Fig. 13.1b, which collects light from the object over a larger area and directs it to the image. A convex lens can form an image just like a pinhole, and the fundamental geometry of image formation for a thin lens is shown in ◻ Fig. 13.2. The positive $z$-axis is the camera's optical axis.

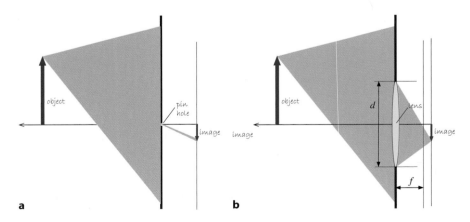

◻ **Fig. 13.1** Light gathering ability of **a** pinhole camera and **b** a lens

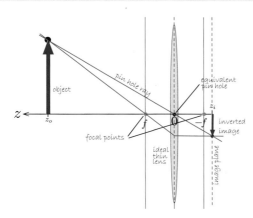

☐ **Fig. 13.2** Image formation geometry for a thin convex lens shown in 2-dimensional cross section. A lens has two focal points at a distance of $f$ on each side of the lens. By convention the camera's optical axis is the $z$-axis

The $z$-coordinate of the object and its image, with respect to the lens center, are related by the thin lens equation

$$\frac{1}{|z_o|} + \frac{1}{|z_i|} = \frac{1}{f} \tag{13.1}$$

where $z_o$ is the $z$-coordinate of the object, $z_i$ is the $z$-coordinate of the image, and $f$ is the focal length of the lens. ▶ For $z_o > f$ an inverted image is formed on the image plane at $z_i < -f$.

The downside of using a lens is the need to focus. The image plane in a camera is the surface of the sensor chip, so the focus ring of the camera moves the lens along the optical axis so that it is a distance $z_i$ from the image plane – for an object at infinity $z_i = -f$. Our own eye has a single convex lens made from transparent crystallin proteins, and focus is achieved by muscles which change its shape – a process known as accommodation. A high-quality camera lens is a compound lens comprising multiple glass or plastic lenses.

In computer vision it is common to use the central perspective imaging model shown in ☐ Fig. 13.3. A ray from the world point P to the origin of the camera frame {C} intersects the image plane, located at $z = f$, at the projected point p. Unlike the pinhole or lens model, the projected image is noninverted. The $z$-axis intersects the image plane at the principal point which is the origin of the 2D image coordinate frame. Using similar triangles we can show that the world point with a coordinate vector $\boldsymbol{P} = (X, Y, Z)$ is projected to a point on the image-plane with a

The inverse of focal length is known as diopter. For thin lenses placed close together their combined diopter is close to the sum of their individual diopters.

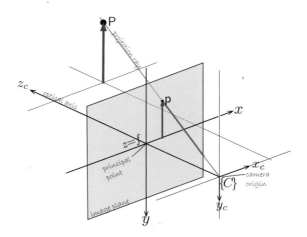

**Fig. 13.3** The central-projection model. The image plane is a distance $f$ in front of the camera's origin, and a noninverted image is formed. The camera's coordinate frame is right-handed with the $z$-axis defining the center of the field of view

coordinate vector $\boldsymbol{p} = (x, y)$ by

$$x = f\frac{X}{Z}, \quad y = f\frac{Y}{Z} \tag{13.2}$$

which are the *retinal* image-plane coordinates. For the case where $f = 1$ the coordinates are referred to as the normalized or canonical image-plane coordinates. Equation (13.2) is a projective transformation, or more specifically, a perspective projection.

This mapping from the 3-dimensional world to a 2-dimensional image has consequences that we can see in ◻ Fig. 13.4 – parallel lines converge and circles become ellipses. More formally we can say that the transformation, from the world to the image plane, has the following characteristics:

- It performs a mapping from 3-dimensional space to the 2-dimensional image plane: $\mathcal{P} : \mathbb{R}^3 \mapsto \mathbb{R}^2$.

**Fig. 13.4** The effect of perspective transformation. **a** Parallel lines converge; **b** circles become ellipses

### Excurse 13.2: Lens Aperture

The *f-number* of a lens, typically marked on the rim, is a dimensionless quantity $F = f/d$ where $d$ is the diameter of the lens (often denoted $\phi$ on the lens rim). The $f$-number is *inversely* related to the light gathering ability of the lens. To reduce the amount of light falling on the image plane, the effective diameter is reduced by a mechanical aperture, or iris, which *increases* the $f$-number. Illuminance on the image plane is inversely proportional to $F^2$ since it depends on the light gathering area. Increasing the $f$-number by a factor of $\sqrt{2}$ or one "stop" will halve the illuminance at the sensor – photographers refer to this as "stopping down". The $f$-number graduations on the lens increase by $\sqrt{2}$ at each stop. An $f$-number is conventionally written in the form $f/1.4$ for $F = 1.4$.

### Excurse 13.3: Focus and Depth of Field

Ideally a group of light rays from a point in the scene meet at a point on the image plane. With imperfect focus, the rays form a finite-sized spot called the circle of confusion which is the point-spread function of the optical system. By convention, if the size of the circle is around that of a pixel then the image is *acceptably* focused.

A pinhole camera has no focus control and always creates a focused image of objects irrespective of their distance. A lens does not have this property – the focus ring changes the distance between the lens and the image plane and must be adjusted so that the object of interest is *acceptably* focused. Photographers refer to depth of field, which is the range of object distances for which *acceptably* focused images are formed. Depth of field is high for small aperture settings where the lens is more like a pinhole, but this means less light and noisier images or longer exposure time and motion blur. This is the photographer's dilemma!

---

- Straight lines in the world are projected to straight lines on the image plane.
- Parallel lines in the world are projected to lines that intersect at a vanishing point as seen in ◨ Fig. 13.4a. In drawing, this effect is known as foreshortening. The exception are fronto-parallel lines – lines lying in a plane parallel to the image plane – which always remain parallel.
- Conics ▸ in the world are projected to conics on the image plane. For example, a circle is projected as a circle or an ellipse as shown in ◨ Fig. 13.4b.

    *Conic sections, or conics, are a family of curves obtained by the intersection of a plane with a cone and are discussed in ▸ App. C.2.1.2. They include circles, ellipses, parabolas and hyperbolas.*

- The size (area) of a shape is not preserved and depends on its distance from the camera.
- The mapping is not one-to-one and no unique inverse exists. That is, given $(x, y)$ we cannot uniquely determine $(X, Y, Z)$. All that can be said is that the world point P lies somewhere along the red projecting ray shown in ◨ Fig. 13.3. This is an important topic that we will return to in ▸ Chap. 14.
- The transformation is not conformal – it does not preserve shape since internal angles are not preserved. Translation, rotation and scaling are examples of conformal transformations.

### 13.1.2 Modeling a Perspective Camera

We can write the coordinates of the image-plane point p as a homogeneous vector $\tilde{p} = (\tilde{x}, \tilde{y}, \tilde{z})^\top \in \mathbb{P}^2$ where

$$\tilde{x} = fX, \quad \tilde{y} = fY, \quad \tilde{z} = Z$$

and the tilde indicates homogeneous quantities. In compact matrix form this becomes ◄

▶ App. C.2 provides a refresher on homogeneous coordinates.

$$\tilde{p} = \begin{pmatrix} f & 0 & 0 \\ 0 & f & 0 \\ 0 & 0 & 1 \end{pmatrix} \begin{pmatrix} X \\ Y \\ Z \end{pmatrix} \tag{13.3}$$

where the nonhomogeneous image-plane coordinates are

$$x = \frac{\tilde{x}}{\tilde{z}}, \quad y = \frac{\tilde{y}}{\tilde{z}}$$

If we also write the coordinate of the world point P as a homogeneous vector $^C\tilde{P} = (X, Y, Z, 1)^\top \in \mathbb{P}^3$, then the perspective projection can be written in *linear* form as

$$\tilde{p} = \begin{pmatrix} f & 0 & 0 & 0 \\ 0 & f & 0 & 0 \\ 0 & 0 & 1 & 0 \end{pmatrix} {}^C\tilde{P} \tag{13.4}$$

or

$$\tilde{p} = \mathbf{C} \, {}^C\tilde{P} \tag{13.5}$$

where **C** is a $3 \times 4$ matrix known as the camera matrix and $^C\tilde{P}$ is the coordinate of the world point with respect to the camera frame {C}.

The camera matrix can be factorized as

$$\tilde{p} = \begin{pmatrix} f & 0 & 0 \\ 0 & f & 0 \\ 0 & 0 & 1 \end{pmatrix} \underbrace{\begin{pmatrix} 1 & 0 & 0 & 0 \\ 0 & 1 & 0 & 0 \\ 0 & 0 & 1 & 0 \end{pmatrix}}_{\Pi} {}^C\tilde{P}$$

where the matrix $\Pi$ is the projection matrix that maps three dimensions into two.

Using the Toolbox, we can create a model of a central-perspective camera with a 15 mm lens

```
>>> camera = CentralCamera(f=0.015);
```

which is an instance of the `CentralCamera` class − a subclass of the `CameraBase` class. By default, the camera is at the origin of the world frame with its optical axis pointing in the world $z$-direction as shown in ◻ Fig. 13.3. We define a world point

```
>>> P = [0.3, 0.4, 3.0];
```

in units of meters and the corresponding retinal image-plane coordinates are

```
>>> camera.project_point(P)
array([[ 0.0015],
       [ 0.002]])
```

When projecting multiple points, the world points are the columns of a $3 \times N$ array and the projected points are the corresponding columns of a $2 \times N$ array.

The point on the image plane is at $(1.5, 2.0)$ mm with respect to the principal point. ◄ This is a very small displacement but it is commensurate with the size of a typical image sensor.

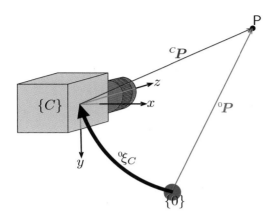

**◻ Fig. 13.5** Camera and point geomery, showing the world and camera frames and the world point P with respect to each frame

In general the camera will have an arbitrary pose $\xi_C$ with respect to the world frame {0} as shown in ◻ Fig. 13.5. The coordinate vector of the world point P with respect to the camera frame {C} is

$$^C P = {}^C \xi_0 \cdot {}^0 P \tag{13.6}$$

or using homogeneous coordinates

$$^C P = \left( {}^0 \mathbf{T}_C \right)^{-1} {}^0 P$$

where ${}^0 \mathbf{T}_C \in \mathbf{SE}(3)$.

We can easily demonstrate this by moving our camera 0.5 m to the left

```
>>> camera.project_point(P, pose=SE3.Tx(-0.5))
array([[   0.004],
       [   0.002]])
```

where the pose of the camera $\xi_C$ is provided as an SE3 object. We see that the $x$-coordinate has increased from 1.5 mm to 4.0 mm, that is, the image point has moved in the opposite direction – to the *right*.

### 13.1.3  Discrete Image Plane

In a digital camera the image plane is a $W \times H$ grid of light-sensitive elements called photosites that correspond directly to the picture elements (or pixels) of the image as shown in ◻ Fig. 13.6. The pixel coordinates are a 2-vector $(u, v) \subset \mathbb{N}_0^2$ and by convention the origin $(0, 0)$ is at the top-left corner of the image plane. The pixels are uniform in size and centered on a regular grid. The pixel coordinate $(u, v)$ is related to the image-plane coordinate $(x, y)$ by

$$u = \frac{x}{\rho_w} + u_0, \quad v = \frac{y}{\rho_h} + v_0 \tag{13.7}$$

where $\rho_w$ and $\rho_h$ are the width and height of each photosite respectively, and $(u_0, v_0)$ is the principal point – the pixel coordinate of the point where the optical axis intersects the image plane with respect to the origin of the $uv$-coordinate frame. In matrix form, the homogeneous relationship between image plane coordinates and pixel coordinates is

$$\begin{pmatrix} \tilde{u} \\ \tilde{v} \\ \tilde{w} \end{pmatrix} = \begin{pmatrix} 1/\rho_w & s & u_0 \\ 0 & 1/\rho_h & v_0 \\ 0 & 0 & 1 \end{pmatrix} \begin{pmatrix} \tilde{x} \\ \tilde{y} \\ \tilde{z} \end{pmatrix} .$$

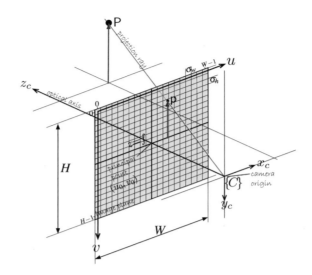

☐ **Fig. 13.6** Central projection model showing image plane and discrete pixels

With precise semiconductor fabrication processes it is unlikely the axes on the sensor chip are not orthogonal, but this term may help model other imperfections in the optical system.

where $s = 0$ but is sometimes written with a nonzero value to account for skew, that is, the $u$- and $v$-axes being nonorthogonal. ◄ Now (13.4) can be written in terms of pixel coordinates

$$\tilde{p} = \underbrace{\begin{pmatrix} 1/\rho_w & 0 & u_0 \\ 0 & 1/\rho_h & v_0 \\ 0 & 0 & 1 \end{pmatrix} \begin{pmatrix} f & 0 & 0 \\ 0 & f & 0 \\ 0 & 0 & 1 \end{pmatrix}}_{\mathbf{K}} \underbrace{\begin{pmatrix} 1 & 0 & 0 & 0 \\ 0 & 1 & 0 & 0 \\ 0 & 0 & 1 & 0 \end{pmatrix}}_{\mathbf{\Pi}} {}^C\tilde{P} \tag{13.8}$$

where $\tilde{p} = (\tilde{u}, \tilde{v}, \tilde{w})$ is the homogeneous coordinate, in units of pixels, of the projected world point P. The product of the first two terms is the $3 \times 3$ camera intrinsic matrix $\mathbf{K}$. The nonhomogeneous image-plane pixel coordinates are

$$u = \frac{\tilde{u}}{\tilde{w}}, \quad v = \frac{\tilde{v}}{\tilde{w}} . \tag{13.9}$$

### Excurse 13.4: Image Sensors

The light-sensitive cells in a camera chip, the photosites (see ► Exc. 11.3), are arranged in a dense array. There is a 1:1 relationship between the photosites on the sensor and the pixels in the output image. Photosites are typically square with a side length in the range $1$–$10\,\mu$m. Professional cameras have large photosites for increased light sensitivity, whereas cellphone cameras have small sensors and therefore small less-sensitive photosites. The ratio of the number of horizontal to vertical pixels is the aspect ratio and is commonly $4 : 3$ or $16 : 9$ (see ► Exc. 11.6). The dimension of the sensor is measured diagonally across the array and is commonly expressed in inches, e.g. $\frac{1}{3}$, $\frac{1}{4}$ or $\frac{1}{2}$ inch. However the active sensing area of the chip has a diagonal that is typically around $\frac{2}{3}$ of the given dimension.

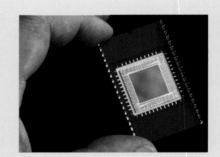

We can model a camera with $1280 \times 1024$ pixels, its principal point at image-plane coordinate $(640, 512)$ and $10\,\mu$m square photosites by

```
>>> camera = CentralCamera(f=0.015, rho=10e-6, name="mycamera",
...                imagesize=[1280, 1024], pp=[640, 512])
          Name: mycamera [CentralCamera]
    pixel size: 1e-05 x 1e-05
    image size: 1280 x 1024
          pose: t = 0, 0, 0; rpy/yxz = 0°, 0°, 0°
  principal pt: [      640        512]
  focal length: [    0.015      0.015]
```

which displays the parameters of the camera model including the camera pose. ▶ The nonhomogeneous image-plane coordinates of the previously defined world point are

```
>>> camera.project_point(P)
array([[      790],
       [      712]])
```

in units of pixels.

We have displayed a different roll-pitch-yaw rotation order $YXZ$. Given the way we have defined the camera axes, the camera orientation with respect to the world frame is a yaw about the vertical or $y$-axis, followed by a pitch about the $x$-axis followed by a roll about the optical axis or $z$-axis.

### 13.1.4 Camera Matrix

Combining (13.6) and (13.8) we can write the camera projection in general form as

$$
\tilde{\boldsymbol{p}} = \begin{pmatrix} f/\rho_w & 0 & u_0 \\ 0 & f/\rho_h & v_0 \\ 0 & 0 & 1 \end{pmatrix} \begin{pmatrix} 1 & 0 & 0 & 0 \\ 0 & 1 & 0 & 0 \\ 0 & 0 & 1 & 0 \end{pmatrix} (^0\mathbf{T}_C)^{-1} \tilde{\boldsymbol{P}}
$$

and substituting $^C\mathbf{T}_0$ for $(^0\mathbf{T}_C)^{-1}$ we can write

$$
\begin{aligned}
\tilde{\boldsymbol{p}} &= \begin{pmatrix} f/\rho_w & 0 & u_0 \\ 0 & f/\rho_h & v_0 \\ 0 & 0 & 1 \end{pmatrix} \begin{pmatrix} 1 & 0 & 0 & 0 \\ 0 & 1 & 0 & 0 \\ 0 & 0 & 1 & 0 \end{pmatrix} \begin{pmatrix} ^C\mathbf{R}_0 & ^C\boldsymbol{t}_0 \\ \mathbf{0} & 1 \end{pmatrix} \tilde{\boldsymbol{P}} \\
&= \underbrace{\mathbf{K}}_{\text{intrinsic}} \underbrace{\begin{pmatrix} ^C\mathbf{R}_0 & ^C\boldsymbol{t}_0 \end{pmatrix}}_{\text{extrinsic}} \tilde{\boldsymbol{P}} \\
&= \mathbf{C}\tilde{\boldsymbol{P}}
\end{aligned} \tag{13.10}
$$

where all the terms are rolled up into the camera matrix $\mathbf{C}$. ▶ This is a $3 \times 4$ homogeneous transformation which performs scaling, translation and perspective projection. It is often also referred to as the projection matrix or the camera calibration matrix. It can be factored into a $3 \times 3$ camera intrinsic matrix $\mathbf{K}$, and a $3 \times 4$ camera extrinsic matrix.

The terms $f/\rho_w$ and $f/\rho_h$ are the focal length expressed in units of pixels.

🛈 It is important to note that the rotation and translation extrinsic parameters describe the world frame with respect to the camera.

The projection is often written in functional form as

$$
\boldsymbol{p} = \mathcal{P}(\boldsymbol{P}, \mathbf{K}, \boldsymbol{\xi}_C) \tag{13.11}
$$

where $\boldsymbol{P}$ is the coordinate vector of the point P in the world frame. $\mathbf{K}$ is the camera intrinsic matrix which comprises the intrinsic characteristics of the camera and

**Excurse 13.5: Ambiguity of Perspective Projection**

We have already mentioned the fundamental ambiguity with perspective projection, that we cannot distinguish between a large distant object and a smaller closer object. We can rewrite (13.10) as

$$\tilde{p} = \mathbf{C}\tilde{P} = \mathbf{C}(\mathbf{H}^{-1}\mathbf{H})\tilde{P} = (\mathbf{CH}^{-1})(\mathbf{H}\tilde{P}) = \mathbf{C}'\tilde{P}'$$

where $\mathbf{H} \in \mathbb{R}^{3\times3}$ is an arbitrary nonsingular matrix. This implies that an infinite number of camera $\mathbf{C}'$ and world-point coordinate $\tilde{P}'$ combinations will result in the same image-plane projection $\tilde{p}$.

This illustrates the essential difficulty in determining 3-dimensional world coordinates from 2-dimensional projected coordinates. It can only be solved if we have information about the camera or the 3-dimensional object.

sensor such as $f, \rho_w, \rho_h, u_0$ and $v_0$. $\xi_C$ is the 3-dimensional pose of the camera and comprises a minimum of six parameters – the extrinsic parameters – that describe camera translation and orientation and is generally represented by an **SE**(3) matrix.

There are 5 intrinsic and 6 extrinsic parameters – a total of 11 independent parameters to describe a camera. The camera matrix has 12 elements so one degree of freedom, the overall scale factor, is unconstrained and can be arbitrarily chosen. Typically, the matrix is normalized such that the lower-right element is one.

The camera intrinsic parameter matrix $\mathbf{K}$ for this camera is

```
>>> camera.K
array([[   1500,       0,     640],
       [      0,    1500,     512],
       [      0,       0,       1]])
```

and the camera matrix $\mathbf{C}$ is

```
>>> camera.C()
array([[   1500,       0,     640,       0],
       [      0,    1500,     512,       0],
       [      0,       0,       1,       0]])
```

Hence `C` is a method, not a property.

and depends on the camera pose. ◄ If pose is not specified, it defaults to the `pose` property of the camera object.

The field of view of a camera is a function of its focal length $f$. A wide-angle lens has a small focal length, a telephoto lens has a large focal length, and a zoom lens has an adjustable focal length. The field of view can be determined from the geometry of ◘ Fig. 13.6. In the horizontal direction the half-angle of view is

$$\frac{\theta_h}{2} = \tan^{-1}\frac{W\rho_w}{2f}$$

where $W$ is the number of pixels in the horizontal direction. We can then write

$$\theta_h = 2\tan^{-1}\frac{W\rho_w}{2f}, \quad \theta_v = 2\tan^{-1}\frac{H\rho_h}{2f} . \tag{13.12}$$

We note that the field of view is also a function of the dimensions of the camera chip which is $W\rho_w \times H\rho_h$. The field of view is computed by the `fov` method of the camera object

```
>>> np.rad2deg(camera.fov())
array([   46.21,     37.69])
```

in degrees in the horizontal and vertical directions respectively.

The camera object can perform visibility checking for points

```
>>> P = np.column_stack([[0, 0, 10], [10, 10, 10]])
array([[ 0, 10],
       [ 0, 10],
       [10, 10]])
>>> p, visible = camera.project_point(P, visibility=True)
>>> visible
array([ True, False])
```

where the argument `visibility` causes an extra return value which is a boolean array indicating which of the world points is visible, that is, is projected within the bounds of the discrete image plane.

### 13.1.5 Projecting Points

The `CentralCamera` class can project multiple world points or lines to the image plane. Using the Toolbox we create a $3 \times 3$ grid of points in the $xy$-plane with overall side length 0.2 m and centered at $(0, 0, 1)$

```
>>> P = mkgrid(n=3, side=0.2, pose=SE3.Tz(1.0));
>>> P.shape
(3, 9)
```

which returns a $3 \times 9$ matrix with one column per grid point where each column comprises the coordinates in $X, Y, Z$ order. The first four columns are

```
>>> P[:, :4]
array([[    -0.1,    -0.1,    -0.1,        0],
       [    -0.1,       0,     0.1,     -0.1],
       [       1,       1,       1,        1]])
```

By default `mkgrid` generates a grid in the $xy$-plane that is centered at the origin. The optional last argument is an **SE**(3) matrix that transforms the points and allows the plane to be arbitrarily positioned and oriented.

The image-plane coordinates of these grid points are

```
>>> camera.project_point(P)
array([[ 490, 490, 490, 640, 640, 640, 790, 790, 790],
       [ 362, 512, 662, 362, 512, 662, 362, 512, 662]])
```

which can be plotted

```
>>> camera.plot_point(P);
```

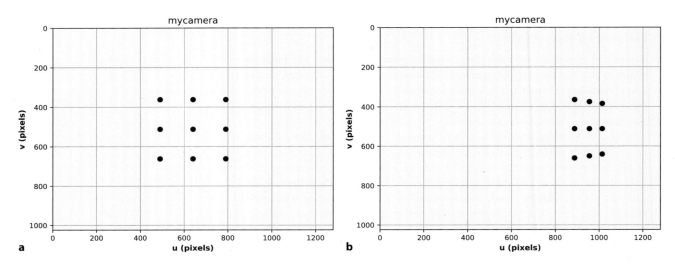

**◘ Fig. 13.7** Two views of a planar grid of points. **a** Frontal view, **b** oblique view

to give the virtual camera view shown in ◘ Fig. 13.7a. The camera pose

```
>>> T_camera = SE3.Trans(-1, 0, 0.5) * SE3.Ry(0.9);
```

results in an oblique view of the plane

```
>>> camera.clf() # clear the virtual image plane
>>> camera.plot_point(P, pose=T_camera);
```

shown in ◘ Fig. 13.7b. We can clearly see the effect of perspective projection which has distorted the shape of the grid – the top and bottom edges, which are parallel lines, have been projected to lines that converge at a vanishing point.

"Ideal" in this context means special rather than perfect. This is a homogeneous line, see ▶ App. C.2.1.1, whose third element is zero, effectively a line at infinity. See also ideal point.

The vanishing point for a line can be determined from the projection of its ideal line. The points in each row of the grid define lines that are parallel to the world $x$-axis, or the vector $(1, 0, 0)$. The corresponding ideal line ◀ is the homogeneous vector $(1, 0, 0, 0)$ and the vanishing point is the projection of this vector

```
>>> camera.project_point([1, 0, 0, 0], pose=T_camera)
array([[    1830],
       [     512]])
```

which is $(1830, 512)$ and just to the right of the visible image plane. ◀

If `project_point` is passed a 4-vector it is assumed to be a coordinate in $\mathbb{P}^3$.

The `plot_point` method also returns image-plane coordinates

```
>>> p = camera.plot_point(P, pose=T_camera)
array([[ 887.8,  887.8,  887.8,  955.2,  955.2,  955.2,  1014,
        1014,  1014],
       [ 364.3,    512,  659.7,  374.9,    512,  649.1,  384.1,
         512,  639.9]])
```

and in this case, the image-plane coordinates have a fractional component which means that the point is not projected to the center of the pixel. However a photosite responds to light equally ◀ over its surface area, so the discrete pixel coordinate can be obtained by rounding.

This is not strictly true for CMOS sensors where transistors reduce the light-sensitive area by the fill factor – the fraction of each photosite's area that is light sensitive.

A 3-dimensional object, a cube, can be defined and projected in a similar fashion. The vertices of a cube with side length $0.2\,\mathrm{m}$ and centered at $(0, 0, 1)$ can be defined by

```
>>> cube = mkcube(0.2, pose=SE3.Tz(1));
>>> cube.shape
(3, 8)
```

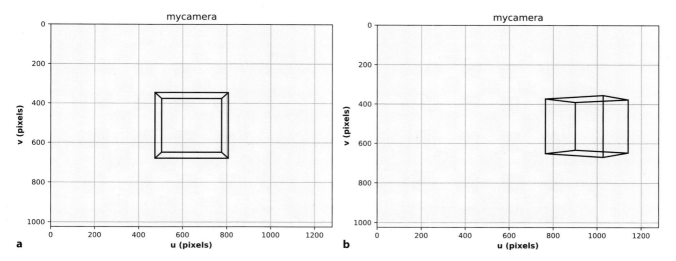

**◻ Fig. 13.8** Line segment representation of a cube. **a** Frontal view, **b** oblique view

which returns a $3 \times 8$ array with one column per vertex. The image-plane points can be plotted as before by

```
>>> camera.clf()  # clear the virtual image plane
>>> camera.plot_point(cube);
```

Alternatively we can create an *edge* representation of the cube by

```
>>> X, Y, Z = mkcube(0.2, pose=SE3.Tz(1), edge=True)
>>> X.shape
(2, 5)
```

which returns three 2D arrays ▶ that define the endpoints of the 12 edges of the cube and display that

The edges are in the same 3-dimensional mesh format used by the Matplotlib method `plot_wireframe`.

```
>>> camera.plot_wireframe(X, Y, Z)
```

as shown in ◻ Fig. 13.8, along with an oblique view generated by moving the camera

```
>>> camera.clf()  # clear the virtual image plane
>>> camera.plot_wireframe(X, Y, Z, pose=T_camera);
```

Successive calls to `plot_wireframe` provides a simple method of animation. For example, to show a cube tumbling in space is

```
>>> X, Y, Z = mkcube(0.2, edge=True)
>>> for theta in np.linspace(0, 2 * pi, 100):
...     T_cube = SE3.Tz(1.5)
...             * SE3.RPY(theta * np.array([1.1, 1.2, 1.3]))
...     camera.clf()
...     camera.plot_wireframe(X, Y, Z, objpose=T_cube)
...     plt.pause(0.1)
```

The cube is defined with its center at the origin, and its vertices are transformed at each time step.

### 13.1.6 Lens Distortion

No lenses are perfect, and the low-cost lenses used in many webcams are far from perfect. Lens imperfections result in a variety of distortions including chromatic aberration (color fringing), spherical aberration or astigmatism (variation in focus across the scene), and geometric distortions where points on the image plane are

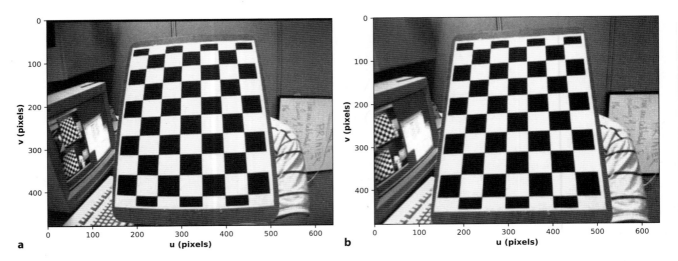

**Fig. 13.9** Lens distortion. **a** Distorted image, the curvature of the vertical lines on the checkerboard is quite pronounced, **b** undistorted image. This is calibration image #12 shipped with OpenCV and included in the Toolbox as `"calibration/left12.jpg"`

displaced from where they should be according to (13.3). An example of geometric distortion is shown in ▫ Fig. 13.9a, and is generally the most problematic effect that we encounter for robotic applications. It has two components: radial distortion, and tangential distortion.

Radial distortion causes image points to be translated along radial lines from the principal point and is well approximated by a polynomial

$$\delta r = k_1 r^3 + k_2 r^5 + k_3 r^7 + \cdots \tag{13.13}$$

where $r$ is the distance of the image-plane point from the principal point in retinal coordinates. Radial distortion can appear in one of two ways, as barrel distortion or pincushion distortion. Barrel distortion occurs when magnification decreases with distance from the principal point, and causes straight lines near the edge of the image to curve outward. Pincushion distortion occurs when magnification increases with distance from the principal point, and causes straight lines near the edge of the image to curve inward. Misalignment of the lens's optical elements leads to tangential distortion at right angles to the radii, but it is generally less significant than radial distortion.

A point $(u_d, v_d)$ in the distorted image is related to its true coordinate $(u, v)$ by

$$u_d = u + \delta_u, \quad v_d = v + \delta_v \tag{13.14}$$

where the displacement is

$$\begin{pmatrix} \delta_u \\ \delta_v \end{pmatrix} = \underbrace{\begin{pmatrix} u\left(k_1 r^2 + k_2 r^4 + k_3 r^6 + \cdots\right) \\ v\left(k_1 r^2 + k_2 r^4 + k_3 r^6 + \cdots\right) \end{pmatrix}}_{\text{radial}} + \underbrace{\begin{pmatrix} 2p_1 uv + p_2(r^2 + 2u^2) \\ p_1(r^2 + 2v^2) + 2p_2 uv \end{pmatrix}}_{\text{tangential}} .$$

$$\tag{13.15}$$

This displacement vector can be plotted for different values of $(u, v)$ as shown in ▫ Fig. 13.12b.

In practice three coefficients are sufficient to describe the radial distortion and the distortion model is often parameterized by $(k_1, k_2, k_3, p_1, p_2)$ which are considered as additional intrinsic parameters. Distortion can be modeled by the `CentralCamera` class using the `distortion` option, for example

```
>>> camera = CentralCamera(f=0.015, rho=10e-6,
...     imagesize=[1280, 1024], pp=[512, 512],
...     distortion=[k1, k2, k3, p1, p2])
```

## Excurse 13.7: Understanding the Geometry of Vision

It has taken humankind a long time to understand light, color and human vision. The Ancient greeks had two schools of thought. The emission theory, supported by Euclid and Ptolemy, held that sight worked by the eye emitting rays of light that interacted with the world somewhat like the sense of touch. The intromission theory, supported by Aristotle and his followers, had physical forms entering the eye from the object.

Euclid of Alexandria (325–265) arguably got the geometry of image formation correct, but his rays emanated from the eye, not the object. Claudius Ptolemy (100–170) wrote Optics and discussed reflection, refraction, and color but today there remains only a poor Arabic translation of his work.

The Arab philosopher Hasan Ibn al-Haytham (aka Alhazen, 965–1040) wrote a seven-volume treatise Kitab al-Manazir (Book of Optics) around 1020. He combined the mathematical rays of Euclid, the medical knowledge of Galen, and the intromission theories of Aristotle. He wrote that "from each point of every colored body, illuminated by any light, issue light and color along every straight line that can be drawn from that point". He understood refraction, but believed the eye's lens, not the retina, received the image – like many early thinkers he struggled with the idea of an inverted image on the retina. A Latin translation of his work was a great influence on later European scholars.

It was not until 1604 that geometric optics and human vision came together when the German astronomer and mathematician Johannes Kepler (1571–1630) published Astronomiae Pars Optica (The Optical Part of Astronomy). He was the first to recognize that images are projected inverted and reversed by the eye's lens onto the retina – the image being corrected later "in the hollows of the brain". (Image from Astronomiae Pars Optica, Johannes Kepler, 1604; reprinted from Felix Platter, De corporis humani structura et usu. Libri III, König, 1583.)

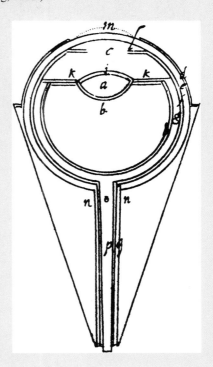

## 13.2 Camera Calibration

The camera projection model (13.10) has a number of parameters that in practice are unknown, as well as lens disortion parameters. In general the principal point is *not* at the center of the photosite array. The focal length written on a lens is only accurate ▶ to 4% of what it is stated to be, and is only correct if the lens is focused at infinity. It is also common experience that the intrinsic parameters change if a lens is detached and reattached, or adjusted for focus or aperture. ▶ The only intrinsic parameters that it may be possible to obtain are the photosite dimensions $\rho_w$ and $\rho_h$, from the sensor manufacturer's data sheet. The extrinsic parameters, the camera's pose, raises the question of which point in the camera should be considered its reference point?

Camera calibration is the process of determining the camera's intrinsic parameters, and the extrinsic parameters with respect to the world coordinate system. Calibration techniques rely on sets of world points whose *relative* coordinates are known, and whose corresponding image-plane coordinates are also known.

We will present two approaches. The first is easy to understand but ignores lens distortion. The second, based on a planar checkerboard, is used by many contempory camera calibration tools and also estimates lens distortion.

According to ANSI Standard PH3.13-1958 "Focal Length Marking of Lenses".

Changing the focus of a lens shifts the lens along the optical axis. In some designs, changing focus rotates the lens so if it is not perfectly symmetric this will move the distortions with respect to the image plane. Changing the aperture alters the parts of the lens that light rays pass through and hence the distortion that they incur.

### 13.2.1 Calibrating with a 3D Target

We expand (13.10), substitute it into (13.9), set $\tilde{p} = (u, v, 1)$ and rearrange as

$$c_{0,0}X + c_{0,1}Y + c_{0,2}Z + c_{0,3} - c_{2,0}uX - c_{2,1}uY - c_{2,2}uZ - c_{2,3}u = 0$$
$$c_{1,0}X + c_{1,1}Y + c_{1,2}Z + c_{1,3} - c_{2,0}vX - c_{2,1}vY - c_{2,2}vZ - c_{2,3}v = 0 \quad (13.16)$$

where $c_{i,j}$ are elements of the unknown camera matrix $\mathbf{C}$, $(X, Y, Z)$ are the coordinates of a point on the calibration target, and $(u, v)$ are the corresponding image plane coordinates.

Calibration requires a 3-dimensional target such as shown in ◻ Fig. 13.10. The position of the center of each marker $(X_i, Y_i, Z_i)$, $i = 0, ..., N-1$ with respect to the target frame {T} must be known, but {T} itself is not known. An image is captured and the *corresponding* image-plane coordinates $(u_i, v_i)$ are determined. As discussed in ▶ Sect. 13.1.4 we can set $c_{2,3} = 1$, and then stack the two equations of (13.16) for each of the $N$ markers to form the matrix equation

$$\begin{pmatrix} X_0 & Y_0 & Z_0 & 1 & 0 & 0 & 0 & 0 & -u_0X_0 & -u_0Y_0 & -u_0Z_0 \\ 0 & 0 & 0 & 0 & X_0 & Y_0 & Z_0 & 1 & -v_0X_0 & -v_0Y_0 & -v_0Z_0 \\ \vdots & \vdots & \vdots & \vdots & \vdots & \vdots & \vdots & \vdots & \vdots & \vdots & \vdots \\ X_{N-1} & Y_{N-1} & Z_{N-1} & 1 & 0 & 0 & 0 & 0 & -u_{N-1}X_{N-1} & -u_{N-1}Y_{N-1} & -u_{N-1}Z_{N-1} \\ 0 & 0 & 0 & 0 & X_{N-1} & Y_{N-1} & Z_{N-1} & 1 & -v_{N-1}X_{N-1} & -v_{N-1}Y_{N-1} & -v_{N-1}Z_{N-1} \end{pmatrix}$$

$$\times \begin{pmatrix} c_{0,0} \\ c_{0,1} \\ \vdots \\ c_{2,2} \end{pmatrix} = \begin{pmatrix} u_0 \\ v_0 \\ \vdots \\ u_{N-1} \\ v_{N-1} \end{pmatrix} \quad (13.17)$$

which can be solved for the camera matrix elements $(c_{0,0} \cdots c_{2,2})$. This approach is known as the direct linear transform. Equation (13.17) has 11 unknowns and requires $N \geq 6$ for solution. Often more than six points will be used, leading

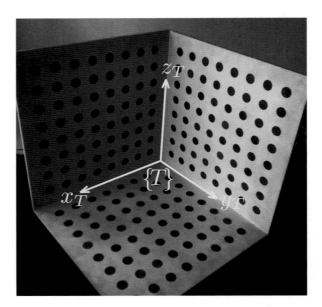

◻ **Fig. 13.10** A 3D calibration target where the circular markers are situated on three planes. The position of each circle with respect to {T} is accurately known. The image-plane coordinates of the markers are generally considered to be the centroids of the circles in the image. (image courtesy of Fabien Spindler)

> **Excurse 13.8: Where is the Camera's Center?**
>
> A compound lens has many cardinal points including focal points, nodal points, principal points and planes, and entry and exit pupils. The *entrance pupil* is a point on the optical axis of a compound lens system that is its center of perspective or its *no-parallax point*. We could consider it to be the *virtual pinhole*. Rotating the camera and lens about this point will not change the relative geometry of targets at different distances in the perspective image.
>
> Rotating about the entrance pupil is important in panoramic photography to avoid parallax errors in the final, stitched panorama. A number of web pages are devoted to discussion of techniques for determining the position of this point, and some even tabulate the position of the entrance pupil for popular lenses. Some of these sites refers to this point incorrectly as the nodal point, even though the techniques they provide do identify the entrance pupil.
>
> Depending on the lens design, the entrance pupil may be behind, within, or in front of the lens system.

to an overdetermined set of equations which can be solved using a least squares approach.

If the points are coplanar then the left-hand matrix of (13.17) becomes rank deficient. This is why the calibration target must be 3-dimensional, typically an array of dots or squares on two or three planes as shown in ◘ Fig. 13.10.

We will illustrate this with an example where the calibration target is a cube. The calibration points are its vertices, its coordinate frame {T} is centered in the cube, and its axes are normal to the faces of the cube faces. The coordinates of the markers, with respect to {T}, are

```
>>> P = mkcube(0.2);
```

The calibration target is at some "unknown pose" $^C\xi_T$ with respect to the camera which we choose to be

```
>>> T_unknown = SE3.Trans(0.1, 0.2, 1.5) * SE3.RPY(0.1, 0.2, 0.3);
```

Next we create a perspective camera whose parameters we will attempt to estimate

```
>>> camera_unknown = CentralCamera(f=0.015, rho=10e-6,
                     imagesize=[1280, 1024], noise=0.05, seed=0)
          Name: perspective [CentralCamera]
    pixel size: 1e-05 x 1e-05
    image size: 1280 x 1024
          pose: t = 0, 0, 0; rpy/yxz = 0°, 0°, 0°
  principal pt: [     640        512]
  focal length: [   0.015      0.015]
```

We have also specified that zero-mean Gaussian noise with $\sigma = 0.05$ pix is added to the $(u, v)$ coordinates to model camera noise and errors in the computer vision algorithms. The image-plane coordinates of the calibration target points at its "unknown" pose and observed by a camera with "unknown" parameters are

```
>>> p = camera_unknown.project_point(P, objpose=T_unknown);
```

Now, using just the object model P and the observed image-plane points p we can estimate the camera matrix

```
>>> C, resid = CentralCamera.points2C(P, p)
>>> C
array([[   852.8,    -233.1,     633.6,       740],
       [    223,       990,      294.8,       712],
       [  -0.131,   0.06549,    0.6488,         1]])
```

which is function of the intrinsic and extrinsic parameters as described by (13.10). Extracting these parameters from the matrix elements is covered in ▶ Sect. 13.2.3.

The residual, the worst-case error between the projection of a world point using the camera matrix $\mathbf{C}$ and the actual image-plane location is

```
>>> resid
0.03767
```

which is a very small fraction of a pixel. It is not zero because of the noise we added to the camera projections.

Linear techniques such as this cannot estimate lens distortion parameters. Distortion will lead to a larger residual and error in the elements of the camera matrix. However, for many situations this might be acceptably low. Distortion parameters are often estimated using a nonlinear optimization over all parameters, typically 16 or more, and this approximate linear solution can be used as the initial parameter estimate.

### 13.2.2 Calibrating with a Checkerboard

Many popular tools for calibrating cameras with lens distortion use a planar checkerboard target which can be easily printed. A number of images, typically ten to twenty, are taken of the target at different distances and orientations as shown in ◻ Fig. 13.11.

A set of calibration target images from OpenCV are shipped with the Toolbox and we can create an image collection

```
>>> images = ImageCollection("calibration/*.jpg");
>>> len(images)
14
```

which we then pass to the calibration routine

```
>>> K, distortion, frames = CentralCamera.images2C(images,
...                              gridshape=(7,6), squaresize=25e-3)
```

The size of the squares on this board is not known but 25 mm is assumed. Errors in this will effect the estimated focal length.

along with information about the shape of the corner-point grid, see ◻ Fig. 13.11, and the dimensions of each square. ◀

The intrinsic matrix

```
>>> K
array([[   534.1,        0,    341.5],
       [      0,    534.1,    232.9],
       [      0,        0,       1]])
```

indicates that the principal point is at $(341.5, 232.9)$ compared to the center of the image which is at

```
>>> images[0].centre
(320, 240)
```

The return value `frames` is a list of objects that hold results for each input image that was successfully processed, in this case just 11 out of the 14 input images. It has properties `image` for the annotated input image, as shown in ◻ Fig. 13.11, and `pose` for the camera pose relative to the checkerboard as an `SE3` object. We can visualize the estimated camera pose for each input image camera

```
>>> for frame in frames:
...     CentralCamera.plot(pose=frame.pose, scale=0.05)
```

which is shown in ◻ Fig. 13.12a.

The estimated lens distortion is returned in the 1D array `distortion` which contains the parameters of (13.15) in the order $(k_1, k_2, p_1, p_2, k_3)$ – note that $k_3$ is

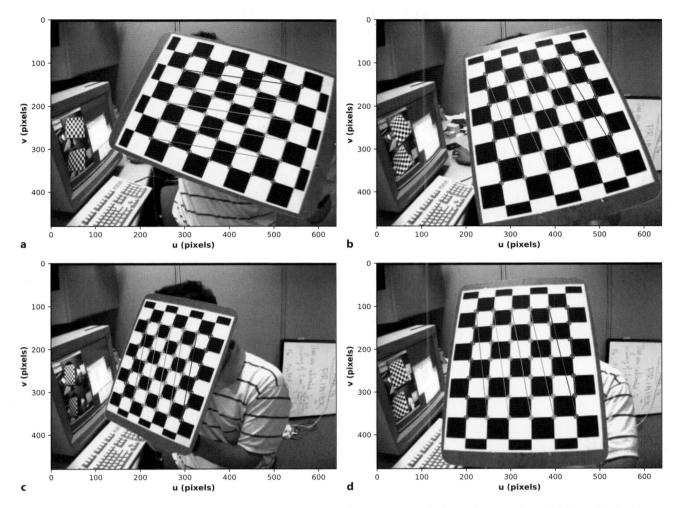

■ **Fig. 13.11** Example frames from the OpenCV calibration example, showing the automatically detected corner points which form a 6 × 7 grid

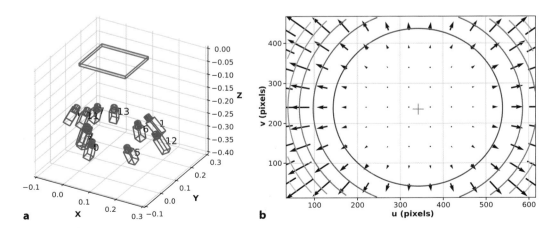

■ **Fig. 13.12** Calibration results from the example checkerboard images. **a** The estimated camera poses relative to the target (shown in blue) for each calibration image; **b** The estimated radial distortion function showing the displacement $(-\delta_u, -\delta_v)$ from distorted coordinate to true coordinate (exaggerated scale), contour lines of constant distortion magnitude, and the principal point out of sequence

```
>>> distortion
array([  -0.293,    0.1077,   0.00131,  -3.109e-05,   0.04348])
```

### 13.2.2.1  Correcting for Lens Distortion

A common problem is to remove the effect of lens distortion and create the image as seen by a perfect lens. To achieve this, we will use the distortion model from ► Sect. 13.1.6 and image warping from ► Sect. 11.7.4.

We first unpack the estimated camera parameters from the previous section

```
>>> u0 = K[0, 2]; v0 = K[1, 2]; fpix_u = K[0, 0]; fpix_v = K[1,1];
>>> k1, k2, p1, p2, k3 = distortion;
```

We choose the undistorted image to be the same size as the distorted image, and the coordinate matrices are

```
>>> U, V = images[12].meshgrid()
```

and using (13.7) we convert the pixel coordinates to retinal image-plane coordinates ◄

In units of meters with respect to the camera's principal point.

```
>>> u = (U - u0) / fpix_u;
>>> v = (V - v0) / fpix_v;
```

The radial distance of the pixels from the principal point is

```
>>> r = np.sqrt(u**2 + v**2);
```

and the retinal coordinate errors due to distortion are given by (13.15)

```
>>> delta_u = u * (k1*r**2 + k2*r**4 + k3*r**6) + p1*u*v
...           + p2*(r**2 + 2*u**2);
>>> delta_v = v * (k1*r**2 + k2*r**4 + k3*r**6)
...           + p1*(r**2 + 2*v**2) + p2*u*v;
```

and the distorted retinal coordinates are

```
>>> ud = u + delta_u; vd = v + delta_v;
```

We convert these back to pixel coordinates in the distorted image

```
>>> Ud = ud * fpix_u + u0;
>>> Vd = vd * fpix_v + v0;
```

Applying the warp

```
>>> undistorted = images[12].warp(Ud, Vd)
Image: 640 x 480 (uint8)
```

gives the results shown in ◘ Fig. 13.9b. The change is quite subtle, but is most pronounced at the edges and corners of the image where $r$ is the greatest.

The distortion-correction vector field $(-\delta_u, -\delta_v)$ can be displayed as a sparse quiver plot

```
>>> plt.clf()    # clear 3D plot
>>> plt.quiver(Ud[::50, ::50], Vd[::50, ::50], -delta_u[::50, ::50],
...            -delta_v[::50, ::50]);
```

along with the magnitude of the distortion as contour lines

```
>>> magnitude = np.sqrt(delta_u**2 + delta_v**2);
>>> plt.contour(U, V, magnitude);
```

and these are shown in ◘ Fig. 13.12b.

### 13.2.3  Decomposing the Camera Calibration Matrix

The elements of the camera matrix are functions of the intrinsic and extrinsic parameters. However, given a camera matrix most of the camera parameters can be recovered.

Continuing the example from ▶ Sect. 13.2.1, we compute the null space of $\mathbf{C}$ which is the world origin in the camera frame

```
>>> o = linalg.null_space(C);
>>> o.T
array([[ 0.08068,  -0.1712,   -0.8136,    0.5497]])
```

which is expressed in homogeneous coordinates that we can convert to nonhomogeneous, or Euclidean, form

```
>>> h2e(o).T
array([[ 0.1468,   -0.3114,    -1.48]])
```

which is close to the true pose of the "unknown" target with respect to the camera

```
>>> T_unknown.inv().t
array([ 0.1464,  -0.3105,   -1.477])
```

To recover orientation as well as the intrinsic parameters we can *decompose* the previously estimated camera matrix

```
>>> est = CentralCamera.decomposeC(C)
         Name: invC [CentralCamera]
   pixel size: 1.0 x 0.999937
         pose: t = 0.147, -0.311, -1.48;
               rpy/yxz = -16.1°, -5.65°, -11.4°
 principal pt: [  642.1     512.9]
 focal length: [   1504     1504]
```

which returns a `CentralCamera` object with its parameters set to values that result in the same camera matrix. We note some differences compared to our true camera model. The focal length is very large, the true value is 0.015 m, and that the pixel sizes are very large, the true value is $10 \times 10^{-6}$. From (13.10) we see that focal length and pixel dimensions always appear together as factors $f/\rho_w$ and $f/\rho_h$. ▶ The `decomposeC` method has set $\rho_w = 1$ but the ratios of the estimated parameters

```
>>> est.f / est.rho[0]
array([   1504,      1504])
```

These quantities have units of pixels since $\rho$ has units of m pixel$^{-1}$. It is quite common in the literature to consider $\rho = 1$ and the focal length is given in pixels. If the pixels are not square then different focal lengths $f_u$ and $f_v$ must be used for the horizontal and vertical directions respectively.

---

**Excurse 13.9: Properties of the Camera Matrix**

The camera matrix $\mathbf{C} \in \mathbb{R}^{3\times4}$ has some important structure and properties:

- It can be partitioned $\mathbf{C} = (\mathbf{M}|\mathbf{c}_3)$ into a nonsingular matrix $\mathbf{M} \subset \mathbb{R}^{3\times3}$ and a vector $\mathbf{c}_3 \in \mathbb{R}^3$, where $\mathbf{c}_3 = -\mathbf{M}\mathbf{c}$ and $\mathbf{c}$ is the world origin in the camera frame. We can recover this by $\mathbf{c} = -\mathbf{M}^{-1}\mathbf{c}_3$.
- The null space of $\mathbf{C}$ is $\tilde{\mathbf{c}}$.
- A pixel at coordinate $\mathbf{p}$ corresponds to a ray in space parallel to the vector $\mathbf{M}^{-1}\tilde{\mathbf{p}}$.
- The matrix $\mathbf{M} = \mathbf{K}\,^C\mathbf{R}_0$ is the product of the camera intrinsics and the camera inverse orientation. We can perform an $RQ$-decomposition of $\mathbf{M} = \bar{\mathbf{R}}\bar{\mathbf{Q}}$ where $\bar{\mathbf{R}}$ is an upper-

triangular matrix (which is $\mathbf{K}$) and an orthogonal matrix $\bar{\mathbf{Q}}$ (which is $^C\mathbf{R}_0$).
- The bottom row of $\mathbf{C}$ defines the principal plane, which is parallel to the image plane and contains the camera origin.
- If the rows of $\mathbf{M}$ are vectors $\mathbf{m}_i$, $i \in [0, 2]$ then:
  - $\mathbf{m}_2^\top$ is a vector normal to the principal plane and parallel to the optical axis and $\mathbf{M}\mathbf{m}_2^\top$ is the principal point in homogeneous form.
  - if the camera has zero skew, that is $k_{0,1} = 0$, then $(\mathbf{m}_0 \times \mathbf{m}_2) \cdot (\mathbf{m}_1 \times \mathbf{m}_2) = 0$
  - and, if the camera has square pixels, that is $\rho_u = \rho_v = \rho$ then the focal length can be recovered by $\|\mathbf{m}_0 \times \mathbf{m}_2\| = \|\mathbf{m}_1 \times \mathbf{m}_2\| = f/\rho$

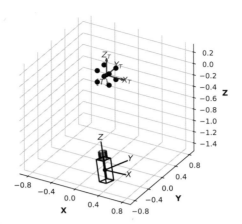

**◘ Fig. 13.13** Calibration target points and estimated camera pose with respect to the target frame {T} which is assumed to be at the origin

are very close to the ratio for the true parameters of the camera

```
>>> camera.f / camera.rho[0]
array([   1500,    1500])
```

The small error in the estimated parameter values is due to the noisy image-plane coordinate values that we used in the calibration process.

The pose of the estimated camera is with respect to the calibration target {T} and is therefore ${}^T\hat{\boldsymbol{\xi}}_C$. The true pose of the target with respect to the camera is ${}^C\boldsymbol{\xi}_T$. If our estimation is accurate then ${}^C\boldsymbol{\xi}_T \oplus {}^T\hat{\boldsymbol{\xi}}_C$ will be $\varnothing$. We earlier set the variable T_unknown equal to ${}^C\boldsymbol{\xi}_T$ and for our example the relative pose between the true and estimated camera pose is quite small

```
>>> (T_unknown * est.pose).printline(orient="camera")
t = -0.000106, -0.000605, -0.00303;
rpy/yxz = -0.0159°, -0.0615°, 0.0872°
```

The camera pose has been estimated to a few millimeters in position and a fraction of a degree in orientation.

To visualize this scenario we plot the calibration markers as small red spheres

```
>>> plotvol3([-0.9, 0.9, -0.9, 0.9, -1.5, 0.3]);
>>> plot_sphere(0.03, P, color="r");
>>> SE3().plot(frame="T", color="b", length=0.3);
```

as well as frame {T} which we have set at the world origin. The estimated pose of the camera is superimposed by

```
>>> est.plot(scale=0.3, color="black", frame=True);
```

and the result is shown in ◘ Fig. 13.13. The problem of determining the pose of a camera with respect to a calibration object is an important problem in photogrammetry known as the camera-location-determination problem.

### 13.2.4 Pose Estimation with a Calibrated Camera

The pose-estimation problem is to determine the pose ${}^C\boldsymbol{\xi}_B$ of an object's body coordinate frame {B} with respect to a calibrated camera. The geometry of the target is known, that is, we know the position of a number of points $(X_i, Y_i, Z_i)$, $i = 0, ..., N-1$ on the target with respect to its coordinate frame {B}. The camera's intrinsic parameters are assumed to be known. An image is captured and the *corresponding* image-plane coordinates $(u_i, v_i)$ are determined using computer vision algorithms.

Estimating the object's pose using corresponding points $(u_i, v_i)$ and $(X_i, Y_i, Z_i)$, and the camera intrinsic parameters is known as the Perspective-$n$-Point problem or PnP for short. It is a simpler problem than camera calibration and decomposition because there are fewer parameters to estimate – camera calibration has 11 parameters while PnP has just 6.

To illustrate pose estimation we will create a calibrated camera with known parameters

```
>>> camera_calib = CentralCamera.Default(noise=0.1, seed=0);
```

and which will add zero-mean Gaussian noise with $\sigma = 0.1$ pix to the projected image plane coordinates.

The object whose pose we wish to determine is a cube with side lengths of 0.2 m and a coordinate frame {B} centered in the cube and with its axes normal to the faces of the cube. The "model" of the object is the coordinates of its vertices with respect to frame {B}

```
>>> P = mkcube(0.2);
```

which is a $3 \times 8$ array.

The object is at some arbitrary but unknown pose ${}^C\xi_B$ pose with respect to the camera

```
>>> T_unknown = SE3.Trans(0.1, 0.2, 1.5) * SE3.RPY(0.1, 0.2, 0.3);
```

The image-plane coordinates of the object's points, at its unknown pose, are

```
>>> p = camera_calib.project_point(P, objpose=T_unknown);
```

where the object points are transformed before being projected to the image plane.

Now using just the object model `P`, the observed image-plane points `p` and the calibrated camera `camera_calib` we estimate the relative pose ${}^C\xi_B$ of the object

```
>>> T_est = camera_calib.estpose(P, p).printline()
t = 0.1, 0.2, 1.5; rpy/zyx = 5.77°, 11.5°, 17.2°
>>> T_unknown.printline()
t = 0.1, 0.2, 1.5; rpy/zyx = 5.73°, 11.5°, 17.2°
```

which is almost identical (within display precision) to the unknown pose `T_unknown` of the object.

In reality the image features coordinates will be imperfectly estimated by the vision system and we have modeled this by adding zero-mean Gaussian noise to the image feature coordinates. It is also important that point correspondence is established, that is, each image plane point is associated with the correct 3D point in the object model.

## 13.3 Wide Field-of-View Cameras

We have discussed perspective imaging in quite some detail since it is the model of our own eyes and most cameras that we encounter. However, perspective cameras fundamentally constrain us to a limited field of view. The thin lens equation (13.1) is singular for points with $Z = f$ which limits the field of view to at most one hemisphere – real lenses achieve far less. As the focal length decreases, radial distortion is increasingly difficult to eliminate, and eventually a limit is reached, beyond which lenses cannot practically be built. The only way forward is to drop the constraint of perspective imaging. In ▶ Sect. 13.3.1 we describe the geometry of image formation with wide-angle lens systems.

An alternative to refractive optics is to use a reflective surface to form an image as shown in ❑ Fig. 13.14. Newtonian telescopes are based on reflection from concave mirrors rather than refraction by lenses. Mirrors are free of color fringing and are easier to scale up in size than a lens. Nature has also evolved reflective optics

◘ **Fig. 13.14** Images formation by reflection from a curved surface (*Cloud Gate*, Chicago, Anish Kapoor, 2006). Note that straight lines have become curves

– the spookfish and some scallops (see ◘ Fig. 1.8) have eyes based on reflectors formed from guanine crystals. In ▶ Sect. 13.3.2 we describe the geometry of image formation with a combination of lenses and mirrors.

### 13.3.1    Fisheye Lens Camera

A fisheye lens image is shown in ◘ Fig. 13.15 and we see that straight lines in the world are curved, and that the field of view is warped into a circle on the image plane. Image formation is modeled using the notation shown in ◘ Fig. 13.16 where the camera is positioned at the origin of the world frame O and its optical axis is the $z$-axis. The world point P is represented in spherical coordinates $(R, \theta, \phi)$, where $\theta$ is the angle outward from the optical axis in the red plane, and $\phi$ is the angle of rotation of the red plane about the optical axis. We can write

$$R = \sqrt{X^2 + Y^2 + Z^2}, \quad \theta = \cos^{-1}\frac{R}{Z}, \quad \phi = \tan^{-1}\frac{Y}{X} \ .$$

On the image plane of the camera, we represent the projection p in polar coordinates $(r, \phi)$ with respect to the principal point, where $r = r(\theta)$. The Cartesian image-plane coordinates are

$$u = r(\theta)\cos\phi, \quad v = r(\theta)\sin\phi$$

and the exact nature of the function $r(\theta)$ depends on the type of fisheye lens. Some common projection models are listed in ◘ Table 13.1 and all have a scaling parameter $k$.

Using the Toolbox we can create a fisheye camera model

```
>>> camera = FishEyeCamera(
...              projection="equiangular",
...              rho=10e-6,
...              imagesize=[1280, 1024]
...              );
```

**Fig. 13.15** Fisheye lens image. Note that straight lines in the world are no longer projected as straight lines. Note also that the field of view is mapped to a circular region on the image plane

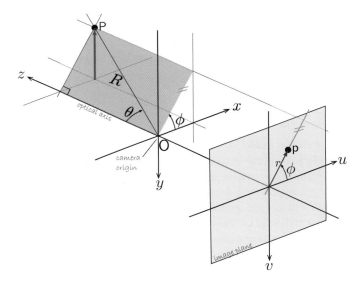

**Fig. 13.16** Image formation for a fisheye lens camera. The world point P is represented in spherical coordinates $(R, \theta, \phi)$ with respect to the camera's origin

**Table 13.1** Fisheye lens projection models

| Mapping | Equation |
| --- | --- |
| Equiangular | $r = k\theta$ |
| Stereographic | $r = k\tan\left(\frac{\theta}{2}\right)$ |
| Equisolid | $r = k\sin\left(\frac{\theta}{2}\right)$ |
| Polynomial | $r = k_1\theta + k_2\theta^2 + \cdots$ |

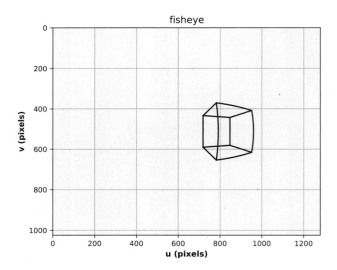

**◻ Fig. 13.17** A cube projected using the `FishEyeCamera` class. The straight edges of the cube are curves on the image plane

which is an instance of the `FishEyeCamera` class – a subclass of the Toolbox's `CameraBase` class and polymorphic with the `CentralCamera` class discussed earlier. If $k$ is not specified, as in this example, then it is computed such that a hemispheric field of view is projected into the maximal circle on the image plane. As is the case for perspective cameras, the parameters such as principal point and pixel dimensions are generally not known and must be estimated using some calibration procedure.

We create an edge-based model of a cube with side length 1 m

```
>>> X, Y, Z = mkcube(side=1, centre=[1, 1, 0.8], edge=True);
```

and project it to the fisheye camera's image plane

```
>>> camera.plot_wireframe(X, Y, Z, color="k");
```

and the result is shown in ◻ Fig. 13.17. We see that straight lines in the world are no longer straight lines in the image.

Wide angle lenses are available with 180° and even 190° field of view, however they have some practical drawbacks. Firstly, the spatial resolution is lower since the camera's pixels are spread over a wider field of view. We also note from ◻ Fig. 13.15 that the field of view is a circular region which means that nearly 25% of the rectangular image plane is wasted. Secondly, outdoors images are more likely to include a lot of bright sky so the camera will automatically reduce its exposure. As a consequence, some non-sky parts of the scene could be underexposed.

### 13.3.2 Catadioptric Camera

From the Greek for curved mirrors (catoptrics) and lenses (dioptrics).

A catadioptric imaging system comprises both reflective and refractive elements, ◄ a mirror and a lens, as shown in ◻ Fig. 13.18a. An example catadioptric image is shown in ◻ Fig. 13.18b.

Image formation is modeled using the geometry shown in ◻ Fig. 13.19. A ray is constructed from the point P to the focal point of the mirror at O which is the origin of the camera system. The ray has an elevation angle of

$$\theta = \tan^{-1} \frac{Z}{X^2 + Y^2} + \frac{\pi}{2}$$

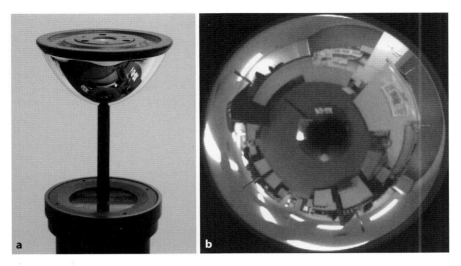

**Fig. 13.18** Catadioptric camera. **a** Catadioptric camera system comprising a conventional perspective camera is looking upward at the mirror; **b** Catadioptric image. Note the dark spot in the center which is the support that holds the mirror above the lens. The floor is in the center of the image and the ceiling is at the edge (photos by Michael Milford)

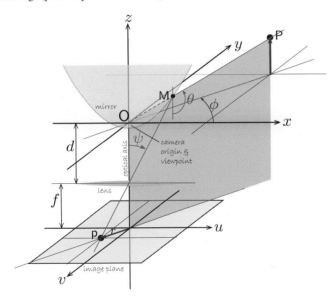

**Fig. 13.19** Catadioptric image formation. A ray from point P at elevation angle $\theta$ and azimuth $\phi$ toward O is reflected from the mirror surface at M and is projected by the lens on to the image plane at p

upward from the optical axis and intersects the mirror at the point M. The reflected ray makes an angle $\psi$ with respect to the optical axis which is a function of the incoming ray angle, that is $\psi(\theta)$. The relationship between $\theta$ and $\psi$ is determined by the tangent to the mirror at the point M and is a function of the shape of the mirror. Many different mirror shapes are used for catadioptric imaging including spherical, parabolic, elliptical and hyberbolic. In general the function $\psi(\theta)$ is nonlinear but an interesting class of mirror is the equiangular mirror for which

$$\theta = \alpha \psi \ .$$

The reflected ray enters the camera lens at angle $\psi$ from the optical axis, and from the lens geometry we can write

$$r = \lambda \tan \psi$$

which is the distance from the principal point. The image-plane point p can be represented in polar coordinates by $p = (r, \phi)$ and the corresponding Cartesian coordinate is

$$u = r \cos \phi, \quad v = r \sin \phi,$$

where $\phi$ is the azimuth angle in the horizontal plane

$$\phi = \tan^{-1} \frac{Y}{X} \, .$$

In ◼ Fig. 13.19 we have assumed that all rays pass through a single focal point or viewpoint – O in this case. This is referred to as central imaging and the resulting image can be transformed perfectly to a perspective image. The equiangular mirror does not meet this constraint and is therefore a noncentral imaging system – the focal point varies with the angle of the incoming ray and lies along a short locus within the mirror known as the caustic. Conical, spherical and equiangular mirrors are all noncentral. In practice the variation in the viewpoint is very small compared to the world scale and many such mirrors are well approximated by the central model.

Using the Toolbox we can model a catadioptric camera, in this case for an equiangular mirror

```
>>> camera = CatadioptricCamera(
...         projection="equiangular",
...         rho=10e-6,
...         imagesize=[1280, 1024],
...         maxangle=pi/4
...         );
```

**13**

---

### Excurse 13.10: Specular Reflection

Specular reflection occurs with a mirror-like surface. Incoming rays are reflected such that the angle of incidence equals the angle of reflection or $\theta_r = \theta_i$. This is in contrast to diffuse or Lambertian reflection which scatters incoming rays over a range of angles.

Speculum is Latin for mirror and speculum metal ($\frac{2}{3}$ copper, $\frac{1}{3}$ tin) is an alloy that can be highly polished. It was used by Newton and Herschel for the curved mirrors in their reflecting telescopes. The image is of the 48 inch speculum mirror from Herschel's 40 foot telescope, completed in 1789, which is now in the British Science Museum. (Image by Mike Peel (► http://mikepeel.net) licensed under CC-BY-SA)

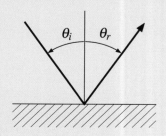

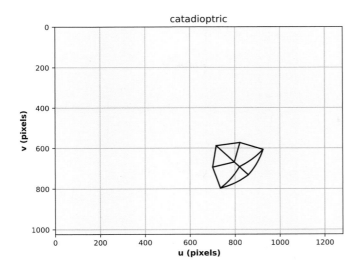

**Fig. 13.20** A cube projected with an equiangular catadioptric camera

which is an instance of a `CatadioptricCamera` object which is a subclass of the Toolbox's `CameraBase` class and polymorphic with the `CentralCamera` class discussed earlier. The option `maxangle` specifies the maximum elevation angle $\theta$ from which the parameters $\alpha$ and $f$ are determined such that the maximum elevation angle corresponds to a circle that maximally fits the image plane. The parameters can be individually specified using the options `alpha` and `focal`. Other supported projection models include parabolic and spherical, and each camera type has different options as described in the online documentation.

We create an edge-based cube model

```
>>> X, Y, Z = mkcube(1, centre=[1, 1, 0.8], edge=True)
```

which we project onto the image plane

```
>>> camera.plot_wireframe(X, Y, Z, color="k");
```

and the result is shown in ◘ Fig. 13.20.

Catadioptric cameras have the advantage that they can view 360° in azimuth, but they also have some practical drawbacks. They share many of the problems of fisheye lenses such as reduced spatial resolution, wasted image-plane pixels and exposure control. In some designs there is also a blind spot due to the mirror support, either a central stalk as seen in ◘ Fig. 13.18, or a number of side supports.

### 13.3.3 Spherical Camera

The fisheye lens and catadioptric systems guide the light rays from a large field of view onto an image plane. Ultimately the 2-dimensional image plane is a limiting factor and it is advantageous to consider an image *sphere* as shown in ◘ Fig. 13.21.

The world point P is projected by a ray to the origin of a unit sphere O. The projection is the point p where the ray intersects the surface of the sphere, and can be represented by the coordinate vector $\boldsymbol{p} = (x, y, z)$ where

$$x = \frac{X}{R}, \quad y = \frac{Y}{R}, \quad \text{and} \quad z = \frac{Z}{R}, \tag{13.18}$$

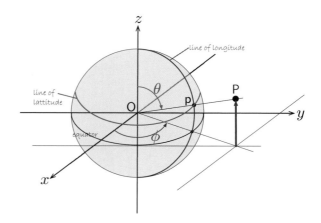

**Fig. 13.21** Spherical image formation. The world point P is projected to p on the surface of the unit sphere and represented by the angles of colatitude $\theta$ and longitude $\phi$

and $R = \sqrt{X^2 + Y^2 + Z^2}$ is the radial distance to the world point. The surface of the sphere is defined by $x^2 + y^2 + z^2 = 1$ so one of the three Cartesian coordinates is redundant. A minimal two-parameter representation for a point on the surface of a sphere $\boldsymbol{p} = (\phi, \theta)$ comprises the angle of colatitude measured down from the North pole

$$\theta = \sin^{-1} r, \quad \theta \in [0, \pi] \tag{13.19}$$

where $r = \sqrt{x^2 + y^2}$, and the azimuth angle (or longitude)

$$\phi = \tan^{-1} \frac{y}{x}, \quad \phi \in [-\pi, \pi) . \tag{13.20}$$

The polar and Cartesian coordinates of the point p are related by

$$x = \sin\theta \cos\phi, \quad y = \sin\theta \sin\phi, \quad z = \cos\theta . \tag{13.21}$$

Using the Toolbox we can create a spherical camera model

```
>>> camera = SphericalCamera()
         Name: spherical [SphericalCamera]
   pixel size: 1.0 x 1.0
         pose: t = 0, 0, 0; rpy/yxz = 0°, 0°, 0°
```

which is an instance of the `SphericalCamera` class – a subclass of the Toolbox's `CameraBase` class and polymorphic with the `CentralCamera` class discussed earlier.

As in the previous examples, we can create an edge-based cube model

```
>>> X, Y, Z = mkcube(1, centre=[2, 3, 1], edge=True)
```

and project it onto the sphere

```
>>> camera.plot_wireframe(X, Y, Z, color="k");
```

and this is shown in ◘ Fig. 13.22. To aid visualization, the spherical image plane has been unwrapped into a rectangle – lines of longitude and latitude are displayed as vertical and horizontal lines respectively. The top and bottom edges correspond to the north and south poles respectively.

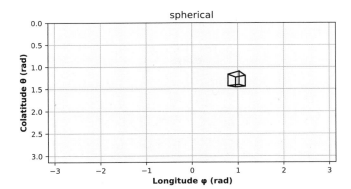

■ **Fig. 13.22**  Cube projected by a spherical camera. The spherical image plane is represented in Cartesian coordinates

It is not yet possible to buy a spherical camera, but prototypes have been demonstrated in several laboratories. The spherical camera is more useful as a conceptual construct to simplify the discussion of wide-angle imaging. As we show in the next section we can transform images from perspective, fisheye or catadioptric camera onto the sphere where we can treat them in a unified manner.

## 13.4  Unified Imaging Model

We have introduced a number of different imaging models in this chapter. Now we will discuss how to transform an image captured with one type of camera, to the image that would have been captured with a different type of camera. For example, given a fisheye lens projection we will generate the corresponding projection for a spherical camera or a perspective camera. The unified imaging model provides a powerful framework to consider very different types of cameras such as standard perspective, catadioptric and many types of fisheye lens.

The unified imaging model is a two-step process and the notation is shown in ■ Fig. 13.23. The first step is spherical projection of the world point P to the surface of the unit sphere p′ as discussed in the previous section and described by (13.18) and (13.19). The view point O is the center of the sphere which is a distance $m$ from the image plane along its normal $z$-axis. The single view point implies a *central* camera.

In the second step, the point p′ is reprojected to the image plane p using the view point F which is at a distance $\ell$ along the $z$-axis above O. The image-plane point p is described in polar coordinates as $\boldsymbol{p} = (r, \phi)$ where

$$r = \frac{(\ell + m) \sin \theta}{\ell - \cos \theta} \ .  \tag{13.22}$$

The unified imaging model has only two parameters $m$ and $\ell$ and these are a function of the type of camera as listed in ■ Table 13.2. For a perspective camera, the two view points O and F are coincident and the geometry becomes the same as the central perspective model shown in ■ Fig. 13.3.

For catadioptric cameras with mirrors that are conics, the focal point F lies between the center of the sphere and the north pole, that is, $0 < \ell < 1$. This projection model is somewhat simpler than the catadioptric camera geometry shown in ■ Fig. 13.19. The imaging parameters are written in terms of the conic parameters eccentricity $\varepsilon$ and latus rectum $4p$. ▶

The projection with F at the north pole is known as  stereographic projection and is used in many fields to project the surface of a sphere onto a plane. Many fisheye lenses are extremely well approximated by F above the north pole.

The length of a chord parallel to the directrix and passing through the focal point.

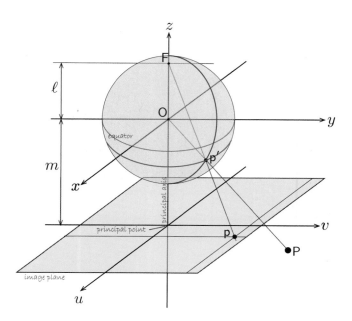

**Fig. 13.23** Unified imaging model. The world point P is projected to p′ on the surface of the unit sphere using the viewpoint O, and then reprojected to point p on the image plane using the viewpoint F

**Table 13.2** Unified imaging model parameters $\ell$ and $m$ according to camera type. $\varepsilon$ is the eccentricity of the conic and $4p$ is the latus rectum

| Imaging | $\ell$ | $m$ |
|---|---|---|
| Perspective | 0 | $f$ |
| Stereographic | 1 | $f$ |
| Fisheye | $> 1\,f$ | |
| Catadioptric (elliptical, $0 < \varepsilon < 1$) | $\frac{2\varepsilon}{1+\varepsilon^2}$ | $\frac{2\varepsilon(2p-1)}{1+\varepsilon^2}$ |
| Catadioptric (parabolic, $\varepsilon = 1$) | 1 | $2p - 1$ |
| Catadioptric (hyperbolic, $\varepsilon > 1$) | $\frac{2\varepsilon}{1+\varepsilon^2}$ | $\frac{2\varepsilon(2p-1)}{1+\varepsilon^2}$ |

### 13.4.1    Mapping Wide-Angle Images to the Sphere

We can use the unified imaging model in reverse. Consider an image captured by a wide field of view camera, such as the fisheye image shown in ◘ Fig. 13.24a. If we know the location of F, then we can project each point p from the fisheye image onto the sphere at p′ to create a spherical image, even though we do not have a spherical camera.

In order to achieve this inverse mapping we need to know some parameters of the camera that captured the image. A common feature of images captured with a fisheye lens or catadioptric camera is that the outer bound of the image is a circle. This circle can be found and its center estimated quite precisely – this is the principal point. A variation of the camera calibration procedure of ▶ Sect. 13.2.2 is applied, which uses corresponding world and image-plane points from the planar calibration target shown in ◘ Fig. 13.24a. This particular camera has a field of view of 190° and its calibration parameters have been estimated to be

```
>>> u0 = 528.1214; v0 = 384.0784; l = 2.7899; m = 996.4617;
```

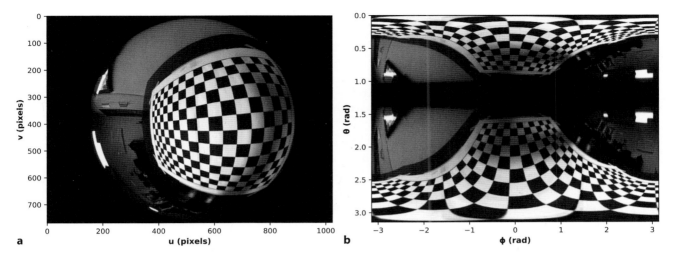

**Fig. 13.24** Fisheye image of a planar calibration target. **a** Fisheye image (image courtesy of Peter Hansen); **b** Image warped to $(\phi, \theta)$ coordinates

We will illustrate the mapping using the image shown in ● Fig. 13.24a

```
>>> fisheye = Image.Read("fisheye_target.png", dtype="float",
...                      mono=True)
Image: 1024 x 768 (float32) [.../images/fisheye_target.png]
```

The spherical image will have 500 pixels in the colatitude and longitude directions, and the coordinate matrices

```
>>> n = 500;
>>> phi_range = np.linspace(-pi, pi, n);   # longitude
>>> theta_range = np.linspace(0, pi, n);   # colatitude
>>> Phi, Theta = np.meshgrid(phi_range, theta_range);
```

span the entire sphere with longitude from $-\pi$ to $+\pi$ radians and colatitude from 0 to $\pi$ radians with 500 steps in each direction.

To map the fisheye image to the sphere we require a mapping from the coordinates of a point in the output spherical image to the coordinates of a point in the input image. This function is the second step of the unified imaging model (13.22) which we implement as

```
>>> r = (1 + m) * np.sin(Theta) / (1 - np.cos(Theta));
```

from which the corresponding Cartesian coordinates in the input image are

```
>>> U = r * np.cos(Phi) + u0;
>>> V = r * np.sin(Phi) + v0;
```

Using image warping from ▶ Sect. 11.7.4 we map the fisheye image to a spherical image

```
>>> spherical = fisheye.warp(U, V, domain=(phi_range, theta_range))
Image: 500 x 500 (float32)
```

where the third argument is the domain of the spherical image and contains a vector that spans its horizontal and vertical axes. The result

```
>>> spherical.disp(axes=(r"$\phi$", r"$\theta$"));
```

is shown in ● Fig. 13.24b. The image appears reflected about the equator and this is because the mapping from a point on the image plane to the sphere is double valued – F is above the north pole so the ray intersects the sphere twice. The top and bottom row of this image corresponds to the principal point, while the dark band above the equator corresponds to the circular outer edge of the input image.

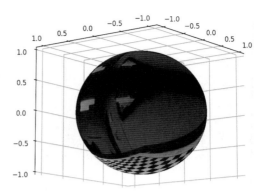

**◨ Fig. 13.25** Fisheye image mapped to the unit sphere. We can see the planar grid lying on a table, and looking closely we can also see ceiling lights, a door and a whiteboard

The image is extremely distorted but this coordinate system is very convenient to texture map onto a sphere

```
>>> ax = plotvol3();
>>> plot_sphere(radius=1, ax=ax, filled=True, resolution=n,
...    facecolors=spherical.colorize().A, cstride=1, rstride=1);
```

and this is shown in ◨ Fig. 13.25. Using the Matplotlib figure toolbar we can rotate the sphere and view it from any direction.

*Any* wide-angle image that can be expressed in terms of central imaging parameters can be similarly projected onto a sphere. So too can multiple perspective images obtained from a camera array, such as shown in ◨ Fig. 13.27.

## 13.4.2    Mapping from the Sphere to a Perspective Image

Given a spherical image we now want to reconstruct a perspective view in a particular direction. We can think of this as being at viewpoint O, inside the sphere, and looking outward at a small surface area which is close to flat and approximates a perspective camera view. This is the second step of the unified imaging model, but with F now at the center of the sphere – this makes the geometry of ◨ Fig. 13.23 similar to the central perspective geometry of ◨ Fig. 13.3. The perspective camera's optical axis is the $-z$-axis of the sphere.

For this example we will use the spherical image created in the previous section. We wish to create a perspective image of $1000 \times 1000$ pixels and with a field-of-view of 45°. The field of view can be written in terms of the image width $W$ and the unified imaging parameter $m$ as

$$\theta_{\text{FOV}} = 2\tan^{-1}\frac{W}{2m}$$

For a 45° field-of-view we require

```
>>> W = 1000;
>>> m = W / 2 / np.tan(np.deg2rad(45 / 2))
1207
```

and for  perspective projection we require

```
>>> l = 0;
```

We also require the principal point to be in the center of the image

```
>>> u0, v0 = W / 2, W / 2;
```

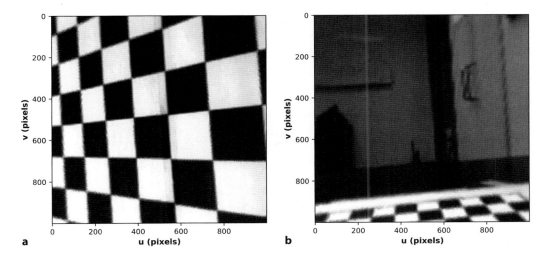

**◻ Fig. 13.26** Perspective projection of spherical image ◻ Fig. 13.25 with a field of view of 45°. Note that the lines on the checkerboard are now straight. **a** looking down through the south pole; **b** looking toward the door and whiteboard

The domain of the output image will be

```
>>> U, V = Image.meshgrid(width=W, height=W);
```

The polar coordinate $(r, \phi)$ of each point in the output image is

```
>>> U0 = U - u0; V0 = V - v0;
>>> r = np.sqrt(U0**2 + V0**2);
>>> phi = np.arctan2(V0, U0);
```

and the corresponding spherical coordinates $(\phi, \theta)$ are

```
>>> Phi = phi;
>>> Theta = pi - np.arctan(r / m);
```

We now warp from spherical coordinates to the perspective image plane

```
>>> spherical.warp(Phi, Theta).disp();
```

and the result is shown in ◻ Fig. 13.26a. This is the view from a perspective camera at the center of the sphere looking down through the south pole. We see that the lines on the checkerboard calibration target are now straight as we would expect from a perspective image.

Of course we are not limited to just looking along the $-z$-axis of the sphere. In ◻ Fig. 13.25 we can see some other features of the room such as a door, a whiteboard and some ceiling lights. We can point our virtual perspective camera in their direction by first rotating the spherical image

```
>>> spherical2 = spherical.rotate_spherical(SO3.Ry(0.9)
...                                     * SO3.Rz(-1.5))
Image: 500 x 500 (float32)
```

so that the $-z$-axis now points toward the distant wall. Repeating the warp process we obtain the result shown in ◻ Fig. 13.26b in which we can clearly see a door and a whiteboard. ▶

```
>>> spherical2.warp(Phi, Theta).disp();
```

The original wide-angle image contains a lot of detail though it can be hard to see because of the distortion. After mapping the image to the sphere we can create a virtual perspective camera view along any line of sight. This is only possible if the original image was taken with a central camera that has a single viewpoint. In theory we cannot create a perspective image from a noncentral wide-angle image but in practice, if the caustic is small, the parallax errors introduced into the perspective image will be negligible.

From a single wide-angle image we can create a perspective view in any direction without having any mechanical pan/tilt mechanism – it's just computation. In fact multiple users could look in different directions simultaneously from a live feed from a single wide-angle camera.

## 13.5    Novel Cameras

### 13.5.1    Multi-Camera Arrays

The cost of cameras and computation continues to fall making it feasible to warp and stitch images from multiple perspective cameras onto a cylindrical or spherical image plane. One such camera is shown in ◘ Fig. 13.27a and uses five cameras to capture a 360° panoramic view as shown in ◘ Fig. 13.27c. The camera in ◘ Fig. 13.27b uses six cameras to achieve an almost spherical field of view.

These camera arrays are not central cameras since light rays converge on the focal points of the individual cameras, not the center of the camera assembly. This can be problematic when imaging objects at short range but in typical use the distance between camera focal points, the caustic, is small compared to distances in the scene. The different viewpoints do have a real advantage however when it comes to capturing the light field.

### 13.5.2    Light-Field Cameras

As we discussed in the early part of this chapter a traditional perspective camera captures a representation of a scene using the two dimensions of the film or sensor. We can think of the captured image as a 2-dimensional function $\mathcal{L}(X, Y)$ that describes the light emitted by the 3D scene. The function is scalar-valued $\mathcal{L}(X, Y) \in \mathbb{R}$ for the monochrome case and vector-valued $\mathcal{L}(X, Y) \in \mathbb{R}^3$ for a tristimulus color representation. ◄

> We could add an extra dimension to represent polarization of the light.

The pin-hole camera of ◘ Fig. 13.1 allows only a very small number of light rays to pass through the aperture, yet space is filled with innumerable light rays

◘ **Fig. 13.27** Omnidirectional camera array. **a** Five perspective cameras provide a 360° panorama with a 72° vertical field of view (camera by Occam Vision Group). **b** Panoramic camera array uses six perspective cameras to provide 90% of a spherical field of view. **c** A seamless panoramic image (3760 × 480 pixels) as output by the camera **a** (images **a** and **c** by Edward Pepperell; image **b** courtesy of Point Grey Research)

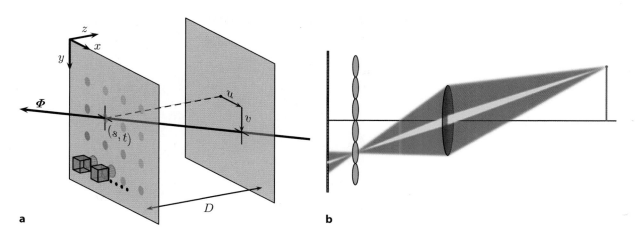

☐ **Fig. 13.28** **a** The light ray Φ passes through the image plane at point $(u, v)$ and the center of the camera at $(s, t)$. This is similar to the central projection model shown in ☐ Fig. 13.3. Any ray can be described by two points, in this case $(u, v)$ and $(s, t)$. **b** Path of light rays from object through main objective lens and lenslet array to the photosite array (images courtesy Donald G. Dansereau)

that provide a richer and more complete description of the world. An objective lens directs multiple rays from a world point to a photosite – increasing the brightness – but the direction information associated with those rays is lost since the photosite responds to all light rays equally, irrespective of their angle to the surface.

The geometric description of *all* the light rays in the scene is called the plenoptic function. ▶ Each ray has a position and direction in 3-dimensional space and could be represented by $\mathcal{L}(X, Y, Z, \theta, \phi)$. ▶ However, a more convenient representation is $\mathcal{L}(s, t, u, v)$ using the 2-plane parameterization shown in ☐ Fig. 13.28a which we can think of as an array of cameras, each with a 2-dimensional image plane. The conventional perspective camera image is the view from a single camera, or a 2-dimensional slice of the full plenoptic function.

The ideas behind the light field have been around for decades, but it is only in recent years that the technology to capture light fields has become widely available. Early light-field cameras were arrays of regular cameras arranged in a plane, such as shown in ☐ Fig. 13.29, or on a sphere surrounding the scene, but these tended to be physically large, complex and expensive to construct. More recently, low-cost and compact light-field cameras based on microlens arrays have come on to the market. The selling point for early consumer light-field cameras was the ability to refocus the image *after* taking the picture. However, the light-field image has many other virtues including synthesizing novel views, 3D reconstruction, low-light imaging and seeing through particulate obscurants.

The microlens or lenslet array is a regular grid of tiny lenses, typically comprising hundreds of thousands of lenses, which is placed a fraction of a millimeter above the surface of the camera's photosite array. The main objective lens focuses an image onto the surface of the microlens array as shown in ☐ Fig. 13.28b. The microlens directs incoming light to one of a small, perhaps $8 \times 8$, patch of photosites according to its direction. The resulting image captures information about both the origin of the ray (the lenslet) and its direction (the particular photosite beneath the lenslet). By contrast, in a standard perspective camera all the rays, irrespective of direction, contribute to the output of the photosite. The light-field camera pixels are sometimes referred to as *raxels* and the resolution of these cameras is typically expressed in megarays.

The raw image from the sensor array looks like ☐ Fig. 13.30a but can be *decoded* into a 4-dimensional light field, as shown in ☐ Fig. 13.30b, and used to render novel views.

The word plenoptic comes from the Latin word plenus meaning full or complete.

Lines in 3D-space have four parameters, see Plücker lines in ▶ App. C.1.2.2.

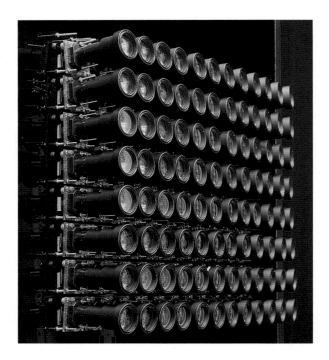

**◻ Fig. 13.29** An 8 × 12 camera array as described in Wilburn et al. (2005) (image courtesy of Marc Levoy, Stanford University)

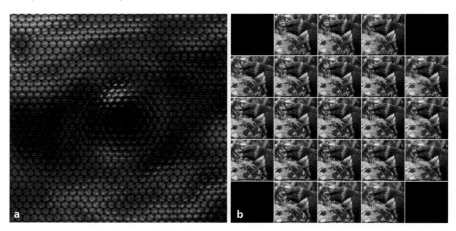

**◻ Fig. 13.30 a** Closeup of image formed on the sensor by the lenslet array. The lenslets are circular but packed hexagonally, and they project hexagonal images because of the hexagonal iris which is designed to avoid overlap of the lenslet images. **b** array of images rendered from the light field for different camera view points (figures courtesy Donald G. Dansereau)

## 13.6  Applications

### 13.6.1  Fiducial Markers

◻ Fig. 13.31 shows a scene that contains a number of artificial marker objects, often referred to as fiducial markers. In robotics two families of fiducial markers are commonly used: ArUco markers, originally developed for augmented reality applications, and AprilTags.

The markers have high-contrast patterns that make them easy to detect in a scene, and OpenCV has functions to determine their pose with respect to the cam-

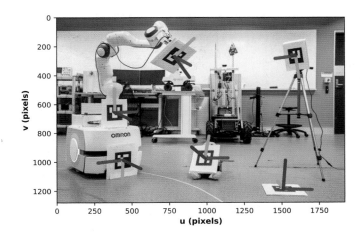

**◘ Fig. 13.31** A laboratory scene with a number of Aruco 4×4_50 markers. This is the same scene as ◘ Fig. 2.6 but with the markers present

era. The pattern of squares indicates the orientation of the marker but also encodes some binary data bits which can be used to uniquely identify the marker. If the camera intrinsics and the dimensions of the marker are known, then we can estimate the pose of the marker with respect to the camera.

Both types of marker are grouped into families. For example the ArUco 4×4_ families have a 4 × 4 grid of black or white squares. These contain a total of 16 bits of data, but only a subset of the $2^{16}$ possible codes are used in order to be able to robustly detect the tag, and determine its orientation. The 4×4_50 family uses just 50 unique codes whereas the 4×4_250 family uses 250 codes.

A scene with ArUco markers is provided with the Toolbox

```
>>> scene = Image.Read("lab-scene.png", rgb=False)
Image: 1920 x 1280 (uint8), B:G:R [.../images/lab-scene.png]
>>> scene.disp();
```

and is shown in ◘ Fig. 13.31. This is similar to ◘ Fig. 2.6 but with the ArUco tags present.

The intrinsic parameters of the camera were found using a little detective work ▶ allowing us to create a calibrated camera model

```
>>> camera = CentralCamera(f=3045, imagesize=scene.shape,
...                         pp=(2016, 1512), rho=1.4e-6);
```

Taken with a Sony ILCE-7RM3, focal length from the EXIF metadata in the image file and sensor data from the camera datasheet.

that matches the camera that took the picture.

We then use the `fidicual` method of the image

```
>>> markers = scene.fiducial(dict="4x4_50", K=camera.K, side=67e-3);
```

where we also pass in the name of the marker family, the camera intrinsic matrix and the side length of the markers. The result is a list of `Fiducial` objects, for example

```
>>> markers[2]
id=5: t = 0.182, 0.0068, 1.14e+06; rpy/zyx = 145°, 29.5°, -3.62°
```

displays the marker's identity and its pose with respect to the camera, which are the attributes `id` and `pose` respectively. The image coordinates of the corners of the marker are

```
>>> markers[2].corners
array([[    947,    1048,    1031,     927],
       [    922,     921,    1025,    1024]], dtype=float32)
```

The underlying OpenCV function that draws the axes into the image requires that the image has BGR color order. We achieve this by the `rgb=False` option to `Image.Read`.

which is the marker at the center bottom of the image held by the teddy bear. The object also has a method to render a coordinate frame into the image ◀

```
>>> for marker in markers:
...     marker.draw(scene, length=0.10, thick=20)
```

and the result is shown in ◘ Fig. 13.31.

### 13.6.2 Planar Homography

Consider a coordinate frame attached to a planar surface such that the $z$-axis is normal to the plane and the $x$- and $y$-axes lie within the plane. We view the marker with a calibrated camera and recalling (13.10)

$$\begin{pmatrix} \tilde{u} \\ \tilde{v} \\ \tilde{w} \end{pmatrix} = \begin{pmatrix} c_{0,0} & c_{0,1} & c_{0,2} & c_{0,3} \\ c_{1,0} & c_{1,1} & c_{1,2} & c_{1,3} \\ c_{2,0} & c_{2,1} & c_{2,2} & 1 \end{pmatrix} \begin{pmatrix} X \\ Y \\ Z \\ 1 \end{pmatrix}.$$

For all points in the plane $Z = 0$, so we can remove $Z$ from the equation

$$\begin{pmatrix} \tilde{u} \\ \tilde{v} \\ \tilde{w} \end{pmatrix} = \begin{pmatrix} c_{0,0} & c_{0,1} & \cancel{c_{0,2}} & c_{0,3} \\ c_{1,0} & c_{1,1} & \cancel{c_{1,2}} & c_{1,3} \\ c_{2,0} & c_{2,1} & \cancel{c_{2,2}} & 1 \end{pmatrix} \begin{pmatrix} X \\ Y \\ \cancel{Z} \\ 1 \end{pmatrix}$$

$$= \begin{pmatrix} c_{0,0} & c_{0,1} & c_{0,2} \\ c_{1,0} & c_{1,1} & c_{1,2} \\ c_{2,0} & c_{2,1} & 1 \end{pmatrix} \begin{pmatrix} X \\ Y \\ 1 \end{pmatrix} = \mathbf{H} \begin{pmatrix} X \\ Y \\ 1 \end{pmatrix}$$

resulting in an invertible linear relationship between homogeneous coordinate vectors in the world plane and in the image plane. This relationship is called a homography, a planar homography or a projective homographyand the matrix $\mathbf{H} \in \mathbb{R}^{3 \times 3}$ is nonsingular.

We define a central perspective camera that is up high and looking obliquely downward

```
>>> T_camera = SE3.Tz(8) * SE3.Rx(-2.8);
>>> camera = CentralCamera.Default(f=0.012, pose=T_camera);
```

at a ground plane as shown in ◘ Fig. 13.32. A shape on the ground plane is defined by a set of 2-dimensional coordinates

```
>>> P = np.column_stack([[-1, 1], [-1, 2], [2, 2], [2, 1]])
array([[-1, -1,  2,  2],
       [ 1,  2,  2,  1]])
```

where each column is the $(X, Y)$ coordinate of a point.

We can project the ground plane points to the image plane in the familiar way using the `CentralCamera` object, after first adding $Z = 0$ for each world point

```
>>> camera.project_point(np.vstack([P, np.zeros((4,))]))
array([[  347.6,    353.8,    792.4,    804.8],
       [  764.9,    616.3,    616.3,    764.9]])
```

The homography is computed from the camera matrix by deleting column two

```
>>> H = np.delete(camera.C(), 2, axis=1)
array([[  1200,    167.5,     3769],
       [     0,   -963.2,     6985],
       [     0,    0.335,    7.538]])
```

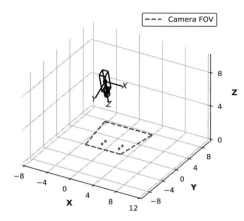

**□ Fig. 13.32** An overhead camera looking obliquely downward to a ground plane on which a shape is defined by four red markers

and we can use that to transform the point coordinates from the ground plane directly to the image plane

```
>>> homtrans(H, P)
array([[  347.6,    353.8,    792.4,    804.8],
       [  764.9,    616.3,    616.3,    764.9]])
```

The Toolbox function `homtrans` performs the appropriate conversions to and from homogeneous coordinates, and the results are the same as those obtained by perspective projection using the camera object.

Since **H** is invertible we can also perform the inverse mapping. The camera has a $1000 \times 1000$ image plane so the coordinates of its corners are

```
>>> p = np.column_stack([[0, 0], [0, 1000], [1000, 1000], [1000, 0]])
array([[   0,     0, 1000, 1000],
       [   0, 1000, 1000,    0]])
```

and on the ground plane these are the points

```
>>> Pi = homtrans(np.linalg.inv(H), p)
array([[  -4.153,   -3.081,    3.081,    4.153],
       [   7.252,   -0.426,   -0.426,    7.252]])
```

which are shown in blue in □ Fig. 13.32. The figure was created by

```
>>> camera.plot(scale=2, color="black");
>>> plot_sphere(radius=0.1, centre=np.vstack((P, np.zeros((4,)))),
...             color="red");
>>> plot_sphere(radius=0.1, centre=np.vstack((Pi, np.zeros((4,)))),
...             color="blue");
```

## 13.7 Advanced Topics

### 13.7.1 Projecting 3D Lines and Quadrics

In ▶ Sect. 13.1 we projected 3D points to the image plane, and we projected 3D line segments by simply projecting their endpoints and joining them on the image plane. To project a continuous line in 3-dimensional space we must first decide how to represent it, and there are many possibilities which are discussed in ▶ App. C.2.2.3. One useful parameterization is Plücker coordinates – a 6-vector with many similarities to twists.

We can easily create a line in Plücker coordinates using the Toolbox. A line that passes through the points $(0, 0, 1)$ and $(1, 1, 1)$ is

```
>>> L = Line3.Join((0, 0, 1), (1, 1, 1))
{ -1 1 0; -1 -1 0}
```

which is an instance of a `Line3` object that is represented as a 6-vector with two components: a moment vector and a direction vector. Other class methods allow a line to be specified using a point and a direction, or the intersection of two planes. The direction of the Plücker line is parallel to the vector

```
>>> L.w
array([ -1,    -1,    0])
```

The `Line3` object also has methods for plotting, as well as determining its intersection point with another `Line3` object or a `Plane3` object. There are many representations of a Plücker line including the 6-vector used above, a minimal 4-vector, and a skew-symmetric $4 \times 4$ matrix ◄ computed using the `skew` method. The latter is used to project the line by

$$\boldsymbol{\ell} = \bigvee{}_{x}\left(\mathbf{C}\,[\mathbf{L}]_{\times}\mathbf{C}^{\top}\right) \in \mathbb{R}^{3}$$

where $\mathbf{C} \in \mathbb{R}^{3 \times 4}$ is the camera matrix, and results in a 2-dimensional line expressed in homogeneous coordinates. Observing this line with the default camera

```
>>> camera = CentralCamera.Default();
>>> l = camera.project_line(L)
array([[      1],
       [     -1],
       [      0]])
```

results in a diagonal line across the image plane. We could plot this using `plot_homline` or on the camera's virtual image plane by

```
>>> camera.plot_line2(l)
```

or in a single step by

```
>>> camera.plot_line3(L)
```

Quadrics, short for quadratic surfaces, are a rich family of 3-dimensional surfaces. There are 17 standard types including spheres, ellipsoids, hyperboloids, paraboloids, cylinders and cones all described by points $\tilde{x} \in \mathbb{P}^3$ such that

$$\tilde{x}^{\top}\mathbf{Q}\,\tilde{x} = 0$$

where $\mathbf{Q} \in \mathbb{R}^{4 \times 4}$ is symmetric. The *outline* of the quadric is projected to the image plane by

$$c^{*} = \mathbf{C}\mathbf{Q}^{*}\mathbf{C}^{\top} \in \mathbb{R}^{3 \times 3}$$

where $(\cdot)^{*}$ represents the adjugate operation, ◄ and $c$ is a matrix representing a conic section on the image plane and the outline is the set of points $p$ such that

$$\tilde{p}^{\top}c\,\tilde{p} = 0$$

and for $\tilde{p} = (u, v, 1)$ can be expanded as

$$Au^2 + Buv + Cv^2 + Du + Ev + F = 0$$

where

$$c = \left( \begin{array}{cc|c} A & B/2 & D/2 \\ B/2 & C & E/2 \\ \hline D/2 & E/2 & F \end{array} \right).$$

The determinant of the top-left submatrix indicates the type of conic: negative for a hyperbola, 0 for a parabola and positive for an ellipse.

$2 \times 2$ and $3 \times 3$ skew-symmetric matrices were discussed in ► Chap. 2. The $4 \times 4$ skew-symmetric matrix has six unique elements.

$\mathbf{A}^{*} = \det(\mathbf{A})\mathbf{A}^{-1}$ which is the transpose of the cofactor matrix. If $\mathbf{B} = \mathbf{A}^{*}$ then $\mathbf{A} = \mathbf{B}^{*}$. See ► App. B for more details.

To demonstrate this, we define a camera looking toward the origin

```
>>> T_camera = pose=SE3.Trans(0.2, 0.1, -5) * SE3.Rx(0.2);
>>> camera = CentralCamera.Default(f=0.015, pose=T_camera);
```

and define a unit sphere at the origin

```
>>> Q = np.diag([1, 1, 1, -1]);
```

then compute its projection to the image plane

```
>>> adj = lambda A: np.linalg.det(A) * np.linalg.inv(A);
>>> C = camera.C();
>>> c = adj(C @ adj(Q) @ C.T)
array([[-5.402e+07, -4.029e+05,  2.389e+10],
       [-4.029e+05, -5.231e+07,  4.122e+10],
       [ 2.389e+10,  4.122e+10, -3.753e+13]])
```

which is a $3 \times 3$ matrix describing a 2-dimensional conic. The determinant of the top-left submatrix

```
>>> np.linalg.det(c[:2, :2])
2.826e+15
```

is positive indicating an ellipse, and a simple way to visualize this is to create a symbolic implicit equation and plot it

```
>>> from sympy import symbols, Matrix, Eq, plot_implicit
>>> x, y = symbols("x y")
>>> X = Matrix([[x, y, 1]]);
>>> ellipse = X * Matrix(c) * X.T;
>>> plot_implicit(Eq(ellipse[0], 1), (x, 0, 1_000), (y, 0, 1_000));
```

for $x, y \in [0, 1000]$.

### 13.7.2 Nonperspective Cameras

The camera matrix, introduced in ▶ Sect. 13.1.4, is a $3 \times 4$ matrix. Any $3 \times 4$ matrix corresponds to some type of camera, but most would result in wildly distorted images. The camera matrix in (13.10) has a special structure – it is a subset of all possible $3 \times 4$ matrices – and corresponds to a perspective camera. The camera projection matrix **C** from (13.10) can be written generally as

$$\mathbf{C} = \begin{pmatrix} c_{0,0} & c_{0,1} & c_{0,2} & c_{0,3} \\ c_{1,0} & c_{1,1} & c_{1,2} & c_{1,3} \\ c_{2,0} & c_{2,1} & c_{2,2} & c_{2,3} \end{pmatrix}$$

which has arbitrary scale so one element, typically $c_{23}$ is set to one – this matrix has 11 unique elements or 11DoF.

Orthographic or parallel projection is a simple perspective-free projection of 3D points onto a plane, like a "plan view". For small objects close to the camera this projection can be achieved using a telecentric lens. The apparent size of an object is independent of its distance.

For the case of an aerial robot flying high over relatively flat terrain the variation of depth, the vertical relief, $\Delta_Z$ is small compared to the average depth of the scene $\overline{Z}$, that is $\Delta_Z \ll \overline{Z}$. We can use a scaled-orthographic projection which is an orthographic projection followed by uniform scaling $m = f/\overline{Z}$.

These two nonperspective cameras are special cases of the more general affine camera model which is described by a matrix of the form

$$\mathbf{C} = \begin{pmatrix} c_{0,0} & c_{0,1} & c_{0,2} & c_{0,3} \\ c_{1,0} & c_{1,1} & c_{1,2} & c_{1,3} \\ 0 & 0 & 0 & 1 \end{pmatrix}$$

that can be factorized as

$$
\mathbf{C} = \underbrace{\begin{pmatrix} m_x & s & 0 \\ 0 & m_y & 0 \\ 0 & 0 & 1 \end{pmatrix}}_{\text{intrinsic}} \underbrace{\begin{pmatrix} 1 & 0 & 0 & 0 \\ 0 & 1 & 0 & 0 \\ 0 & 0 & 0 & 1 \end{pmatrix}}_{\boldsymbol{\Pi}} \underbrace{\begin{pmatrix} \mathbf{R} & t \\ 0 & 1 \end{pmatrix}}_{\text{extrinsic}} .
$$

It can be shown that the principal point is undefined for such a camera model which simplifies the intrinsic matrix. A skew parameter $s$ is commonly introduced to handle the case of nonorthogonal sensor axes. The projection matrix $\boldsymbol{\Pi}$ is different compared to the perspective case in (13.10) – the last two columns are swapped. We can delete the zero column of that matrix and compensate by deleting the third row of the extrinsic matrix resulting in

$$
\mathbf{C} = \underbrace{\begin{pmatrix} m_x & s & 0 \\ 0 & m_y & 0 \\ 0 & 0 & 1 \end{pmatrix}}_{\text{3 DoF}} \underbrace{\begin{pmatrix} r_{0,0} & r_{0,1} & r_{0,2} & t_x \\ r_{1,0} & r_{1,1} & r_{1,2} & t_y \\ 0 & 0 & 0 & 1 \end{pmatrix}}_{\text{5 DoF}} .
$$

The $2 \times 3$ submatrix of the rotation matrix has 6 elements but 3 constraints – the two rows have unit norms and are orthogonal – and therefore has 3 DoF.

This has at most 8 DoF ◄ and the independence from depth is very clear since $t_z$ does not appear. The case where skew $s = 0$ and $m_x = m_y = 1$ is orthographic projection and has only 5 DoF, while the scaled-orthographic case when $s = 0$ and $m_x = m_y = m$ has 6 DoF. The case where $m_x \neq m_y$ is known as weak perspective projection, although this term is sometimes also used to describe scaled-orthographic projection.

## 13.8  Wrapping Up

This Chapter has introduced the image formation process which connects the physical world of light rays to an array of pixels which comprise a digital image. The images that we are familiar with are perspective projections of the world in which 3 dimensions are compressed into 2 dimensions. This leads to ambiguity about object size – a large object in the distance looks the same as a small object that is close. Straight lines and conics are unchanged by this projection but shape distortion occurs – parallel lines can appear to converge and circles can appear as ellipses. We have modeled the perspective projection process and described it in terms of eleven parameters – intrinsic and extrinsic. Geometric lens distortion adds additional lens parameters. Camera calibration is the process of estimating these parameters and two approaches were introduced. We also discussed pose estimation where the pose of an object with known geometry can be estimated from a perspective projection obtained using a calibrated camera.

Perspective images are limited in their field of view and we discussed several wide-angle imaging systems based on the fisheye lens, catadioptrics and multiple cameras. We also discussed the ideal wide-angle camera, the spherical camera, which is currently still a theoretical construct. However, it can be used as an intermediate representation in the unified imaging model which provides one model for almost all camera geometries. We used the unified imaging model to convert a fisheye camera image to a spherical image and then to a perspective image along a specified view axis. We also covered some more recent camera developments such as panoramic camera arrays and light-field cameras. We used two application examples to introduce fiducial markers, and a linear mapping between a plane in the world and the image plane which is applicable to many problems. Finally, we covered some advanced topics such as projecting lines and quadrics from the world to the image plane, and some generalizations of the perspective camera matrix.

In this chapter we treated imaging as a problem of pure geometry with a small number of world points or line segments. In the next chapter we will discuss the geometric relationships between different images of the same scene.

## 13.8.1 Further Reading and Resources

Computer vision textbooks such as Davies (2017) and Klette (2014), Gonzalez and Woods (2018), Forsyth and Ponce (2012) and Szeliski (2022) all provide coverage of the topics introduced in this chapter. Hartley and Zisserman (2003) provide very detailed coverage of image formation using geometric and mathematical approaches, while Ma et al. (2003) provide a mathematical approach. Many topics in geometric computer vision have also been studied by the photogrammetric community, but different language is used. For example, camera calibration is known as camera resectioning, and pose estimation is known as space resectioning. The updated Manual of Photogrammetry (McGlone 2013) provides comprehensive and definitive coverage of the field including history, theory and applications of aircraft and satellite imagery. The revised classic textbook by DeWitt and Wolf (2014) is a thorough and readable introduction to photogrammetry. Wade (2007) reviews the progression, over many centuries, of humankind's understanding of the visual process.

#### ▪▪ Camera Calibration

The homogeneous transformation calibration (Sutherland 1974) approach of ▶ Sect. 13.2.1 is known as the direct linear transform (DLT) in the photogrammetric literature. Wolf (1974) describes extensions to the linear camera calibration with models that include up to 18 parameters and suitable nonlinear optimization estimation techniques. A more concise description of nonlinear calibration is provided by Forsyth and Ponce (2012). Hartley and Zisserman (2003) describe how the linear calibration model can be obtained using features such as lines within the scene.

Tsai (1986) introduced calibration using a planar target, and this is the basis of most current camera calibration tools. Corners on a checkerboard can be determined to a very high precision, whereas for grids of dots the center of the ellipsoidal projection is not the center of the circular marker and will introduce errors (Heikkila and Silven 1997).

There are many good camera calibration toolboxes available on the web. An early and popular tool was the Camera Calibration Toolbox for MATLAB by Jean-Yves Bouguet (2010) and is still available at ▶ http://www.vision.caltech. edu/bouguetj/calib_doc. Many of the modern tools were inspired by this, and have added functionality like automatically finding the checkerboard target which is otherwise tedious to locate in every image. This Toolbox uses OpenCV to perform the underlying calibration. OpenCV can also perform calibration for fisheye lenses, though that is not exposed by the Toolbox.

Pose estimation is a classic problem in computer vision and for which there exists a very large literature. The approaches can be broadly divided into analytic and iterative solutions. Assuming that lens distortion has been corrected the analytic solutions for three and four noncollinear points are given by Fischler and Bolles (1981), DeMenthon and Davis (1992) and Horaud et al. (1989). Typically multiple solutions exist, but for four coplanar points there is a unique solution. Six or more points always yield unique solutions, as well as the intrinsic camera calibration parameters. Iterative solutions were described by Rosenfeld (1959) and Lowe (1991). The pose estimation in the Toolbox is a wrapper around the OpenCV function `solvePnP` which implements iterative and non-iterative algorithms, the latter based on Lepetit et al. (2009). Pose estimation requires a geometric model of the object

**Excurse 13.11: Photogrammetry**

Photogrammetry is the science of understanding the geometry of the world from images. The techniques were developed by the French engineer Aimé Laussedat (1819–1907) working for the Army Corps of Engineers in the 1850s. He produced the first measuring camera and developed a mathematical analysis of photographs as perspective projections. He pioneered the use of aerial photography as a surveying tool to map Paris – using rooftops as well as uncrewed balloons and kites.

Photogrammetry is normally concerned with making maps from images acquired at great distance, but the subfield of close-range or terrestrial photogrammetry is concerned with camera to object distances less than 100 m which is directly relevant to robotics. (Image from La Métrophotographie, Aimé Laussedat, 1899)

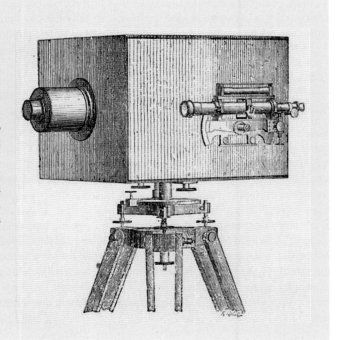

and such computer vision approaches are known as *model-based vision*. An interesting historical perspective on model-based vision is the 1987 video by the late Joe Mundy which is available at ▶ https://www.archive.org/details/JosephMu1987.

■■ **Wide Field-of-View Cameras**

There is recent and growing interest in this type of camera and today, good quality lightweight fisheye lenses and catadioptric camera systems are available. Nayar (1997) provides an excellent motivation for, and introduction to, wide-angle imaging. A very useful online resource is the catadioptric sensor design page at ▶ https://www.math.drexel.edu/~ahicks/design and a page of links to research groups, companies and workshops at ▶ https://www.cis.upenn.edu/~kostas/omni. html. Equiangular mirror systems were described by Chahl and Srinivasan (1997) and Ollis et al. (1999). Nature's solution, the reflector-based scallop eye, is described in Colicchia et al. (2009). The book of Daniilidis and Klette (2006) is a collection of papers on nonperspective imaging and Benosman and Kang (2001) is another, earlier, published collection of papers. Some information is available through CVonline at ▶ https://homepages.inf.ed.ac.uk/rbf/CVonline in the section *Image Physics*.

A Toolbox for calibrating wide-angle cameras by Scaramuzza et al. (2006) can be found online and has MATLAB and C++ code. OpenCV has support for fisheye camera calibration, projection and image undistortion. Both model the fisheye projection as a radial distortion polynomial $r(\theta)$.

The unified imaging model was introduced by Geyer and Daniilidis (2000) in the context of catadioptric cameras. Later it was shown (Ying and Hu 2004) that many fisheye cameras can also be described by this model. The fisheye calibration of ▶ Sect. 13.4.1 was described by Hansen et al. (2010) who estimate $\ell$ and $m$ rather than a polynomial function $r(\theta)$ as does Scaramuzza's Toolbox.

There is a huge and growing literature on light-field imaging but as yet no textbook. A great introduction to light fields and its application to robotics is the thesis by Dansereau (2014). The same author has a MATLAB Toolbox available at ▶ https://mathworks.com/matlabcentral/fileexchange/75250-light-field-toolbox. An interesting description of an early camera array is given by Wilburn et al. (2005) and the associated video demonstrates many capabilities. Light-field imaging is a subset of the larger, and growing, field of computational photography.

■■ **Fiducial Markers**

AprilTags were described in Olson (2011) and there is an implementation at ▶ https://github.com/AprilRobotics/apriltag. ArUco markers are described in Romero-Ramirez et al. (2018) and also at ▶ https://www.uco.es/investiga/grupos/ava/node/26. Both types of marker are detected by the OpenCV function `detectMarkers`.

### 13.8.2 Exercises

1. Create a central camera and a cube target and visualize it for different camera and cube poses. Create and visualize different 3D mesh shapes such as created by the Spatial Math Toolbox functions `cylinder`, `cuboid` and `sphere`.
2. Write code to fly the camera in an orbit around the cube, always facing toward the center of the cube.
3. Write code to fly the camera through the cube.
4. Create a central camera with lens distortion and which is viewing a $10 \times 10$ planar grid of points. Vary the distortion parameters and see the effect this has on the shape of the projected grid. Create pincushion and barrel distortion.
5. Repeat the homogeneous camera calibration exercise of ▶ Sect. 13.2.1 and the decomposition of ▶ Sect. 13.2.3. Investigate the effect of the number of calibration points, noise and camera distortion on the calibration residual and estimated target pose.
6. Determine the solid angle for a rectangular pyramidal field of view that subtends angles $\theta_h$ and $\theta_v$.
7. Run the camera calibration example yourself. Use the provided images or capture your own checkerboard images.
8. Calibrate the camera on your computer.
9. Derive (13.15).
10. For the camera calibration matrix decomposition example (▶ Sect. 13.2.3) determine the roll-pitch-yaw orientation error between the true and estimated camera pose.
11. Pose estimation (▶ Sect. 13.2.4):
    a) Repeat the pose estimation exercise for different object poses (closer, further away).
    b) Repeat for different levels of camera noise.
    c) What happens as the number of points is reduced?
    d) Does increasing the number of points counter the effects of increased noise?
    e) Change the intrinsic parameters of the camera `cam` before invoking the `estpose` method. What is the effect of changing the focal length and the principal point by say 5%.
    f) Introduce a correspondence error `p[:, [1,2]] = p[:, [2,1]]`. What happens?
12. Repeat exercises 2 and 3 for the fisheye camera and the spherical camera.
13. With reference to ◘ Fig. 13.19 derive the function $\psi(\theta)$ for a parabolic mirror.
14. With reference to ◘ Fig. 13.19 derive the equation of the equiangular mirror $z(x)$ in the $xz$-plane.
15. Quadrics (▶ Sect. 13.7.1):
    a) Write a function that generates a 3D plot of a quadric given a $4 \times 4$ matrix. Hint: for complex 3D graphics try using pyvista, mayavi or plotly.

b) Write code to compute the quadric matrix for a sphere at arbitrary location and of arbitrary radius.

c) Write code to compute the quadric matrix for an arbitrary circular cylinder.

d) Write code to plot the planar conic section described by a $3 \times 3$ matrix.

16. Project an ellipsoidal or spherical quadric to the image plane. The result will be the implicit equation for a conic – write code to plot the implicit equation.

13

# Using Multiple Images

**Contents**

© The Author(s), under exclusive license to Springer Nature Switzerland AG 2023
P. Corke, *Robotics, Vision and Control*, Springer Tracts in Advanced Robotics 146,
https://doi.org/10.1007/978-3-031-06469-2_14

Almost! We can determine the translation of the camera only up to an unknown scale factor, that is, the translation is $\lambda \hat{t} \in \mathbb{R}^3$ where the direction $\hat{t}$ is known but $\lambda$ is not.

**14**

In ▶ Chap. 12 we learned about point features which are distinctive *points* in an image. They correspond to visually distinctive physical features in the world that can be reliably detected in different views of the same scene, irrespective of viewpoint or lighting conditions. They are characterized by high image gradients in orthogonal directions and frequently occur on the corners of objects. However the 3-dimensional coordinate of the corresponding world point was lost in the perspective projection process which we discussed in ▶ Chap. 13 – we mapped a 3-dimensional world point to a 2-dimensional image coordinate. All we know is that the world point lies along some ray in space corresponding to the pixel coordinate, as shown in ◨ Fig. 13.6. To recover the missing third dimension we need additional information. In ▶ Sect. 13.2.4 the additional information was camera calibration parameters plus a geometric object model, and this allowed us to estimate the object's 3-dimensional pose from 2-dimensional image data.

In this chapter we consider an alternative approach in which the additional information comes from *multiple* views of the same scene. As already mentioned, the pixel coordinates from a single view constrains the world point to lie along some ray. If we can locate the same world point in another image, taken from a different but known pose, we can determine another ray along which that world point must lie. The world point lies at the intersection of these two rays – a process known as triangulation or 3D reconstruction. Even more powerfully, if we observe sufficient points, we can estimate the 3D motion of the camera between the views as well as the 3D structure of the world. ◀

The underlying challenge is to find the same world point in multiple images. This is the *correspondence problem*, an important but nontrivial problem that we will discuss in ▶ Sect. 14.1. In ▶ Sect. 14.2 we revisit the fundamental geometry of image formation developed in ▶ Chap. 13 for the case of a single camera. If you haven't yet read that chapter, or it's been a while since you read it, it would be helpful to (re)acquaint yourself with that material. We extend the geometry to encompass multiple image planes and show the geometric relationship between pairs of images. Stereo vision is an important technique for robotics where information from two images of a scene, taken from different viewpoints, is combined to determine the 3-dimensional structure of the world. We discuss sparse and dense approaches to stereo vision in ▶ Sects. 14.3 and 14.4 respectively. Bundle adjustment is an advanced concept, but a very general approach to combining information from many cameras, and is introduced in ▶ Sect. 14.3.2.

For some applications we might use RGBD cameras which return depth as well as color information and the underlying principles of such cameras are introduced in ▶ Sect. 14.6. The 3-dimensional information is typically represented as a *point cloud*, a set of 3D points, and techniques for plane fitting and alignment of such data are introduced in ▶ Sect. 14.7.

We finish this chapter, and this part of the book, with three application examples that put the concepts we have learned into practice. ▶ Sect. 14.8.1 describes how we can transform an image with obvious perspective distortion into one without, effectively synthesizing the view from a virtual camera at a different location. ▶ Sect. 14.8.2 describes mosaicing which is the process of taking multiple overlapping images from a moving camera and *stitching* them together to form one large virtual image. ▶ Sect. 14.8.3 describes how we can process a sequence of images from a moving camera to locate consistent world points and to estimate the camera motion and 3-dimensional world structure.

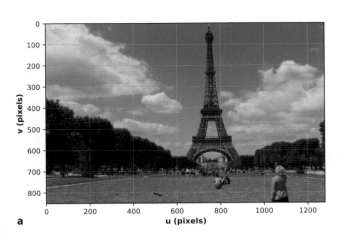

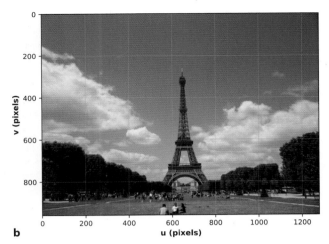

**Fig. 14.1** Two views of the Eiffel tower. The images were captured approximately simultaneously using two different handheld digital cameras. **a** 7 Mpix camera with $f = 7.4$ mm; **b** 10 Mpix camera with $f = 5.2$ mm (image by Lucy Corke). The images have quite different scale and the tower is 700 and 600 pixels tall in **a** and **b** respectively. The camera that captured image **b** is held by the person in the bottom-right corner of **a**

## 14.1  Point Feature Correspondence

Correspondence is the problem of finding the pixel coordinates in two different images that correspond to the same point in the world. ▶ Consider the pair of real images

This is another example of the data association problem that we have encountered several times in this book.

```
>>> view1 = Image.Read("eiffel-1.png")
Image: 1280 x 851 (uint8), R:G:B [.../images/eiffel-1.png]
>>> view2 = Image.Read("eiffel-2.png")
Image: 1280 x 960 (uint8), R:G:B [.../images/eiffel-2.png]
```

shown in ◻ Fig. 14.1. They show the same scene viewed from two different positions using two different cameras – the pixel size, focal length and number of pixels for each image are quite different. The scenes are complex and we see immediately that determining correspondence is not trivial. More than half the pixels in each scene correspond to blue sky and it is impossible to match a blue pixel in one image to the corresponding blue pixel in the other – these pixels are insufficiently distinct. This situation is common and can occur with homogeneous image regions such as dark shadows, expanses of water or snow, or smooth human-made objects such as walls or the bodies of cars.

The solution is to choose only those points that are distinctive, and we can use the point feature detectors that were introduced in the last chapter. Harris point features can be detected

```
>>> hf = view1.Harris(nfeat=150)
Harris features, 150 points
>>> view1.disp(darken=True); hf.plot();
```

and are shown in ◻ Fig. 14.2a. Alternatively, we can detect SIFT features

```
>>> sf = view1.SIFT().sort().filter(minscale=10)[:150]
SIFT features, 150 points
>>> view1.disp(darken=True);
>>> sf.plot(filled=True, color="y", alpha=0.3)
```

which are shown in ◻ Fig. 14.2b. The SIFT features were sorted by descending feature strength, filtered to select those with a scale greater than 10 and then the top 150 selected. Using either type of feature point, we have simplified the problem

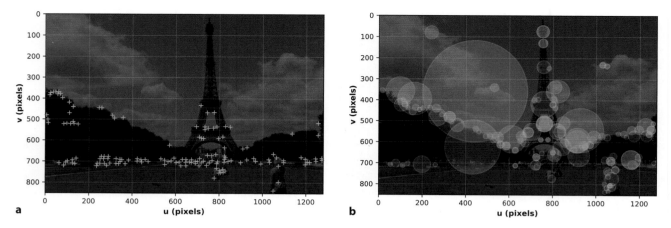

**◘ Fig. 14.2** 150 point features computed for ◘ Fig. 14.1a. **a** Harris point features; **b** SIFT point features showing scale

– instead of having to deal with millions of pixels we now have to handle just 150 distinctive point features.

Consider the general case of two sets of features points: $\{\,^1\boldsymbol{p}_i \in \mathbb{R}^2, i = 0, \ldots, N_1-1\}$ in the first image and $\{\,^2\boldsymbol{p}_j \in \mathbb{R}^2, j = 0, \ldots, N_2-1\}$ in the second image. ◀ Since these are distinctive image points we would expect a significant number of points in image one would correspond to points found in image two. The problem is to determine which $(\,^2u_j, \,^2v_j)$, if any, corresponds to each $(\,^1u_i, \,^1v_i)$.

We cannot use the feature coordinates alone to determine correspondence – the features will have different coordinates in each image. For example in ◘ Fig. 14.1 we see that most features are lower in the right-hand image. We cannot use the intensity or color of the pixels either. Variations in white balance, illumination and exposure setting make it highly unlikely that corresponding pixels will have the same value. Even if intensity variation was eliminated there are likely to be tens of thousands of pixels in the other image with exactly the same intensity value – it is not sufficiently unique. We need some richer way of *describing* each feature.

In practice we consider a region of pixels around the feature point called the *support region*. We describe this region with a *feature descriptor* which is a distinctive and unique numerical description of the feature point and its support region. For the Toolbox implementation of the Harris detector, the descriptor

```
>>> hf[0].descriptor.shape
(121,)
```

is a unit vector containing the contents of an $11 \times 11$ window around the feature point, from which the mean value has been subtracted. Feature descriptors are typically vectors and are therefore often referred to as feature vectors.

Determining the similarity of two descriptors using normalized cross-correlation is simply the dot product of two descriptors and the resulting similarity measure $s \in [-1, 1]$ has some meaning – a perfect match is $s = 1$ and $s \geq 0.8$ is typically considered a good match. For the example above

```
>>> hf[0].distance(hf[1], metric="ncc")
array([[ 0.5908]])
```

and the correlation score indicates a poor match.

This window-based descriptor is distinctive and invariant to changes in image intensity but it is not invariant to scale or rotation. Other descriptors of the surrounding region that we could use include census and rank values, as well as histograms of intensity or color. Histograms have the advantage of being invariant to rotation but they say nothing about the spatial relationship between the pixels,

The feature coordinates are assumed to be real numbers, determined to subpixel precision. The Harris detector is the only one that does not provide subpixel precision.

that is, the same pixel values in a completely different spatial arrangement have the same histogram.

The SIFT algorithm computes a different type of descriptor

```
>>> sf[0].descriptor.shape
(128,)
```

which is a 128-element vector that describes the pixels within the feature's support region in a way which is scale and rotationally invariant. It is based on the image in the scale-space sequence corresponding to the feature's scale, rotated according to the feature's orientation, and then normalized ▶ to increase its invariance to changes in image intensity. Similarity between descriptors is based on Euclidean distance.

```
>>> sf[0].distance(sf[1], metric="L2")
array([[   484.1]])
```

The Toolbox uses the OpenCV implementation where the descriptor is a vector of 128 `float32` values, which have integer values, and the norm is approximately 512.

which is a large value indicating poor similarity. The SIFT descriptor is quite invariant to image intensity, scale and rotation. SIFT is both a point feature detector and a descriptor, whereas the Harris operator is just a point feature detector which must be used with with some descriptor and associated distance measure – in the Toolbox this is a normalized local window and the ZNCC distance norm is simply the dot product of two descriptors.

For the remainder of this chapter we will use SIFT features. They are computationally more expensive but pay for themselves in terms of the quality of matches between widely different views of the same scene. We compute SIFT features for each image

```
>>> sf1 = view1.SIFT()
SIFT features, 3181 points
>>> sf2 = view2.SIFT()
SIFT features, 3028 points
```

which results in two sets of `SIFTFeature` objects each containing a few thousand point features.

Next we match the two sets of SIFT features based on the distance between the SIFT descriptors

```
>>> matches = sf1.match(sf2);
>>> len(matches)
860
```

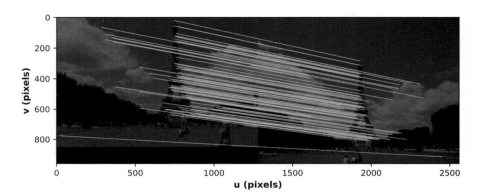

**◻ Fig. 14.3** Feature correspondence shown by yellow lines joining putative corresponding points in the two images, which are stacked horizontally. Subset of 100 matches based on SIFT descriptor similarity. We can see that at least one proposed correspondence is clearly incorrect

We refer to them as putative or candidate because, although they are very likely to correspond, this has not yet been confirmed.

which results in a `FeatureMatch` object that represents 860 *putative* or *candidate* correspondences. ◀ The first five putative correspondences are

```
>>> matches[:5].list()
   0:    24.00 (1118.6, 178.7) <--> (952.5, 417.9)
   1:    24.12 (900.2, 636.6) <--> (775.6, 802.3)
   2:    25.81 (760.3, 125.0) <--> (656.2, 369.1)
   3:    27.33 (820.5, 519.0) <--> (708.0, 701.6)
   4:    28.12 (801.1, 632.4) <--> (694.1, 800.3)
```

which shows the $L_2$ distance between the two feature descriptors, and the corresponding feature coordinate in the first and second image. The matches are ordered by increasing distance – decreasing similarity.

We can overlay a subset of these matches on the original image pair

```
>>> matches.subset(100).plot(color="yellow")
```

The `FeatureMatch` object has a reference to the original images and displays them horizontally concatenated.

and the result is shown in ◻ Fig. 14.3. ◀ Yellow lines connect the matched features in each image and the lines show a consistent pattern. Most of these correspondences seem quite sensible, but at least one of this small subset is obviously incorrect and we will deal with that problem shortly. The `subset` method of the `FeatureMatch` object returns another `FeatureMatch` object with the specified number of features sampled evenly from the original. If all correspondences were shown we would just see a solid yellow mass.

The correspondences are

```
>>> c = matches.correspondence();
>>> c[:, :5]
array([[2890, 2467, 1661, 2218, 2081],
       [2258, 1945, 1412, 1773, 1705]])
```

which is an array with one column per putative correspondence. The first column indicates that feature number 2890 in image one is most similar to feature number 2258 in image two and so on. In terms of the feature objects computed above, this says that SIFT feature `sf1[2890]` corresponds with SIFT feature `sf2[2258]`.

The Euclidean distance between the matched feature descriptors is given by the `distance` property and the distribution of these, with no thresholding applied, is

```
>>> plt.hist(matches.distance, cumulative=True, density=True);
```

shown in ◻ Fig. 14.4. It shows that 25% of all matches have descriptor distances below 75 whereas the maximum distance can be much larger – such matches are less likely to be valid.

A classical way to select good matches is by some distance threshold

```
>>> m = sf1.match(sf2, thresh=20);
```

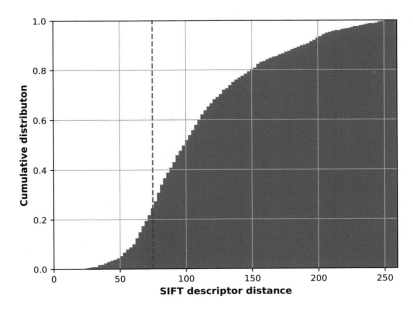

☐ **Fig. 14.4** Cumulative distribution of the distance between putative matched features

but choosing a suitable threshold value is always challenging. Alternatively, we could sort the matches by decreasing match quality and choose the 10 best matches

```
>>> m = sf1.match(sf2, sort=True)[:10];
```

For SIFT features we commonly employ two heuristics. The first is the ratio test, and for each feature in the first image we find the top two matches in the second image. If the ratio of the distances between the best and second best match exceeds a threshold we deem the match to be ambiguous and discard it. The ratio test is enabled by

```
>>> m = sf1.match(sf2, ratio=0.8)
951 matches
```

and the default threshold is 0.75. The second is consistency checking or cross checking where we match a feature in the first image to one in the second image, and then match that back to the first image. If we don't end up with the starting feature we discard the match. We can enable this check by

```
>>> m = sf1.match(sf2, crosscheck=True)
1372 matches
```

but it is off by default, since it doubles the matching time.

Feature matching is computationally expensive – it is an $O(N^2)$ problem since every feature descriptor in one image must be compared with every feature descriptor in the other image. The OpenCV algorithms used by the Toolbox employ a kd-tree so that similar descriptors – nearest neighbors in feature space – can be found cheaply.

Although the quality of matching shown in ☐ Fig. 14.3 looks quite good there is at least one obviously incorrect match in this small subset. We can discern a pattern in the lines joining the corresponding points, they are slightly converging and sloping down to the right. This pattern is a function of the relative pose between the two camera views, and understanding this is key to determining which of the putative matches are correct. That is the topic of the next section.

## 14.2 **Geometry of Multiple Views**

Assuming the world point does not move.

We start by studying the geometric relationships between images of a single point P observed from two different viewpoints and this is shown in ◘ Fig. 14.5. This geometry could represent the case of two cameras simultaneously viewing the same scene, or one moving camera taking a picture from two different viewpoints. ◄ The center of each camera, the origins of {1} and {2}, plus the world point P defines a plane in space – the epipolar plane. The world point P is projected onto the image planes of the two cameras at points $^1$p and $^2$p respectively, and these points are known as conjugate points. The intersection of the epipolar plane and a camera's image plane is an epipolar line.

Consider image one. The image-plane point $^1$p is a function of the world point P, and point $^1$e is a function of the relative pose of camera two. These two points, plus the camera center define the epipolar plane, which in turn defines the epipolar line $^2\ell$ in image two. By definition, the conjugate point $^2$p must lie on that line. Conversely $^1$p must lie along the epipolar line in image one $^1\ell$ that is defined by $^2$p in image two.

This is a very fundamental and important geometric relationship – given a point in one image we know that its conjugate is constrained to lie along a line in the other image – the epipolar constraint. We illustrate this with a simple example that mimics the geometry of ◘ Fig. 14.5

```
>>> camera1 = CentralCamera(name="camera 1", f=0.002, imagesize=1000,
...                         rho=10e-6, pose=SE3.Tx(-0.1)*SE3.Ry(0.4))
          Name: camera 1 [CentralCamera]
     pixel size: 1e-05 x 1e-05
     image size: 1000 x 1000
           pose: t = -0.1, 0, 0; rpy/yxz = 0°, 0°, 22.9°
   principal pt: [     500       500]
   focal length: [   0.002     0.002]
```

which returns an instance of the `CentralCamera` class as discussed previously in ► Sect. 13.1.2. Similarly for the second camera

```
>>> camera2 = CentralCamera(name="camera 2", f=0.002, imagesize=1000,
...                         rho=10e-6, pose=SE3.Tx(0.1)*SE3.Ry(-0.4))
          Name: camera 2 [CentralCamera]
     pixel size: 1e-05 x 1e-05
     image size: 1000 x 1000
           pose: t = 0.1, 0, 0; rpy/yxz = 0°, 0°, -22.9°
```

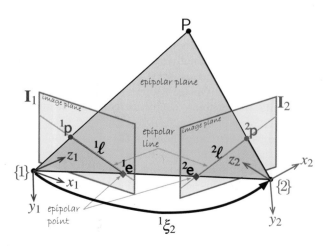

◘ **Fig. 14.5** Epipolar geometry showing the two cameras with associated coordinate frames {1} and {2} and image planes. Three black dots, the world point P and the two camera centers, define the epipolar plane. The intersection of this plane with the image-planes are the epipolar lines

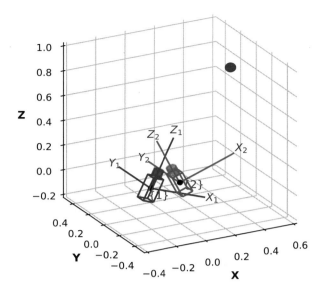

**◻ Fig. 14.6** Visualization of two cameras and a target point. The origins of the two cameras are offset along the *x*-axis and the cameras are *verged*, that is, their optical axes intersect

```
principal pt:   [       500        500]
focal length:   [     0.002      0.002]
```

and the pose of the two cameras is visualized by

```
>>> ax = plotvol3([-0.4, 0.6, -0.5, 0.5, -0.2, 1]);
>>> camera1.plot(ax=ax, scale=0.15, shape="camera", frame=True,
...              color="blue");
>>> camera2.plot(ax=ax, scale=0.15, shape="camera", frame=True,
...              color="red");
```

which is shown in ◻ Fig. 14.6. We define an arbitrary world point

```
>>> P=[0.5, 0.1, 0.8];
```

which we display as a small sphere

```
>>> plot_sphere(0.03, P, color="blue");
```

which is also shown in ◻ Fig. 14.6. We project this point to both cameras

```
>>> p1 = camera1.plot_point(P)
array([[   549.7],
       [   520.6]])
>>> p2 = camera2.plot_point(P)
array([[    734],
       [   534.4]])
```

and this is shown in ◻ Fig. 14.7. The epipoles are computed by projecting the center of each camera to the other camera's image plane

```
>>> e1 = camera1.plot_point(camera2.centre, "kd")
array([[    973],
       [    500]])
>>> e2 = camera2.plot_point(camera1.centre, "kd")
array([[  26.96],
       [    500]])
```

and these are shown in ◻ Fig. 14.7 as a black ◆-marker.

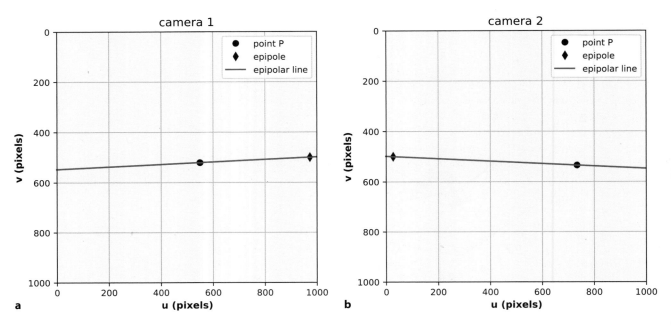

**◘ Fig. 14.7** The virtual image planes of two Toolbox `CentralCamera` objects with the perspective projection of point P, the projection of the other camera's center, and the epipolar line

### 14.2.1 The Fundamental Matrix

The epipolar relationship shown graphically in ◘ Fig. 14.5 can be expressed concisely and elegantly as

$$^2\tilde{\boldsymbol{p}}^{\top}\,\mathbf{F}\,^1\tilde{\boldsymbol{p}} = 0 \tag{14.1}$$

where $^1\tilde{\boldsymbol{p}}$ and $^2\tilde{\boldsymbol{p}}$ are homogeneous coordinate vectors describing the image points $^1\mathsf{p}$ and $^2\mathsf{p}$ respectively, and $\mathbf{F} \subset \mathbb{R}^{3\times3}$ is known as the fundamental matrix. We can rewrite this as

$$^2\tilde{\boldsymbol{p}}^{\top}\,^2\tilde{\ell} = 0 \tag{14.2}$$

where

$$^2\tilde{\ell} \simeq \mathbf{F}\,^1\tilde{\boldsymbol{p}} \tag{14.3}$$

describes a 2-dimensional line, the epipolar line, along which conjugate points in image two must lie. This line is a function of the point $^1\mathsf{p}$ in image one and (14.2) is a powerful test as to whether or not a point in image two is a possible conjugate.

Taking the transpose of both sides of (14.1) yields

$$^1\tilde{\boldsymbol{p}}^{\top}\,\mathbf{F}^{\top}\,^2\tilde{\boldsymbol{p}} = 0 \tag{14.4}$$

from which we can write the epipolar line for camera one

$$^1\tilde{\ell} \simeq \mathbf{F}^{\top}\,^2\tilde{\boldsymbol{p}} \tag{14.5}$$

in terms of a point viewed by camera two.

The fundamental matrix is a function of the camera parameters and the relative camera pose between the views

$$\mathbf{F} \simeq \mathbf{K}_2^{-\top} \left[ {}^2t_1 \right]_\times {}^2\mathbf{R}_1 \mathbf{K}_1^{-1} \tag{14.6}$$

where $\mathbf{K}_1, \mathbf{K}_2 \in \mathbb{R}^{3\times3}$ are the camera intrinsic matrices defined in (13.8) ▶. ${}^2\mathbf{R}_1 \in \mathbf{SO}(3)$ and ${}^2t_1 \in \mathbb{R}^3$ are the relative orientation and translation of camera one with respect to camera two or ${}^2\xi_1$.

If both images were captured with the same camera then $\mathbf{K}_1 = \mathbf{K}_2$.

❗ This might be the inverse of what you expect, it is camera one with respect to camera two, but the mathematics are expressed more simply this way. Toolbox functions always describe camera pose with respect to the world frame, or camera two with respect to camera one.

The fundamental matrix that relates the two views is returned by the method `F` of the `CentralCamera` class, for example

```
>>> F = camera1.F(camera2)
array([[        0, -1.947e-06, 0.0009735],
       [-1.947e-06,         0, 0.001895],
       [0.0009735, 5.248e-05,  -0.9735]])
```

and for the two image points computed earlier we can evaluate (14.1)

```
>>> e2h(p2).T @ F @ e2h(p1)
array([[      0]])
```

which is zero ▶ indicating that the points are conjugates.

The fundamental matrix has some important properties. It is singular with a rank of two

The result is a scalar, but the NumPy result is a $1 \times 1$ array which is equivalent.

```
>>> np.linalg.matrix_rank(F)
2
```

and has seven degrees of freedom. ▶ The epipoles are *encoded* in the null space of the matrix. The epipole for camera one is the right null space of $\mathbf{F}$

The matrix $\mathbf{F} \subset \mathbb{R}^{3\times3}$ has seven underlying parameters so its nine elements are not independent. The overall scale is not defined, and it has a constraint that $\det(\mathbf{F}) = 0$.

```
>>> e1h = linalg.null_space(F);
>>> e1h.T
array([[ 0.8894,    0.457, 0.0009141]])
```

in homogeneous coordinates or

```
>>> e1 = h2e(e1h)
array([[     973],
       [     500]])
```

This is the right null space of the matrix transpose, see ▶ App. B.

in Euclidean coordinates – as shown in ◘ Fig. 14.7. The epipole for camera two is the left null space ◀ of the fundamental matrix

```
>>> e2h = linalg.null_space(F.T);
>>> e2 = h2e(e2h)
array([[  26.96],
       [    500]])
```

The Toolbox can display epipolar lines using the `plot_epiline` methods of the `CentralCamera` class

```
>>> camera2.plot_epiline(F, p1, color="red")
```

which is shown in ◘ Fig. 14.7 as a red line in the camera two image plane. We see, as expected, that the projection of P lies on this epipolar line. The epipolar line for camera one is

```
>>> camera1.plot_epiline(F.T, p2, color="red");
```

### 14.2.2    The Essential Matrix

The epipolar geometric constraint can also be expressed as

$$^2\tilde{\boldsymbol{x}}^\top \mathbf{E}\ ^1\tilde{\boldsymbol{x}} = 0 \tag{14.7}$$

For a camera with a focal length of 1 and the coordinate origin at the principal point as shown in ◘ Fig. 13.3.

where $\mathbf{E} \subset \mathbb{R}^{3\times3}$ is the essential matrix and $^2\tilde{\boldsymbol{x}}$ and $^1\tilde{\boldsymbol{x}}$ are conjugate points in homogeneous normalized image coordinates. ◀ This matrix is a simple function of the relative camera pose

$$\mathbf{E} \simeq [^2\boldsymbol{t}_1]_\times\ ^2\mathbf{R}_1 \tag{14.8}$$

See ▶ App. B.2.2.

A 3-dimensional translation $(x, y, z)$ with unknown scale can be considered as $(x', y', 1)$.

where $(^2\mathbf{R}_1, {}^2\boldsymbol{t}_1)$ represent $^2\boldsymbol{\xi}_1$, the relative pose of camera one with respect to camera two. The essential matrix is singular, has a rank of two, and has two equal nonzero singular values ◀ and one of zero. The essential matrix has only 5 degrees of freedom and is completely defined by 3 rotational and 2 translational ◀ parameters. For pure rotation, when $\boldsymbol{t} = 0$, the essential matrix is not defined.

We recall from (13.8) that $\tilde{\boldsymbol{p}} \simeq \mathbf{K}\tilde{\boldsymbol{x}}$ and substituting into (14.7) we can write

$$^2\tilde{\boldsymbol{p}}^\top \underbrace{\mathbf{K}_2^{-\top}\mathbf{E}\mathbf{K}_1^{-1}}_{\mathbf{F}}\ ^1\tilde{\boldsymbol{p}} = 0 \ . \tag{14.9}$$

Equating terms with (14.1) yields a relationship between the two matrices

$$\mathbf{E} \simeq \mathbf{K}_2^\top \mathbf{F}\mathbf{K}_1 \tag{14.10}$$

If both images were captured with the same camera then $\mathbf{K}_1 = \mathbf{K}_2$.

in terms of the intrinsic parameters of the two cameras involved. ◀ This is implemented by the `E` method of the `CentralCamera` class

```
>>> E = camera1.E(F)
array([[       0, -0.07788,        0],
       [-0.07788,        0,   0.1842],
       [       0,  -0.1842,        0]])
```

where the intrinsic parameters of camera one (which is the same as camera two) are used.

Like the camera matrix in ▶ Sect. 13.2.3, the essential matrix can be decomposed to yield the relative pose $^1\xi_2$ as an **SE**(3) matrix. ▶ The inverse is not unique and the underlying OpenCV function returns four solutions

```
>>> T_1_2 = camera1.decomposeE(E);
>>> T_1_2.printline(orient="camera")
t = -0.921, 0, -0.389; rpy/yxz = 0°, 0°, -45.8°
t = 0.921, 0, 0.389; rpy/yxz = 0°, 0°, -45.8°
t = -0.921, 0, -0.389; rpy/yxz = 180°, 0°, 180°
t = 0.921, 0, 0.389; rpy/yxz = 180°, 0°, 180°
```

where the translations are given as unit vectors $\hat{t}$ and the true translation is $s\hat{t}$ where $s$ is the unknown scale. If we consider that $s \in \mathbb{R}$, that is, it is a *signed* real number, then there are really only two solutions. The first and second pair are really just one solution, differing by the sign of $s$. ▶ The true relative pose from camera one to camera two is ▶

```
>>> T_1_2_true = camera1.pose.inv() * camera2.pose;
>>> T_1_2_true.printline(orient="camera")
t = 0.184, 0, 0.0779; rpy/yxz = 0°, 0°, -45.8°
```

and the translation, as a unit vector, is

```
>>> T_1_2_true.t / np.linalg.norm(T_1_2_true.t)
array([ 0.9211,       0,    0.3894])
```

which indicates that, in this case, solution two is the correct one.

In the general case we do not know the pose of the two cameras, so how do we determine the correct solution? One approach is to place the first camera at the origin and use the estimated relative pose as the pose of the second camera. A distant point on the optical axis of the first camera

```
>>> Q = [0, 0, 10];
```

is projected to the first camera at

```
>>> camera1.project_point(Q).T
array([[   417.8,       500]])
```

which is, as expected, in the middle of the image plane. This point should also be visible to the second camera. We can check the visibility of the estimated relative poses in `pose_1_2`

```
>>> for T in T_1_2:
...     print(camera1.project_point(Q, pose=T).T)
[[   746.1       500]]
[[    670       500]]
[[    nan       nan]]
[[    nan       nan]]
```

The `nan` values indicate that the point `Q` is not visible from the last two estimated camera poses – the camera is actually facing away from the point. We can perform this more compactly by providing a test point

```
>>> T = camera1.decomposeE(E, Q);
>>> T.printline(orient="camera")
t = -0.921, 0, -0.389; rpy/yxz = 0°, 0°, -45.8°
```

in which case only the first valid solution, of those given above, is returned.

In summary these $3 \times 3$ matrices, the fundamental and the essential matrix, encode the parameters and relative pose of the two cameras. The fundamental matrix and a point in one image defines an epipolar line in the other image along which its conjugate point must lie. The essential matrix encodes the relative pose of the two camera's centers and the pose can be extracted, with four possible values, and with translation scaled by an unknown factor. In this example, the fundamental matrix was computed from known camera motion and intrinsic parameters. The real

Although (14.8) is written in terms of $^2\xi_1 \sim (^2\mathbf{R}_1, {}^2t_1)$ the Toolbox function returns $^1\xi_2$.

As observed by Hartley and Zisserman (2003, p 259), not even the sign of $t$ can be determined when decomposing an essential matrix.

We have specified a different roll-pitch-yaw rotation order $YXZ$. Given the way we have defined the camera axes, the camera orientation with respect to the world frame is a yaw about the vertical or $y$-axis, followed by a pitch about the $x$-axis followed by a roll about the optical axis or $z$-axis.

world isn't generally like this – camera motion is difficult to measure and the camera may not be calibrated. Instead we can estimate the fundamental matrix directly from corresponding image points.

### 14.2.3  Estimating the Fundamental Matrix from Real Image Data

Assume that we have $N$ pairs of corresponding points in two views of the same scene ($^1p_i$, $^2p_i$), $i = 0, \ldots, N-1$. To demonstrate this we create a set of ten random point features (within a $2 \times 2 \times 2$ m cube) whose center is located 3 m in front of the cameras

```
>>> P = np.random.uniform(low=-1, high=1, size=(3, 10))
...        + np.c_[0, 0, 3].T;
```

and project these points onto both camera image planes

```
>>> p1 = camera1.project_point(P);
>>> p2 = camera2.project_point(P);
```

The fundamental matrix can be estimated from just seven points, but in that case there will be three possible solutions.

If $N \geq 8$ the fundamental matrix can be estimated from these two sets of *corresponding* points ◄

```
>>> F, resid = CentralCamera.points2F(p1, p2)
>>> resid
1.31e-07
```

where the residual is the maximum value of the left-hand side of (14.1) and is ideally zero. The value here is not quite zero, and this is due to the accumulation of errors from finite-precision arithmetic. The estimated matrix has the required rank property

```
>>> np.linalg.matrix_rank(F)
2
```

For camera two we can plot the projected points

```
>>> camera2.plot_point(P);
```

and overlay the epipolar lines generated by each point in image one

```
>>> camera2.plot_epiline(F, p1, color="red")
```

which is shown in ◘ Fig. 14.8. We see a family or *pencil* of epipolar lines, and that every point in image two lies on an epipolar line. Note how the epipolar lines all converge on the epipole which is possible in this case ◄ because the two cameras are verged as shown in ◘ Fig. 14.6.

The example has been contrived so that the epipoles lie within the images, that is, that each camera can see the center of the other camera. A common imaging geometry is for the optical axes to be parallel, such as shown in ◘ Fig. 14.22 in which case the epipoles are at infinity (the third element of the homogeneous coordinate is zero) and all the epipolar lines are parallel.

To demonstrate the importance of correct point correspondence we will repeat the example above but introduce two *bad* data associations by swapping two elements in `p2`

```
>>> p2[:,[5, 6]] = p2[:,[6, 5]];
```

The fundamental matrix estimation

```
>>> _, resid = CentralCamera.points2F(p1, p2);
>>> resid
0.007143
```

now has a residual that is over 4 orders of magnitude larger than previously. This means that the point correspondence cannot be well *explained* by the epipolar relationship of (14.1).

If we knew the fundamental matrix, we could test whether a pair of putative corresponding points are in fact conjugates by measuring how far one is from the

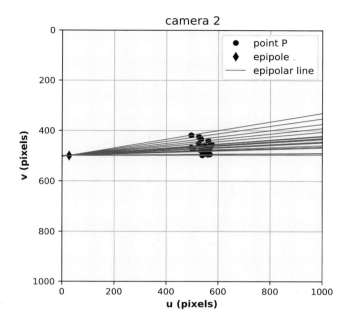

**◻ Fig. 14.8** A pencil of epipolar lines on the camera two image plane. Note how all epipolar lines pass through the epipole which is the projection of camera one's center

epipolar line defined by the other

```
>>> CentralCamera.epidist(F, p1[:, 0], p2[:,0])
array([[4.448e-05]])
>>> CentralCamera.epidist(F, p1[:, 5], p2[:,5])
array([[  34.82]])
```

which shows that point 0 is a good fit, but point 5 (which we swapped with point 6), is a poor fit. However we have to first estimate the fundamental matrix, and that requires that point correspondence is known. We break this deadlock with an ingenious algorithm called Random Sampling and Consensus or RANSAC.

The underlying principle is delightfully simple. Estimating a fundamental matrix requires eight points so we randomly choose eight putative corresponding points (the sample), from the set of ten points, and estimate **F** to create a *model*. This model is tested against all the other putative pairs and those that fit ▶ vote for this model. The process is repeated a number of times and the model that had the most supporters (the consensus) is returned. Since the sample is small the chance that it contains all valid corresponding pairs is high. The point pairs that support the model are termed inliers, while those that do not are outliers.

RANSAC is remarkably effective and efficient at finding the inlier set, even in the presence of large numbers of outliers (more than 50%), and is applicable to a wide range of problems. Within the Toolbox we can select RANSAC as an *option* when computing the fundamental matrix

To within a settable reprojection threshold `ransacReprojThreshold` in units of pixels.

```
>>> F, resid, inliers = CentralCamera.points2F(p1, p2,
...                          method="ransac", confidence=0.99, seed=0);
>>> resid
1.54e-07
```

and we obtain an excellent final residual. The set of inliers is also returned

```
>>> inliers
array([True, True, True, True, True, False, False, True, True, True])
```

and the two incorrect associations, points 5 and 6, are flagged as False – they are outliers. The fourth parameter is the desirable level of confidence that the estimated matrix is correct. If this parameter is chosen to be too close to one, then RANSAC

---

### Excurse 14.3: Robust Estimation with RANSAC

We create a set of ten discrete points on the line $y = 3x - 10$

```
>>> x = np.arange(11);
>>> y = 3 * x - 10;
```

and add a random number to four randomly chosen points

```
>>> nbad = 4;
>>> np.random.seed(1)  # set random seed
>>> bad = np.random.choice(len(x), nbad,
...                         replace=False)
array([2,  3,  4,  9])
>>> y[bad] = y[bad] + np.random.rand(nbad)
... * 10
```

and fitting a line using standard least squares

```
>>> from scipy import stats
>>> m, c, *_ = stats.linregress(x, y)
>>> plt.plot(x, m * x + c, 'r--');
```

results in the red-dashed line which is clearly biased away from the true line by the outlier data points.

In clear contrast, the RANSAC algorithm (see examples/ransac_line)

```
>>> params, inliers = ransac_line(x, y)
>>> params
```

```
(3.0, -10.0)
>>> inliers
[1, 7, 0, 5, 6, 8, 10]
```

has found the true line parameters and returned a list of the "good" data points, despite 40% of the points not fitting the model. RANSAC has found the parameters of the consensus line, the line that the largest number of points agree on.

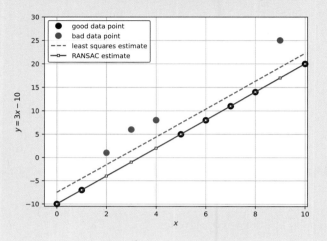

---

will perform more trials at the expense of computation time. That may in turn lead to the maximum iteration count being exceeded, requiring that parameter to be increased. Keep in mind also that the results of RANSAC will vary from run to run due to the random subsampling performed, but in this example we have set the random number seed. Using RANSAC may require some trial and error to choose good parameters for a specific problem.

We return now to the pair of images of the Eiffel tower shown in ■ Fig. 14.3. At the end of ▶ Sect. 14.1 we had found *putative* correspondences based on descriptor similarity but there were a number of clearly incorrect matches. If we apply RANSAC while computing the fundamental matrix then incorrect correspondences will be detected and only point pairs that meet the epipolar constraint of the fundamental matrix will be included in the inlier set

```
>>> F, resid, inliers = CentralCamera.points2F(matches.p1,
...                            matches.p2, method="ransac",
...                            confidence=0.99);
>>> resid
0.0324
>>> sum(inliers) / len(inliers)
0.8244
```

which shows that just over 80% of the points were inliers. Lense distortion has been ignored in this example which means that the epipolar relationship (14.1) cannot be exactly satisfied. ◄

In this case RANSAC has identified

```
>>> sum(~inliers)
151
```

Lens distortion causes points to be displaced on the image plane and this violates the epipolar geometry. Images can be corrected by warping as discussed in ▶ Sect. 11.7.4.

outliers, or incorrect data associations, from the SIFT feature matching stage – the putative matching was worse than it looked.

An alternative way to express this is to have the `FeatureMatch` object perform this operation

```
>>> F, resid = matches.estimate(CentralCamera.points2F,
...                             method="ransac", confidence=0.99,
...                             seed=0);
```

and invoke the function given by the first argument. The `FeatureMatch` object retains the returned information about the inliers, and can now display the status of each match

```
>>> matches
860 matches, with 709 (82.4%) inliers
>>> matches[:10].list()
  0:  +   24.00 (1118.6, 178.7) <--> (952.5, 417.9)
  1:  +   24.12 (900.2, 636.6) <--> (775.6, 802.3)
  2:  +   25.81 (760.3, 125.0) <--> (656.2, 369.1)
  3:  +   27.33 (820.5, 519.0) <--> (708.0, 701.6)
  4:  +   28.12 (801.1, 632.4) <--> (694.1, 800.3)
  5:  +   29.33 (1094.0, 184.7) <--> (932.9, 423.0)
  6:  +   31.06 (781.0, 214.4) <--> (672.8, 443.8)
  7:  +   33.60 (1094.0, 184.7) <--> (932.9, 423.0)
  8:  +   33.70 (526.5, 484.2) <--> (462.4, 673.6)
  9:  +   33.79 (759.7, 332.0) <--> (655.9, 543.0)
```

with a plus or minus sign to indicate where the match is an inlier or outlier respectively. We can plot some of the inliers

```
>>> matches.inliers.subset(100).plot(color="g");
```

or some of the outliers

```
>>> matches.outliers.subset(100).plot(color="red")
```

and these are shown in ◨ Fig. 14.9.

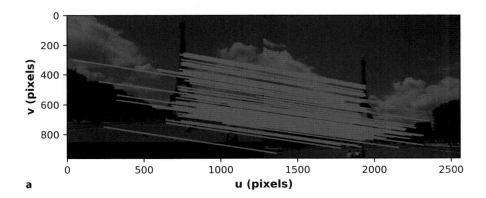

a

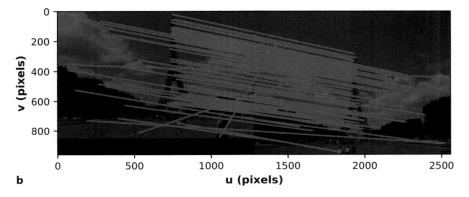

b

◨ **Fig. 14.9** Results of SIFT feature matching after RANSAC. **a** Subset of all inlier matches; **b** subset of the outlier matches, some are visibly incorrect while others are more subtly wrong

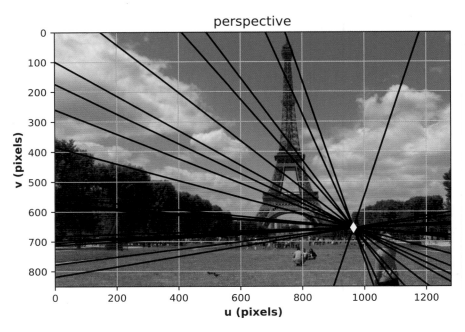

□ **Fig. 14.10**   Image from □ Fig. 14.1a showing epipolar lines converging on the projection of the second camera's center. In this case, the second camera is visible in the bottom right of the image

To overlay the epipolar lines on an image it is convenient to create a `CentralCamera` object and display the image in its virtual image plane

```
>>> camera = CentralCamera();
>>> camera.disp(view1);
```

Now we can overlay the epipolar lines, computed from the corresponding points found in the second image

```
>>> camera.plot_epiline(F.T, matches.inliers.subset(20).p2,
...                     color="black");
```

and the result is shown in □ Fig. 14.10. The epipolar lines intersect at the epipolar point which we can clearly see is the projection of the second camera in the first image. ◀ The epipole at

We only plot a small subset of the epipolar lines since they are too numerous and would obscure the image.

```
>>> epipole = h2e(linalg.null_space(F))
array([[  964.3],
       [  654.4]])
>>> camera.plot_point(epipole, "wd");
```

is also superimposed on the plot as a white diamond. With two handheld cameras and an overlapping field of view, we have been able to pinpoint the second camera in the first image. The result is not quite perfect – there is a horizontal offset of about 80 pixels which is likely to be due to a small orientation error in one or both cameras which were handheld and only approximately synchronized.

### 14.2.4 **Planar Homography**

In ▶ Sect. 13.6.2 we considered a single camera viewing world points that lie on a plane. In this section we will consider two different cameras viewing a set of world points $P_i$ that lie on a plane. The camera image plane projections ${}^1p_i$ and ${}^2p_i$ are

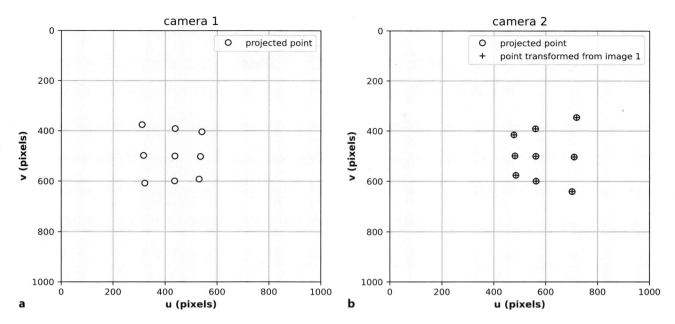

■ **Fig. 14.11** Views of the oblique planar grid of points from two different view points. The grid points are projected as open circles, while the plus signs in **b** indicate points transformed from the camera one image plane by the homography

related by

$$^2\tilde{p}_i \simeq \mathbf{H}\,^1\tilde{p}_i \qquad\qquad (14.11)$$

where $\mathbf{H} \subset \mathbb{R}^{3\times3}$ is a nonsingular matrix known as a homography, a planar homography, or the homography *induced* by the plane. ▶

For example consider again the pair of cameras from ■ Fig. 14.6 now observing a $3 \times 3$ grid of points

```
>>> T_grid = SE3.Tz(1) * SE3.Rx(0.1) * SE3.Ry(0.2);
>>> P = mkgrid(3, 1.0, pose=T_grid);
```

where `T_grid` is the pose of the grid coordinate frame {G} and the grid points are centered in the frame's $xy$-plane. The points are projected to both cameras

```
>>> p1 = camera1.plot_point(P, "o");
>>> p2 = camera2.plot_point(P, "o");
```

and the projections are shown in ■ Fig. 14.11a, b.

Just as we did for the fundamental matrix, we can estimate the matrix H from two sets of corresponding points

```
>>> H, resid = CentralCamera.points2H(p1, p2)
>>> H
array([[ -0.4187, -0.0003935,    397.8],
       [ -0.6981,    0.3738,    309.5],
       [-0.001396, -1.459e-05,        1]])
```

which requires $N \geq 4$ corresponding point pairs.

According to (14.11) we can predict the position of the grid points in image two from the corresponding image one coordinates

```
>>> p2b = homtrans(H, p1);
```

An homography matrix has arbitrary scale and therefore 8 degrees of freedom. With respect to (14.13) the rotation, translation and normal have 3, 3 and 2 degrees of freedom respectively. Homographies form a group: the product of two homographies is another homography, the identity homography is a unit matrix and an inverse operation is the matrix inverse.

which we can can superimpose on image two as +-symbols

```
>>> camera2.plot_point(p2b, "+");
```

as shown in ◨ Fig. 14.11b. We see that the predicted points are perfectly aligned with the actual projection of the world points. The inverse of the homography matrix

$$^{1}\tilde{\boldsymbol{p}}_{i} \simeq \mathbf{H}^{-1}\ ^{2}\tilde{\boldsymbol{p}}_{i} \tag{14.12}$$

performs the inverse mapping, from image two coordinates to image one

```
>>> p1b = homtrans(np.linalg.inv(H), p1);
```

The fundamental matrix constrains the conjugate point to lie along a line but the homography tells us *exactly* where the conjugate point will be in the other image – provided that the points lie on a plane.

We can use this proviso to our advantage as a test for whether or not points lie on a plane. We will add some extra world points ◄ to our example

These points lie along the ray from the camera one center to an extra row of points in the grid plane. However their $z$-coordinates have been chosen to be 0.4, 0.5 and 0.6 m respectively.

```
>>> Q = np.array([
...    [-0.2302,    -0.0545,    0.2537],
...    [ 0.3287,    0.4523,    0.6024],
...    [ 0.4000,    0.5000,    0.6000] ]);
```

which we plot in 3D

```
>>> plotvol3([-1, 1, -1, 1, 0, 2]);
>>> plot_sphere(0.05, P, color="blue");
>>> plot_sphere(0.05, Q, color="red");
>>> camera1.plot(color="blue", frame=True);
>>> camera2.plot(color="red", frame=True);
```

and this is shown in ◨ Fig. 14.12. The new points, shown in red, are clearly not in the same plane as the original blue points. Viewed from camera one

```
>>> p1 = camera1.plot_point(np.hstack((P, Q)), "o");
```

as shown in ◨ Fig. 14.13a, these new points *appear* as an extra row in the grid of points we used above. However in the second view

```
>>> p2 = camera2.plot_point(np.hstack((P, Q)), "o");
```

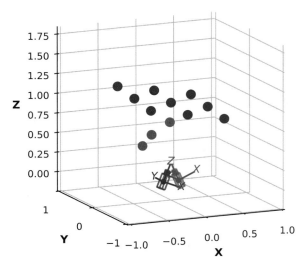

◨ **Fig. 14.12** World view of target points and two camera poses. Blue points lie in a planar grid, while the red points appear to lie in the grid from the viewpoint of camera one

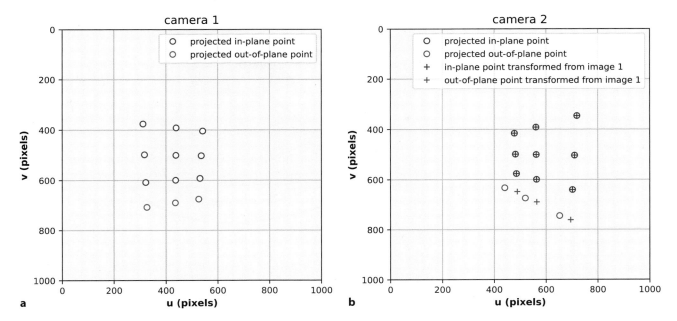

The points observed from two different view points. The grid points are projected as open circles. Plus signs in **b** indicate points transformed from the camera one image plane by the homography. Blue indicates points on the planar grid and red indicates the non-coplanar points

as shown in □ Fig. 14.13b these *out of plane* points no longer form a regular grid. If we apply the homography to the camera one image points

```
>>> p2h = homtrans(H, p1);
```

we find where they should be in the camera two image, *if* they belonged to the plane implicit in the homography

```
>>> camera2.plot_point(p2h, "+");
```

The original nine points overlap, but the three new points do not. We could use this as the basis for an automated test based on the prediction error. In this case

```
>>> np.linalg.norm(homtrans(H, p1) - p2, axis=0)
array([1.262e-05, 9.673e-06, 6.544e-07, 8.302e-06, 4.34e-06,
       1.189e-05, 1.679e-05, 5.884e-06, 3.927e-05,
          50.6,     46.44,     45.38])
```

it is clear that the last three points do not belong to the plane that induced the homography.

In this example we estimated the homography based on two sets of corresponding points which were projections of known planar points. In practice we do not know in advance which points belong to the plane and RANSAC comes to our aid again

```
>>> H, resid, inliers = CentralCamera.points2H(p1, p2,
...                                 method="ransac");
>>> resid
4.838e-05
>>> inliers
array([ True,  True,  True,  True,  True,  True,  True,  True,  True,
       False, False, False])
```

which finds the homography that best explains the relationship between the sets of image points. It has also identified those points which support the homography, and the last three points do not – they are outliers.

The geometry related to the homography is shown in □ Fig. 14.14. We can express the homography in normalized image coordinates ▶

See ▶ Sect. 13.1.2.

$$^2\tilde{x} \simeq \mathbf{H}_E \, ^1\tilde{x}$$

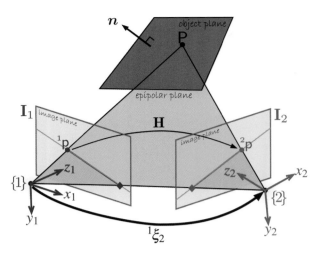

**◘ Fig. 14.14** Geometry of homography showing two cameras with associated coordinate frames {1} and {2} and image planes. The world point P belongs to a plane with surface normal $\boldsymbol{n}$. **H** is the homography, a $3 \times 3$ matrix that maps $^1\mathsf{p}$ to $^2\mathsf{p}$

where $\mathbf{H}_E$ is the Euclidean homography which is written

$$\mathbf{H}_E \simeq {}^2\mathbf{R}_1 + \frac{{}^2\boldsymbol{t}_1}{d}\boldsymbol{n}^\top \tag{14.13}$$

in terms of the relative camera pose $^2\boldsymbol{\xi}_1$ and the plane $\boldsymbol{n}^\top \boldsymbol{P} + d = 0$ with respect to frame {1}. The Euclidean and projective homographies are related by

$$\mathbf{H}_E \simeq \mathbf{K}^{-1}\mathbf{H}\mathbf{K}$$

where $\mathbf{K}$ is the camera intrinsic parameter matrix.

Just as we did for the essential matrix, we can decompose a projective homography to yield the relative pose $^1\boldsymbol{\xi}_2$ in homogeneous transformation form ◄ as well as the normal to the plane

Although (14.13) is written in terms of $(^2\mathbf{R}_1, {}^2\boldsymbol{t}_1)$ the Toolbox function returns the inverse which is $^1\boldsymbol{\xi}_2$.

```
>>> T, normals = camera1.decomposeH(H);
>>> T.printline(orient="camera")
t = -0.185, 0, -0.0783; rpy/yxz = -8.39e-06°, 4.61e-06°, -45.8°
t = 0.185, 0, 0.0783; rpy/yxz = -8.39e-06°, 4.61e-06°, -45.8°
t = 0.0197, 0.0192, -0.199; rpy/yxz = 1.09°, 0.338°, -34.4°
t = -0.0197, -0.0192, 0.199; rpy/yxz = 1.09°, 0.338°, -34.4°
```

Again there are multiple solutions, and we need to apply additional information to determine the correct one. As usual, the translational component of the transformation matrix has an unknown scale factor. We know from ◘ Fig. 14.12 that the camera motion is predominantly in the $x$-direction and that the plane normal is approximately parallel to the camera's optical- or $z$-axis and this knowledge leads us to choose the second solution. The true relative pose from camera one to two is

```
>>> (camera1.pose.inv() * camera2.pose).printline(orient="camera")
t = 0.184, 0, 0.0779; rpy/yxz = 0°, 0°, -45.8°
```

and supports our choice. The pose of the grid with respect to camera one is

```
>>> camera1.pose.inv() * T_grid
   0.9797   -0.03888   -0.1968   -0.2973
```

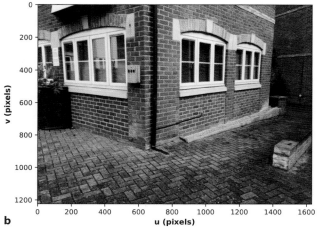

**a**        **u (pixels)**        **b**        **u (pixels)**

■ **Fig. 14.15** Two views of a courtyard taken from different viewpoints which we refer to as left and right. Image **b** was taken approximately 30 cm to the right of image **a**. Image **a** has superimposed dots that are features that fit the first three dominant planes. The camera was handheld

```
0.01983    0.995     -0.09784    0
0.1996     0.09195    0.9756      0.96
0          0          0           1
```

and the third column of the rotation submatrix is the grid's normal ▶ which matches the estimated normal associated with solution two

```
>>> normals[1].T
array([[ -0.1968, -0.09784,  0.9756]])
```

We can just as easily apply this technique to a pair of real images, a left and right view of a courtyard and walls

```
>>> walls_l = Image.Read("walls-l.png", reduce=2);
>>> walls_r = Image.Read("walls-r.png", reduce=2);
```

which are shown in ■ Fig. 14.15. The images have been downsized by a factor of 2 in each dimension to reduce computation time later. We start by finding the SIFT features

```
>>> sf_l = walls_l.SIFT();
>>> sf_r = walls_r.SIFT();
```

and the 1000 best putative correspondences are

```
>>> matches = sf_l.match(sf_r);
```

We use RANSAC to find the subset of correspondences points that best fits a plane in the world

```
>>> H, resid = matches.estimate(CentralCamera.points2H,
...                    confidence=0.9, seed=0)
>>> matches
2796 matches, with 1053 (37.7%) inliers
```

and the confidence was reduced to allow for lens distortion and the planes being not perfectly smooth. In this case the majority of point pairs do not fit the model, that is they do not belong to the plane that induces the homography **H**. The good news is that 1053 points *do* belong to some plane, and we can superimpose those points as red dots on the image

```
>>> walls_l.disp();
>>> plot_point(matches.inliers.p1, "r.");
```

as shown in ■ Fig. 14.15a. The RANSAC consensus plane is the right-hand wall. If we remove the matches corresponding to this plane, that is take the outlier points

```
>>> not_plane = matches.outliers;
```

Since the points are in the $xy$-plane of the grid frame {G} the normal is the $z$-axis.

and repeat the RANSAC homography estimation step with `not_plane` we will find the next most dominant plane, the green points in ◘ Fig. 14.15a, which is the left-hand wall. We can repeat this process until we have found all the planes in the scene, terminating when the residual exceeds some threshold which indicates the remaining points do not fit a plane. Planes are very common in human-made environments and we will revisit homographies and their decomposition in ▶ Sect. 14.8.1.

## 14.3 Sparse Stereo

Stereo vision is a technique for estimating the 3-dimensional structure of the world from two images taken from different viewpoints, as for example shown in ◘ Fig. 14.15. Our eyes are separated by 50–80 mm and the difference between these two viewpoints is an important, but not the only, part of how we sense distance. This section introduces sparse stereo, a natural extension of what we have learned about feature matching, which recovers the world coordinate $(X, Y, Z)$ for each corresponding point pair. Dense stereo, covered in ▶ Sect. 14.4 attempts to recover the world coordinate $(X, Y, Z)$ for *every pixel* in the image.

### 14.3.1 3D Triangulation

To illustrate sparse stereo we will return to the pair of images shown in ◘ Fig. 14.15. We have already found the SIFT features, so we will find putative correspondences, and then estimate the fundamental matrix

```
>>> matches = sf_l.match(sf_r)
2796 matches
>>> F, resid = matches.estimate(CentralCamera.points2F,
...                             confidence=0.99, seed=0);
```

which captures the relative geometry of the two views, and we retain only the inlier set

```
>>> matches = matches.inliers  # keep only the inliers
2459 matches, with 2459 (100.0%) inliers
```

We can display the epipolar lines for a subset of right-hand image points overlaid on the left-hand image

```
>>> camera = CentralCamera();
>>> camera.disp(walls_l);
>>> camera.plot_epiline(F.T, matches.subset(40).p2, "yellow");
```

which is shown in ◘ Fig. 14.16. In this case the epipolar lines are approximately horizontal and parallel, which is expected for camera motion that was a translation in the camera's $x$-direction. ◘ Fig. 14.17 shows the epipolar geometry for stereo vision. It is clear that as the point moves away from the camera, from P to P′, the conjugate point in the right-hand image moves to the right along the epipolar line.

The origin of {1} and the image plane point $^1$p defines a ray in space, as does the origin of {2} and $^2$p. These two rays intersect at the world point P – the process of triangulation – but to determine these rays we need to know the pose and intrinsic parameters of each camera. It is convenient to consider that the camera one frame {1} is the world frame {0}, and only the pose of camera two $\xi_2$ is unknown. However, we can estimate the relative pose $^1\xi_2$ by decomposing the essential matrix computed between the two views. We already have the fundamental matrix, but to determine the essential matrix according to (14.10) we need the camera's intrinsic parameters. With a little sleuthing we can find them!

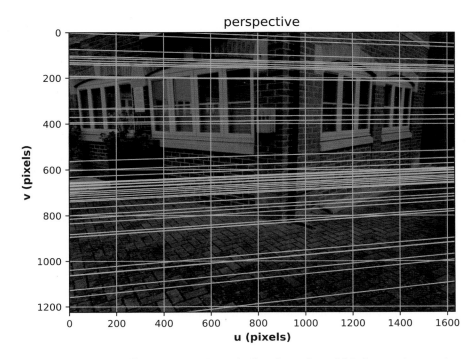

**Fig. 14.16** Image of **Fig. 14.15a with epipolar lines for a subset of right image points superimposed

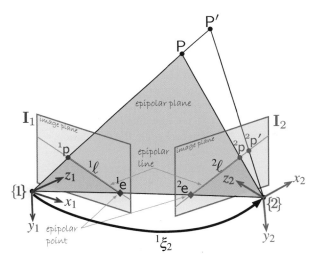

**Fig. 14.17** Epipolar geometry for stereo vision. We can see clearly that as the depth of the world point increases, from P to P′, the projection moves rightward along the epipolar line in the second image plane

The camera focal length is stored in the metadata of the image, as we discussed in ▶ Sect. 11.1.1, and its value is

```
>>> f = walls_1.metadata("FocalLength")
4.15
```

in units of mm. The dimensions of the pixels $\rho_w \times \rho_h$ are not included in the image metadata but some web-based research on this model of camera

```
>>> name = walls_1.metadata("Model")
'iPhone 5s'
```

We have doubled the pixel dimensions to account for halving the image resolution when we loaded the images. A low-resolution image effectively has larger photosites.

suggests that the camera has an image sensor with $1.5\,\mu m$ pixels. We create a `CentralCamera` object based on the known focal length, pixel size and image dimension ◄

```
>>> camera = CentralCamera(name=name, imagesize=walls_1.shape,
...                         f=f/1000, rho=2*1.5e-6)
            Name: iPhone 5s [CentralCamera]
       pixel size: 3e-06 x 3e-06
       image size: 1224 x 1632
             pose: t = 0, 0, 0; rpy/yxz = 0°, 0°, 0°
    principal pt: [     612        816]
    focal length: [ 0.00415   0.00415]
```

In the absence of any other information, we make the reasonable assumption that the principal point is at the center of the image.

The essential matrix is obtained by applying the camera intrinsic parameters to the fundamental matrix

```
>>> E = camera.E(F)
array([[-0.09511,    -8.169,    -2.394],
       [   7.763,     1.671,    -41.98],
       [   2.583,      42.8,      1.11]])
```

which we decompose to determine the camera motion

```
>>> T_1_2 = camera.decomposeE(E, [0, 0, 10]);
>>> T_1_2.printline(orient="camera")
t = -0.982, 0.0639, -0.179; rpy/yxz = -0.39°, 1.92°, 0.474°
```

We chose a test point P at $(0, 0, 10)$, a distant point along the optical axis, to determine the correct solution for the relative camera motion. The camera orientation was kept as constant as possible between the two views, so the roll-pitch-yaw angles should be close to zero – the estimated rotation is less than two degrees of rotation about any axis.

The estimated translation $t$ from {1} to {2} has an unknown scale factor. Once again we bring in an extra piece of information – when we took the images the camera position changed by approximately $0.3\,m$ in the positive $x$-direction. The estimated translation has the correct direction, dominant $x$-axis motion, but the magnitude is quite wrong. We therefore scale the translation

```
>>> t = T_1_2.t;
>>> s = 0.3 / t[0]   # estimate of translation scale factor
-0.3056
>>> T_1_2.t = s * t  # scaled translation
>>> T_1_2.printline(orient="camera")
t = 0.3, -0.0195, 0.0547; rpy/yxz = -0.39°, 1.92°, 0.474°
```

and we have an estimate of $^{1}\xi_{2}$ – the relative pose of camera two with respect to camera one represented as an **SE**(3) matrix.

Each image-plane point corresponds to a ray in space, see ► Exc. 13.7 and sometimes called a raxel, which we represent by a 3-dimensional line. In the first corresponding point pair, the ray from camera one is

```
>>> ray1 = camera.ray(matches[0].p1)
{ 0 0 0; -0.027349 -0.4743 1}
```

which is an instance of a `Line3` object which is represented using Plücker coordinates. The corresponding ray from the second camera is

```
>>> ray2 = camera.ray(matches[0].p2, pose=T_1_2)
{ -0.0074528 0.30579 0.15004; -0.18397 -0.48815 0.98576}
```

where we have specified the relative pose of camera two that we just estimated. The two rays intersect at ◄

The `closest_to_line` method returns the point on `ray1` that is closest to `ray2`.

```
>>> P, e = ray1.closest_to_line(ray2);
>>> P
array([-0.05306,   -0.9203,      1.94])
```

which is a point with a $z$-coordinate, or depth, of almost 2 m. Due to errors in the estimate of camera two's pose the two rays do not actually intersect, but their closest point is returned. At that closest point, the lines are

```
>>> e
0.02978
```

nearly 30 mm apart. Considering the lack of rigor in this exercise, two handheld camera shots and only approximate knowledge of the magnitude of the camera displacement, the recovered depth information is quite remarkable. ▶

We can do this for all matched points, creating a set of 3-dimensional lines in world space for each camera

```
>>> ray1 = camera.ray(matches.p1);
>>> ray2 = camera.ray(matches.p2, pose=T_1_2);
```

where the return values are `Line3` objects that each containing multiple lines

```
>>> len(ray1)
2459
```

Then we find the closest intersection point for every line in `r1` with the corresponding line in `r2`

```
>>> P, e = ray1.closest_to_line(ray2);
>>> P.shape
(3, 2459)
```

where `P` is an array of closest points, one per column, and the last row

```
>>> z = P[2, :];
>>> z.mean()
2.116
```

is the depth coordinate. The elements of the array `e` contains the distance between the lines at their closest points and some simple statistics

```
>>> np.median(e)
0.02008
>>> e.max()
0.1383
```

show that the median error is of the order of centimeters, very small compared to the scale of the scene. However there are some significant outliers as indicated by the maximum error value.

We can plot these points using Matplotlib

```
>>> plotvol3();
>>> plt.plot(P[0,:], P[1,:], P[2,:], '.', markersize=2);
```

but the result is rather crude. A much better alternative is to create a `PointCloud` object ▶

```
>>> walls_pcd = PointCloud(P)
PointCloud with 2459 points.
>>> walls_pcd.transform(SE3.Rx(pi));  # make y-axis upward
```

which we can display

```
>>> walls_pcd.disp()
```

and the result is shown in ◻ Fig. 14.18a. The point cloud display is interactive, allowing control of viewpoint and zoom. Type the h-key to print a list of the supported mouse and keyboard shortcuts.

There are a few obvious outlier points which are characterized by having few neighbors. We can reject all points that have fewer than 10 points within a sphere

Even small errors in the estimated orientation between the camera poses will lead to large closing errors at distances of several meters. The closing error observed here would be induced by a rotational error of less than 1°.

This requires that you have Open3D installed, see ▶ App. A.

▶ go.sn.pub/3XlqAO

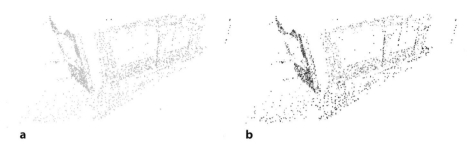

**a**                                    **b**

**◻ Fig. 14.18** Point clouds reconstructed from the image pair of ◻ Fig. 14.15. **a** Point cloud where color indicates depth (warmer colored points are nearer), note some incorrect distant (dark blue) points; **b** Colored point cloud

of radius 20 cm

```
>>> walls_pcd = walls_pcd.remove_outlier(nb_points=10, radius=0.2)
PointCloud with 2341 points.
```

which is a copy of the original point cloud but with over 100 points removed.

For extra realism we can set the color of the points from the original left-hand RGB image

```
>>> colors = []
>>> for m in matches:
...     colors.append(walls_1.image[int(m.p1[1]), int(m.p1[0]), :])
>>> pcd = SE3.Rx(pi) * PointCloud(P, colors=np.array(colors).T)
>>> pcd.disp()
```

Left multiplying a `PointCloud` object by an `SE3` object transforms the points according to $\boldsymbol{\xi} \cdot \boldsymbol{P}_i$.

creating what is known as a colored point cloud which is shown in ◻ Fig. 14.18b. ◀ We will discuss point clouds in more detail in ▶ Sect. 14.7.

This is an example of stereopsis where we have used information from two overlapping images to infer the 3-dimensional position of points in the world. For obvious reasons, the approach used here is referred to as sparse stereo because we only compute distance at a tiny subset of pixels in the image. More commonly the relative pose between the cameras would be known, as would the camera intrinsic parameters.

### 14.3.2    Bundle Adjustment (advanced topic)

In the previous section we used triangulation to estimate the 3D coordinates of a sparse set of landmark points in the world, but this was an approximation based on a guesstimate of the relative pose between the cameras. To assess the quality of our solution we can *reproject* the estimated 3D landmark points onto the image planes based on the estimated camera poses and the known camera model. The reprojection error is the image-plane distance between the back-projected landmark and its observed position on the image plane.

Continuing the example above, we can reproject the first triangulated point to each of the cameras

```
>>> p1_reproj = camera.project_point(P[:, 0]);
>>> p2_reproj = camera.project_point(P[:, 0], pose=T_1_2);
```

and the resulting image plane errors are

```
>>> (p1_reproj - matches[0].p1).T
array([[       0,        0]])
>>> (p2_reproj - matches[0].p2).T
array([[  -1.154,    23.49]])
```

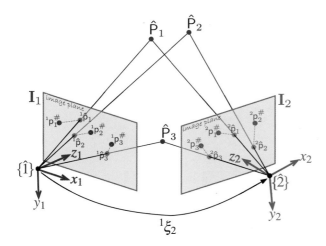

☐ **Fig. 14.19** Bundle adjustment notation illustrated with a simple problem comprising only two cameras and three landmark points $P_i$ in the world. The estimated camera poses and point positions are indicated, as are the estimated and measured image-plane coordinates. The reprojection errors are shown as dashed-gray lines. The problem is solved when the variables are adjusted so that the total reprojection error is as small as possible

The point in the first camera has zero error, since the point P lies on the ray defined by that image-plane point, but for camera two there is an error of over 20 pixels. This implies there is an error in the estimate of the world point, the estimate of camera two's pose, or both. To reduce this error we can use a technique called bundle adjustment.

Bundle adjustment is an optimization process that simultaneously adjusts the camera poses and the landmark coordinates so as to minimize the total back-projection error. It uses 2D measurements from a set of images of the same scene to recover information related to the 3D geometry of the imaged scene, as well as the poses of the cameras. This is also called Structure from Motion (SfM) or Structure and Motion Estimation (SaM) – *structure* being the 3D landmarks in the world and *motion* being a sequence of camera poses. It is also called visual SLAM (VSLAM) since it is very similar to the pose-graph SLAM problem discussed in ▶ Sect. 6.5. That was a planar problem solved in the three dimensions of **SE**(2) whereas bundle adjustment involves camera poses in **SE**(3) and points in $\mathbb{R}^3$.

To formalize the problem, consider a camera with known intrinsic parameters at $N$ different poses $\boldsymbol{\xi}_i \in \textbf{SE}(3), i = 0, \ldots, N-1$ and a set of $M$ landmark points $P_j$ represented by coordinate vectors $\boldsymbol{P}_j \in \mathbb{R}^3, j = 0, \ldots, M-1$. The camera with pose $\boldsymbol{\xi}_i$ observes $P_j$, and the image-plane projection $p_j$ is represented by the coordinate vector $^i\boldsymbol{p}_j^{\#} \in \mathbb{R}^2$. The notation is shown in ☐ Fig. 14.19 for two cameras.

In general only a subset of landmarks is visible from any camera, and this visibility information can be represented elegantly using a graph as shown in ☐ Fig. 14.20, where each camera pose $\boldsymbol{\xi}_i$ and each landmark coordinate is a vertex. ▶ Edges between camera and landmark vertices represent observations, and the value of the edge is the observed image-plane coordinate. The estimated value of the image-plane projection of landmark $j$ on the image plane of camera $i$ is

The visibility information can also be represented by a visibility matrix, an $N \times M$ array where element $(i, j)$ is 1 if landmark $j$ is visible from camera $i$, otherwise 0.

$$^i\hat{\boldsymbol{p}}_j = \mathcal{P}\left(\hat{\boldsymbol{\xi}}_i, \ \hat{\boldsymbol{P}}_j; \ \mathbf{K}\right)$$

and the reprojection error – the difference between the estimated and observed projection – is $^i\hat{\boldsymbol{p}}_j - {}^i\boldsymbol{p}_j^{\#}$.

Using the Toolbox we start by creating a `BundleAdjust` object

```
>>> bundle = BundleAdjust(camera)
```

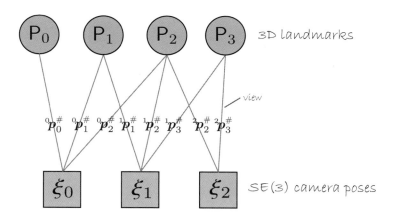

**Fig. 14.20** A visibility graph showing camera vertices (red) and landmark vertices (blue). Lines connecting vertices represent a view of that landmark from that camera, and the edge value is the observed image-plane coordinate. Landmark $P_0$ is viewed by only one camera, $P_1$ and $P_3$ are viewed by two cameras, and $P_2$ is viewed by three cameras

In this case we only estimated the relative pose $^1\xi_2$ but we can consider the first camera pose as the reference coordinate frame $\xi_1 = 0$ and $\xi_2 = {}^1\xi_2$.

which contains a graph of camera poses and landmarks. The argument is a camera model with known intrinsic parameters. Next we add the camera pose vertices to the graph ◄

```
>>> view0 = bundle.add_view(SE3(), fixed=True);
>>> view1 = bundle.add_view(SE3.Tx(0.3));
```

and indicate that the first camera pose is known and that we do not need to optimize for it. The second camera's pose was estimated in ► Sect. 14.3.1 from the essential matrix, but for this example we will use a cruder approximation to illustrate the power of bundle adjustment. The `add_view` method returns a `ViewPoint` instance that describes a particular camera pose vertex within the bundle adjustment graph, and which we will use below.

Next, for a subset of the landmarks estimated previously using triangulation and the feature match objects

```
>>> for (Pj, mj) in zip(P[:, ::4].T, matches[::4]):
...     landmark = bundle.add_landmark(Pj)           # add vertex
...     bundle.add_projection(view0, landmark, mj.p1)  # add edge
...     bundle.add_projection(view1, landmark, mj.p2)  # add edge
```

we create a landmark vertex `landmark` which is a `LandMark` instance, and then add the measurements by specifying the camera vertex, the landmark vertex and its projection on the image plane. The problem is now fully defined and a summary can be displayed

```
>>> bundle
Bundle adjustment problem:  2 views
   1 locked views: [0]
  615 landmarks
  1230 projections
  1857 total states
  1851 variable states
  2460 equations
  landmarks per view: min=615, max=615, avg=615.0
  views per landmark: min=2, max=2, avg=2.0
```

The graph contained in the `BundleAdjust` object can be plotted by

```
>>> bundle.plot()
```

and an example is shown in ◻ Fig. 14.21.

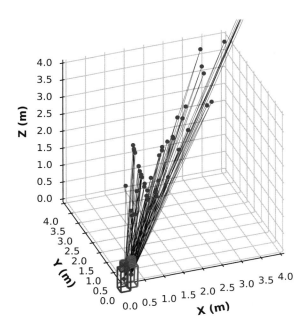

■ **Fig. 14.21** Bundle adjustment problem, for a subset of points, shown as an embedded graph. Dots represent landmark positions, camera icons represent camera pose, and lines denote observations. Camera 1 is blue and camera 2 is red

To solve this optimization problem we put all the variables we wish to adjust into a single state vector that contains camera poses and landmark coordinates

$$x = \{\xi_0, \xi_1, \ldots, \xi_{N-1} | P_0, P_1, \ldots, P_{M-1}\} \in \mathbb{R}^{6N+3M}$$

where the **SE**(3) camera pose is represented in a vector format $\xi_i \sim (t, r) \in \mathbb{R}^6$ comprising translation $t \in \mathbb{R}^3$ and rotation $r \in \mathbb{R}^3$; and $P_j \in \mathbb{R}^3$. Possible representations of rotation include Euler angles, roll-pitch-yaw angles, angle-axis or exponential coordinate representations. For bundle adjustment it is common to use the vector component of a unit quaternion which is singularity free and has only three parameters – we will use this representation here. ▶

The number of unknowns in this system is $6N + 3M$: 6 unknowns for each camera pose and 3 unknowns for the position of each landmark point. However we have up to $2NM$ equations due to the measured $(u, v)$ projections of each world point on the image planes. Typically, the pose of one camera is assumed to be the reference coordinate frame, and this reduces the number of unknowns to $6(N - 1) + 3M$.

In the problem we are discussing $N = 2$, but one camera is locked, and $M = 615$ so we have $6 \times (2 - 1) + 3 \times 615 = 1857$ unknowns and $2 \times 2 \times 615 = 2460$ equations – an overdetermined set of equations for which a solution should be possible. For our problem, the state vector

```
>>> x = bundle.getstate();
>>> x.shape
(1857,)
```

includes the pose of the fixed camera, although that will remain constant. The pose of camera two is stored in the second block of 6 elements

```
>>> x[6:12]
array([    0.3,        0,        0,        0,        0,        0])
```

as translation followed by rotation. The first landmark is stored in

```
>>> x[12:15]
array([-0.05306,    -0.9203,      1.94])
```

The double mapping property of unit quaternions means that any unit quaternion can be written with a nonnegative scalar component. By definition the unit quaternion has a unit norm, so the scalar component can be easily recovered $s = \sqrt{1 - v_x^2 - v_y^2 - v_z^2}$ given the vector component, see ▶ Sect. 2.3.1.7.

Bundle adjustment is a minimization problem – it finds the camera poses and landmark positions that minimize the total reprojection error across all the edges

$$\boldsymbol{x}^* = \arg \min_{\boldsymbol{x}} \sum_k F_k(\boldsymbol{x})$$

where $F_k(\cdot) > 0$ is a nonnegative scalar cost associated with the graph edge $k$ from camera $i$ to landmark $j$. The reprojection error of a landmark at $\mathsf{P}_j$ onto the camera at pose $\xi_i$ is

$$\boldsymbol{f}_k(\boldsymbol{x}) = \mathcal{P}\!\left(\hat{\boldsymbol{\xi}}_i, \; \hat{\boldsymbol{P}}_j; \; \mathbf{K}\right) - {}^i\boldsymbol{p}_j^{\#} \in \mathbb{R}^2$$

and the scalar cost is the squared Euclidean reprojection error

$$F_k(\boldsymbol{x}) = \boldsymbol{f}_k^{\top}(\boldsymbol{x}) \; \boldsymbol{f}_k(\boldsymbol{x}) \;.$$

Although written as a function of the entire state vector, $F_k(\boldsymbol{x})$ depends on only two elements of that vector: $\boldsymbol{\xi}_i$ and $P_j$ where $i$ and $j$ are the vertices connected by edge $k$. The total error, the sum of the squared back-projection error for all edges, can be computed for any value of the state vector and for the initial conditions is

```
>>> bundle.errors(x)
1.407e+06
```

The bundle adjustment task is to adjust the camera and landmark parameters to reduce this value. We have framed bundle adjustment as a sparse nonlinear least squares problem, and this can be solved numerically if we have a sufficiently good initial estimate of $\boldsymbol{x}$.

The first step in solving this problem is to linearize it. The reprojection error $\boldsymbol{f}_k(\boldsymbol{x})$ can be linearized about the current state $\boldsymbol{x}_0$ of the system

$$\boldsymbol{f}_k'(\boldsymbol{\Delta}) \approx \boldsymbol{f}_{0,k} + \mathbf{J}_k \boldsymbol{\Delta}$$

where $\boldsymbol{f}_{0,k} = \boldsymbol{f}_k(\boldsymbol{x}_0)$ and

$$\mathbf{J}_k = \frac{\partial \boldsymbol{f}_k(\boldsymbol{x})}{\partial \boldsymbol{x}} \in \mathbb{R}^{2 \times (6N+3M)}$$

Linearization and Jacobians are discussed in ▶ App. E, and solution of sparse nonlinear equations in ▶ App. F.2.4.

is a Jacobian matrix ◀ which depends only on the camera pose $\boldsymbol{\xi}_i$ and the landmark position $\boldsymbol{P}_j$ so is therefore mostly zeros

$$\mathbf{J}_k = \begin{pmatrix} \mathbf{0}_{2\times6} & \cdots & \mathbf{A}_i & \cdots & \mathbf{B}_j & \cdots & \mathbf{0}_{2\times3} \end{pmatrix}, \quad \text{where} \quad \mathbf{A}_i = \frac{\partial \boldsymbol{f}_k(\boldsymbol{x})}{\partial \boldsymbol{\xi}_i} \in \mathbb{R}^{2\times6},$$

$$\mathbf{B}_j = \frac{\partial \boldsymbol{f}_k(\boldsymbol{x})}{\partial \boldsymbol{P}_j} \in \mathbb{R}^{2\times3} \;.$$

The CentralCamera class has a method to compute the two Jacobians and the image-plane projection in a single call

```
>>> p, A, B = camera.derivatives(t, r, P);
```

The structure of the Jacobian matrix $\mathbf{A}_i$ is specific to the chosen representation of camera orientation r. The Jacobians, particularly $\mathbf{A}_i$, are quite complex to derive but can be automatically generated using SymPy and the script symbolic/bundle-adjust.py in the Machine Vision Toolbox.

where t and r are the camera pose as a translation and rotation vector, and P is the landmark coordinate vector. ◀ Translating the camera by $\boldsymbol{d}$ or translating the point by $-\boldsymbol{d}$ have an equivalent effect on the image, and therefore $\mathbf{B}_j$ is the negative of the first three columns of $\mathbf{A}_j$.

Now everything is in place to allow us to solve the bundle adjustment problem

```
>>> x_new, resid = bundle.optimize(x);
Bundle adjustment cost 1.41e+06 -- initial
Bundle adjustment cost 868 (solved in 2.46 sec)
Bundle adjustment cost 238 (solved in 2.44 sec)
Bundle adjustment cost 238 (solved in 2.46 sec)
Bundle adjustment cost 238 (solved in 2.46 sec)
Bundle adjustment cost 238 (solved in 2.46 sec)
 * 5 iterations in 13.0 seconds
 * Final RMS error is 0.44 pixels
```

At each step the derivatives are computed by the `derivatives` method and used to update the state estimate. The displayed messages shows how the total cost (squared reprojection error) decreases at each iteration, reducing by several orders of magnitude. The final result has an RMS reprojection error ▶ better than half a pixel for each landmark, which is impressive given that the images were captured with a handheld phone camera and we have completely ignored lens distortion.

The square root of the sum of the squared reprojection errors, divided by the number of points.

The result is a new state vector that contains the updated camera poses and landmark positions which we can use to update the state of all viewpoint and landmark vertices in the graph

```
>>> bundle.setstate(x_new);
```

and the refined camera pose is

```
>>> bundle.views[1].pose.printline(orient="camera")
t = 0.31, -0.0268, 0.074; rpy/yxz = -0.585°, 1.25°, 0.336°
```

compared to the value from the essential matrix decomposition which was

```
>>> T_1_2.printline(orient="camera")
t = 0.3, -0.0195, 0.0547; rpy/yxz = -0.39°, 1.92°, 0.474°
```

The refined estimate of the first landmark is

```
>>> bundle.landmarks[0].P
array([-0.05306, -0.9203, 1.94])
```

While the overall RMS error is low we can look at the final reprojection error in more detail

```
>>> e = np.sqrt(bundle.getresidual());
>>> e.shape
(2, 615)
```

where element $(i, j)$ is the reprojection error in pixels for camera $i$ and landmark $j$. The median error for cameras one and two

```
>>> np.median(e, axis=1)
array([ 0.2741, 0.2625])
```

are around a quarter of a pixel, while the maximum errors

```
>>> np.max(e, axis=1)
array([ 1.614, 1.61])
```

are less than 2 pixels.

Bundle adjustment finds the optimal *relative* pose and positions – not absolute pose. For example, if all the cameras and landmarks moved 1 m in the $x$-direction, the total reprojection error would be the same. To remedy this we can fix or *anchor* one or more cameras or landmarks – in this example, we fixed the first camera. The values of the fixed poses and positions are kept in the state vector but they are not updated during the iterations – their Jacobians do not need to be computed and the Hessian matrix used to solve the update at each iteration is smaller since the rows and columns corresponding to those fixed parameters can be deleted.

The fundamental issue of scale ambiguity with monocular cameras, as we discussed for the essential matrix in ▶ Sect. 14.2.2, applies here as well. A scaled model of the same world with a similarly scaled camera translation is indistinguishable from the real thing. More formally, if the whole problem was scaled so that $P'_j = \lambda P_j$, $[\xi'_i]_t = \lambda [\xi_i]_t$ and $\lambda \neq 0$, the total reprojection error would be the same. The solution we obtained above has an arbitrary scale or value of $\lambda$ – changing the initial condition for the camera poses or landmark coordinates will lead to a solution with a different scale. We can remedy this by anchoring the pose of at least two cameras, one camera and one landmark, or two landmarks.

The bundle adjustment technique, but not this implementation, allows for constraints between cameras. For example, a multi-camera rig moving through space would use constraints to ensure the fixed relative pose of the cameras at each time step. Odometry from wheels or inertial sensing could be used to constrain the distance between camera coordinate frames to enforce the correct scale, or orientation from an IMU could be used to constrain the camera attitude. In the underlying graph representation of the problem as shown in ◘ Fig. 14.20 this would involve adding additional edges between the camera vertices. Constraints could also be added between landmarks that had a known relative position, for example the corners of a window – this would involve adding additional edges between the relevant landmark vertices. This is now very similar to the posegraph SLAM solution we introduced in ▶ Sect. 6.5.

The particular problem we studied is unusual in that every camera views every landmark. In a more common situation the camera might be moving in a very large environment so any one camera will only see a small subset of landmarks. In a real-time system, a limited bundle adjustment might be performed with respect to occasional frames known as keyframes, and a bundle adjustment over all frames, or all keyframes, performed at a lower rate in the background.

In this example we have assumed the camera intrinsic parameters are known and constant. Theoretically, bundle adjustment can solve for intrinsic as well as extrinsic parameters. We simply add additional parameters for each camera in the state vector and adjust the Jacobian $\mathbf{A}$ accordingly. However, given the coupling between intrinsic and extrinsic parameters this may lead to poor performance. If we chose to estimate the elements of the camera matrix ($c_{0,0}, \ldots, c_{2,2}$) directly, then the state vector would contain 11 ◄ rather than 6 elements for each camera. If $\mathbf{C}_i$ is the true camera matrix, then bundle adjustment will estimate an arbitrary linear transformation $\mathbf{C}_i \mathbf{Q}$ where $\mathbf{Q} \in \mathbb{R}^{4 \times 4}$ is some nonsingular matrix. Correspondingly, the estimated world points will be $\mathbf{Q}^{-1} \tilde{\boldsymbol{P}}_j$ where $\tilde{\boldsymbol{P}}_j$ is their true value. Fortunately, projection matrices for realistic cameras have well defined structure (13.10) and properties as described in ▶ Exc. 13.7, and these provide constraints that allow us to estimate $\mathbf{Q}$. Estimating an arbitrary $\mathbf{C}_i$ is referred to as a projective reconstruction. This can be *upgraded* to an affine reconstruction (using an affine camera model) or a metric reconstruction (using a perspective camera model) by suitable choice of $\mathbf{Q}$.

The camera matrix has an arbitrary scale factor. Changes in focal length and $z$-axis translation have similar image-plane effects as do change in principal point and camera $x$- and $y$-axis translation.

## 14.4 Dense Stereo

A stereo image pair is commonly taken simultaneously using two cameras, generally with parallel optical axes, and separated by a known distance referred to as the camera baseline. ◘ Fig. 14.22 shows stereo camera systems which simultaneously capture images from both cameras, perform dense stereo matching and send the results to a host computer for action. Stereo cameras are a common sensor for mobile robotics, particularly outdoor robots, and can be seen in ◘ Figs. 1.5a, 1.9 and 4.16.

To illustrate stereo image processing, we load the left and right images comprising a stereo pair

```
>>> rocks_l = Image.Read("rocks2-l.png", reduce=2)
Image: 638 x 555 (uint8), R:G:B [.../images/rocks2-l.png]
>>> rocks_r = Image.Read("rocks2-r.png", reduce=2)
Image: 638 x 555 (uint8), R:G:B [.../images/rocks2-r.png]
```

and we can display these two images side by side

```
>>> rocks_l.stdisp(rocks_r)
```

as shown in ◘ Fig. 14.23. This is an interactive tool, and clicking on a point in the left-hand image updates a pair of cross hairs that mark the *same* coordinate relative to the right-hand image. Clicking in the right-hand image sets another vertical cross hair and displays the difference between the horizontal coordinate of the two crosshairs. The cross hairs as shown are set to a point on the digit "5", painted on a foreground rock, and we observe several things. Firstly the spot has the same vertical coordinate in both images, and this implies that the epipolar lines are horizontal.

**Fig. 14.22** Stereo cameras that compute disparity onboard using FPGA hardware. Camera baselines of 100 and 250 mm are shown (image courtesy of Nerian Vision GmbH)

Secondly, in the right-hand image the spot has moved to the left by 73.18 pixels. If we probed more points we would see this horizontal shift decreases for conjugate points that are further from the camera.

As shown in ■ Fig. 14.17 the conjugate point in the right-hand image moves rightward along the epipolar line as the point depth increases. For the parallel-axis camera geometry the epipolar lines are parallel and horizontal, so conjugate points must have the same $v$-coordinate. If the coordinates of two corresponding points are $(^Lu, ^Lv)$ and $(^Ru, ^Rv)$ then $^Rv = ^Lv$. The displacement along the horizontal epipolar line $d = ^Lu - ^Ru$, where $d \geq 0$, is called *disparity*.

The dense stereo process is illustrated in ■ Fig. 14.24. For the pixel at $(^Lu, ^Lv)$ in the left-hand image we know that its corresponding pixel is at some coordinate $(^Lu - d, ^Lv)$ in the right-hand image where $d \in \{d_{\min}, \ldots, d_{\max}\}$. To reliably find

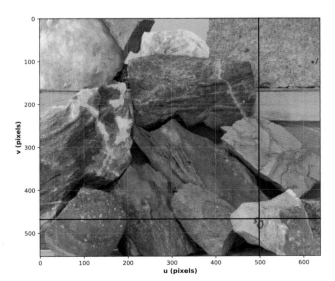
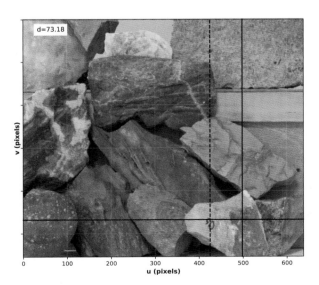

**Fig. 14.23** The stereo image browsing window created by `stdisp`. The black cross hair in the left-hand image has been positioned at the top right of the digit 5 on the rightmost foreground rock. Another black cross hair is automatically positioned at the same coordinate in the right-hand image. Clicking on the corresponding point in the right-hand image sets the dashed-black cross-hair, and the panel at the top indicates a horizontal shift of 73.18 pixels to the left. This stereo image pair is from the Middlebury 2006 stereo dataset "Rocks2", fullsize (Scharstein and Pal 2007). The focal length $f/\rho$ is 3740 pixels, and the baseline is 160 mm. The images have been cropped so that the actual disparity should be offset by 274 pixels

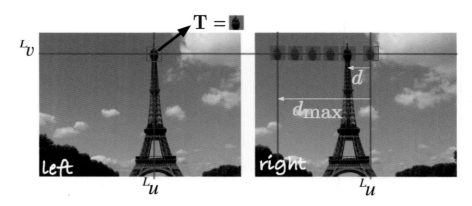

**◘ Fig. 14.24** Stereo matching. A search window in the right image, starting at $u = {}^L u$, is moved leftward along the horizontal epipolar line $v = {}^L v$ until it best matches the template window **T** from the left image

the corresponding point for a pixel in the left-hand image we create an $N \times N$ pixel *template* region **T** about that pixel – shown as a red square. We *slide* the template window horizontally leftward across the right-hand image. The position at which the template is most similar is considered to be the corresponding point from which disparity is calculated. Compared to the matching problem we discussed in ▶ Sect. 11.5.2, this one is much simpler because there is no change in relative scale or orientation between the two images.

The epipolar constraint means that we only need to perform a 1-dimensional search for the corresponding point. The template is moved horizontally in $D$ steps of 1 pixel from $d_{\min}$ to $d_{\max}$. At each template position we perform a template matching operation, as discussed in ▶ Sect. 11.5.2. For an $N \times N$ template these have a computational cost of $O(N^2)$, and for a $W \times H$ image the total cost of dense stereo matching is $O(DWHN^2)$. This is high, but feasible in real time using optimized algorithms, SIMD instructions sets, and possibly GPU hardware.

To perform stereo matching for the image pair in ◘ Fig. 14.23 using the Toolbox is quite straightforward ◀

This is a simple pure Python implementation for pedagogical purposes only.

```
>>> disparity, *_ = rocks_l.stereo_simple(rocks_r, hw=3,
...                                       drange=[40, 90]);
```

The result is a 2D array **D**, the same size as `rocks_l`, whose elements $d_{u,v}$ are the disparity at pixel $(u, v)$ in the *left* image – the conjugate pixel would be at $(u - d_{u,v}, v)$ in the right image. We can display the disparity as an image – a disparity image

```
>>> disparity.disp(colorbar=True);
```

which is shown in ◘ Fig. 14.25. Disparity images have a distinctive ghostly appearance since all surface color and texture is absent. The second argument to `stereo_simple` is the half-width of the template, in this case we are using a $7 \times 7$ window. The third argument is the range of disparities to be searched, in this case from 40 to 90 pixels so the pixel values in the disparity image lie in the range [40, 90]. The disparity range was determined by examining some far and near points using `stdisp`. ◀ Template matching is performed using the zero-mean normalized cross correlation (ZNCC) similarity measure introduced in ▶ Sect. 11.5.2.

We could chose a range such as [0, 90] but this increases the search time: 91 disparities would have to be evaluated instead of 51. It also increases the possibility of matching errors as discussed in ▶ Sect. 14.4.2.

In the disparity image we can clearly see that the rocks at the bottom of the pile have a larger disparity than those at the top – the bottom rocks are closer to the camera. There are also some errors, such as the anomalous bright values around the edges of some rocks. These pixels are indicated as being nearer than they really are. The disparity is set to `nan` around the edge of the image where the similarity matching template extends beyond the boundary of the image, and wherever the

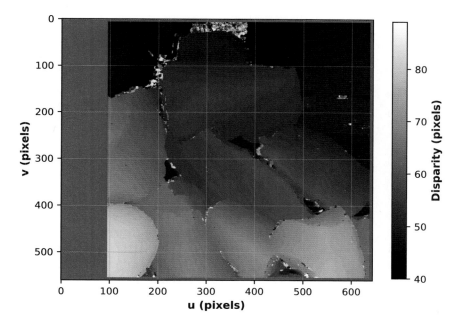

**Fig. 14.25** Disparity image for the rock pile stereo pair, where brighter means higher disparity or shorter range. Pixels with a value of nan, shown in red, are where disparity could not be computed. Note the quantization in gray levels since we search for disparity in steps of one pixel

denominator of the ZNCC similarity metric (■ Table 11.1) is equal to zero. ► The nan values are displayed as red.

This occurs if all the pixels in either template have exactly the same value.

The stereo_simple method returns two additional values

```
>>> disparity, similarity, DSI = \
...     rocks_l.stereo_simple(rocks_r, hw=3, drange=[40, 90])
```

The disparity-space image (DSI) is a 3D array

```
>>> DSI.shape
(561, 644, 50)
```

denoted $\mathcal{D}$ and illustrated in ■ Fig. 14.26. Its elements $\mathcal{D}(u, v, d)$ are the similarity between the templates centered at $(u, v)$ in the left image and $(u - d, v)$ in the right image. ► The disparity image we saw earlier is simply the position of the maximum value in the $d$-direction evaluated at every pixel

This is a large matrix (144 Mbyte) which is why the images were reduced in size when loaded.

```
>>> np.argmax(DSI, axis=2);
```

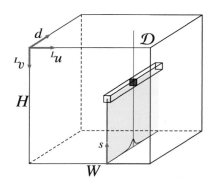

**Fig. 14.26** The disparity space image (DSI) is a 3-dimensional image where element $\mathcal{D}(u, v, d)$ is the similarity between the template window centered at $({}^{L}u, {}^{L}v)$ in the left image and the same sized window centered at $({}^{L}u - d, {}^{L}v)$ in the right image

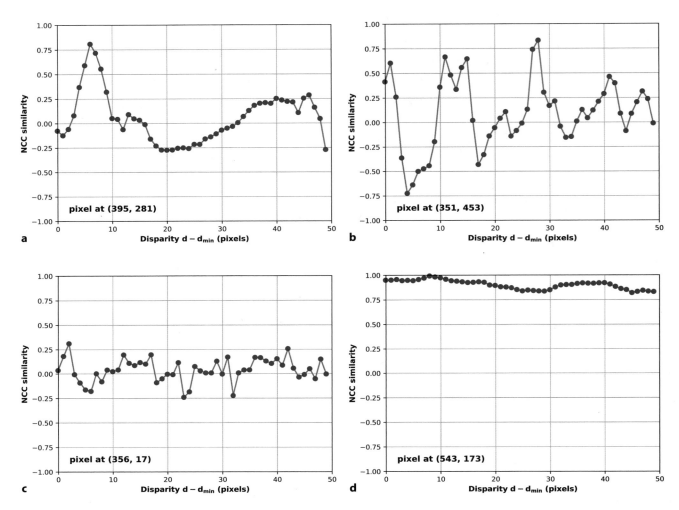

**☐ Fig. 14.27** Template ZNCC similarity score versus disparity at four selected points in the image. **a** Single strong peak; **b** multiple peaks; **c** weak peak; **d** broad peak

The similarity image `similarity` is the same size as `rocks_1`, whose pixel values are the peak similarity scores, the maxima in the $d$-direction evaluated at every pixel

```
>>> similarity_values = np.max(DSI, axis=2);
```

Each fibre of the DSI in the $d$-direction, as shown in ☐ Fig. 14.26, is the ZNCC similarity measure versus disparity for the corresponding pixel in the left image. For the pixel at $(395, 281)$ we can plot this

```
>>> plt.plot(DSI[281, 395, :], "o-");
```

which is shown in ☐ Fig. 14.27a. A strong match occurs at $d - d_{min} = 6$, and since $d_{min} = 40$ the disparity at this peak is 46 pixels.

### 14.4.1  Peak Refinement

The disparity at each pixel is an integer value $d = d_{min}, \ldots, d_{max}$ at which the greatest similarity was found. ☐ Fig. 14.27a shows a single unambiguous strong peak and we can use the peak and adjacent points to refine the estimate of the peak's position. ◄ We model similarity around the peak as a second-order function

This 1-dimensional peak refinement is discussed in ► App. J.1.

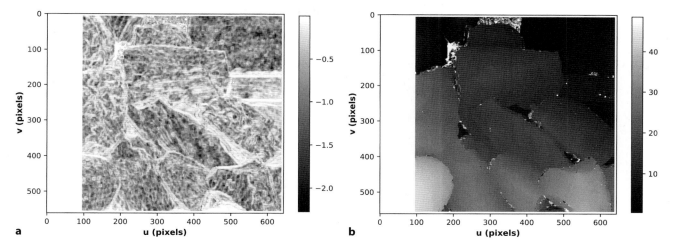

**Fig. 14.28** Peak refinement. **a** Value of the $A$ term, peak sharpness increases with lower (darker) values; **b** subpixel interpolated disparity value

of disparity

$$s = Ad^2 + Bd + C \qquad (14.14)$$

and the coefficients $(A, B, C)$ can be estimated from three points and we choose the peak value and its two neighbors. For the ZNCC similarity measure, the best match is a maxima which means that the parabola is inverted and $A < 0$. The $A$ coefficient will have a large magnitude for a sharp peak.

The maximum value of the fitted parabola occurs when its derivative is zero, from which we can obtain a more precise estimate of the position of the similarity peak which occurs at the disparity

$$\hat{d} = \frac{-B}{2A} .$$

We can compute the refined disparity values, and the $A$ coefficient, for every pixel by

```
>>> disparity_refined, A = Image.DSI_refine(DSI)
```

and these are shown in ■ Fig. 14.28.

## 14.4.2 Stereo Failure Modes

While the similarity versus disparity curve in ■ Fig. 14.27a has a single sharp peak, the results shown in ■ Fig. 14.27b–c exhibit various common failure modes.

### 14.4.2.1 Multiple Peaks

■ Fig. 14.27b shows several peaks of almost similar amplitude, and this means that the template pattern was found multiple times in the search. This occurs when there are regular vertical features in the scene as is often the case in human-made scenes: brick walls, rows of windows, architectural features or a picket fence. The problem, illustrated in ■ Fig. 14.29, is commonly known as the picket fence effect and more properly as spatial aliasing.

There is no real cure for this problem ▶ but we can detect its presence. The ambiguity ratio is the ratio of the height of the second highest peak to the height of the highest peak – a high-value indicate that the result is uncertain and should not be used.

Multi-camera stereo, using more than two cameras, is a powerful method to solve this ambiguity.

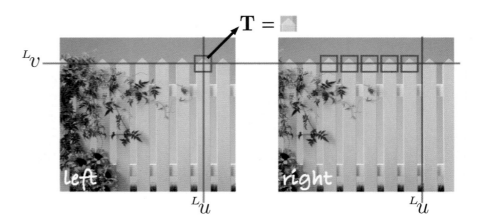

**Fig. 14.29** Picket fence effect. The template will match well at a number of different disparities. This problem occurs in any scene with repeating patterns

The chance of detecting incorrect peaks can be reduced by ensuring that the disparity search range is as small as possible but this requires some knowledge of the expected range of objects.

#### 14.4.2.2 Weak Matching

A weak match is shown in ◘ Fig. 14.27c. This typically occurs when the corresponding scene point is not visible in the right-hand view due to occlusion – also known as the missing parts problem. Occlusion is illustrated in ◘ Fig. 14.30 and it is clear that point 3 is only visible to the left camera. If the conjugate point is occluded, the stereo matching algorithm will match the left-image template to the most similar, but wrong, template in the right image. This figure is an exaggerated depiction, but real images do suffer this problem where the depth changes rapidly. In our rock pile scenario this will occur at the edges of the rocks, which is exactly where we observe the incorrect disparities in ◘ Fig. 14.25. The problem becomes more prevalent as the baseline increases. The problem also occurs when the corresponding point does not lie within the disparity search range, that is, the disparity search range is too small.

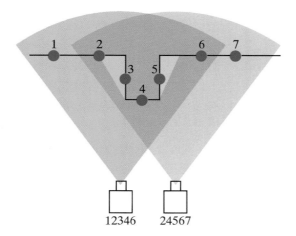

**Fig. 14.30** Occlusion in stereo vision. The field of view of the two cameras are shown as colored sectors. Points 1 and 7 fall outside the overlapping view area and are seen by only one camera each. Point 5 is occluded from the left camera, and point 3 is occluded from the right camera. The order of points seen by each camera is given underneath it

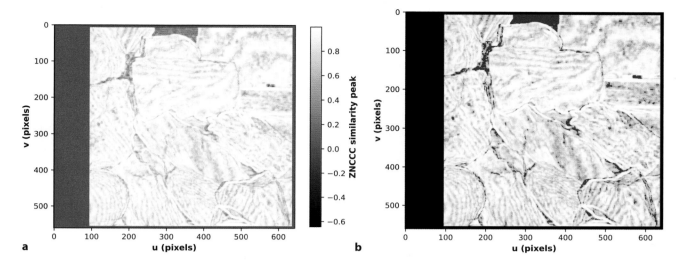

**Fig. 14.31** Stereo template similarity. **a** Similarity image where brighter means higher similarity and red indicates `nan` values in the result where disparity could not be computed; **b** disparity image with pixels having low similarity score marked in blue

The occlusion problem cannot be cured but it can be detected. The simplest method is to consider the returned similarity score

```
>>> similarity.disp();
```

as shown in ◘ Fig. 14.31a. We see that the erroneous disparity values correspond to low similarity scores. Disparity results where similarity is low can be discarded

```
>>> similarity.choose("blue", similarity < 0.6).disp();
```

and this is shown in ◘ Fig. 14.31b where pixels with similarity $s < 0.6$ are displayed in blue, and could be excluded from subsequent calculations. The distribution of maximum similarity scores

```
>>> plt.hist(similarity.view1d(), 100, (0, 1), cumulative=True,
...          density=True);
```

is shown in ◘ Fig. 14.32. We see that only 5% of pixels have a similarity score less than 0.6, and that more than half of all pixels have a similarity score greater than 0.9. However some of these pixels may suffer from the problem of multiple peaks.

A simple but effective way to test for occlusion is to perform the matching in two directions – left-right consistency checking. Starting with a pixel in the left-hand image, the strongest match in the right-image is found. Then the strongest match to that pixel is found in the left-hand image. If this is where we started, then the match is considered valid. However if the corresponding point was occluded in the right image the first match will be a weak one to a different feature, and there is a high probability that the second match will be to a different pixel in the left image.

From ◘ Fig. 14.30 it is clear that pixels on the left-side of the left-hand image cannot overlap at all with the right-hand image – point 1 for example is outside the field of view of the right-hand camera. This is the reason for the large number of incorrect matches on the left-hand side of the disparity image in ◘ Fig. 14.25. It is common practice to discard the $d_{max}$ left-most columns (90 in this case) of the disparity image.

### 14.4.2.3 Broad Peak

The final problem that can arise is a similarity function with a very broad peak as shown in ◘ Fig. 14.27d. The breadth makes it difficult to precisely estimate the maxima. This generally occurs when the template region has very low texture,

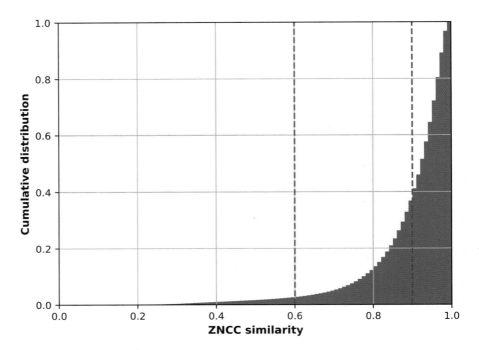

□ **Fig. 14.32** Cumulative probability of ZNCC template similarities. The probability of a similarity above $\sim 0.9$ is $\sim 45\%$

for example corresponding to the sky, dark shadows, sheets of water, snow, ice or smooth human-made objects. Simply put, in a region that is all gray, a gray template matches equally well with any number of possible gray regions.

One approach to detect this is to apply a threshold to the similarity peak sharpness $|A|$ from ► Sect. 14.4.1. Another approach is to quantify the variation of pixel values in the template using measures such as the difference between the maximum and minimum value or the variance of the pixel values. If the template has too little variation it is less likely to result in a strong peak.

For indoor scenes, where textureless surfaces are common, we can project artificial texture onto the surface using an infrared pattern projector. This approach is used by the Intel RealSense D400 series of depth cameras.

### 14.4.2.4 Quantifying Failure Modes

For the various problem cases just discussed, disparity cannot be determined but the problem can be detected. This is important since it allows those pixels to be marked as having no known range, and the robot can be prudent about regions whose 3-dimensional structure is not reliably known – it is important to know what we don't know. Where reliable depth information from stereo vision is missing, a robot should be cautious and not asssume it is free space.

We use a number of simple measures to mark elements of the disparity image as being invalid or unreliable. We start by creating an array `status` the same size as `disparity` and initialized to one

```
>>> status = np.ones(disparity.shape);
```

The elements are set to different values if they correspond to specific failure conditions

```
>>> U, V = disparity.meshgrid()
>>> status[np.isnan(disparity.image)] = 5    # no similarity computed
>>> status[U <= 90] = 2                        # no overlap
>>> status[similarity.image < 0.6] = 3         # weak match
>>> status[A.image >= -0.1] = 4                # broad peak
```

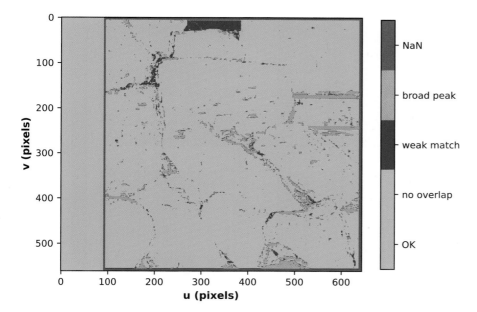

■ **Fig. 14.33** Stereo matching status on a per pixel basis

We can display this array as an image

```
>>> plt.imshow(status);
```

which is shown in ■ Fig. 14.33 with a labeled colormap. While there are quite a number of problem pixels, the good news is that there are a lot of light green pixels! In fact

```
>>> (status == 1).sum() / status.size * 100
77.01
```

nearly 80% of disparity values pass our battery of quality tests. The blue pixels, indicating weak similarity, occur around the edges of rocks and are due to occlusion. The orange pixels, indicating a broad peak, occur in areas that are fairly smooth, either deep shadow between rocks or the flat wall behind.

The valid disparity is

```
>>> disparity_valid = disparity.choose(0, status!=1)
Image: 644 x 561 (float32)
```

where pixels that have any of the defects just discussed are set to zero. This information is now in a useful form for a robot – unreliable disparity values are clearly marked.

### 14.4.2.5 Slicing the DSI

An alternative way to gain insight into the matching process, and what can go wrong, is to consider how disparity varies along a row of the image – effectively a depth cross-section of the scene. We can obtain this by taking a slice of the DSI in the $v$-direction, for example the slice for $v = 100$ is

```
>>> Image(DSI[100, :, :].T).disp();
```

A number of slices are shown in ■ Fig. 14.34 for selected values of $v$ and within each of these $ud$-planes we see a bright path (high similarity values) that represents disparity $D(u)$. Note the significant discontinuities in the path for the plane at $v = 100$ which correspond to sudden changes in depth at $u = 200$. The path also *fades away* in places where the maximum similarity is low because there is no strong match in the right-hand image – the most likely cause is occlusion or lack of texture. A class of dense-stereo matching algorithms – semi-global block matching – attempt to find a path across the image that maximizes similarity and continuity.

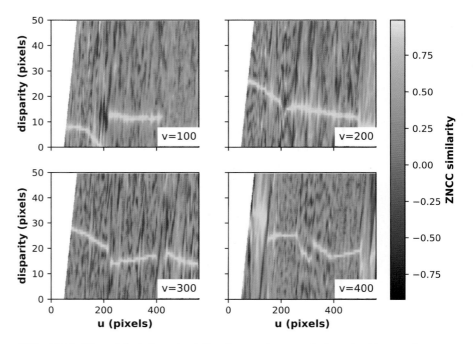

■ **Fig. 14.34** Slices of the 3-dimensional disparity space image are 2-dimensional images of template similarity as a function of disparity and $u$. Shown here for four rows of the image

### 14.4.2.6  Summary

The design of a stereo-vision system has three degrees of freedom. The first is the baseline distance between the cameras. As this increases, the disparities become larger making it possible to estimate depth to greater precision, but the occlusion problem becomes worse. Second, the disparity search range needs to be set carefully. If the maximum is too large the chance of spatial aliasing increases but if too small then points close to the camera will generate incorrect and weak matches. A large disparity range also increases the computation time. Third, template size involves a tradeoff between computation time and quality of the disparity image. A small template size can pick up fine depth structure but tends to give results that are much noisier since a small template is more susceptible to ambiguous matches. A large template gives a smoother disparity image, but requires greater computation. It also increases the chance that the template will contain pixels belonging to objects at different depths which is referred to as the mixed pixel problem. This can cause poor quality matching at the edges of objects, and the resulting disparity image appears blurred. One solution is to use a nonparametric local  transform such as the rank or census transform prior to performing correlation. Since these rely on the ordering of intensity values, not the values themselves, they give better performance at object boundaries.

### 14.4.2.7  Advanced Stereo Matching

In the previous sections we used a simple stereo matching algorithm coded in Python and shipped with the Toolbox. It has instructional value but is slow and far from the state-of-the-art. OpenCV includes a number of more advanced stereo matching algorithms.

The block matching stereo algorithm is similar in principle to `stereo_simple` but it is fast and provides disparity estimated to sub-pixel accuracy. Repeating the rock pile example

```
>>> disparity_BM = rocks_l.stereo_BM(rocks_r, hw=3, drange=[40, 90],
...                                              speckle=(200, 2))
```

```
Image: 638 x 555 (float32)
>>> disparity_BM.disp();
```

the results are shown in ◼ Fig. 14.35b. The anomalous black and white spots, referred to as speckle, indicate pixels where disparity was computed incorrectly. There are two approaches to reducing speckle: increase the block matching window size, or introduce a speckle filter. The effect of increasing the window size can be seen in ◼ Fig. 14.35c–d where the disparity image becomes smoother but discontinuities are lost. Speckles are small regions with consistent disparity that are different to their surrounds, and a variant of connected component analysis is used to detect them. The speckle filter has two parameters: the maximum area of the speckle region in pixels, and the maximum variation in disparity within the region. The effect of the speckle filter can be seen in the difference between ◼ Fig. 14.35b, with the filter enabled, and ◼ Fig. 14.35a with it disabled.

In the approaches so far, disparity computed at each pixel is independent of other pixels, but for real scenes adjacent pixels typically belong to the same surface and disparity will be quite similar – this is referred to as the *smoothness constraint*. Disparity will be discontinuous at the edges of surfaces. In the context of ◼ Fig. 14.34 we can think of this as finding the best-possible path of strongly similarity values across the image that also enforces the smoothness constraint in the horizontal direction. This is implemented by the semi-global block matching algorithm

```
>>> rocks_1.stereo_SGBM(rocks_r, hw=3, drange=[40, 90],
...                     speckle=(200, 2)).disp();
```

and the result is shown in ◼ Fig. 14.35e, but at the cost of increased computation.

### 14.4.2.8 3D Reconstruction

The final step of the stereo vision process is to convert the disparity values, in units of pixels, to world coordinates in units of meters – a process known as 3D reconstruction. In the earlier discussion on sparse stereo we determined the world point from the intersection of two lines in 3-dimensional space. For a stereo camera with parallel optical axes as shown in ◼ Fig. 14.22, the geometry is much simpler as illustrated in ◼ Fig. 14.36. ▶ For the red and blue triangles we can write

$$X = Z \tan \theta_1, \quad \text{and} \quad X - b = Z \tan \theta_2$$

where $b$ is the baseline and the angles of the rays correspond to the horizontal image coordinate $^i u, i = \{L, R\}$

$$\tan \theta_i = \frac{\rho_u (^i u - u_0)}{f} .$$

where $\rho_u$ is the pixel width, $u_0$ is the horizontal coordinate of the principal point, and $f$ is the focal length. Substituting and eliminating $X$ gives

$$Z = \frac{fb}{\rho_u (^L u - ^R u)} = \frac{fb}{\rho_u d}$$

which shows that depth is inversely proportional to disparity. We can also recover the $X$- and $Y$-coordinates so the 3D point coordinate is

$$\boldsymbol{P} = \frac{b}{d} \left( ^L u - u_0, ^L v - v_0, \frac{f}{\rho_u} \right) \tag{14.15}$$

which can be computed for every pixel.

The rock pile stereo pair has been rectified to account for minor alignment errors in the stereo cameras.

14

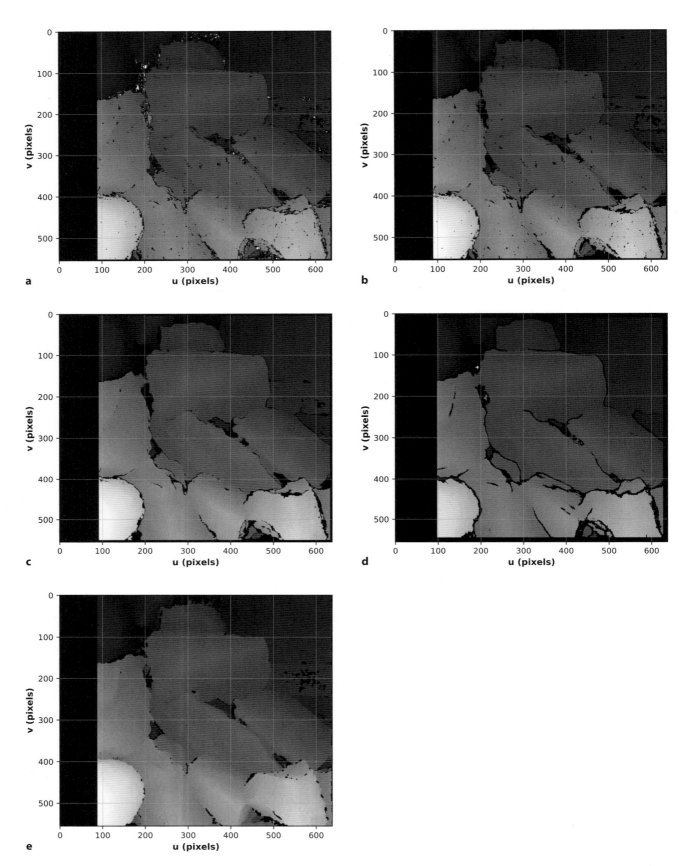

■ **Fig. 14.35** Disparity images for OpenCV stereo matchers. **a** `stereo_BM` blocksize $7 \times 7$, no speckle filter; **b** `stereo_BM` blocksize $7 \times 7$; **c** `stereo_BM` blocksize $11 \times 11$; **d** `stereo_BM` blocksize $23 \times 23$; **e** `stereo_SGBM` blocksize $7 \times 7$. **b–d** have the speckle filter set to (200, 2)

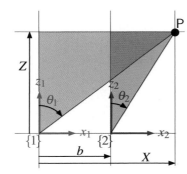

**Fig. 14.36** Stereo geometry for parallel camera axes. $X$ and $Z$ are measured with respect to camera one, $b$ is the baseline

A good stereo system can estimate disparity with an accuracy of 0.2 pixels. Distant points have a small disparity and the error in the estimated 3D coordinate will be significant. A rule of thumb is that stereo systems typically have a maximum range of $50b$.

> ❗ The images shown in ☐ Fig. 14.23, from the Middlebury dataset, were taken with a very-wide camera baseline. The left edge of the left-image and the right edge of the right-image have no overlap and have been cropped. Cropping $N$ pixels from the left side of the left-hand image, reduces the disparity by $N$.

The true disparity is

```
>>> di = disparity_BM.image * 2 + 274;
```

where the factor of two accounts for the fact that disparity was computed for a half-resolution image, and 274 is to compensate for image cropping. We compute the $X$-, $Y$- and $Z$-coordinate of each pixel as separate arrays

```
>>> U, V = disparity_BM.meshgrid();
>>> u0, v0 = disparity.centre;
>>> f = 3740;   # pixels, according to Middlebury website
>>> b = 0.160;  # m, according to Middlebury website
>>> X = b * (U - u0) / di; Y = b * (V - v0) / di; Z = f * b / di;
```

which can be displayed as a surface

```
>>> fig, ax = plt.subplots(subplot_kw={"projection": "3d"})
>>> ax.plot_surface(X, Y, Z)
>>> ax.view_init(-100, -100)
```

but the visual quality is rather poor.

We will use the popular package Open3D to display a dense colored point cloud. We pass in the depth image z, the original color image, and the projection model as a `CentralCamera` object

```
>>> cam = CentralCamera(f=f, imagesize=rocks_1.shape);
>>> pcd = PointCloud(Z, image=rocks_1, camera=cam, depth_trunc=1.9)
PointCloud with 354090 points.
>>> pcd *= SE3.Rx(pi);  # make y-axis upward
```

where the `depth_trunc` parameter specifies that points greater than this distance will be excluded, and this includes the wall behind the rock pile. ▶ The result is displayed by

```
>>> pcd.disp()
```

and shown in ☐ Fig. 14.37. Interactive controls allow the viewpoint and zoom to be changed.

Inplace multiplication of a `PointCloud` object by an `SE3` object transforms the points according to $\xi \cdot P_i$.

▶ go.sn.pub/pjm5oA

**Fig. 14.37** 3-dimensional reconstruction of the rock pile stereo pair as a colored point cloud. This plot was created using the `PointCloud` class which wraps the Open3D package. The images in this case were full, rather than half, resolution

### 14.4.3 Image Rectification

The rock pile stereo pair of ◉ Fig. 14.23 has conjugate points on the same row in the left- and right-hand images – they are an epipolar-aligned image pair. Stereo cameras, such as shown in ◉ Fig. 14.22, are built with precision to ensure that the optical axes of the cameras are parallel and that the $u$- and $v$-axes of the two sensor chips are parallel. However there are limits to the precision of mechanical alignment, and lens distortion will also introduce error. Typically one or both images are warped to correct for these errors – a process known as rectification.

We will illustrate rectification using the courtyard stereo pair from ◉ Fig. 14.15

```
>>> walls_l = Image.Read('walls-l.png', reduce=2)
Image: 1632 x 1224 (uint8), R:G:B [.../images/walls-l.png]
>>> walls_r = Image.Read('walls-r.png', reduce=2)
Image: 1632 x 1224 (uint8), R:G:B [.../images/walls-r.png]
```

which ◉ Fig. 14.16 showed were far from being epipolar aligned. We first find the SIFT features

```
>>> sf_l = walls_l.SIFT()
SIFT features, 20426 points
>>> sf_r = walls_r.SIFT()
SIFT features, 18941 points
```

and determine the putative matches

```
>>> matches = sf_l.match(sf_r);
```

then determine the epipolar relationship

```
>>> F, resid = matches.estimate(CentralCamera.points2F,
...                             method="ransac", confidence=0.95);
```

The rectification step requires the set of corresponding points, which is contained in `matches`, and the fundamental matrix

```
>>> H_l, H_r = walls_l.rectify_homographies(matches, F)
```

and returns a pair of homography matrices that are used to warp the input images

```
>>> walls_l_rect = walls_l.warp_perspective(H_l)
Image: 1632 x 1224 (uint8), R:G:B
>>> walls_r_rect = walls_r.warp_perspective(H_r)
Image: 1632 x 1224 (uint8), R:G:B
```

These rectified images

```
>>> walls_l_rect.stdisp(walls_r_rect)
```

are shown in ◉ Fig. 14.38. Corresponding points in the scene now have the same vertical coordinate. As we have observed previously when warping images, not all

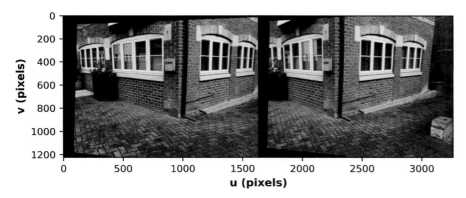

**Fig. 14.38** Rectified images of the courtyard scene from **Fig. 14.15. The black pixels at the left and bottom of each image correspond to pixels beyond the bounds of the original image

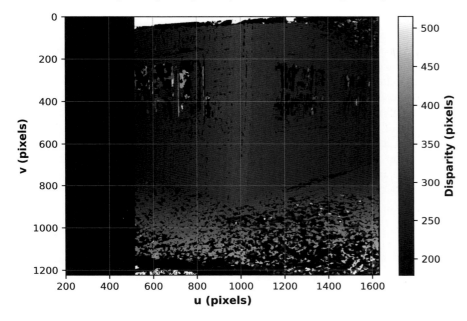

**Fig. 14.39** Dense stereo disparity image for the courtyard scene from **Fig. 14.15. The walls and ground show a clear depth gradient

of the output pixels are mapped to the input images and this results in undefined pixels which are displayed as black.

We can think of these images as having come from a virtual stereo camera with parallel axes and aligned pixel rows, and they can now be used for dense stereo matching

```
>>> walls_l_rect.stereo_SGBM(walls_r_rect, hw=7, drange=[180, 530],
...                          speckle=(50, 2)).disp();
```

and the result is shown in **Fig. 14.39. The disparity range parameters were determined interactively using `stdisp` to check the disparity for near and far points in the rectified image pair. The window half size of 7 was arrived at with a little trial and error, this value corresponding to a $15 \times 15$ window and produces a reasonably smooth result, at the expense of computation time. Matching of points in the foreground is poor, possibly due to spatial aliasing or insufficient disparity search range. Nevertheless this is quite an impressive result – using only two images taken from a handheld camera we have been able to create a dense 3-dimensional representation of the scene.

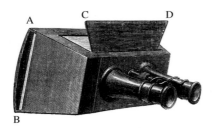

## 14.5 **Anaglyphs**

Human stereo perception relies on each eye providing a different viewpoint of the scene. However even if we look at a 2D photograph of a 3D scene we still get some sense of depth, albeit reduced, because our brain uses many visual cues, in addition to stereo, to infer depth. Since the invention of photography in the 19th century people have been fascinated by 3D photographs and movies, and the enduring popularity of 3D movies is further evidence of this.

The key to most 3D display technologies is to take the image from two cameras, with a similar baseline to the human eyes (50–80 mm) and present those images again to the corresponding eyes. Old fashioned stereograms required a binocular viewing device or could, with difficulty, be viewed by squinting at the stereo pair and crossing your eyes. More modern and convenient means of viewing stereo pairs are LCD shutter (gaming) glasses or polarized glasses which allow full-color stereo movie viewing, or head mounted displays.

An old, but inexpensive, method of viewing and distributing stereo information is through anaglyph images where the left and right images are overlaid in different colors. Typically red is used for the left eye and cyan (greeny blue) for the right eye, but many other color combinations are commonly used. The anaglyph is viewed through glasses with different colored lenses. The red lens allows only the red part of the anaglyph image to enter the left eye, while the cyan lens allows only the cyan parts of the image to enter the right eye. The disadvantage is that only the scene intensity, not its color, can be portrayed. The big advantage of anaglyphs is that they can be printed on paper or imaged onto ordinary movie film and viewed with simple and cheap glasses such as those shown in ◘ Fig. 14.40a.

The rock pile stereo pair can be displayed as an anaglyph

```
>>> walls_l.anaglyph(walls_r, "rc").disp();
```

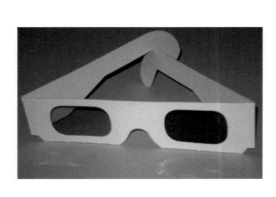

**a**                    **b**

◘ **Fig. 14.40** Anaglyphs for stereo viewing. **a** Anaglyph glasses shown with red and blue lenses, **b** anaglyph rendering of the rock scene from ◘ Fig. 14.23 with the left image in red and the right image in cyan

which is shown in ◨ Fig. 14.40b. The argument `"rc"` indicates that left and right images are encoded in **r**ed and **c**yan respectively. Other color options include: **b**lue, **g**reen, **m**agenta and **o**range.

## 14.6 Other Depth Sensing Technologies

There are a number of alternative approaches to obtaining 3-dimensional information about the world around a robot.

### 14.6.1 Depth from Structured Light

A classic, yet effective, method of estimating the 3D structure of a scene is structured light. This is conceptually similar to stereo vision, but we replace the left camera with a projector that imposes a pattern of light on the scene.

The simplest example of structured light, shown in ◨ Fig. 14.41a, uses a vertical plane of light. ▶ This is equivalent, in a stereo system, to a left-hand image that is a vertical line. The image of the line projected onto the surface viewed from the right-hand camera will be a distorted version of the line, as shown in ◨ Fig. 14.41b. The disparity between the virtual left-hand image and the actual right-hand image is a function of the depth of points along the line.

Finding the light stripe on the scene is a relatively simple vision problem. In each image row we search for the pixel corresponding to the projected stripe, based on intensity or color. If the camera coordinate frames are parallel, then depth is computed by (14.15).

To achieve depth estimates over the whole scene, we need to move the light plane horizontally across the scene and there are many ways to achieve this: mechanically rotating the laser stripe projector, using a moving mirror to deflect the stripe or using a data projector and software to create the stripe image. However, sweeping the light plane across the scene is slow and fundamentally limited by the rate at which we can acquire successive images of the scene. One way to speed up the process is to project multiple lines on the scene, but then we have to solve the correspondence problem which is not simple if parts of some lines are occluded. Many solutions have been proposed but generally involve coding the lines in some way – using different colors or using a sequence of binary or Gray-coded line patterns which can match $2^N$ lines in just $N$ frames.

Laser-based line projectors, so called "laser stripers" or "line lasers", are available for just a few dollars. They comprise a low-power solid-state laser and a cylindrical lens or diffractive optical element.

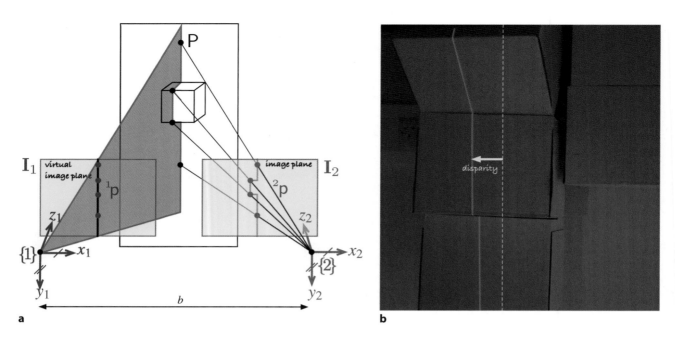

**◘ Fig. 14.41** **a** Geometry of structured light showing a light projector on the left and a camera on the right; four conjugate points are marked with dots on the left and right images and the scene; **b** a real structured light scenario showing the light stripe falling on a stack of carboard boxes. The superimposed dashed line represents the stripe position for a plane at infinity. Disparity, the left shift of the projected line relative to the dashed line, is inversely proportional to depth

This approach was used in the original Kinect for Xbox 360 introduced in 2010, and still available in the Asus Xtion sensor. It was developed by the Israeli company PrimeSense which was bought by Apple in 2013. The newer Azure Kinect uses per pixel time-of-flight measurement.

A related approach is to project a known but random pattern of *dots* onto the scene as shown in ◘ Fig. 14.42a. Each dot can be identified by the unique pattern of dots in its surrounding window. ◄ One lens projects an infrared dot pattern using a laser with a diffractive optical element which is viewed, see ◘ Fig. 14.42a, by an infrared sensitive camera behind another lens from which the depth image shown in ◘ Fig. 14.42c is computed. The shape of the dots also varies with distance, due to imperfect focus, and this provides additional cues about the distance of a point. A third lens is a regular color camera which provides the view shown in ◘ Fig. 14.42b. This is an example of an RGBD camera, returning RGB color values as well as depth (D) at every pixel.

Structured light approaches work well for ranges of a few meters indoors, for textureless surfaces, and in the dark. However outdoors, the projected pattern is overwhelmed by ambient illumination from the sun.

Some stereo systems, such as the Intel RealSense D400 series, also employ a dot pattern projector, sometimes known as a speckle or texture projector. This provides artificial texture that helps the stereo vision system when it is looking at textureless surfaces where matching is frequently weak and ambiguous, as discussed in ▶ Sect. 14.4.2. Such a sensor has the advantage of working on textureless surfaces which are common indoors where the sun is not a problem, and outdoors using pure stereo where scene texture is usually rich.

### 14.6.2 Depth from Time-of-Flight

An emerging alternative to stereo vision are cameras based on time-of-flight measurement. The camera emits a pulse of infrared light that illuminates the entire scene, and every pixel records the intensity and time of return of the reflected energy. The time measurement problem is challenging since a depth resolution of 1 cm requires timing precision of the order of 70 picoseconds and needs to be performed at every pixel.

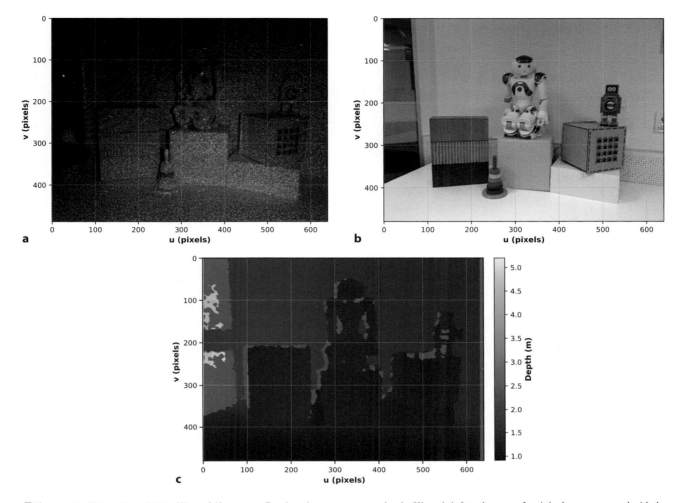

**◨ Fig. 14.42** 3D imaging with the Kinect 360 sensor. **a** Random dot pattern as seen by the Kinect's infrared camera; **b** original scene captured with the Kinect's color camera; **c** computed depth image. Red pixels indicate nan values where depth could not be computed due to occlusion or the maximum range being exceeded, as for example through the window on the left side of the scene (images courtesy William Chamberlain)

This type of camera works well indoors and even in complete darkness, but outdoors under full sun the maximum range is limited, just as it is for structured light. In fact it is worse than structured light. The illumination energy is limited by eye-safety considerations and structured light concentrates that energy over a sparse dot field, whereas time-of-flight cameras spread it over an area.

## 14.7 Point Clouds

A point cloud is a set of coordinates $\{(X_i, Y_i, Z_i), i \in 0, \ldots, N-1\}$ of points in the world. A colored point cloud is a set of tuples $\{(X_i, Y_i, Z_i, R_i, G_i, B_i), i \in 0, \ldots, N-1\}$ comprising the coordinates of a world point and its tristimulus value. Point clouds can be derived from all sorts of 3-dimensional sensors including stereo vision, RGBD cameras and LiDARs. A point cloud is a useful way to combine 3-dimensional data from multiple sensors (of different types) or sensors moving over time.

For a robotics application we need to extract some concise meaning from the thousands or millions of points in the point cloud. Common data reduction strategies include finding dominant planes, and registering points to known 3-dimensional models of objects.

▶ go.sn.pub/bosWgD

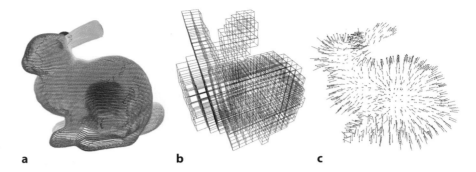

**◱ Fig. 14.43** **a** Point cloud of the Stanford bunny with 35,947 points, nearer points have a warmer color; **b** voxelized point cloud shown as a wire mesh; **c** a sparse needle plot of surface normals (Stanford bunny model courtesy Stanford Graphics Laboratory)

This model is well known in the computer graphics community. It was created by Greg Turk and Marc Levoy in 1994 at Stanford University, using a Cyberware 3030 MS scanner and a ceramic rabbit figurine.

The Polygon file format is a simple file format that contains $(X, Y, Z)$ tuples for the points, as well as faces and color information. Open3D can read PLY, STL, OBJ, OFF and GLTF file formats as well as lists of vertices provided as simple ASCII text.

"voxel" is a contraction of volume element, the 3D analog of a "pixel" or 2D picture element. A related approach is an octree where cubes are clustered, where possible, into larger cubes so as to further reduce storage requirements.

The order of the points is generally not important, and a kd-tree is typically constructed to reduce the computation involved in finding the neighbors of a point. The exception is an organized point cloud where the points correspond to pixels in an image and are given in image row-major order.

We will illustrate this by loading a point cloud of the famous Stanford bunny. ◀ Just as for 2-dimensional images, there are many common file formats, and here we will load a PLY file ◀

```
>>> bunny_pcd = PointCloud.Read('data/bunny.ply')
PointCloud with 35947 points.
>>> bunny_pcd.disp(block=False)
```

which is shown in ◱ Fig. 14.43a. The mouse can be used to rotate the object, and typing H in the display window generates a list of keyboard and mouse options.

Point clouds can be manipulated in various ways, for example they can be converted to a voxel grid ◀

```
>>> pcd = bunny_pcd.voxel_grid(voxel_size=0.01).disp()
```

which is shown as a wiremesh in ◱ Fig. 14.43b. Each cube has a side length of 1 cm and is included in the 3D grid if it contains one or more point cloud points. This is useful representation for collision detection, but it is also a way to reduce resolution

```
>>> pcd = bunny_pcd.downsample_voxel(voxel_size=0.01)
PointCloud with 795 points.
```

where each point in the new point cloud is the centroid of the points that fall within each voxel element. We can also compute the normal at each point, based on the surrounding points to which a local surface is fitted

```
>>> pcd.normals(radius=0.1, max_nn=30)
>>> pcd.disp(block=False)
```

and this is shown in ◱ Fig. 14.43b as a sparse needle map. Point clouds can also be cropped or triangulated into surface meshes for rendering as solid 3D objects, and the points can be transformed by appling a rigid-body transformation.

### 14.7.1 Fitting a Plane

Planes are common in our built world and for robotics, planes can be used to model the ground (for wheeled mobile robot driving or UAV landing) and walls. Given a set of 3-dimensional coordinates, a point cloud, there are two common approaches to finding a plane of best fit. The first, and simplest, approach for finding the plane is to fit the data to an ellipsoid. The ellipsoid will have one very small radius in the direction normal to the plane – that is, it will be an elliptical plate. The inertia tensor of the ellipsoid is equal to the moment matrix which is calculated directly from the points

$$\mathbf{J} = \sum_{i=0}^{N-1} \boldsymbol{x}_i \boldsymbol{x}_i^\top \in \mathbb{R}^{3\times3} \tag{14.16}$$

where $\boldsymbol{x} = \boldsymbol{P}_i - \overline{\boldsymbol{P}}$ are the coordinates of the points, as column vectors, with respect to the centroid of the points $\overline{\boldsymbol{P}} = \frac{1}{N} \sum_{i=0}^{N-1} \boldsymbol{P}_i$. The radii of the ellipsoid are the square root of the eigenvalues of $\mathbf{J}$, and the eigenvector corresponding to the smallest eigenvalue is the direction of the minimum radius which is the normal to the plane.

Outlier data points are problematic with this simple estimator since they significantly bias the solution. A number of approaches are commonly used but a simple one is to modify (14.16) to include a weight

$$\mathbf{J} = \sum_{i=0}^{N-1} w_i \boldsymbol{x}_i \boldsymbol{x}_i^\top$$

which is inversely related to the distance of $\boldsymbol{x}_i$ from the plane and solve iteratively. Initially all weights are set to $w_i = 1$, and on subsequent iterations the weights ▶ are set according to the distance of point $\mathsf{P}_i$ from the plane estimated at the previous step. ▶ App. C.1.4 has more details about ellipses.

Alternatively we could apply RANSAC by taking samples of three points to estimate the parameters of a plane $aX + bY + cZ + d = 0$. We can illustrate this with the earlier courtyard scene

Alternatively a Cauchy-Lorentz function $w = \beta^2/(d^2 + \beta^2)$ is used where $d$ is the distance of the point from the plane and $\beta$ is the half-width. The function is smooth for $d = [0, \infty)$ and has a value of $\frac{1}{2}$ when $d = \beta$.

```
>>> pcd = walls_pcd
PointCloud with 2341 points.
>>> plane, plane_pcd, pcd =
...    pcd.segment_plane(distance_threshold=0.05, seed=0)
>>> plane
array([ 0.7343,  -0.1192,   0.6682,   0.8823])
```

which has returned the equation of the first plane found, and two point clouds. The first returned point cloud contains all the points within the plane

```
>>> plane_pcd
PointCloud with 1245 points.
```

which has 1245 points that lie with 5 cm of the estimated plane. The second returned point cloud contains all the other points in the scene, and we can repeat the process

```
>>> plane, plane_pcd, pcd =
...    pcd.segment_plane(distance_threshold=0.05, seed=0)
>>> plane
array([ 0.6998,   0.1586,  -0.6965,  -1.413])
```

to find the next plane, and so on.

## 14.7.2 Matching Two Sets of Points

Consider the problem shown in ◻ Fig. 14.44a where we have a model of the bunny at a known pose shown as a blue point cloud. The red point cloud is the result of either the bunny moving, or the 3D camera moving. By aligning the two point clouds we can determine the rigid-body motion of the bunny or the camera.

Formally, this is the point cloud registration problem. Given two sets of $n$-dimensional points: the model $\{M_i \in \mathbb{R}^n, i = 0, \ldots, N_M-1\}$ and some noisy observed data $\{D_j \in \mathbb{R}^n, j = 0, \ldots, N_D-1\}$ determine the rigid-body motion from the data coordinate frame to the model frame

$$^D\hat{\xi}_M = \arg\min_{\xi} \sum_{i,j} \|D_j - \xi \cdot M_i\| .$$

A well known solution to this problem is the iterated closest point algorithm or ICP. At each iteration, the first step is to compute a translation that makes the centroids of the two point clouds coincident ◄

> We consider the general case where the two points clouds have different numbers of points, that is, $N_D \neq N_M$.

$$\overline{M} = \frac{1}{N_M} \sum_{i=0}^{N_M-1} M_i, \ \ \overline{D} = \frac{1}{N_D} \sum_{j=0}^{N_D-1} D_j$$

from which we compute a displacement

$$t = \overline{D} - \overline{M} .$$

Next we compute approximate correspondence. For each data point $D_j$ we find the closest model point $M_i$. ◄ Correspondence is not unique and quite commonly several points in one set can be associated with a single point in the other set, and consequently some points will be unpaired. Often the sensor returns only a subset of points in the model, for instance a laser scanner can see the front but not the back of an object. This approach to correspondence is far from perfect but it is surprisingly good in practice and *improves* the alignment of the point clouds so that in the next iteration the estimated correspondences will be a little more accurate.

> This is computationally expensive, but organizing the points in a kd-tree improves the efficiency.

The corresponding points are used to compute, via an outer product, the moment matrix

$$\mathbf{J} = \sum_{i,j} (M_i - \overline{M})(D_j - \overline{D})^\top \in \mathbb{R}^{n \times n}$$

which encodes the rotation between the two point sets. The singular value decomposition is $\mathbf{J} = \mathbf{U}\boldsymbol{\Sigma}\mathbf{V}^\top$ and the rotation matrix is given by (F.2).

The estimated relative pose between the two point clouds is $\xi^\Delta \sim (\mathbf{R}, t)$ and the model points are transformed so that they are closer to the data points

$$M_i \leftarrow \xi^\Delta \cdot M_i, \ i = 0, \ldots, N_M-1$$
$$\xi \leftarrow \xi \oplus \xi^\Delta$$

a          b

☐ **Fig. 14.44** Iterated closest point (ICP) matching of two point clouds: model (blue) and data (red) **a** before registration, **b** after registration; observed data points have been transformed to the model coordinate frame (Stanford bunny model courtesy Stanford Graphics Laboratory)

and the process is repeated until it converges. The correspondences used are unlikely to have all been correct, and therefore the estimate of the relative orientation between the sets is only an approximation.

We will illustrate this with the bunny point cloud. The model will be a random subsample of the bunny comprising 10% of the points

```
>>> model = bunny_pcd.downsample_random(0.1, seed=0)
PointCloud with 3594 points.
```

The data that we try to fit to the model is another random subsample, this time only 5% of the points, and we apply a rigid-body transformation to those points ▶

```
>>> data = SE3.Trans(0.3, 0.4, 0.5) * SE3.Rz(50, unit="deg")
...            * bunny_pcd.downsample_random(0.05, seed=-1);
```

Random sampling means that the chance of the same bunny point being in both point clouds is low. Despite this ICP does a very good job of matching and estimating the transformation.

We will color each of the point clouds and display them together

```
>>> model.paint([0, 0, 1])  # blue
PointCloud with 3594 points.
>>> data.paint([1, 0, 0])   # red
PointCloud with 1797 points.
>>> (model + data).disp(block=False)
```

which is shown in ◘ Fig. 14.44a. The + operator concatenates two `PointCloud` objects into a new `PointCloud` object.

```
>>> T, status = model.ICP(data, max_correspondence_distance=1,
...                       max_iteration=2_000, relative_fitness=0,
...                       relative_rmse=0)
>>> T.printline()
t = 0.3, 0.4, 0.5; rpy/zyx = 0.0976°, -0.0385°, 49.9°
```

which is the "unknown" relative pose of the data point cloud that we chose above. Now we can transform the data point cloud to the model coordinate frame and overlay it on the model

```
>>> (model + T.inv() * data).disp(block=False)
```

which is shown in ◘ Fig. 14.44b. We see that the red and blue point clouds are well aligned.

## 14.8 Applications

### 14.8.1 Perspective Correction

Consider the image

```
>>> notredame = Image.Read("notre-dame.png");
>>> notredame.disp();
```

shown in ◘ Fig. 14.45. The shape of the building is significantly distorted because the camera's optical axis was not normal to the plane of the building – we see evidence of perspective foreshortening or keystone distortion. We manually pick four points, clockwise from the bottom left, that are the corners of a large rectangle on the planar face of the building

```
>>> picked_points = plt.ginput(4);
```

which is a list of tuples which are the $u$- and $v$-coordinate of each selected point and this can be converted to a $2 \times 4$ array,

```
>>> p1 = np.array(picked_points).T;
```

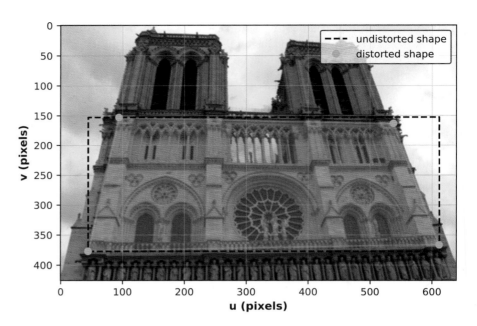

**◻ Fig. 14.45** Image taken from the ground shows the effect of foreshortening which gives the building a trapezoidal appearance (also known as keystone distortion). Four points on the approximately planar face of the building have been manually picked as indicated by the yellow dots and shading (Notre Dame de Paris)

which adopts the Toolbox standard for a set of points, with one column per point. To complete this example we will use some previously picked points

```
>>> p1 = np.array([
...         [ 44.1364,   94.0065,   537.8506,   611.8247],
...         [377.0654,  152.7850,   163.4019,   366.4486]]);
```

and overlay them on the image of the cathedral along with a translucent yellow trapezoid

```
>>> plot_polygon(p1, filled=True, color="y", alpha=0.4, linewidth=2);
>>> plot_point(p1, "yo");
```

We use the extrema of these points to define the vertices of a rectangle in the image

```
>>> mn = p1.min(axis=1);
>>> mx = p1.max(axis=1);
>>> p2 = np.array([[mn[0], mn[0], mx[0], mx[0]],
...                [mx[1], mn[1], mn[1], mx[1]]]);
```

which we overlay on the image as a dashed-black rectangle

```
>>> plot_polygon(p2, "k--", close=True, linewidth=2);
```

The sets of points `p1` and `p2` are projections of world points that lie approximately in a plane, so we can compute an homography

```
>>> H, _ = CentralCamera.points2H(p1, p2, method="leastsquares")
>>> H
array([[      1.4,     0.3827,    -136.6],
       [  -0.07853,     1.805,    -83.11],
       [-0.0002687,  0.001557,         1]])
```

that will transform the vertices of the yellow trapezoid to the vertices of the dashed-black rectangle ◄

$$\tilde{\boldsymbol{p}}_2 \simeq \mathbf{H}\tilde{\boldsymbol{p}}_1 \ .$$

That is, the homography maps image coordinates from the distorted keystone shape to an undistorted rectangular shape.

An homography can also be computed from four lines in the plane, for example the building's edges, but this is not supported by the Toolbox.

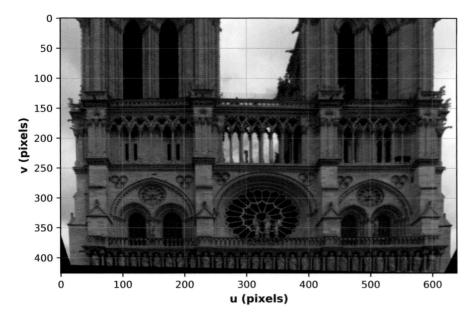

● **Fig. 14.46** A fronto-parallel view synthesized from ● Fig. 14.45. The image has been transformed so that the marked points become the corners of a rectangle in the image. The black pixels at the bottom edge have no corresponding pixels in the input image

We can apply this homography to the coordinate of every pixel in an output image in order to warp the input image. We use the Toolbox generalized image warping function

```
>>> notredame.warp_perspective(H).disp();
```

and the result shown in ● Fig. 14.46 is a synthetic fronto-parallel view. This is equivalent to the view that would be seen by a drone camera, with its optical axis normal to the facade of the cathedral. However points that are not in the plane, such as the left-hand side of the right bell tower, have been distorted.

In addition to creating this synthetic view, we can decompose the homography to recover the camera motion from the actual to the virtual viewpoint, and also the surface normal of the cathedral. The coordinate frames for this example are sketched in ● Fig. 14.47 and shows the actual and virtual camera poses. As in ▶ Sect. 14.2.4 we need to determine the camera calibration matrix so that we can convert the projective homography to a Euclidean homography. We obtain the focal length from the metadata in the image file

```
>>> f = notredame.metadata("FocalLength")
7.4
```

which is in units of millimeters, and the sensor of this camera is known to be $7.18 \times 5.32$ mm. We create a calibrated camera

```
>>> cam = CentralCamera(imagesize=notredame.shape, f=f/1000,
...                      sensorsize=[7.18e-3, 5.32e-3])
          Name: perspective [CentralCamera]
    pixel size: 1.121875e-05 x 1.2488262910798123e-05
    image size: 426 x 640
          pose: t = 0, 0, 0; rpy/yxz = 0°, 0°, 0°
  principal pt: [     213      320]
  focal length: [  0.0074    0.0074]
```

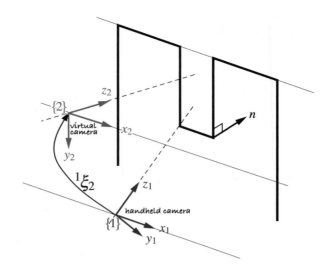

**◻ Fig. 14.47** Notre-Dame example showing the two camera coordinate frames. Frame {1} is the viewpoint of the camera that took the image, and {2} is the viewpoint of the synthetic fronto-parallel view

and we use that to compute and decompose the Euclidean homography

```
>>> pose, normals = cam.decomposeH(H)
>>> pose.printline(orient="camera")
t = 0.139, -0.614, 0.238; rpy/yxz = 3.69°, -36.9°, -8.32°
t = -0.139, 0.614, -0.238; rpy/yxz = 3.69°, -36.9°, -8.32°
t = -0.0308, 0.119, -0.661; rpy/yxz = 1.06°, 0.94°, 1.36°
t = 0.0308, -0.119, 0.661; rpy/yxz = 1.06°, 0.94°, 1.36°
```

We could automate this decision making by computing a homography from each possible camera pose and testing if it maps p1 to p2.

which returns an SE3 instance holding four possible camera poses, and a list of plane normals in the original camera frame. In this case the first solution is the correct one, since it represents a translation in the $-y$-direction and a significant negative pitch (rotation about the $x$-axis) angle. ◀ The camera needs to be pitched downward to make the camera's optical axis horizontal, and elevated to achieve the pose of the virtual camera that captures the fronto-parallel view. The estimated camera translation is not to scale.

The frontal plane of the cathedral $n$ is defined with respect to the original camera pose {1}

```
>>> normals[0].T
array([[-0.02788,    0.1523,    0.9879]])
```

and lies in the $yz$-plane as implied by ◻ Fig. 14.47.

## 14.8.2  Image Mosaicing

Mosaicing or image stitching is the process of creating a large-scale composite image from a number of overlapping images. It is commonly applied to drone and satellite images to create a continuous single image of the Earth's surface. It can also be applied to images of the ocean floor captured from downward looking cameras on an underwater robot. The panorama generation software supplied with, or built into, digital cameras and smart phones is another example of mosaicing.

As a rule of thumb images should overlap by 60% of area in the forward direction and 30% sideways.

The input to the mosaicing process is a sequence of overlapping images. ◀ It is not necessary to know the camera calibration parameters or the pose of the camera where the images were taken – the camera can rotate arbitrarily between images and the scale can change.

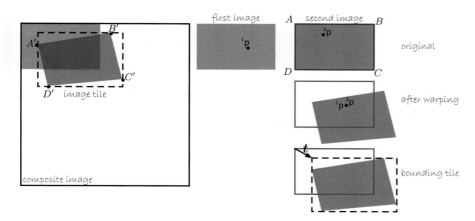

☐ **Fig. 14.48** The first image in the sequence is shown in red, the second in blue. The second image is warped into the image tile and then blended into the composite image

For this example we load a set of overlapping images

```
>>> images = ImageCollection("mosaic/aerial2-*.png", mono=True);
```

which are each $1280 \times 1024$. We start by creating an empty composite image that is $2000 \times 2000$ pixels

```
>>> composite = Image.Zeros(2_000, 2_000)
Image: 2000 x 2000 (uint8)
```

that will hold the mosaic. The essentials of the mosaicing process are shown in ☐ Fig. 14.48.

The first image is easy, and we simply paste it into the top left corner

```
>>> composite.paste(images[0], (0, 0));
```

of the composite image as shown in red in ☐ Fig. 14.48. The next image, shown in blue, requires more work – it needs to be rotated, scaled and translated so that it correctly overlays the red image.

We will assume the scene is planar which is reasonable for high-altitude photography where the vertical relief ▶ is small. This means that we can use a homography to map points between the different camera views. The first step is to identify common feature points, also known as tie points, and we use now familiar tools

The ratio of the height of points above the plane to the distance of the camera from the plane.

```
>>> next_image = images[1]
Image: 1113 x 825 (uint8), id=1 [.../mosaic/aerial2-002.png]
>>> sf_c = composite.SIFT()
SIFT features, 30281 points
>>> sf_next= next_image.SIFT()
SIFT features, 20521 points
>>> match = sf_c.match(sf_next);
```

and then estimate the homography which maps $^1$p to $^2$p

```
>>> H, _ = match.estimate(CentralCamera.points2H, "ransac",
...                       confidence=0.99);
>>> H
array([[ 0.9725,    0.137,    -245.6],
       [ -0.137,    0.9725,    -95.92],
       [2.668e-09, 1.639e-09,        1]])
```

using RANSAC to eliminate outlier matches. We map a point $^2$p in the new image to its corresponding coordinate in the composite image by

$$^1p \simeq \mathbf{H}^{-1}\,{}^2p$$

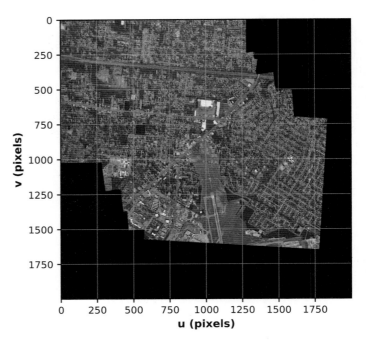

**□ Fig. 14.49** Example image mosaic. At the bottom of the frame we can clearly see three overlapping views of the airport runway which shows good alignment between the frames

and we can do this for every pixel in the new image by warping

```
>>> tile, topleft, corners =
...    next_image.warp_perspective(H, inverse=True, tile=True)
```

As shown in ◻ Fig. 14.48 the warped blue image falls outside the bounds of the original blue image, and the option `tile=True` specifies that the returned image is the minimum containing rectangle ◄ of the warped image. This image is referred to as a *tile* and shown with a dashed-black line. The return value `topleft` is the coordinate of the tile's coordinate frame with respect to the original image – the vector *t* in ◻ Fig. 14.48. In general not every pixel in the tile has a corresponding point in the input image and those pixels are set to zero.

Now the tile has to be *blended* into the composite mosaic image

```
>>> composite.paste(tile, topleft, method="blend");
```

and the result is shown in ◻ Fig. 14.49. We can clearly see that the images are overlaid with good alignment. The process is simply repeated for each of the input images.

The `"blend"` option sets the composite pixel to the new pixel value if the composite pixel was zero, otherwise it sets it to the average of the composite and new pixel value. This is a simple and cheap strategy, and if the images were taken with the same exposure then the edges of the tiles would not be visible. If the exposures were different, the two sets of overlapping pixels have to be analyzed to determine the average intensity offset and scale factor which can be used to correct the tile before blending – a process known as tone matching.

We also need to consider the effect of points in the image that are not in the ground plane such as those on a tall building. An image taken from directly overhead will show just the roof of the building, but an image taken from further away will be an oblique view that shows the side of the building. In a mosaic we want to create the illusion that we are directly above every point in the image so we should not see the sides of any building. This type of image is known as an orthophoto, and unlike a perspective view, where rays converge on the camera's focal point, the rays are all parallel which implies a viewpoint at infinity. ◄ At every pixel in the

The bounding box of the tile is computed by applying the homography to the image corners $A = (0, 0)$, $B = (W − 1, 0)$, $C = (W − 1, H − 1)$ and $D = (0, H − 1)$, where $W$ and $H$ are the width and height respectively, and finding the bounds in the $u$- and $v$-directions.

Google Earth sometimes provides an imperfect orthophoto. When looking at cities we might see oblique views of buildings.

composite image we can choose a pixel from any of the overlapping tiles. To best approximate an orthophoto we should choose the pixel that is closest to overhead, that is, prior to warping the pixel that was closest to the principal point.

In photogrammetry the type of mosaic just introduced is referred to as an uncontrolled digital mosaic since it does not use explicit control points – manually identified corresponding features in the images. The principles illustrated here can also be applied to the problem of image stabilization. The homography is used to map features in the new image to the location they had in the previous image. The full code for this example is `examples/mosaic`.

### 14.8.3  Visual Odometry

A common problem in robotics is to estimate the distance a robot has traveled, and this is a key input to all of the localization algorithms discussed in ▶ Chap. 6. For a wheeled robot we can use information from the wheel encoders, but these are subject to random errors (slippage) as well as systematic errors (imprecisely known wheel radius). However for an aerial or underwater robot the problem of odometry is much more difficult. Visual odometry (VO) is the process of using information from consecutive images to estimate the robot's relative motion from one camera image to the next.

We load a sequence of images taken from a car driving along a road

```
>>> left = ZipArchive("bridge-l.zip", filter="*.pgm", mono=True,
...                    dtype="uint8", maxintval=4095,
...                    roi=[20, 750, 20, 480]);
>>> len(left)
251
```

and the options trim out a central region of the images which have been rectified, and convert the `uint16` pixels containing 12-bit values to 8-bit integer values. ▶ The image sequence can be displayed as an animation

```
>>> for image in images:
...     image.disp(reuse=True, block=0.05)
```

at around 20 frames per second.

For each frame we can detect point features and overlay these as an animation

```
>>> fig, ax = plt.subplots()
>>> for image in left:
...     ax.clear()                              # clear the axes
...     image.disp(ax=ax)                       # display the image
...     features = image.ORB(nfeatures=20)      # compute ORB features
...     features.plot();                        # display ORB features
...     plt.pause(0.05)                         # small delay
```

and a single frame of this sequence is shown in ◘ Fig. 14.50. For variety, and computational efficiency, we use ORB features which are commonly used for this type of online application. We see that the point features *stick* reliably to points in the world over many frames, and show a preference for the corners of signs and cars, as well as the edges of trees. The motion of features in the image is known as optical flow, and is a function of the camera's motion through the world and the 3-dimensional structure of the world. ▶

To understand the 3-dimensional world structure we will use sparse stereo as discussed in ▶ Sect. 14.3. The vehicle in this example was fitted with a stereo camera and we load the corresponding right-camera images ▶

```
>>> right = ZipArchive("bridge-r.zip", mono=True, dtype="uint8",
...                     maxintval=4095, roi=[20, 750, 20, 480]);
```

This is the left image sequence from EISATS Bridge sequence in dataset 4 (Klette 2011) at ▶ http://www.mi.auckland.ac.nz/EISATS.

We will discuss optical flow in more detail in the next chapter. The magnitude of optical flow – the speed of a world point on the image plane – is proportional to camera velocity divided by the distance of the world point from the camera. Optical flow therefore has a scale ambiguity – a camera moving quickly through a world with distant points yields the same flow magnitude as a slower camera moving past closer points. To resolve this we need to use additional information. For example if we knew that the points were on the road surface, that the road was flat, and the height of the camera above the road then we can resolve this unknown scale. However this assumption is quite strict and would not apply for something like a drone moving over unknown terrain.

Both sets of images are rectified, the image cropping is to exclude the warping artefacts.

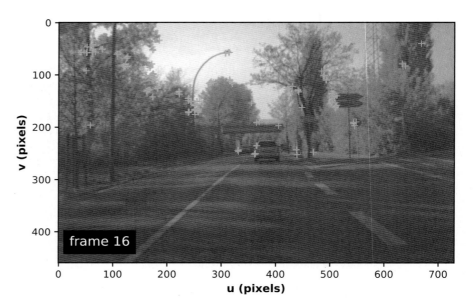

**⬛ Fig. 14.50** Frame number 16 from the `bridge-1` image sequence with overlaid features (image from .enpeda. project, Klette et al. 2011)

For each pair of left and right images we extract features, and determine correspondence by robustly matching features using descriptor similarity and the epipolar constraint implied by a fundamental matrix. Next, we compute horizontal disparity between corresponding features and, assuming the cameras are fully calibrated, we triangulate the image-plane coordinates to determine the world coordinates of the landmark points with respect to the left-hand camera on the vehicle.

We could match the 3D point clouds at the current and previous time step using a technique like iterated closest point (ICP) in order to determine the camera pose change. This is the so-called 3D-3D approach to visual odometry and, while the principle is sound, it works poorly in practice. Firstly, some of the 3D points may be on other moving objects and this violates the assumption of ICP that the sensor or the object moves, but not both. Secondly, the estimated range to distant points is quite inaccurate since errors in estimated disparity become significant when disparity is small.

An alternative approach, 3D-2D matching, projects the 3D points at the current time step into the previous image and finds the camera pose that minimizes the error with respect to the observed feature coordinates – this is bundle adjustment. Typically this is done for just one image and we will choose the left image. To establish correspondence of features over time we find correspondences between left-image features that had a match with the right image, and a match with features from the previous left image – again enforcing an epipolar constraint. We now know the correspondence between points in the three views of the scene as shown in ⬛ Fig. 14.51.

We could also frame this as a bundle adjustment problem with three cameras: left and right cameras at the current time step, and the left camera at the previous time step.

At each time step we set up a bundle adjustment problem that has two cameras and a number of landmarks determined from stereo triangulation. ◄ The first camera is associated with the previous time step and is fixed at the reference frame origin. The second camera is associated with the current time step and would be expected to have a translation in the positive $z$-axis direction. We could obtain an initial estimate of the second camera's pose by estimating and decomposing an essential matrix, but we will instead set it to the origin.

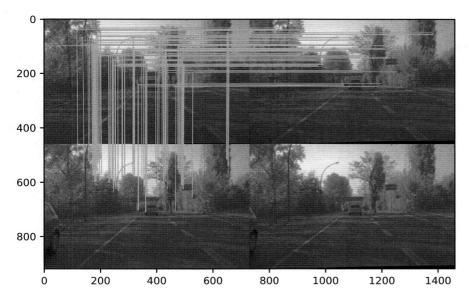

**☐ Fig. 14.51** Feature correspondence for visual odometry. The top row is a stereo pair at the current time step, and the bottom row is a stereo pair at the previous time step. Epiplolar consistent correspondences between three of the image images are shown in yellow (images from .enpeda. project, Klette et al. 2011)

For brevity, the full example is not included here but the details, with comments, can be found in the example script

```
>>> %run -m visodom
```

which displays graphics like ☐ Fig. 14.51 for every frame.

The final results for camera translation are shown in ☐ Fig. 14.52a, and we notice a value of around 0.5 m at each time step. There are also some anomalously high values and these have two causes. Firstly, the bundle adjustment process has failed to converge properly as indicated by the per-frame final squared error shown in ☐ Fig. 14.52b. The median error is around $0.07$ pixels$^2$ and 80% of the frames have a bundle-adjustment error of less than $0.1$ pixels$^2$. However the maximum is around $0.45$ pixels$^2$ and we could exclude such bad results and perhaps infer the translation from the previous value. The likely source of error is incorrect point correspondences. Bundle adjustment assumes that all points in the world are fixed but in this sequence there are numerous moving objects. We used the epipolar constraint between current and previous frame to ensure that only features points consistent with a moving camera and a fixed world are in the inlier set. There is also the motion of cars on the overpass bridges that is not consistent with the motion induced by the car's own motion. A more sophisticated bundle adjustment algorithm would detect and reject such points. Finally, there is the issue of the number of matching features points and their distribution over the image.

The number of inlier matches is plotted in ☐ Fig. 14.52c and varies considerably over the sequence. However, we can see that this is not strongly correlated with error. A bigger problem is that a preponderance of the matched feature points lie in the top part of the frame and correspond to world points that are quite distant from the cameras, and this leads to poor translation estimation. A more sophisticated approach to feature detection would choose features more uniformly spread over the image.

The second cause of error is a common one when using video data for robots. The clue is that a number of camera displacements are suspiciously close to exactly twice the median value. Each image in the sequence was assigned a timestamp

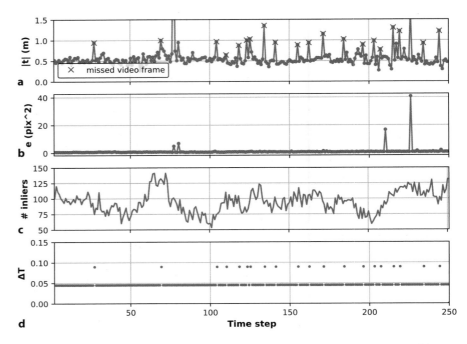

**Fig. 14.52**   Visual odometry results. **a** Estimated norm of camera displacement; **b** bundle adjustment final error per frame; **c** number of inlier matches; **d** image time stamp

when it was received by the computer and those timestamps can be loaded from the left-image archive

```
>>> ts = np.loadtxt(left.open("timestamps.dat"));
```

and the difference between timestamps

```
>>> plt.plot(np.diff(ts));
```

is shown in ⬛ Fig. 14.52d. We see that while the median time between images is 44.6 ms, there are many spikes where the interval is twice that. The computer logging the images has skipped a frame, perhaps it was unable to write image data to memory or disk as quickly as it was arriving. So the interval between the frames was twice as long, the vehicle traveled twice as far, and the spikes on our estimated displacement are in fact correct. This is not an uncommon situation – in a robot system all data should be timestamped, and timestamps should be checked to detect problems like this. The displacements in ⬛ Fig. 14.52a that have this timestamp issue are annotated with a red cross. The median velocity over these frames is 11.53 m s$^{-1}$ or around 40 km h$^{-1}$.

For a vehicle or ground robot the estimated displacements in 3-dimensional space over time are not independent – they are constrained by the vehicle's kinodynamic model as discussed in ▶ Chap. 4. We could use this model to smooth the results and discount erroneous velocity estimates. If the bundle adjuster included constraints on camera pose we could set the weighting to penalize infeasible motion in the lateral and vertical directions as well as roll and pitch motion.

## 14.9 Wrapping Up

This chapter has covered many topics with the aim to demonstrate a multiplicity of concepts that are of use in real robotic vision systems. There have been two common threads through this chapter. The first was the use of point features which correspond to distinctive points in the world, and allow matching of these between images taken from different viewpoints. The second thread was the loss of scale in the perspective projection process, and techniques based on additional sources of information to recover scale, for example stereo vision, structured light or bundle adjustment.

We extended the geometry of single camera imaging to the case of two cameras, and showed how corresponding points in the two images are constrained by the fundamental matrix. We showed how the fundamental matrix can be estimated from image data, the effect of incorrect data association, and how to overcome this using the RANSAC algorithm. Using camera intrinsic parameters the essential matrix can be computed and then decomposed to give the camera motion between the two views, but the translation has an unknown scale factor. With some extra information such as the magnitude of the translation, the camera motion can be estimated completely. Given the camera motion, then the 3-dimensional coordinates of points in the world can be estimated.

For the special case where world points lie on a plane they *induce* a homography that is a linear mapping of image points between images. The homography can be used to detect points that do not lie in the plane, and can be decomposed to give the camera motion between the two views (translation again has an unknown scale factor) and the normal to the plane.

If the fundamental matrix is known, then a pair of overlapping images can be rectified to create an epipolar-aligned stereo pair, and dense stereo matching can be used to recover the world coordinates for every point. Errors due to effects such as occlusion and lack of texture were discussed, as were techniques to detect these situations.

We used bundle adjustment to solve the structure and motion estimation problem – using 2D measurements from a set of images of the scene to recover information related to the 3D geometry of the scene as well as the locations of the cameras. Stereo vision is a simple case where the motion is known – fixed by the stereo baseline – and we are interested only in structure. The visual odometry problem is complementary and we are interested only in the motion of the camera, not the scene structure.

We used point clouds, including colored point clouds, to represent and visualize the 3-dimensional world. We also demonstrated some operations on point cloud data such as outlier removal, downsampling, normal estimation, plane fitting and registration.

Finally, these multi-view techniques were then used for application examples such as perspective correction, mosaic creation, and visual odometry.

### 14.9.1 **Further Reading**

3-dimensional reconstruction and camera pose estimation has been studied by the photogrammetry community since the mid nineteenth century, see ▶ Exc. 13.11. 3-dimensional computer vision or *robot vision* has been studied by the computer vision and artificial intelligence communities since the 1960s. This book follows the language and nomenclature associated with the computer vision literature, but the photogrammetric literature can be comprehended with only a little extra difficulty. The similarity of a stereo camera to our own two eyes is very striking, and while we do make strong use of stereo vision it is not the only technique we use to infer distance (Cutting 1997).

Significant early work on multi-view geometry was conducted at laboratories such as Stanford, SRI International, MIT AI laboratory, CMU, JPL, INRIA, Oxford and ETL Japan in the 1980s and 1990s and led to a number of text books being published in the early 2000s. The definitive references for multiple-view geometry are Hartley and Zisserman (2003) and Ma et al. (2003). These books present quite dif-

**Table 14.1** Rosetta stone. Summary of notational differences between two other popular textbooks and this book

| Object | Hartley & Zisserman 2003 | Ma et al. 2003 | This book |
|---|---|---|---|
| World point | $\mathbf{X}$ | $P$ | P |
| Image plane point | $\mathbf{x}, \mathbf{x}'$ | $x_1, x_2$ | $^1\mathsf{p}, {}^2\mathsf{p}$ |
| $i^{\text{th}}$ image plane point | $\mathbf{x}_i, \mathbf{x}'_i$ | $x_1^i, x_2^i$ | $^1\mathsf{p}_i, {}^2\mathsf{p}_i$ |
| Camera motion | $R, \mathbf{t}$ | $R, T$ | $\mathbf{R}, t$ |
| Normalized coordinates | $\mathbf{x}, \mathbf{x}'$ | $x_1, x_2$ | $(\bar{u}, \bar{v})$ |
| Camera matrix | P | $\Pi$ | $\mathbf{C}$ |
| Homogeneous quantities | $\mathbf{x}, \mathbf{X}$ | $x, P$ | $\tilde{p}, \tilde{P}$ |
| Homogeneous equivalence | $\mathbf{x} = \mathbf{PX}$ | $\lambda x = \Pi P$ $x \sim \Pi P$ | $\tilde{p} \simeq \mathbf{C}\tilde{P}$ |

ferent approaches to the same body of material. The former takes a more geometric approach while the latter is more mathematical. Unfortunately they use quite different notation, and each differs from the notation used in this book – a summary of the important notational elements is given in ◨ Table 14.1. These books all cover feature extraction (using Harris point features, since they were published before scale invariant feature detectors such as SIFT were developed); the geometry of one, two and $N$ views; fundamental and essential matrices; homographies; and the recovery of 3-dimensional scene structure and camera motion through offline batch techniques. Both provide the key algorithms in pseudo-code and have some supporting MATLAB code on their associated web sites. The slightly earlier book by Faugeras et al. (2001) covers much of the same material using a fairly mathematical approach and with different notation again. The older book by Faugeras (1993) focuses on sparse stereo from line features. The book by Szeliski (2022) provides a very readable and deeper discussion of the topics in this chapter and a PDF version is available for personal use from ▶ https://szeliski.org/Book.

SIFT and other feature detectors were previously discussed in ▶ Sects. 14.1 and 12.3.2.1. The performance of feature detectors and their matching performance is covered in Mikolajczyk and Schmid (2005) which reviews a number of different feature descriptors. Arandjelović and Zisserman (2012) discuss some important points when matching feature descriptors.

The RANSAC algorithm described by Fischler and Bolles (1981) is the workhorse of all the feature-based methods discussed in this chapter but fails with very small inlier ratios. Subsequent work includes vector field consensus (VFC) by Ma et al. (2014) and progressive sample consensus (PROSAC) by Chum and Matas (2005).

The term fundamental matrix was defined in the thesis of Luong (1992). The book by Xu and Zhang (1996) is a readable introduction to epipolar geometry. Epipolar geometry can also be formulated for nonperspective cameras in which case the epipolar line becomes an epipolar curve (Mičušík and Pajdla 2003; Svoboda and Pajdla 2002). For three views, the geometry is described by the trifocal tensor $\mathcal{T}$ which is a $3 \times 3 \times 3$ tensor with 18 degrees of freedom that relates a point in one image to epipolar lines in two other images (Hartley and Zisserman 2003; Ma et al. 2003). An important early paper on epipolar geometry for an image sequence is Bolles et al. (1987).

The essential matrix was first described a decade earlier in a letter to Nature (Longuet-Higgins 1981) by the theoretical chemist and cognitive scientist Christopher Longuet-Higgins (1923–2004). The paper describes a method of estimating

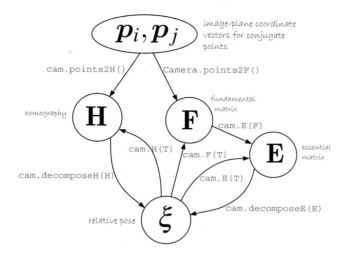

**Fig. 14.53** Toolbox functions and camera object methods, and their inter-relationship. `cam.` denotes an instance method, and `Camera.` denotes a class method

the essential matrix from eight corresponding point pairs. The decomposition of the essential matrix was first described in Faugeras (1993, § 7.3.1) but is also covered in the texts Hartley and Zisserman (2003) and Ma et al. (2003). In this chapter we have estimated camera motion by first computing the essential matrix and then decomposing it. The first step requires at least eight pairs of corresponding points but algorithms such as Nistér (2003), Li and Hartley (2006) compute the motion directly from just five pairs of points. Decomposition of a homography is described by Faugeras and Lustman (1988), Hartley and Zisserman (2003), Ma et al. (2003), and the comprehensive technical report by Malis and Vargas (2007). The relationships between these matrices, camera motion, and the relevant Toolbox functions are summarized in ■ Fig. 14.53.

Stereo cameras and stereo matching software are available today from many sources and can provide high-resolution depth maps at more than 10 Hz on standard computers. In the 1990s this was challenging, and custom hardware including FPGAs was required to achieve real-time operation (Woodfill and Von Herzen 1997; Corke et al. 1999). The application of stereo vision for planetary rover navigation is discussed by Matthies (1992). More than two cameras can be used, and multi-camera stereo was introduced by Okutomi and Kanade (1993) and provides robustness to problems such as the picket fence effect.

Brown et al. (2003) provide a readable review of stereo vision techniques with a focus on real-time issues. An old, but clearly written, book on the principles of stereo vision is Shirai (1987). Scharstein and Szeliski (2002) consider the stereo process as four steps: matching, aggregation, disparity computation and refinement. The cost and performance of different algorithms for each step are compared. The `stereo_simple` dense stereo matching method presented in ▶ Sect. 14.4 is a very conventional correlation-based stereo algorithm, and would be described as: NCC matching, box filter aggregation, and winner takes all. The disparity is computed independently at each pixel, but for real scenes adjacent pixels belong to the same surface and disparity will be quite similar – this is referred to as the *smoothness constraint*. Ensuring smoothness can be achieved using Markov random fields (MRFs), total variation with regularizers (Pock 2008), or more efficient semi-global matching (SGM) algorithms (Hirschmüller 2008). The very popular library for efficient large-scale stereo matching (LIBELAS) by Geiger et al. (2010) uses an alternative to global optimization that provides fast and accurate results for a variety of indoor and outdoor scenes. Stereo vision involves a significant amount of computation but there is considerable scope for parallelization using multiple cores, MIMD instruction sets, GPUs, custom chips and FPGAs. The use of non-parametric local transforms is described by Zabih and Woodfill (1994) and Banks and Corke (2001).

Deep learning has been applied to stereo matching (Luo et al. 2016) as well as for estimating depth directly from monocular image sequences (Garg et al. 2016), perhaps implicitly using the sort of monocular cues used by humans (Cutting 1997). Deep networks have also been applied to problems such as visual odometry (Zhan et al. 2018).

The ICP algorithm (Besl and McKay 1992) is used for a wide range of applications from robotics to medical imaging. ICP is fast but determining the correspondences via nearest neighbors is an expensive $O(N^2)$ operation. This can be speeded up by using a kd-tree to organize the data points. Many variations have been developed that make the approach robust to outlier data and to improve computational speed for large datasets. Salvi et al. (2007) provide a recent review and comparison of some different algorithms. Determining the relative orientation between two sets of points is a classical problem and the SVD approach used here is described by Arun et al. (1987). Solutions based on quaternions and orthonormal rotation matrices have been described by Horn (Horn et al. 1988; Horn 1987).

Structure from motion (SfM), the simultaneous recovery of world structure and camera motion, is a classical problem in computer vision. Two useful review papers are by Huang and Netravali (1994) which provides a taxonomy of approaches, and Jebara et al. (1999). Broida et al. (1990) describe an early recursive SfM technique for a monocular camera sequence using an EKF where each world point is represented by its $(X, Y, Z)$ coordinate. McLauchlan provides a detailed description of a variable-length state estimator for SfM (McLauchlan 1999). Azarbayejani and Pentland (1995) present a recursive approach where each world point is parameterized by a scalar, its depth with respect to the first image. A more recent algorithm with bounded estimation error is described by Chiuso et al. (2002) and also discusses the problem of scale variation. The MonoSLAM system by Davison et al. (2007) is an impressive monocular SfM system that maintains a local map that includes features even when they are not currently in the field of view. Newcombe et al. (2011) performed camera tracking and dense 3D reconstruction from a single moving RGB camera. The application of SfM to large-scale urban mapping is becoming increasing popular and Pollefeys et al. (2008) describe a system for offline processing of large image sets. The ORB-SLAM family of visual SLAM algorithms (Campos et al. 2021) is powerful and popular way to compute the 3D path of a camera given a sequence of RGB, RGBD or stereo images.

Bundle adjustment or structure from motion (SfM) is a big field with a large literature that cover many variants of the problem, for example robustness to outliers, and specific applications and camera types. Classical introductions include Triggs et al. (2000) and Hartley and Zisserman (2003). Theses by Warren (2015), Sünderhauf (2012) and Strasdat (2012) are comprehensive and readable. Unfortunately, every reference uses different notation. Estimating the camera matrix for each view, computing a projective reconstruction, and then upgrading it to a Euclidean reconstruction is described by Hartley and Zisserman (2003) and Ma et al. (2003).

The SfM problem can be simplified by using stereo rather than monocular image sequences (Molton and Brady 2000; Zhang et al. 1992), or by incorporating inertial data (Strelow and Singh 2004). A readable two-part tutorial introduction to visual odometry (VO) is Scaramuzza and Fraundorfer (2011) and Fraundorfer and Scaramuzza (2012). Visual odometry is discussed by Nistér et al. (2006) using point features and monocular or stereo vision. Maimone et al. (2007) describe experience with stereo-camera VO on the Mars rover, and Corke et al. (2004) describe monocular catadioptric VO for a prototype planetary rover.

Mosaicing is a process as old as photography. In the past it was highly skilled and labor intensive requiring photographs, scalpels and sandpaper. The surfaces of the Moon and nearby planets were mosaiced manually in the 1960s using imagery sent back by robotic spacecraft. High-quality offline mosaicing tools are available for creating panoramas, for example Hugin, PhotoStitcher and AutoStitch.

Image sequence analysis is the core of many real-time robotic vision systems. Early work on real-time feature tracking across frames, such as Hager and Toyama (1998) and Lucas and Kanade (1981), was limited by available computation. It was typically based on the computationally cheaper Harris detectors or the pyramidal Kanade-Lucas-Tomasi (KLT) tracker. Today, efficient implementations of complex feature detectors like SIFT and SURF, perhaps using GPU hardware, are capable of real-time performance.

### 14.9.2 Resources

The field of computer vision has progressed through the availability of standard datasets. These have enabled researchers to quantitatively compare the performance of different algorithms on the same data. One of the earliest collections of stereo image pairs was the JISCT dataset (Bolles et al. 1993). The more recent Middlebury dataset (Scharstein and Szeliski 2002) at ▶ https://vision.middlebury. edu/stereo provides an extensive collection of stereo images, at high resolution, taken at different exposure settings and including ground truth data. Stereo images from various NASA Mars rovers are available online as left+right pairs or encoded in anaglyphs. Motion datasets include people moving inside a building ▶ https:// homepages.inf.ed.ac.uk/rbf/CAVIARDATA1, traffic scenes ▶ http://i21www.ira. uka.de/image_sequences, and from a moving vehicle ▶ http://www.mi.auckland. ac.nz/EISATS.

The popular LIBELAS library (▶ http://www.cvlibs.net/software/libelas) for large-scale stereo matching supports parallel processing using OpenMP and has MATLAB and ROS interfaces. Various stereo-vision algorithms are compared for speed and accuracy at the KITTI (▶ http://www.cvlibs.net/datasets/kitti/eval_ scene_flow.php) and Middlebury (▶ http://vision.middlebury.edu/stereo/eval3) benchmark sites.

The Epipolar Geometry Toolbox (Mariottini and Prattichizzo 2005) for MAT-LAB by Mariottini and Prattichizzo is available at ▶ http://egt.dii.unisi.it and handles perspective and catadioptric cameras. Andrew Davison's monocular visual SLAM system (MonoSLAM) for C and MATLAB is available at ▶ https://www. doc.ic.ac.uk/~ajd/software.html.

The sparse bundle adjustment software (SBA) by Lourakis (▶ https://users.ics. forth.gr/~lourakis/sba) is an efficient C implementation that is widely used and has a MATLAB and OpenCV wrapper. The Toolbox `BundleAdjust` class can import SBA example files. One application is Bundler (▶ https://www.cs.cornell.edu/ ~snavely/bundler) which can perform matching of points from thousands of cameras over city scales and has enabled reconstruction of cities such as Rome (Agarwal et al. 2011), Venice and Dubrovnik. Some of these large-scale datasets are available from ▶ https://grail.cs.washington.edu/projects/bal and ▶ https://www. cs.cornell.edu/projects/p2f. Other open source solvers that can be used for sparse bundle adjustment include $g^2o$, SSBA and CERES, all implemented in C++. $g^2o$ by Kümmerle et al. (2011) (▶ https://github.com/RainerKuemmerle/g2o) can also be used to solve SLAM problems. SSBA by Christopher Zach is available at ▶ https:// github.com/chzach/SSBA. The CERES solver from Google (▶ http://ceres-solver. org) is a library for modeling and solving large complex optimization problems on desktop and mobile platforms.

Open3D (▶ http://www.open3d.org) is a large-scale, open and standalone package for 2D/3D image and point cloud processing with support for feature detectors and descriptors, 3D registration, kd-trees, shape segmentation, surface meshing, and visualization (Zhou et al. 2018). The Toolbox `PointCloud` class is a lightweight wrapper of Open3D. The Point Data Abstraction Library (PDAL) (▶ http://www.pdal.io) is a library and set of Unix command line tools for manipulating point cloud data.

Point clouds can be stored in a number of common formats that can include optional color information as well as point coordinates. Point Cloud Data (PCD) files were defined by the earlier Point Cloud library (PCL) project (▶ https://pointclouds.org). Polygon file format (PLY) files are designed to describe meshes but can be used to represent an unmeshed point cloud, and there are a number of great visualizers such as MeshLab and potree. Open3D and PDAL allowing reading and writing many point cloud file formats from Python. LAS format is widely used for large point clouds and colored point clouds, it is a binary format and LAZ format is a compressed version. Cloud Compare (▶ https://www.danielgm.net/cc/) is a cross-platform open-source GUI-based tool for displaying, comparing and editing massive point clouds stored in a variety of formats.

A song about the fundamental matrix can be found at ▶ https://danielwedge.com/fmatrix/.

### 14.9.3 Exercises

1. Point features and matching (▶ Sect. 14.1). Examine the cumulative distribution of feature strength for Harris and SIFT features. What is an appropriate way to choose strong features for feature matching?
2. Feature matching. We could define the quality of descriptor-based feature matching in terms of the percentage of inliers after applying RANSAC.
   a) Take any image. We will match this image against various transforms of itself to explore the robustness of SIFT and Harris features. The transforms are: *(a)* scale the intensity by 70%; *(b)* add Gaussian noise with standard deviation of 0.05, 0.5 and 2 gray values; *(c)* scale the size of the image by 0.9, 0.8, 0.7, 0.6 and 0.5; *(d)* rotate by 5, 10, 15, 20, 30, 40°.
   b) For the Harris detector compare the matching performance for patch descriptor sizes of $3 \times 3$, $7 \times 7$ and $11 \times 11$ and $15 \times 15$.
   c) Try increasing the suppression radius for Harris point features. Does the lower density of matches improve the matching performance?
   d) Is there any correlation between outlier matches and feature strength?
3. Write the equation for the epipolar line in image two, given a point in image one.
4. Show that the epipoles are the null space of the fundamental matrix.
5. Can you determine the camera matrix **C** for camera two given the fundamental matrix and the camera matrix for camera one?
6. Estimating the fundamental matrix (▶ Sect. 14.2.3)
   a) For the synthetic data example vary the number of points and the additive Gaussian noise and observe the effect on the residual.
   b) For the Eiffel tower example observe the effect of varying the parameters to RANSAC. Repeat this with just the top 50% of SIFT features after sorting by feature strength.
   c) What is the probability of drawing 8 inlier points in a random sample (without replacement) from $N$ inliers and $M$ outliers?
7. Epipolar geometry (▶ Sect. 14.2)
   a) Create two central cameras, one at the origin and the other translated in the $x$-direction. For a sparse fronto-parallel grid of world points display the family of epipolar lines in image two that correspond to the projected points in image one. Describe these epipolar lines? Repeat for the case where camera two is translated in the $y$- and $z$-axes and rotated about the $x$-, $y$- and

$z$-axes. Repeat this for combinations of motion such as $x$- and $z$-translation or $x$-translation and $y$-rotation.

b) The example of ◼ Fig. 14.16 has epipolar lines that slope slightly upward. What does this indicate about the two camera viewpoints?

8. Essential matrix (▶ Sect. 14.2.2)

a) Create a set of corresponding points for a camera undergoing pure rotational motion, and compute the fundamental and essential matrix. Can you recover the rotational motion?

b) For a case of translational and rotational motion visualize both poses that result from decomposing the essential matrix. Sketch it or alternatively use `CentralCamera.plot`.

9. Homography (▶ Sect. 14.2.4)

a) Compute Euclidean homographies for translation in the $x$-, $y$- and $z$-directions and for rotation about the $x$-, $y$- and $z$-axes. Convert these to projective homographies and apply to a fronto-parallel grid of points. Is the resulting image motion what you would expect? Apply these homographies as a warp to a real image such as Mona Lisa.

b) Decompose the homography of ◼ Fig. 14.15, the courtyard image, to determine the plane of the wall with respect to the camera. You will need the camera intrinsic parameters.

c) Reverse the order of the points given to `points2H` and show that the result is the inverse homography.

10. Load a reference image of this book's cover from `rvc3_cover.png`. Next, capture an image that includes the book's front cover, compute SIFT features, match them and use RANSAC to estimate a homography between the two views of the book cover. Decompose the homography to estimate rotation and translation. Put all of this into a real-time loop and continually display the pose of the book relative to the camera.

11. Sparse stereo (▶ Sect. 14.3)

a) The line intersection method can return the closest distance between the lines (which is ideally zero). Plot a histogram of the closing error and compute the mean and maximum error.

b) The assumed camera translation magnitude was 30 cm. Repeat for 25 and 35 cm. Are the closing error statistics changed? Can you determine what translation magnitude minimizes this error?

12. Bundle adjustment (▶ Sect. 14.3.2)

a) Vary the initial condition for the second camera, for example, set it to the identity matrix.

b) Set the initial camera translation to 3 m in the $x$-direction, and scale the landmark coordinates by 10×. What is the final value of the back-projection error and the second camera pose.

c) Experiment with anchoring landmarks and cameras.

d) Derive the two Jacobians **A** (hard) and **B**.

13. Derive a relationship for depth in terms of disparity for the case of verged cameras. That is, cameras with their optical axes intersecting similar to the cameras shown in ◼ Fig. 14.6.

14. Stereo vision. Using the rock piles example (◼ Fig. 14.23)

a) Zoom in on the disparity image and examine pixel values on the boundaries of the image and around the edges of rocks.

b) Experiment with different window sizes. What effects do you observe in the disparity image and computation time?

c) Experiment with changing the disparity range. Try [50,90], [30,90], [40,80] and [40,100]. What happens to the disparity image and why?

d) Display the epipolar lines on image two for selected points in image one.

e) Obtain the disparity space image **D**
   i. For selected pixels $(u, v)$ plot $\mathbf{D}(u, v, d)$ versus $d$. Look for pixels that have a sharp peak, broad peak and weak peak. For a selected row $v$ display $\mathbf{D}(u, v, d)$ as an image. What does this represent?
   ii. For a particular pixel plot $s$ versus $d$, fit a parabola around the maxima and overlay this on the plot.
   iii. Use raw data from the DSI, and compute the peak location to subpixel resolution for every pixel.
   iv. Use raw data from the DSI, find the second peak at each pixel and compute the ambiguity ratio.

15. Anaglyphs (▶ Sect. 14.5)
   a) Download an anaglyph image and convert it into a pair of grayscale images, then compute dense stereo.
   b) Create your own anaglyph from a stereo pair.
   c) Write code to plot shapes, like squares, that appear at different depths. Or, a single shape that moves in and out of the screen.
   d) Write code to plot a 3-dimensional shape like a cube.

16. Stereo vision. For a pair of identical cameras with a focal length of 8 mm, $1000 \times 1000$ pixels that are $10\,\mu$m square on an 80 mm baseline and with parallel optical axes:
   a) Sketch, or write code to display, the fields of views of the camera in a plan view. If the cameras are viewing a plane surface normal to the principal axes how wide is the horizontal overlapping field of view in units of pixels?
   b) Assuming that disparity error is normally distributed with $\sigma = 0.1$ pixels compute and plot the distribution of error in the $z$-coordinate of the reconstructed 3D points which have a mean disparity of 0.5, 1, 2, 5, 10 and 20 pixels. Draw 1000 random values of disparity, convert these to $Z$ and plot a histogram (distribution) of their values.

17. A da Vinci on your wall. Acquire an image of a room in your house and display it using the Toolbox. Select four points, using `ginput`, to define the corners of the virtual frame on your wall. Perhaps use the corners of an existing rectangular feature in your room such as a window, poster or picture. Estimate the appropriate homography, warp the Mona Lisa image and insert it into the original image of your room.

18. Plane fitting (▶ Sect. 14.7.1)
   a) Test the robustness of the plane fitting algorithm to additive noise and outlier points.
   b) Implement an iterative approach with weighting to minimize the effect of outliers.
   c) Create a RANSAC-based plane fit algorithm that takes random samples of three points to solve for the equation of a plane ▶ App. C.1.3.

19. ICP (▶ Sect. 14.7.2)
   a) Change the initial relative pose between the point clouds. Try some very large rotations.
   b) Explore the robustness of ICP by simulating some realistic sensor errors. For example, add Gaussian or impulse noise to the data points. Or add an increasing number of spurious data points.
   c) How does matching performance vary as the number of data points increases and decreases.
   d) Discover and explore all the ICP options provided by the underlying Open3D function.

20. Perspective correction (▶ Sect. 14.8.1)
   a) Create a virtual view looking downward at 45° to the front of the cathedral.
   b) Create a virtual view from the original camera viewpoint but with the camera rotated 20° to the left.
   c) Find another real picture with perspective distortion and attempt to correct it.

21. Mosaicing (▶ Sect. 14.8.2)
    a) Run the example file `mosaic` and watch the whole mosaic being assembled.
    b) Modify the way the tile is pasted into the composite image to use pixel setting rather than averaging.
    c) Modify the way the tile is pasted into the composite image so that pixels closest to the principal point are used.
    d) Run the software on a set of your own overlapping images and create a panorama.
22. Image stabilization can be used to virtually stabilize an unsteady camera, perhaps one that is handheld, on a drone or on a mobile robot traversing rough terrain. Capture a short image sequence $\mathbf{I}_0, \mathbf{I}_1 \cdots \mathbf{I}_{N-1}$ from an unsteady camera. For frame $i$, $i \geq 1$ estimate a homography with respect to frame 0, warp the image appropriately, and store it in an array. Animate the stabilized image sequence.
23. Visual odometry (▶ Sect. 14.8.2). Modify the example script to:
    a) use Harris or SIFT features instead of ORB. What happens to accuracy and execution time?
    b) track the features from frame to frame and display the displacement (optical flow) vectors at each frame
    c) using a scale-invariant point feature, track points across multiple frames. You will need to create a a datastructure to manage the life cycle of each point: it is first seen, it is seen again, then it is lost.
    d) ensure that features are more uniformly spread over the scene, investigate the `gridify` method for `Feature` objects.
    e) plot the fundamental matrix residuals at each time step (there are two of them). Is there a pattern here? Adjust the RANSAC parameters so as to reduce the number of times bundle adjustment fails.
    f) use a robust bundle adjuster, either find one or implement one (hard).
    g) use a Kalman filter with simple vehicle dynamics to smooth the velocity estimates or optionally skip the update step if the bundle adjustment result is poor.
    h) explore the statistics of the bundle adjustment error over all the frames. Does increasing the number of iterations help?
    i) the bundle adjustment initializes the second camera at the origin. Does setting its displacment to that determined for the previous frame improve performance?
    j) redo the bundle adjustment with three cameras: left and right cameras at the current time step, and the left camera at the previous time step.
24. Learn about kd-trees and create your own implementation.

# Vision-Based Control

Contents

It is common to talk about a robot moving to an object, but in reality the robot is only moving to a pose at which it expects the object to be. This is a subtle but deep distinction. A consequence of this is that the robot will fail to grasp the object if it is not at the expected pose. It will also fail if imperfections in the robot mechanism or controller result in the end-effector not actually achieving the end-effector pose that was specified. In order for this conventional approach to work successfully we need to solve two quite difficult problems: determining the pose of the object and ensuring the robot achieves that pose.

The first problem, determining the pose of an object, is typically avoided in manufacturing applications by ensuring that the object is always precisely placed. This requires mechanical jigs and fixtures which are expensive, and have to be built and set up for every different part the robot needs to interact with, somewhat negating the flexibility of robotic automation.

» *the root cause of the problem is that the robot cannot "see" what it is doing.*

Consider if the robot could see the object and its end-effector, and could use that information to guide the end-effector toward the object. This is what humans call hand-eye coordination and what we will call vision-based control or visual servo control – the use of information from one or more cameras to guide a robot in order to achieve a task.

The pose of the target does not need to be known a priori; the robot moves toward the observed target wherever it might be in the workspace. There are numerous advantages of this approach: part position tolerance can be relaxed, the ability to deal with parts that are moving comes almost for free, and any errors in the robot's intrinsic accuracy will be compensated for.

A vision-based control system involves continuous measurement of the target and the robot using vision to create a feedback signal and moves the robot arm until the visually observed error between the robot and the target is zero. Vision-based control is quite different to taking an image, determining where the target is and then reaching for it. The advantage of continuous measurement and feedback is that it provides great robustness with respect to any errors in the system.

In this part of the book we bring together much that we have learned previously: kinematics and dynamics for robot arms and mobile robots; geometric aspects of image formation; and feature extraction. The part comprises two chapters. ► Chap. 15 discusses the two classical approaches to visual servoing which are known as position-based and image-based visual servoing. The image coordinates of world features are used to move the robot toward a desired pose relative to the observed object. The first approach requires explicit estimation of object pose from image features, but because it is performed in a closed-loop fashion any errors in pose estimation are compensated for. The second approach involves no pose estimation and uses image-plane information directly. Both approaches are discussed in the context of a perspective camera which is free to move in three dimensions, and their respective advantages and disadvantages are described. The chapter also includes a discussion of the problem of determining object depth, and the use of line and ellipse image features.

► Chap. 16 extends the discussion to hybrid visual-servo algorithms which overcome the limitations of the position- and image-based visual servoing by using the best features of both. The discussion is then extended to nonperspective cameras such as fisheye lenses and catadioptric optics as well as arm robots, holonomic and nonholonomic ground robots, and a aerial robot.

This part of the book is pitched at a higher level than earlier parts. It assumes a good level of familiarity with the rest of the book, and the increasingly complex examples are sketched out rather than described in detail. The text introduces the essential mathematical and algorithmic principles of each technique, but the full details are to be found in the source code of the Python classes that implement the controllers, or in the details of the block diagrams. The results are also increasingly hard to depict in a book and are best understood by running the supporting Python bdsim code and plotting the results or watching the animations.

# Vision-Based Control

**Contents**

© The Author(s), under exclusive license to Springer Nature Switzerland AG 2023
P. Corke, *Robotics, Vision and Control*, Springer Tracts in Advanced Robotics 146,
https://doi.org/10.1007/978-3-031-06469-2_15

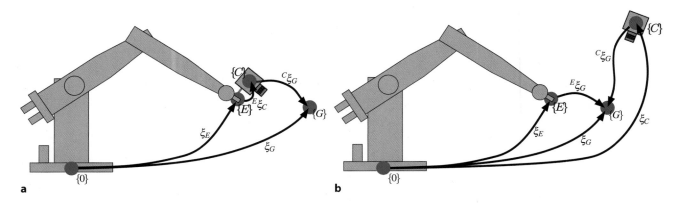

**Fig. 15.1** Visual servo configurations as a pose graph. The coordinate frames are: world {0}, end effector {E}, camera {C} and goal {G}. **a** End-point closed-loop configuration (eye-in-hand); **b** end-point open-loop configuration

`chap15.ipynb`

▶ go.sn.pub/WETNOx

The task in visual servoing is to control the pose of the robot's end effector, relative to some goal object, using visual features extracted from an image of the goal object. The features might be centroids of points, equations of lines or ellipses, or even brightness patterns of pixels. The image of the goal is a function of its camera-relative pose $^C\xi_G$, and so to are the features extracted from the image.

There are two possible configurations of camera and robot. The configuration of ☐ Fig. 15.1a has the camera mounted on the robot's end effector observing the goal, and is referred to as end-point closed-loop or eye-in-hand. The configuration of ☐ Fig. 15.1b has the camera at a fixed point in the world, observing both the goal and the robot's end effector, and is referred to as end-point open-loop. In the

**15**

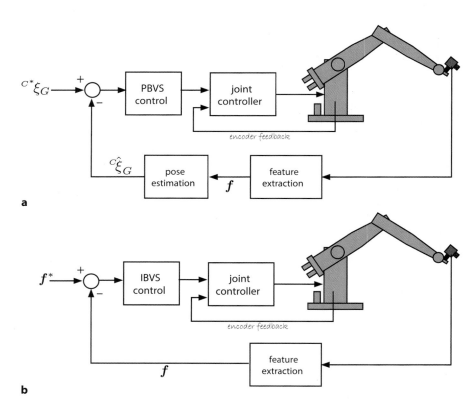

**Fig. 15.2** The two distinct classes of visual servo system. **a** Position-based visual servo; **b** Image-based visual servo

remainder of this book we will discuss only the eye-in-hand configuration. Vision-based control can work with other types of robots besides manipulator arms – the camera could be attached to a mobile ground robot or a aerial robot, and these are discussed in the next chapter.

For either configuration, there are two fundamentally different approaches to control: Position-Based Visual Servo (PBVS) and Image-Based Visual Servo (IBVS). Position-based visual servoing, shown in ◻ Fig. 15.2a, uses observed visual features, a calibrated camera and a known geometric model of the goal object to determine its pose with respect to the camera. The robot then moves toward the desired pose and the control is performed in task space. Good algorithms exist for pose estimation – see ▶ Sect. 13.2.4 – but it relies critically on the accuracy of the camera calibration and the model of the object's geometry. PBVS is discussed in ▶ Sect. 15.1.

Image-based visual servoing, shown in ◻ Fig. 15.2b, omits the pose estimation step, and uses the image features directly. The control is performed in image-coordinate space $\mathbb{R}^2$. The desired camera pose with respect to the goal is defined *implicitly* by the desired image feature values. IBVS is a challenging control problem since the image features are a highly nonlinear function of camera pose. IBVS is discussed in ▶ Sect. 15.2.

## 15.1 Position-Based Visual Servoing

For a PBVS system the relationships between the relevant poses is shown as a pose graph in ◻ Fig. 15.3. We wish to move the camera from its initial pose {C} to a desired pose {C*} which is defined with respect to the goal frame {G} by a relative pose $^{C^*}\xi_G$. The critical loop of the pose graph, indicated by the dashed-black, arrow is

$$\xi^\Delta \oplus {}^{C^*}\xi_G = {}^C\xi_G \qquad (15.1)$$

where $^C\xi_G$ is the current relative pose of the goal with respect to the camera. However, we cannot directly determine this since the pose of the goal in the world frame, $\xi_G$, shown by the gray arrow is unknown. Instead, we will use a vision-based estimate $^C\hat{\xi}_G$ of the goal pose relative to the camera. Pose estimation was discussed in ▶ Sect. 13.2.4 for a general object with known geometry, and in ▶ Sect. 13.6.1 for a fiducial marker. In both cases it requires a calibrated camera.

We rearrange (15.1) as

$$\xi^\Delta = {}^C\hat{\xi}_G \ominus {}^{C^*}\xi_G$$

which is the camera motion, in the camera frame, required to achieve the desired camera pose relative to the goal. The change in pose might be quite large so we do

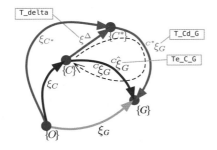

◻ **Fig. 15.3** Pose graph for PBVS example. Frame {C} is the current camera pose, frame {C*} is the desired camera pose, and frame {G} is the goal object. The Python variable names are shown in the gray boxes: an estimate (^) is indicated by the e suffix and a desired value (*) by the d suffix

not attempt to make this movement in a single time step, rather we move to a point closer to $\{C^*\}$ by

$$\boldsymbol{\xi}_{C\langle k+1\rangle} \leftarrow \boldsymbol{\xi}_{C\langle k\rangle} \oplus \Lambda\big(\boldsymbol{\xi}^{\Delta}\langle k\rangle, \lambda\big)$$

which is a fraction $\lambda \in (0, 1)$ of the translation and rotation required – $\Lambda(\cdot)$ is an interpolation as discussed in ▶ Sect. 3.3.5.

Using the Toolbox, we start by defining a camera with default intrinsic parameters

```
>>> camera = CentralCamera.Default(pose=SE3.Trans(1, 1, -2));
```

and an initial pose. The goal comprises four points that form a square of side length 0.5 m that lies in the *xy*-plane and is centered at the origin of the goal frame {G}

```
>>> P = mkgrid(2, side=0.5)
array([[    -0.25,     -0.25,      0.25,       0.25],
       [    -0.25,      0.25,      0.25,      -0.25],
       [        0,         0,         0,          0]])
```

and we assume that {G} is unknown to the control system. The image-plane projections of the world points are

```
>>> p = camera.project_point(P, objpose=SE3.Tz(1))
array([[   166.7,     166.7,       300,        300],
       [   166.7,       300,       300,      166.7]])
```

In code we represent $^X\boldsymbol{\xi}_Y$ as T_X_Y, an SE3 instance, which we read as the transformation from X to Y.

from which the pose of the goal with respect to the camera $^C\hat{\boldsymbol{\xi}}_G$ is estimated ◄

```
>>> Te_C_G = camera.estpose(P, p, frame="camera");
>>> Te_C_G.printline()
t = -1, -1, 3; rpy/zyx = 0°, 0°, 0°
```

indicating that the frame {G} is a distance of 3 m in front of the camera. The required relative pose is 1 m in front of the camera and fronto-parallel to it

```
>>> T_Cd_G = SE3.Tz(1);
```

so the required change in camera pose $\xi^{\Delta}$ is

```
>>> T_delta = Te_C_G * T_Cd_G.inv();
>>> T_delta.printline()
t = -1, -1, 2; rpy/zyx = 0°, 0°, 0°
```

and the new value for camera pose, given $\lambda = 0.05$, is

```
>>> camera.pose = camera.pose * T_delta.interp1(0.05);
```

We repeat the process, at each time step moving a fraction of the required relative pose until the goal is achieved. In this way, even if the goal moves or the robot has errors and does not move exactly as requested, the motion computed at the next time step will account for that error. ◄

The Toolbox provides a class to simulate a PBVS system, display results graphically and also save the time history of the camera motion and related quantities. We start as before by defining a camera with some initial pose

```
>>> camera = CentralCamera.Default(pose = SE3.Trans(1, 1, -2));
```

and the desired pose of the goal with respect to the camera is

```
>>> T_Cd_G = SE3.Tz(1);
```

which has the goal 1 m in front of the camera and fronto-parallel to it. We create an instance of the PBVS class

To ensure that the updated pose is a proper **SE**(3) matrix it should be normalized at each iteration, or the composition performed using the @ operator.

```
>>> pbvs = PBVS(camera, P=P, pose_g=SE3.Trans(-1, -1, 2),
...             pose_d=T_Cd_G, plotvol=[-1, 2, -1, 2, -3, 2.5])
Visual servo object: camera=default perspective camera
  100 iterations, 0 history
cdTg: t = 0, 0, 1; rpy/yxz = 0°, 0°, 0°
```

15

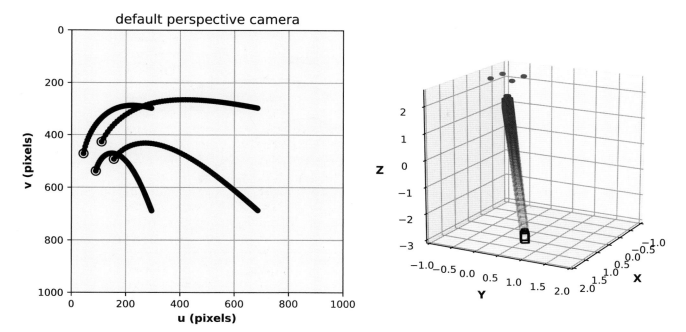

**◻ Fig. 15.4** Snapshot from the PBVS servo simulation. The left axis is the camera's image plane, showing the path followed by the projected world points. The right axis is the world view showing the 3D world points and the camera poses

which is a subclass of the `VisualServo` class and implements the controller outlined above. The arguments to the object constructor are a `CentralCamera` object, which has an initial pose, the object points relative to frame {G}, the pose of frame {G} which is required to project the points for pose estimation, and the desired relative pose of the camera with respect to {G}. The `run` method implements the PBVS control algorithm that drives the camera and

```
>>> pbvs.run(200);
```

will run the simulation for 200 time steps – repeatedly calling the object's `step` method to simulate the motion for a single time step. ► The simulation animates the features moving on the image plane of the camera and a 3-dimensional visualization of the camera and the world points – as shown in ◻ Fig. 15.4. The simulation completes after a defined number of iterations or when $\|\boldsymbol{\xi}^\Delta\|$ falls below some threshold. In this case, the simulation stops after 120 time steps when the velocity demand falls below the default threshold.

The simulation results are stored within the object for later analysis. We can plot the path of the goal features in the image, the camera velocity versus time or camera pose versus time

```
>>> pbvs.plot_p();      # plot image plane trajectory
>>> pbvs.plot_vel();    # plot camera velocity
>>> pbvs.plot_pose();   # plot camera trajectory
```

which are shown in ◻ Fig. 15.5. We see that the feature points have followed a curved path in the image, and that the camera's translation and orientation have converged smoothly on the desired values.

The pose of the `camera` object is updated at each time step. The initial pose of the camera is stashed when the object is constructed, and is restored every time `run` is called using the camera's `reset` method.

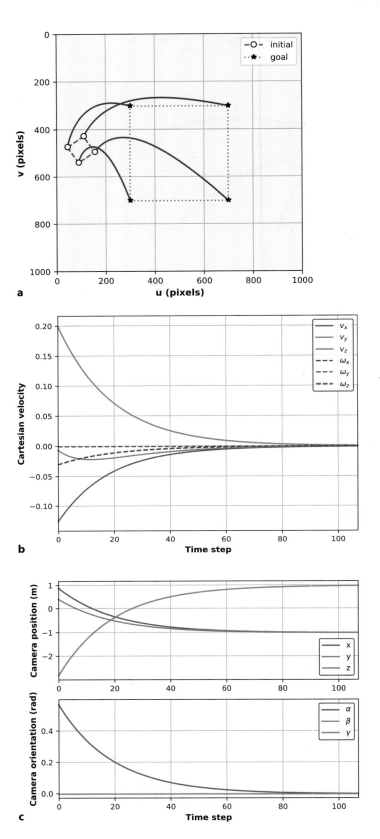

**Fig. 15.5** Results of PBVS simulation. **a** Image-plane feature motion; **b** Cartesian velocity; **c** camera pose where roll-pitch-yaw angles are in camera YXZ order

## 15.2 Image-Based Visual Servoing

IBVS differs fundamentally from PBVS by not estimating the relative pose of the goal. The relative pose is actually implicit in the values of the image features. ◘ Fig. 15.6 shows two views of a square object defined by its four corner points. The view from the initial camera pose is shown in red, and it is clear that the camera is viewing the object obliquely. The desired, or goal, view is shown in blue where the camera is further from the object, and its optical axis is normal to the plane of the object – a fronto-parallel view.

The control problem can be expressed in terms of image coordinates. The task is to move the feature points indicated by the round markers, to the points indicated by the star-shaped markers. The points may, but do not have to, follow the straight line paths indicated by the arrows. Moving the feature points in the image *implicitly* changes the camera pose – we have changed the problem from pose estimation to direct control of points on the image plane.

### 15.2.1 Camera and Image Motion

Consider the default camera

```
>>> camera = CentralCamera.Default();
```

and a world point at

```
>>> P = [1, 1, 5];
```

which has image coordinates

```
>>> p0 = camera.project_point(P)
array([[      660],
       [      660]])
```

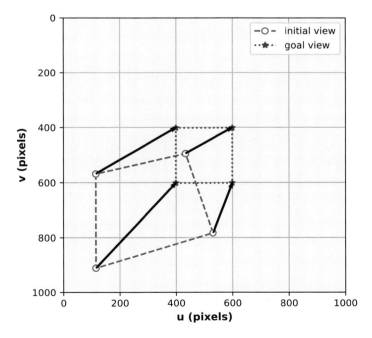

◘ **Fig. 15.6** Two different views of a square object: initial and goal

If we displace the camera slightly in the $x$-direction the pixel coordinates will become

```
>>> p_dx = camera.project_point(P, pose=SE3.Tx(0.1))
array([[     644],
       [     660]])
```

Using the camera coordinate conventions of ◘ Fig. 13.5, the camera has moved to the right so the image point has moved to the left. The *sensitivity* of image motion to camera motion $\Delta \boldsymbol{p}/\Delta x$ is

```
>>> (p_dx - p0) / 0.1
array([[    -160],
       [       0]])
```

which is an approximation to the derivative $\partial \boldsymbol{p}/\partial x$. We can repeat this for $z$-axis translation

```
>>> (camera.project_point(P, pose=SE3.Tz(0.1) ) - p0) / 0.1
array([[   32.65],
       [   32.65]])
```

which shows equal image-plane motion in the $u$- and $v$-directions. For $x$-axis rotation

```
>>> (camera.project_point(P, pose=SE3.Rx(0.1)) - p0) / 0.1
array([[   40.96],
       [   851.9]])
```

the image motion is predominantly in the $v$-direction. It is clear that camera motion along, and about, the different degrees of freedom in 3-dimensional space causes quite different motion of image points. Earlier, in (13.11), we expressed perspective projection in functional form

$$\boldsymbol{p} = \mathcal{P}(\boldsymbol{P}, \mathbf{K}, \xi_C) \tag{15.2}$$

and its derivative with respect to time is

$$\dot{\boldsymbol{p}} = \frac{\mathrm{d}\boldsymbol{p}}{\mathrm{d}t} = \frac{\partial \boldsymbol{p}}{\partial \boldsymbol{\xi}} \frac{\mathrm{d}\boldsymbol{\xi}}{\mathrm{d}t} = \mathbf{J}_p(\boldsymbol{P}, \mathbf{K}, \boldsymbol{\xi}_C) \, \boldsymbol{v} \tag{15.3}$$

where $\mathbf{J}_p$ is a Jacobian-like object, but because we have taken the derivative with respect to a pose $\boldsymbol{\xi} \in \mathbf{SE}(3)$, rather than a vector, it is technically called an *interaction matrix*. However in the visual servoing world it is more commonly called an *image Jacobian* or a *feature sensitivity matrix*. This is a local linearization ◄ of the highly nonlinear function (15.2). The camera moves in 3-dimensional space and its velocity is a *spatial velocity* $\boldsymbol{v} = (v_x, v_y, v_z, \omega_x, \omega_y, \omega_z) \in \mathbb{R}^6$, a concept introduced in ► Sect. 3.1, and comprises translational and rotational velocity components.

See ► App. E.

Consider a camera moving with a body velocity $\boldsymbol{v} = (\boldsymbol{v}, \boldsymbol{\omega})$ in the world frame and observing a world point P described by a camera relative coordinate vector $^C\boldsymbol{P} = (X, Y, Z)$. The velocity of the point relative to the camera frame is

$$^C\dot{\boldsymbol{P}} = -\left(\boldsymbol{\omega} \times {}^C\boldsymbol{P} + \boldsymbol{v}\right) \tag{15.4}$$

which we can write in scalar form as

$$\dot{X} = Y\omega_z - Z\omega_y - v_x$$
$$\dot{Y} = Z\omega_x - X\omega_z - v_y$$
$$\dot{Z} = X\omega_y - Y\omega_x - v_z \ . \tag{15.5}$$

The perspective projection (13.2) for normalized image-plane coordinates is

$$x = \frac{X}{Z}, \quad y = \frac{Y}{Z}$$

and the temporal derivative, using the quotient rule, is

$$\dot{x} = \frac{\dot{X}Z - X\dot{Z}}{Z^2}, \quad \dot{y} = \frac{\dot{Y}Z - Y\dot{Z}}{Z^2}.$$

Substituting in (15.5), $X = xZ$ and $Y = yZ$, we can write these in matrix form as

$$
\begin{pmatrix} \dot{x} \\ \dot{y} \end{pmatrix} =
\begin{pmatrix}
-\frac{1}{Z} & 0 & \frac{x}{Z} & xy & -(1+x^2) & y \\
0 & -\frac{1}{Z} & \frac{y}{Z} & 1+y^2 & -xy & -x
\end{pmatrix}
\begin{pmatrix} v_x \\ v_y \\ v_z \\ \omega_x \\ \omega_y \\ \omega_z \end{pmatrix}
\tag{15.6}
$$

which relates camera spatial velocity to feature velocity in normalized image co-ordinates. The normalized image-plane coordinates are related to the pixel coordinates by (13.8)

$$u = \frac{f}{\rho_u} x + u_0, \quad v = \frac{f}{\rho_v} y + v_0$$

which we rearrange as

$$x = \frac{\rho_u}{f}\bar{u}, \quad y = \frac{\rho_v}{f}\bar{v} \tag{15.7}$$

where $\bar{u} = u - u_0$ and $\bar{v} = v - v_0$ are the pixel coordinates relative to the principal point. The temporal derivative is

$$\dot{x} = \frac{\rho_u}{f}\dot{\bar{u}}, \quad \dot{y} = \frac{\rho_v}{f}\dot{\bar{v}} \tag{15.8}$$

and substituting (15.7) and (15.8) into (15.6) leads to

$$
\begin{pmatrix} \dot{u} \\ \dot{v} \end{pmatrix} =
\begin{pmatrix}
-\frac{f}{\rho_u Z} & 0 & \frac{\bar{u}}{Z} & \frac{\rho_v \bar{u}\,\bar{v}}{f} & -\frac{f^2 + \rho_u^2 \bar{u}^2}{\rho_u f} & \frac{\rho_v \bar{v}}{\rho_u} \\
0 & -\frac{f}{\rho_v Z} & \frac{\bar{v}}{Z} & \frac{f^2 + \rho_v^2 \bar{v}^2}{\rho_v f} & -\frac{\rho_u \bar{u}\,\bar{v}}{f} & -\frac{\rho_u \bar{u}}{\rho_v}
\end{pmatrix}
\begin{pmatrix} v_x \\ v_y \\ v_z \\ \omega_x \\ \omega_y \\ \omega_z \end{pmatrix}
$$

noting that $\dot{u} = \dot{\bar{u}}$ and $\dot{v} = \dot{\bar{v}}$.

For the typical case where $\rho_u = \rho_v = \rho$ we can express the focal length in pixels $f' = f/\rho$ and write

$$
\begin{pmatrix} \dot{u} \\ \dot{v} \end{pmatrix} =
\underbrace{
\begin{pmatrix}
-\frac{f'}{Z} & 0 & \frac{\bar{u}}{Z} & \frac{\bar{u}\,\bar{v}}{f'} & -\frac{f'^2 + \bar{u}^2}{f'} & \bar{v} \\
0 & -\frac{f'}{Z} & \frac{\bar{v}}{Z} & \frac{f'^2 + \bar{v}^2}{f'} & -\frac{\bar{u}\,\bar{v}}{f'} & -\bar{u}
\end{pmatrix}
}_{\mathbf{J}_p(\boldsymbol{p}, Z)}
\begin{pmatrix} v_x \\ v_y \\ v_z \\ \omega_x \\ \omega_y \\ \omega_z \end{pmatrix}
\tag{15.9}
$$

in terms of pixel coordinates *with respect to the principal point*. We can write this in concise matrix form as

$$\dot{p} = \mathbf{J}_p(p, Z)\, v \qquad\qquad (15.10)$$

where $\mathbf{J}_p$ is the $2 \times 6$ image Jacobian matrix for a point feature with coordinate vector $p = (\overline{u}, \overline{v})$ ◀ and $Z$ which is the $z$-coordinate of the point in the camera frame and often referred to as the point's depth.

This is commonly written in terms of $u$ and $v$ rather than $\overline{u}$ and $\overline{v}$ but we use the overbar notation to emphasize that the coordinates are with respect to the principal point, not the image origin which is typically in the top-left corner.

The Toolbox `CentralCamera` class provides the method `visjac_p` to compute the image Jacobian, and for the example above it is

```
>>> J = camera.visjac_p(p0, depth=5)
array([[  -160,       0,      32,      32,    -832,     160],
       [     0,    -160,      32,     832,     -32,    -160]])
```

where the first argument is the pixel coordinate of the point of interest, and the second argument is the depth of the point. The approximate numerical derivatives computed above correspond to the first, third and fourth columns respectively. Image Jacobians can also be derived for line and circle features and these are discussed in ▶ Sect. 15.3.

For a given camera velocity, the velocity of the point is a function of the point's coordinate, its depth and the camera's intrinsic parameters. Each column of the Jacobian indicates the velocity of an image feature point caused by one unit of the corresponding component of the velocity vector. The `flowfield` method of the `CentralCamera` class shows, for a particular camera velocity, the image-plane velocity for a grid of world points projected to the image plane. For camera translational velocity in the $x$-direction the flow field

```
>>> camera.flowfield([1, 0, 0, 0, 0, 0]);
```

is shown in ◻ Fig. 15.7a. As expected, moving the camera to the right causes all the projected points to move to the left. The motion of points on the image plane is known as *optical flow* and can be computed from image sequences, by tracking point features as discussed in ▶ Sect. 14.8.3. Eq. (15.9) is often referred to as the optical flow equation.

For translation in the $z$-direction

```
>>> camera.flowfield([0, 0, 1, 0, 0, 0]);
```

the points radiate outward from the principal point – the Star Trek warp effect – as shown in ◻ Fig. 15.7b. Rotation about the $z$-axis

```
>>> camera.flowfield([0, 0, 0, 0, 0, 1]);
```

causes the points to rotate about the principal point as shown in ◻ Fig. 15.7c.

Rotational motion about the $y$-axis

```
>>> camera.flowfield([0, 0, 0, 0, 1, 0]);
```

is shown in ◻ Fig. 15.7d and is very similar to the case of $x$-axis translation, with some small curvature for points far from the principal point. This similarity is because the first and fifth column of the image Jacobian are approximately equal in this case.

Any point on the optical axis will project to the the principal point, and at a depth of 1 m, the image Jacobian is

```
>>> camera.visjac_p(camera.pp, depth=1)
array([[  -800,       0,       0,       0,    -800,       0],
       [     0,    -800,       0,     800,       0,       0]])
```

and we see that the first and fifth columns are equal. This implies that translation in the $x$-direction causes the same image motion as rotation about the $y$-axis. You can easily demonstrate this equivalence by watching how the world moves as you

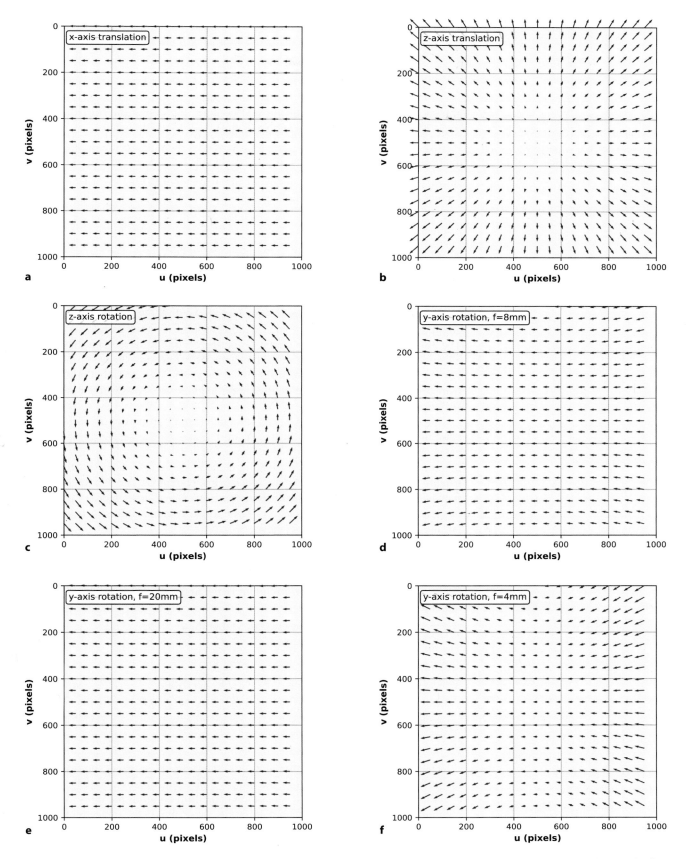

**Fig. 15.7** Image-plane velocity vectors for canonic camera velocities where all corresponding world points lie in a fronto-parallel plane. The flow vectors are normalized – they are shown with correct relative scale within each plot, but not between plots

translate your head to the right or rotate your head to the right – in both cases the world appears to move to the left. As the focal length increases, the fifth column

$$\lim_{f' \to \infty} \begin{pmatrix} -\frac{f'^2 + \bar{u}^2}{f'} \\ -\frac{\bar{u}\,\bar{v}}{f'} \end{pmatrix} = \begin{pmatrix} -f' \\ 0 \end{pmatrix}$$

approaches a scalar multiple of the first column. Increasing the focal length to $f = 20\,\text{mm}$ (the default focal length is $8\,\text{mm}$) leads to the flow field

```
>>> camera.f = 20e-3;
>>> camera.flowfield([0, 0, 0, 0, 1, 0]);
```

shown in ◘ Fig. 15.7e which is almost identical to that of ◘ Fig. 15.7a. Conversely, for small focal lengths (wide-angle cameras) the flow field

```
>>> camera.f = 4e-3;
>>> camera.flowfield([0, 0, 0, 0, 1, 0]);
```

shown in ◘ Fig. 15.7f has much more pronounced curvature. The same principle applies for the second and fourth columns except for a difference of sign – there is an equivalence between translation in the $y$-direction and rotation about the $-x$-axis.

The curvature in the wide-angle flow field shown in ◘ Fig. 15.7f means that the ambiguity between translational and rotation motion, discussed earlier, is resolved. The edges of the field of view contain motion cues that differentiate these types of motion. As the field of view gets larger, this effect becomes more pronounced. For narrow field of view, such as for our own eyes, we need additional information from sensors to resolve this ambiguity and the angular rate sensors in our vestibular system serve that purpose.

The Jacobian matrix of (15.9) has some interesting properties. It does not depend at all on the world coordinates $X$ or $Y$, only on the image-plane coordinates $(\bar{u}, \bar{v})$. However, the first three columns depend on the point's depth $Z$ and this reflects the fact that, for a translating camera, the image-plane velocity is inversely proportional to depth. Again, you can easily demonstrate this to yourself – translate your head sideways and observe that near objects move more in your field of view than distant objects. However, if you rotate your head all objects, near and far, move equally in your field of view.

The matrix has a rank of two, ◄ and therefore has a null space of dimension four. The null space comprises a set of spatial velocity vectors that individually, or in any linear combination, cause the world point to have *no motion* in the image. Consider the simple case of a world point lying on the optical axis which projects to the principal point

```
>>> J = camera.visjac_p(camera.pp, depth=1);
```

The null space of the Jacobian is

```
>>> linalg.null_space(J)
array([[      0,        0,   -0.7071,        0],
       [      0,   0.7071,        0,        0],
       [      1,        0,        0,        0],
       [      0,   0.7071,        0,        0],
       [      0,        0,   0.7071,        0],
       [      0,        0,        0,        1]])
```

The first column indicates that motion in the $z$-direction, along the ray toward the point, results in no motion in the image. Nor does rotation about the $z$-axis, as indicated by the last column. The second and third columns are more complex, combining rotation and translation. Essentially these exploit the image motion ambiguity mentioned above. Since $x$-axis translation causes the same image motion as $y$-axis rotation, the third column indicates that if one is positive and the other negative the resulting image motion will be zero – that is translating left while rotating to the right.

**15**    The rank cannot be less than 2, even if $Z \to \infty$.

We can consider the motion of two points by stacking their Jacobians

$$
\begin{pmatrix} \dot{u}_0 \\ \dot{v}_0 \\ \dot{u}_1 \\ \dot{v}_1 \end{pmatrix} = \begin{pmatrix} \mathbf{J}_p(\boldsymbol{p}_0,\ Z_0) \\ \mathbf{J}_p(\boldsymbol{p}_1,\ Z_1) \end{pmatrix} \boldsymbol{v}
$$

to give a $4 \times 6$ matrix which will have a null space with just two columns. One of these null-space camera motions corresponds to rotation around a line joining the two points.

For three points

$$
\begin{pmatrix} \dot{u}_0 \\ \dot{v}_0 \\ \dot{u}_1 \\ \dot{v}_1 \\ \dot{u}_2 \\ \dot{v}_2 \end{pmatrix} = \begin{pmatrix} \mathbf{J}_p(\boldsymbol{p}_0,\ Z_0) \\ \mathbf{J}_p(\boldsymbol{p}_1,\ Z_1) \\ \mathbf{J}_p(\boldsymbol{p}_2,\ Z_2) \end{pmatrix} \boldsymbol{v} \tag{15.11}
$$

the matrix will be full rank, nonsingular, so long as the points are not coincident or collinear.

## 15.2.2  Controlling Feature Motion

So far we have shown how points move in the image plane as a consequence of camera motion. As is often the case, it is the inverse problem that is more useful – what camera motion is needed in order to move the image features at a desired velocity?

For the case of three points $\{(u_i, v_i), i = 0, 1, 2\}$ and corresponding velocities $\{(\dot{u}_i, \dot{v}_i)\}$, we can invert (15.11)

$$
\boldsymbol{v} = \begin{pmatrix} \mathbf{J}_p(\boldsymbol{p}_0,\ Z_0) \\ \mathbf{J}_p(\boldsymbol{p}_1,\ Z_1) \\ \mathbf{J}_p(\boldsymbol{p}_2,\ Z_2) \end{pmatrix}^{-1} \begin{pmatrix} \dot{u}_0 \\ \dot{v}_0 \\ \dot{u}_1 \\ \dot{v}_1 \\ \dot{u}_2 \\ \dot{v}_2 \end{pmatrix} \tag{15.12}
$$

and solve for the required camera spatial velocity. The remaining question is how to determine the point velocity? Typically, we use a simple linear controller

$$
\dot{\boldsymbol{p}}^* = \lambda(\boldsymbol{p}^* - \boldsymbol{p}) \tag{15.13}
$$

that *drives* the image plane points toward their desired values $\boldsymbol{p}^*$. Combined with (15.12) we write ▶

$$
\boldsymbol{v} = \lambda \begin{pmatrix} \mathbf{J}_p(\boldsymbol{p}_0,\ Z_0) \\ \mathbf{J}_p(\boldsymbol{p}_1,\ Z_1) \\ \mathbf{J}_p(\boldsymbol{p}_2,\ Z_2) \end{pmatrix}^{-1} (\boldsymbol{p}^* - \boldsymbol{p})
$$

That's it! This controller computes a velocity that moves the camera in such a way that the feature points move toward their desired position in the image. It is important to note that nowhere have we required the pose of the camera or of the object – everything has been computed in terms of what can be measured on the image plane. ▶ As the controller runs, the image-plane points $\boldsymbol{p}_i$ move so $\mathbf{J}_p(\cdot)$ needs to be updated frequently, typically for every new camera image.

Note that papers based on the task function approach (Espiau et al. 1992) write this as actual position minus demanded position and use $-\lambda$ in (15.14) to ensure negative feedback. We use the common convention from control theory where the error signal is "demand minus actual".

We do require the point depth $Z$, but we will deal that issue shortly.

**❗ Correspondence between the observed and desired points is essential**
It is critically important when computing (15.13) that the points $p_i^*$ and $p_i$ correspond. In a contrived scene, the points could be the centroids of shapes with uniques colors, sizes or shapes. In a more general case, correspondence can be solved if one point is unique and the ordering of the points, often with respect to the centroid of the points, is known.

For the general case where $N > 3$ points we can stack the Jacobians for all features and solve for camera velocity using the pseudoinverse

$$
v = \lambda \begin{pmatrix} \mathbf{J}_p(\boldsymbol{p}_0,\, Z_0) \\ \vdots \\ \mathbf{J}_p(\boldsymbol{p}_{N-1},\, Z_{N-1}) \end{pmatrix}^+ (\boldsymbol{p}^* - \boldsymbol{p}) \tag{15.14}
$$

Note that it is possible to specify a set of feature point velocities which are inconsistent, that is, there is no possible camera motion that will result in the required image motion. In such a case the pseudoinverse will find a solution that minimizes the norm of the feature velocity error.

The Jacobian is a first-order approximation of the relationship between camera motion and image-plane motion. Faster convergence is achieved by using a second-order approximation and it has been shown that this can be obtained very simply

$$
v = \frac{\lambda}{2} \left[ \begin{pmatrix} \mathbf{J}_p(\boldsymbol{p}_0,\, Z_0) \\ \vdots \\ \mathbf{J}_p(\boldsymbol{p}_{N-1},\, Z_{N-1}) \end{pmatrix}^+ + \begin{pmatrix} \mathbf{J}_p(\boldsymbol{p}_0^*,\, Z_0^*) \\ \vdots \\ \mathbf{J}_p(\boldsymbol{p}_{N-1}^*,\, Z_{N-1}^*) \end{pmatrix}^+ \right] (\boldsymbol{p}^* - \boldsymbol{p}) \tag{15.15}
$$

by taking the mean of the pseudoinverse of the image Jacobians at the current and desired states.

For $N \geq 3$ the matrix can be poorly conditioned if the points are nearly co-incident or collinear. In practice, this means that some camera motions will cause very small image motions, that is, the motion has low perceptibility. There is strong similarity with the concept of manipulability that we discussed in ▶ Sect. 8.3.2 and we take a similar approach in formalizing perceptibility. Consider a camera spatial velocity of unit magnitude

$$
v^\top v = 1
$$

and from (15.10) we can write the camera velocity in terms of the pseudoinverse

$$
v = \mathbf{J}^+ \dot{\boldsymbol{p}}
$$

where $\mathbf{J} \in \mathbb{R}^{2N \times 6}$ is the Jacobian stack and the point velocities are $\dot{\boldsymbol{p}} \in \mathbb{R}^{2N}$. Substituting yields

$$
\dot{\boldsymbol{p}}^\top \mathbf{J}^{+\top} \mathbf{J}^+ \dot{\boldsymbol{p}} = 1
$$
$$
\dot{\boldsymbol{p}}^\top \left( \mathbf{J} \mathbf{J}^\top \right)^{-1} \dot{\boldsymbol{p}} = 1
$$

which is the equation of an ellipse in the image-plane velocity space. The eigenvectors of $\mathbf{J}\mathbf{J}^\top$ define the principal axes of the ellipse and the singular values of $\mathbf{J}$ are the radii. The ratio of the maximum to minimum radius is given by the condition number of $\mathbf{J}\mathbf{J}^\top$ and indicates the anisotropy of the feature motion. A high value indicates that some of the points have low velocity in response to some camera motions. An alternative to stacking all the point feature Jacobians in (15.14) is to select just three that, when stacked, result in the best conditioned square matrix which can then be inverted.

Using the Toolbox we start by defining a camera with default intrinsic parameters

```
>>> camera = CentralCamera.Default(pose=SE3.Trans(1, 1, -2));
```

and an initial pose. The goal comprises four points that form a square of side length 0.5 m that lies in the $xy$-plane and is centered at $(0, 0, 3)$

```
>>> P = mkgrid(2, side=0.5, pose=SE3.Tz(3));
```

The desired position of the goal features on the image plane are a $400 \times 400$ square centered on the principal point

```
>>> pd = 200 * np.array([[-1, -1, 1, 1], [-1, 1, 1, -1]])
...       + np.c_[camera.pp]
array([[   300,    300,    700,    700],
       [   300,    700,    700,    300]])
```

which implicitly has the square goal fronto-parallel to the camera. The camera's initial projection of the world points is

```
>>> p = camera.project_point(P)
array([[   300,    300,    380,    380],
       [   300,    380,    380,    300]])
```

and p and pd each have one column per point.

We compute the image-plane error

```
>>> e = pd - p
array([[     0,      0,    320,    320],
       [     0,    320,    320,      0]])
```

and the stacked image Jacobian

```
>>> J = camera.visjac_p(p, depth=1);
```

is an $8 \times 6$ matrix in this case, since p contains four points. The Jacobian does require the point depth which we do not know, so for now we will just choose a constant value $Z = 1$. ▶ This is an important topic that we will address in ▶ Sect. 15.2.3.

Here we provide a single value which is taken as the depth of all the points. Alternatively we could provide a 1D array to specify the depth of each point individually.

The control law determines the required translational and angular velocity of the camera ▶

```
>>> lmbda = 0.1;
>>> v = lmbda * np.linalg.pinv(J) @ e.flatten(order="F")
array([   -0.1,    -0.1,     0.4,       0,       0,       0])
```

To create the error column vector as described in (15.12) we need to flatten the 2D array e in column-major order.

where lmbda is the gain, a positive number, and we take the pseudoinverse of the nonsquare Jacobian to implement (15.14). The resulting velocity is expressed in the camera coordinate frame, and integrating it over a unit time step results in a spatial displacement of the same magnitude. The camera pose is updated by

$$\xi_{C\langle k+1\rangle} \leftarrow \xi_{C\langle k\rangle} \oplus \Delta^{-1}(\nu_{\langle k\rangle})$$

where $\Delta^{-1}(\cdot)$ is described in ▶ Sect. 3.1.5. Using the Toolbox this is implemented as

```
>>> camera.pose = camera.pose @ SE3.Delta(v);
```

where we ensure that the transformation remains a proper **SE**(3) matrix by using the @ operator.

Similar to the PBVS example, we create an instance of the IBVS class

```
>>> camera = CentralCamera.Default(pose=SE3.Trans(1, 1, -3)
...             * SE3.Rz(0.6));
>>> ibvs = IBVS(camera, P=P, p_d=pd);
```

which is a subclass of the `VisualServo` class and which implements the controller outlined above. The object constructor takes a `CentralCamera` object, with a specified initial pose, as its argument. The controller then drives the camera to achieve the desired image-plane point configuration specified by `p_d`. The simulation is run for 25 time steps

```
>>> ibvs.run(25);
```

which repeatedly calls the object's `step` method which simulates motion for a single time step. The simulation animates the image plane of the camera as well as a 3-dimensional visualization of the camera and the world points.

The simulation results are stored within the object for later analysis. We can plot the path of the goal features on the image plane, the camera velocity versus time or camera pose versus time

```
>>> ibvs.plot_p();     # plot image plane trajectory
>>> ibvs.plot_vel();   # plot camera velocity
>>> ibvs.plot_pose();  # plot camera trajectory
```

which are shown in ◨ Fig. 15.8. We see that the feature points have followed an approximately straight-line path in the image, and the camera pose has changed smoothly toward some final value. The condition number of the image Jacobian

```
>>> ibvs.plot_jcond();
```

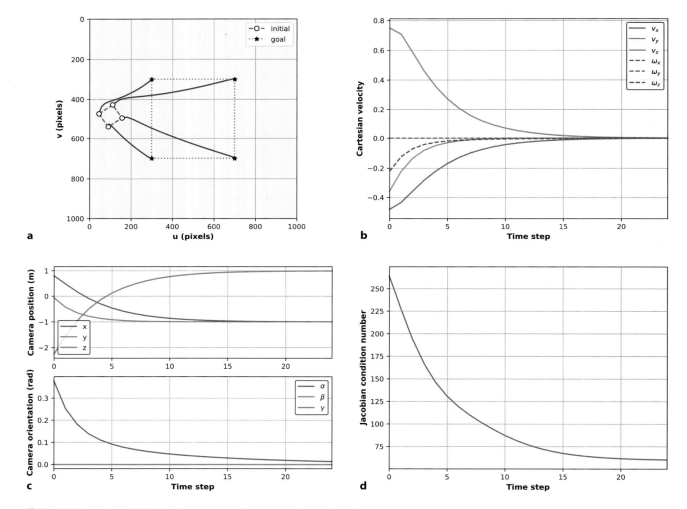

◨ **Fig. 15.8**  Results of IBVS simulation, created by `IBVS`. **a** Image-plane feature motion; **b** spatial velocity components; **c** camera pose where roll-pitch-yaw angles are in camera YXZ order; **d** image Jacobian condition number (lower is better)

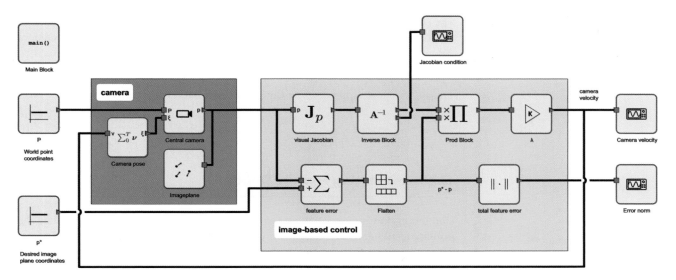

**Fig. 15.9** Block diagram model `models/IBVS`. The `Main Block` points to the Python file `IBVS-main.py` which creates an instance of the `CentralCamera` object and the clock required for discrete-time system integration

decreases over the motion indicating that the Jacobian is becoming better conditioned, and this is a consequence of the features moving further apart – becoming less coincident.

How is $p^*$ determined? The image points can be found by demonstration, by moving the camera to the desired pose and recording the observed image coordinates. Alternatively, if the camera calibration parameters and the goal geometry are known, the desired image coordinates can be computed for any specified goal pose. This calculation, world point point projection, is computationally cheap and performed only once before visual servoing commences.

The IBVS system can also be expressed in terms of a block diagram model as shown in ◘ Fig. 15.9. That model can be simulated by ▶

```
>>> %run -m IBVS-main -H
```

and the scope blocks animate the evolution of the image plane features, camera velocity and feature error against time. The initial pose of the camera is set by a parameter of the `Camera pose` block, and the world points are parameters of a constant block. The `CentralCamera` object is defined in `IBVS-main.py` and is a parameter to the `Central camera`, `visual Jacobian` and `ImagePlane` blocks. Other parameters such as the camera model, world points and gains are defined in the properties of the various blocks and in `IBVS-main.py`.

Running the simulation adds a simulation results object `out` to the global namespace

```
>>> out
results:
t          | ndarray (1001,)
x          | ndarray (0,)
xnames     | list = []
clock0     : Struct
           > t           | ndarray (1001,)
           > x           | ndarray (1001, 6)
y0         | ndarray (1001,)
y1         | ndarray (1001,)
y2         | ndarray (1001, 6)
ynames     | list = ['total feature error[0]', 'Inverse Block[1]',
                     'lambda[0]']
```

Block diagram models were introduced in ▶ Sect. 4.1.1.1. The `-H` option stops the function from blocking until all the figures are dismissed.

The element `t` contains simulation time and the elements `y0`, `y1` and `y2` correspond to the output ports of the blocks listed in `ynames` respectively. We can plot camera velocity, the output of the $\lambda$ block against time by

```
>>> plt.plot(out.t, out.y2)
```

The camera pose is the discrete-time state associated with `clock0` and can be plotted

```
>>> plt.plot(out.clock0.t, out.clock0.x)
```

### 15.2.3  Estimating Feature Depth

Computing the image Jacobian requires knowledge of the camera intrinsics, the principal point and focal length, but in practice it is quite tolerant to errors in these. The Jacobian also requires knowledge of $Z_i$, the $z$-coordinate, or the depth of, each point. In the simulations just discussed, we have assumed that depth is known – this is easy in simulation but not so in reality.

A number of approaches have been proposed to deal with the problem of unknown depth. The simplest is to just assume a constant value for the depth, and this is quite reasonable if the required camera velocity is approximately in a plane parallel to the plane of the object points. To evaluate the performance of different constant estimates of point depth, we can compare the effect of choosing $Z = 1$ and $Z = 10$ for the example above where the true depth is initially $Z = 3$

```
>>> ibvs = IBVS(camera, P=P, p_d=pd, depth=1);
>>> ibvs.run(50)
>>> ibvs = IBVS(camera, P=P, p_d=pd, depth=10);
>>> ibvs.run(50)
```

and the results are shown in ◻ Fig. 15.10. We see that the image-plane paths are no longer straight, because the Jacobian is now a poor approximation of the relationship between the camera motion and image feature motion. We also see that for $Z = 1$ the convergence is much slower than for the $Z = 10$ case. The Jacobian for $Z = 1$ overestimates the optical flow, so the inverse Jacobian underestimates the required camera velocity. For the $Z = 10$ case, the camera displacement at each time step is large leading to a very jagged path. Nevertheless, for quite significant errors in depth, IBVS has converged, and it is widely observed in practice that IBVS is remarkably tolerant to errors in $Z$.

A second approach is to assume a calibrated camera and use standard computer vision techniques such as pose estimation or sparse stereo to estimate the value of $Z$. However, this does defeat an important advantage of IBVS which is to avoid explicit pose estimation and 3D models.

A third approach, which we will expand on here, is to estimate depth online using observed frame-to-frame feature motion, known camera intrinsics, and knowledge of the camera spatial velocity $(\boldsymbol{v}, \boldsymbol{\omega})$ which the controller computes. We can create a depth estimator by rewriting (15.9)

$$
\begin{pmatrix} \dot{u} \\ \dot{v} \end{pmatrix} = \begin{pmatrix} -\frac{f}{\rho_u Z} & 0 & \frac{\bar{u}}{Z} & \frac{\rho_u \overline{uv}}{f} & -\frac{f^2 + \rho_u^2 \bar{u}^2}{\rho_u f} & \bar{v} \\ 0 & -\frac{f}{\rho_v Z} & \frac{\bar{v}}{Z} & \frac{f^2 + \rho_v^2 \bar{v}^2}{\rho_v f} & -\frac{\rho_v \overline{uv}}{f} & -\bar{u} \end{pmatrix} \begin{pmatrix} \boldsymbol{v} \\ \boldsymbol{\omega} \end{pmatrix}
$$

$$
= \left( \tfrac{1}{Z} \mathbf{J}_t \mid \mathbf{J}_\omega \right) \begin{pmatrix} \boldsymbol{v} \\ \boldsymbol{\omega} \end{pmatrix}
$$

$$
= \frac{1}{Z} \mathbf{J}_t \boldsymbol{v} + \mathbf{J}_\omega \boldsymbol{\omega}
$$

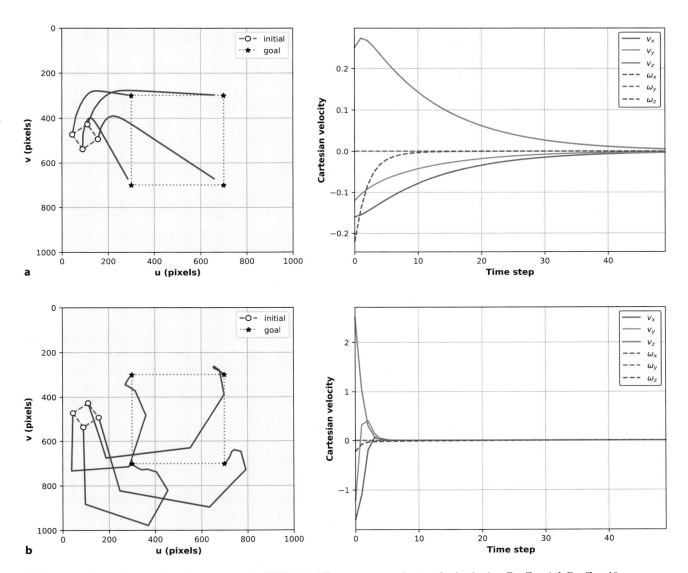

**Fig. 15.10** Image feature paths and camera pose for IBVS with different constant estimates of point depth: **a** For $Z = 1$; **b** For $Z = 10$

and then rearranging into estimation form

$$(\mathbf{J}_t v)\frac{1}{Z} = \begin{pmatrix} \dot{u} \\ \dot{v} \end{pmatrix} - \mathbf{J}_\omega \boldsymbol{\omega} \qquad (15.16)$$

The right-hand side is the observed optical flow from which the expected optical flow, due to camera rotation, is subtracted – a process referred to as derotating optical flow. The remaining optical flow, after subtraction, is only due to translation. Writing (15.16) in compact form

$$\mathbf{A}\boldsymbol{\theta} = \boldsymbol{b} \qquad (15.17)$$

we have a simple linear equation with one unknown parameter $\boldsymbol{\theta} = 1/Z$ which can be solved using least-squares.

In our example we can enable this by

```
>>> ibvs = IBVS(camera, P=P, p_d=pd, depthest=True);
>>> ibvs.run()
```

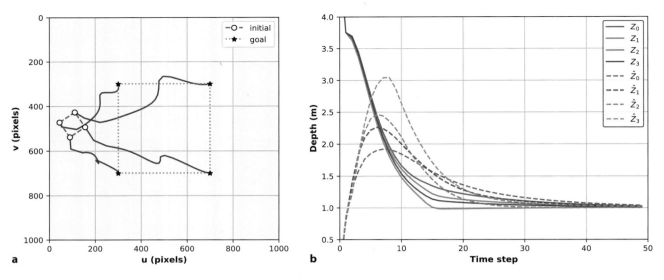

■ **Fig. 15.11**  IBVS with online depth estimator. **a** Feature paths; **b** comparison of estimated and true depth for all four points

and the result is shown in ■ Fig. 15.11. ■ Fig. 15.11b shows the estimated and true point depth versus time. The estimated depth was initially zero, a poor choice, but it has risen rapidly and then tracked the actual goal depth as the controller converges. ■ Fig. 15.11a shows the feature motion, and we see that the features initially move in the wrong direction because the gross error in depth has led to an image Jacobian that predicts poorly how feature points will move.

### 15.2.4  Performance Issues

The control law for PBVS is defined in terms of 3-dimensional pose so there is no mechanism by which the motion of the image features is directly regulated. For the PBVS example shown in ■ Fig. 15.5, the feature points followed a curved path on the image plane, and therefore it is possible that they could leave the camera's field of view. For a different initial camera pose

```
>>> pbvs.pose_0 = SE3.Trans(-2.1, 0, -3) * SE3.Rz(5*pi/4);
>>> pbvs.run()
```

the result is shown in ■ Fig. 15.12a and we see that two of the points move beyond the image boundary which would cause the PBVS control to fail. ◄

In this simulation the image plane coordinates are still computed and used, even though they fall outside the image bounds. Visibility checking can be enabled by the `visibility` option to the camera's `project_point` method.

By contrast, the IBVS control for the same initial camera pose

```
>>> ibvs.pose_0 = pbvs.pose_0;
>>> ibvs.run()
>>> ibvs.plot_p();
```

gives the feature trajectories shown in ■ Fig. 15.12b, but in this case there is no direct control over the task-space camera velocity. This can sometimes result in surprising motion, particularly when the goal is rotated about the optical axis

```
>>> ibvs.pose_0 = SE3.Tz(-1) * SE3.Rz(2);
>>> ibvs.run(50)
```

where the image features and camera pose are shown in ■ Fig. 15.13a. We see that the camera has performed an unnecessary translation along the $z$-axis – initially away from the goal, and then back again. This phenomenon is termed camera retreat. The resulting motion is not time optimal and can require large and possibly

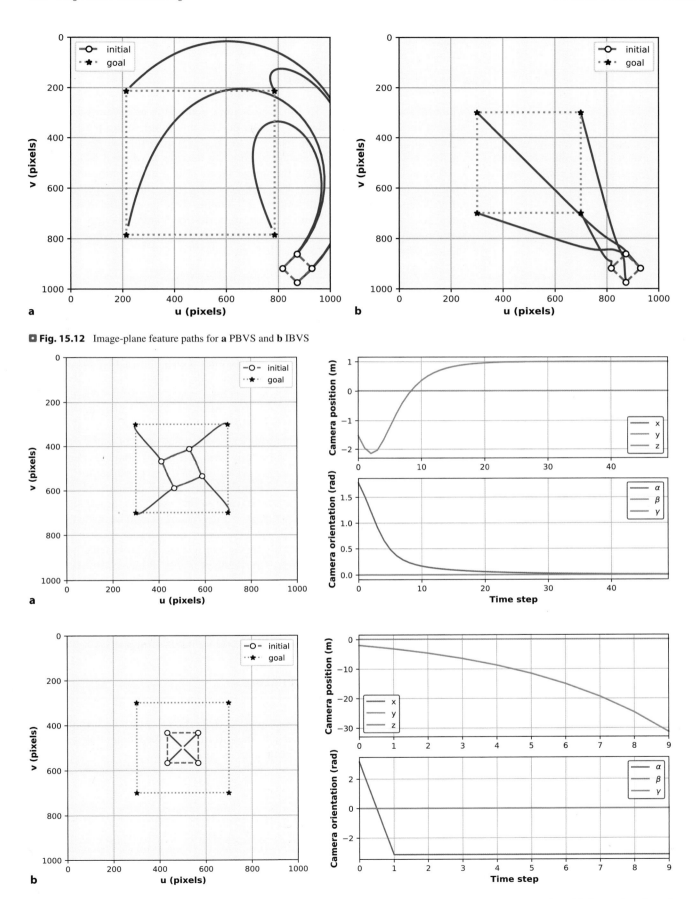

**Fig. 15.12** Image-plane feature paths for **a** PBVS and **b** IBVS

**Fig. 15.13** Image feature paths and camera pose for IBVS with pure goal rotation about the optical axis. **a** for rotation of 1 rad; **b** for rotation of π rad

unachievable camera motion. An extreme example arises for a pure rotation about the optical axis by $\pi$ rad

```
>>> ibvs.pose_0 = SE3.Tz(-1) * SE3.Rz(pi);
>>> ibvs.run(10)
```

As the image points converge on the principal point the Jacobian condition number increases and it will eventually become rank deficient.

which is shown in ◘ Fig. 15.13b. The feature points are, as usual, moving in a straight line toward their desired values, but for this problem the paths all pass through the principal point. The only way the goal feature points can be at the principal point is if the camera is at negative infinity, and that is where it is headed! ◄

A final consideration is that the image Jacobian is a linearization of a highly nonlinear system. If the motion at each time step is large, then the linearization is not valid and the features will follow curved rather than linear paths in the image, as we saw in ◘ Fig. 15.10. This can occur if the desired feature positions are a long way from the initial positions and/or the gain $\lambda$ is too high. One solution is to limit the maximum norm of the commanded velocity

$$
\nu = \begin{cases} \nu_{max} \frac{\nu}{|\nu|} & \text{if } |\nu| > \nu_{max} \\ \nu & \text{if } |\nu| \le \nu_{max} \end{cases}
$$

and this is enabled by the `vmax` option to the constructor. We can see from all the camera velocity plots that the highest velocity is commanded at the first time step when the feature error is largest. We can also use a technique called *smooth start* to modify the image-plane error $e = p^* - p$

$$
e' = e - e_0 \, e^{-\mu t}
$$

before it is used to compute the control, where $e_0$ is the value of $e$ at time $t = 0$. The second term ensures that $e'$ is initially zero and grows gradually with time. This can be enabled using the `smoothstart`=$\mu$ option to the constructor.

The feature paths do not have to be straight lines and nor do the features have to move with asymptotic velocity – we have used these only for simplicity. Using the trajectory planning methods of ► Sect. 3.3 the features could be made to follow any arbitrary trajectory in the image.

In summary, IBVS is a remarkably robust approach to vision-based control. We have seen that it is tolerant to errors in the depth of points. We have also shown that it can produce less than optimal Cartesian paths for the case of large rotations about the optical axis, and we will discuss remedies to these problems in the next chapter. There are of course some practical complexities. If the camera is on the end of the robot it might interfere with the task, or when the robot is close to the target the camera might be unable to focus, or the target might be obscured by the gripper.

## 15.3 Using Other Image Features

So far we have considered only point features but IBVS can also be formulated to work with other image features such as lines, as found by the Hough transform, the shape of an ellipse or even pixel intensities.

### 15.3.1 Line Features

We represent a line using the $(\theta, \rho)$ parameterization that was introduced for the Hough transform in ► Sect. 12.2

$$
u \cos \theta + v \sin \theta = \rho \ .
$$

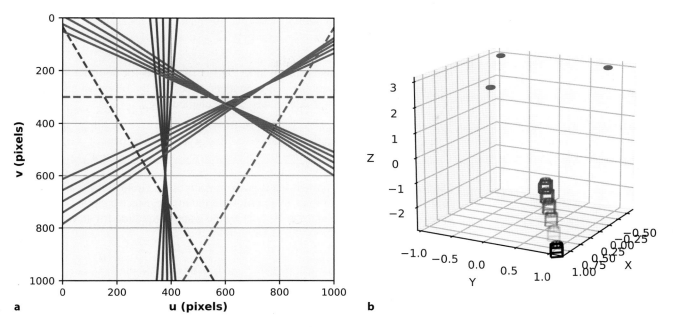

**◻ Fig. 15.14** IBVS using line features. **a** The image plane showing three line features (red, green and blue) over time (solid) and the desired line features (dashed); **b** the camera and three points that form the three world lines

The rate of change of the line parameters is related to camera velocity by

$$\begin{pmatrix} \dot{\theta} \\ \dot{\rho} \end{pmatrix} = \mathbf{J}_l \, \boldsymbol{v}$$

and the Jacobian is

$$\mathbf{J}_l = \begin{pmatrix} \lambda_\theta \cos\theta & \lambda_\theta \sin\theta & -\lambda_\theta\rho & -\rho\cos\theta & -\rho\sin\theta & -1 \\ \lambda_\rho \cos\theta & \lambda_\rho \sin\theta & -\lambda_\rho\rho & (1+\rho^2)\sin\theta & -(1+\rho^2)\cos\theta & 0 \end{pmatrix}.$$

Just as the point-feature Jacobian required some partial 3-dimensional knowledge (the point depth $Z$), the line-feature Jacobian requires the 3-dimensional plane that contains the line $aX + bY + cZ + d = 0$ and $\lambda_\theta = (a\sin\theta - b\cos\theta)/d$ and $\lambda_\rho = (a\rho\cos\theta + b\rho\sin\theta + c)/d$. There are an infinite number of planes that contain the line and we choose one for which $d \neq 0$. This is computed by the `visjac_l` method of a `CentralCamera` object.

Like a point feature, a line feature provides a Jacobian with two rows, so we require a minimum of three lines in order to have a Jacobian of full rank. ▶

We illustrate this with an example comprising three lines that all lie in the plane $Z = 3$, and we can conveniently construct three points in that plane using the `circle` function with just three boundary points

Interestingly a line feature provides two rows of the stacked Jacobian, yet two points which define a line segment would provide four rows.

```
>>> P = circle([0, 0, 3], 0.5, resolution=3);
```

and use the familiar `CentralCamera` class methods to project these to the image. For each pair of points $\{(u_i, v_i), (u_j, v_j), i \neq j\}$ we compute the equations of the line

$$\tan\theta_i = \frac{u_j - u_i}{v_i - v_j}, \quad \rho_i = u_i \cos\theta + v_i \sin\theta$$

The simulation is run in familiar fashion

```
>>> ibvs = IBVS_l.Example(camera);  # quick problem setup
>>> ibvs.run()
```

and a snapshot of results is shown in ◼ Fig. 15.14. Just as for point-feature visual servoing, we need to establish correspondence between the observed and desired lines and for real-world problems this can be challenging.

### 15.3.2 Ellipse Features

Circles are common in human-made environments and a circle in the world will be projected, in the general case, to an ellipse in the image. Since a circle is a special case of an ellipse, we can describe both by the general equation for an ellipse ◀

Ellipses, and fitting ellipses to data, are described in more detail in ▶ App. C.1.4.

$$u^2 + E_0 v^2 - 2E_1 uv + 2E_2 u + 2E_3 v + E_4 = 0 \tag{15.18}$$

and the rate of change of the ellipse parameters $E_i$ is related to camera velocity by

$$\begin{pmatrix} \dot{E}_0 \\ \dot{E}_1 \\ \vdots \\ \dot{E}_4 \end{pmatrix} = \mathbf{J}_e(\mathbf{E},\ \rho)\boldsymbol{v}$$

and the Jacobian is

$$\mathbf{J}_e(\mathbf{E}, \rho)$$
$$= \begin{pmatrix} 2\beta E_1 - 2\alpha E_0 & 2E_0(\beta - \alpha E_1) & 2\beta E_3 - 2\alpha E_0 E_2 & 2E_3 & 2E_0 E_2 & -2E_1(E_0 + 1) \\ \beta - \alpha E_1 & \beta E_1 - \alpha(2E_1^2 - E_0) & \alpha(E_3 - 2E_1 E_2) + \beta E_2 & E_2 & 2E_1 E_2 - E_3 & E_0 - 2E_1^2 - 1 \\ \gamma - \alpha E_2 & \alpha(E_3 - 2E_1 E_2) + \gamma E_1 & \gamma E_2 - \alpha(2E_2^2 - E_4) & -E_1 & 1 + 2E_2^2 - E_4 & E_3 - 2E_1 E_2 \\ E_2\beta + E_1\gamma - 2\alpha E_3 & E_3\beta + E_0\gamma - 2\alpha E_1 E_3 & \beta E_4 + \gamma E_3 - 2\alpha E_2 E_3 & E_4 - E_0 & 2E_2 E_3 + E_1 & -2E_1 E_3 - E_2 \\ 2\gamma E_2 - 2\alpha E_4 & 2\gamma E_3 - 2\alpha E_1 E_4 & 2\gamma E_4 - 2\alpha E_2 E_4 & -2E_3 & 2E_2 E_4 + 2E_2 & -2E_1 E_4 \end{pmatrix}.$$

Once again, some partial 3-dimensional knowledge is required and the ellipse-feature Jacobian requires the 3-dimensional plane that contains the ellipse $aX + bY + cZ + d = 0$ and $\alpha = -a/d$, $\beta = -b/d$ and $\gamma = -c/d$. This Jacobian has a maximum rank of five, but this drops to three when the projection is of a circle centered in the image plane, and a rank of two if the circle has zero radius. The image Jacobian for an ellipse feature is computed by the `visjac_e` method of the `CentralCamera` class.

An advantage of the ellipse feature is that the ellipse can be computed from the set of boundary points without needing to solve the correspondence problem. The ellipse feature can also be computed from the moments of all the points within the ellipse boundary.

We illustrate this with an example of a circle of ten points in the world

```
>>> P = circle([0, 0, 3], 0.5, resolution=10);
```

and the `CentralCamera` class projects these to the image plane

```
>>> p = camera.project_point(P, pose=camera.pose, retinal=True);
```

at the current camera pose and we convert to retinal image coordinates. The parameters of an ellipse are calculated using the methods of ▶ App. C.1.4.2

```
>>> x, y = p
>>> A = np.column_stack([y**2, -2*x*y, 2*x, 2*y, np.ones(x.shape)]);
>>> b = -(x**2);
>>> E, *_ = np.linalg.lstsq(A, b, rcond=None)
```

which returns a 5-vector of ellipse parameters. The Jacobian

```
>>> plane = [0, 0, 1, -3];  # plane Z=3
>>> J = camera.visjac_e(E, plane);
>>> J.shape
(5, 6)
```

is a $5 \times 6$ matrix which has a maximum rank of only 5 so we cannot uniquely solve
for the camera velocity. We have at least two options. Firstly, if our final view is of
a circle then we may not be concerned about rotation around the circle's normal,
and in this case we can delete the sixth column of the Jacobian to make it square
and set $\omega_z$ to zero. Secondly, and the approach taken in this example, is to combine
the features for the ellipse and a single point ▶

Here we arbitrarily choose the first
point, any one will do, but we need to
establish correspondence in every
frame.

$$\begin{pmatrix} \dot{E}_0 \\ \dot{E}_1 \\ \vdots \\ \dot{E}_4 \\ \dot{u}_0 \\ \dot{v}_0 \end{pmatrix} = \begin{pmatrix} \mathbf{J}_e(\mathbf{E}) \\ \mathbf{J}_p(\boldsymbol{p}_0) \end{pmatrix} \boldsymbol{v}$$

and the stacked Jacobian is now $7 \times 6$ and we can solve for camera velocity using the pseudoinverse. As for the previous IBVS examples, the desired velocity is
proportional to the difference between the current and desired feature values

$$\begin{pmatrix} \dot{E}_0 \\ \dot{E}_1 \\ \vdots \\ \dot{E}_4 \\ \dot{u}_0 \\ \dot{v}_0 \end{pmatrix} = \lambda \begin{pmatrix} E_0^* - E_0 \\ E_1^* - E_1 \\ \vdots \\ E_4^* - E_4 \\ u_0^* - u_0 \\ v_0^* - v_0 \end{pmatrix}$$

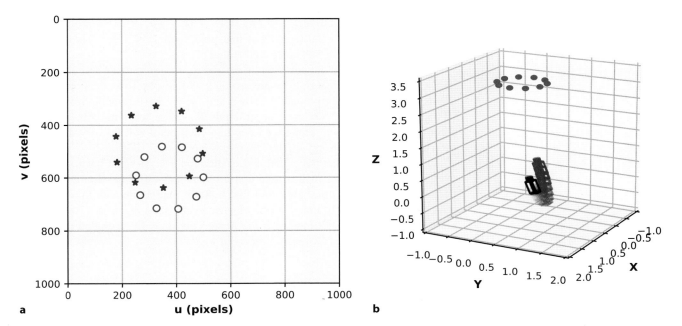

**□ Fig. 15.15** IBVS using ellipse features, showing the image plane with current points (circle) and demanded (star); **b** a world view showing the
points and the camera

The simulation is run in the now familiar fashion

```
>>> ibvs = IBVS_e.Example();  # quick problem setup
>>> ibvs.run()
```

and a snapshot of results is shown in ◘ Fig. 15.15.

### 15.3.3  Photometric Features

When servoing using point or line features we have to determine the error between current and desired features, and this requires determining correspondence between features in the current and the desired images. Correspondence is a complex task which can be complicated by occlusions or features leaving the camera's field of view. In photometric visual servoing we work directly with the pixel values and no correspondences are required.

The image feature is a vector that contains the intensity of every pixel in the image – current or desired – stacked into a very tall vector of height $N = W \times H$. The rate of change of the pixel values $\dot{\mathbf{I}}$ is related to the camera velocity by

$$
\begin{pmatrix} \dot{i}_0 \\ \dot{i}_1 \\ \vdots \\ \dot{i}_{N-1} \end{pmatrix} = \mathbf{J}_{\mathrm{I}}(\mathbf{I})\boldsymbol{v}
$$

where the Jacobian is

$$
\mathbf{J}_I(\mathbf{I}) = \begin{pmatrix} \nabla_{\mathrm{I}}(\boldsymbol{p}_0)\,\mathbf{J}_p(\boldsymbol{p}_0, Z_0) \\ \nabla_{\mathrm{I}}(\boldsymbol{p}_1)\,\mathbf{J}_p(\boldsymbol{p}_1, Z_1) \\ \vdots \\ \nabla_{\mathrm{I}}(\boldsymbol{p}_{N-1})\,\mathbf{J}_p(\boldsymbol{p}_{N-1}, Z_{N-1}) \end{pmatrix} \in \mathbb{R}^{N \times 6}
$$

and where $\boldsymbol{p}_i$ is the image-plane coordinate of the pixel corresponding to the $i^{\mathrm{th}}$ element of the feature vector, $\nabla_{\mathrm{I}}(\boldsymbol{p}) = \big(\nabla_u(\boldsymbol{p}), \nabla_v(\boldsymbol{p})\big) \in \mathbb{R}^{1\times 2}$ are the image gradients in the $u$- and $v$-directions at that pixel, and $\mathbf{J}_p(\cdot) \in \mathbb{R}^{2\times 6}$ is the image point-feature Jacobian (15.9) computed at that pixel.

Once again we need the point depth $Z_i$, and we should use our best estimate of that, as previously discussed. If we are servoing with respect to a planar image then the depth might be approximately known and, as we have remarked previously, IBVS is quite robust to errors in point depth. If we are servoing with respect to a complex 3-dimensional scene then depth at each pixel will be very difficult to determine and we again rely on the inherent robustness of IBVS.

In order to converge, the actual and destination images must have significant overlap. The derivation makes assumptions that the scene has Lambertian reflectance, no specular highlights, and that lighting magnitude and direction does not change over time. In practice photometric visual servoing works well even if these assumptions are not met, or if the images are partially occluded during the camera motion.

### 15.4  Wrapping Up

In this chapter we have learned about the fundamentals of vision-based robot control, and the fundamental techniques developed over two decades up to the mid 1990s. There are two distinct configurations. The camera can be attached to the robot observing the goal (eye-in-hand) or fixed in the world observing both robot and goal. Another form of distinction is the control structure: Position-Based Visual Servo (PBVS) and Image-Based Visual Servo (IBVS). The former involves pose estimation based on a calibrated camera and a geometric model of the goal object, while the latter performs the control directly in the image plane. Each approach has certain advantages and disadvantages. PBVS performs efficient straight-line Cartesian camera motion in the world, but may cause image features to leave the image plane. IBVS always keeps features in the image plane, but may result in trajectories that exceed the reach of the robot, particularly if it requires a large amount of rotation about the camera's optical axis. IBVS also requires a touch of 3-dimensional information (the depth of the feature points) but is quite robust to errors in depth and it is quite feasible to estimate the depth as the camera moves. IBVS can be formulated to work not only with point features, but also with lines, ellipses and pixel values. An arbitrary number of features (which can be any mix of points, lines or ellipses) from an arbitrary number of cameras can be combined simply by stacking the relevant Jacobian matrices.

So far in our simulations we have determined the required camera velocity and moved the camera accordingly, without consideration of the mechanism or robot that moves it. In the next chapter we consider cameras attached to arm-type robots, mobile ground robots and aerial robots.

## 15.4.1 Further Reading

The tutorial paper by Hutchinson et al. (1996) was the first comprehensive articulation and taxonomy of the field, and Chaumette and Hutchinson (2006 and 2007) provide a more recent tutorial introduction. Chapters on visual servoing are included in Siciliano and Khatib (2016, § 34) and Spong et al. (2006, § 12).

It is well known that IBVS is very tolerant to errors in depth and its effect on control performance is examined in detail in Marey and Chaumette (2008). Feddema and Mitchell (1989) performed a partial 3D reconstruction to determine point depth, based on observed features and known goal geometry. Papanikolopoulos and Khosla (1993) described adaptive control techniques to estimate depth, as used in this chapter. Hosoda and Asada (1994), Jägersand et al. (1996) and Piepmeier et al. (1999) have shown how the image Jacobian matrix itself can be estimated online from measurements of robot and image motion. The second-order visual servoing technique was introduced by Malis (2004).

The most common image Jacobian is based on the motion of points in the image, but it can also be derived for the parameters of lines in the image plane (Chaumette 1990; Espiau et al. 1992) and the parameters of an ellipse in the image plane (Espiau et al. 1992). Mahoney et al. (2002) describe how the choice of features effects the closed-loop dynamics. Moments of binary regions have been proposed for visual servoing of planar scenes (Chaumette 2004; Tahri and Chaumette 2005). More recently the ability to servo directly from image pixel values, without segmentation or feature extraction, has been described by Collewet et al. (2008) and subsequent papers, and more recently by Bakthavatchalam et al. (2015) and Crombez et al. (2015). Deep learning has been applied to visual servoing by Bateux et al. (2018). The literature on PBVS is much smaller, but the paper by Westmore and Wilson (1991) is a good introduction. They use an EKF to implicitly perform pose estimation – the goal pose is the filter state and the innovation between predicted and observed feature coordinates updates the goal pose state. Hashimoto et al. (1991) present simulations to compare position-based and image-based approaches.

15

## ▪▪ History and Background

Visual servoing has a very long history – the earliest reference is by Shirai and Inoue (1973) who describe how a visual feedback loop can be used to correct the position of a robot to increase task accuracy. They demonstrated a system with a servo cycle time of 10 s, and this highlights the early challenges with real-time feature extraction. Up until the late 1990s this required bulky and expensive special-purpose hardware such as that shown in ▢ Fig. 15.16. Significant early work on industrial applications occurred at SRI International during the late 1970s (Hill and Park 1979; Makhlin 1985).

In the 1980s Weiss et al. (1987) introduced the classification of visual servo structures as either position-based or image-based. They also introduced a distinction between visual servo and dynamic look and move, the former uses only visual feedback whereas the latter uses joint feedback and visual feedback. This latter distinction is no longer in common usage and most visual servo systems today make use of joint-position and visual feedback, commonly encoder-based joint velocity loops as discussed in ▶ Sect. 9.1.6 with an outer vision-based position loop. Weiss (1984) applied adaptive control techniques for IBVS of a robot arm without joint-level feedback, but the results were limited to low degree of freedom arms due to the low-sample rate vision processing available at that time. Others have looked at incorporating the manipulator dynamics (9.11) into controllers that command motor torque directly (Kelly 1996; Kelly et al. 2002a, 2002b) but all still require joint angles in order to evaluate the manipulator Jacobian, and the joint rates to provide damping. Feddema (Feddema and Mitchell 1989; Feddema 1989) used closed-loop joint control to overcome problems due to low visual sampling rate and demonstrated IBVS for 4-DoF. Chaumette, Rives and Espiau (Chaumette et al. 1991; Rives et al. 1989) describe a similar approach using the task function method (Samson et al. 1990) and show experimental results for robot positioning using a goal object with four features. Feddema et al. (1991) describe an algorithm to select which subset of the available features give the best conditioned square Jacobian. Hashimoto et al. (1991) have shown that there are advantages in using a larger number of features and using a pseudoinverse to solve for velocity. Control and stability in closed-loop visual control systems was addressed by several researchers (Corke and Good 1992; Espiau et al. 1992; Papanikolopoulos et al. 1993) and feedforward predictive, rather than feedback, controllers were proposed by Corke (1994) and Corke and Good (1996).

▢ **Fig. 15.16** In the early 1990s (Corke 1994) it took a 19 inch VMEbus rack of hardware image processing cards, capable of 10 Mpixs$^{-1}$ throughput or 50 Hz framerate for $512 \times 512$ images, to do real-time visual servoing

The 1993 book edited by Hashimoto (1993) was the first collection of papers covering approaches and applications in visual servoing. The 1996 book by Corke (1996b) is now out of print but available free online and covers the fundamentals of robotics and vision for controlling the dynamics of an image-based visual servoing system. It contains an extensive, but dated, collection of references to visual servoing applications including industrial applications, camera control for tracking, high-speed planar micromanipulators, road vehicle guidance, aircraft refueling, and fruit picking. Another important collection of papers (Kriegman et al. 1998) stems from a 1998 workshop on the synergies between control and vision: how vision can be used for control and how control can be used for vision. Another workshop collection by Chesi and Hashimoto (2010) covers more recent algorithmic developments and application.

Visual servoing has been applied to a diverse range of problems that normally require human hand-eye skills such as ping-pong (Andersson 1989), juggling (Rizzi and Koditschek 1991) and inverted pendulum balancing (Dickmanns and Graefe 1988a; Andersen et al. 1993), catching (Sakaguchi et al. 1993; Buttazzo et al. 1993; Bukowski et al. 1991; Skofteland and Hirzinger 1991; Skaar et al. 1987; Lin et al. 1989), and controlling a labyrinth game (Andersen et al. 1993).

### 15.4.2 Exercises

1. Position-based visual servoing (▶ Sect. 15.1)
   a) Run the PBVS example. Experiment with varying parameters such as the initial camera pose, the path fraction $\lambda$ and adding pixel noise to the output of the camera.
   b) Create a PBVS system that servos with respect to a fiducial marker from ▶ Sect. 13.6.1.
   c) Create a block diagram model for PBVS.
   d) Use a different camera model for the pose estimation (slightly different focal length or principal point) and observe the effect on final end-effector pose.
   e) Implement an EKF-based PBVS system as described in Westmore and Wilson (1991).
2. Optical flow fields (▶ Sect. 15.2.1)
   a) Plot the optical flow fields for cameras with different focal lengths.
   b) Plot the flow field for some composite camera motions such as $x$- and $y$-translation, $x$- and $z$-translation, and $x$-translation and $z$-rotation.
3. For the case of two points the image Jacobian is $4 \times 6$ and the null space has two columns. What camera motions do they correspond to?
4. Image-based visual servoing (▶ Sect. 15.2.2)
   a) Run the IBVS example, either the Python script or block diagram version. Experiment with varying the gain $\lambda$. Remember that $\lambda$ can be a scalar or a diagonal matrix which allows different gain settings for each degree of freedom.
   b) Implement the function to limit the maximum norm of the commanded velocity.
   c) Experiment with adding pixel noise to the output of the camera.
   d) Experiment with different initial camera poses and desired image-plane coordinates.
   e) Experiment with different number of goal points, from three up to ten. For the cases where $N > 3$ compare the performance of the pseudoinverse with just selecting a subset of three points (first three or random three). Design an algorithm that chooses a subset of points which results in the stacked Jacobian with the best condition number?

f) Create a set of desired image-plane points that form a rectangle rather than a square. There is no perspective viewpoint from which a square appears as a rectangle (why is this?). What does the IBVS system do?

g) Create a set of desired image-plane points that cannot be reached, for example swap two adjacent world or image points. What does the IBVS system do?

h) Use a different camera model for the image Jacobian (slightly different focal length or principal point) and observe the effect on final end-effector pose.

i) Implement second-order IBVS using (15.15).

j) For IBVS we generally force points to move in straight lines but this is just a convenience. Use a trajectory generator to move the points from initial to desired position with some sideways motion, perhaps a half or full cycle of a sine wave. What is the effect on camera Cartesian motion?

k) Implement stereo IBVS. Hint: stack the point feature Jacobians for both cameras and determine the desired feature positions on each camera's image plane.

l) Simulate the output of a camera at arbitrary pose observing a planar target. Use knowledge of homographies and image warping to achieve this.

m) Implement an IBVS controller using SIFT features. Use the SIFT descriptors to ensure robust correspondence.

5. Derive the image Jacobian for the case where the camera is limited to just pan and tilt motion. What happens if the camera pans and tilts about a point that is not the camera center?

6. When discussing motion perceptibility we used the identity $\mathbf{J}_p^{+\top}\mathbf{J}_p^+ = (\mathbf{J}_p\mathbf{J}_p^\top)^{-1}$. Prove this. Hint, use the singular value decomposition $\mathbf{J} = \mathbf{U}\mathbf{\Sigma}\mathbf{V}^\top$ and remember that $\mathbf{U}$ and $\mathbf{V}$ are orthogonal matrices.

7. End-point open-loop visual servo systems have not been discussed in this book. Consider a group of goal points on the robot end effector as well as the those on the goal object, both being observed by a single camera (challenging).

a) Create an end-point open-loop PBVS system.

b) Use a different camera model for the pose estimation (slightly different focal length or principal point) and observe the effect on final end-effector relative pose.

c) Create an end-point open-loop IBVS system.

d) Use a different camera model for the image Jacobian (slightly different focal length or principal point) and observe the effect on final end-effector relative pose.

8. Run the line-based visual servo example (▶ Sect. 15.3.1).

9. Ellipse-based visual servo (▶ Sect. 15.3.2)

a) Run the ellipse-based visual servo example.

b) Modify to servo five degrees of camera motion using just the ellipse parameters (without the point feature).

c) For an arbitrary shape we can compute its equivalent ellipse which is expressed in terms of an inertia tensor and a centroid. Determine the ellipse parameters of (15.18) from the inertia tensor and centroid. Create an ellipse feature visual servo to move to a desired view of the arbitrary shape (challenging).

d) Try servoing on the equivalent ellipse of a blob, perhaps the shark image from ◨ Fig. 12.8c.

10. Implement a simulation of photometric visual servoing. You could use homographies and perspective warping to simulate a camera moving with respect to a planar image, and the derivative of Gaussian kernel to compute the image gradients. Investigate the performance when servoing from different initial camera translations and rotations. Vary the assumed depth, and vary the parameters of the derivative kernel.

# Advanced Visual Servoing

## Contents

© The Author(s), under exclusive license to Springer Nature Switzerland AG 2023
P. Corke, *Robotics, Vision and Control*, Springer Tracts in Advanced Robotics 146,
https://doi.org/10.1007/978-3-031-06469-2_16

This chapter builds on the previous one and introduces some advanced visual servo techniques and applications. ► Sect. 16.1 introduces a hybrid visual servo method that avoids some of the limitations of the IBVS and PBVS schemes described previously.

Wide-angle cameras such as fisheye lenses and catadioptric cameras have significant advantages for visual servoing. ► Sect. 16.2 shows how IBVS can be reformulated for polar rather than Cartesian image-plane coordinates. This is directly relevant to fisheye lenses but also gives improved rotational control when using a perspective camera. The unified imaging model from ► Sect. 13.4 allows most cameras (perspective, fisheye and panoramic) to be represented by a spherical projection model, and ► Sect. 16.3 shows how IBVS can be reformulated for spherical cameras.

► Sect. 16.4 presents a number of brief application examples. These illustrate, using block diagrams, how visual servoing can be used with different types of cameras (perspective and spherical) and different types of robots (arm-type robots, mobile ground robots and aerial robots). Examples include a 6-axis robot manipulator with an eye-in-hand camera; a mobile robot moving to a specific pose which could be used for navigating through a doorway or docking; and a quadrotor moving to, and hovering at, a fixed pose with respect to a goal on the ground.

## 16.1 XY/Z-Partitioned IBVS

In ► Sect. 15.2.4, we encountered the problem of camera retreat in an IBVS system. This phenomenon can be explained intuitively by the fact that the IBVS control law causes feature points to move in straight lines on the image plane, but for a camera rotating about its optical axis the points will naturally move along circular arcs. The linear IBVS controller dynamically changes the overall image scale so that motion along an arc appears as motion along a straight line. The scale change is achieved by $z$-axis translation which we observe as camera retreat.

Partitioned methods eliminate camera retreat by using IBVS to control some degrees of freedom, while using a different controller for the remaining degrees of freedom. The XY/Z hybrid schemes consider the $x$- and $y$-axes as one group, and the $z$-axes as another group. The approach is based on a couple of insights. Firstly, and intuitively, the camera retreat problem is a $z$-axis phenomenon: $z$-axis rotation leads to unwanted $z$-axis translation. Secondly, from ◻ Fig. 15.7, the image-plane motion due to $x$- and $y$-axis translational and rotational motion are all quite similar, whereas the optical flow due to $z$-axis rotation and translation are radically different.

We partition the point-feature optical flow of (15.10) so that

$$\dot{\boldsymbol{p}} = \mathbf{J}_{xy}\boldsymbol{v}_{xy} + \mathbf{J}_z\boldsymbol{v}_z \tag{16.1}$$

where $\boldsymbol{v}_{xy} = (v_x, v_y, \omega_x, \omega_y)$, $\boldsymbol{v}_z = (v_z, \omega_z)$, and $\mathbf{J}_{xy}$ and $\mathbf{J}_z$ are respectively columns 0,1,3,4 and columns 2, 5 of $\mathbf{J}_p$. We compute $\boldsymbol{v}_z$ using a separate controller, so we can write (16.1) as

$$\boldsymbol{v}_{xy} = \lambda\mathbf{J}_{xy}^{+}(\dot{\boldsymbol{p}}^* - \mathbf{J}_z\boldsymbol{v}_z) \tag{16.2}$$

where $\dot{\boldsymbol{p}}^*$ is the desired feature point velocity that we have used previously for IBVS control as in (15.13).

The $z$-axis velocities $\boldsymbol{v}_z = (v_z, \omega_z)$ are computed directly from two scalar image features $\theta$ and $A$ shown in ◻ Fig. 16.1. The first image feature $\theta \in [-\pi, \pi)$, is the angle between the $u$-axis and the directed line segment joining feature points $i$ and $j$. ◄ For numerical conditioning, it is advantageous to select the longest line segment that can be constructed from the feature points, and allowing that this may

$\theta$ in this case has a different meaning compared to its use in ► Sects. 12.2 and 15.3.1. The line here is $u\sin\theta + v\cos\theta + d = 0$.

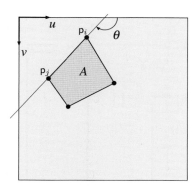

**◻ Fig. 16.1** Image features for $XY/Z$ partitioned IBVS control. As well as the coordinates of the four points which are used directly for IBVS, we use the angle of the longest line segment $\theta$ and the polygon area $A$

change during the motion as the feature point configuration changes. The desired rotational rate is obtained using a simple proportional control law

$$\omega_z^* = \lambda_{\omega_z}(\theta^* \ominus \theta)$$

where the operator $\ominus$ indicates modulo-$2\pi$ subtraction which is implemented by the Toolbox functions `angdiff`, `wrap_mpi_pi` or `wrap_0_2pi`. As always for motion on a circle, there are two directions to move to achieve the goal. If the rotation is limited, for instance by a mechanical stop on a robot joint, then the sign of $\omega_z$ should be chosen so as to avoid motion through that stop.

The second image feature that we use is a function of the area $A \in \mathbb{R}_{>0}$ of the regular polygon whose vertices are the image feature points. The advantages of this measure are: it is strongly dependent on $z$-axis translation; it is a scalar; it is invariant to rotation about the $z$-axis, thus decoupling camera rotation from $z$-axis translation; and it can be cheaply computed. The area of the polygon is just the zero-order moment, $m_{00}$ which can be computed from the vertices using the Toolbox function `Polygon2(p).area()`. The feature we will use for control is the square root of the area

$$\sigma = \sqrt{m_{00}}$$

which has units of length, in pixels. The desired camera $z$-axis translation rate is obtained using a simple proportional control law

$$v_z^* = \lambda_{v_z}(\sigma^* - \sigma) \tag{16.3}$$

The features discussed above for $z$-axis translation and rotation control are simple and inexpensive to compute, but work best when the goal's normal is within $\pm40°$ of the camera's optical axis. When the goal plane is not orthogonal to the optical axis its area will appear diminished, due to perspective, which causes the camera to initially approach the goal. Perspective will also change the perceived angle of the line segment which can cause small, but unnecessary, $z$-axis rotational motion.

A block diagram model is shown in ◻ Fig. 16.2. The simulation is run by

```
>>> %run -m IBVS-partitioned-main -H
```

and the image-plane features are animated, while scope blocks plot the camera velocity and feature error against time. ▶ The parameters such as the camera model, initial camera pose and world points are defined in the properties of the various blocks and in the source code of `IBVS-partitioned-main`.

Unlike the standard IBVS, this controller can drive feature points out of the field of view. If points are moving toward the edge of the field of view, the simplest

The `-H` option stops the function from blocking until all the figures are dismissed.

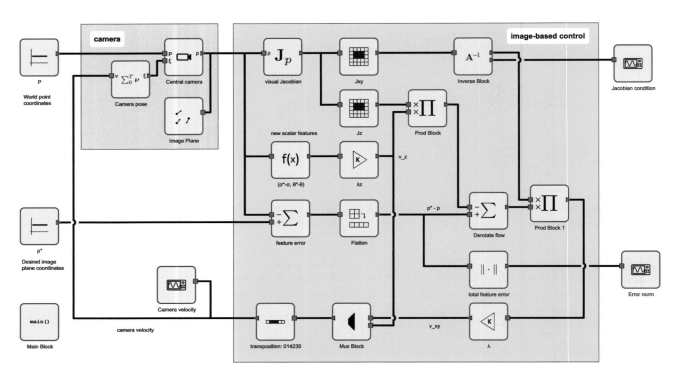

**◻ Fig. 16.2** Block diagram model `IBVS-partitioned` of a $XY/Z$-partitioned visual servo scheme, an extension of the IBVS system shown in ◻ Fig. 15.9. The initial camera pose is set in the `camera pose` block and the desired image-plane points $p^*$ are set in the lower left red block

way to keep them in view is to move the camera away from the goal. We define a repulsive force that acts on the camera, pushing it away as a point approaches the boundary of the image plane

$$
F_z(p) = \begin{cases} \left(\frac{1}{d(p)} - \frac{1}{d_0}\right)\frac{1}{d^2(p)} & \text{if } d(p) \le d_0 \\ 0 & \text{if } d(p) > d_0 \end{cases}
$$

where $d(p)$ is the shortest distance to the edge of the image plane from the image point coordinate $p$, and $d_0$ is the width of the image zone in which the repulsive force acts. For a $W \times H$ image

$$
d(p) = \min(u, v, W - u, H - v) \tag{16.4}
$$

Such a repulsion force could be incorporated into the $z$-axis translation controller

$$
v_z^* = \lambda_{v_z}(\sigma^* - \sigma) - \eta \sum_{i=1}^{N} F_z(p_i)
$$

where $\eta$ is a gain constant with units of damping. The repulsion force is discontinuous and may lead to chattering where the feature points oscillate in and out of the repulsive force – this could be remedied by introducing smoothing filters and velocity limiters or velocity dampers.

## 16.2 IBVS Using Polar Coordinates

In ▶ Sect. 15.3 we showed image feature Jacobians for point features expressed in terms of their Cartesian coordinates $(u, v)$, but other coordinate systems could be used. In polar coordinates the image point is represented by $p = (\phi, r)$ where $\phi$ is the angle

$$\phi = \tan^{-1} \frac{\overline{v}}{\overline{u}} \tag{16.5}$$

and $\overline{u}$ and $\overline{v}$ are the image coordinates with respect to the principal point rather than the image origin. The distance of the point from the principal point is

$$r = \sqrt{\overline{u}^2 + \overline{v}^2} \; . \tag{16.6}$$

The Cartesian and polar representations are related by

$$\overline{u} = r \cos \phi, \quad \overline{v} = r \sin \phi \tag{16.7}$$

Taking the derivatives with respect to time

$$\begin{pmatrix} \dot{\overline{u}} \\ \dot{\overline{v}} \end{pmatrix} = \begin{pmatrix} -r \sin \phi & \cos \phi \\ r \cos \phi & \sin \phi \end{pmatrix} \begin{pmatrix} \dot{\phi} \\ \dot{r} \end{pmatrix}$$

and inverting

$$\begin{pmatrix} \dot{\phi} \\ \dot{r} \end{pmatrix} = \begin{pmatrix} -\frac{1}{r} \sin \phi & \frac{1}{r} \cos \phi \\ \cos \phi & \sin \phi \end{pmatrix} \begin{pmatrix} \dot{\overline{u}} \\ \dot{\overline{v}} \end{pmatrix}$$

and substituting into (15.9) along with (16.7), we can write

$$\begin{pmatrix} \dot{\phi} \\ \dot{r} \end{pmatrix} = \mathbf{J}_{p,p} \begin{pmatrix} v_x \\ v_y \\ v_z \\ \omega_x \\ \omega_y \\ \omega_z \end{pmatrix} \tag{16.8}$$

where the feature Jacobian is

$$\mathbf{J}_{p,p} = \begin{pmatrix} \frac{f}{rZ} \sin \phi & -\frac{f}{rZ} \cos \phi & 0 & \frac{f}{r} \cos \phi & \frac{f}{r} \sin \phi & -1 \\ -\frac{f}{Z} \cos \phi & -\frac{f}{Z} \sin \phi & \frac{r}{Z} & \frac{f^2+r^2}{f} \sin \phi & -\frac{f^2+r^2}{f} \cos \phi & 0 \end{pmatrix} \tag{16.9}$$

This Jacobian is unusual in that it has three constant elements. In the first row, the zero indicates that polar angle is invariant to translation along the optical axis (points move along radial lines), and the negative one indicates that the polar angle of a feature (with respect to the $u$-axis) decreases with positive camera rotation. In the second row, the zero indicates that radius $r$ is invariant to rotation about the $z$-axis. As for the Cartesian point features, the translational part of the Jacobian (the first three columns) are proportional to $1/Z$. Note also that the Jacobian is undefined for $r = 0$, that is for a point on the optical axis. The image Jacobian is computed by the `visjac_p_polar` method of the `CentralCamera` class.

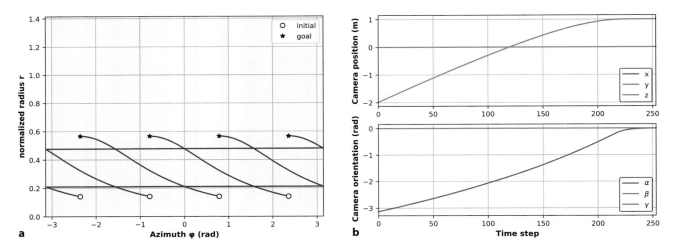

**◘ Fig. 16.3** IBVS using polar coordinates. **a** Feature motion in polar $\phi r$-space; **b** camera motion in Cartesian space

The desired feature velocity is a function of feature error

$$\dot{\boldsymbol{p}}^* = \lambda \begin{pmatrix} \phi^* \ominus \phi \\ r^* - r \end{pmatrix}$$

where $\ominus$ is modulo-$2\pi$ subtraction for the angular component. The choice of units (pixels and radians) means that $|r| \gg |\phi|$ and radius should be normalized

$$r = \frac{\sqrt{\bar{u}^2 + \bar{v}^2}}{\sqrt{W^2 + H^2}}$$

so that $r$ and $\phi$ are of approximately the same order of magnitude.

An example of IBVS using polar coordinates is implemented by the class `IBVS_polar`. We first create a default camera model

```
>>> camera = CentralCamera.Default(pose=SE3.Tz(-2)*SE3.Rz(pi))
>>> P = mkgrid(2, 0.5, pose=SE3.Tz(2))
>>> ibvs = IBVS_polar(camera, lmbda=0.1, P=P, pose_d=SE3.Tz(1),
...                   depth=2, graphics=False)
```

which is the case shown in ◘ Fig. 15.13, for a rotation of $\pi$ about the optical axis, where the standard IBVS algorithm failed. When we run the simulation

```
>>> ibvs.run()
```

we see the camera moving directly toward the goal and rotating as required, and the features take direct paths on the polar image plane. The time history of feature and camera motion are

```
>>> ibvs.plot_p()
>>> ibvs.plot_pose()
```

◘ Fig. 16.3a shows that the features have moved upward and to the left on the $\phi r$-plane, which could be considered as a cylinder, the left- and right-hand sides are joined, and we see that the polar angle of two features have wrapped around from $-\pi$ to $\pi$. ◘ Fig. 16.3b shows that the camera has undergone monotonic translational motion in the $z$-direction and shows no sign of camera retreat.

It has been observed that the performance of polar IBVS is the complement of Cartesian IBVS – it generates good camera motion for the case of large rotation, but poorer motion for the case of large translation. To gain the benefit of both representations we could, for every feature point, stack the $2 \times 6$ Jacobian for the Cartesian point coordinates (15.9) *and* the $2 \times 6$ Jacobian for the polar coordinates (16.9). The leads to a $4N \times 6$ Jacobian which can be inverted using the pseudoinverse as in previous examples.

16

## 16.3 IBVS for a Spherical Camera

In ▶ Sect. 13.3 we introduced several nonperspective cameras such as the fisheye lens camera and the catadioptric camera. Given the particular projection equations for any camera, we can derive an image feature Jacobian from first principles. However the many different lens and mirror shapes leads to many different projection models and image Jacobians. In ▶ Sect. 13.4 we showed that feature points from any type of camera can be projected to a sphere, so all we need to derive is an image Jacobian for a spherical camera.

We proceed in a similar manner to the perspective camera in ▶ Sect. 15.2.1. Referring to ◻ Fig. 13.21, the world point P is represented by the coordinate vector $P = (X, Y, Z)$ in the camera frame, and is projected onto the surface of the sphere at the point p represented by the coordinate vector $p = (x, y, z)$ where the projection ray pierces the sphere. We can write

$$x = \frac{X}{R}, \quad y = \frac{Y}{R}, \quad \text{and } z = \frac{Z}{R} \tag{16.10}$$

where $R = \sqrt{(X^2 + Y^2 + Z^2)}$ is the distance from the camera origin to the world point.

For a unit image sphere the surface constraint $x^2 + y^2 + z^2 = 1$ means that one of the Cartesian coordinates is redundant. We will use a minimal spherical coordinate system comprising the azimuth angle (or longitude)

$$\phi = \tan^{-1} \frac{y}{x} \in [-\pi, \pi) \tag{16.11}$$

and the angle of colatitude ▶

$$\theta = \sin^{-1} r \in [0, \pi] \tag{16.12}$$

Colatitude is zero at the north pole and increases to π at the south pole.

where $r = \sqrt{x^2 + y^2}$, and this yields a point feature vector $p = (\phi, \theta)$.

Taking the derivatives of (16.12) and (16.11) with respect to time, and substituting (15.5) as well as

$$X = R \sin\theta \cos\phi, \quad Y = R \sin\theta \sin\phi, \quad Z = R \cos\theta \tag{16.13}$$

we obtain, in matrix form, the spherical optical flow equation

$$\begin{pmatrix} \dot{\phi} \\ \dot{\theta} \end{pmatrix} = \mathbf{J}_{p,s}(\phi, \theta, R) \begin{pmatrix} v_x \\ v_y \\ v_z \\ \omega_x \\ \omega_y \\ \omega_z \end{pmatrix} \tag{16.14}$$

where the image feature Jacobian is

$$\mathbf{J}_{p,s} = \begin{pmatrix} \frac{\sin\phi}{R\sin\theta} & -\frac{\cos\phi}{R\sin\theta} & 0 & \frac{\cos\phi\,\cos\theta}{\sin\theta} & \frac{\sin\phi\,\cos\theta}{\sin\theta} & -1 \\ -\frac{\cos\phi\,\cos\theta}{R} & -\frac{\sin\phi\,\cos\theta}{R} & \frac{\sin\theta}{R} & \sin\phi & -\cos\phi & 0 \end{pmatrix} \tag{16.15}$$

Like the polar coordinate case, this Jacobian also has three constant values. In the first row, the zero indicates that azimuth angle is invariant to translation along the optical axis, while the negative one indicates that the azimuth changes in the opposite sense to camera rotation about the optical axis. The zero in the second row indicates that colatitude is invariant to rotation about the optical axis. As for all image Jacobians, the translational submatrix (the first three columns) is a function of point depth $R$. This Jacobian is computed by the `visjac_p` method of the `SphericalCamera` class.

The Jacobian is not defined at the north and south poles where $\sin\theta = 0$ and azimuth also has no meaning at these points. This is a singularity, and as we remarked in ▶ Sect. 2.3.1.3, in the context of Euler angle representation of orientation, this is a consequence of using a minimal representation. However, in general the benefits outweigh the costs for this application.

For control purposes we follow the normal procedure of computing one $2 \times 6$ Jacobian, (16.15), for each of $N$ feature points and stacking them to form a $2N \times 6$ matrix

$$\begin{pmatrix} \dot{\phi}_0 \\ \dot{\theta}_0 \\ \vdots \\ \dot{\phi}_{N-1} \\ \dot{\theta}_{N-1} \end{pmatrix} = \begin{pmatrix} \mathbf{J}_0 \\ \vdots \\ \mathbf{J}_{N-1} \end{pmatrix} v \qquad (16.16)$$

The control law is

$$v = \mathbf{J}^+ \dot{p}^* \qquad (16.17)$$

where $\dot{p}^*$ is the desired velocity of the features in $\phi\theta$-space. Typically we choose this to be proportional to feature error

$$\dot{p}^* = \lambda(p^* \ominus p) \qquad (16.18)$$

where $\lambda$ is a positive gain, $p$ is the current point in $\phi\theta$-coordinates, and $p^*$ the desired value. This results in locally linear motion of features within the feature space. $\ominus$ denotes modulo subtraction and returns the smallest angular distance given that $\phi = [-\pi, \pi)$ and $\theta \in [0, \pi]$.

An example of IBVS using spherical coordinates (◻ Fig. 16.4) is implemented by the class `IBVS_sph`. We first create a spherical camera and a set of world points

```
>>> camera = SphericalCamera(pose=SE3.Trans(0.3, 0.3, -2)
...                                   * SE3.Rz(0.4))
>>> P = mkgrid(2, side=1.5, pose=SE3.Tz(0.5))
```

and then a spherical IBVS object

```
>>> ibvs = IBVS_sph(camera, P=P, pose_d=SE3.Tz(-1.5), verbose=False,
...                     graphics=False)
```

where `pose_d` is the camera pose used to determine the desired image sphere points – it is not used for control. When we run run the simulation

```
>>> ibvs.run()
```

we see feature motion on the $\phi\theta$-plane and the camera and world points in a world view.

Spherical imaging has many advantages for visual servoing. Firstly, a spherical camera eliminates the need to explicitly keep features in the field of view which we have seen is a problem with both position-based visual servoing and some hybrid schemes. Secondly, we previously observed an ambiguity between the optical flow fields for $y$-axis rotation and $x$-axis translation ◀ for a small field of view. For IBVS with a long focal length camera this can lead to slow convergence and/or sensitivity to noise in feature coordinates. For a spherical camera, with the largest possible field of view, this ambiguity is reduced. ◀

Spherical cameras do not yet exist ◀ but we can can project feature coordinates from one or more cameras, of any type, onto the image sphere. Then the camera velocity can be computed using the control law in spherical coordinates.

As well as for $x$-axis rotation and $-y$-axis translation.

Provided that the world points are well distributed around the sphere.

The camera of ◻ Fig. 13.27b comes close with 90% of a spherical field of view.

**16**

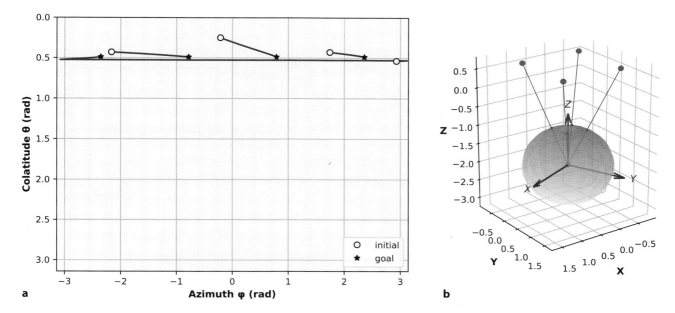

▣ **Fig. 16.4** IBVS using spherical camera and coordinates. **a** Feature motion in $\phi\theta$-space; **b** four goal points projected onto the sphere in its initial pose

## 16.4  Applications

This section briefly introduces a number of advanced control application that bring together all that we have learnt about robotics, vision and control. Each is described by a detailed block diagram, which you can run and experiment with, and a succinct overview.

### 16.4.1  Arm-Type Robot

In this example, the camera is carried by a 6-axis robot manipulator which can control all six degrees of freedom of the camera's pose. We will assume that the robot's joints are ideal velocity sources, that is, they move at precisely the velocity that was commanded. A modern robot is very close to this ideal, typically having high-performance joint controllers using velocity and position feedback from encoders on the joints.

The *nested* control structure for a robot joint was discussed in ▶ Sect. 9.1.7. The inner velocity loop uses joint velocity feedback to ensure that the joint moves at the desired speed. The outer position loop uses joint position feedback to determine the joint speed required to follow the trajectory. These loops frequently operate at a 1 kHz sample rate. By contrast, in this visual servo system, the position loop is provided by the vision system which has a low sample rate, typically only 25 or 30 Hz, and often with a high latency of one or two sample times.

The block diagram model of this eye-in-hand system is shown in ▣ Fig. 16.5 and simulates the camera, IBVS control, and the robot. ▶ The desired camera velocity is input to a resolved-rate motion controller based on the manipulator Jacobian. The joint angle rates are input to a discrete-time integrator which represents the robot's velocity loops. The resulting joint angles are input to a forward kinematics block which outputs the end-effector pose. A perspective camera with default parameters is mounted on the robot's end effector, and its axes are aligned with the end-effector coordinate frame. The camera block has one parameter which is a

In this case the venerable Puma 560 from Part III of this book, but you can change this to any robot supported by the Toolbox.

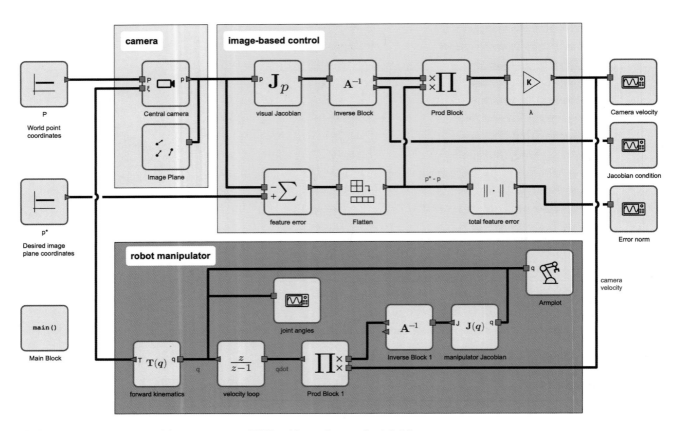

■ **Fig. 16.5**  Block diagram model `IBVS-arm` uses IBVS to drive a robot arm that is holding a camera

`CentralCamera` object, and its inputs are the coordinates of the goal points, which are the corners of a square in the *yz*-plane, and the camera pose in the world frame. The output image features are used to compute an image Jacobian with an assumed $Z$ value for every point, and also to determine the feature error in image space. The image Jacobian is inverted and a gain applied to determine the spatial velocity of the camera. The control loop is now closed, and the system will drive the robot and camera to achieve the desired image-plane point configuration.

We run this model by

```
>>> %run -m IBVS-arm-main -H
```

which displays an animation of the robot moving and the features moving on the image plane of a virtual camera.

The simulation adds a results object called `out` to the global namespace. The robot joint angles at each time step are the states of the discrete-time integrator connected to `clock0` and can be plotted by

```
>>> plt.plot(out.clock0.t, out.clock0.x)
```

with one row per time step. The parameters such as the robot and camera model, initial robot configuration and world points are defined in the properties of the various blocks and in the source code of `IBVS-arm-main.py`.

Note that this model does not include any dynamics of the robot arm or the vision system – the joints are modeled as perfect velocity control devices, and the vision system is modeled as having no delay. For high-performance visual control these two dynamic effects are important and need to be taken into consideration. Nevertheless, this model could form the basis of a higher fidelity model that incorporates these real-world effects.

## 16.4.2 Mobile Robot

In this section we consider a camera mounted on a mobile robot moving in a planar environment. We will first consider a holonomic robot, that is one that has an omni-directional base and can move in any direction. Then we extend the solution to a nonholonomic car-like base which touches on some of the issues discussed in ▶ Chap. 4. The camera observes a number of landmark points with known 3-dimensional coordinates. ▶ The visual servo controller will drive the robot until its view of the landmarks matches the desired view.

These points can be above the ground plane where the robot operates.

### 16.4.2.1 Holonomic Mobile Robot

For this problem we assume an upward-looking central perspective camera fixed to the robot, and a number of ceiling-mounted landmarks that are continuously visible to the camera as the robot moves along the path. The vehicle's coordinate frame is such that the $x$-axis is forward and the $z$-axis is upward.

We define a perspective camera

```
>>> camera = CentralCamera.Default(f=0.002);
```

with a wide field of view so that it can keep the landmarks in view as it moves. With respect to the robot's body frame {B}, the camera is mounted at a constant relative pose $^B\xi_C$

```
>>> T_B_C = SE3.Trans(0.2, 0.1, 0.3);
```

which is to the front left of the vehicle, 30 cm above ground level, with its optical axis upward, and its $x$-axis pointing forward. The two landmarks are 3 m above the ground and situated at $x = 0$ and $y = \pm 1$ m in the world frame

```
>>> P = np.array([[0, 1, 3], [0, -1, 3]]).T;
```

The desired vehicle position is with the center of its rear axle at $(-2, 0)$.

The robot operates in the $xy$-plane and can rotate only about the $z$-axis, so we will remove the columns of the image Jacobian from (15.9) that correspond to nonpermissible motion and write

$$\begin{pmatrix} \dot{u} \\ \dot{v} \end{pmatrix} = \begin{pmatrix} -\frac{f}{\rho_u Z} & 0 & \bar{v} \\ 0 & -\frac{f}{\rho_v Z} & \bar{u} \end{pmatrix} \begin{pmatrix} v_x \\ v_y \\ \omega_z \end{pmatrix} \tag{16.19}$$

As for standard IBVS case, we stack these Jacobians, one per landmark, and then invert the equation to solve for the vehicle velocity. Since there are only three unknown components of velocity, and each landmark contributes two equations, we need two or more feature points in order to solve for velocity.

The block diagram model is shown in ◪ Fig. 16.6 and is a hybrid of earlier models such as ◪ Figs. 4.7 and 15.9. The model is simulated by

```
>>> %run -m IBVS-holonomic-main -H
```

and displays an animation of the vehicle's path in the $xy$-plane and the camera view. Results are stored in the simulation results object `out` and can be displayed as for previous examples. The parameters such as the robot and camera model, initial robot configuration and world points are defined in the properties of the various blocks and in the source code of `IBVS-holonomic-main.py`.

### 16.4.2.2 Nonholonomic Mobile Robot

The challenge of driving a nonholonomic mobile robot to a pose were discussed in ▶ Chap. 4, and a nonlinear pose controller was introduced in ▶ Sect. 4.1.1.4. The notation for our problem is shown in ◪ Fig. 16.7 and once again we use a controller based on the polar coordinates $(\rho, \alpha, \beta)$. For this control example we will use PBVS techniques to estimate the variables needed for control. We assume

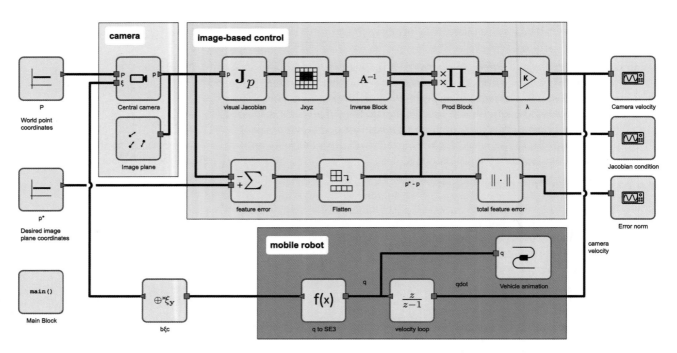

**◻ Fig. 16.6** Block diagram model `IBVS-holonomic` drives a holonomic mobile robot to a pose using IBVS control

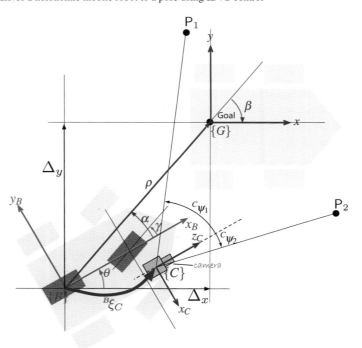

**◻ Fig. 16.7** PBVS for nonholonomic vehicle (bicycle model) vehicle moving toward a goal pose: $\rho$ is the distance to the goal, $\beta$ is the angle of the goal vector with respect to the world frame, and $\alpha$ is the angle of the goal vector with respect to the vehicle frame. $P_1$ and $P_2$ are landmarks which are at bearing angles of $^C\psi_1$ and $^C\psi_2$ with respect to the camera

a central perspective camera that is fixed to the robot body frame {B} with a relative pose $^B\xi_C$; a number of landmarks with known locations relative to a goal frame {G}; the landmarks are continuously visible to the camera; and that the vehicle's heading $\theta$ is also known, perhaps using a compass or some other sensor.

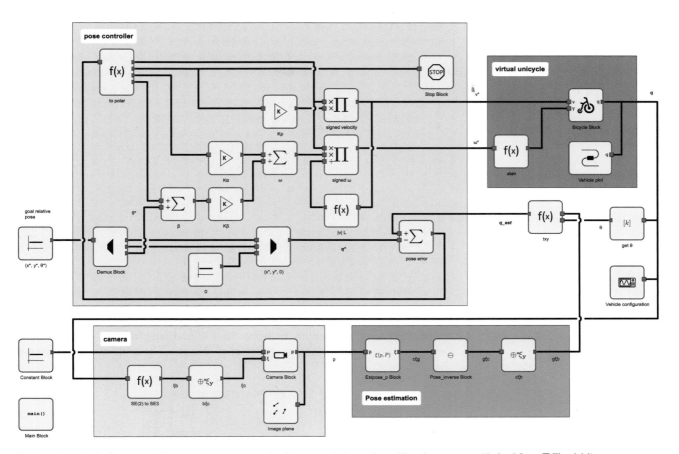

**Fig. 16.8** Block diagram model `IBVS-nonholonomic` drives a nonholonomic mobile robot to a pose (derived from Fig. 4.14)

The block diagram model is shown in Fig. 16.8 and the view of the landmarks is simulated by the camera block and its output, the projected points, are input to a pose estimation block and the known locations of the landmarks are set as parameters.

As discussed in ▶ Sect. 13.2.4 at least three landmarks are needed, and in this example five landmarks are used. The pose estimator outputs the estimated pose of the landmarks with respect to the camera $^C\hat{\boldsymbol{\xi}}_G$ which are transformed to the robot's pose with respect to the goal. The $x$- and $y$-translational components of this pose are combined with estimated heading angle ▶ to yield an estimate of the vehicle's configuration $(\hat{\boldsymbol{x}}, \hat{\boldsymbol{y}}, \hat{\theta})$ which is input to the pose controller. The remainder of the system is essentially the same as the example from Fig. 4.14.

The simulation is run by

In a real system, heading angle would come from a compass, but in this simulation we "cheat" and use the true heading angle.

```
>>> %run -m IBVS-nonholonomic-main -H
```

and displays an animation of the vehicle's path in the $xy$-plane and the camera view. Results are stored in the simulation results object `out` and can be displayed as for previous examples. The parameters such as the robot and camera model, initial robot configuration and world points are defined in the properties of the various blocks and in the source code of `IBVS-nonholonomic-main.py`.

### 16.4.3 Aerial Robot

A spherical camera is particularly suitable for aerial and underwater robots that move in 3-dimensional space. In this example we consider a spherical camera attached to a quadrotor, and we will use IBVS to servo the quadrotor to a particular pose with respect to four goal points on the ground.

As discussed in ▶ Sect. 4.3 the quadrotor is underactuated and we cannot independently control all 6 degrees of freedom in task space – we can only control position $(X, Y, Z)$ and yaw angle. Roll and pitch angle are the control inputs used to achieve translation in the horizontal plane, and must be zero when the vehicle is in equilibrium. The block diagram model is shown in ◼ Fig. 16.9. This controller attempts to keep the quadrotor at a constant relative pose with respect to the goal points. If the goal moves, so too will the quadrotor – we could imagine a scheme like this being used to land a quadrotor on a moving vehicle – a car, mobile robot or boat.

The model is a hybrid of the quadrotor controller from ◼ Fig. 4.24 and the underactuated IBVS system of ◼ Fig. 16.6. There are however a number of key differences. Firstly, in the quadrotor control of ◼ Fig. 4.24 we used a rotation matrix to map $xy$-error in the world frame to the pitch and roll demand of the vehicle. This is not needed for the visual servo case since the $xy$-error is observed directly by the camera in the body frame. Secondly, like the mobile robot case, the vehicle is underactuated and here the Jacobian comprises only the columns corresponding to $(v_x, v_y, v_z, \omega_z)$. Thirdly, we are using a spherical camera, so a `SphericalCamera` object is passed to the camera and visual Jacobian blocks.

Fourthly, there is coupling between the roll and pitch motion of the quadrotor and the image-plane feature coordinates. We recall how the quadrotor cannot

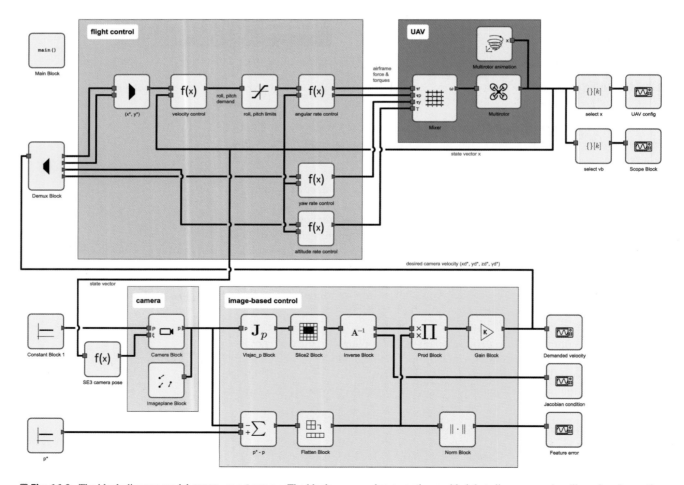

◼ **Fig. 16.9** The block diagram model `IBVS-quadrotor`. The block `p* - p` has an option enabled that allows proper handling of angles on the sphere during subtraction

translate without first tilting into the direction it wishes to translate, and this will cause the features to move in the image and increase the image feature error. For small amounts of roll and pitch this can be ignored but for aggressive maneuvers it should be taken into account. While not implemented in this model, the principle is to use the image Jacobian to approximate ▶ the feature displacements $(\Delta\phi, \Delta\theta)$ as a function of displacements in camera roll and pitch angle which are rotations about the $x$- and $y$-axes respectively

$$\begin{pmatrix} \Delta\phi \\ \Delta\theta \end{pmatrix} = \begin{pmatrix} \sin\psi & -\cos\phi \\ \frac{\cos\phi\cos\theta}{\sin\phi} & \frac{\sin\phi\cos\theta}{\sin\phi} \end{pmatrix} \begin{pmatrix} \Delta\alpha \\ \Delta\beta \end{pmatrix}$$

and these can be subtracted from the features observed by the camera to give the features that would be observed by a camera at the origin of frame {B} but with its axes parallel to those of frame {0}. This scheme is sometimes referred to as feature derotation since it mimics in software, the effect of a nonrotating or gimbal-stabilized camera.

Comparing ◩ Fig. 16.9 to ◩ Fig. 4.24 we see the visual controller performs the function of the outermost *position loops* for $x$- and $y$-position, altitude and yaw. It generates the required inputs for the velocity loops of those degrees of freedom directly. Note that rate information is still required as input to the velocity loops and in a real aerial robot this would be derived from an inertial measurement unit.

The simulation is run by

```
>>> %run -m IBVS-quadrotor-main -H
```

and displays an animation of the vehicle's path in three dimensions and the camera view while various scope blocks plot the camera velocity and feature error against time. Results are stored in the simulation results object `out` and can be displayed as for previous examples. The parameters such as the quadrotor and camera model, initial quadrotor configuration and world points are defined in the properties of the various blocks and in the source code of `IBVS-quadrotor-main.py`.

This is a first-order approximation to the feature motion.

## 16.5  Wrapping Up

### 16.5.1  Further Reading

A good introduction to advanced visual servo techniques is the tutorial article by Chaumette and Hutchinson (2007) and also the visual servoing chapter in Siciliano and Khatib (2016, § 34). Much of the interest in so-called hybrid techniques was sparked by Chaumette's paper (Chaumette 1998) which introduced the specific example that drives the camera of a point-based IBVS system to infinity, for the case of goal rotation by π about the optical axis. One of the first methods to address this problem was 2.5D visual servoing, proposed by Malis et al. (1999), which augments the image-based point features with a minimal Cartesian feature. Other notable early hybrid methods were proposed by Morel et al. (2000) and Deguchi (1998) which partitioned the image Jacobian into a translational and rotational part. An homography is computed between the initial and final view (so the goal points must be planar) and then decomposed to determine a rotation and translation. Morel et al. combine this rotational information with translational control based on IBVS of the point features. Conversely, Deguchi et al. combine this translational information with rotational control based on IBVS. Since translation is only determined up to an unknown scale factor some additional means of determining scale is required.

Corke and Hutchinson (2001) presented an intuitive geometric explanation for the problem of the camera moving away from the goal during servoing, and proposed a partitioning scheme split by axes: $x$- and $y$-translation and rotation in one group, and $z$-translation and rotation in the other. Another approach to hybrid visual servoing is to switch rapidly between IBVS and PBVS approaches (Gans et al. 2003). The performance of several partitioned schemes is compared by Gans et al. (2003).

The polar form of the image Jacobian for point features (Iwatsuki and Okiyama 2002a, 2002b; Chaumette and Hutchinson 2007) handles the IBVS failure case nicely, but results in somewhat suboptimal camera translational motion (Corke et al. 2009) – the converse of what happens for the Cartesian formulation.

The Jacobian for a spherical camera is similar to the polar form. The two angle parameterization was first described in Corke (2010) and was used for control and structure-from-motion estimation. There has been relatively little work on spherical visual servoing. Fomena and Chaumette (2007) consider the case for a single spherical object from which they extract features derived from the projection to the spherical imaging plane such as the center of the circle and its apparent radius. Tahri et al. (2009) consider spherical image features such as lines and moments. Hamel and Mahony (2002) describe kino-dynamic control of an underactuated aerial robot using point features.

The robot manipulator dynamics (9.11) and the perspective projection (13.2) are highly nonlinear, and a function of the state of the manipulator and the goal. Almost all visual servo systems consider that the robot is velocity controlled, and that the underlying dynamics are suppressed and linearized by tight control loops. As we learned in ▶ Sect. 9.1, this is the case for arm-type robots and in the quadrotor example we used a similar nested control structure. This approach is necessitated by the short time constants of the underlying mechanical dynamics, and the slow sample rate and latency of any visual control loop. Modern computers and high-speed cameras make it theoretically possible to do away with axis-level velocity loops but it is far simpler to use them. High-frame-rate cameras are increasingly available but have lower resolution and may require higher levels of scene illumination, and the latency of the vision system becomes important.

Visual servoing of nonholonomic robots is nontrivial since Brockett's theorem (1983) shows that no linear time-invariant controller can control it. The approach used in this chapter was position based and is a minor extension of the pose controller introduced in ▶ Sect. 4.1.1.4. IBVS approaches have been proposed (Tsakiris et al. 1998; Masutani et al. 1994) but require that the camera is attached to the mobile base by a robot with a small number of degrees of freedom. Mariottini et al. (2007) describe a two-step servoing approach where the camera is rigidly attached to the base and the epipoles of the geometry, defined by the current and desired camera views, are explicitly servoed. Usher (Usher et al. 2003; Usher 2005) describes a switching control law that takes the robot onto a line that passes through the desired pose, and then along the line to the pose – experimental results on an outdoor vehicle are presented. The similarity between mobile robot navigation and visual servoing problem is discussed in Corke (2001).

## 16.5.2 Resources

The controllers demonstrated in this chapter have all worked with simulated robotic systems, and have executed much slower than real time. In order to put visual control into practice we need to have fast image processing and feature extraction algorithms, as well as means of communicating with the robot hardware. Fortunately there are lots of tools and technologies to help with this: the Robot Operating System (aka, ROS ▶ www.ros.org) is a comprehensive robot software framework for creating robots, OpenCV for image processing (▶ www.opencv.org), and ViSP for creating visual trackers and controllers ▶ https://visp.inria.fr.

## 16.5.3 Exercises

1. XY/Z-partitioned IBVS (▶ Sect. 16.1)
   a) Investigate the generated motion for different combinations of initial camera translation and rotation, and compare to the classical IBVS scheme of the last chapter.
   b) Create a scenario where the features leave the image.
   c) Add a repulsion field to ensure that the features remain within the image.
   d) Investigate variations of (16.3). Instead of driving the difference of area to zero, try driving the ratio of current and desired area to one, or the logarithm of this ratio to zero. See Mahony et al. (2002).
2. Investigate the performance of polar and spherical IBVS for different combinations of initial camera translation and rotation, and compare to the classical IBVS scheme of the last chapter.
3. Arm-robot IBVS example (▶ Sect. 16.4.1)
   a) Add an offset (rotation and/or translation) between the end effector and the camera. Your controller will need to incorporate an additional Jacobian (see ▶ Sect. 3.1.3) to account for this.
   b) Add a delay after the camera block to model the image processing time. Investigate the response as the delay is increased, and the tradeoff between gain and delay. You might like to plot a discrete-time root locus diagram for this dynamic system.
   c) Model a moving goal. Show the tracking error, that is, the distance between the camera and the goal.
   d) Investigate feedforward techniques to improve the control (Corke 1996b). Hint, instead of basing the control on where the goal was seen by the camera, base it on where it will be some short time into the future. How far into the future? What is a good model for this estimation? Investigate $\alpha - \beta$ filters as one option for a simple predictor. (challenging)
   e) An eye-in-hand camera for a docking task might have problems as the camera gets really close to the goal. How might you configure the goal points and camera to avoid this?
4. Mobile robot visual servo (▶ Sect. 16.4.2)
   a) For the holonomic and nonholonomic cases replace the perspective camera with a catadioptric camera.
   b) For the holonomic case with a catadioptric camera, move the robot through a series of via points, each defined in terms of a set of desired feature coordinates.
   c) For the nonholonomic case, implement the pure pursuit and line following controllers from ▶ Chap. 4 but in this case using visual features. For pure pursuit, consider the object being pursued carries one or two point features. For the line following case, consider using one or two line features.
5. Display the feature flow fields, like ◘ Fig. 15.7, for the polar $r - \phi$ and spherical $\theta - \phi$ projections (▶ Sects. 16.2 and 16.3). For the spherical case can you plot the flow vectors on the surface of a sphere?
6. Quadrotor visual servo (▶ Sect. 16.4.3)
   a) Replace the spherical camera with a perspective camera.
   b) Create a controller to follow a series of point features rather than hover over a single point (challenging).
   c) Add image feature derotation to minimize the effect of vehicle roll and pitch on the visual control.
7. Implement the 2.5D visual servo scheme by Malis et al. (1999) (challenging).

# Supplementary Information

# A Installing the Toolboxes

The most up-to-date instructions on installing the required software, and getting started, can always be found at

 https://github.com/petercorke/RVC3-python

## A.1  Installing the Packages

This book depends on the following open-source Python packages:
- Robotics Toolbox for Python `roboticstoolbox`
- Machine Vision Toolbox for Python `machinevisiontoolbox`

which in turn have dependencies on other packages created by the author and third parties.

The Python package `rvc3python` provides a simple one-step installation of these Toolboxes and their dependencies

```
$ pip install rvc3python
```

or

```
$ conda install rvc3python
```

The package provides additional resources for readers of the book including:
- `rvctool`, a command line script that is an IPython wrapper. It imports the above mentioned packages using `import *` and then provides an interactive computing environment. By default `rvctool` has prompts like the regular Python REPL not IPython, and it automatically displays the results of expressions like MATLAB® does – put a semicolon on the end of the line to suppress that. `rvctool` allows cutting and pasting in lines from the book, and prompt characters are ignored.
- Jupyter notebooks, one per chapter, containing all the code examples.
- The code to produce every Python-generated figure in the book.
- All example scripts.
- All block diagram models.

The GitHub repository `RVC3-python` also contains online renderable versions of selected 3D figures, and all line drawings in EPS format.

## A.2  Block Diagram Models

Block diagram models are simulated using the Python package `bdsim` which can run models:
- written in Python using `bdsim` block classes and wiring methods.
- created graphically using `bdedit` and saved as a `.bd` (JSON format) file.

## A.3  Getting Help

The GitHub repository `RVC3-python` hosts a number of other resources for users:
- *Wiki pages* provides answers to frequently asked questions.
- *Discussions* between users on aspects of the book and the core toolboxes.
- *Issues* can be used for reporting and discussing issues and bugs.

# B Linear Algebra

## B.1  Vectors

The term *vector* has multiple meanings which can lead to confusion:
- In computer science, a vector is an array of numbers or a tuple.
- In physics and engineering, a vector is a physical quantity like force or velocity which has a magnitude, direction and unit.
- In mathematics, a vector is an object that belong to a vector space.
- In linear algebra, a vector is a one-dimensional matrix organized as either a row or a column.

In this book, we use all these interpretations and rely on the context to disambiguate, but this section is concerned with the last two meanings. We consider only real vectors which are an ordered *n-tuple* of real numbers which is usually written as

$$\boldsymbol{v} = (v_0, v_1, \ldots, v_{n-1}) \in \mathbb{R}^n$$

where $v_0$, $v_1$, etc. are called the scalar components of $\boldsymbol{v}$, and $v_i$ is called the $i^{\text{th}}$ component of $\boldsymbol{v}$. The symbol $\mathbb{R}^n = \mathbb{R} \times \mathbb{R} \times \cdots \times \mathbb{R}$ is a Cartesian product that denotes the set of ordered $n$-tuples of real numbers. For a 3-vector, we often write the elements as $\boldsymbol{v} = (v_x, v_y, v_z)$.

The coordinate of a point in an $n$-dimensional space is also represented by an *n-tuple* of real numbers. A coordinate vector is that same tuple but interpreted as a linear combination $\boldsymbol{v} = v_0 \boldsymbol{e}_0 + v_1 \boldsymbol{e}_1 + \cdots + v_{n-1} \boldsymbol{e}_{n-1}$ of the orthogonal basis vectors $\{\boldsymbol{e}_0, \boldsymbol{e}_1, \ldots, \boldsymbol{e}_{n-1}\}$ of the space – this is a vector from the origin to the point. In this book, the basis vectors are denoted by $\{\hat{\boldsymbol{x}}, \hat{\boldsymbol{y}}\}$ or $\{\hat{\boldsymbol{x}}, \hat{\boldsymbol{y}}, \hat{\boldsymbol{z}}\}$ for 2 or 3 dimensions respectively.

In mathematics, an $n$-dimensional *vector space* (also called a *linear space*) is a group-like object that contains a collection of objects called *vectors* $\boldsymbol{v} \in \mathbb{R}^n$. It supports the operations of vector addition $\boldsymbol{a} + \boldsymbol{b} = (a_0 + b_0, a_1 + b_1, \ldots, a_{n-1} + b_{n-1})$, and multiplication ("scaling") by a number $s$ called a scalar $s\boldsymbol{a} = (sa_0, sa_1, \ldots, sa_{n-1})$. The negative of a vector, scaling by $-1$, is obtained by negating each element of the vector $-\boldsymbol{a} = (-a_0, -a_1, \ldots, -a_{n-1})$.

The symbol $\mathbb{R}^n$ is used to denote a space of points or vectors, and context is needed to resolve the ambiguity. We need to be careful to distinguish points and vectors because the operations of addition and scalar multiplication, while valid for vectors, are meaningless for points. We can add a vector to the coordinate vector of a point to obtain the coordinate vector of another point, and we can subtract one coordinate vector from another, and the result is the displacement between the points.

For many operations in linear algebra, it is important to distinguish between column and row vectors

$$\boldsymbol{v} = \begin{pmatrix} v_0 \\ v_1 \\ \vdots \\ v_{n-1} \end{pmatrix} \quad \text{or} \quad \boldsymbol{v} = (v_0, v_1, \ldots, v_{n-1})$$

which are equivalent to an $n \times 1$ and a $1 \times n$ matrix (see next section) respectively. Most vectors in this book are column vectors which are sometimes written compactly as $(v_0, v_1, \ldots, v_{n-1})^\top$, where $\cdot^\top$ denote a matrix transpose. Both can be denoted by $\mathbb{R}^n$, or distinguished by $\mathbb{R}^{n \times 1}$ or $\mathbb{R}^{1 \times n}$ for column and row vectors respectively.

The magnitude or length of a vector is a nonnegative scalar given by its $p$-norm

`np.linalg.norm(v)`

$$\|\boldsymbol{v}\|_p = \left(\sum_{i=0}^{n-1} |v_i|^p\right)^{1/p} \in \mathbb{R}_{\geq 0}\ .$$

The Euclidean length of a vector is given by $\|\boldsymbol{v}\|_2$ which is also referred to as the $L_2$ norm and is generally assumed when $p$ is omitted, for example $\|\boldsymbol{v}\|$. A unit vector ◄ is one where $\|\boldsymbol{v}\|_2 = 1$ and is denoted as $\hat{\boldsymbol{v}}$. The $L_1$ norm is the sum of the absolute value of the elements of the vector, and is also known as the Manhattan distance, it is the distance traveled when confined to moving along the lines in a grid. The $L_\infty$ norm is the maximum element of the vector.

The unit vector corresponding to the vector $\boldsymbol{v}$ is $\hat{\boldsymbol{v}} = \frac{1}{\|\boldsymbol{v}\|_2}\boldsymbol{v}$.

The dot or inner product of two vectors is a scalar

`np.dot(a, b)`

$$\boldsymbol{a} \cdot \boldsymbol{b} = \boldsymbol{b} \cdot \boldsymbol{a} = \sum_{i=0}^{n-1} a_i b_i = \|\boldsymbol{a}\|\,\|\boldsymbol{b}\|\cos\theta$$

where $\theta$ is the angle between the vectors. $\boldsymbol{a} \cdot \boldsymbol{b} = 0$ when the vectors are orthogonal. If $\boldsymbol{a}$ and $\boldsymbol{b}$ are column vectors, $\boldsymbol{a}, \boldsymbol{b} \in \mathbb{R}^{n\times 1}$, the dot product can be written as

$$\boldsymbol{a} \cdot \boldsymbol{b} = \boldsymbol{a}^\top \boldsymbol{b} = \boldsymbol{b}^\top \boldsymbol{a}$$

The outer product $\boldsymbol{a}\boldsymbol{b}^\top \in \mathbb{R}^{n\times n}$ has a maximum rank of one, and if $\boldsymbol{b} = \boldsymbol{a}$ is a symmetric matrix.

For 3-vectors, the cross product

`np.cross(a, b)`

$$\boldsymbol{a} \times \boldsymbol{b} = -\boldsymbol{b} \times \boldsymbol{a} = \det\begin{pmatrix} \hat{\boldsymbol{x}} & \hat{\boldsymbol{y}} & \hat{\boldsymbol{z}} \\ a_0 & a_1 & a_2 \\ b_0 & b_1 & b_2 \end{pmatrix} = \|\boldsymbol{a}\|\,\|\boldsymbol{b}\|\sin\theta\,\hat{\boldsymbol{n}}$$

where $\hat{\boldsymbol{x}}$ is a unit vector parallel to the $x$-axis, etc., and $\hat{\boldsymbol{n}}$ is a unit vector normal to the plane containing $\boldsymbol{a}$ and $\boldsymbol{b}$ whose direction is given by the right-hand rule. If the vectors are parallel, $\boldsymbol{a} \times \boldsymbol{b} = 0$.

## B.2  Matrices

Real matrices are a subset of all matrices. For the general case of complex matrices, the term Hermitian is the analog of symmetric, and unitary is the analog of orthogonal. $\mathbf{A}^H$ denotes the Hermitian transpose, the complex conjugate transpose of the complex matrix $\mathbf{A}$. Matrices are rank-2 tensors.

In this book we are concerned, almost exclusively, with real $m \times n$ matrices ◄

$$\mathbf{A} = \begin{pmatrix} a_{0,0} & a_{0,1} & \cdots & a_{0,n-1} \\ a_{1,0} & a_{1,1} & \cdots & a_{1,n-1} \\ \vdots & \vdots & \ddots & \vdots \\ a_{m-1,0} & a_{m-1,1} & \cdots & a_{m-1,n-1} \end{pmatrix} \in \mathbb{R}^{m\times n}$$

with $m$ rows and $n$ columns. If $n = m$, the matrix is square.

Matrices of the same size $\mathbf{A} \in \mathbb{R}^{m\times n}$ and $\mathbf{B} \in \mathbb{R}^{m\times n}$ can be added

`A+B`

$$\mathbf{C} = \mathbf{A} + \mathbf{B}, \quad c_{i,j} = a_{i,j} + b_{i,j}, \quad i = 0, ..., m-1, \ j = 0, ..., n-1$$

or multiplied element-wise

`A*B`

$$\mathbf{C} = \mathbf{A} \circ \mathbf{B}, \quad c_{i,j} = a_{i,j} b_{i,j}, \quad i = 0, ..., m-1, \ j = 0, ..., n-1$$

which is also called the Hadamard product. Two matrices with *conforming* dimensions $\mathbf{A} \in \mathbb{R}^{m\times n}$ and $\mathbf{B} \in \mathbb{R}^{n\times p}$ can be multiplied

`A@B`

$$\mathbf{C} = \mathbf{A}\mathbf{B}, \quad c_{i,j} = \sum_{k=0}^{n-1} a_{i,k} b_{k,j}, \quad i = 0, ..., m-1, \ j = 0, ..., p-1$$

which is matrix multiplication and $\mathbf{C} \in \mathbb{R}^{m\times p}$.

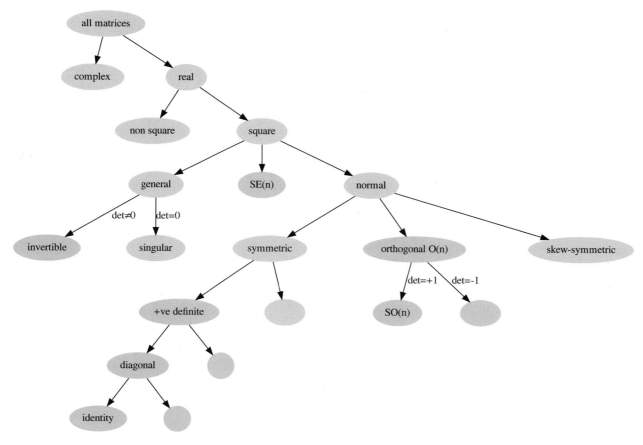

**□ Fig. B.1** Taxonomy of matrices. Matrices shown in blue are never singular

The transpose of a matrix is

A.T

$$\mathbf{B} = \mathbf{A}^\top, \quad b_{i,j} = a_{j,i}, \quad \forall i, j$$

and it can be shown that

$$(\mathbf{AB})^\top = \mathbf{B}^\top \mathbf{A}^\top, \quad (\mathbf{ABC})^\top = \mathbf{C}^\top \mathbf{B}^\top \mathbf{A}^\top, \quad \text{etc.}$$

The elements of a matrix can also be matrices, creating a block or partitioned matrix

$$\mathbf{A} = \begin{pmatrix} \mathbf{B}_{0,0} & \mathbf{B}_{0,1} & \cdots & \mathbf{B}_{0,n-1} \\ \mathbf{B}_{1,0} & \mathbf{B}_{1,1} & \cdots & \mathbf{B}_{1,n-1} \\ \vdots & \vdots & \ddots & \vdots \\ \mathbf{B}_{m-1,0} & \mathbf{B}_{m-1,1} & \cdots & \mathbf{B}_{m-1,n-1} \end{pmatrix}$$

where all matrices in a column must have the same width, and all matrices in a row must have the same height. A block diagonal matrix has blocks, not necessarily of the same size, arranged along the diagonal

$$\mathbf{A} = \begin{pmatrix} \mathbf{B}_0 & & & \\ & \mathbf{B}_1 & & 0 \\ & & \ddots & \\ 0 & & & \mathbf{B}_{p-1} \end{pmatrix}$$

A taxonomy of matrices is shown in □ Fig. B.1.

## B.2.1  Square Matrices

`np.linalg.inv(A)`

A square matrix may have an inverse $\mathbf{A}^{-1}$ in which case

$$\mathbf{A}\mathbf{A}^{-1} = \mathbf{A}^{-1}\mathbf{A} = \mathbf{1}_{n\times n}$$

where

`np.eye(n)`

$$\mathbf{1}_{n\times n} = \begin{pmatrix} 1 & & & 0 \\ & 1 & & \\ & & \ddots & \\ 0 & & & 1 \end{pmatrix} \in \mathbb{R}^{n\times n}$$

is the identity matrix, a unit diagonal matrix, sometimes written as $\mathbf{I}$. The inverse $\mathbf{A}^{-1}$ exists provided that the matrix $\mathbf{A}$ is non-singular, that is, its determinant $\det(\mathbf{A}) \neq 0$. The inverse can be computed from the matrix of cofactors. If $\mathbf{A}$ and $\mathbf{B}$ are square and non-singular, then

`np.linalg.det(A)`

$$(\mathbf{A}\mathbf{B})^{-1} = \mathbf{B}^{-1}\mathbf{A}^{-1}, \quad (\mathbf{A}\mathbf{B}\mathbf{C})^{-1} = \mathbf{C}^{-1}\mathbf{B}^{-1}\mathbf{A}^{-1}, \quad \text{etc.}$$

and also

$$\left(\mathbf{A}^\top\right)^{-1} = \left(\mathbf{A}^{-1}\right)^\top = \mathbf{A}^{-\top}.$$

The inverse can be written as

$$\mathbf{A}^{-1} = \frac{1}{\det(\mathbf{A})}\mathrm{adj}(\mathbf{A})$$

Confusingly, sometimes also referred to as the adjoint matrix, but in this book that term is reserved for the matrix introduced in ▶ Sect. 3.1.3.

where $\mathrm{adj}(\mathbf{A})$ is the transpose of the matrix of cofactors and known as the adjugate matrix and sometimes denoted by $\mathbf{A}^*$. ◄ If $\mathbf{A}$ is non-singular, the adjugate can be computed by

$$\mathrm{adj}(\mathbf{A}) = \det(\mathbf{A})\mathbf{A}^{-1}.$$

For a square $n \times n$ matrix, if:

$\mathbf{A} = \mathbf{A}^\top$ – the matrix is **symmetric**. The inverse of a symmetric matrix is also symmetric. Many matrices that we encounter in robotics are symmetric, for example, covariance matrices and manipulator inertia matrices.

$\mathbf{A}^{-1} = \mathbf{A}^\top$ – the matrix is **orthogonal**. The matrix is also known as **orthonormal** since its column (and row) vectors must be of unit length, and the columns (and rows) are orthogonal (normal) to each other, that is, their dot products are zero. The product of two orthogonal matrices is also an orthogonal matrix. The set of $n \times n$ orthogonal matrices forms a group $\mathbf{O}(n)$ under the operation of matrix multiplication known as the orthogonal group. The determinant of an orthogonal matrix is either $+1$ or $-1$. The subgroup of orthogonal matrices with determinant $+1$ is called the special orthogonal group denoted by $\mathbf{SO}(n)$.

$\mathbf{A} = -\mathbf{A}^\top$ – the matrix is **skew symmetric** or **anti symmetric**. Such a matrix has a zero diagonal, and is always singular if $n$ is odd. Any matrix can be written as the sum of a symmetric matrix and a skew-symmetric matrix.

There is a mapping from a vector to a skew-symmetric matrix which, for the case $\boldsymbol{v} \in \mathbb{R}^3$, is

`skew(v)`

$$\mathbf{A} = [\boldsymbol{v}]_\times = \begin{pmatrix} 0 & -v_z & v_y \\ v_z & 0 & -v_x \\ -v_y & v_x & 0 \end{pmatrix} \tag{B.1}$$

The inverse mapping is

$$v = \vee_{\times}(\mathbf{A}) \ .$$

vex(A)

If $v \in \mathbb{R}^n$ and $\mathbf{A} \in \mathbb{R}^{n \times n}$ then $[\mathbf{A}v]_{\times} = \det(\mathbf{A})\mathbf{A}^{-\top}[v]_{\times}\mathbf{A}^{-1}$.

For the 3-dimensional case, the cross product can be written as the matrix-vector product $v_1 \times v_2 = [v_1]_{\times}v_2$, and $v^{\top}[v]_{\times} = [v]_{\times}v = 0$, $\forall v$. If $\mathbf{R} \in \mathbf{SO}(3)$ then $[\mathbf{R}v]_{\times} = \mathbf{R}[v]_{\times}\mathbf{R}^{\top}$.

$\mathbf{A}^{\top}\mathbf{A} = \mathbf{A}\mathbf{A}^{\top}$ – the matrix is **normal** and can be diagonalized by an orthogonal matrix $\mathbf{U}$ so that $\mathbf{U}^{\top}\mathbf{A}\mathbf{U}$ is a diagonal matrix. All symmetric, skew-symmetric and orthogonal matrices are normal matrices as are matrices of the form $\mathbf{A} = \mathbf{B}^{\top}\mathbf{B} = \mathbf{B}\mathbf{B}^{\top}$ where $\mathbf{B}$ is an arbitrary matrix.

The square matrix $\mathbf{A} \in \mathbb{R}^{n \times n}$ can be applied as a linear transformation to a vector $x \in \mathbb{R}^n$

$$x' = \mathbf{A}x$$

which results in another vector, generally with a change in its length and direction. For example, in two dimensions, if $x$ is the set of all points lying on a circle, then $x'$ defines points that lie on some ellipse. However, there are some important special cases. If $\mathbf{A} \in \mathbf{SO}(n)$, the transformation is isometric and the vector's length is unchanged, $\|x'\| = \|x\|$. The (right) eigenvectors of the matrix are those vectors $x$ such that

np.linalg.eig(A)

$$\mathbf{A}x = \lambda_i x, \quad i = 0, \ldots, n-1 \tag{B.2}$$

that is, their direction is unchanged when transformed by the matrix. They are simply scaled by $\lambda_i$, the corresponding eigenvalue. The matrix $\mathbf{A}$ has $n$ eigenvalues (the *spectrum* of the matrix) which can be real or complex pairs. For an orthogonal matrix, the eigenvalues lie on a unit circle in the complex plane, $|\lambda_i| = 1$, and the eigenvectors are all orthogonal to one another.

The Toolbox demo

```
>>> %run -m eigdemo 1 2 3 4
```

shows an animation of a rotating 2D coordinate vector $x$ and the transformed vector $\mathbf{A}x$. The arguments are the row-wise elements of $\mathbf{A}$. Twice per revolution $x$ and $\mathbf{A}x$ are parallel – these are the eigenvectors of $\mathbf{A}$.

If $\mathbf{A}$ is non-singular, then the eigenvalues of $\mathbf{A}^{-1}$ are the reciprocal of those of $\mathbf{A}$, and the eigenvectors of $\mathbf{A}^{-1}$ are parallel to those of $\mathbf{A}$. The eigenvalues of $\mathbf{A}^{\top}$ are the same as those of $\mathbf{A}$ but the eigenvectors are different.

The trace of a matrix is the sum of the diagonal elements

$$\mathrm{tr}(\mathbf{A}) = \sum_{i=0}^{n-1} a_{i,i}$$

np.trace(A)

which is also the sum of the eigenvalues

$$\mathrm{tr}(\mathbf{A}) = \sum_{i=0}^{n-1} \lambda_i \ .$$

For a rotation matrix, either 2D or 3D, the trace is related to the angle of rotation

$$\theta = \cos^{-1}\frac{\mathrm{tr}(\mathbf{R}) - 1}{2}$$

about the rotation axis, but due to the limited range of $\cos^{-1}$, values of $\theta$ above $\pi$ cannot be properly determined.

The determinant of the matrix is equal to the product of the eigenvalues

$$\det(\mathbf{A}) = \prod_{i=0}^{n-1} \lambda_i$$

thus a matrix with one or more zero eigenvalues will be singular.

The eigenvalues of a real **symmetric** matrix $\mathbf{A}$ are all real and we classify such a matrix according to the sign of its eigenvalues:

- $\lambda_i > 0, \forall i$, positive-definite, never singular, written as $\mathbf{A} \succ 0$
- $\lambda_i \geq 0, \forall i$, positive-semi-definite, possibly singular, written as $\mathbf{A} \succeq 0$
- $\lambda_i < 0, \forall i$, negative-definite, never singular, written as $\mathbf{A} \prec 0$
- otherwise, indefinite.

The diagonal elements of a positive-definite matrix are positive, $a_{i,i} > 0$, and its determinant is positive $\det(\mathbf{A}) > 0$. The inverse of a positive-definite matrix is also positive-definite.

We will frequently encounter the matrix quadratic form

$$s = \boldsymbol{x}^\top \mathbf{A} \boldsymbol{x} \tag{B.3}$$

which is a scalar. If $\mathbf{A}$ is positive-definite, then $s > 0, \forall \boldsymbol{x} \neq 0$. For the case that $\mathbf{A}$ is diagonal this can be written

$$s = \sum_{i=0}^{n-1} a_{i,i} x_i^2$$

which is a weighted sum of squares. If $\mathbf{A} = \mathbf{1}$ then $s = (\|\boldsymbol{v}\|_2)^2$. If $\mathbf{A}$ is symmetric then

$$s = \sum_{i=0}^{n-1} a_{i,i} x_i^2 + 2 \sum_{i=0}^{n-1} \sum_{j=i+1}^{n-1} a_{i,j} x_i x_j$$

and the result also includes products or correlations between elements of $\boldsymbol{x}$.

The Mahalanobis distance is a weighted distance or norm

$$s = \sqrt{\boldsymbol{x}^\top \mathbf{P}^{-1} \boldsymbol{x}}$$

where $\mathbf{P} \in \mathbb{R}^{n \times n}$ is a covariance matrix which down-weights elements of $\boldsymbol{x}$ where uncertainty is high.

The matrices $\mathbf{A}^\top \mathbf{A}$ and $\mathbf{A}\mathbf{A}^\top$ are always symmetric and positive-semi-definite. This implies than any symmetric matrix $\mathbf{A}$ can be written as

$$\mathbf{A} = \mathbf{L}\mathbf{L}^\top$$

`np.linalg.cholesky(A)`
`sp.linalg.sqrtm(A)`

where $\mathbf{L}$ is the Cholesky decomposition of $\mathbf{A}$.

Other matrix factorizations of $\mathbf{A}$ include the matrix square root

$$\mathbf{A} = \mathbf{S}\mathbf{S}$$

where $\mathbf{S}$ is the square root of $\mathbf{A}$ or $\mathbf{A}^{\frac{1}{2}}$ which is positive definite (and symmetric) if $\mathbf{A}$ is positive definite, and QR-decomposition

`np.linalg.qr(A)`

$$\mathbf{A} = \mathbf{Q}\mathbf{R}$$

where $\mathbf{Q}$ is an orthogonal matrix and $\mathbf{R}$ is an upper triangular matrix.

If $\mathbf{T}$ is any non-singular matrix, then

$$\mathbf{A} = \mathbf{T}\mathbf{B}\mathbf{T}^{-1}$$

is known as a similarity transformation or conjugation. $\mathbf{A}$ and $\mathbf{B}$ are said to be similar, and it can be shown that the eigenvalues are unchanged by this transformation.

The matrix form of (B.2) is

$$\mathbf{A}\mathbf{X} = \mathbf{X}\mathbf{\Lambda}$$

where $\mathbf{X} \in \mathbb{R}^{n \times n}$ is a matrix of eigenvectors of $\mathbf{A}$, arranged column-wise, and $\mathbf{\Lambda}$ is a diagonal matrix of corresponding eigenvalues. If $\mathbf{X}$ is non-singular, we can rearrange this as

$$\mathbf{A} = \mathbf{X}\mathbf{\Lambda}\mathbf{X}^{-1}$$

which is the eigen or spectral decomposition of the matrix. This implies that the matrix can be diagonalized by a similarity transformation

$$\mathbf{\Lambda} = \mathbf{X}^{-1}\mathbf{A}\mathbf{X} .$$

If $\mathbf{A}$ is symmetric, then $\mathbf{X}$ is orthogonal and we can instead write

$$\mathbf{A} = \mathbf{X}\mathbf{\Lambda}\mathbf{X}^{\top} . \tag{B.4}$$

The determinant of a square matrix $\mathbf{A} \in \mathbb{R}^{n \times n}$ is the factor by which the transformation scales volumes in an $n$-dimensional space. For two dimensions, imagine a shape defined by points $x_i$ with an enclosed area $\Lambda$. The shape formed by the points $\mathbf{A}x_i$ would have an enclosed area of $\Lambda \det(\mathbf{A})$. If $\mathbf{A}$ is singular, the points $\mathbf{A}x_i$ would be coincident or collinear and have zero enclosed area. In a similar way for three dimensions, the determinant is a scale factor applied to the volume of a set of points transformed by $\mathbf{A}$.

The columns of $\mathbf{A} = (c_0, c_1, \ldots, c_{n-1})$ can be considered as a set of vectors that define a space – the column space. Similarly, the rows of $\mathbf{A}$ can be considered as a set of vectors that define a space – the row space. The column rank of a matrix is the number of linearly independent columns of $\mathbf{A}$. Similarly, the row rank is the number of linearly independent rows of $\mathbf{A}$. The column rank and the row rank are always equal and are simply called the rank of $\mathbf{A}$, and this has an upper bound of $n$. A square matrix for which $\text{rank}(\mathbf{A}) < n$ is said to be rank deficient or not of full rank. The rank shortfall $n - \text{rank}(\mathbf{A})$ is the nullity of $\mathbf{A}$. In addition, $\text{rank}(\mathbf{AB}) \leq \min(\text{rank}(\mathbf{A}), \text{rank}(\mathbf{B}))$ and $\text{rank}(\mathbf{A} + \mathbf{B}) \leq \text{rank}(\mathbf{A}) + \text{rank}(\mathbf{B})$. If $x$ is a column vector, the matrix $xx^{\top}$, the outer product of $x$, has rank 1 for all $x \neq \mathbf{0}$.

`np.linalg.matrix_rank(A)`

## B.2.2 Non-Square and Singular Matrices

Singular square matrices can be treated as a special case of a non-square matrix. For a non-square matrix $\mathbf{A} \in \mathbb{R}^{m \times n}$ the rank is the dimension of the largest non-singular square submatrix that can be formed from $\mathbf{A}$, and has an upper bound of $\min(m, n)$.

A non-square or singular matrix cannot be inverted, but we can determine the left generalized inverse or pseudoinverse or Moore-Penrose pseudoinverse

$$\mathbf{A}^{+}\mathbf{A} = \mathbf{1}_{n \times n}$$

where $\mathbf{A}^{+} = (\mathbf{A}^{\top}\mathbf{A})^{-1}\mathbf{A}^{\top}$. The right generalized inverse is

$$\mathbf{A}\mathbf{A}^{+} = \mathbf{1}_{m \times m}$$

where $\mathbf{A}^+ = \mathbf{A}^\top (\mathbf{A}\mathbf{A}^\top)^{-1}$. The qualifier left or right denotes which side of $\mathbf{A}$ the pseudoinverse appears.

If the matrix $\mathbf{A}$ is not of full rank, then it has a finite null space or kernel. A vector $\boldsymbol{x}$ lies in the null space of the matrix if

`sp.linalg.null_space(A)`

$$\mathbf{A}\boldsymbol{x} = \mathbf{0} \ .$$

More precisely, this is the right-null space. A vector lies in the left-null space if

$$\boldsymbol{x}^\top \mathbf{A} = \mathbf{0} \ .$$

The left-null space is equal to the right-null space of $\mathbf{A}^\top$.

The null space is defined by a set of orthogonal basis vectors whose cardinality is the nullity of $\mathbf{A}$. Any linear combination of these null-space basis vectors lies in the null space.

For a non-square matrix $\mathbf{A} \in \mathbb{R}^{m \times n}$, the analog to (B.2) is

$$\mathbf{A}\boldsymbol{v}_i = \sigma_i \boldsymbol{u}_i$$

where $\boldsymbol{u}_i \in \mathbb{R}^m$ and $\boldsymbol{v}_i \in \mathbb{R}^n$ are respectively the right and left singular vectors of $\mathbf{A}$, and $\sigma_i$ its singular values. The singular values are nonnegative real numbers that are the square root of the eigenvalues of $\mathbf{A}\mathbf{A}^\top$ and $\boldsymbol{u}_i$ are the corresponding eigenvectors. $\boldsymbol{v}_i$ are the eigenvectors of $\mathbf{A}^\top \mathbf{A}$.

The singular value decomposition or SVD of the matrix $\mathbf{A}$ is

`np.linalg.svd(A)`

$$\mathbf{A} = \mathbf{U}\boldsymbol{\Sigma}\mathbf{V}^\top$$

where $\mathbf{U} \in \mathbb{R}^{m \times m}$ and $\mathbf{V} \in \mathbb{R}^{n \times n}$ are both orthogonal matrices comprising, as columns, the corresponding singular vectors $\boldsymbol{u}_i$ and $\boldsymbol{v}_i$. $\boldsymbol{\Sigma} \in \mathbb{R}^{m \times n}$ is a diagonal matrix of the singular values $\sigma_i$ in *decreasing* magnitude

$$\boldsymbol{\Sigma} = \begin{pmatrix} \sigma_0 & & & & & \\ & \ddots & & & & 0 \\ & & \sigma_{r-1} & & & \\ & & & 0 & & \\ & 0 & & & \ddots & \\ & & & & & 0 \end{pmatrix}$$

`np.linalg.cond(A)`

where $r = \text{rank}(\mathbf{A})$ is the rank of $\mathbf{A}$ and $\sigma_{i+1} \le \sigma_i$. For the case where $r < n$, the diagonal will have $n - r$ zero elements as shown. Columns of $\mathbf{V}$ corresponding to the zero columns of $\boldsymbol{\Sigma}$ define the null space of $\mathbf{A}$. The condition number of a matrix $\mathbf{A}$ is $\max(\boldsymbol{\sigma})/\min(\boldsymbol{\sigma})$ and a high value means the matrix is close to singular or "poorly conditioned".

# C Geometry

Geometric concepts such as points, lines, ellipses and planes are critical to the fields of robotics and robotic vision. We briefly summarize key representations in both Euclidean and projective (homogeneous coordinate) space.

## C.1 Euclidean Geometry

### C.1.1 Points

A point in an $n$-dimensional space is represented by an $n$-tuple, an ordered set of $n$ numbers $(x_0, x_1, \ldots, x_{n-1})^\top$ which define the coordinates of the point. The tuple can also be interpreted as a column vector – a coordinate vector – from the origin to the point. A point in 2-dimensions is written as $p = (x, y)^\top$, and in 3-dimensions is written as $p = (x, y, z)^\top$.

### C.1.2 Lines

#### C.1.2.1 Lines in 2D

A line is defined by $\ell = (a, b, c)^\top$ such that

$$ax + by + c = 0 \tag{C.1}$$

which is a generalization of the line equation we learned in school $y = mx + c$, but which can easily represent a vertical line by setting $b = 0$. $v = (a, b)$ is a vector normal to the line, and $v = (-b, a)$ is a vector parallel to the line. The line that joins two points $p_1$ and $p_2$, $p_i = (x_i, y_i)$, is given by the solution to

`Line2.Join`

$$\begin{pmatrix} x_1 & y_1 & 1 \\ x_2 & y_2 & 1 \end{pmatrix} \begin{pmatrix} a \\ b \\ c \end{pmatrix} = \mathbf{0}$$

which is found from the right null space of the leftmost term. The intersection point of two lines $\ell_1$ and $\ell_2$ is

`l1.intersects(l2)`

$$\begin{pmatrix} a_1 & b_1 \\ a_2 & b_2 \end{pmatrix} \begin{pmatrix} x \\ y \end{pmatrix} = - \begin{pmatrix} c_1 \\ c_2 \end{pmatrix}$$

which has no solution if the lines are parallel – the leftmost term is singular.
We can also represent the line in polar form

$$x \cos \theta + y \sin \theta + \rho = 0$$

where $\theta$ is the angle from the $x$-axis to the line and $\rho$ is the normal distance between the line and the origin, as shown in ◻ Fig. 12.20.

#### C.1.2.2 Lines in 3D and Plücker Coordinates

We can define a line by two points, **p** and **q**, as shown in ◻ Fig. C.1, which would require a total of six parameters $\ell = (p_x, p_y, p_z, q_x, q_y, q_z)$. However, since these points can be arbitrarily chosen, there would be an infinite set of parameters that represent the same line making it hard to determine the equivalence of two lines.

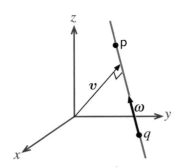

◻ **Fig. C.1**  Describing a line in three dimensions

There are advantages in representing a line as

$$\boldsymbol{\ell} = (\boldsymbol{\omega} \times \boldsymbol{q}, \boldsymbol{p} - \boldsymbol{q}) = (\boldsymbol{v}, \boldsymbol{\omega}) \in \mathbb{R}^6$$

where $\boldsymbol{\omega}$ is the direction of the line and $\boldsymbol{v}$ is the moment of the line – a vector from the origin to a point on the line and which is normal to the line. This is a Plücker coordinate vector – a 6-dimensional quantity subject to two constraints: the coordinates are homogeneous and thus invariant to overall scale factor; and $\boldsymbol{v} \cdot \boldsymbol{\omega} = 0$. Lines therefore have 4 degrees-of-freedom ◀ and the Plücker coordinates lie on a 4-dimensional manifold – the Klein quadric – in 6-dimensional space. Lines with $\boldsymbol{\omega} = \boldsymbol{0}$ lie at infinity and are known as ideal lines. ◀

> This is not intuitive but consider two parallel planes and an arbitrary 3D line passing through them. The line can be described by the 2-dimensional coordinates of its intersection point on each plane – a total of four coordinates.

> Ideal as in imaginery, not as in perfect.

We will first define two points

```
>>> P = [2, 3, 4]; Q = [3, 5, 7];
```

and then create a Plücker line object

```
>>> L = Line3.Join(P, Q)
{ 1   -2   1; -1   -2   -3}
```

which displays the $\boldsymbol{v}$ and $\boldsymbol{\omega}$ components. These can be accessed as properties

```
>>> L.v.T
array([      1,        -2,        1])
>>> L.w.T
array([     -1,        -2,       -3])
```

A Plücker line can also be represented as a skew-symmetric matrix

```
>>> L.skew()
array([[      0,        1,        2,       -1],
       [     -1,        0,        1,       -2],
       [     -2,       -1,        0,       -3],
       [      1,        2,        3,        0]])
```

> Since lines have infinite length, we need to specify a finite volume in which to draw it.

To plot this line, we first define a region of 3D space ◀ then plot it in blue

```
>>> plotvol3([-5, 5]);
>>> L.plot("b");
```

The line is the set of all points

$$\boldsymbol{p}(\lambda) = \frac{\boldsymbol{v} \times \boldsymbol{\omega}}{\boldsymbol{\omega} \cdot \boldsymbol{\omega}} + \lambda \boldsymbol{\omega}, \quad \lambda \in \mathbb{R}$$

which can be generated parametrically in terms of the scalar parameter $\lambda$

```
>>> L.point([0, 1, 2])
array([[  0.5714,     0.3042,    0.03691],
       [  0.1429,    -0.3917,    -0.9262],
       [ -0.2857,     -1.087,     -1.889]])
```

where the columns are points on the line corresponding to $\lambda = 0, 1, 2$.

A point $x$ is closest to the line when

$$\lambda = \frac{(x - q) \cdot \omega}{\omega \cdot \omega} \ .$$

For the point $(1, 2, 3)^\top$, the closest point on the line, and its distance, is given by

```
>>> [x, d] = L.closest_to_point([1, 2, 3])
>>> x
array([    1.571,     2.143,     2.714])
>>> d
0.6547
```

The line intersects the plane $n^\top x + d = 0$ at the point coordinate

$$x = \frac{v \times n - d\omega}{\omega \cdot n} \ .$$

For the $xy$-plane, the line intersects at

```
>>> p, _ = L.intersect_plane([0, 0, 1, 0])
>>> p
array([   0.6667,     0.3333,         0])
```

Two lines can be identical, coplanar or skewed. Identical lines have linearly dependent Plücker coordinates, that is, $\ell_1 = \lambda \ell_2$ or $\hat{\ell}_1 = \hat{\ell}_2$. If coplanar, they can be parallel or intersecting, and, if skewed they can be intersecting or not. If lines have $\omega_1 \times \omega_2 = 0$ they are parallel, otherwise, they are skewed.

The minimum distance between two lines is

$$d = \omega_1 \cdot v_2 + \omega_2 \cdot v_1$$

and is zero if they intersect.

For two lines $\ell^1$ and $\ell^2$, the side operator is a permuted dot product

$$\text{side}(\ell^1, \ \ell^2) = \ell_0^1 \ell_4^2 + \ell_1^1 \ell_5^2 + \ell_2^1 \ell_3^2 + \ell_3^1 \ell_2^2 + \ell_4^1 \ell_0^2 + \ell_5^1 \ell_1^2$$

which is zero if the lines intersect or are parallel, and is computed by the `side` method. For `Line3` objects `L1` and `L2`, the operators `L1 ^ L2` and `L1 | L2` are true if the lines are respectively intersecting or parallel.

### Excurse C.1: Julius Plücker

Plücker (1801–1868) was a German mathematician and physicist who made contributions to the study of cathode rays and analytical geometry. He was born at Elberfeld and studied at Düsseldorf, Bonn, Heidelberg and Berlin, and went to Paris in 1823 where he was influenced by the French geometry movement. In 1825, he returned to the University of Bonn, was made professor of mathematics in 1828 (at age 27), and professor of physics in 1836. In 1858, he proposed that the lines of the spectrum, discovered by his colleague Heinrich Geissler (of Geissler tube fame), were characteristic of the chemical substance which emitted them. In 1865, he returned to geometry and invented what was known as line geometry. He was the recipient of the Copley Medal from the Royal Society in 1866, and is buried in the Alter Friedhof (Old Cemetery) in Bonn.

We can transform a Plücker line between frames by

$$^{B}\boldsymbol{\ell}' = \mathrm{Ad}(\,^{B}\boldsymbol{\xi}_{A})\,^{A}\boldsymbol{\ell}$$

where the adjoint of the rigid-body motion is described by (D.2). This can be computed by premultiplying a `Line3` instance by an `SE3` instance.

### C.1.3  Planes

A plane is defined by a 4-vector $\boldsymbol{\pi} = (a, b, c, d)^{\top}$ and is the set of all points $\boldsymbol{x} = (x, y, z)^{\top}$ such that

$$ax + by + cz + d = 0$$

which can be written in point-normal form as

$$\boldsymbol{n}^{\top}(\boldsymbol{x} - \boldsymbol{p}) = 0$$

where $\boldsymbol{n} = (a, b, c)$ is the normal to the plane, and $\boldsymbol{p} \in \mathbb{R}^{3}$ is a point in the plane.

`Plane.PointNormal(p, n)`

A plane with the normal $\boldsymbol{n}$ and containing the point with coordinate vector $\boldsymbol{p}$ is $\boldsymbol{\pi} = (n_{x}, n_{y}, n_{z}, \boldsymbol{n} \cdot \boldsymbol{p})^{\top}$. A plane can also be defined by 3 points $\boldsymbol{p}_{1}$, $\boldsymbol{p}_{2}$ and $\boldsymbol{p}_{3}$, where $\boldsymbol{p}_{i} = (x_{i}, y_{i}, z_{i})$

`Plane.ThreePoints(p)`

$$\begin{pmatrix} x_{1} & y_{1} & z_{1} & 1 \\ x_{2} & y_{2} & z_{2} & 1 \\ x_{3} & y_{3} & z_{3} & 1 \end{pmatrix} \boldsymbol{\pi} = \boldsymbol{0}$$

and solved for using the right null space of the leftmost term, or by two nonparallel lines $\boldsymbol{\ell}_{1}$ and $\boldsymbol{\ell}_{2}$

`Plane.TwoLines(l1, l2)`

$$\boldsymbol{\pi} = (\boldsymbol{\omega}_{1} \times \boldsymbol{\omega}_{2},\ \boldsymbol{v}_{1} \cdot \boldsymbol{\omega}_{2})$$

or by a Plücker line $(\boldsymbol{v}, \boldsymbol{w})$ and a point with coordinate vector $\boldsymbol{p}$

`Plane.LinePoint(l, p)`

$$\boldsymbol{\pi} = (\boldsymbol{\omega} \times \boldsymbol{p} - \boldsymbol{v},\ \boldsymbol{v} \cdot \boldsymbol{p})\ .$$

A point can also be defined as the intersection point of three planes $\boldsymbol{\pi}_{1}$, $\boldsymbol{\pi}_{2}$ and $\boldsymbol{\pi}_{3}$

`Plane.intersection()`

$$\begin{pmatrix} a_{1} & b_{1} & c_{1} \\ a_{2} & b_{2} & c_{2} \\ a_{3} & b_{3} & c_{3} \end{pmatrix} \begin{pmatrix} x \\ y \\ z \end{pmatrix} = - \begin{pmatrix} d_{1} \\ d_{2} \\ d_{3} \end{pmatrix}\ .$$

If the left-hand matrix is singular, then two or more planes are parallel and have either zero or infinitely many intersection points.

The Plücker line formed by the intersection of two planes $\boldsymbol{\pi}_{1}$ and $\boldsymbol{\pi}_{2}$ is

$$\boldsymbol{\ell} = (\boldsymbol{n}_{1} \times \boldsymbol{n}_{2},\ d_{2}\boldsymbol{n}_{1} - d_{1}\boldsymbol{n}_{2})\ .$$

### C.1.4  Ellipses and Ellipsoids

An ellipse belongs to the family of planar curves known as conics. The simplest form of an ellipse, centered at $(0, 0)$, is defined implicitly by the points $(x, y)$ such that

$$\frac{x^{2}}{a^{2}} + \frac{y^{2}}{b^{2}} = 1$$

and is shown in ▢ Fig. C.2a. ► This canonical ellipse is centered at the origin and has its major and minor axes aligned with the $x$- and $y$-axes. The radius in the $x$-direction is $a$ and in the $y$-direction is $b$. The longer of the two radii is known as the semi-major axis length, and the other is the semi-minor axis length.

A circle is a special case of an ellipse where $a = b = r$ and $r$ is the radius.

We can write the ellipse in matrix quadratic form (B.3) as

$$(x \quad y)\begin{pmatrix} 1/a^2 & 0 \\ 0 & 1/b^2 \end{pmatrix}\begin{pmatrix} x \\ y \end{pmatrix} = 1$$

and more compactly as

$$\boldsymbol{x}^{\top}\mathbf{E}\boldsymbol{x} = 1 \tag{C.2}$$

where $\boldsymbol{x} = (x, y)^{\top}$ and $\mathbf{E}$ is a symmetric matrix

$$\mathbf{E} = \begin{pmatrix} \alpha & \gamma \\ \gamma & \beta \end{pmatrix}. \tag{C.3}$$

Its determinant $\det(\mathbf{E}) = \alpha\beta - \gamma^2$ defines the type of conic

$$\det(\mathbf{E})\begin{cases} > 0 & \text{ellipse} \\ = 0 & \text{parabola} \\ < 0 & \text{hyperbola} \end{cases}$$

An ellipse can therefore be represented by a positive-definite symmetric matrix $\mathbf{E}$. Conversely, any positive-definite symmetric $2 \times 2$ matrix, such as the inverse of an inertia tensor or covariance matrix, has an equivalent ellipse.

Nonzero values of $\gamma$ change the orientation of the ellipse. The ellipse can be arbitrarily centered at $\boldsymbol{x}_c = (x_c, y_c)$ by writing it in the form

$$(\boldsymbol{x} - \boldsymbol{x}_c)^{\top}\mathbf{E}(\boldsymbol{x} - \boldsymbol{x}_c) = 1$$

which leads to the general ellipse shown in ▢ Fig. C.2b.

Since $\mathbf{E}$ is symmetric, it can be diagonalized by (B.4)

$$\mathbf{E} = \mathbf{X}\boldsymbol{\Lambda}\mathbf{X}^{\top}$$

where $\mathbf{X}$ is an orthogonal matrix comprising the eigenvectors of $\mathbf{E}$, and the diagonal elements of $\boldsymbol{\Lambda}$ are the eigenvalues of $\mathbf{E}$. The quadratic form (C.2) becomes

$$\boldsymbol{x}^{\top}\mathbf{X}\boldsymbol{\Lambda}\mathbf{X}^{\top}\boldsymbol{x} = 1$$
$$(\mathbf{X}^{\top}\boldsymbol{x})^{\top}\boldsymbol{\Lambda}(\mathbf{X}^{\top}\boldsymbol{x}) = 1$$
$$\boldsymbol{x}'^{\top}\boldsymbol{\Lambda}\boldsymbol{x}' = 1$$

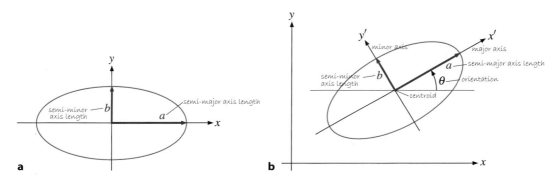

▢ **Fig. C.2** Ellipses. **a** Canonical ellipse centered at the origin and aligned with the $x$- and $y$-axes; **b** general form of ellipse which is similar to (C.2) but with the ellipse defined by the diagonal matrix $\mathbf{\Lambda}$ with respect to the rotated coordinated frame $\mathbf{x}' = \mathbf{X}^\top \mathbf{x}$. The major and minor ellipse axes are aligned with the eigenvectors of $\mathbf{E}$.

The radii of the ellipse are related to the inverse square root of the eigenvalues

$$r_i = \frac{1}{\sqrt{\lambda_i}} \tag{C.4}$$

so the major and minor radii of the ellipse, $a$ and $b$ respectively, are determined by the smallest and largest eigenvalues respectively. The area of an ellipse is

$$\Lambda = \pi r_1 r_2 = \frac{\pi}{\sqrt{\det(\mathbf{E})}} \tag{C.5}$$

since $\det(\mathbf{E}) = \prod \lambda_i$. The eccentricity is

$$\varepsilon = \frac{\sqrt{a^2 - b^2}}{a} \ .$$

Alternatively, the ellipse can be represented in polynomial form by writing as

$$(\mathbf{x} - \mathbf{x}_c)^\top \begin{pmatrix} \alpha & \gamma \\ \gamma & \beta \end{pmatrix} (\mathbf{x} - \mathbf{x}_c) = 1$$

and expanding to

$$e_0 x^2 + e_1 y^2 + e_2 xy + e_3 x + e_4 y + e_5 = 0$$

where $e_0 = \alpha$, $e_1 = \beta$, $e_2 = 2\gamma$, $e_3 = -2(\alpha x_c + \gamma y_c)$, $e_4 = -2(\beta y_c + \gamma x_c)$ and $e_5 = \alpha x_c^2 + \beta y_c^2 + 2\gamma x_c y_c - 1$. The ellipse has only five degrees of freedom, its center coordinate and the three unique elements in $\mathbf{E}$. For a nondegenerate ellipse where $e_1 \neq 0$, we can rewrite the polynomial in normalized form

$$x^2 + \epsilon_0 y^2 + \epsilon_1 xy + \epsilon_2 x + \epsilon_3 y + \epsilon_4 = 0 \tag{C.6}$$

with five unique parameters $\epsilon_i = e_{i+1}/e_0$, $i = 0, ..., 4$.

Consider the ellipse

$$\mathbf{x}^\top \begin{pmatrix} 1 & 1 \\ 1 & 2 \end{pmatrix} \mathbf{x} = 1$$

which is represented in Python by

```
>>> E = np.array([[1, 1], [1, 2]])
array([[1, 1],
       [1, 2]])
```

We can plot this by

```
>>> plot_ellipse(E)
```

which is shown in ◼ Fig. C.3. The eigenvectors and eigenvalues of $\mathbf{E}$ are respectively

```
>>> e, v = np.linalg.eig(E)
>>> e
array([   0.382,    2.618])
>>> v
array([[ -0.8507,  -0.5257],
       [  0.5257,  -0.8507]])
```

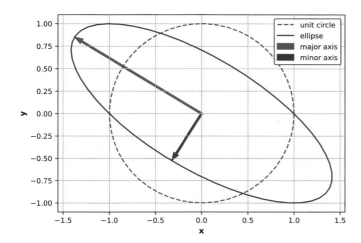

**□ Fig. C.3** Ellipse corresponding to a symmetric 2 × 2 matrix. The arrows indicate the major and minor axes of the ellipse

The ellipse radii are

```
>>> r = 1 / np.sqrt(e)
array([  1.618,    0.618])
```

If either radius is equal to zero the ellipse is degenerate and becomes a line. If both radii are zero the ellipse is a point.

The eigenvectors are unit vectors in the major and minor axis directions ▶ and we will scale them by the radii to yield radius vectors which we can plot

Note that np.linalg.eig does not sort the eigenvalues

```
>>> plot_arrow((0, 0), v[:,0]*r[0], color="r", width=0.02);
>>> plot_arrow((0, 0), v[:,1]*r[1], color="b", width=0.02);
```

The orientation of the ellipse is the angle of the major axis with respect to the horizontal axis and is

$$\theta = \tan^{-1} \frac{v_y}{v_x}$$

where $\boldsymbol{v} = (v_x, v_y)$ is the eigenvector corresponding to the smallest eigenvalue (the largest radius). For our example, this is

```
>>> np.rad2deg(np.arctan2(v[1, 0], v[0, 0]))
148.3
```

in units of degrees, counter-clockwise from the $x$-axis.

The Toolbox function plot_ellipsoid will draw an ellipsoid and **E** is a 3 × 3 matrix.

### C.1.4.1 Drawing an Ellipse

In order to draw an ellipse, we first define a set of points on the unit circle $\boldsymbol{y} = (x, y)^\top$ such that

$$\boldsymbol{y}^\top \boldsymbol{y} = 1 \ . \tag{C.7}$$

We rewrite (C.2) as

$$\boldsymbol{x}^\top \mathbf{E}^{\frac{1}{2}} \mathbf{E}^{\frac{1}{2}} \boldsymbol{x} = 1 \tag{C.8}$$

where $\mathbf{E}^{\frac{1}{2}}$ is the matrix square root, and equating (C.7) and (C.8), we can write

$$\boldsymbol{x}^\top \mathbf{E}^{\frac{1}{2}} \, \mathbf{E}^{\frac{1}{2}} \boldsymbol{x} = \boldsymbol{y}^\top \boldsymbol{y} \ .$$

which leads to

$$y = \mathbf{E}^{\frac{1}{2}} x$$

which we can rearrange as

$$x = \mathbf{E}^{-\frac{1}{2}} y$$

that transforms a point on the unit circle to a point on an ellipse, and $\mathbf{E}^{-\frac{1}{2}}$ is the inverse of the matrix square root. If the ellipse is centered at $x_c$, rather than the origin, we can perform a change of coordinates

$$(x - x_c)^{\top} \mathbf{E}^{\frac{1}{2}} \mathbf{E}^{\frac{1}{2}} (x - x_c) = 1$$

from which we write the transformation as

$$x = \mathbf{E}^{-\frac{1}{2}} y + x_c$$

Drawing an ellipsoid is tackled in an analogous fashion.

Continuing the code example above

```
>>> E = np.array([[1, 1], [1, 2]]);
```

We define a set of points on the unit circle

```
>>> th = np.linspace(0, 2*pi, 50);
>>> y = np.vstack([np.cos(th), np.sin(th)]);
```

which we transform to points on the perimeter of the ellipse

```
>>> x = linalg.sqrtm(np.linalg.inv(E)) @ y;
>>> plt.plot(x[0, :], x[1, :], '.');
```

which is encapsulated in the Toolbox function

```
>>> plot_ellipse(E, centre=[0, 0])
```

In many cases we wish to plot an ellipse given by $\mathbf{A}^{-1}$. Computing the inverse of $\mathbf{A}$ is inefficient because `plot_ellipse` will invert it again, so in this case we write `plot_ellipse(A, inverted=True)`.

### C.1.4.2  Fitting an Ellipse to Data

A common problem is to find the equation of an ellipse that best fits a set of points that lie within the ellipse boundary, or that lie on the ellipse boundary.

#### From a Set of Interior Points

To fit an ellipse to a set of points within the ellipse boundary, we find the ellipse that has the same mass properties as the set of points. From the set of $N$ points $x_i = (x_i, y_i)^{\top}$, we can compute the moments

$$m_{00} = N, \quad m_{10} = \sum_{i=0}^{N-1} x_i, \quad m_{01} = \sum_{i=0}^{N-1} y_i \ .$$

The center of the ellipse is taken to be the centroid of the set of points

$$x_c = (x_c, \ y_c)^{\top} = \left( \frac{m_{10}}{m_{00}}, \ \frac{m_{01}}{m_{00}} \right)^{\top}$$

which allows us to compute the central second moments

$$\mu_{20} = \sum_{i=0}^{N-1} (x_i - x_c)^2$$

$$\mu_{02} = \sum_{i=0}^{N-1} (y_i - y_c)^2$$

$$\mu_{11} = \sum_{i=0}^{N-1} (x_i - x_c)(y_i - y_c) .$$

The inertia tensor for a general ellipse is the symmetric matrix

$$\mathbf{J} = \begin{pmatrix} \mu_{20} & \mu_{11} \\ \mu_{11} & \mu_{02} \end{pmatrix}$$

where the diagonal terms are the moments of inertia and the off-diagonal terms are the products of inertia. Inertia can be computed more directly as the summation of $N$ rank-1 matrices

$$\mathbf{J} = \sum_{i=0}^{N-1} (\boldsymbol{x}_i - \boldsymbol{x}_c)(\boldsymbol{x}_i - \boldsymbol{x}_c)^\top .$$

The inertia tensor and its equivalent ellipse are inversely related by

$$\mathbf{E} = \frac{m_{00}}{4} \mathbf{J}^{-1} .$$

To demonstrate this, we can create a set of points that lie within the ellipse used in the example above

```
>>> rng = np.random.default_rng(0);
>>> # create 200 random points inside the ellipse
>>> x = [];
>>> while len(x) < 200:
...     p = rng.uniform(low=-2, high=2, size=(2,1))
...     if np.linalg.norm(p.T @ E @ p) <= 1:
...         x.append(p)
>>> x = np.hstack(x);  # create 2 x 50 array
>>> plt.plot(x[0, :], x[1, :], "k."); # plot them
>>>
>>> # compute the moments
>>> m00 = mpq_point(x, 0, 0);
>>> m10 = mpq_point(x, 1, 0);
>>> m01 = mpq_point(x, 0, 1);
>>> xc = np.c_[m10, m01] / m00;
>>>
>>> # compute the central second moments
>>> x0 = x - xc.T;
>>> u20 = mpq_point(x0, 2, 0);
>>> u02 = mpq_point(x0, 0, 2);
>>> u11 = mpq_point(x0, 1, 1);
>>>
>>> # compute inertia tensor and ellipse matrix
>>> J = np.array([[u20, u11], [u11, u02]]);
>>> E_est = m00 / 4 * np.linalg.inv(J);
```

which results in an estimate

```
>>> E_est
array([[  1.053,    1.038],
       [  1.038,    1.98]])
```

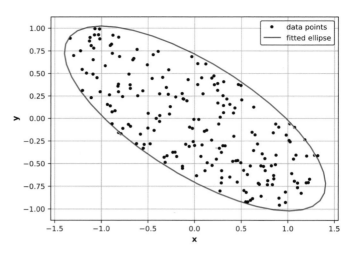

**◘ Fig. C.4**  200 random data points with a fitted ellipse

that is similar to the original value of E. The point data is shown in ◘ Fig. C.4. We can overlay the estimated ellipse on the point data

```
>>> plot_ellipse(E_est, "r")
```

and the result is shown in red in ◘ Fig. C.4.

### From a Set of Perimeter Points

Given a set of points $(x_i, y_i)$ that lie on the perimeter of an ellipse, we use the polynomial form of the ellipse (C.6) for each point. We write this in matrix form with one row per point

$$
\begin{pmatrix}
y_0^2 & x_0 y_0 & x_0 & y_0 & 1 \\
y_1^2 & x_1 y_2 & x_1 & y_1 & 1 \\
\vdots & \vdots & \vdots & \vdots & \vdots \\
y_{N-1}^2 & x_{N-1} y_{N-1} & x_{N-1} & y_{N-1} & 1
\end{pmatrix}
\begin{pmatrix}
\epsilon_0 \\
\epsilon_1 \\
\epsilon_2 \\
\epsilon_3 \\
\epsilon_4
\end{pmatrix}
=
\begin{pmatrix}
-x_0^2 \\
-x_1^2 \\
\vdots \\
-x_{N-1}^2
\end{pmatrix}
$$

and for $N \geq 5$ we can solve for the ellipse parameter vector using least squares.

## C.2  Homogeneous Coordinates

A point in homogeneous coordinates, or the projective space $\mathbb{P}^n$, is represented by a coordinate vector $\tilde{x} = (\tilde{x}_0, \tilde{x}_1, \ldots, \tilde{x}_n)$ and the tilde is used to indicate that the quantity is homogeneous. The Euclidean coordinates are related to the projective or homogeneous coordinates by

$$
x_i = \frac{\tilde{x}_i}{\tilde{x}_{n+1}}, \quad i = 0, \ldots, n-1
$$

Conversely, a homogeneous coordinate vector can be constructed from a Euclidean coordinate vector by

$$
\tilde{x} = (x_0, \ x_1, \ \ldots, \ x_{n-1}, \ 1) \ .
$$

The extra *degree of freedom* offered by projective coordinates has several advantages. It allows points and lines at infinity, known as ideal points and lines, to be

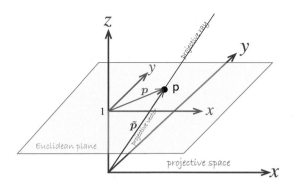

**⬛ Fig. C.5** A point p on the Euclidean plane is described by a coordinate vector $\boldsymbol{p} \in \mathbb{R}^2$ which is equivalent to the three-dimensional vector in the projective space $\tilde{\boldsymbol{p}} \in \mathbb{P}^2$

represented using only finite numbers. It also means that scale is unimportant, that is, $\tilde{\boldsymbol{x}}$ and $\tilde{\boldsymbol{x}}' = \lambda \tilde{\boldsymbol{x}}$ both represent the same Euclidean point for all $\lambda \neq 0$. We express this as $\tilde{\boldsymbol{x}} \simeq \tilde{\boldsymbol{x}}'$. We can apply a rigid-body transformation to points $\tilde{\boldsymbol{x}} \in \mathbb{P}^n$ by multiplying the homogeneous coordinate by an $n \times n$ homogeneous transformation matrix.

Homogeneous vectors are important in computer vision when we consider points and lines that exist in a plane, for example, a camera's image plane. We can also consider that the homogeneous form represents a ray in projective space as shown in ⬛ Fig. C.5. Any point on the projective ray is equivalent to p and this relationship between points and rays is at the core of the projective transformation. You can experiment with this by running the notebook `examples/homogeneous-coords.ipynb`.

### C.2.1 Two Dimensions

#### C.2.1.1 Points and Lines

In two dimensions, there is a duality between points and lines. In $\mathbb{P}^2$, a line is represented by a vector $\tilde{\boldsymbol{\ell}} = (\ell_0, \ell_1, \ell_2)^\top$, not all zero, and the equation of the line is the set of all points $\tilde{\boldsymbol{x}} \in \mathbb{P}^2$ such that

$$\tilde{\boldsymbol{\ell}}^\top \tilde{\boldsymbol{x}} = 0$$

which is the point equation of a line. This expands to $\ell_0 x + \ell_1 y + \ell_2 = 0$ and can be manipulated into the more familiar representation of a line $y = -\frac{\ell_0}{\ell_1} x - \frac{\ell_2}{\ell_1}$. Note that this form can represent a vertical line which the familiar form $y = mx + c$ cannot. The nonhomogeneous vector $(\ell_0, \ell_1)^\top$ is normal to the line, ▸ and $(-\ell_1, \ell_0)^\top$ is parallel to the line.

> Hence, this line representation is referred to as normal form.

A point is defined by the intersection of two lines. If we write the point equations for two lines $\tilde{\boldsymbol{\ell}}_1^\top \tilde{\boldsymbol{x}} = 0$ and $\tilde{\boldsymbol{\ell}}_2^\top \tilde{\boldsymbol{x}} = 0$, their intersection is the point with homogeneous coordinates

$$\tilde{\boldsymbol{p}} = \tilde{\boldsymbol{\ell}}_1 \times \tilde{\boldsymbol{\ell}}_2$$

and is known as the line equation of a point. Similarly, a line passing through two points $\tilde{\boldsymbol{p}}_1, \tilde{\boldsymbol{p}}_2 \in \mathbb{P}^2$, said to be *joining* the points, is given by the cross-product

$$\tilde{\boldsymbol{\ell}}_{12} = \tilde{\boldsymbol{p}}_1 \times \tilde{\boldsymbol{p}}_2 \; .$$

Consider the case of two parallel lines at 45° to the horizontal axis

```
>>> l1 = [1, -1, 0];
>>> l2 = [1, -1, -1];
```

which we can plot

```
>>> plot_homline(l1, "b");
>>> plot_homline(l2, "r");
```

The intersection point of these parallel lines is

```
>>> np.cross(l1, l2)
array([1, 1, 0])
```

This is an *ideal point* since the third coordinate is zero – the equivalent Euclidean point would be at infinity. Projective coordinates allow points and lines at infinity to be simply represented and manipulated without special logic.

The distance from a line $\tilde{\ell}$ to a point $\tilde{p}$ is

$$d = \frac{\tilde{\ell}^\top \tilde{p}}{\tilde{p}_2 \sqrt{\tilde{\ell}_0^2 + \tilde{\ell}_1^2}} \ . \tag{C.9}$$

### C.2.1.2  Conics

Conic sections are an important family of planar curves that includes circles, ellipses, parabolas and hyperbolas. They can be described generally as the set of points $\tilde{x} \in \mathbb{P}^2$ such that

$$\tilde{x}^\top \mathbf{\Omega} \tilde{x} = 0$$

where $\mathbf{\Omega}$ is a symmetric matrix

$$\mathbf{\Omega} = \left( \begin{array}{cc|c} a & b/2 & d/2 \\ b/2 & c & e/2 \\ \hline d/2 & e/2 & f \end{array} \right) \ . \tag{C.10}$$

The determinant of the top-left submatrix indicates the type of conic: positive for an ellipse, zero for a parabola, and negative for a hyperbola.

### C.2.2  Three Dimensions

In three dimensions, there is a duality between points and planes.

### C.2.2.1  Lines

For two points $\tilde{p}, \tilde{q} \in \mathbb{P}^3$ in homogeneous form, the line that joins them is defined by a $4 \times 4$ skew-symmetric matrix

$$\begin{aligned} \mathbf{L} &= \tilde{q}\tilde{p}^\top - \tilde{p}\tilde{q}^\top \\ &= \begin{pmatrix} 0 & v_3 & -v_2 & -\omega_1 \\ -v_3 & 0 & v_1 & -\omega_2 \\ v_2 & -v_1 & 0 & -\omega_3 \\ \omega_1 & \omega_2 & \omega_3 & 0 \end{pmatrix} \end{aligned} \tag{C.11}$$

whose six unique elements comprise the Plücker coordinate representation of the line. This matrix is rank 2 and the determinant is a quadratic in the Plücker coordinates – a 4-dimensional quadric hypersurface known as the Klein quadric. All points that lie on this manifold are valid lines. Many of the relationships in

▶ App. C.1.2.2 (between lines and points and planes) can be expressed in terms of this matrix. This matrix is returned by the `skew` method of the `Line3` class.

For a perspective camera with a camera matrix $\mathbf{C} \in \mathbb{R}^{3 \times 4}$, the 3-dimensional Plücker line represented as a $4 \times 4$ skew-symmetric matrix $\mathbf{L}$ is projected onto the image plane as

$$\tilde{\ell} = \mathbf{C}\mathbf{L}\mathbf{C}^\top \in \mathbb{P}^2$$

which is a homogeneous 2-dimensional line. This is computed by the `project_line` method of the `CentralCamera` class.

### C.2.2.2 Planes

A plane $\tilde{\pi} \in \mathbb{P}^3$ is described by the set of points $\tilde{x} \in \mathbb{P}^3$ such that $\tilde{\pi}^\top \tilde{x} = 0$. A plane can be defined by a line $\mathbf{L} \in \mathbb{R}^{4 \times 4}$ and a homogeneous point $\tilde{p}$

$$\tilde{\pi} = \mathbf{L}\tilde{p} \qquad\qquad\qquad \text{Plane.LinePoint()}$$

or three points

$$\begin{pmatrix} \tilde{p}_1^\top \\ \tilde{p}_2^\top \\ \tilde{p}_3^\top \end{pmatrix} \tilde{\pi} = \mathbf{0} \qquad\qquad\qquad \text{Plane.ThreePoints()}$$

and the solution is found from the right-null space of the matrix.

The intersection, or incidence, of three planes is the dual

$$\begin{pmatrix} \tilde{\pi}_1^\top \\ \tilde{\pi}_2^\top \\ \tilde{\pi}_3^\top \end{pmatrix} \tilde{p} = \mathbf{0} \qquad\qquad\qquad \text{Plane.intersection()}$$

and is an ideal point, zero last component, if the planes do not intersect at a point.

### C.2.2.3 Quadrics

Quadrics, short for quadratic surfaces, are a rich family of 3-dimensional *surfaces*. There are 17 standard types including spheres, ellipsoids, hyperboloids, paraboloids, cylinders and cones, all described by the set of points $\tilde{x} \in \mathbb{P}^3$ such that

$$\tilde{x}^\top \mathbf{Q} \tilde{x} = 0$$

and $\mathbf{Q} \in \mathbb{R}^{4 \times 4}$ is symmetric.

For a perspective camera with a camera matrix $\mathbf{C} \in \mathbb{R}^{3 \times 4}$, the *outline* of the 3-dimensional quadric is projected to the image plane by

$$\boldsymbol{\Omega}^* = \mathbf{C}\mathbf{Q}^*\mathbf{C}^\top$$

where $\boldsymbol{\Omega}$ is given by (C.10), and $(\cdot)^*$ represents the adjugate operation, see ▶ App. B.1.

## C.3 Geometric Transformations

A linear transformation is

$$y = \mathbf{A}x \qquad\qquad\qquad (\text{C.12})$$

Scaling about an arbitrary point.

while an affine transformation

$$y = \mathbf{A}x + b \tag{C.13}$$

comprises a linear transformation *and* a change of origin. Examples of affine transformations include translation, scaling, homothety, ◄ reflection, rotation, shearing, and any arbitrary composition of these. Every linear transformation is affine, but not every affine transformation is linear.

In homogeneous coordinates, we can write (C.13) as

$$\tilde{y} = \mathbf{H}\tilde{x}, \quad \text{where } \mathbf{H} = \begin{pmatrix} \mathbf{A} & b \\ \mathbf{0} & 1 \end{pmatrix}$$

and the transformation operates on a point with homogeneous coordinates $\tilde{x}$.

Projective space is a generalization of affine space which, in turn, is a generalization of Euclidean space. An affine space has no distinguished point that serves as an origin and hence no vector can be uniquely associated to a point. An affine space has only displacement vectors between two points in the space. Subtracting two points results in a displacement vector, and adding a displacement vector to a point results in a new point. If a displacement vector is defined as the difference between two homogeneous points $\tilde{p}$ and $\tilde{q}$, then the difference $\tilde{p} - \tilde{q}$ is a 4-vector whose last element will be zero, distinguishing a point from a displacement vector.

In two dimensions, the most general transformation is a projective transformation, also known as a collineation

$$\mathbf{H} = \begin{pmatrix} h_{0,0} & h_{0,1} & h_{0,2} \\ h_{1,0} & h_{1,1} & h_{1,2} \\ h_{2,1} & h_{2,1} & 1 \end{pmatrix}$$

which is unique up to scale and one element has been normalized to one. It has 8 degrees of freedom.

The affine transformation is a subset where the elements of the last row are fixed

$$\mathbf{H} = \begin{pmatrix} a_{0,0} & a_{0,1} & t_x \\ a_{1,0} & a_{1,1} & t_y \\ 0 & 0 & 1 \end{pmatrix}$$

and has 6 degrees of freedom, two of which are the translation $(t_x, t_y)$.

The similarity transformation is a more restrictive subset

$$\mathbf{H} = \begin{pmatrix} s\mathbf{R} & t \\ \mathbf{0} & 1 \end{pmatrix} = \begin{pmatrix} sr_{0,0} & sr_{0,1} & t_x \\ sr_{1,0} & sr_{1,1} & t_y \\ 0 & 0 & 1 \end{pmatrix}$$

where $\mathbf{R} \in \mathbf{SO}(2)$ resulting in only 4 degrees of freedom, and $s < 0$ causes a reflection. Similarity transformations, without reflection, are sometimes referred to as a Procrustes transformation.

Finally, the Euclidean or rigid-body transformation

$$\mathbf{H} = \begin{pmatrix} \mathbf{R} & t \\ \mathbf{0} & 1 \end{pmatrix} = \begin{pmatrix} r_{0,0} & r_{0,1} & t_x \\ r_{1,0} & r_{1,1} & t_y \\ 0 & 0 & 1 \end{pmatrix} \in \mathbf{SE}(2)$$

is the most restrictive and has only 3 degrees of freedom. Some graphical examples of the effect of the various transformations on a square are shown in ◘ Fig. C.6.

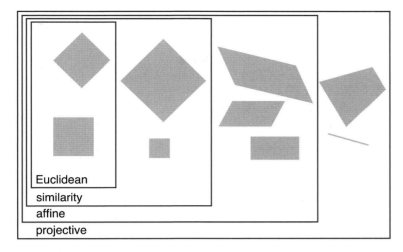

◘ **Fig. C.6** A 2-dimensional dark gray square with various transformations applied: from the most limited (Euclidean) to the most general (projective)

◘ **Table C.1** For various planar transformation families, the possible geometric transformations and the geometric properties which are preserved are listed

|  | Euclidean | Similarity | Affine | Projective |
|---|---|---|---|---|
| **Geometric transformation** | | | | |
| Rotation | ✓ | ✓ | ✓ | ✓ |
| Translation | ✓ | ✓ | ✓ | ✓ |
| Reflection | | ✓ | ✓ | ✓ |
| Uniform scaling | | ✓ | ✓ | ✓ |
| Nonuniform scaling | | | ✓ | ✓ |
| Shear | | | ✓ | ✓ |
| Perspective projection | | | | ✓ |
|  | **Euclidean** | **Similarity** | **Affine** | **Projective** |
| **Preserved geometric properties (invariants)** | | | | |
| Length | ✓ | | | |
| Angle | ✓ | ✓ | | |
| Ratio of lengths | ✓ | ✓ | | |
| Parallelism | ✓ | ✓ | ✓ | |
| Incidence | ✓ | ✓ | ✓ | ✓ |
| Cross ratio | ✓ | ✓ | ✓ | ✓ |

The possible geometric transformations for each type of transformation are summarized in ◘ Table C.1 along with the geometric properties which are unchanged, or invariant, under that transformation. We see that while Euclidean is most restrictive in terms of the geometric transformations it can perform, it is able to preserve important properties such as length and angle.

# D Lie Groups and Algebras

We cannot go very far in the study of rotations or rigid-body motion without coming across the terms Lie groups, Lie algebras or Lie brackets – all named in honor of the Norwegian mathematician Sophus Lie. Rotations and rigid-body motion in two and three dimensions can be represented by matrices that have special structure, they form Lie groups, and they have Lie algebras.

We will start simply by considering the set of all real $2 \times 2$ matrices $\mathbf{A} \in \mathbb{R}^{2 \times 2}$

$$\mathbf{A} = \begin{pmatrix} a_{0,0} & a_{0,1} \\ a_{1,0} & a_{1,1} \end{pmatrix}$$

which we could write as a linear combination of basis matrices

$$\mathbf{A} = a_{0,0} \begin{pmatrix} 1 & 0 \\ 0 & 0 \end{pmatrix} + a_{0,1} \begin{pmatrix} 0 & 1 \\ 0 & 0 \end{pmatrix} + a_{1,0} \begin{pmatrix} 0 & 0 \\ 1 & 0 \end{pmatrix} + a_{1,1} \begin{pmatrix} 0 & 0 \\ 0 & 1 \end{pmatrix}$$

where each basis matrix represents a *direction* in a 4-dimensional space of $2 \times 2$ matrices. That is, the four axes of this space are *parallel* with each of these basis matrices. Any $2 \times 2$ matrix can be represented by a point in this space – this particular matrix is a point with the coordinates $(a_{0,0}, a_{0,1}, a_{1,0}, a_{1,1})$.

All proper rotation matrices, those belonging to $\mathbf{SO}(2)$, are a *subset* of points within the space of all $2 \times 2$ matrices. For this example, the points lie in a 1-dimensional subset, a closed curve, in the 4-dimensional space. This is an instance of a manifold, a lower-dimensional smooth *surface* embedded within a space.

The notion of a curve in the 4-dimensional space makes sense when we consider that the $\mathbf{SO}(2)$ rotation matrix

$$\mathbf{A} = \begin{pmatrix} \cos \theta & \sin \theta \\ -\sin \theta & \cos \theta \end{pmatrix}$$

has only one free parameter, and varying that parameter moves the point along the manifold.

Invoking mathematical formalism, we say that rotations $\mathbf{SO}(2)$ and $\mathbf{SO}(3)$, and rigid-body motions $\mathbf{SE}(2)$ and $\mathbf{SE}(3)$ are matrix Lie groups and this has two implications. Firstly, they are an *algebraic group*, a mathematical structure comprising

### Excurse D.2: Sophus Lie

Lie, pronounced lee, (1842–1899) was a Norwegian mathematician who obtained his Ph.D. from the University of Christiania in Oslo in 1871. He spent time in Berlin working with Felix Klein, and later contributed to Klein's Erlangen program to characterize geometries based on group theory and projective geometry. On a visit to Milan during the Franco-Prussian war, he was arrested as a German spy and spent one month in prison. He is best known for his discovery that continuous transformation groups (now called Lie groups) can be understood by linearizing them and studying their generating vector spaces. He is buried in the Vår Frelsers gravlund in Oslo. (Image by Ludwik Szacinski)

elements and a single operator. In simple terms, a group $\mathbf{G}$ has the following properties:

1. If $g_1$ and $g_2$ are elements of the group, that is, $g_1, g_2 \in \mathbf{G}$, then the result of the group's operator $\diamond$ is also an element of the group: $g_1 \diamond g_2 \in \mathbf{G}$. In general, groups are not commutative, so $g_1 \diamond g_2 \neq g_2 \diamond g_1$. For rotations and rigid-body motions, the group operator $\diamond$ represents composition. ▶

2. The group operator is associative, that is, $(g_1 \diamond g_2) \diamond g_3 = g_1 \diamond (g_2 \diamond g_3)$.

3. There is an identity element $I \in \mathbf{G}$ such, for every $g \in \mathbf{G}$, $g \diamond I = I \diamond g = g$. ▶

4. For every $g \in \mathbf{G}$, there is a unique inverse $h \in \mathbf{G}$ such that $g \diamond h = h \diamond g = I$. ▶

In this book's notation, $\oplus$ is the group operator for relative pose.

The second implication of being a Lie group is that there is a smooth (differentiable) manifold structure. At any point on the manifold, we can construct tangent vectors. The set of all tangent vectors at that point form a vector space – the tangent space. This is the multidimensional equivalent to a tangent line on a curve, or a tangent plane on a solid. We can think of this as the set of all possible derivatives of the manifold at that point.

In this book's notation, the identity for relative pose (implying null motion) is denoted by $\varnothing$ so we can say that
$$\xi \oplus \varnothing = \varnothing \oplus \xi = \xi.$$

The tangent space *at the identity* is described by the Lie algebra of the group, and the basis directions of the tangent space are called the generators of the group. Points in this tangent space map to elements of the group via the exponential function. If $\mathbf{g}$ is the Lie algebra for group $\mathbf{G}$, then

In this book's notation, we use the operator $\ominus \xi$ to form the inverse of a relative pose.

$$e^X \in \mathbf{G}, \ \forall X \in \mathbf{g}$$

where the elements of $\mathbf{g}$ and $\mathbf{G}$ are matrices of the same size and each of them has a specific structure.

The surface of a sphere is a manifold in a 3-dimensional space and at any point on that surface we can create a tangent vector. In fact, we can create an infinite number of them and they lie within a plane which is a 2-dimensional vector space – the tangent space. We can choose a set of basis directions and establish a 2-dimensional coordinate system and we can map points on the plane to points on the sphere's surface.

Now, consider an arbitrary real $3 \times 3$ matrix $\mathbf{A} \in \mathbb{R}^{3 \times 3}$

$$\mathbf{A} = \begin{pmatrix} a_{0,0} & a_{0,1} & a_{0,2} \\ a_{1,0} & a_{1,1} & a_{1,2} \\ a_{2,0} & a_{2,1} & a_{2,2} \end{pmatrix}$$

which we could write as a linear combination of basis matrices

$$\mathbf{A} = a_{0,0} \begin{pmatrix} 1 & 0 & 0 \\ 0 & 0 & 0 \\ 0 & 0 & 0 \end{pmatrix} + a_{0,1} \begin{pmatrix} 0 & 1 & 0 \\ 0 & 0 & 0 \\ 0 & 0 & 0 \end{pmatrix} + \cdots + a_{2,2} \begin{pmatrix} 0 & 0 & 0 \\ 0 & 0 & 0 \\ 0 & 0 & 1 \end{pmatrix}$$

where each basis matrix represents a *direction* in a 9-dimensional space of $3 \times 3$ matrices. Every possible $3 \times 3$ matrix is represented by a point in this space.

Not all matrices in this space are proper rotation matrices belonging to $\mathbf{SO}(3)$, but those that do will lie on a manifold since $\mathbf{SO}(3)$ is a Lie group. The null rotation, represented by the identity matrix, is one point in this space. At that point, we can construct a tangent space which has only 3 dimensions. Every point in the tangent space can be expressed as a linear combination of basis matrices

$$\Omega = \omega_1 \underbrace{\begin{pmatrix} 0 & 0 & 0 \\ 0 & 0 & -1 \\ 0 & 1 & 0 \end{pmatrix}}_{\mathbf{G}_1} + \omega_2 \underbrace{\begin{pmatrix} 0 & 0 & 1 \\ 0 & 0 & 0 \\ -1 & 0 & 0 \end{pmatrix}}_{\mathbf{G}_2} + \omega_3 \underbrace{\begin{pmatrix} 0 & -1 & 0 \\ 1 & 0 & 0 \\ 0 & 0 & 0 \end{pmatrix}}_{\mathbf{G}_3} \tag{D.1}$$

which is the Lie algebra of the $\mathbf{SO}(3)$ group. The bases of this space: $\mathbf{G}_1$, $\mathbf{G}_2$ and $\mathbf{G}_3$ are called the generators of $\mathbf{SO}(3)$ and belong to $\mathbf{so}(3)$. ▶

The equivalent algebra is denoted using lowercase letters and is a set of matrices.

Equation (D.1) can be written as a skew-symmetric matrix

$$\mathbf{\Omega} = [\boldsymbol{\omega}]_\times = \begin{pmatrix} 0 & -\omega_3 & \omega_2 \\ \omega_3 & 0 & -\omega_1 \\ -\omega_2 & \omega_1 & 0 \end{pmatrix} \in \mathbf{so}(3)$$

parameterized by the vector $\boldsymbol{\omega} = (\omega_1, \omega_2, \omega_3) \in \mathbb{R}^3$ which is known as the Euler vector or exponential coordinates. This reflects the 3 degrees of freedom of the $\mathbf{SO}(3)$ group embedded in the space of all $3 \times 3$ matrices. The 3DOF is consistent with our intuition about rotations in 3D space and also Euler's rotation theorem.

Mapping between vectors and skew-symmetric matrices is frequently required and the following shorthand notation will be used

$$[\cdot]_\times \colon \mathbb{R} \mapsto \mathbf{so}(2), \ \mathbb{R}^3 \mapsto \mathbf{so}(3),$$
$$\vee_\times(\cdot) \colon \mathbf{so}(2) \mapsto \mathbb{R}, \ \mathbf{so}(3) \mapsto \mathbb{R}^3.$$

The first mapping is performed by the Toolbox function `skew` and the second by `vex` (which is named after the $\vee_\times$ operator).

The exponential of *any* matrix in $\mathbf{so}(3)$ is a valid member of $\mathbf{SO}(3)$

$$\mathbf{R}(\theta\hat{\boldsymbol{\omega}}) = e^{[\theta\hat{\boldsymbol{\omega}}]_\times} \in \mathbf{SO}(3)$$

and an efficient closed-form solution is given by Rodrigues' rotation formula

$$\mathbf{R}(\theta\hat{\boldsymbol{\omega}}) = \mathbf{1} + \sin\theta[\hat{\boldsymbol{\omega}}]_\times + (1 - \cos\theta)[\hat{\boldsymbol{\omega}}]_\times^2$$

Finally, consider an arbitrary real $4 \times 4$ matrix $\mathbf{A} \in \mathbb{R}^{4\times4}$

$$\mathbf{A} = \begin{pmatrix} a_{0,0} & a_{0,1} & a_{0,2} & a_{0,3} \\ a_{1,0} & a_{1,1} & a_{1,2} & a_{1,3} \\ a_{2,0} & a_{2,1} & a_{2,2} & a_{2,3} \\ a_{3,0} & a_{3,1} & a_{3,2} & a_{3,3} \end{pmatrix}$$

which we could write as a linear combination of basis matrices

$$\mathbf{A} = a_{0,0}\begin{pmatrix} 1 & 0 & 0 & 0 \\ 0 & 0 & 0 & 0 \\ 0 & 0 & 0 & 0 \\ 0 & 0 & 0 & 0 \end{pmatrix} + a_{0,1}\begin{pmatrix} 0 & 1 & 0 & 0 \\ 0 & 0 & 0 & 0 \\ 0 & 0 & 0 & 0 \\ 0 & 0 & 0 & 0 \end{pmatrix} + \cdots + a_{3,3}\begin{pmatrix} 0 & 0 & 0 & 0 \\ 0 & 0 & 0 & 0 \\ 0 & 0 & 0 & 0 \\ 0 & 0 & 0 & 1 \end{pmatrix}$$

where each basis matrix represents a *direction* in a 16-dimensional space of all possible $4 \times 4$ matrices. Every $4 \times 4$ matrix is represented by a point in this space.

Not all matrices in this space are proper rigid-body transformation matrices belonging to $\mathbf{SE}(3)$, but those that do lie on a smooth manifold. The null motion (zero rotation and translation), which is represented by the identity matrix, is one point in this space. At that point, we can construct a tangent space, which has 6 dimensions in this case, and points in the tangent space can be expressed as a linear combination of basis matrices

$$\mathbf{\Sigma} = \omega_1\begin{pmatrix} 0 & 0 & 0 & 0 \\ 0 & 0 & -1 & 0 \\ 0 & 1 & 0 & 0 \\ 0 & 0 & 0 & 0 \end{pmatrix} + \omega_2\begin{pmatrix} 0 & 0 & 1 & 0 \\ 0 & 0 & 0 & 0 \\ -1 & 0 & 0 & 0 \\ 0 & 0 & 0 & 0 \end{pmatrix} + \omega_3\begin{pmatrix} 0 & -1 & 0 & 0 \\ 1 & 0 & 0 & 0 \\ 0 & 0 & 0 & 0 \\ 0 & 0 & 0 & 0 \end{pmatrix}$$

$$+ v_1\begin{pmatrix} 0 & 0 & 0 & 1 \\ 0 & 0 & 0 & 0 \\ 0 & 0 & 0 & 0 \\ 0 & 0 & 0 & 0 \end{pmatrix} + v_2\begin{pmatrix} 0 & 0 & 0 & 0 \\ 0 & 0 & 0 & 1 \\ 0 & 0 & 0 & 0 \\ 0 & 0 & 0 & 0 \end{pmatrix} + v_3\begin{pmatrix} 0 & 0 & 0 & 0 \\ 0 & 0 & 0 & 0 \\ 0 & 0 & 0 & 1 \\ 0 & 0 & 0 & 0 \end{pmatrix}$$

and these generator matrices belong to the Lie algebra of the group $\mathbf{SE}(3)$ which is enoted by $\mathbf{se}(3)$. This can be written in general form as

$$\Sigma = [S] = \left( \begin{array}{ccc|c} 0 & -\omega_3 & \omega_2 & v_1 \\ \omega_3 & 0 & -\omega_1 & v_2 \\ -\omega_2 & \omega_1 & 0 & v_3 \\ \hline 0 & 0 & 0 & 0 \end{array} \right) \in \mathbf{se}(3)$$

which is an augmented skew-symmetric matrix parameterized by $S = (v, \omega) \in \mathbb{R}^6$ which is referred to as a twist and has physical interpretation in terms of a screw axis direction and position. The sparse matrix structure and this concise parameterization reflects the 6 degrees of freedom of the $\mathbf{SE}(3)$ group embedded in the space of all $4 \times 4$ matrices. We extend our earlier shorthand notation

$$[\cdot] \colon \mathbb{R}^3 \mapsto \mathbf{se}(2), \ \mathbb{R}^6 \mapsto \mathbf{se}(3),$$
$$\vee(\cdot) \colon \mathbf{se}(2) \mapsto \mathbb{R}^3, \ \mathbf{se}(3) \mapsto \mathbb{R}^6 \,.$$

We can use these operators to convert between a twist representation which is a 6-vector and a Lie algebra representation which is a $4 \times 4$ augmented skew-symmetric matrix. We convert the Lie algebra to the Lie group representation using

$$\mathbf{T}(\theta \hat{S}) = \mathrm{e}^{[\theta \hat{S}]} \in \mathbf{SE}(3)$$

or the inverse using the matrix logarithm. The exponential and the logarithm each have an efficient closed-form solution given by `trexp` and `trlog`.

## Transforming a Twist – the Adjoint Representation

We have seen in ▶ Sect. 2.4.7 that rigid-body motions can be described by a twist which represents motion in terms of a screw axis direction and position. For example, in ◘ Fig. D.1, the twist $^A S$ can be used to transform points on the body. If the screw is rigidly attached to the body which undergoes some rigid-body motion $^A\xi_B$ the new twist is

$$^B S = \mathrm{Ad}\big(^B\xi_A\big)\,^A S$$

where

$$\mathrm{Ad}\big(^A\xi_B\big) = \left( \begin{array}{cc} ^A\mathbf{R}_B & \big[^A t_B\big]_\times {}^A\mathbf{R}_B \\ \mathbf{0} & ^A\mathbf{R}_B \end{array} \right) \in \mathbb{R}^{6\times 6} \tag{D.2}$$

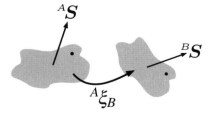

◘ **Fig. D.1** Points in the body (gray cloud) can be transformed by the twist $^A S$. If the body and the screw axis undergo a rigid-body transformation $^A\xi_B$, the new twist is $^B S$

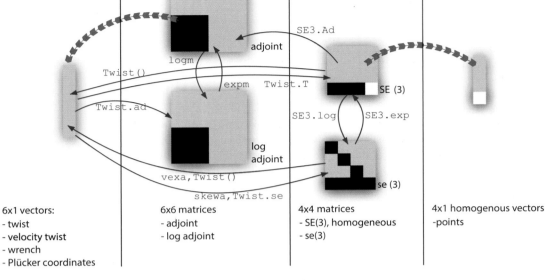

**Fig. D.2** The menagerie of **SE**(3) related quantities. Matrix values are coded as: 0 (black), 1 (white), other values (gray). Transformations between types are indicated by blue arrows with the relevant class plus method name. Operations are indicated by red arrows: the tail-end object operates on the head-end object and results in another object of the head-end type

is the adjoint representation of the rigid-body motion. Alternatively, we can write

$$\mathrm{Ad}(e^{[S]}) = e^{\mathrm{ad}(S)}$$

where $\mathrm{ad}(S)$ is the logarithm of the adjoint and defined in terms of the twist parameters as

$$\mathrm{ad}(S) = \begin{pmatrix} [\boldsymbol{\omega}]_\times & [\boldsymbol{v}]_\times \\ \mathbf{0} & [\boldsymbol{\omega}]_\times \end{pmatrix} \in \mathbb{R}^{6\times6} \ .$$

The relationship between the various mathematical objects discussed are shown in ◾ Fig. D.2.

# E Linearization, Jacobians and Hessians

In robotics and computer vision, the equations we encounter are often nonlinear. To apply familiar and powerful analytic techniques, we must work with linear or quadratic approximations to these equations. The principle is illustrated in ◻ Fig. E.1 for the 1-dimensional case, and the analytical approximations shown in red are made at $x = x_0$. The approximation equals the nonlinear function at $x_0$ but is increasingly inaccurate as we move away from that point. We call this a *local approximation* since it is valid in a region local to $x_0$ – the size of the valid region depends on the severity of the nonlinearity. This approach can be extended to an arbitrary number of dimensions.

## E.1 Scalar Function of a Scalar

The function $f\colon \mathbb{R} \mapsto \mathbb{R}$ can be expressed as a Taylor series

$$f(x_0 + \Delta) = f(x_0) + \frac{\mathrm{d}f}{\mathrm{d}x}\Delta + \frac{1}{2}\frac{\mathrm{d}^2 f}{\mathrm{d}x^2}\Delta^2 + \cdots$$

which we truncate to form a first-order or linear approximation

$$f'(\Delta) \approx f(x_0) + J(x_0)\Delta$$

or a second-order approximation

$$f'(\Delta) \approx f(x_0) + J(x_0)\Delta + \frac{1}{2}H(x_0)\Delta^2$$

where $\Delta \in \mathbb{R}$ is an infinitesimal change in $x$ relative to the linearization point $x_0$, and the first and second derivatives are given by

$$J(x_0) = \left.\frac{\mathrm{d}f}{\mathrm{d}x}\right|_{x_0}, \quad H(x_0) = \left.\frac{\mathrm{d}^2 f}{\mathrm{d}x^2}\right|_{x_0}$$

respectively.

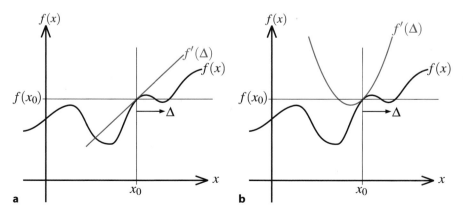

a                    b

◻ **Fig. E.1** The nonlinear function $f(x)$ is approximated at the point $x = x_0$ by **a** the red line – a linear or first-order approximation, **b** the red parabola – a second-order approximation. At the linearization point, both curves are equal and tangent to the function while for **b** the second derivatives also match

**Excurse E.3: Ludwig Otto Hesse**

Hesse (1811–1874) was a German mathematician, born in Königsberg, Prussia, who studied under Jacobi and Bessel at the University of Königsberg. He taught at Königsberg, Halle, Heidelberg and finally at the newly established Polytechnic School in Munich. In 1869, he joined the Bavarian Academy of Sciences. He died in Munich, but was buried in Heidelberg which he considered his second home.

## E.2    Scalar Function of a Vector

The scalar field $f(\boldsymbol{x}) \colon \mathbb{R}^n \mapsto \mathbb{R}$ can be expressed as a Taylor series

$$f(\boldsymbol{x}_0 + \boldsymbol{\Delta}) = f(\boldsymbol{x}_0) + \mathbf{J}(\boldsymbol{x}_0)\boldsymbol{\Delta} + \frac{1}{2}\boldsymbol{\Delta}^\top \mathbf{H}(\boldsymbol{x}_0)\boldsymbol{\Delta} + \cdots$$

which we can truncate to form a first-order or linear approximation

$$f'(\boldsymbol{\Delta}) \approx f(\boldsymbol{x}_0) + \mathbf{J}(\boldsymbol{x}_0)\boldsymbol{\Delta}$$

or a second-order approximation

$$f'(\boldsymbol{\Delta}) \approx f(\boldsymbol{x}_0) + \mathbf{J}(\boldsymbol{x}_0)\boldsymbol{\Delta} + \frac{1}{2}\boldsymbol{\Delta}^\top \mathbf{H}(\boldsymbol{x}_0)\boldsymbol{\Delta}$$

where $\boldsymbol{\Delta} \in \mathbb{R}^n$ is an infinitesimal change in $\boldsymbol{x} \in \mathbb{R}^n$ relative to the linearization point $\boldsymbol{x}_0$, $\mathbf{J} \in \mathbb{R}^{1 \times n}$ is the vector version of the first derivative, and $\mathbf{H} \in \mathbb{R}^{n \times n}$ is the Hessian – the matrix version of the second derivative.

The derivative of the function $f(\cdot)$ with respect to the vector $\boldsymbol{x}$ is

$$\mathbf{J}(\boldsymbol{x}) = \nabla f(\boldsymbol{x}) = \begin{pmatrix} \frac{\partial f}{\partial x_0} \\ \frac{\partial f}{\partial x_1} \\ \vdots \\ \frac{\partial f}{\partial x_{n-1}} \end{pmatrix} \in \mathbb{R}^n$$

and is a column vector that points in the direction at which the function $f(\boldsymbol{x})$ has maximal increase. It is often written as $\nabla_x f$ to make explicit that the differentiation is with respect to $\boldsymbol{x}$.

The Hessian is an $n \times n$ symmetric matrix of second derivatives

$$\mathbf{H}(\boldsymbol{x}) = \begin{pmatrix} \frac{\partial^2 f}{\partial x_0^2} & \frac{\partial^2 f}{\partial x_0 \partial x_1} & \cdots & \frac{\partial^2 f}{\partial x_0 \partial x_{n-1}} \\ \frac{\partial^2 f}{\partial x_0 \partial x_1} & \frac{\partial^2 f}{\partial x_1^2} & \cdots & \frac{\partial^2 f}{\partial x_1 \partial x_{n-1}} \\ \vdots & \vdots & \ddots & \vdots \\ \frac{\partial^2 f}{\partial x_0 \partial x_{n-1}} & \frac{\partial^2 f}{\partial x_1 \partial x_{n-1}} & \cdots & \frac{\partial^2 f}{\partial x_{n-1}^2} \end{pmatrix} \in \mathbb{R}^{n \times n} .$$

The function is at a critical point when $\|\mathbf{J}(x)\| = 0$. If $\mathbf{H}(x)$ is positive-definite then the function is at a local minimum, if negative-definite then a local maximum, otherwise the result is inconclusive.

For functions which are quadratic in $x$, as is the case for least-squares problems, it can be shown that the Hessian is

$$\mathbf{H}(x) = \mathbf{J}^\top(x)\,\mathbf{J}(x) + \sum_{i=0}^{m-1} f_i(x)\frac{\partial^2 f_i}{\partial x^2} \approx \mathbf{J}^\top(x)\,\mathbf{J}(x)$$

which is frequently approximated by just the first term and this is key to Gauss-Newton least-squares optimization discussed in ▶ App. F.2.3.

## E.3 Vector Function of a Vector

The vector field $f(x)\colon \mathbb{R}^n \mapsto \mathbb{R}^m$ can also be written as

$$f(x) = \begin{pmatrix} f_0(x) \\ f_1(x) \\ \vdots \\ f_{m-1}(x) \end{pmatrix} \in \mathbb{R}^m$$

where $f_i\colon \mathbb{R}^m \to \mathbb{R}$, $i = 0, \ldots, m-1$. The derivative of $f$ with respect to the vector $x$ can be expressed in matrix form as a Jacobian matrix

$$\mathbf{J}(x) = \begin{pmatrix} \frac{\partial f_0}{\partial x_0} & \cdots & \frac{\partial f_0}{\partial x_{n-1}} \\ \vdots & \ddots & \vdots \\ \frac{\partial f_{m-1}}{\partial x_0} & \cdots & \frac{\partial f_{m-1}}{\partial x_{n-1}} \end{pmatrix} \in \mathbb{R}^{m \times n}$$

which can also be written as

$$\mathbf{J}(x) = \begin{pmatrix} \nabla f_0^\top \\ \nabla f_1^\top \\ \vdots \\ \nabla f_{m-1}^\top \end{pmatrix}$$

This derivative is also known as the tangent map of $f$, denoted by $\mathrm{T}f$, or the differential of $f$ denoted by $\mathrm{D}f$. To make explicit that the differentiation is with respect to $x$ this can be denoted as $\mathbf{J}_x$, $\mathrm{T}_x f$, $\mathrm{D}_x f$ or even $\partial f / \partial x$.

The Hessian in this case is a rank-3 tensor which we can think of as a list, of length $m$, of $n \times n$ Hessians $\mathbf{H} = \big(\mathbf{H}_0(x), \ldots, \mathbf{H}_{m-1}(x)\big)$. In Python this could be represented by an $n \times n \times m$ array.

## E.4 Deriving Jacobians

Jacobians of functions are required for many optimization algorithms, for example pose-graph optimization, as well as for the extended Kalman filter. They can be computed either numerically or symbolically.

Consider (6.9) for the range and bearing angle of a landmark given the pose of the vehicle and the position of the landmark. We can express this as a lambda function

```
>>> zrange = lambda xi, xv, w: np.array([
...             np.linalg.norm(xi - xv[:2]) + w[0],
...             np.arctan((xi[1] - xv[1]) / (xi[0] - xv[0]))
...                   -xv[2] + w[1]]);
```

To estimate the Jacobian $\mathbf{H}_{\boldsymbol{x}_v} = \partial \boldsymbol{h} / \partial \boldsymbol{x}_v$ for $\boldsymbol{x}_v = (1, 2, \frac{\pi}{3})$ and $\boldsymbol{x}_i = (10, 8)$, we can compute a first-order numerical difference

```
>>> xv = np.r_[1, 2, pi/3]; xi = np.r_[10, 8]; w = np.r_[0, 0];
>>> h0 = zrange(xi, xv, w)
array([   10.82,  -0.4592])
>>> d = 0.001;
>>> J = np.column_stack([
...         zrange(xi, xv + [d, 0, 0], w) - h0,
...         zrange(xi, xv + [0, d, 0], w) - h0,
...         zrange(xi, xv + [0, 0, d], w) - h0
...                     ]) / d
array([[ -0.832,  -0.5547,       0],
       [ 0.05129, -0.07693,     -1]])
```

which shares the characteristic last column with the Jacobian shown in (6.14). Note that in computing this Jacobian we have set the measurement noise `w` to zero. The principal difficulty with this approach is choosing `d`, the difference used to compute the finite-difference approximation to the derivative. Too large and the results will be quite inaccurate if the function is nonlinear, too small and numerical problems will lead to reduced accuracy.

The Toolbox provides a convenience function to perform this

```
>>> numjac(lambda x: zrange(xi, x, w), xv)
array([[ -0.8321,  -0.5547,       0],
       [ 0.05128, -0.07692,     -1]])
```

where the arguments are a scalar function of a vector and the value of the vector about which to linearize.

Alternatively, we can perform the differentiation symbolically. This particular function is relatively simple and the derivatives can be determined easily by hand using differential calculus. The numerical derivative can be used as a quick check for correctness. To avoid the possibility of error, or for more complex functions, we can perform the symbolic differentiation using any of a large number of computer algebra packages. Using SymPy we define some symbolic column vectors for $\boldsymbol{x}_i$ and $\boldsymbol{x}_v$ which are 2D arrays

```
>>> from sympy import Matrix, MatrixSymbol, sqrt, atan, simplify,
... pycode
>>> xi = MatrixSymbol("xi", 2, 1)
xi
>>> xv = MatrixSymbol("xv", 3, 1)
xv
>>> w = Matrix([0, 0])
Matrix([
[0],
[0]])
```

Now we have to rewrite the function in a SymPy friendly way, removing NumPy functions and replacing them with equivalent SymPy functions ◄

NumPy functions will not accept symbolic variables.

```
>>> zrange = lambda xi, xv, w: Matrix([
...         sqrt((xi[0] - xv[0])**2 + (xi[1] - xv[1])**2) + w[0],
...         atan((xi[1] - xv[1]) / (xi[0] - xv[0]))
...                 -xv[2] + w[1]]);
>>> z = zrange(xi, xv, w)
Matrix([[sqrt((xi[0, 0] - xv[0, 0])**2 + (xi[1, 0] - xv[1, 0])**2)],
        [atan((xi[1, 0] - xv[1, 0])/(xi[0, 0] - xv[0, 0]))
        - xv[2, 0]]])
```

which is simply (6.9) in SymPy symbolic form and using SymPy `Matrix` rather than NumPy array. The result is less compact because SymPy supports only 2D arrays, so we have to provide two indices to reference an element of a vector. The Jacobian is computed by

```
>>> J = z.jacobian(xv)
Matrix([
    [(-xi[0, 0] + xv[0, 0])/sqrt((xi[0, 0] - xv[0, 0])**2
                                + (xi[1, 0] - xv[1, 0])**2),
     (-xi[1, 0] + xv[1, 0])/sqrt((xi[0, 0] - xv[0, 0])**2
                                + (xi[1, 0] - xv[1, 0])**2),
     0],
    [(xi[1, 0] - xv[1, 0])/((xi[0, 0] - xv[0, 0])**2
                                + (xi[1, 0] - xv[1, 0])**2),
     (-xi[0, 0] + xv[0, 0])/((xi[0, 0] - xv[0, 0])**2
                                + (xi[1, 0] - xv[1, 0])**2),
     -1]
    ])
```

which has the required shape

```
>>> J.shape
(2, 3)
```

and the characteristic last column. We could cut and paste this code into our program or automatically create a Python callable function using pycode.

Another approach is automatic differentiation (AD) of code. The package ADOL-C is an open-source tool that can differentiate C and C++ programs, that is, given a function written in C, it will return a Jacobian or higher-order derivative function written in C. It is available at ▶ https://github.com/coin-or/ADOL-C. For Python code, there is an open-source tool called Tangent available at ▶ https://github.com/google/tangent.

# F Solving Systems of Equations

Solving systems of linear and nonlinear equations, particularly over-constrained systems, is a common problem in robotics and computer vision.

## F.1  Linear Problems

### F.1.1  Nonhomogeneous Systems

These are equations of the form

$$\mathbf{A}x = b$$

where we wish to solve for the unknown vector $x \in \mathbb{R}^n$ where $\mathbf{A} \in \mathbb{R}^{m \times n}$ and $b \in \mathbb{R}^m$ are known constants.

If $n = m$ then $\mathbf{A}$ is square, and if $\mathbf{A}$ is non-singular then the solution is

$$x = \mathbf{A}^{-1}b \ .$$

In practice, we often encounter systems where $m > n$, that is, there are more equations than unknowns. In general, there will not be an exact solution, but we can attempt to find the *best* solution, in a least-squares sense, which is

$$x^* = \arg \min_x ||\mathbf{A}x - b|| \ .$$

That solution is given by

$$x^* = \left(\mathbf{A}^\top \mathbf{A}\right)^{-1} \mathbf{A}^\top b = \mathbf{A}^+ b$$

which is known as the pseudoinverse or more formally the left-generalized inverse. Using SVD where $\mathbf{A} = \mathbf{U}\mathbf{\Sigma}\mathbf{V}^\top$, this is

$$x = \mathbf{V}\mathbf{\Sigma}^{-1}\mathbf{U}^\top b$$

where $\mathbf{\Sigma}^{-1}$ is simply the element-wise inverse of the diagonal elements of $\mathbf{\Sigma}$.

If the matrix is singular, or the system is under constrained $n < m$, then there are infinitely many solutions. We can again use the SVD approach

$$x = \mathbf{V}\mathbf{\Sigma}^{-1}\mathbf{U}^\top b$$

where this time $\mathbf{\Sigma}^{-1}$ is the element-wise inverse of the *nonzero* diagonal elements of $\mathbf{\Sigma}$, all other zeros are left in place.

All these problems can be solved using SciPy

```
>>> x = linalg.spsolve(A, b)
```

### F.1.2  Homogeneous Systems

These are equations of the form

$$\mathbf{A}x = 0 \tag{F.1}$$

and always have the trivial solution $x = 0$. If $\mathbf{A}$ is square and non-singular, that is the only solution. Otherwise, if $\mathbf{A}$ is not of full rank, that is, the matrix is non-square, or square and singular, then there are an infinite number of solutions which are linear combinations of vectors in the right null space of $\mathbf{A}$ which is computed by the SciPy function `sp.linalg.null_space`.

### F.1.3 Finding a Rotation Matrix

Consider two sets of points in $\mathbb{R}^n$ related by an unknown rotation $\mathbf{R} \in \mathbf{SO}(n)$. The points are arranged column-wise as $\mathbf{P} = \{\boldsymbol{p}_0, \ldots, \boldsymbol{p}_{m-1}\} \in \mathbb{R}^{n \times m}$ and $\mathbf{Q} = \{\boldsymbol{q}_0, \ldots, \boldsymbol{q}_{m-1}\} \in \mathbb{R}^{n \times m}$ such that

$$\mathbf{RP} = \mathbf{Q} \ .$$

To solve for $\mathbf{R}$, we first compute the moment matrix

$$\mathbf{M} = \sum_{i=0}^{m-1} \boldsymbol{q}_i \boldsymbol{p}_i^\top$$

and then take the SVD such that $\mathbf{M} = \mathbf{U}\boldsymbol{\Sigma}\mathbf{V}^\top$. The least squares estimate of the rotation matrix is

$$\mathbf{R} = \mathbf{U}\begin{pmatrix} 1 & 0 & 0 \\ 0 & 1 & 0 \\ 0 & 0 & s \end{pmatrix}\mathbf{V}^\top, \ \ s = \det(\mathbf{U})\det(\mathbf{V}^\top) \tag{F.2}$$

and is guaranteed to be an $\mathbf{SO}(3)$ matrix (Horn 1987; Umeyama 1991).

## F.2 Nonlinear Problems

Many problems in robotics and computer vision involve sets of nonlinear equations. Solution of these problems requires linearizing the equations about an estimated solution, solving for an improved solution and iterating. Linearization is discussed in ▶ App. E.

### F.2.1 Finding Roots

Consider a set of equations expressed in the form

$$\boldsymbol{f}(\boldsymbol{x}) = \mathbf{0} \tag{F.3}$$

where $\boldsymbol{f} : \mathbb{R}^n \mapsto \mathbb{R}^m$. This is a nonlinear version of (F.1), and we first linearize the equations about our best estimate of the solution $\boldsymbol{x}_0$

$$\boldsymbol{f}(\boldsymbol{x}_0 + \boldsymbol{\Delta}) = \boldsymbol{f}_0 + \mathbf{J}(\boldsymbol{x}_0)\boldsymbol{\Delta} + \frac{1}{2}\boldsymbol{\Delta}^\top \mathbf{H}(\boldsymbol{x}_0)\boldsymbol{\Delta} + \text{H.O.T.} \tag{F.4}$$

where $\boldsymbol{f}_0 = \boldsymbol{f}(\boldsymbol{x}_0) \in \mathbb{R}^{m \times 1}$ is the function value and $\mathbf{J} = \mathbf{J}(\boldsymbol{x}_0) \in \mathbb{R}^{m \times n}$ the Jacobian, both evaluated at the linearization point, $\boldsymbol{\Delta} \in \mathbb{R}^{n \times 1}$ is an infinitesimal change in $\boldsymbol{x}$ relative to $\boldsymbol{x}_0$ and H.O.T. denotes higher-order terms. $\mathbf{H}(\boldsymbol{x}_0)$ is the $n \times n \times m$ Hessian tensor. ▶

We take the first two terms of (F.4) to form a linear approximation

To evaluate $\boldsymbol{\Delta}^\top \mathbf{H}(\boldsymbol{x}_0)\boldsymbol{\Delta}$ we write it as $\boldsymbol{\Delta}^\top (\mathbf{H}(\boldsymbol{x}_0)\bar{\times}_2\boldsymbol{\Delta})$ which uses a mode-2 tensor-vector product.

$$\boldsymbol{f}'(\boldsymbol{\Delta}) \approx \boldsymbol{f}_0 + \mathbf{J}(\boldsymbol{x}_0)\boldsymbol{\Delta} \tag{F.5}$$

and solve an approximation of the original problem $\boldsymbol{f}'(\boldsymbol{\Delta}) = \mathbf{0}$

$$\boldsymbol{f}_0 + \mathbf{J}(\boldsymbol{x}_0)\boldsymbol{\Delta} = \mathbf{0} \ \Rightarrow \ \boldsymbol{\Delta} = -\mathbf{J}^{-1}(\boldsymbol{x}_0)\boldsymbol{f}_0 \ .$$

If $n \neq m$, then $\mathbf{J}$ is non-square and we can use the pseudoinverse `np.linalg.pinv(J)`, or a least squares solver `scipy.linalg.spsolve(J, f0)`. The computed step $\boldsymbol{\Delta}$ is based on an approximation to the original nonlinear function so $\boldsymbol{x}_0 + \boldsymbol{\Delta}$ will generally not be the solution but it will be closer. This leads to an iterative solution – the Newton-Raphson method:

**repeat**
  compute $\boldsymbol{f}_0 = \boldsymbol{f}(\boldsymbol{x}_0)$, $\mathbf{J} = \mathbf{J}(\boldsymbol{x}_0)$
  $\boldsymbol{\Delta} = -\mathbf{J}^{-1}\boldsymbol{f}_0$
  $\boldsymbol{x}_0 \leftarrow \boldsymbol{x}_0 + \boldsymbol{\Delta}$
**until** $\|\boldsymbol{f}_0\| < \epsilon$

### F.2.2  Nonlinear Minimization

A very common class of problems involves finding the *minimum* of a scalar function $f(\boldsymbol{x}): \mathbb{R}^n \mapsto \mathbb{R}$ which can be expressed as

$$\boldsymbol{x}^* = \arg\min_{\boldsymbol{x}} f(\boldsymbol{x}) \ .$$

The derivative of the linearized system of equations (F.5) is

$$\frac{\mathrm{d}\boldsymbol{f}'}{\mathrm{d}\boldsymbol{\Delta}} = \mathbf{J}(\boldsymbol{x}_0)$$

and if we consider the function to be a multi-dimensional surface, then $\mathbf{J}(\boldsymbol{x}_0) \in \mathbb{R}^{n \times 1}$ is a vector indicating the direction and magnitude of the *slope* at $\boldsymbol{x} = \boldsymbol{x}_0$ so an update of

$$\boldsymbol{\Delta} = -\beta\mathbf{J}(\boldsymbol{x}_0)$$

will move the estimate *down hill* toward the minimum. This leads to an iterative solution called gradient descent:

**repeat**
  compute $\mathbf{J} = \mathbf{J}(\boldsymbol{x}_0)$
  $\boldsymbol{\Delta} = -\beta\mathbf{J}$
  $\boldsymbol{x}_0 \leftarrow \boldsymbol{x}_0 + \boldsymbol{\Delta}$
**until** $\|\boldsymbol{\Delta}\| < \epsilon$

and the challenge is to choose the appropriate value of $\beta$ which controls the step size. If too small, then many iterations will be required. If too large, the solution may be unstable and not converge.

If we include the second-order term from (F.4), the approximation becomes

$$\boldsymbol{f}'(\boldsymbol{\Delta}) \approx \boldsymbol{f}_0 + \mathbf{J}(\boldsymbol{x}_0)\boldsymbol{\Delta} + \frac{1}{2}\boldsymbol{\Delta}^\top\mathbf{H}(\boldsymbol{x}_0)\boldsymbol{\Delta}$$

and includes the $n \times n$ Hessian matrix. To find its minima we take the derivative and set it to zero

$$\frac{\mathrm{d}\boldsymbol{f}'}{\mathrm{d}\boldsymbol{\Delta}} = \mathbf{0} \ \Rightarrow \ \mathbf{J}(\boldsymbol{x}_0) + \mathbf{H}(\boldsymbol{x}_0)\boldsymbol{\Delta} = \mathbf{0}$$

and the update is

$$\boldsymbol{\Delta} = -\mathbf{H}^{-1}(\boldsymbol{x}_0)\mathbf{J}(\boldsymbol{x}_0) \ .$$

This leads to another iterative solution – Newton's method. The challenge is determining the Hessian of the nonlinear system, either by numerical approximation or symbolic manipulation.

## F.2.3  Nonlinear Least-Squares Minimization

Very commonly, the scalar function we wish to optimize is a quadratic cost function

$$F(x) = ||f(x)||^2 = f(x)^\top f(x)$$

where $f(x): \mathbb{R}^n \mapsto \mathbb{R}^m$ is some vector-valued nonlinear function which we can linearize as

$$f'(\Delta) \approx f_0 + \mathbf{J}\Delta$$

and the scalar cost is

$$\begin{aligned} F(\Delta) &\approx (f_0 + \mathbf{J}\Delta)^\top (f_0 + \mathbf{J}\Delta) \\ &\approx f_0^\top f_0 + \underline{f_0^\top \mathbf{J}\Delta} + \underline{\Delta^\top \mathbf{J}^\top f_0} + \Delta^\top \mathbf{J}^\top \mathbf{J}\Delta \\ &\approx f_0^\top f_0 + 2f_0^\top \mathbf{J}\Delta + \Delta^\top \mathbf{J}^\top \mathbf{J}\Delta \end{aligned}$$

where $\mathbf{J}^\top \mathbf{J} \in \mathbb{R}^{n \times n}$ is the *approximate* Hessian from ▶ App. E.2. ▶

To minimize the error of this linearized least squares system, we take the derivative with respect to $\Delta$ and set it to zero

One of the underlined terms is the transpose of the other, but since both result in a scalar, the transposition doesn't matter.

$$\frac{\mathrm{d}F}{\mathrm{d}\Delta} = 0 \Rightarrow 2f_0^\top \mathbf{J} + \Delta^\top \mathbf{J}^\top \mathbf{J} = 0$$

which we can solve for the locally optimal update

$$\Delta = -(\mathbf{J}^\top \mathbf{J})^{-1} \mathbf{J}^\top f_0 \tag{F.6}$$

which is the pseudoinverse or left generalized-inverse of $\mathbf{J}$. Once again, we iterate to find the solution – a Gauss-Newton iteration.

### Numerical Issues

When solving (F.6), we may find that the Hessian $\mathbf{J}^\top \mathbf{J}$ is poorly conditioned or singular and we can add a damping term $\lambda$

$$\Delta = -(\mathbf{J}^\top \mathbf{J} + \lambda \mathbf{1}_{n \times n})^{-1} \mathbf{J}^\top f_0$$

which makes the system more positive-definite. Since $\mathbf{J}^\top \mathbf{J} + \mathbf{1}_{n \times n}$ is effectively in the denominator, increasing $\lambda$ will decrease $\|\Delta\|$ and slow convergence.

How do we choose $\lambda$? We can experiment with different values, but a better way is the Levenberg-Marquardt algorithm (▣ Fig. F.1) which adjusts $\lambda$ to ensure convergence. If the error increases compared to the last step, then the step is repeated with increased $\lambda$ to reduce the step size. If the error decreases, then $\lambda$ is reduced to increase the convergence rate. The updates vary continuously between Gauss-Newton (low $\lambda$) and gradient descent (high $\lambda$).

For problems where $n$ is large, inverting the $n \times n$ approximate Hessian is expensive. Typically, $m < n$ which means the Jacobian is not square and (F.6) can be rewritten as

$$\Delta = -\mathbf{J}^\top (\mathbf{J}\mathbf{J}^\top)^{-1} f_0$$

which is the right pseudoinverse and involves inverting a smaller matrix. We can reintroduce a damping term

$$\Delta = -\mathbf{J}^\top (\mathbf{J}\mathbf{J}^\top + \lambda \mathbf{1}_{m \times m})^{-1} f_0$$

```
 1  initialize λ
 2  repeat
 3  │   compute f_0 = f(x_0), J = J(x_0), H = J^T J
 4  │   Δ = -(H + λ1_{n×n})^{-1} J^T f_0
 5  │   if  f(x_0 + Δ) < f(x_0) then
 6  │   │   – error decreased: reduce damping
 7  │   │   x_0 ← x_0 + Δ
 8  │   │   λ ← λ/c
 9  │   else
10  │   │   – error increased: discard and raise damping
11  │   │   λ ← cλ
12  │   end
13  until ‖Δ‖ < ε
```

◻ **Fig. F.1**  Levenberg-Marquadt algorithm, $c$ is typically chosen in the range 2 to 10

and if $\lambda$ is large this becomes simply

$$\Delta \approx -\beta \mathbf{J}^\top \boldsymbol{f}_0$$

but exhibits very slow convergence.

If $\boldsymbol{f}_k(\cdot)$ has additive noise that is zero mean, normally distributed and time invariant, we have a maximum likelihood estimator of $\boldsymbol{x}$. Outlier data has a significant impact on the result since errors are squared. Robust estimators minimize the effect of outlier data and in an M-estimator

$$F(\boldsymbol{x}) = \rho(\boldsymbol{f}_k(\boldsymbol{x}))$$

the squared norm is replaced by a loss function $\rho(\cdot)$ which models the likelihood of its argument. Unlike the squared norm, these functions flatten off for large values, and some common examples include the Huber loss function and the Tukey biweight function.

### F.2.4  Sparse Nonlinear Least Squares

For a large class of problems, the overall cost is the sum of quadratic costs

$$F(\boldsymbol{x}) = \sum_k \| \boldsymbol{f}_k(\boldsymbol{x}) \|^2 = \sum_k \boldsymbol{f}_k(\boldsymbol{x})^\top \boldsymbol{f}_k(\boldsymbol{x}) \ . \tag{F.7}$$

Consider the problem of fitting a model $\boldsymbol{z} = \phi(\boldsymbol{w}; \boldsymbol{x})$ where $\phi \colon \mathbb{R}^p \mapsto \mathbb{R}^m$ with parameters $\boldsymbol{x} \in \mathbb{R}^n$ to a set of data points $(\boldsymbol{w}_k, \boldsymbol{z}_k)$. The error vector associated with the $k^{\text{th}}$ data point is

$$\boldsymbol{f}_k(\boldsymbol{x}) = \boldsymbol{z}_k - \phi(\boldsymbol{w}_k; \boldsymbol{x}) \in \mathbb{R}^m$$

and minimizing (F.7) gives the optimal model parameters $\boldsymbol{x}$.

Another example is pose-graph optimization as used for pose-graph SLAM and bundle adjustment. Edge $k$ in the graph connects vertices $i$ and $j$ and has an associated cost $\boldsymbol{f}_k(\cdot) \colon \mathbb{R}^n \mapsto \mathbb{R}^m$

$$\boldsymbol{f}_k(\boldsymbol{x}) = \hat{\boldsymbol{e}}_k(\boldsymbol{x}) - \boldsymbol{e}_k^{\#} \tag{F.8}$$

where $e_k^\#$ is the observed value of the edge parameter and $\hat{e}_k(x)$ is the estimate based on the state $x$ of the pose graph. This is linearized

$$f_k'(\Delta) \approx f_{0,k} + \mathbf{J}_k \Delta$$

and the squared error for the edge is

$$F_k(x) = f_k^\top(x) \Omega_k f_k(x)$$

where $\Omega_k \in \mathbb{R}^{m \times m}$ is a positive-definite constant matrix ▶ which we combine as

$$
\begin{aligned}
F_k(\Delta) &\approx (f_{0,k} + \mathbf{J}_k \Delta)^\top \Omega_k (f_{0,k} + \mathbf{J}_k \Delta) \\
&\approx f_{0,k}^\top \Omega_k f_{0,k} + f_{0,k}^\top \Omega_k \mathbf{J}_k \Delta + \Delta^\top \mathbf{J}_k^\top \Omega_k f_{0,k} + \Delta^\top \mathbf{J}_k^\top \Omega_k \mathbf{J}_k \Delta \\
&\approx c_k + 2 b_k^\top \Delta + \Delta^\top \mathbf{H}_k \Delta
\end{aligned}
$$

This can be used to specify the significance of the edge $\det(\Omega_k)$ with respect to other edges, as well as the relative significance of the elements of $f_k(\cdot)$.

where $c_k = f_{0,k}^\top \Omega_k f_{0,k}$, $b_k^\top = f_{0,k}^\top \Omega_k \mathbf{J}_k$ and $\mathbf{H}_k = \mathbf{J}_k^\top \Omega_k \mathbf{J}_k$. The total cost is the sum of all edge costs

$$
\begin{aligned}
F(\Delta) &= \sum_k F_k(\Delta) \\
&\approx \sum_k \left( c_k + 2 b_k^\top \Delta + \Delta^\top \mathbf{H}_k \Delta \right) \\
&\approx \sum_k c_k + 2 \left( \sum_k b_k^\top \right) \Delta + \Delta^\top \left( \sum_k \mathbf{H}_k \right) \Delta \\
&\approx c + 2 b^\top \Delta + \Delta^\top \mathbf{H} \Delta
\end{aligned}
$$

where

$$b^\top = \sum_k f_{0,k}^\top \Omega_k \mathbf{J}_k, \quad \mathbf{H} = \sum_k \mathbf{J}_k^\top \Omega_k \mathbf{J}_k$$

are summations over the edges of the graph. Once they are computed, we proceed as previously, taking the derivative with respect to $\Delta$ and setting it to zero, solving for the update $\Delta$ and iterating using ◻ Fig. F.2.

## State Vector

The state vector is a concatenation of all poses and coordinates in the optimization problem. For pose-graph SLAM, it takes the form

$$x = \{\xi_0, \xi_1, \ldots, \xi_{N-1}\} \in \mathbb{R}^{N_x} .$$

Poses must be represented in a vector form and preferably one that is compact and singularity free. For **SE**(2), this is quite straightforward and we use $\xi \sim (x, y, \theta) \in \mathbb{R}^3$, and $N_\xi = 3$. For **SE**(3), we will use $\xi \sim (t, r) \in \mathbb{R}^6$ which comprises translation $t \in \mathbb{R}^3$ and rotation $r \in \mathbb{R}^3$, and $N_\xi = 6$. Rotation $r$ can be triple angles (Euler or roll-pitch-yaw), axis-angle (Euler vector), exponential coordinates or the vector part of a unit quaternion as discussed in ▶ Sect. 2.3.1.7. The state vector $x$ has length $N_x = N N_\xi$, and comprises a sequence of subvectors one per pose – the $i^{\text{th}}$ subvector is $x_i \in \mathbb{R}^{N_\xi}$.

For pose-graph SLAM with landmarks, or bundle adjustment, the state vector comprises poses and coordinate vectors

$$x = \{\xi_0, \xi_1, \ldots, \xi_{N-1} | P_0, P_1, \ldots, P_{M-1}\} \in \mathbb{R}^{N_x}$$

and we denote the $i^{\text{th}}$ and $j^{\text{th}}$ subvectors of $x$ as $x_i \in \mathbb{R}^{N_\xi}$ and $x_j \in \mathbb{R}^{N_P}$ that correspond to $\xi_i$ and $P_j$ respectively, where $N_P = 2$ for **SE**(2) and $N_P = 3$ for **SE**(3). The state vector $x$ length is $N_x = N N_\xi + M N_P$.

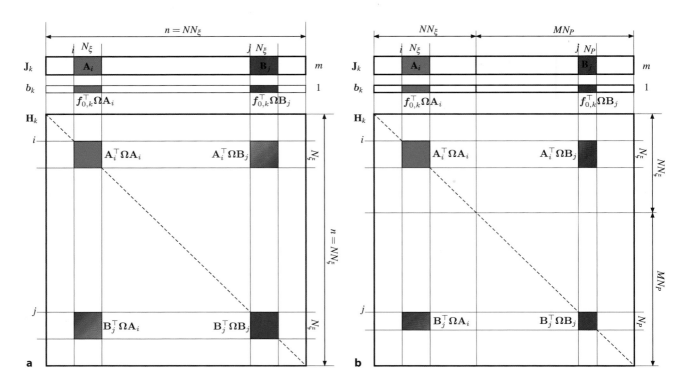

**Fig. F.2** Inherent structure of the error vector, Jacobian and Hessian matrices for graph-based least-squares problems. **a** Pose-graph SLAM with $N$ vertices representing robot pose as $\mathbb{R}^{N_\xi}$; **b** bundle adjustment with $N$ vertices representing camera pose as $\mathbb{R}^{N_\xi}$ and $M$ vertices representing landmark position as $\mathbb{R}^{N_P}$. The indices $i$ and $j$ denote the $i^{\text{th}}$ and $j^{\text{th}}$ block – not the $i^{\text{th}}$ and $j^{\text{th}}$ row or column. Zero values are indicated by white

## Inherent Structure

A key observation is that the error vector $\boldsymbol{f}_k(\boldsymbol{x})$ for edge $k$ depends only on the associated vertices $i$ and $j$, and this means that the Jacobian

$$\mathbf{J}_k = \frac{\partial \boldsymbol{f}_k(\boldsymbol{x})}{\partial \boldsymbol{x}} \in \mathbb{R}^{m \times N_x}$$

is mostly zeros

$$\mathbf{J}_k = \begin{pmatrix} \mathbf{0} & \cdots & \mathbf{A}_i & \cdots & \mathbf{B}_j & \cdots & \mathbf{0} \end{pmatrix}, \quad \mathbf{A}_i = \frac{\partial \boldsymbol{f}_k(\boldsymbol{x})}{\partial \boldsymbol{x}_i}, \quad \mathbf{B}_j = \frac{\partial \boldsymbol{f}_k(\boldsymbol{x})}{\partial \boldsymbol{x}_j}$$

where $\mathbf{A}_i \in \mathbb{R}^{m \times N_\xi}$ and $\mathbf{B}_j \in \mathbb{R}^{m \times N_\xi}$ or $\mathbf{B}_j \in \mathbb{R}^{m \times N_P}$ according to the state vector structure.

This sparse block structure means that the vector $\boldsymbol{b}_k$ and the Hessian $\mathbf{J}_k^\top \boldsymbol{\Omega}_k \mathbf{J}_k$ also have a sparse block structure as shown in ◩ Fig. F.2. The Hessian has just four small nonzero blocks, so rather than compute the product $\mathbf{J}_k^\top \boldsymbol{\Omega}_k \mathbf{J}_k$, which involves many multiplications by zero, we can just compute the four nonzero blocks and add them into the Hessian for the least-squares system. All blocks in a row have the same height, and in a column have the same width. For pose-graph SLAM with landmarks, or bundle adjustment, the blocks are of different sizes as shown in ◩ Fig. F.2b.

If the value of an edge represents pose, then (F.8) must be replaced with $\boldsymbol{f}_k(\boldsymbol{x}) = \hat{\boldsymbol{e}}_k(\boldsymbol{x}) \ominus \boldsymbol{e}_k^{\#}$. We generalize this with the $\boxminus$ operator to indicate the use of $-$ or $\ominus$, subtraction or relative pose, as appropriate. Similarly, when updating the state vector at the end of an iteration, the poses must be compounded $\boldsymbol{x}_0 \leftarrow \boldsymbol{x}_0 \oplus \boldsymbol{\Delta}$ and positions incremented by $\boldsymbol{x}_0 \leftarrow \boldsymbol{x}_0 + \boldsymbol{\Delta}$ which we generalize with the $\boxplus$ operator. The pose-graph optimization is solved by the iteration in ◩ Fig. F.3.

```
 1  repeat
 2  │   H ← 0, b ← 0
 3  │   foreach k do
 4  │   │   f_{0,k}(x_0) = ê_k(x) ⊟ e_k^#
 5  │   │   (i, j) = vertices(k)
 6  │   │   compute A_i(x_i), B_j(x_j)
 7  │   │   b_i ← b_i + f_{0,k}^⊤ Ω_k A_i
 8  │   │   b_j ← b_j + f_{0,k}^⊤ Ω_k B_j
 9  │   │   H_{i,i} ← H_{i,i} + A_i^⊤ Ω_k A_i
10  │   │   H_{i,j} ← H_{i,j} + A_i^⊤ Ω_k B_j
11  │   │   H_{j,i} ← H_{j,i} + B_j^⊤ Ω_k A_i
12  │   │   H_{j,j} ← H_{j,j} + B_j^⊤ Ω_k B_j
13  │   end
14  │   Δ = −H^{-1} b
15  │   x_0 ← x_0 ⊞ Δ
16  until ‖Δ‖ < ε
```

**▢ Fig. F.3** Pose graph optimization. For Levenberg-Marquardt optimization, replace line 14 with lines 4–12 from ▢ Fig. F.1

## Large Scale Problems

For pose-graph SLAM with thousands of poses or bundle adjustment with thousands of cameras and millions of landmarks, the Hessian matrix will be massive leading to computation and storage challenges. The overall Hessian is the summation of many edge Hessians structured as shown in ▢ Fig. F.2, and the total Hessian for two problems we have discussed are shown in ▢ Fig. F.4. They have clear structure which we can exploit.

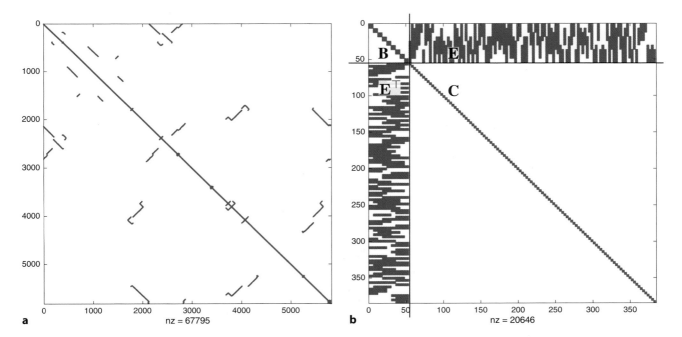

**▢ Fig. F.4** Hessian sparsity maps produced using the `plt.spy` function, the number of nonzero elements is shown beneath the plot. **a** Hessian for the pose-graph SLAM problem of ▢ Fig. 6.18, the diagonal elements represent pose constraints between successive vertices due to odometry, the off-diagonal terms represent constraints due to revisiting locations (loop closures); **b** Hessian for a bundle adjustment problem with 10 cameras and 110 landmarks (`examples/bademo`)

Firstly, in both cases the Hessian is sparse – that is, it contains mostly zeros. Instead of storing all those zeros (at 8 bytes each) we can employ sparse matrix techniques that maintain a list of the nonzero elements. SciPy provides classes in `scipy.sparse` to create sparse matrices which have similar behavior to nonsparse NumPy arrays, as well as `spsolve` to solve $\mathbf{A}x = b$ where the matrices and vectors are sparse.

Secondly, for the bundle adjustment case in ◼ Fig. F.2b, we see that the Hessian has a block structure, so we partition the system as

$$
\begin{pmatrix} \mathbf{B} & \mathbf{E} \\ \mathbf{E}^\top & \mathbf{C} \end{pmatrix} \begin{pmatrix} \mathbf{\Delta}_\xi \\ \mathbf{\Delta}_P \end{pmatrix} = \begin{pmatrix} b_\xi \\ b_P \end{pmatrix}
$$

> A block diagonal matrix is inverted by simply inverting each of the nonzero blocks along its diagonal.

where $\mathbf{B}$ and $\mathbf{C}$ are block diagonal. ◄ The subscripts $\xi$ and $P$ denote the blocks of $\mathbf{\Delta}$ and $b$ associated with camera poses and landmark positions respectively. We solve first for the camera pose updates $\mathbf{\Delta}_\xi$

$$
\mathbf{S}\mathbf{\Delta}_\xi = b_\xi - \mathbf{E}\mathbf{C}^{-1}b_P
$$

where $\mathbf{S} = \mathbf{B} - \mathbf{E}\mathbf{C}^{-1}\mathbf{E}^\top$ is the Schur complement of $\mathbf{C}$ and is a symmetric positive-definite matrix that is also block diagonal. Then we solve for the update to landmark positions

$$
\mathbf{\Delta}_P = \mathbf{C}^{-1}\left(b_P - \mathbf{E}^\top \mathbf{\Delta}_\xi\right) .
$$

More sophisticated techniques exploit the fine-scale block structure to further reduce computational time, for example GTSAM (► https://github.com/borglab/gtsam) and SLAM++ (► https://sourceforge.net/projects/slam-plus-plus).

### Anchoring

Optimization provides a solution where the *relative* poses and positions give the lowest overall cost, and the solution will have an arbitrary transformation with respect to a global reference frame. To obtain absolute poses and positions, we must anchor or fix some vertices – assign them values with respect to the global frame and prevent the optimization from adjusting them. The appropriate way to achieve this is to remove from $\mathbf{H}$ and $b$ the rows and columns corresponding to the anchored poses and positions. We then solve a lower dimensional problem for $\mathbf{\Delta}'$ which will be shorter than $x$. Careful bookkeeping is required to add the subvectors of $\mathbf{\Delta}'$ into $x$ for the update.

# G Gaussian Random Variables

The 1-dimensional Gaussian function

$$g(x) = \frac{1}{\sqrt{\sigma^2 2\pi}} e^{-\frac{1}{2\sigma^2}(x-\mu)^2} \tag{G.1}$$

is completely described by the position of its peak $\mu$ and its width $\sigma$. The total area under the curve is unity and $g(x) > 0$, $\forall x$. The function can be plotted using the RVC Toolbox function

```
>>> x = np.linspace(-6, 6, 500);
>>> plt.plot(x, gauss1d(0, 1, x), "r");
>>> plt.plot(x, gauss1d(0, 2**2, x), "b--");
```

and ◻ Fig. G.1 shows two Gaussians with zero mean and $\sigma = 1$ and $\sigma = 2$. Note that the second argument to `gauss1d` is $\sigma^2$.

If the Gaussian is considered to be a probability density function (PDF), then this is the well known normal distribution and the peak position $\mu$ is the mean value and the width $\sigma$ is the standard deviation. A random variable drawn from a normal distribution is often written as $X \sim N(\mu, \sigma^2)$, and $\sigma^2$ is the variance. $N(0, 1)$ is referred to as the standard normal distribution – the NumPy function `random.normal` draws random numbers from this distribution. To draw one hundred Gaussian random numbers with mean `mu` and standard deviation `sigma` is

```
>>> g = np.random.normal(loc=mu, scale=sigma, size=(100,));
```

The probability that a random value falls within an interval $x \in [x_1, x_2]$ is obtained by integration

$$P = \int_{x_1}^{x_2} g(x)\,dx = \Phi(x_2) - \Phi(x_1), \text{ where } \Phi(x) = \int_{-\infty}^{x} g(x)\,dx$$

or evaluation of the cumulative normal distribution function $\Phi(x)$ returned by `scipy.stats.norm.cdf`. The marked points in ◻ Fig. G.1 at $\mu \pm 1\sigma$ delimit the $1\sigma$ confidence interval. The area under the curve over this interval is 0.68, so the probability of a random value being drawn from this interval is 68%.

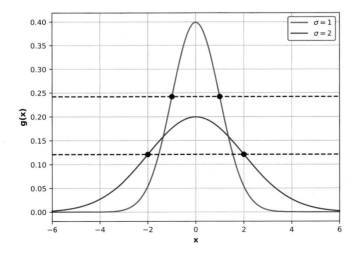

◻ **Fig. G.1** Two Gaussian functions, both with mean $\mu = 0$, and with standard deviation $\sigma = 1$, and $\sigma = 2$. The markers indicate the points $x = \mu \pm 1\sigma$. The area under both curves is unity

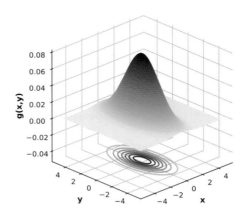

■ **Fig. G.2** The 2-dimensional Gaussian with covariance $\mathbf{P} = \mathrm{diag}(1^2, 2^2)$. Contour lines of constant probability density are shown beneath

The Gaussian can be extended to an arbitrary number of dimensions. The $n$-dimensional Gaussian, or multivariate normal distribution, is

$$g(\boldsymbol{x}) = \frac{1}{\sqrt{\det(\mathbf{P})(2\pi)^n}} e^{-\frac{1}{2}(\boldsymbol{x}-\boldsymbol{\mu})^\top \mathbf{P}^{-1}(\boldsymbol{x}-\boldsymbol{\mu})} \tag{G.2}$$

and compared to the scalar case of (G.1) $\boldsymbol{x} \in \mathbb{R}^n$ and $\boldsymbol{\mu} \in \mathbb{R}^n$ have become vectors, the squared term in the exponent has been replaced by a matrix quadratic form, and $\sigma^2$, the variance, has become a symmetric positive-definite matrix $\mathbf{P} \in \mathbb{R}^{n \times n}$. We can plot a 2-dimensional Gaussian

```
>>> x, y = np.meshgrid(np.linspace(-5, 5, 100),
...         np.linspace(-5, 5, 100));
>>> P = np.diag([1, 2])**2;
>>> g = gauss2d([0, 0], P, x, y);
>>> ax = plotvol3();
>>> ax.plot_surface(x, y, g);
>>> ax.contour(x, y, g, zdir="z", offset=-0.05);
```

as a surface which is shown in ■ Fig. G.2. The contour lines are ellipses and in this example, the radii in the $y$- and $x$-directions are in the ratio 2 : 1 as defined by the ratio of the standard deviations. If the covariance matrix is diagonal, as in this case, then the ellipses are aligned with the $x$- and $y$-axes as we saw in ▶ App. C.1.4, otherwise they are rotated.

If the 2-dimensional Gaussian is considered to be a PDF, then $\mathbf{P}$ is the covariance matrix where the diagonal elements $p_{i,i}$ represent the variance of $x_i$ and the off-diagonal elements $p_{i,j}$ are the correlationss between $x_i$ and $x_j$. The inverse of the covariance matrix is known as the *information matrix*. If the variables are independent or uncorrelated, the matrix $\mathbf{P}$ would be diagonal. In this case, $\boldsymbol{\mu} = (0, 0)$ and $\mathbf{P} = \mathrm{diag}(1^2, 2^2)$ which corresponds to uncorrelated variables with standard deviation of 1 and 2 respectively. If the PDF represented the position of a vehicle, it is most likely to be at coordinates $(0, 0)$. The probability of being within any region of the $xy$-plane is the volume under the surface of that region, and involves a non-trivial integration to determine.

The connection between Gaussian probability density functions and ellipses can be found in the quadratic exponent of (G.2) which is the equation of an ellipse or ellipsoid. ◄ All the points that satisfy

It is also the definition of Mahalanobis distance, the covariance weighted distance between $\boldsymbol{x}$ and $\boldsymbol{\mu}$.

$$(\boldsymbol{x} - \boldsymbol{\mu})^\top \mathbf{P}^{-1}(\boldsymbol{x} - \boldsymbol{\mu}) = s$$

result in a constant probability density value, that is, a contour of the 2-dimensional Gaussian. If $p$ is the probability that the point $\boldsymbol{x}$ lies inside the ellipse, then $s$ is given by the inverse-cumulative $\chi^2$ distribution function ▶ with $n$ degrees of freedom, 2 in this case. For example, the 50% confidence interval is

```
>>> from scipy.stats.distributions import chi2
>>> chi2.ppf(0.5, 2)
1.386
```

where the first argument is the probability and the second is the number of degrees of freedom.

To draw a 2-dimensional covariance ellipse, we use the general approach for ellipses outlined in ▶ App. C.1.4, but the right-hand side of the ellipse equation is $s$ not 1, and $\mathbf{E} = \mathbf{P}^{-1}$. Using the Toolbox this is `plot_ellipse(P, confidence=0.5, inverted=True)`.

If we draw a vector of length $n$ from the multivariate Gaussian, each element is normally distributed. The sum of squares of independent normally distributed values has a $\chi^2$ (chi-squared) distribution with $n$ degrees of freedom.

# H Kalman Filter

> **»** *All models are wrong. Some models are useful.*
> – George Box

Consider the system shown in ■ Fig. H.1. The physical robot is a "black box" which has a true state or pose $x$ that evolves over time according to the applied inputs. We cannot directly measure the state, but sensors on the robot have outputs which are a function of that true state. Our challenge is: given the system inputs and sensor outputs, *estimate* the unknown true state $x$ and how certain we are of that estimate.

At face value, this might seem hard or even impossible, but we will assume that the system has the model shown inside the black box, and we know how this system behaves. Firstly, we know how the state evolves over time as a function of the inputs – this is the state transition ◄ model $f(\cdot)$, and we know the inputs $u$ that are applied to the system. Our model is unlikely to be perfect ◄ and it is common to represent this uncertainty by an imaginary random number generator which is corrupting the system state – process noise. Secondly, we know how the sensor output depends on the state – this is the sensor model $h(\cdot)$ and its uncertainty is also modeled by an imaginary random number generator – sensor noise.

The imaginary random number sources $v$ and $w$ are *inside* the black box so the random numbers are also unknowable. However, we can describe the characteristics of these random numbers – their *distribution* which tells us how likely it is that we will *draw* a random number with a particular value. A lot of noise in physical systems can be modeled well by the Gaussian (aka normal) distribution $N(\mu, \sigma^2)$ which is characterized by a mean $\mu$ and a standard deviation $\sigma$, and the Gaussian distribution has some nice mathematical properties that we will rely on. ◄ Typically, we assume that the noise has zero mean, and that the covariance of the process and sensor noise are $\mathbf{V}$ and $\mathbf{W}$ respectively.

Often called the process or motion model.

For example, wheel slippage on a mobile ground robot or wind gusts for a UAV.

There are infinitely many possible distributions that these noise sources could have, which could be asymmetrical or have multiple peaks. We should never assume that noise is Gaussian – we should attempt to determine the distribution by understanding the physics of the process and the sensor, or from careful measurement and analysis.

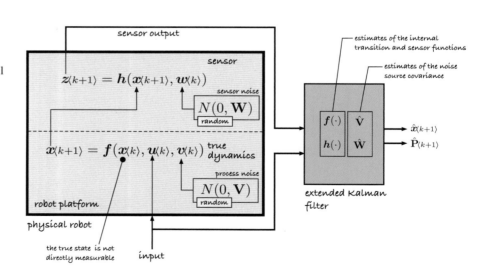

■ **Fig. H.1**  The physical robot on the left is a "black box" whose true state cannot be directly measured. However, it can be inferred from sensor output which is a function of the unknown true state, and the assumed model shown inside the black box

In general terms, the problem we wish to solve is:

> Given a model of the system $f(\cdot)$, $h(\cdot)$, estimates of $\mathbf{V}$ and $\mathbf{W}$; the known inputs applied to the system $u$; and some noisy sensor measurements $z$, find an estimate $\hat{x}$ of the system state and the uncertainty $\hat{\mathbf{P}}$ in that estimate.

In a robotic localization context, $x$ is the unknown position or pose of the robot, $u$ is the commands sent to the motors, and $z$ is the output of various sensors on the robot. For a ground robot, $x$ would be the pose in $\mathbf{SE}(2)$, $u$ would be the motor commands, and $z$ might be the measured odometry or range and bearing to landmarks. For an aerial robot, $x$ would be the pose in $\mathbf{SE}(3)$, $u$ are the known forces applied to the airframe by the propellers, and $z$ might be the measured accelerations and angular velocities. ▶

The state is a vector and there are many approaches to mapping pose to a vector, especially the rotational component – Euler angles, quaternions, and exponential coordinates are commonly used, see ▶ App. F.2.4.

## H.1 Linear Systems – Kalman Filter

Consider the transition model described as a discrete-time linear time-invariant system

$$x_{\langle k+1 \rangle} = \mathbf{F}x_{\langle k \rangle} + \mathbf{G}u_{\langle k \rangle} + v_{\langle k \rangle} \tag{H.1}$$

$$z_{\langle k \rangle} = \mathbf{H}x_{\langle k \rangle} + w_{\langle k \rangle} \tag{H.2}$$

where $k$ is the time step, $x \in \mathbb{R}^n$ is the state vector, and $u \in \mathbb{R}^m$ is a vector of inputs to the system at time $k$, for example, a velocity command, or applied forces and torques. The matrix $\mathbf{F} \in \mathbb{R}^{n \times n}$ describes the dynamics of the system, that is, how the states evolve with time. The matrix $\mathbf{G} \in \mathbb{R}^{n \times m}$ describes how the inputs are coupled to the system states. The vector $z \in \mathbb{R}^p$ represents the outputs of the system as measured by sensors. The matrix $\mathbf{H} \in \mathbb{R}^{p \times n}$ describes how the system states are mapped to the system outputs which we can observe.

To account for errors in the motion model ($\mathbf{F}$ and $\mathbf{G}$) or unmodeled disturbances, we introduce a Gaussian random variable $v \in \mathbb{R}^n$ termed the process noise. $v_{\langle k \rangle} \sim N(0, \mathbf{V})$, that is, it has zero mean and covariance $\mathbf{V} \in \mathbb{R}^{n \times n}$. Covariance is a matrix quantity which is the variance for a multi-dimensional distribution – it is a symmetric positive-definite matrix. The sensor measurement model $\mathbf{H}$ is not perfect either, and this is modeled by sensor measurement noise, a Gaussian random variable $w \in \mathbb{R}^p$, $w_{\langle k \rangle} \sim N(0, \mathbf{W})$ and covariance $\mathbf{W} \in \mathbb{R}^{p \times p}$.

The Kalman filter is an optimal estimator for the case where the process and measurement noise are both zero-mean Gaussian noise. The filter has two steps: prediction and update. The prediction is based on the previous state and the inputs that were applied

$$\hat{x}^+_{\langle k+1 \rangle} = \mathbf{F}\hat{x}_{\langle k \rangle} + \mathbf{G}u_{\langle k \rangle} \tag{H.3}$$

$$\hat{\mathbf{P}}^+_{\langle k+1 \rangle} = \underline{\mathbf{F}\hat{\mathbf{P}}_{\langle k \rangle}\mathbf{F}^\top} + \hat{\mathbf{V}} \tag{H.4}$$

where $\hat{x}^+$ is the predicted estimate of the state and $\hat{\mathbf{P}} \in \mathbb{R}^{n \times n}$ is the predicted estimate of the covariance, or uncertainty, in $\hat{x}^+$. The notation $^+$ makes explicit that the left-hand side is a prediction at time $k + 1$ based on information from time $k$. $\hat{\mathbf{V}}$ is a constant and our best estimate of the covariance of the process noise.

The underlined term in (H.4) *projects* the estimated covariance from the current time step to the next. Consider a one dimensional example where $F$ is a scalar and the state estimate $\hat{x}_{\langle k \rangle}$ has a PDF which is Gaussian with a mean $\mu_{\langle k \rangle}$ and a variance $\sigma^2 \langle k \rangle$. The prediction equation maps the state and its Gaussian distribution to a new Gaussian distribution with a mean $F\mu\langle k \rangle$ and a variance $F^2\sigma^2\langle k \rangle$. The term $\mathbf{F}\langle k \rangle \mathbf{P}\langle k \rangle \mathbf{F}\langle k \rangle^\top$ is the matrix form of this since

$$\text{cov}(\mathbf{F}\boldsymbol{x}) = \mathbf{F}\,\text{cov}(\boldsymbol{x})\mathbf{F}^\top \tag{H.5}$$

which scales the covariance appropriately.

The prediction of $\hat{\mathbf{P}}$ in (H.4) involves the addition of two positive-definite matrices so the uncertainty will increase – this is to be expected since we have used an uncertain model to predict the future value of an already uncertain estimate. $\hat{\mathbf{V}}$ must be a reasonable estimate of the covariance of the actual process noise. If we overestimate it, that is our estimate of process noise is larger than it really is, then we will have a larger than necessary increase in uncertainty at this step, leading to a pessimistic estimate of our certainty.

To counter this growth in uncertainty, we need to introduce new information such as measurements made by the sensors since they depend on the state. The difference between what the sensors measure and what the sensors are predicted to measure is

$$\boldsymbol{v} = \boldsymbol{z}^{\#}\langle k+1 \rangle - \mathbf{H}\hat{\boldsymbol{x}}^{+}\langle k+1 \rangle \in \mathbb{R}^p \ .$$

Some of this difference is due to noise in the sensor, the measurement noise, but the remainder provides valuable information related to the error between the actual and the predicted value of the state. Rather than considering this as error, we refer to it more positively as *innovation* – new information.

The second step of the Kalman filter, the *update* step, maps the innovation into a correction for the predicted state, optimally tweaking the estimate based on what the sensors observed

$$\hat{\boldsymbol{x}}\langle k+1 \rangle = \hat{\boldsymbol{x}}^{+}\langle k+1 \rangle + \mathbf{K}\boldsymbol{v}, \tag{H.6}$$

$$\hat{\mathbf{P}}\langle k+1 \rangle = \hat{\mathbf{P}}^{+}\langle k+1 \rangle - \mathbf{K}\mathbf{H}\hat{\mathbf{P}}^{+}\langle k+1 \rangle . \tag{H.7}$$

Uncertainty is now *decreased* or *deflated*, since new information, from the sensors, is being incorporated. The matrix

$$\mathbf{K} = \mathbf{P}^{+}\langle k+1 \rangle \mathbf{H}^\top \left( \underline{\mathbf{H}\mathbf{P}^{+}\langle k+1 \rangle \mathbf{H}^\top} + \hat{\mathbf{W}} \right)^{-1} \in \mathbb{R}^{n \times p} \tag{H.8}$$

is known as the Kalman gain. The term in parentheses is the estimated covariance of the innovation, and comprises the uncertainty in the state and the estimated measurement noise covariance. If the innovation has high uncertainty in some dimensions, then the Kalman gain will be correspondingly small, that is, if the new information is uncertain, then only small changes are made to the state vector. The underlined term *projects* the covariance of the state estimate into the space of sensor values. $\hat{\mathbf{W}}$ is a constant and our best estimate of the covariance of the sensor noise.

The covariance matrix must be symmetric, but after many updates the accumulated numerical errors may result in the matrix no longer being symmetric. The symmetric structure can be enforced by using the Joseph form of (H.7)

$$\hat{\mathbf{P}}\langle k+1 \rangle = (\mathbf{1}_{n \times n} - \mathbf{K}\mathbf{H})\hat{\mathbf{P}}^{+}\langle k+1 \rangle (\mathbf{1}_{n \times n} - \mathbf{K}\mathbf{H})^\top + \mathbf{K}\hat{\mathbf{V}}\mathbf{K}^\top$$

but this is computationally more costly.

The equations above constitute the classical Kalman filter which is widely used in robotics, aerospace and econometric applications. The filter has a number of important characteristics. Firstly, it is optimal, but only if the noise is truly Gaussian with zero mean and time-invariant parameters. This is often a good assumption but not always. Secondly, it is recursive, the output of one iteration is the input to the next. Thirdly, it is asynchronous. At a particular iteration, if no sensor information is available, we only perform the prediction step and not the update. In the case that there are different sensors, each with their own $\mathbf{H}$, and different sample rates, we just apply the update with the appropriate $z$ and $\mathbf{H}$ whenever sensor data becomes available.

The filter must be initialized with some reasonable value of $\hat{x}$ and $\hat{\mathbf{P}}$, as well as good choices of the estimated covariance matrices $\hat{\mathbf{V}}$ and $\hat{\mathbf{W}}$. As the filter runs, the estimated covariance $\|\hat{\mathbf{P}}\|$ decreases but never reaches zero – the minimum value can be shown to be a function of $\hat{\mathbf{V}}$ and $\hat{\mathbf{W}}$. The Kalman-Bucy filter is a continuous-time version of this filter.

The covariance matrix $\hat{\mathbf{P}}$ is rich in information. The diagonal elements $\hat{p}_{i,i}$ are the variance, or uncertainty, in the state $x_i$. The off-diagonal elements $\hat{p}_{i,j}$ are the correlations between states $x_i$ and $x_j$ and indicate that the errors in the states are not independent. The correlations are critical in allowing any piece of new information to *flow through* to adjust all the states that affect a particular process output.

If elements of the state vector represent 2-dimensional position, the corresponding $2 \times 2$ submatrix of $\hat{\mathbf{P}}$ is the inverse of the equivalent uncertainty ellipse $\mathbf{E}^{-1}$, see ▶ App. C.1.4.

## H.2 Nonlinear Systems – Extended Kalman Filter

For the case where the system is not linear, it can be described generally by two functions: the state transition (the motion model in robotics) and the sensor model

$$x_{\langle k+1\rangle} = f(x_{\langle k\rangle}, \ u_{\langle k\rangle}, \ v_{\langle k\rangle}) \tag{H.9}$$

$$z_{\langle k\rangle} = h(x_{\langle k\rangle}, \ w_{\langle k\rangle}) \tag{H.10}$$

and as before we represent model uncertainty, external disturbances and sensor noise by Gaussian random variables $v$ and $w$.

We linearize the state transition function about the current state estimate $\hat{x}_{\langle k\rangle}$ as shown in ◼ Fig. H.2 resulting in

$$x'_{\langle k+1\rangle} \approx \mathbf{F}_x \, x'_{\langle k\rangle} + \mathbf{F}_u \, u_{\langle k\rangle} + \mathbf{F}_v \, v_{\langle k\rangle} \tag{H.11}$$

$$z'_{\langle k\rangle} \approx \mathbf{H}_x \, x'_{\langle k\rangle} + \mathbf{H}_w \, w_{\langle k\rangle} \tag{H.12}$$

where $\mathbf{F}_x = \partial f / \partial x \in \mathbb{R}^{n \times n}$, $\mathbf{F}_u = \partial f / \partial u \in \mathbb{R}^{n \times m}$, $\mathbf{F}_v = \partial f / \partial v \in \mathbb{R}^{n \times n}$, $\mathbf{H}_x = \partial h / \partial x \in \mathbb{R}^{p \times n}$ and $\mathbf{H}_w = \partial h / \partial w \in \mathbb{R}^{p \times p}$ are Jacobians, see ▶ App. E, of the functions $f(\cdot)$ and $h(\cdot)$. Equating coefficients between (H.1) and (H.11) gives $\mathbf{F} \sim \mathbf{F}_x$, $\mathbf{G} \sim \mathbf{F}_u$ and $v_{\langle k\rangle} \sim \mathbf{F}_v v_{\langle k\rangle}$; and between (H.2) and (H.12) gives $\mathbf{H} \sim \mathbf{H}_x$ and $w_{\langle k\rangle} \sim \mathbf{H}_w w_{\langle k\rangle}$.

Taking the prediction equation (H.9) with $v_{\langle k\rangle} = 0$, and the covariance equation (H.4) with the linearized terms substituted, we can write the prediction step as

$$\hat{x}^+_{\langle k+1\rangle} = f(\hat{x}_{\langle k\rangle}, \ u_{\langle k\rangle})$$

$$\hat{\mathbf{P}}^+_{\langle k+1\rangle} = \mathbf{F}_x \hat{\mathbf{P}}_{\langle k\rangle} \mathbf{F}_x^\top + \mathbf{F}_v \hat{\mathbf{V}} \mathbf{F}_v^\top$$

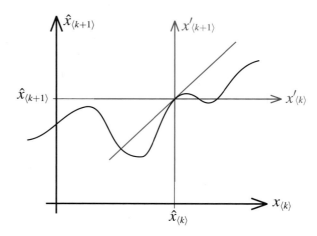

**◻ Fig. H.2** One dimensional example illustrating how the nonlinear state transition function $f : x_k \mapsto x_{k+1}$ shown in black is linearized about the point $(\hat{x}_{\langle k \rangle}, \hat{x}_{\langle k+1 \rangle})$ shown in red

and the update step as

$$\hat{\boldsymbol{x}}_{\langle k+1 \rangle} = \hat{\boldsymbol{x}}^{+}_{\langle k+1 \rangle} + \mathbf{K} \boldsymbol{\nu}$$

$$\hat{\mathbf{P}}_{\langle k+1 \rangle} = \hat{\mathbf{P}}^{+}_{\langle k+1 \rangle} - \mathbf{K} \mathbf{H}_x \hat{\mathbf{P}}^{+}_{\langle k+1 \rangle}$$

where the Kalman gain is now

$$\mathbf{K} = \mathbf{P}^{+}_{\langle k+1 \rangle} \mathbf{H}_x^{\top} \left( \mathbf{H}_x \mathbf{P}^{+}_{\langle k+1 \rangle} \mathbf{H}_x^{\top} + \mathbf{H}_w \hat{\mathbf{W}} \mathbf{H}_w^{\top} \right)^{-1} . \tag{H.13}$$

Properly, these matrices should be denoted as depending on the time step, i.e. $\mathbf{F}_{x \langle k \rangle}$, but this has been dropped in the interest of readability.

These equations are only valid at the linearization point $\hat{\boldsymbol{x}}_{\langle k \rangle}$ – the Jacobians $\mathbf{F}_x$, $\mathbf{F}_v$, $\mathbf{H}_x$, $\mathbf{H}_w$ must be computed at every iteration. ◄ The full procedure is summarized in ◻ Fig. H.3.

A fundamental problem with the extended Kalman filter is that PDFs of the random variables are no longer Gaussian after being operated on by the nonlinear functions $\boldsymbol{f}(\cdot)$ and $\boldsymbol{h}(\cdot)$. We can illustrate this by considering a nonlinear scalar function $y = (x + 2)^2 / 4$. We will draw a million Gaussian random numbers from the normal distribution $N(5, 4)$ which has a mean of 5 and a standard deviation of 2

```
>>> x = np.random.normal(loc=5, scale=2, size=(1_000_000,));
```

then apply the function

```
>>> y = (x + 2)**2 / 4;
```

and plot the probability density function of $y$

```
>>> plt.hist(y, bins=200, density=True, histtype="step");
```

which is shown in ◻ Fig. H.4. We see that the PDF of $y$ is substantially changed and no longer Gaussian. It has lost its symmetry, so the mean value is greater than the mode. The Jacobians that appear in the EKF equations appropriately scale the covariance but the resulting non-Gaussian distribution breaks the assumptions which guarantee that the Kalman filter is an optimal estimator. Alternative approaches to dealing with system nonlinearity include the iterated EKF described

**Input:** $\hat{\boldsymbol{x}}_{\langle k \rangle} \in \mathbb{R}^n$, $\hat{\mathbf{P}}_{\langle k \rangle} \in \mathbb{R}^{n \times n}$, $\boldsymbol{u}_{\langle k \rangle} \in \mathbb{R}^m$, $\boldsymbol{z}_{\langle k+1 \rangle} \in \mathbb{R}^p$; $\hat{\mathbf{V}} \in \mathbb{R}^{n \times n}$,
     $\hat{\mathbf{W}} \in \mathbb{R}^{p \times p}$

**Output:** $\hat{\boldsymbol{x}}_{\langle k+1 \rangle} \in \mathbb{R}^n$, $\hat{\mathbf{P}}_{\langle k+1 \rangle} \in \mathbb{R}^{n \times n}$

*– linearize about* $\boldsymbol{x} = \hat{\boldsymbol{x}}_{\langle k \rangle}$

  compute Jacobians: $\mathbf{F}_x \in \mathbb{R}^{n \times n}$, $\mathbf{F}_v \in \mathbb{R}^{n \times n}$, $\mathbf{H}_x \in \mathbb{R}^{p \times n}$, $\mathbf{H}_w \in \mathbb{R}^{p \times p}$

*– the prediction step*

$\hat{\boldsymbol{x}}^+_{\langle k+1 \rangle} = \boldsymbol{f}(\hat{\boldsymbol{x}}_{\langle k \rangle}, \boldsymbol{u}_{\langle k \rangle})$ // *predict state at next time step*

$\hat{\mathbf{P}}^+_{\langle k+1 \rangle} = \mathbf{F}_x \hat{\mathbf{P}}_{\langle k \rangle} \mathbf{F}_x^\top + \mathbf{F}_v \hat{\mathbf{V}} \mathbf{F}_v^\top$ // *predict covariance at next time step*

*– the update step*

$\boldsymbol{\nu} = \boldsymbol{z}_{\langle k+1 \rangle} - \mathbf{h}(\hat{\boldsymbol{x}}^+_{\langle k+1 \rangle})$    // *innovation: measured - predicted sensor value*

$\mathbf{K} = \mathbf{P}^+_{\langle k+1 \rangle} \mathbf{H}_x^\top (\mathbf{H}_x \, \mathbf{P}^+_{\langle k+1 \rangle} \mathbf{H}_x^\top + \mathbf{H}_w \hat{\mathbf{W}} \mathbf{H}_w^\top)^{-1}$ // *Kalman gain*

$\hat{\boldsymbol{x}}_{\langle k+1 \rangle} = \hat{\boldsymbol{x}}^+_{\langle k+1 \rangle} + \mathbf{K}\boldsymbol{\nu}$ // *update state estimate*

$\hat{\mathbf{P}}_{\langle k+1 \rangle} = \hat{\mathbf{P}}^+_{\langle k+1 \rangle} - \mathbf{K}\mathbf{H}_x \hat{\mathbf{P}}^+_{\langle k+1 \rangle}$ // *update covariance estimate*

**◘ Fig. H.3**  Procedure EKF

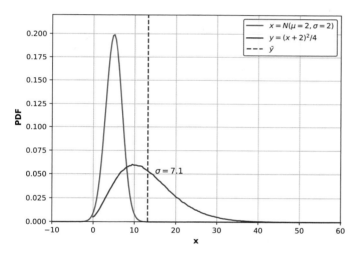

**◘ Fig. H.4**  PDF of the state $x$ which is Gaussian $N(5, 4)$ and the PDF of the nonlinear function $y = (x + 2)^2/4$. The distribution of $y$ is no longer Gaussian and the mean no longer corresponds to the peak

by Jazwinski (2007), the Unscented Kalman Filter (UKF) (Julier and Uhlmann 2004), or the sigma-point filter which uses discrete sample points (sigma points) to approximate the PDF.

# I Graphs

A graph is an abstract representation of a set of objects connected by links and depicted visually as shown in ◘ Fig. I.1. Mathematically, a graph is denoted by $G(V, E)$ where $V$ are the vertices or nodes, and $E$ are the links that connect pairs of vertices and are called edges or arcs. Edges can be directed (shown as arrows) or undirected (shown as line segments) as in this case. Edges can have an associated weight or cost associated with moving from one vertex to another. A sequence of edges from one vertex to another is a path, and a sequence that starts and ends at the same vertex is a cycle. An edge from a vertex to itself is a loop. Graphs can be used to represent transport, communications or social networks, and this branch of mathematics is graph theory.

We will illustrate graphs using the Python package `pgraph` that supports embedded graphs where the vertices are associated with a point in an $n$-dimensional space. To create a new graph

```
>>> import pgraph
>>> g = pgraph.UGraph()
```

and, by default, the vertices of the graph exist in 2-dimensional space. We can add five randomly-placed vertices to the graph

```
>>> np.random.seed(0)  # ensure repeatable results
>>> for i in range(5):
...    g.add_vertex(np.random.rand(2));
```

The method `add_vertex` returns a reference to the newly created vertex object, but we can obtain the reference by indexing the graph object. For example vertex 1 is

```
>>> g[1]
UVertex[#1, coord=(0.6028, 0.5449)]
```

and is an instance of `UVertex`, a vertex in an undirected graph; it has the coordinate $(0.6028, 0.5449)$; and has been assigned a default name `"#1"` but we could specify a different name. ◄ We can also reference a vertex by its name

*Vertex names must be unique within the graph.*

```
>>> g["#1"]
UVertex[#1, coord=(0.6028, 0.5449)]
```

We can create edges between pairs of vertex objects

```
>>> g.add_edge(g[0], g[1]);
>>> g.add_edge(g[0], g[2]);
```

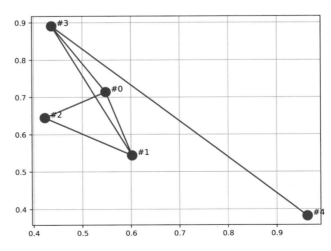

◘ **Fig. I.1**    An example graph generated by the `UGraph` class

```
>>> g.add_edge(g[0], g[3]);
>>> g.add_edge(g[1], g[2]);
>>> g.add_edge(g[1], g[3]);
>>> g.add_edge(g[3], g[4]);
```

and the `add_edge` method returns a reference to the edge object just created. By default, the edge cost is the Euclidean distance between the vertices, but this can be overridden by a third argument to `add_edge`.

A summary of the graph can be displayed

```
>>> print(g)
UGraph: 5 vertices, 6 edges, 1 component
```

This graph has one component, that is, all the vertices are connected into one network. The graph can be plotted by

```
>>> g.plot()
```

as shown in ◫ Fig. I.1. The vertices are shown as blue circles, and the edges are shown as lines joining vertices. Many options exist to change default plotting behavior. Note that only graphs embedded in 2- and 3-dimensional space can be plotted.

The neighbors of vertex 1 are

```
>>> g[1].adjacent()
[UVertex[#0, coord=(0.5488, 0.7152)],
 UVertex[#2, coord=(0.4237, 0.6459)],
 UVertex[#3, coord=(0.4376, 0.8918)]]
```

which is a list of neighboring vertex objects, those connected to it by an edge. Each edge is an object and the edges connecting to vertex 1 are

```
>>> g[1].edges()
[Edge{[#0] -- [#1], cost=0.1786},
 Edge{[#1] -- [#2], cost=0.2056},
 Edge{[#1] -- [#3], cost=0.3842}]
```

and each edge has references to the vertices it connects

```
>>> g[1].edges()[0].endpoints
[UVertex[#0, coord=(0.5488, 0.7152)],
 UVertex[#1, coord=(0.6028, 0.5449)]]
```

The cost or length of an edge is

```
>>> g[1].edges()[0].cost
0.1786
```

Arbitrary data can be attached to any vertex or edge by adding attributes to the vertex or edge objects, or by subclassing the `UVertex` and `Edge` classes.

The vertex closest to the coordinate $(0.5, 0.5)$ is

```
>>> g.closest((0.5, 0.5))
(UVertex[#1, coord=(0.6028, 0.5449)], 0.1121)
```

The minimum-distance path between any two vertices in the graph can be computed using well-known algorithms such as A* (Hart et al. 1968)

```
>>> path, length, _ = g.path_Astar(g[2], g[4])
>>> path
[UVertex[#2, coord=(0.4237, 0.6459)],
 UVertex[#0, coord=(0.5488, 0.7152)],
 UVertex[#3, coord=(0.4376, 0.8918)],
 UVertex[#4, coord=(0.9637, 0.3834)]]
```

which is a list of vertices representing a total path length of

```
>>> length
1.083
```

Methods exist to compute various other representations of the graph such as adjacency, incidence, degree and Laplacian matrices.

# J Peak Finding

A commonly encountered problem is finding, or refining an estimate of, the position of the peak of some discrete 1- or 2-dimensional signal.

## J.1    1D Signal

Consider the discrete 1-dimensional signal $y(k)$, $k \in \mathbb{N}_0$

```
>>> y = mvtb_load_matfile("data/peakfit.mat")["y"];
>>> plt.plot(y, "-o");
```

shown in ◘ Fig. J.1a.

Finding the position of the peak to the nearest integer $k$ is straightforward using the NumPy function `argmax`

```
>>> k = np.argmax(y)
7
```

which indicates the peak occurs at $k = 7$, and the peak value is

```
>>> y[k]
0.9905
```

In this case, there is more than one peak and we can use the RVC Toolbox function `findpeaks` instead

```
>>> k, ypk = findpeaks(y)
>>> k
array([ 7,   0, 24, 15])
>>> ypk
array([  0.9905,    0.873,    0.6718,   -0.5799])
```

which has returned four maxima ordered by descending value. A common test of the quality of a peak is its magnitude, and the ratio of the height of the second peak to the first peak

```
>>> ypk[1] / ypk[0]
0.8813
```

which is called the ambiguity ratio and is ideally small.

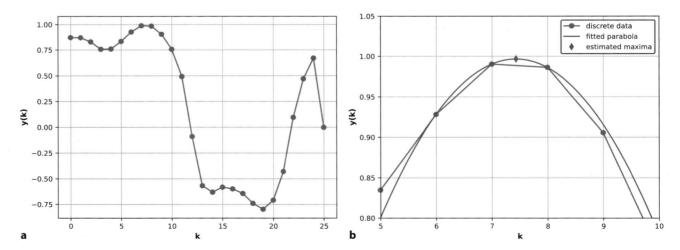

**a**                                    **b**

◘ **Fig. J.1**    Peak fitting. **a** A signal with several local maxima; **b** close-up view of the first maximum with the fitted parabola

The signal $y(k)$ in ◻ Fig. J.1a is a sampled representation of a continuous underlying signal $y(x)$ and the real peak might actually lie between the samples. If we look at a zoomed version of the signal, ◻ Fig. J.1b, we can see that although the maximum is at $k = 7$, the value at $k = 8$ is only slightly lower, so the peak lies somewhere between these two points. A common approach is to fit a parabola

$$y = a\delta_x^2 + b\delta_x + c, \; \delta_x \in \mathbb{R} \tag{J.1}$$

to the points surrounding the peak: $(-1, y_{k-1})$, $(0, y_k)$ and $(1, y_{k+1})$ where $\delta_x = 0$ when $k = 7$. Substituting these into (J.1), we can write three equations

$$y_{k-1} = a - b + c$$
$$y_k = c$$
$$y_{k+1} = a + b + c$$

or in compact matrix form as

$$\begin{pmatrix} y_{k-1} \\ y_k \\ y_{k+1} \end{pmatrix} = \begin{pmatrix} 1 & -1 & 1 \\ 0 & 0 & 1 \\ 1 & 1 & 1 \end{pmatrix} \begin{pmatrix} a \\ b \\ c \end{pmatrix}$$

and then solve for the parabolic coefficients

$$\begin{pmatrix} a \\ b \\ c \end{pmatrix} = \begin{pmatrix} 1 & -1 & 1 \\ 0 & 0 & 1 \\ 1 & 1 & 1 \end{pmatrix}^{-1} \begin{pmatrix} y_{k-1} \\ y_k \\ y_{k+1} \end{pmatrix} = \frac{1}{2} \begin{pmatrix} 1 & -2 & 1 \\ -1 & 0 & 1 \\ 0 & 2 & 0 \end{pmatrix} \begin{pmatrix} y_{k-1} \\ y_k \\ y_{k+1} \end{pmatrix}. \tag{J.2}$$

The maximum of the parabola occurs when its derivative is zero

$$2a\delta_x + b = 0$$

and substituting the values of $a$ and $b$ from (J.2), we find the displacement of the peak of the fitted parabola with respect to the discrete maximum

$$\delta_x = \frac{1}{2} \frac{y_{k-1} - y_{k+1}}{y_{k-1} - 2y_k + y_{k+1}}, \quad \delta_x \in (-1, 1)$$

so the refined, or interpolated, position of the maximum is at

$$\hat{x} = k + \delta_x \in \mathbb{R}$$

and the estimated value of the maximum is obtained by substituting $\delta_x$ into (J.1).

The coefficient $a$ will be negative for a maximum. If $|a|$ is large, it indicates a well-defined sharp peak, whereas a low value indicates a very broad peak for which estimation of a refined peak may not be so accurate.

Continuing the earlier example, we can use the function `findpeaks` to estimate the refined peak positions

```
>>> findpeaks(y, interp=True)
(array([   7.439,    23.73,    15.24]), array([  0.9969,    0.7037,
 -0.578]))
```

where the argument `interp` uses a second-order polynomial by default, but this can be overridden by setting `interp` to the desired polynomial order. The fitted parabola for the first peak is shown in red in ◻ Fig. J.1b.

If the signal has superimposed noise, then there are likely to be multiple peaks, many of which are quite minor, and this can be overcome by specifying the *scale* of the peak. For example, the peaks that are greater than all other values within $\pm 5$ values in the horizontal direction are

```
>>> findpeaks(y, scale=5)
(array([ 7,  0, 24]), array([  0.9905,    0.873,    0.6718]))
```

This technique is called nonlocal maxima suppression and requires that an appropriate scale is chosen. In this case, the result is unchanged since the signal is fairly smooth.

## J.2  2D Signal

An example 2-dimensional discrete signal can be loaded by

```
>>> img = mvtb_load_matfile("data/peakfit.mat")["image"]
array([[ -0.0696,  0.03483,   0.1394,   0.2436,    0.348],
       [    0.08,   0.3754,   0.3202,     0.44,     0.56],
       [ 0.04003,   0.1717,   0.3662,   0.4117,     0.52],
       [0.000214,   0.2062,   0.8766,   0.4462,   0.4802],
       [-0.03997,  0.09166,   0.2862,   0.3317,     0.44],
       [   -0.08,  0.04003,   0.1602,     0.28,      0.4]])
```

and the maximum value of $0.8766$ is at [3, 2] using NumPy indexing conventions. We can find this by

```
>>> k = np.argmax(img)
17
```

Starting with 0 at the top-left, across each row, then back to the left side of the next lower row, and so on.

and the maximum is at element 17 in row-major order. ◄ We can think of this as a 1D index into a flattened version of the image

```
>>> img.ravel()[k]
0.8766
```

Alternatively, we convert to array subscripts

```
>>> np.unravel_index(k, img.shape)
(3, 2)
```

We can find this concisely using the RVC Toolbox function `findpeaks2d`

```
>>> xy = findpeaks2d(img)
array([[       2,        3,    0.8766],
       [       1,        1,    0.3754]])
```

which has returned two local maxima, one per row, using the image coordinate indexing convention. The third column is the peak value. This function will return all nonlocal maxima where the size of the local region is given by the `scale` option that is the half width of the 2D window in which the maximum is found.

We can refine the estimate of the 2D peak by following a similar procedure to the 1D case. We fit a *paraboloid*

$$z = ax^2 + by^2 + cx + dy + e \tag{J.3}$$

which has five coefficients that can be calculated from the center value (the discrete maximum) and its four neighbors (north, south, east and west) using a similar procedure to above. ◄ The displacement of the estimated peak with respect to the central point is

A paraboloid normally has an $xy$ term giving a total of 6 parameters. This would require using one additional point in order to solve the equation, which breaks the simplicity and symmetry of this scheme. All neighbors could be used, requiring a least-squares solution.

$$\delta_x = \frac{1}{2}\frac{z_e - z_w}{2z_c - z_e - z_w}, \quad \delta_x \in (-1, 1)$$

$$\delta_y = \frac{1}{2}\frac{z_n - z_a}{2z_c - z_n - z_s}, \quad \delta_y \in (-1, 1)$$

The coefficients $a$ and $b$ represent the sharpness of the peak in the $x$- and $y$-directions, and the quality of the peak can be considered as being $\min(|a|, |b|)$.

For this example, the refined peaks are given by

```
>>> xy = findpeaks2d(img, interp=True)
array([[    2.109,    2.964,    0.8839,    0.5505],
       [    1.343,    1.126,    0.4003,    0.1753]])
```

which has one row per peak, comprising the coordinates, the interpolated height of the peak, and the sharpness of the peak. The highest peak in this case is at image coordinate $(2.109, 2.964)$ with a height of $0.8839$. When this process is applied to image data, it is referred to as subpixel interpolation.

# References

Achtelik MW (2014) Advanced closed loop visual navigation for micro aerial vehicles. Ph.D. thesis, ETH Zurich

Adorno B, Marinho M (2021) DQ Robotics: A Library for Robot Modeling and Control. IEEE Robotics Automation Magazine, vol. 28, 102–116

Agarwal S, Furukawa Y, Snavely N, Simon I, Curless B, Seitz SM, Szeliski R (2011) Building Rome in a day. Commun ACM 54(10):105–112

Agarwal P, Burgard W, Stachniss C (2014) Survey of geodetic mapping methods: Geodetic approaches to mapping and the relationship to graph-based SLAM. IEEE Robot Autom Mag 21(3):63–80

Agrawal M, Konolige K, Blas M (2008) CenSurE: Center surround extremas for realtime feature detection and matching. In: Forsyth D, Torr P, Zisserman A (eds) Lecture notes in computer science. Computer Vision – ECCV 2008, vol 5305. Springer-Verlag, Berlin Heidelberg, pp 102–115

Albertos P, Mareels I (2010) Feedback and control for everyone. Springer-Verlag, Berlin Heidelberg

Altmann SL (1989) Hamilton, Rodrigues, and the quaternion scandal. Math Mag 62(5):291–308

Alton K, Mitchell IM (2006) Optimal path planning under defferent norms in continuous state spaces. In: Proceedings of the IEEE International Conference on Robotics and Automation (ICRA). pp 866–872

Andersen N, Ravn O, Sørensen A (1993) Real-time vision based control of servomechanical systems. In: Chatila R, Hirzinger G (eds) Lecture notes in control and information sciences. Experimental Robotics II, vol 190. Springer-Verlag, Berlin Heidelberg, pp 388–402

Andersson RL (1989) Dynamic sensing in a ping-pong playing robot. IEEE T Robotic Autom 5(6):728–739

Antonelli G (2014) Underwater robots: Motion and force control of vehicle-manipulator systems, 3rd ed. Springer Tracts in Advanced Robotics, vol 2. Springer-Verlag, Berlin Heidelberg

Arandjelović R, Zisserman A (2012) Three things everyone should know to improve object retrieval. In: IEEE Conference on Computer Vision and Pattern Recognition (CVPR). pp 2911–2918

Arkin RC (1999) Behavior-based robotics. MIT Press, Cambridge, Massachusetts

Armstrong WW (1979) Recursive solution to the equations of motion of an N-link manipulator. In: Proceedings of the 5th World Congress on Theory of Machines and Mechanisms, Montreal, Jul, pp 1343–1346

Armstrong BS (1988) Dynamics for robot control: Friction modelling and ensuring excitation during parameter identification. Stanford University

Armstrong B (1989) On finding exciting trajectories for identification experiments involving systems with nonlinear dynamics. Int J Robot Res 8(6):28

Armstrong B, Khatib O, Burdick J (1986) The explicit dynamic model and inertial parameters of the Puma 560 Arm. In: Proceedings of the IEEE International Conference on Robotics and Automation (ICRA), vol 3. pp 510–518

Armstrong-Hélouvry B, Dupont P, De Wit CC (1994) A survey of models, analysis tools and compensation methods for the control of machines with friction. Automatica 30(7):1083–1138

Arun KS, Huang TS, Blostein SD (1987) Least-squares fitting of 2 3-D point sets. IEEE T Pattern Anal 9(5):699–700

Asada H (1983) A geometrical representation of manipulator dynamics and its application to arm design. J Dyn Syst-T ASME 105:131

Astolfi A (1999) Exponential stabilization of a wheeled mobile robot via discontinuous control. J Dyn Syst-T ASME 121(1):121–126

Azarbayejani A, Pentland AP (1995) Recursive estimation of motion, structure, and focal length. IEEE T Pattern Anal 17(6):562–575

Bailey T, Durrant-Whyte H (2006) Simultaneous localization and mapping: Part II. IEEE Robot Autom Mag 13(3):108–117

Bakthavatchalam M, Chaumette F, Tahri O (2015) An improved modelling scheme for photometric moments with inclusion of spatial weights for visual servoing with partial appearance/disappearance. In: Proceedings of the IEEE International Conference on Robotics and Automation (ICRA). pp 6037–6043

Baldridge AM, Hook SJ, Grove CI, Rivera G (2009) The ASTER spectral library version 2.0. Remote Sens Environ 113(4):711–715

Ball RS (1876) The theory of screws: A study in the dynamics of a rigid body. Hodges, Foster & Co., Dublin

Ball RS (1908) A treatise on spherical astronomy. Cambridge University Press, New York

Ballard DH (1981) Generalizing the Hough transform to detect arbitrary shapes. Pattern Recogn 13(2):111–122

Banks J, Corke PI (2001) Quantitative evaluation of matching methods and validity measures for stereo vision. Int J Robot Res 20(7):512–532

Barfoot T (2017) State Estimation for Robotics. Cambridge University Press

Bar-Shalom Y, Fortmann T (1988) Tracking and data association. Mathematics in science and engineering, vol 182. Academic Press, London Oxford

Bar-Shalom Y, Li XR, Kirubarajan T (2001) Estimation with applications to tracking and navigation. John Wiley & Sons, Inc., Chichester

Bateux Q, Marchand E, Leitner J, Chaumette F, Corke P (2018) Training Deep Neural Networks for Visual Servoing. IEEE International Conference On Robotics And Automation (ICRA). pp 3307–3314

Bauer J, Sünderhauf N, Protzel P (2007) Comparing several implementations of two recently published feature detectors. In: IFAC Symposium on Intelligent Autonomous Vehicles (IAV). Toulouse

Bay H, Ess A, Tuytelaars T, Van Gool L (2008) Speeded-up robust features (SURF). Comput Vis Image Und 110(3):346–359

Benosman R, Kang SB (2001) Panoramic vision: Sensors, theory, and applications. Springer-Verlag, Berlin Heidelberg

Benson KB (ed) (1986) Television engineering handbook. McGraw-Hill, New York

Berns RS (2019) Billmeyer and Saltzman's Principles of Color Technology, 4th ed. Wiley

Bertolazzi E, Frego M (2014) G1 fitting with clothoids. Mathematical Methods in the Applied Sciences 38(5):881–897

Bertozzi M, Broggi A, Cardarelli E, Fedriga R, Mazzei L, Porta P (2011) VIAC expedition: Toward autonomous mobility. IEEE Robot Autom Mag 18(3):120–124

Besl PJ, McKay HD (1992) A method for registration of 3-D shapes. IEEE T Pattern Anal 14(2): 239–256

Bhat DN, Nayar SK (2002) Ordinal measures for image correspondence. IEEE T Pattern Anal 20(4): 415–423

Biber P, Straßer W (2003) The normal distributions transform: A new approach to laser scan matching. In: Proceedings of the IEEE/RSJ International Conference on intelligent robots and systems (IROS), vol 3. pp 2743–2748

Bishop CM (2006) Pattern recognition and machine learning. Information science and statistics. Springer-Verlag, New York

Blewitt M (2011) Celestial navigation for yachtsmen. Adlard Coles Nautical, London

Bolles RC, Baker HH, Marimont DH (1987) Epipolar-plane image analysis: An approach to determining structure from motion. Int J Comput Vision 1(1):7–55, Mar

Bolles RC, Baker HH, Hannah MJ (1993) The JISCT stereo evaluation. In: Image Understanding Workshop: proceedings of a workshop held in Washington, DC apr 18–21, 1993. Morgan Kaufmann, San Francisco, pp 263

Bolton W (2015) Mechatronics: Electronic control systems in mechanical and electrical engineering, 6th ed. Pearson, Harlow

Borenstein J, Everett HR, Feng L (1996) Navigating mobile robots: Systems and techniques. AK Peters, Ltd. Natick, MA, USA, Out of print and available at http://www-personal.umich.edu/~johannb/Papers/pos96rep.pdf

Borgefors G (1986) Distance transformations in digital images. Comput Vision Graph 34(3):344–371

Bostrom N (2016) Superintelligence: Paths, dangers, strategies. Oxford University Press, Oxford, 432 p

Bouguet J-Y (2010) Camera calibration toolbox for MATLAB®. http://www.vision.caltech.edu/bouguetj/calib_doc

Brady M, Hollerbach JM, Johnson TL, Lozano-Pérez T, Mason MT (eds) (1982) Robot motion: Planning and control. MIT Press, Cambridge, Massachusetts

Braitenberg V (1986) Vehicles: Experiments in synthetic psychology. MIT Press, Cambridge, Massachusetts

Bray H (2014) You are here: From the compass to GPS, the history and future of how we find ourselves. Basic Books, New York

Brockett RW (1983) Asymptotic stability and feedback stabilization. In: Brockett RW, Millmann RS, Sussmann HJ (eds) Progress in mathematics. Differential geometric control theory, vol 27. pp 181–191

Broida TJ, Chandrashekhar S, Chellappa R (1990) Recursive 3-D motion estimation from a monocular image sequence. IEEE T Aero Elec Sys 26(4):639–656

Brooks RA (1986) A robust layered control system for a mobile robot. IEEE T Robotic Autom 2(1):14–23

Brooks RA (1989) A robot that walks: Emergent behaviors from a carefully evolved network. MIT AI Lab, Memo 1091

Brown MZ, Burschka D, Hager GD (2003) Advances in computational stereo. IEEE T Pattern Anal 25(8):993–1008

Brynjolfsson E, McAfee A (2014) The second machine age: Work, progress, and prosperity in a time of brilliant technologies. W.W. Norton & Co., New York

Buehler M, Iagnemma K, Singh S (eds) (2007) The 2005 DARPA grand challenge: The great robot race. Springer Tracts in Advanced Robotics, vol 36. Springer-Verlag, Berlin Heidelberg

Buehler M, Iagnemma K, Singh S (eds) (2010) The DARPA urban challenge. Tracts in Advanced Robotics, vol 56. Springer-Verlag, Berlin Heidelberg

Bukowski R, Haynes LS, Geng Z, Coleman N, Santucci A, Lam K, Paz A, May R, DeVito M (1991) Robot hand-eye coordination rapid prototyping environment. In: Proc ISIR, pp 16.15–16.28

Buttazzo GC, Allotta B, Fanizza FP (1993) Mousebuster: A robot system for catching fast moving objects by vision. In: Proceedings of the IEEE International Conference on Robotics and Automation (ICRA). Atlanta, pp 932–937

Calonder M, Lepetit V, Strecha C, Fua P (2010) BRIEF: Binary robust independent elementary features. In: Daniilidis K, Maragos P, Paragios N (eds) Lecture notes in computer science. Computer Vision – ECCV 2010, vol 6311. Springer-Verlag, Berlin Heidelberg, pp 778–792

Campos C, Elvira R, Rodríguez JJG, Montiel JMM, Tardós JD (2021) ORB-SLAM3: An Accurate Open-Source Library for Visual, Visual–Inertial, and Multimap SLAM. IEEE Trans on Robotics, 37(6), pp 1874–1890

Canny JF (1983) Finding edges and lines in images. MIT, Artificial Intelligence Laboratory, AI-TR-720. Cambridge, MA

Canny J (1987) A computational approach to edge detection. In: Fischler MA, Firschein O (eds) Readings in computer vision: Issues, problems, principles, and paradigms. Morgan Kaufmann, San Francisco, pp 184–203

Censi A (2008) An ICP variant using a point-to-line metric. In: Proceedings of the IEEE International Conference on Robotics and Automation (ICRA). pp 19–25

Chahl JS, Srinivasan MV (1997) Reflective surfaces for panoramic imaging. Appl Optics 31(36):8275–8285

Chaumette F (1990) La relation vision-commande: Théorie et application et des tâches robotiques. Ph.D. thesis, Université de Rennes 1

Chaumette F (1998) Potential problems of stability and convergence in image-based and position-based visual servoing. In: Kriegman DJ, Hager GD, Morse AS (eds) Lecture notes in control and information sciences. The confluence of vision and control, vol 237. Springer-Verlag, Berlin Heidelberg, pp 66–78

Chaumette F (2004) Image moments: A general and useful set of features for visual servoing. IEEE T Robotic Autom 20(4):713–723

Chaumette F, Hutchinson S (2006) Visual servo control 1: Basic approaches. IEEE Robot Autom Mag 13(4):82–90

Chaumette F, Hutchinson S (2007) Visual servo control 2: Advanced approaches. IEEE Robot Autom Mag 14(1):109–118

Chaumette F, Rives P, Espiau B (1991) Positioning of a robot with respect to an object, tracking it and estimating its velocity by visual servoing. In: Proceedings of the IEEE International Conference on Robotics and Automation (ICRA). Seoul, pp 2248–2253

Chesi G, Hashimoto K (eds) (2010) Visual servoing via advanced numerical methods. Lecture notes in computer science, vol 401. Springer-Verlag, Berlin Heidelberg

Chiuso A, Favaro P, Jin H, Soatto S (2002) Structure from motion causally integrated over time. IEEE T Pattern Anal 24(4):523–535

Choset HM, Lynch KM, Hutchinson S, Kantor G, Burgard W, Kavraki LE, Thrun S (2005) Principles of robot motion. MIT Press, Cambridge, Massachusetts

Chum O, Matas J (2005) Matching with PROSAC - progressive sample consensus. IEEE Computer Society Conference On Computer Vision And Pattern Recognition (CVPR), pp 220–226

Colicchia G, Waltner C, Hopf M, Wiesner H (2009) The scallop's eye – A concave mirror in the context of biology. Physics Education 44(2):175–179

Collewet C, Marchand E, Chaumette F (2008) Visual servoing set free from image processing. In: Proceedings of IEEE International Conference on Robotics and Automation (ICRA). pp 81–86

Commission Internationale de L'Éclairage (1987) Colorimetry, 2nd ed. Commission Internationale de L'Eclairage, CIE No 15.2

Corke PI (1994) High-performance visual closed-loop robot control. University of Melbourne, Dept. Mechanical and Manufacturing Engineering. https://hdl.handle.net/11343/38847

Corke PI (1996a) In situ measurement of robot motor electrical constants. Robotica 14(4):433–436

Corke PI (1996b) Visual control of robots: High-performance visual servoing. Mechatronics, vol 2. Research Studies Press (John Wiley). Out of print and available at http://www.petercorke.com/bluebook

Corke PI (2001) Mobile robot navigation as a planar visual servoing problem. In: Jarvis RA, Zelinsky A (eds) Springer tracts in advanced robotics. Robotics Research: The 10th International Symposium, vol 6. IFRR, Lorne, pp 361–372

Corke PI (2007) A simple and systematic approach to assigning Denavit-Hartenberg parameters. IEEE T Robotic Autom 23(3):590–594

Corke PI (2010) Spherical image-based visual servo and structure estimation. In: Proceedings of the IEEE International Conference on Robotics and Automation (ICRA). Anchorage, pp 5550–5555

Corke PI, Armstrong-Hélouvry BS (1994) A search for consensus among model parameters reported for the PUMA 560 robot. In: Proceedings of the IEEE International Conference on Robotics and Automation (ICRA). San Diego, pp 1608–1613

References

Corke PI, Armstrong-Hélouvry BS (1995) A meta-study of PUMA 560 dynamics: A critical appraisal of literature data. Robotica 13(3):253–258

Corke PI, Good MC (1992) Dynamic effects in high-performance visual servoing. In: Proceedings of the IEEE International Conference on Robotics and Automation (ICRA). Nice, pp 1838–1843

Corke PI, Good MC (1996) Dynamic effects in visual closed-loop systems. IEEE T Robotic Autom 12(5):671–683

Corke PI, Haviland J (2021) Not your grandmother's toolbox – the Robotics Toolbox reinvented for Python. IEEE International Conference on Robotics and Automation (ICRA). pp 11357–11363

Corke PI, Hutchinson SA (2001) A new partitioned approach to image-based visual servo control. IEEE T Robotic Autom 17(4):507–515

Corke PI, Dunn PA, Banks JE (1999) Frame-rate stereopsis using non-parametric transforms and programmable logic. In: Proceedings of the IEEE International Conference on Robotics and Automation (ICRA). Detroit, pp 1928–1933

Corke PI, Lobo J, Dias J (2007) An introduction to inertial and visual sensing. The International Journal of Robotics Research, 26(6). pp 519–536

Corke PI, Strelow D, Singh S (2004) Omnidirectional visual odometry for a planetary rover. In: Proceedings of the International Conference on Intelligent Robots and Systems (IROS). Sendai, pp 4007–4012

Corke PI, Spindler F, Chaumette F (2009) Combining Cartesian and polar coordinates in IBVS. In: Proceedings of the International Conference on Intelligent Robots and Systems (IROS). St. Louis, pp 5962–5967

Corke PI, Paul R, Churchill W, Newman P (2013) Dealing with shadows: Capturing intrinsic scene appearance for image-based outdoor localisation. In: Proceedings of the IEEE/RSJ International Conference on Intelligent Robots and Systems (IROS). pp 2085–2092

Craig JJ (1986) Introduction to robotics: Mechanics and control, 1st ed. Addison-Wesley

Craig JJ (1987) Adaptive control of mechanical manipulators. Addison-Wesley Longman Publishing Co., Inc. Boston

Craig JJ (2005) Introduction to robotics: Mechanics and control, 3rd ed. Pearson/Prentice Hall

Craig JJ, Hsu P, Sastry SS (1987) Adaptive control of mechanical manipulators. Int J Robot Res 6(2):16–28

Crombez N, Caron G, Mouaddib EM (2015) Photometric Gaussian mixtures based visual servoing. In: Proceedings of the IEEE/RSJ International Conference on Intelligent Robots and Systems (IROS). pp 5486–5491

Crone RA (1999) A history of color: The evolution of theories of light and color. Kluwer Academic, Dordrecht

Cummins M, Newman P (2008) FAB-MAP: Probabilistic localization and mapping in the space of appearance. Int J Robot Res 27(6):647

Cutting JE (1997) How the eye measures reality and virtual reality. Behav Res Meth Ins C 29(1):27–36

Daniilidis K, Klette R (eds) (2006) Imaging beyond the pinhole camera. Computational Imaging, vol 33. Springer-Verlag, Berlin Heidelberg

Dansereau DG (2014) Plenoptic signal processing for robust vision in field robotics. Ph.D. thesis, The University of Sydney

Davies ER (2017) Computer Vision: Principles, Algorithms, Applications, Learning 5th ed. Academic Press

Davison AJ, Reid ID, Molton ND, Stasse O (2007) MonoSLAM: Real-time single camera SLAM. IEEE T Pattern Anal 29(6):1052–1067

Deguchi K (1998) Optimal motion control for image-based visual servoing by decoupling translation and rotation. In: Proceedings of the International Conference on Intelligent Robots and Systems (IROS). Victoria, Canada, pp 705–711

Dellaert F, Kaess M (2006) Square root SAM: Simultaneous localization and mapping via square root information smoothing. Int J Robot Res 25(12):1181–1203

DeMenthon D, Davis LS (1992) Exact and approximate solutions of the perspective-three-point problem. IEEE T Pattern Anal 14(11):1100–1105

Denavit J, Hartenberg RS (1955) A kinematic notation for lower-pair mechanisms based on matrices. J Appl Mech-T ASME 22(1):215–221

Deo AS, Walker ID (1995) Overview of damped least-squares methods for inverse kinematics of robot manipulators. J Intell Robot Syst 14(1):43–68

Deriche R, Giraudon G (1993) A computational approach for corner and vertex detection. Int J Comput Vision 10(2):101–124

Diankov R (2010) Automated Construction of Robotic Manipulation Programs. PhD thesis, Carnegie Mellon University, Robotics Institute, CMU-RI-TR-10-29

Dickmanns ED (2007) Dynamic vision for perception and control of motion. Springer-Verlag, London

Dickmanns ED, Graefe V (1988a) Applications of dynamic monocular machine vision. Mach Vision Appl 1:241–261

Dickmanns ED, Graefe V (1988b) Dynamic monocular machine vision. Mach Vision Appl 1(4):223–240

Dickmanns ED, Zapp A (1987) Autonomous high speed road vehicle guidance by computer vision. In: Tenth Triennial World Congress of the International Federation of Automatic Control, vol 4. Munich, pp 221–226

Dijkstra EW (1959) A note on two problems in connexion with graphs. Numer Math 1(1):269–271

Dougherty ER, Lotufo RA (2003) Hands-on morphological image processing. Society of Photo-Optical Instrumentation Engineers (SPIE)

Dubins LE (1957). On Curves of Minimal Length with a Constraint on Average Curvature, and with Prescribed Initial and Terminal Positions and Tangents. American Journal of Mathematics, 79(3), 497–516

Duda RO, Hart PE (1972) Use of the Hough transformation to detect lines and curves in pictures. Commun ACM 15(1):11–15

Durrant-Whyte H, Bailey T (2006) Simultaneous localization and mapping: Part I. IEEE Robot Autom Mag 13(2):99–110

Espiau B, Chaumette F, Rives P (1992) A new approach to visual servoing in robotics. IEEE T Robotic Autom 8(3):313–326

Everett HR (1995) Sensors for mobile robots: Theory and application. AK Peters Ltd., Wellesley

Faugeras OD (1993) Three-dimensional computer vision: A geometric viewpoint. MIT Press, Cambridge, Massachusetts

Faugeras OD, Lustman F (1988) Motion and structure from motion in a piecewise planar environment. Int J Pattern Recogn 2(3):485–508

Faugeras O, Luong QT, Papadopoulou T (2001) The geometry of multiple images: The laws that govern the formation of images of a scene and some of their applications. MIT Press, Cambridge, Massachusetts

Featherstone R (1987) Robot dynamics algorithms. Springer

Featherstone R (2010a) A Beginner's Guide to 6-D Vectors (Part 1). IEEE Robotics Automation Magazine, vol. 17, 83–94

Featherstone R (2010b) A Beginner's Guide to 6-D Vectors (Part 2). IEEE Robotics Automation Magazine, vol. 17, 88–99

Feddema JT (1989) Real time visual feedback control for hand-eye coordinated robotic systems. Purdue University

Feddema JT, Mitchell OR (1989) Vision-guided servoing with feature-based trajectory generation. IEEE T Robotic Autom 5(5):691–700

Feddema JT, Lee CSG, Mitchell OR (1991) Weighted selection of image features for resolved rate visual feedback control. IEEE T Robotic Autom 7(1):31–47

Felzenszwalb PF, Huttenlocher DP (2004) Efficient graph-based image segmentation. Int J Comput Vision 59(2):167–181

Ferguson D, Stentz A (2006) Using interpolation to improve path planning: The Field D* algorithm. J Field Robotics 23(2):79–101

Fischler MA, Bolles RC (1981) Random sample consensus: A paradigm for model fitting with applications to image analysis and automated cartography. Commun ACM 24(6):381–395

Flusser J (2000) On the independence of rotation moment invariants. Pattern Recogn 33(9):1405–1410

Fomena R, Chaumette F (2007) Visual servoing from spheres using a spherical projection model. In: Proceedings of the IEEE International Conference on Robotics and Automation (ICRA). Rome, pp 2080–2085

Ford M (2015) Rise of the robots: Technology and the threat of a jobless future. Basic Books, New York

Förstner W (1994) A framework for low level feature extraction. In: Ecklundh J-O (ed) Lecture notes in computer science. Computer Vision – ECCV 1994, vol 800. Springer-Verlag, Berlin Heidelberg, pp 383–394

Förstner W, Gülch E (1987) A fast operator for detection and precise location of distinct points, corners and centres of circular features. In: ISPRS Intercommission Workshop. Interlaken, pp 149–155

Forsyth DA, Ponce J (2012) Computer vision: A modern approach, 2nd ed. Pearson, London

Fraundorfer F, Scaramuzza D (2012) Visual odometry: Part II – Matching, robustness, optimization, and applications. IEEE Robot Autom Mag 19(2):78–90

Freeman H (1974) Computer processing of line-drawing images. ACM Comput Surv 6(1):57–97

Friedman DP, Felleisen M, Bibby D (1987) The little LISPer. MIT Press, Cambridge, Massachusetts

Frisby JP, Stone JV (2010) Seeing: The Computational Approach to Biological Vision. MIT Press

Fu KS, Gonzalez RC, Lee CSG (1987) Robotics: Control, Sensing, Vision, and Intelligence. McGraw-Hill

Funda J, Taylor RH, Paul RP (1990) On homogeneous transforms, quaternions, and computational efficiency. IEEE T Robotic Autom 6(3):382–388

Gans NR, Hutchinson SA, Corke PI (2003) Performance tests for visual servo control systems, with application to partitioned approaches to visual servo control. Int J Robot Res 22(10–11):955

Garg R, Kumar B, Carneiro G, Reid I (2016) Unsupervised CNN for Single View Depth Estimation: Geometry to the Rescue. ECCV

Gautier M, Khalil W (1992) Exciting trajectories for the identification of base inertial parameters of robots. Int J Robot Res 11(4):362

Geiger A, Roser M, Urtasun R (2010) Efficient large-scale stereo matching. In: Kimmel R, Klette R, Sugimoto A (eds) Computer vision – ACCV 2010: 10th Asian Conference on Computer Vision, Queenstown, New Zealand, November 8–12, 2010, revised selected papers, part I. Springer-Verlag, Berlin Heidelberg, pp 25–38

Geraerts R, Overmars MH (2004) A comparative study of probabilistic roadmap planners. In: Boissonnat J-D, Burdick J, Goldberg K, Hutchinson S (eds) Springer tracts in advanced robotics. Algorithmic Foundations of Robotics V, vol 7. Springer-Verlag, Berlin Heidelberg, pp 43–58

Gevers T, Gijsenij A, van de Weijer J, Geusebroek J-M (2012) Color in computer vision: Fundamentals and applications. John Wiley & Sons, Inc., Chichester

Geyer C, Daniilidis K (2000) A unifying theory for central panoramic systems and practical implications. In: Vernon D (ed) Lecture notes in computer science. Computer vision – ECCV 2000, vol 1843. Springer-Verlag, Berlin Heidelberg, pp 445–461

Glover A, Maddern W, Warren M, Reid S, Milford M, Wyeth G (2012) OpenFABMAP: An open source toolbox for appearance-based loop closure detection. In: Proceedings of the IEEE International Conference on Robotics and Automation (ICRA). pp 4730–4735

Gonzalez RC, Woods RE (2018) Digital image processing, 4th ed. Pearson

Grassia FS (1998) Practical parameterization of rotations using the exponential map. Journal of Graphics Tools 3(3):29–48

Gregory RL (1997) Eye and brain: The psychology of seeing. Princeton University Press, Princeton, New Jersey

Grey CGP (2014) Humans need not apply. YouTube video, www.youtube.com/watch?v=7Pq-S557XQU

Grisetti G, Kümmerle R, Stachniss C, Burgard W (2010) A Tutorial on Graph-Based SLAM. IEEE Intelligent Transportation Systems Magazine 2(4). pp 31–43

Groves PD (2013) Principles of GNSS, inertial, and multisensor integrated navigation systems, 2nd ed. Artech House, Norwood, USA

Hager GD, Toyama K (1998) X Vision: A portable substrate for real-time vision applications. Comput Vis Image Und 69(1):23–37

Hamel T, Mahony R (2002) Visual servoing of an under-actuated dynamic rigid-body system: An image based approach. IEEE T Robotic Autom 18(2):187–198

Hamel T, Mahony R, Lozano R, Ostrowski J (2002) Dynamic modelling and configuration stabilization for an X4-flyer. IFAC World Congress 1(2), p 3

Hansen P, Corke PI, Boles W (2010) Wide-angle visual feature matching for outdoor localization. Int J Robot Res 29(1–2):267–297

Harris CG, Stephens MJ (1988) A combined corner and edge detector. In: Proceedings of the Fourth Alvey Vision Conference. Manchester, pp 147–151

Hart PE (2009) How the Hough transform was invented [DSP history]. IEEE Signal Proc Mag 26(6):18–22

Hart PE, Nilsson NJ, Raphael B (1968) A Formal Basis for the Heuristic Determination of Minimum Cost Paths. IEEE Trans Systems Science and Cybernetics, 4(2):100–107

Hartenberg RS, Denavit J (1964) Kinematic synthesis of linkages. McGraw-Hill, New York

Hartley R, Zisserman A (2003) Multiple view geometry in computer vision. Cambridge University Press, New York

Hashimoto K (ed) (1993) Visual servoing. In: Robotics and automated systems, vol 7. World Scientific, Singapore

Hashimoto K, Kimoto T, Ebine T, Kimura H (1991) Manipulator control with image-based visual servo. In: Proceedings of the IEEE International Conference on Robotics and Automation (ICRA). Seoul, pp 2267–2272

Heikkila J, and Silven O (1997) A four-step camera calibration procedure with implicit image correction, Proc IEEE Computer Society Conference on Computer Vision and Pattern Recognition (CVPR), pp 1106–1112

Herschel W (1800) Experiments on the refrangibility of the invisible rays of the sun. Phil Trans R Soc Lond 90:284–292

Hill J, Park WT (1979) Real time control of a robot with a mobile camera. In: Proceedings of the 9th ISIR, SME. Washington, DC. Mar, pp 233–246

Hirata T (1996) A unified linear-time algorithm for computing distance maps. Inform Process Lett 58(3):129–133

Hirschmüller H (2008) Stereo processing by semiglobal matching and mutual information. IEEE Transactions on Pattern Analysis and Machine Intelligence 30(2):328–341

Hirt C, Claessens S, Fecher T, Kuhn M, Pail R, Rexer M (2013) New ultrahigh-resolution picture of Earth's gravity field. Geophys Res Lett 40:4279–4283

Hoag D (1963) Consideration of Apollo IMU gimbal lock. MIT Instrumentation Laboratory, E–1344, https://www.hq.nasa.gov/alsj/e-1344.htm

Holland O (2003) Exploration and high adventure: the legacy of Grey Walter. Philosophical Trans of the Royal Society of London. Series A: Mathematical, Physical and Engineering Sciences, 361(1811)

Hollerbach JM (1980) A recursive Lagrangian formulation of manipulator dynamics and a comparative study of dynamics formulation complexity. IEEE T Syst Man Cyb 10(11):730–736, Nov

Hollerbach JM (1982) Dynamics. In: Brady M, Hollerbach JM, Johnson TL, Lozano-Pérez T, Mason MT (eds) Robot motion – Planning and control. MIT Press, Cambridge, Massachusetts, pp 51–71

Horaud R, Canio B, Leboullenx O (1989) An analytic solution for the perspective 4-point problem. Comput Vision Graph 47(1):33–44

Horn BKP (1987) Closed-form solution of absolute orientation using unit quaternions. J Opt Soc Am A 4(4):629–642

Horn BKP, Hilden HM, Negahdaripour S (1988) Closed-form solution of absolute orientation using orthonormal matrices. J Opt Soc Am A 5(7):1127–1135

Hosoda K, Asada M (1994) Versatile visual servoing without knowledge of true Jacobian. In: Proceedings of the International Conference on Intelligent Robots and Systems (IROS). Munich, pp 186–193

Howard TM, Green CJ, Kelly A, Ferguson D (2008) State space sampling of feasible motions for high-performance mobile robot navigation in complex environments. J Field Robotics 25(6–7):325–345

Hu MK (1962) Visual pattern recognition by moment invariants. IRE T Inform Theor 8:179–187

Hua M-D, Ducard G, Hamel T, Mahony R, Rudin K (2014) Implementation of a nonlinear attitude estimator for aerial robotic vehicles. IEEE T Contr Syst T 22(1):201–213

Huang TS, Netravali AN (1994) Motion and structure from feature correspondences: A review. P IEEE 82(2):252–268

Humenberger M, Zinner C, Kubinger W (2009) Performance evaluation of a census-based stereo matching algorithm on embedded and multi-core hardware. In: Proceedings of the 19th International Symposium on Image and Signal Processing and Analysis (ISPA). pp 388–393

Hunt RWG (1987) The reproduction of colour, 4th ed. Fountain Press, Tolworth

Hunter RS, Harold RW (1987) The measurement of appearance. John Wiley & Sons, Inc., Chichester

Hutchinson S, Hager G, Corke PI (1996) A tutorial on visual servo control. IEEE T Robotic Autom 12(5):651–670

Huynh DQ (2009) Metrics for 3D Rotations: Comparison and Analysis. J Math Imaging Vis 35, pp 155–164

Ings S (2008) The Eye: A Natural History. Bloomsbury Publishing

Iwatsuki M, Okiyama N (2002a) A new formulation of visual servoing based on cylindrical coordinate system with shiftable origin. In: Proceedings of the International Conference on Intelligent Robots and Systems (IROS). Lausanne, pp 354–359

Iwatsuki M, Okiyama N (2002b) Rotation-oriented visual servoing based on cylindrical coordinates. In: Proceedings of the IEEE International Conference on Robotics and Automation (ICRA). Washington, DC, May, pp 4198–4203

Izaguirre A, Paul RP (1985) Computation of the inertial and gravitational coefficients of the dynamics equations for a robot manipulator with a load. In: Proceedings of the IEEE International Conference on Robotics and Automation (ICRA). Mar, pp 1024–1032

Jägersand M, Fuentes O, Nelson R (1996) Experimental evaluation of uncalibrated visual servoing for precision manipulation. In: Proceedings of the IEEE International Conference on Robotics and Automation (ICRA). Albuquerque, NM, pp 2874–2880

Jarvis RA, Byrne JC (1988) An automated guided vehicle with map building and path finding capabilities. In: Robotics Research: The Fourth international symposium. MIT Press, Cambridge, Massachusetts, pp 497–504

Jazwinski AH (2007) Stochastic processes and filtering theory. Dover Publications, Mineola

Jebara T, Azarbayejani A, Pentland A (1999) 3D structure from 2D motion. IEEE Signal Proc Mag 16(3):66–84

Julier SJ, Uhlmann JK (2004) Unscented filtering and nonlinear estimation. P IEEE 92(3):401–422

Kaehler A, Bradski G (2017) Learning OpenCV 3: Computer vision in C++ with the OpenCV library. O'Reilly & Associates, Köln

Kaess M, Ranganathan A, Dellaert F (2007) iSAM: Fast incremental smoothing and mapping with efficient data association. In: Proceedings of the IEEE International Conference on Robotics and Automation (ICRA). pp 1670–1677

Kahn ME (1969) The near-minimum time control of open-loop articulated kinematic linkages. Stanford University, AIM-106

Kálmán RE (1960) A new approach to linear filtering and prediction problems. J Basic Eng-T Asme 82(1):35–45

Kane TR, Levinson DA (1983) The use of Kane's dynamical equations in robotics. Int J Robot Res 2(3):3–21

Karaman S, Walter MR, Perez A, Frazzoli E, Teller S (2011) Anytime motion planning using the RRT*. In: Proceedings of the IEEE International Conference on Robotics and Automation (ICRA). pp 1478–1483

Kavraki LE, Svestka P, Latombe JC, Overmars MH (1996) Probabilistic roadmaps for path planning in high-dimensional configuration spaces. IEEE T Robotic Autom 12(4):566–580

Kelly R (1996) Robust asymptotically stable visual servoing of planar robots. IEEE T Robotic Autom 12(5):759–766

Kelly A (2013) Mobile robotics: Mathematics, models, and methods. Cambridge University Press, New York

Kelly R, Carelli R, Nasisi O, Kuchen B, Reyes F (2002a) Stable visual servoing of camera-in-hand robotic systems. IEEE-ASME T Mech 5(1):39–48

Kelly R, Shirkey P, Spong MW (2002b) Fixed-camera visual servo control for planar robots. In: Proceedings of the IEEE International Conference on Robotics and Automation (ICRA). Washington, DC, pp 2643–2649

Kenwright K (2012) Dual-Quaternions: From Classical Mechanics to Computer Graphics and Beyond. https://xbdev.net/misc_demos/demos/dual_quaternions_beyond/paper.pdf

Khalil W, Creusot D (1997) SYMORO+: A system for the symbolic modelling of robots. Robotica 15(2):153–161

Khalil W, Dombre E (2002) Modeling, identification and control of robots. Kogan Page Science, London

Khatib O (1987) A unified approach for motion and force control of robot manipulators: The operational space formulation. IEEE T Robotic Autom 3(1):43–53

King-Hele D (2002) Erasmus Darwin's improved design for steering carriages and cars. Notes and Records of the Royal Society of London 56(1):41–62

Klafter RD, Chmielewski TA, Negin M (1989) Robotic engineering – An integrated approach. Prentice Hall, Upper Saddle River, New Jersey

Klein CA, Huang CH (1983) Review of pseudoinverse control for use with kinematically redundant manipulators. IEEE T Syst Man Cyb 13:245–250

Klein G, Murray D (2007) Parallel tracking and mapping for small AR workspaces. In: Sixth IEEE and ACM International Symposium on Mixed and Augmented Reality (ISMAR 2007). pp 225–234

Klette R, Kruger N, Vaudrey T, Pauwels K, van Hulle M, Morales S, Kandil F, Haeusler R, Pugeault N, Rabe C (2011) Performance of correspondence algorithms in vision-based driver assistance using an online image sequence database. IEEE T Veh Technol 60(5):2012–2026

Klette R (2014) Concise Computer Vision: An Introduction into Theory and Algorithms. Springer

Koenderink JJ (1984) The structure of images. Biol Cybern 50(5):363–370

Koenderink JJ (2010) Color for the sciences. MIT Press, Cambridge, Massachusetts

Koenig S, Likhachev M (2005) Fast replanning for navigation in unknown terrain. IEEE T Robotic Autom 21(3):354–363

Krajník T, Cristóforis P, Kusumam K, Neubert P, Duckett T (2017) Image features for visual teach-and-repeat navigation in changing environments. Robotics And Autonomous Systems. **88** pp 127–141

Kriegman DJ, Hager GD, Morse AS (eds) (1998) The confluence of vision and control. Lecture notes in control and information sciences, vol 237. Springer-Verlag, Berlin Heidelberg

Kuipers JB (1999) Quaternions and rotation sequences: A primer with applications to orbits, aerospace and virtual reality. Princeton University Press, Princeton, New Jersey

Kümmerle R, Grisetti G, Strasdat H, Konolige K, Burgard W (2011) g²o: A general framework for graph optimization. In: Proceedings of the IEEE International Conference on Robotics and Automation (ICRA). pp 3607–3613

Lam O, Dayoub F, Schulz R, Corke P (2015) Automated topometric graph generation from floor plan analysis. In: Proceedings of the Australasian Conference on Robotics and Automation. Australasian Robotics and Automation Association (ARAA)

Lamport L (1994) LATEX: A document preparation system. User's guide and reference manual. Addison-Wesley Publishing Company, Reading

Land EH, McCann J (1971) Lightness and retinex theory. J Opt Soc Am A 61(1):1–11

Land MF, Nilsson D-E (2002) Animal eyes. Oxford University Press, Oxford

LaValle SM (1998) Rapidly-exploring random trees: A new tool for path planning. Computer Science Dept., Iowa State University, TR 98–11

LaValle SM (2006) Planning algorithms. Cambridge University Press, New York

LaValle SM (2011a) Motion planning: The essentials. IEEE Robot Autom Mag 18(1):79–89

LaValle SM (2011b) Motion planning: Wild frontiers. IEEE Robot Autom Mag 18(2):108–118

LaValle SM, Kuffner JJ (2001) Randomized kinodynamic planning. Int J Robot Res 20(5):378–400

Laussedat A (1899) La métrophotographie. Enseignement supérieur de la photographie. Gauthier-Villars, 52 p

Leavers VF (1993) Which Hough transform? Comput Vis Image Und 58(2):250–264

Lee CSG, Lee BH, Nigham R (1983) Development of the generalized D'Alembert equations of motion for mechanical manipulators. In: Proceedings of the 22nd CDC, San Antonio, Texas. pp 1205–1210

Lepetit V, Moreno-Noguer F, Fua P (2009) EPnP: An accurate $O(n)$ solution to the PnP problem. Int J Comput Vision 81(2):155–166

Li H, Hartley R (2006) Five-point motion estimation made easy. In: 18th International Conference on Pattern Recognition ICPR 2006. Hong Kong, pp 630–633

Li Y, Jia W, Shen C, van den Hengel A (2014) Characterness: An indicator of text in the wild. IEEE T Image Process 23(4):1666–1677

Li T, Bolic M, Djuric P (2015) Resampling methods for particle filtering: Classification, implementation, and strategies. IEEE Signal Proc Mag 32(3):70–86

Lin Z, Zeman V, Patel RV (1989) On-line robot trajectory planning for catching a moving object. In: Proceedings of the IEEE International Conference on Robotics and Automation (ICRA). pp 1726–1731

Lindeberg T (1993) Scale-space theory in computer vision. Springer-Verlag, Berlin Heidelberg

Lipkin H (2005) A Note on Denavit-Hartenberg Notation in Robotics. Proceedings of the 29th ASME Mechanisms and Robotics Conference, pp 921–926

Lloyd J, Hayward V (1991) Real-time trajectory generation using blend functions. In: Proceedings of the IEEE International Conference on Robotics and Automation (ICRA). Seoul, pp 784–789

Longuet-Higgins H (1981) A computer algorithm for reconstruction of a scene from two projections. Nature 293:133–135

Lovell J, Kluger J (1994) Apollo 13. Coronet Books

Lowe DG (1991) Fitting parametrized three-dimensional models to images. IEEE T Pattern Anal 13(5): 441–450

Lowe DG (2004) Distinctive image features from scale-invariant keypoints. Int J Comput Vision 60(2):91–110

Lowry S, Sunderhauf N, Newman P, Leonard J, Cox D, Corke P, Milford M (2015) Visual place recognition: A survey. Robotics, IEEE Transactions on (99):1–19

Lu F, Milios E (1997) Globally consistent range scan alignment for environment mapping. Auton Robot 4:333–349

Lucas SM (2005) ICDAR 2005 text locating competition results. In: Proceedings of the Eighth International Conference on Document Analysis and Recognition, ICDAR05. pp 80–84

Lucas BD, Kanade T (1981) An iterative image registration technique with an application to stereo vision. In: International joint conference on artificial intelligence (IJCAI), Vancouver, vol 2. https://www.ijcai.org/Past/%20Proceedings/IJCAI-81-VOL-2/PDF/017.pdf, pp 674–679

Luh JYS, Walker MW, Paul RPC (1980) On-line computational scheme for mechanical manipulators. J Dyn Syst-T ASME 102(2):69–76

Lumelsky V, Stepanov A (1986) Dynamic path planning for a mobile automaton with limited information on the environment. IEEE T Automat Contr 31(11):1058–1063

Luo W, Schwing A, Urtasun R (2016) Efficient Deep Learning for Stereo Matching. IEEE Conference On Computer Vision And Pattern Recognition (CVPR).

Luong QT (1992) Matrice fondamentale et autocalibration en vision par ordinateur. Ph.D. thesis, Université de Paris-Sud, Orsay, France

Lynch KM, Park FC (2017) Modern robotics: Mechanics, planning, and control. Cambridge University Press, New York

Ma Y, Kosecka J, Soatto S, Sastry S (2003) An invitation to 3D. Springer-Verlag, Berlin Heidelberg

Ma J, Zhao J, Tian J, Yuille A, Tu Z (2014) Robust Point Matching via Vector Field Consensus, IEEE Trans on Image Processing, 23(4), pp 1706–1721

Maddern W, Pascoe G, Linegar C, Newman P (2016) 1 Year, 1000km: The Oxford RobotCar Dataset. The International Journal of Robotics Research 36(1). pp 3–15

Magnusson M, Lilienthal A, Duckett T (2007) Scan registration for autonomous mining vehicles using 3D-NDT. J Field Robotics 24(10):803–827

Magnusson M, Nuchter A, Lorken C, Lilienthal AJ, Hertzberg J (2009) Evaluation of 3D registration reliability and speed – A comparison of ICP and NDT. In: Proceedings of the IEEE International Conference on Robotics and Automation (ICRA). pp 3907–3912

Mahony R, Corke P, Chaumette F (2002) Choice of image features for depth-axis control in image based visual servo control. IEEE/RSJ International Conference On Intelligent Robots And Systems, pp 390–395 vol.1

Mahony R, Kumar V, Corke P (2012) Multirotor aerial vehicles: Modeling, estimation, and control of quadrotor. IEEE Robot Autom Mag (19):20–32

Maimone M, Cheng Y, Matthies L (2007) Two years of visual odometry on the Mars exploration rovers. J Field Robotics 24(3):169–186

Makhlin AG (1985) Stability and sensitivity of servo vision systems. In: Proc 5th International Conference on Robot Vision and Sensory Controls – RoViSeC 5. IFS (Publications), Amsterdam, pp 79–89

Malis E (2004) Improving vision-based control using efficient second-order minimization techniques. In: Proceedings of the IEEE International Conference on Robotics and Automation (ICRA). pp 1843–1848

Malis E, Vargas M (2007) Deeper understanding of the homography decomposition for vision-based control. Research Report, RR-6303, Institut National de Recherche en Informatique et en Automatique (INRIA), 90 p, https://hal.inria.fr/inria-00174036v3/document

Malis E, Chaumette F, Boudet S (1999) 2-1/2D visual servoing. IEEE T Robotic Autom 15(2):238–250

Marey M, Chaumette F (2008) Analysis of classical and new visual servoing control laws. In: Proceedings of the IEEE International Conference on Robotics and Automation (ICRA). Pasadena, pp 3244–3249

Mariottini GL, Prattichizzo D (2005) EGT for multiple view geometry and visual servoing: Robotics vision with pinhole and panoramic cameras. IEEE T Robotic Autom 12(4):26–39

Mariottini GL, Oriolo G, Prattichizzo D (2007) Image-based visual servoing for nonholonomic mobile robots using epipolar geometry. IEEE T Robotic Autom 23(1):87–100

Marr D (2010) Vision: A computational investigation into the human representation and processing of visual information. MIT Press, Cambridge, Massachusetts

Martin D, Fowlkes C, Tal D, Malik J (2001) A database of human segmented natural images and its application to evaluating segmentation algorithms and measuring ecological statistics. Proceedings of the 8th International Conference on Computer Vision, vol 2. pp 416–423

Martins FN, Celeste WC, Carelli R, Sarcinelli-Filho M, Bastos-Filho TF (2008) An adaptive dynamic controller for autonomous mobile robot trajectory tracking. Control Eng Pract 16(11):1354–1363

Masutani Y, Mikawa M, Maru N, Miyazaki F (1994) Visual servoing for non-holonomic mobile robots. In: Proceedings of the International Conference on Intelligent Robots and Systems (IROS). Munich, pp 1133–1140

Matarić MJ (2007) The robotics primer. MIT Press, Cambridge, Massachusetts

Matas J, Chum O, Urban M, Pajdla T (2004) Robust wide-baseline stereo from maximally stable extremal regions. Image Vision Comput 22(10):761–767

Matthews ND, An PE, Harris CJ (1995) Vehicle detection and recognition for autonomous intelligent cruise control. Technical Report, University of Southampton

Matthies L (1992) Stereo vision for planetary rovers: Stochastic modeling to near real-time implementation. Int J Comput Vision 8(1):71–91

Mayeda H, Yoshida K, Osuka K (1990) Base parameters of manipulator dynamic models. IEEE T Robotic Autom 6(3):312–321

McGee LA, Schmidt SF (1985) Discovery of the Kalman filter as a Practical Tool for Aerospace and Industry. NASA-TM-86847

McGlone C (ed) (2013) Manual of photogrammetry, 6th ed. American Society of Photogrammetry

McLauchlan PF (1999) The variable state dimension filter applied to surface-based structure from motion. University of Surrey, VSSP-TR-4/99

Merlet JP (2006) Parallel robots. Kluwer Academic, Dordrecht

Mettler B (2003) Identification modeling and characteristics of miniature rotorcraft. Kluwer Academic, Dordrecht

Mičušík B, Pajdla T (2003) Estimation of omnidirectional camera model from epipolar geometry. In: IEEE Conference on Computer Vision and Pattern Recognition, vol 1. Madison, pp 485–490

Middleton RH, Goodwin GC (1988) Adaptive computed torque control for rigid link manipulations. Syst Control Lett 10(1):9–16

Mikolajczyk K, Schmid C (2004) Scale and affine invariant interest point detectors. Int J Comput Vision 60(1):63–86

Mikolajczyk K, Schmid C (2005) A performance evaluation of local descriptors. IEEE T Pattern Anal 27(10):1615–1630

Mindell DA (2008) Digital Apollo. MIT Press, Cambridge, Massachusetts

Molton N, Brady M (2000) Practical structure and motion from stereo when motion is unconstrained. Int J Comput Vision 39(1):5–23

Montemerlo M, Thrun S (2007) FastSLAM: A scalable method for the simultaneous localization and mapping problem in robotics, vol 27. Springer-Verlag, Berlin Heidelberg

Montemerlo M, Thrun S, Koller D, Wegbreit B (2003) FastSLAM 2.0: An improved particle filtering algorithm for simultaneous localization and mapping that provably converges. In: Proceedings of the 18th International Joint Conference on Artificial Intelligence. Morgan Kaufmann, San Francisco, pp 1151–1156

Moravec H (1980) Obstacle avoidance and navigation in the real world by a seeing robot rover. Ph.D. thesis, Stanford University

Morel G, Liebezeit T, Szewczyk J, Boudet S, Pot J (2000) Explicit incorporation of 2D constraints in vision based control of robot manipulators. In: Corke PI, Trevelyan J (eds) Lecture notes in control and information sciences. Experimental robotics VI, vol 250. Springer-Verlag, Berlin Heidelberg, pp 99–108

Muja M, Lowe DG (2009) Fast approximate nearest neighbors with automatic algorithm configuration. International Conference on Computer Vision Theory and Applications (VISAPP), Lisbon, Portugal (Feb 2009), pp 331–340

Murray RM, Sastry SS, Zexiang L (1994) A mathematical introduction to robotic manipulation. CRC Press, Inc., Boca Raton

NASA (1970) Apollo 13: Technical air-to-ground voice transcription. Test Division, Apollo Spacecraft Program Office, https://www.hq.nasa.gov/alsj/a13/AS13_TEC.PDF

Nayar SK (1997) Catadioptric omnidirectional camera. In: Proceedings of the IEEE Conference on Computer Vision and Pattern Recognition. Los Alamitos, CA, pp 482–488

Neira J, Tardós JD (2001) Data association in stochastic mapping using the joint compatibility test. IEEE T Robotic Autom 17(6):890–897

Neira J, Davison A, Leonard J (2008) Guest editorial special issue on Visual SLAM. IEEE T Robotic Autom 24(5):929–931

Newcombe RA, Lovegrove SJ, Davison AJ (2011) DTAM: Dense tracking and mapping in real-time. In: Proceedings of the International Conference on Computer Vision, pp 2320–2327

Ng J, Bräunl T (2007) Performance comparison of bug navigation algorithms. J Intell Robot Syst 50(1):73–84

Niblack W (1985) An introduction to digital image processing. Strandberg Publishing Company Birkeroed, Denmark

Nistér D (2003) An efficient solution to the five-point relative pose problem. In: IEEE Conference on Computer Vision and Pattern Recognition, vol 2. Madison, pp 195–202

Nistér D, Naroditsky O, Bergen J (2006) Visual odometry for ground vehicle applications. J Field Robotics 23(1):3–20

Nixon MS, Aguado AS (2019) Feature extraction and image processing for Computer Vision, 4th ed. Academic Press, London Oxford

Noble JA (1988) Finding corners. Image Vision Comput 6(2):121–128

Okutomi M, Kanade T (1993) A multiple-baseline stereo. IEEE T Pattern Anal 15(4):353–363

Ollis M, Herman H, Singh S (1999) Analysis and design of panoramic stereo vision using equi-angular pixel cameras. Robotics Institute, Carnegie Mellon University, CMU-RI-TR-99-04, Pittsburgh, PA

Olson E (2011) AprilTag: A robust and flexible visual fiducial system. In: Proceedings of the IEEE International Conference on Robotics and Automation (ICRA). pp 3400–3407

Orin DE, McGhee RB, Vukobratovic M, Hartoch G (1979) Kinematics and kinetic analysis of open-chain linkages utilizing Newton-Euler methods. Math Biosci 43(1/2):107–130

Ortega R, Spong MW (1989) Adaptive motion control of rigid robots: A tutorial. Automatica 25(6):877–888

Otsu N (1975) A threshold selection method from gray-level histograms. Automatica 11:285–296

Owen M, Beard RW, McLain TW (2015) Implementing Dubins Airplane Paths on Fixed-Wing UAVs. Handbook of Uncrewed Aerial Vehicles, pp 1677–1701.

Papanikolopoulos NP, Khosla PK (1993) Adaptive robot visual tracking: Theory and experiments. IEEE T Automat Contr 38(3):429–445

Papanikolopoulos NP, Khosla PK, Kanade T (1993) Visual tracking of a moving target by a camera mounted on a robot: A combination of vision and control. IEEE T Robotic Autom 9(1):14–35

Patel S, Sobh T (2015) Manipulator performance measures-a comprehensive literature survey. Journal Of Intelligent and Robotic Systems, vol. 77, pp 547–570

Paul R (1972) Modelling, trajectory calculation and servoing of a computer controlled arm. Ph.D. thesis, technical report AIM-177, Stanford University

Paul R (1979) Manipulator Cartesian path control. IEEE T Syst Man Cyb 9:702–711

Paul RP (1981) Robot manipulators: Mathematics, programming, and control. MIT Press, Cambridge, Massachusetts

Paul RP, Shimano B (1978) Kinematic control equations for simple manipulators. In: IEEE Conference on Decision and Control, vol 17. pp 1398–1406

Paul RP, Zhang H (1986) Computationally efficient kinematics for manipulators with spherical wrists based on the homogeneous transformation representation. Int J Robot Res 5(2):32–44

Piepmeier JA, McMurray G, Lipkin H (1999) A dynamic quasi-Newton method for uncalibrated visual servoing. In: Proceedings of the IEEE International Conference on Robotics and Automation (ICRA). Detroit, pp 1595–1600

Pivtoraiko M, Knepper RA, Kelly A (2009) Differentially constrained mobile robot motion planning in state lattices. J Field Robotics 26(3):308–333

Pock T (2008) Fast total variation for computer vision. Ph.D. thesis, Graz University of Technology

Pollefeys M, Nistér D, Frahm JM, Akbarzadeh A, Mordohai P, Clipp B, Engels C, Gallup D, Kim SJ, Merrell P, et al. (2008) Detailed real-time urban 3D reconstruction from video. Int J Comput Vision 78(2):143–167, Jul

Pomerleau D, Jochem T (1995) No hands across America Journal. https://www.cs.cmu.edu/~tjochem/nhaa/Journal.html

Pomerleau D, Jochem T (1996) Rapidly adapting machine vision for automated vehicle steering. IEEE Expert 11(1):19–27

Posner I, Corke P, Newman P (2010) Using text-spotting to query the world. In: IEEE/RSJ International Conference on Intelligent Robots and Systems (IROS). IEEE, pp 3181–3186

Pounds P (2007) Design, construction and control of a large quadrotor micro air vehicle. Ph.D. thesis, Australian National University

Pounds P, Mahony R, Gresham J, Corke PI, Roberts J (2004) Towards dynamically-favourable quadrotor aerial robots. In: Proceedings of the Australasian Conference on Robotics and Automation. Canberra

Pounds P, Mahony R, Corke PI (2006) A practical quad-rotor robot. In: Proceedings of the Australasian Conference on Robotics and Automation. Auckland

Poynton CA (2012) Digital video and HD algorithms and interfaces. Morgan Kaufmann, Burlington

Press WH, Teukolsky SA, Vetterling WT, Flannery BP (2007) Numerical recipes, 3rd ed. Cambridge University Press, New York

Prince SJ (2012) Computer vision: Models, learning, and inference. Cambridge University Press, New York

Prouty RW (2002) Helicopter performance, stability, and control. Krieger, Malabar FL

Pujol J (2012) Hamilton, Rodrigues, Gauss, Quaternions, and Rotations: A Historical Reassessment. Communications In Mathematical Analysis, vol. 13

Pynchon T (2006) Against the day. Jonathan Cape, London

Reeds J, Shepp L (1990) Optimal paths for a car that goes both forwards and backwards. Pacific Journal Of Mathematics, vol. 145, pp 367–393

Rekleitis IM (2004) A particle filter tutorial for mobile robot localization. Technical report (TR-CIM-04-02), Centre for Intelligent Machines, McGill University

Rives P, Chaumette F, Espiau B (1989) Positioning of a robot with respect to an object, tracking it and estimating its velocity by visual servoing. In: Hayward V, Khatib O (eds) Lecture notes in control and information sciences. Experimental robotics I, vol 139. Springer-Verlag, Berlin Heidelberg, pp 412–428

Rizzi AA, Koditschek DE (1991) Preliminary experiments in spatial robot juggling. In: Chatila R, Hirzinger G (eds) Lecture notes in control and information sciences. Experimental robotics II, vol 190. Springer-Verlag, Berlin Heidelberg, pp 282–298

Roberts LG (1963) Machine perception of three-dimensional solids. MIT Lincoln Laboratory, TR 315, https://dspace.mit.edu/handle/1721.1/11589

Romero-Ramirez FJ, Muñoz-Salinas R, Medina-Carnicer R (2018) Speeded up detection of squared fiducial markers. Image and Vision Computing, vol 76, pp 38–47

Rosenfeld GH (1959) The problem of exterior orientation in photogrammetry. Photogramm Eng 25(4):536–553

Rosten E, Porter R, Drummond T (2010) FASTER and better: A machine learning approach to corner detection. IEEE T Pattern Anal 32:105–119

Russell S, Norvig P (2020) Artificial intelligence: A modern approach, 4th ed. Pearson

Sakaguchi T, Fujita M, Watanabe H, Miyazaki F (1993) Motion planning and control for a robot performer. In: Proceedings of the IEEE International Conference on Robotics and Automation (ICRA). Atlanta, May, pp 925–931

Salvi J, Matabosch C, Fofi D, Forest J (2007) A review of recent range image registration methods with accuracy evaluation. Image Vision Comput 25(5):578–596

Samson C, Espiau B, Le Borgne M (1990) Robot control: The task function approach. Oxford University Press, Oxford

Scaramuzza D, Martinelli A, Siegwart R (2006) A Toolbox for Easy Calibrating Omnidirectional Cameras. Proc IEEE/RSJ Int Conf on Intelligent Robots and Systems, pp 5695–5701

Scaramuzza D, Fraundorfer F (2011) Visual odometry [tutorial]. IEEE Robot Autom Mag 18(4):80–92

Scharstein D, Pal C (2007) Learning conditional random fields for stereo. In: IEEE Computer Society Conference on Computer Vision and Pattern Recognition (CVPR). Minneapolis, MN

Scharstein D, Szeliski R (2002) A taxonomy and evaluation of dense two-frame stereo correspondence algorithms. Int J Comput Vision 47(1):7–42

Selig JM (2005) Geometric fundamentals of robotics. Springer-Verlag, Berlin Heidelberg

Sharf I, Nahon M, Harmat A, Khan W, Michini M, Speal N, Trentini M, Tsadok T, Wang T (2014) Ground effect experiments and model validation with Draganflyer X8 rotorcraft. Int Conf Uncrewed Aircraft Systems (ICUAS). pp 1158–1166

Sharp A (1896) Bicycles & tricycles: An elementary treatise on their design an construction; With examples and tables. Longmans, Green and Co., London New York Bombay

Sheridan TB (2003) Telerobotics, automation, and human supervisory control. MIT Press, Cambridge, Massachusetts, 415 p

Shi J, Tomasi C (1994) Good features to track. In: Proceedings of the Computer Vision and Pattern Recognition. IEEE Computer Society, Seattle, pp 593–593

Shih FY (2009) Image processing and mathematical morphology: Fundamentals and applications, CRC Press, Boca Raton

Shirai Y (1987) Three-dimensional computer vision. Springer-Verlag, New York

Shirai Y, Inoue H (1973) Guiding a robot by visual feedback in assembling tasks. Pattern Recogn 5(2):99–106

Shoemake K (1985) Animating rotation with quaternion curves. In: Proceedings of ACM SIGGRAPH, San Francisco, pp 245–254

Siciliano B, Khatib O (eds) (2016) Springer handbook of robotics, 2nd ed. Springer-Verlag, New York

Siciliano B, Sciavicco L, Villani L, Oriolo G (2009) Robotics: Modelling, planning and control. Springer-Verlag, Berlin Heidelberg

Siegwart R, Nourbakhsh IR, Scaramuzza D (2011) Introduction to autonomous mobile robots, 2nd ed. MIT Press, Cambridge, Massachusetts

Silver WM (1982) On the equivalence of Lagrangian and Newton-Euler dynamics for manipulators. Int J Robot Res 1(2):60–70

Sivic J, Zisserman A (2003) Video Google: A text retrieval approach to object matching in videos. In: Proceedings of the Ninth IEEE International Conference on Computer Vision. pp 1470–1477

Skaar SB, Brockman WH, Hanson R (1987) Camera-space manipulation. Int J Robot Res 6(4):20–32

Skofteland G, Hirzinger G (1991) Computing position and orientation of a freeflying polyhedron from 3D data. In: Proceedings of the IEEE International Conference on Robotics and Automation (ICRA). Seoul, pp 150–155

Smith R (2007) An overview of the Tesseract OCR engine. In: 9th International Conference on Document Analysis and Recognition (ICDAR). pp 629–633

Sobel D (1996) Longitude: The true story of a lone genius who solved the greatest scientific problem of his time. Fourth Estate, London

Soille P (2003) Morphological image analysis: Principles and applications. Springer-Verlag, Berlin Heidelberg

Solà J (2017) Quaternion kinematics for the error-state Kalman filter. arXiv 1711.02508

Solà J, Deray J, Atchuthan D (2018) A micro Lie theory for state estimation in robotics. arXiv 1812.01537

Spong MW (1989) Adaptive control of flexible joint manipulators. Syst Control Lett 13(1):15–21

Spong MW, Hutchinson S, Vidyasagar M (2006) Robot modeling and control, 2nd ed. John Wiley & Sons, Inc., Chichester

Srinivasan VV, Venkatesh S (1997) From living eyes to seeing machines. Oxford University Press, Oxford

Stachniss C, Burgard W (2014) Particle filters for robot navigation. Foundations and Trends in Robotics 3(4):211–282

Steinvall A (2002) English colour terms in context. Ph.D. thesis, Umeå Universitet

Stentz A (1994) The D* algorithm for real-time planning of optimal traverses. The Robotics Institute, Carnegie-Mellon University, CMU-RI-TR-94-37

Stewart A (2014) Localisation using the appearance of prior structure. Ph.D. thesis, University of Oxford

Stone JV (2012) Vision and brain: How we perceive the world. MIT Press, Cambridge, Massachusetts

Strasdat H (2012) Local accuracy and global consistency for efficient visual SLAM. Ph.D. thesis, Imperial College London

Strelow D, Singh S (2004) Motion estimation from image and inertial measurements. Int J Robot Res 23(12):1157–1195

Sünderhauf N (2012) Robust optimization for simultaneous localization and mapping. Ph.D. thesis, Technische Universität Chemnitz

Sussman GJ, Wisdom J, Mayer ME (2001) Structure and interpretation of classical mechanics. MIT Press, Cambridge, Massachusetts

Sutherland IE (1974) Three-dimensional data input by tablet. P IEEE 62(4):453–461

Svoboda T, Pajdla T (2002) Epipolar geometry for central catadioptric cameras. Int J Comput Vision 49(1):23–37

Szeliski R (2022) Computer vision: Algorithms and applications. Springer-Verlag, Berlin Heidelberg

Tahri O, Chaumette F (2005) Point-based and region-based image moments for visual servoing of planar objects. IEEE T Robotic Autom 21(6):1116–1127

Tahri O, Mezouar Y, Chaumette F, Corke PI (2009) Generic decoupled image-based visual servoing for cameras obeying the unified projection model. In: Proceedings of the IEEE International Conference on Robotics and Automation (ICRA). Kobe, pp 1116–1121

Taylor RA (1979) Planning and execution of straight line manipulator trajectories. IBM J Res Dev 23(4):424–436

ter Haar Romeny BM (1996) Introduction to scale-space theory: Multiscale geometric image analysis. Utrecht University

Thrun S, Burgard W, Fox D (2005) Probabilistic robotics. MIT Press, Cambridge, Massachusetts

Tissainayagam P, Suter D (2004) Assessing the performance of corner detectors for point feature tracking applications. Image Vision Comput 22(8):663–679

Titterton DH, Weston JL (2005) Strapdown inertial navigation technology. IEE Radar, Sonar, Navigation and Avionics Series, vol 17, The Institution of Engineering and Technology (IET), 576 p

Tomasi C, Kanade T (1991) Detection and tracking of point features. Carnegie Mellon University, CMU-CS-91-132

Triggs B, McLauchlan P, Hartley R, Fitzgibbon A (2000) Bundle adjustment – A modern synthesis. Lecture notes in computer science. Vision algorithms: theory and practice, vol 1883. Springer-Verlag, Berlin Heidelberg, pp 153–177

Tsai RY (1986) An Efficient and Accurate Camera Calibration Technique for 3D Machine Vision. Proc IEEE Conference on Computer Vision and Pattern Recognition (CVPR), pp 364–374

Tsakiris D, Rives P, Samson C (1998) Extending visual servoing techniques to nonholonomic mobile robots. In: Kriegman DJ, Hager GD, Morse AS (eds) Lecture notes in control and information sciences. The confluence of vision and control, vol 237. Springer-Verlag, Berlin Heidelberg, pp 106–117

Terzakis G, Lourakis MIA, Ait-Boudaoud D (2018) Modified Rodrigues Parameters: An Efficient Representation of Orientation in 3D Vision and Graphics. J Math Imaging Vis. 60:422–442

Uicker JJ (1965) On the dynamic analysis of spatial linkages using 4 by 4 matrices. Dept. Mechanical Engineering and Astronautical Sciences, NorthWestern University

Umeyama, S (1991) Least-Squares Estimation of Transformation Parameters Between Two Point Patterns. IEEE Trans Pattern Analysis and Machine Intellence 13(4). pp 376–380

Usher K (2005) Visual homing for a car-like vehicle. Ph.D. thesis, Queensland University of Technology

Usher K, Ridley P, Corke PI (2003) Visual servoing of a car-like vehicle – An application of omnidirectional vision. In: Proceedings of the IEEE International Conference on Robotics and Automation (ICRA). Taipei, Sep, pp 4288–4293

Valgren C, Lilienthal AJ (2010) SIFT, SURF & seasons: Appearance-based long-term localization in outdoor environments. Robot Auton Syst 58(2):149–156

Vanderborght B, Sugar T, Lefeber D (2008) Adaptable compliance or variable stiffness for robotic applications. IEEE Robot Autom Mag 15(3):8–9

Vince J (2011) Quaternions for Computer Graphics. Springer

Wade NJ (2007) Image, eye, and retina. J Opt Soc Am A 24(5):1229–1249

Walker MW, Orin DE (1982) Efficient dynamic computer simulation of robotic mechanisms. J Dyn Syst-T ASME 104(3):205–211

Walter WG (1950) An imitation of life. Sci Am 182(5):42–45

Walter WG (1951) A machine that learns. Sci Am 185(2):60–63

Walter WG (1953) The living brain. Duckworth, London

Warren M (2015) Long-range stereo visual odometry for unmanned aerial vehicles. Ph.D. thesis, Queensland University of Technology

Weiss LE (1984) Dynamic visual servo control of robots: An adaptive image-based approach. Ph.D. thesis, technical report CMU-RI-TR-84-16, Carnegie-Mellon University

Weiss L, Sanderson AC, Neuman CP (1987) Dynamic sensor-based control of robots with visual feedback. IEEE T Robotic Autom 3(1):404–417

Westmore DB, Wilson WJ (1991) Direct dynamic control of a robot using an end-point mounted camera and Kalman filter position estimation. In: Proceedings of the IEEE International Conference on Robotics and Automation (ICRA). Seoul, Apr, pp 2376–2384

Whitney DE (1969) Resolved motion rate control of manipulators and human prostheses. IEEE T Man Machine 10(2):47–53

Wiener N (1965) Cybernetics or control and communication in the animal and the machine. MIT Press, Cambridge, Massachusetts

Wilburn B, Joshi N, Vaish V, Talvala E-V, Antunez E, Barth A, Adams A, Horowitz M, Levoy M (2005) High performance imaging using large camera arrays. ACM Transactions on Graphics (TOG) – Proceedings of ACM SIGGRAPH 2005 24(3):765–776

Wolf PR, DeWitt BA (2014) Elements of photogrammetry, 4th ed. McGraw-Hill, New York

Woodfill J, Von Herzen B (1997) Real-time stereo vision on the PARTS reconfigurable computer. In: Proceedings of the IEEE Symposium on FPGAs for Custom Computing Machines, Grenoble. pp 201–210

Xu G, Zhang Z (1996) Epipolar geometry in stereo, motion, and object recognition: A unified approach. Springer-Verlag, Berlin Heidelberg

Yi K, Trulls E, Lepetit V, Fua P (2016) LIFT: Learned Invariant Feature Transform. ECCV 2016. pp 467–483

Ying X, Hu Z (2004) Can we consider central catiodioptric cameras and fisheye cameras within a unified imaging model. In: Pajdla T, Matas J (eds) Lecture notes in computer science. Computer vision – ECCV 2004, vol 3021. Springer-Verlag, Berlin Heidelberg, pp 442–455

Yoshikawa T (1984) Analysis and control of robot manipulators with redundancy. In: Brady M, Paul R (eds) Robotics research: The first international symposium. MIT Press, Cambridge, Massachusetts, pp 735–747

Zabih R, Woodfill J (1994) Non-parametric local transforms for computing visual correspondence. In: Ecklundh J-O (ed) Lecture notes in computer science. Computer Vision – ECCV 1994, vol 800. Springer-Verlag, Berlin Heidelberg, pp 151–158

Zarchan P, Musoff H (2005) Fundamentals of Kalman filtering: A practical approach. Progress in Astronautics and Aeronautics, vol 208. American Institute of Aeronautics and Astronautics

Zhan H, Garg R, Weerasekera C, Li K, Agarwal H, Reid I (2018) Unsupervised Learning of Monocular Depth Estimation and Visual Odometry With Deep Feature Reconstruction. IEEE Conference On Computer Vision And Pattern Recognition (CVPR).

Zhang Z, Faugeras O, Kohonen T, Hunag TS, Schroeder MR (1992) Three D-dynamic scene analysis: A stereo based approach. Springer-Verlag, New York

Zhou Q, Park J, Koltun V (2018) Open3D: A Modern Library for 3D Data Processing. ArXiv:1801.09847

Ziegler J, Bender P, Schreiber M, Lategahn H, Strauss T, Stiller C, Thao Dang, Franke U, Appenrodt N, Keller CG, Kaus E, Herrtwich RG, Rabe C, Pfeiffer D, Lindner F, Stein F, Erbs F, Enzweiler M, Knöppel C, Hipp J, Haueis M, Trepte M, Brenk C, Tamke A, Ghanaat M, Braun M, Joos A, Fritz H, Mock H, Hein M, Zeeb E (2014) Making Bertha drive – An autonomous journey on a historic route. IEEE Intelligent Transportation Systems Magazine 6(2):8–20

# Index of People

# Index of Functions, Classes and Methods[1]

[1]   Classes are shown in bold, methods are prefixed by a dot. Other entries are functions.

# Index of Models[2]

---

2    These block diagram models are included in the rvc3python package in the folder models. To run them requires the bdsim package.

# General Index